STALLCUP'S®

Illustrated Code Changes, 2008 Edition

Based on the NEC® and Related Standards

James G. Stallcup
with
James W. Stallcup
Mark C. Ode

JONES AND BARTLETT PUBLISHERS
Sudbury, Massachusetts
BOSTON TORONTO LONDON SINGAPORE

World Headquarters
Jones and Bartlett Publishers
40 Tall Pine Drive
Sudbury, MA 01776
978-443-5000
info@jbpub.com
www.jbpub.com

Jones and Bartlett Publishers Canada
6339 Ormindale Way
Mississauga, ON L5V 1J2
Canada

Jones and Bartlett Publishers International
Barb House, Barb Mews
London W6 7PA
United Kingdom

National Fire Protection Association
1 Batterymarch Park
Quincy, MA 02169
www.NFPA.org

Jones and Bartlett's books and products are available through most bookstores and online booksellers. To contact Jones and Bartlett Publishers directly, call 800-832-0034, fax 978-443-8000, or visit our website www.jbpub.com.

Substantial discounts on bulk quantities of Jones and Bartlett's publications are available to corporations, professional associations, and other qualified organizations. For details and specific discount information, contact the special sales department at Jones and Bartlett Publishers via the above contact information or send an email to specialsales@jbpub.com.

Production Credits
Chief Executive Officer: Clayton E. Jones
Chief Operating Officer: Donald W. Jones, Jr.
President, Higher Education and Professional Publishing: Robert W. Holland, Jr.
V.P., Sales and Marketing: William J. Kane
V.P., Production and Design: Anne Spencer
V.P., Manufacturing and Inventory Control: Therese Connell
Publisher, Public Safety Group: Kimberly Brophy
Acquisitions Editor: Martin Schumacher
Reprints Coordinator/Production Assistant: Amy Browning
Director of Marketing: Alisha Weisman
Design, Graphics, and Layout: Billy G. Stallcup
Cover Design: Anne Spencer
Text Printing and Binding: Courier Kendallville
Cover Printing: Courier Kendallville

Library of Congress Cataloging-in-Publication Data

Stallcup, James G.
Stallcup's illustrated code changes 2008 / James Stallcup.
p. cm.
ISBN-13: 978-0-7637-5149-4
ISBN-10: 0-7637-5149-9
1. National Fire Protection Association. National Electrical Code (2005) 2. Electric engineering--Insurance requirements--United States.
I. National Electrical Code. II. Title.
TK260.S765 2007
621.319'24021873--dc22
6048 2007019290

Printed in the United States of America
12 11 10 09 08 10 9 8 7 6 5 4 3 2

Forward

As in the past, the changes that have been made to the newest edition of the NEC have been numerous and as always prompted a need for a publication that not only designates these changes in a well designed format but provides a detailed and visually accepted presentation for the easy comprehension of the reader.

Now, once again, the 2008 NEC represents numerous changes. Learning what these changes are and the reasoning behind them shouldn't be the only reason for a user to learn them.

As in all Stallcup publications, the correlation and interactivity of various codes and standards is considered mandatory in the electrical industry. Therefore, it is imperative that there be a Code by which other codes and standards can reference as the Code of standard practice. The NEC is recognized as this Code and therefore, for installation and maintenance of electrical systems, as well as for safety issues, keeping oneself aware of changes not only in the NEC but in the electrical industry as a whole, is a must.

As a note of record, the Stallcups would like to thank Mark Ode for his invaluable contributions to the development of this newest code change publication. His tireless efforts were greatly appreciated.

The Stallcup's

Introduction

The **2008 edition of the National Electrical Code® (NEC®)** contains many comprehensive revisions pertaining to specific NEC rules and regulations. Electrical personnel have an immediate and awesome task in not only learning but implementing these revisions in their everyday design, installation and inspection of electrical systems.

The material in this book, if read and studied carefully in a continuous and enthusiastic manner, will provide a proper update on the revisions in the **2008 NEC**. However, even though it is true that only time and discussion among electrical personnel will provide the answers on how to interpret and apply some of these rules, one can use this book to get a head start.

Illustrated Code Changes explains the major changes in the **2008 NEC** and can be used as a guide for fast and easy reference. These changes are presented in numerical order to correlate with the "Articles" and "Sections" as they appear in the **2008 NEC** and are also illustrated to give a more detailed description. Where appropriate, reasons for revisions and new Articles are given and what kind of impact such changes will have on manufactures, designers, installers, and inspectors.

For every proposal made, there was a reason and hopefully, this book will provide some of the reasons why a change resulted.

CORRELATING THE NEC WITH OTHER STANDARDS

Type of Change	Panel Action	UL	UL 508	API 500 - 1997	API 505 - 1997	OSHA - 1994
-	-	-	-	-	-	-
ROP	**ROC**	**NFPA 70E - 2004**	**NFPA 70B - 2006**	**NFPA 79 - 2007**	**NFPA**	**NEMA**
pg. - # -	pg. - # -	-	-	-	-	-
log: - CMP: -	log: - Submitter: -			2005 NEC: -		IEC: -

With each change, references are made to well known standards that are used in the industry. However, the **ROP and ROC**, that are referenced to in the Table, are documents that are directly involved in the code-making process and are explained below along with the different types of changes and **Code Making Panel** actions.

NEC PROCESS

ROP (Report on Proposals)

The information found in this document not only specifies the proposed changes, but also includes who submitted the change and substantiation for such proposal.

ROC (Report on Comments)

All of the actions that were taken by the **Code Making Panels** and the **Technical Correlating Committee** at the meetings prior to the **NFPA** annual meeting and the adoption o the **2008 NEC** are found in this document.

Type of Change:

New Article, Section, Subsection, Subdivision, Exception and FPN
.... Revision
.... Deletion

Panel Action:

.... Accept
.... Accept in Principle
.... Accept in Part
.... Accept in Principle in Part
.... Reject

This Table was developed by the author with the intention of enhancing the perception of the change while correlating the NEC with other standards. Hopefully, the instructor and/or student of the code will find this information useful, if not interesting.

For seminar purposes the authors have listed 100 of what they feel will be the most important code changes. However, this book represents over 500 - 2008 NEC changes and are intended for the reader's review.

1. 90.2(B)(5)b
2. Article 100 - Neutral Conductor
3. Article 100 – Qualified Person
4. 110.12(A)
5. 110.16
6. 110.26(C)(2)
7. 200.2(B)
8. 210.4(D)
9. 210.5(C)
10. 210.8(B)(4)
11. 210.12(B), Exception (Delete)
12. 210.12(B), Exceptions
13. 210.52(E)(3)
14. 225.39
15. 230.44, Exception
16. 230.205
17. 240.4(D)(1) through (D)(7)
18. 240.21(C)(2)
19. 240.21(H)
20. 240.86(A)
21. 240.92(B)
22. 250.8
23. 250.20(D)
24. 250.28(D)(1) through (D)(3)
25. 250.32(B) and Exception
26. 250.35
27. 250.52(A)(2)
28. 250.52(A)(3)
29. 250.52(A)(5)
30. 250.64(D)
31. 250.68
32. 250.119
33. 250.122
34. 250.146
35. 300.4(E)
36. 300.9
37. 300.12, Exception No. 2
38. Table 300.50
39. 310.7, Exception No. 2 and FPN
40. 310.15(B)(2)(c) and Table
41. 310.15(B)(6)
42. 314.16(B)(1)
43. 314.16(B)(4)
44. 314.24(A) through (C)
45. 314.27(A) and (B)
46. 314.27(E) and Exception
47. 334.15(C)
48. 334.15(C)
49. 334.80
50. 342.30(C)
51. 344.10(A) through (C)
52. 348.12(1)
53. Article 355
54. 376.100(A) through (D)
55. 382.6
56. 404.8(C)
57. 406.4(G)
58. 406.8(A)
59. 406.8(B)(1) and Exception
60. 406.11
61. 408.3(F)
62. 408.34, 408.35, and 408.36
63. 408.54
64. 410.10(D)
65. 410.16(C)
66. 410.130(G)
67. 411.3(A) and (B)
68. 422.51
69. 430.52(C)(1), Exception No. 2
70. 445.19
71. 501.10(B)(1)(7)
72. 501.30(B)
73. 504.2
74. 504.30(A), Exceptions No. 3 and 4
75. 504.70
76. 513.2
77. 513.3
78. 517.32(C)(3)
79. 517.160(A)(5)
80. Article 522
81. 551.42(C)
82. 555.21(A) and (B)
83. 600.4(C)
84. 600.21(E)
85. 600.41(B)
86. 620.21(A)(1)(e)
87. 620.51(C)(1)
88. Article 626
89. 680.12
90. 680.23(B)(6)
91. 680.26(A) through (C)
92. 680.31

93. 695.6(B)(2)
94. 700.9(B)(5)
95. 700.23
96. 702.5(A) and (B)
97. Article 708
98. 725.48(B)(4)
99. 800.25 and 800.26
100. 800.156

Table of Contents

1

Administration and Enforcement, Introduction, Definitions and Requirements

Article 90 and **Chapter 1** of the NEC has always been referred to as the "get acquainted" material that every designer, installer, electrician, apprentice, inspector, and maintenance person must review and understand before the other chapters, articles and sections of the NEC can really be understood and applied.

The first article in the NEC is **Article 90**, which contains the Introduction. **Article 90** covers the purpose of the NEC along with other pertinent information that is applicable throughout each chapter of the NEC.

Chapter 1 acquaints the user of the NEC with definitions and clearance rules that are mandatory to ensure the safety of the general public and personnel working in, near, or on wiring methods and equipment.

Users as well as students of the NEC must review and become acquainted with **Article 90** and **Chapter 1** before attempting to study, learn, and apply the other articles and chapters to a particular design or installation.

It is this concept of study that will make interpretations and applications of the many requirements in the NEC much easier to understand.

NEC Ch. - Article 90
Part - 90.2(B)(5)b

Type of Change	Panel Action	UL	UL 508	API 500 - 1997	API 505 - 1997	OSHA - 1994
Revision	Accept	-	-	-	-	1910.269(a)
ROP	**ROC**	**NFPA 70E - 2004**	**NFPA 70B - 2006**	**NFPA 79 - 2007**	**NFPA**	**NEMA**
pg. 3 # 1-5	pg. 3 # 1-5	-	-	-	-	-
log: 2018 CMP: 1	log: 2018	Submitter: Wayne Robinson		2005 NEC: 90.2(B)(5)b		IEC: -

2005 NEC - 90.2(B) Not Covered.

This *Code* does not cover the following:

(5) Installations under the exclusive control of an electric utility where such installations

a. Consist of service drops or service laterals, and associated metering, or

b. Are located in legally established easements, rights-of-way, ~~or by other agreements either~~ designated by or recognized by public service commissions, utility commissions, or other regulatory agencies having jurisdiction for such installations, or

c. Are on property owned or leased by the electric utility for the purpose of communications, metering, generation, control, transformation, transmission, or distribution of electric energy.

2008 NEC - 90.2(B) Not Covered.

This *Code* does not cover the following:

(5) Installations under the exclusive control of an electric utility where such installations

a. Consist of service drops or service laterals, and associated metering, or

b. Are located in legally established easements, or rights-of-way designated by or recognized by public service commissions, utility commissions, or other regulatory agencies having jurisdiction for such installations, or

c. Are on property owned or leased by the electric utility for the purpose of communications, metering, generation, control, transformation, transmission, or distribution of electric energy.

Author's Comment: Deleting the phrase "or by other agreements either," will limit any agreements, other than a legally established easement or right of way, that a utility company can use to make installations that are outside the scope of the NEC. Utility agreements with Forest Service installations, local, state, and Federal facilities, as well as Indian Tribal installations, many of which are not under the jurisdiction of public service utility commissions or not covered by legally established right-of-ways or easements may be affected by this change.

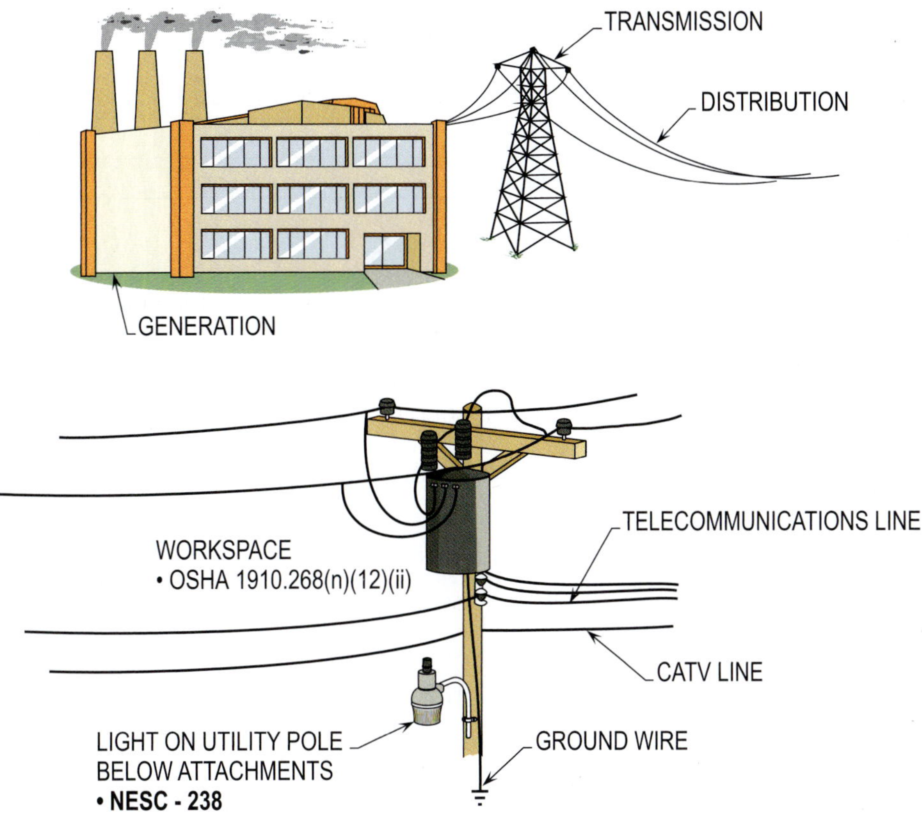

INSTALLATIONS UNDER THE EXCLUSIVE CONTROL OF AN ELECTRIC UTILITY

- SERVICE DROPS OR SERVICE LATERALS, AND ASSOCIATED METERING
- LOCATED IN LEGALLY ESTABLISHED EASEMENTS, OR RIGHT-OF-WAYS DESIGNATED BY OR RECOGNIZED BY PUBLIC SERVICE COMMISSIONS, OR OTHER REGULATORY AGENCIES HAVING JURISDICTION FOR SUCH INSTALLATIONS
- ON PROPERTY OWNED OR LEASED BY THE ELECTRIC UTILITY FOR THE PURPOSE OF COMMUNICATIONS, METERING, GENERATION, CONTROL, TRANSFORMATION, TRANSMISSION OR DISTRIBUTION OF ELECTRIC ENERGY.

NOT COVERED
90.2(B)(5)b

Purpose of Change: This revision clarifies that any agreements will be limited, other than a legally established easement or right-of-way, that a utility company can use to make installations that are outside the scope of the NEC.

NEC Ch. 1 - Article 100
Part I - Bonded (Bonding)

Type of Change	Panel Action	UL	UL 508	API 500 - 1997	API 505 - 1997	OSHA - 1994
Revision	Accept	-	-	-	-	1910.399(a)
ROP	**ROC**	**NFPA 70E - 2004**	**NFPA 70B - 2006**	**NFPA 79 - 2007**	**NFPA**	**NEMA**
pg. 9 # 5-2	pg. - # -	Article 100	3.3.1	-	-	-
log: 1512 CMP: 1	log: -	Submitter: Technical Correlating Committee on NEC		2005 NEC: Article 100		IEC: -

2005 NEC - Article 100

~~**Bonding (Bonded).** The permanent joining of metallic parts to form an electrically conductive path that ensures electrical continuity and the capacity to conduct safely any current likely to be imposed.~~

2008 NEC - Article 100

Bonded (Bonding). Connected to establish electrical continuity and conductivity.

Author's Comment: This revision has been written to apply generally throughout the NEC and simply describe the purpose and function of bonding.

There are conditions in the NEC where specific bonding is required solely to minimize the difference of potential (voltage) between normally energized conductive components.

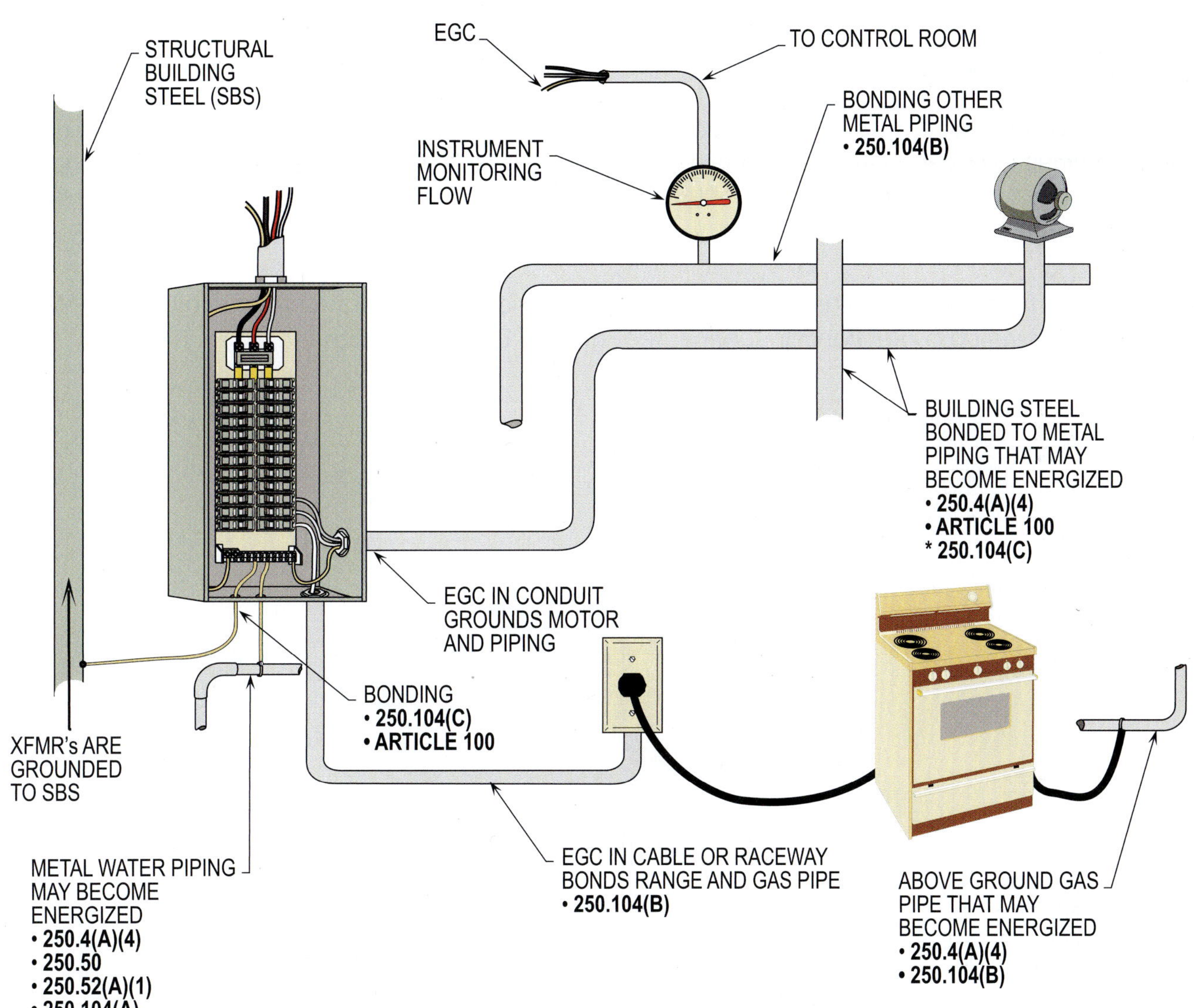

BONDED (BONDING)
ARTICLE 100

Purpose of Change: To revise the definition of bonded (bonding) to apply generally throughout the NEC and describe the purpose and function of bonding.

NEC Ch. 1 - Article 100
Part I - Branch-Circuit Overcurrent Device

Type of Change	Panel Action	UL	UL 508	API 500 - 1997	API 505 - 1997	OSHA - 1994
New Definition	Accept in Principle	248 - 489	-	-	-	-
ROP	**ROC**	**NFPA 70E - 2004**	**NFPA 70B - 2006**	**NFPA 79 - 2007**	**NFPA**	**NEMA - 2002**
pg. 10 / # 10-1a	pg. - / # -	-	-	-	-	AB 1
log: 1259 / CMP: 1	log: -	Submitter: Frank G. Ladonne		2005 NEC: Article 100		IEC: -

2008 NEC - Article 100

Branch-Circuit Overcurrent Device. A device capable of providing protection for service, feeder, and branch circuits and equipment over the full range of overcurrents between its rated current and interrupting rating. Branch circuit overcurrent protective devices are provided with interrupting ratings appropriate for the intended use but no less than 5,000 amperes.

Author's Comment: A new definition has been added to clarify that a branch-circuit overcurrent device is capable of providing protection for service, feeder, and branch circuits.

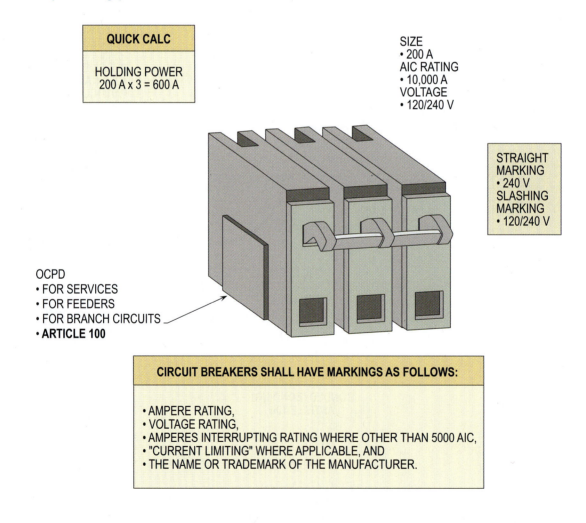

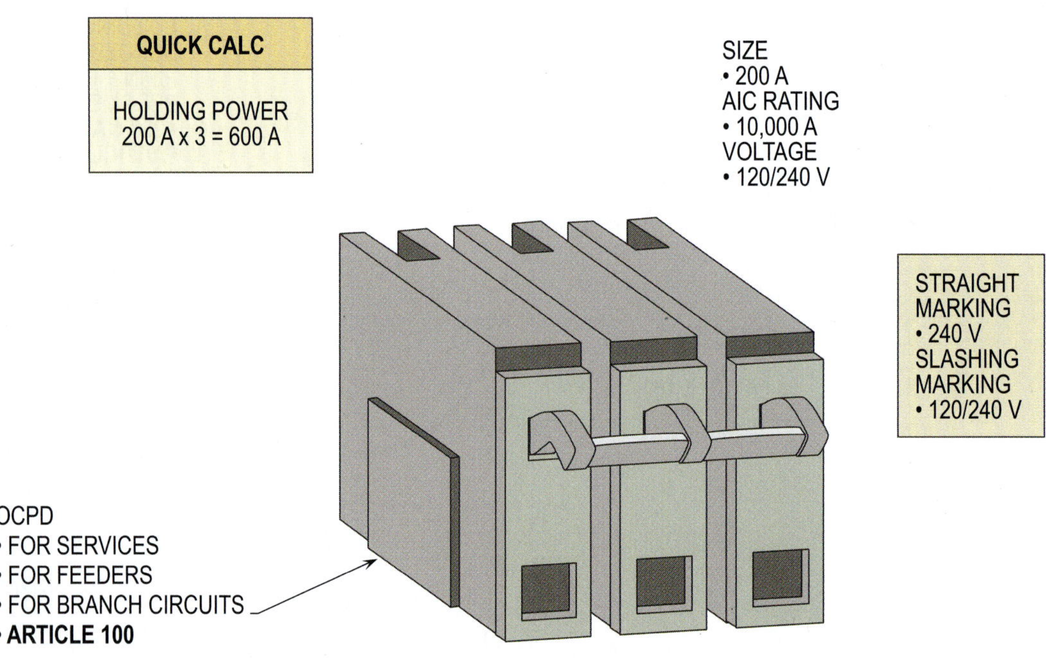

CIRCUIT BREAKERS SHALL HAVE MARKINGS AS FOLLOWS:

- AMPERE RATING,
- VOLTAGE RATING,
- AMPERES INTERRUPTING RATING WHERE OTHER THAN 5000 AIC,
- "CURRENT LIMITING" WHERE APPLICABLE, AND
- THE NAME OR TRADEMARK OF THE MANUFACTURER.

**BRANCH-CIRCUIT OVERCURRENT DEVICE
ARTICLE 100**

Purpose of Change: A new definition has been added to clarify that a branch circuit overcurrent device is capable of providing protection for service, feeder, and branch circuit.

NEC Ch. 1 - Article 100
Part I - Clothes Closet

Type of Change		Panel Action		UL	UL 508	API 500 - 1997	API 505 - 1997	OSHA - 1994
New Definition		Accept in Principle		-	-	-	-	-
ROP		ROC		NFPA 70E - 2004	NFPA 70B - 2006	NFPA 79 - 2007	NFPA	NEMA
pg. 11	# 1-20	pg. 7	# 1-22	-	-	-	-	-
log: 358	CMP: 1	log: 2099	Submitter: Michael J. Johnston			2005 NEC: -		IEC: -

2008 NEC - Article 100

Clothes Closet. A non-habitable room or space intended primarily for storage of garments and apparel.

Author's Comment: A new defintion has been added to clarify what constitutes a "clothes closet."

The Accept in Principle in the ROC stage results in no change to the proposed text for this definition.

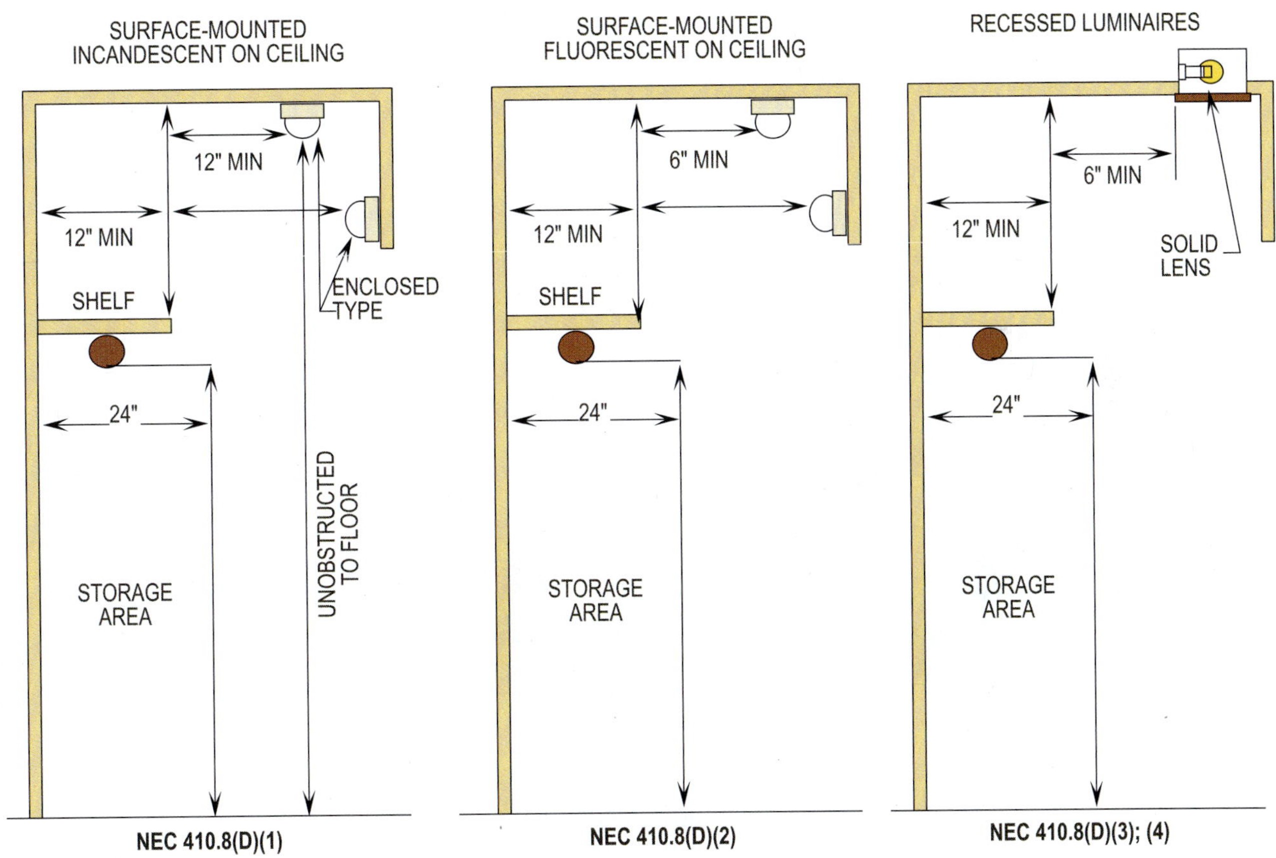

CLOTHES CLOSET
ARTICLE 100

Purpose of Change: A new definition has been added to clarify that a non-habitable room or space intended primarily for storage of garments and apparel constitutes a "clothes closet."

NEC Ch. 1 - Article 100
Part I - Electrical Power Production and Distribution Network

Type of Change		Panel Action		UL	UL 508	API 500 - 1997	API 505 - 1997	OSHA - 1994
New Definition		Accep		-	-	-	-	-
ROP		ROC		NFPA 70E - 2004	NFPA 70B - 2006	NFPA 79 - 2007	NFPA	NEMA
pg. 14	# 1-28	pg. -	# -	-	-	-	-	-
log: 2577	CMP: 1	log: -	Submitter: Timothy M. Croushore			2005 NEC: -		IEC: -

2008 NEC - Article 100

Electrical Power Production and Distribution Network. Power production, distribution, and utilization equipment and facilities, such as electric utility systems that deliver electric power to the connected loads, that are external to and not controlled by an interactive system.

Author's Comment: A new definition has been added to correlate with the definition in Section 2.41 of Underwriter Laboratories Standard 1741 - Inverters, Converters and Controllers for Use in Independent Power Systems.

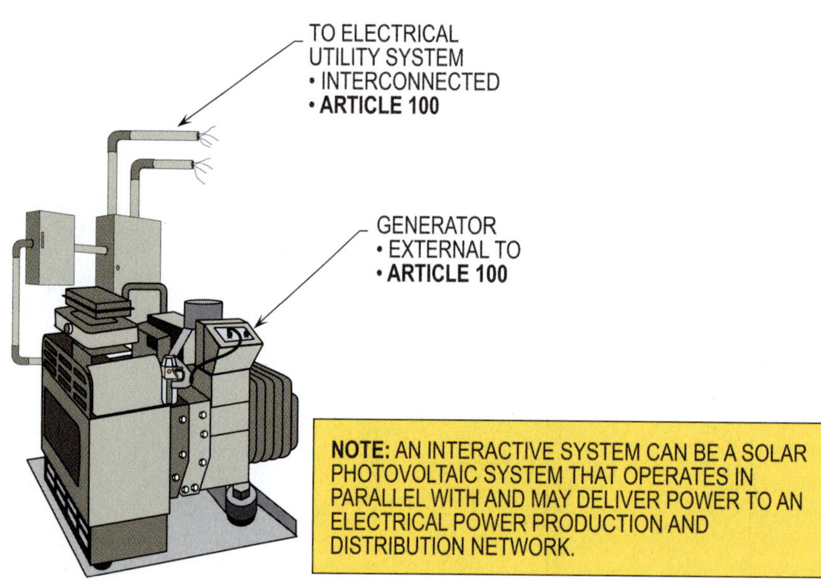
TO ELECTRICAL
UTILITY SYSTEM
• INTERCONNECTED
• ARTICLE 100
GENERATOR
• EXTERNAL TO
• ARTICLE 100
NOTE: AN INTERACTIVE SYSTEM CAN BE A SOLAR PHOTOVOLTAIC SYSTEM THAT OPERATES IN PARALLEL WITH AND MAY DELIVER POWER TO AN ELECTRICAL POWER PRODUCTION AND DISTRIBUTION NETWORK.

**ELECTRICAL POWER PRODUCTION AND DISTRIBUTION NETWORK
ARTICLE 100**

Purpose of Change: A new definition has been added that will be used in **Articles 690**, **692**, and **705**.

NEC Ch. 1 - Article 100
Part I - Equipment

Type of Change	Panel Action	UL	UL 508	API 500 - 1997	API 505 - 1997	OSHA - 1994
Revision	Accept in Principle	-	-	-	-	-
ROP	**ROC**	**NFPA 70E - 2004**	**NFPA 70B - 2006**	**NFPA 79 - 2007**	**NFPA**	**NEMA**
pg. 14 # 1-31	pg. - # -	-	-	-	-	-
log: 2869 CMP: 1	log: -	Submitter: Robert D. Osborne		2005 NEC: Article 100		IEC: -

2005 NEC - Article 100

Equipment. A general term including material, fittings, devices, appliances, luminaires (fixtures), apparatus, and the like used as a part of, or in connection with, an electrical installation.

2008 NEC - Article 100

Equipment. A general term including material, fittings, devices, appliances, luminaires (fixtures), apparatus, machinery, and the like used as a part of, or in connection with, an electrical installation.

Author's Comment: "Machinery" has been added to the definition to ensure that electrical machinery is included in the definition for electrical equipment. Section **110.2** requires conductors and electrical equipment required or permitted by the NEC to be approved. Adding machinery will now make it clear that industrial equipment installations are covered by the NEC.

APPROVAL. THE CONDUCTORS AND EQUIPMENT REQUIRED OR PERMITTED BY THIS CODE SHALL BE ACCEPTABLE ONLY IF APPROVED.

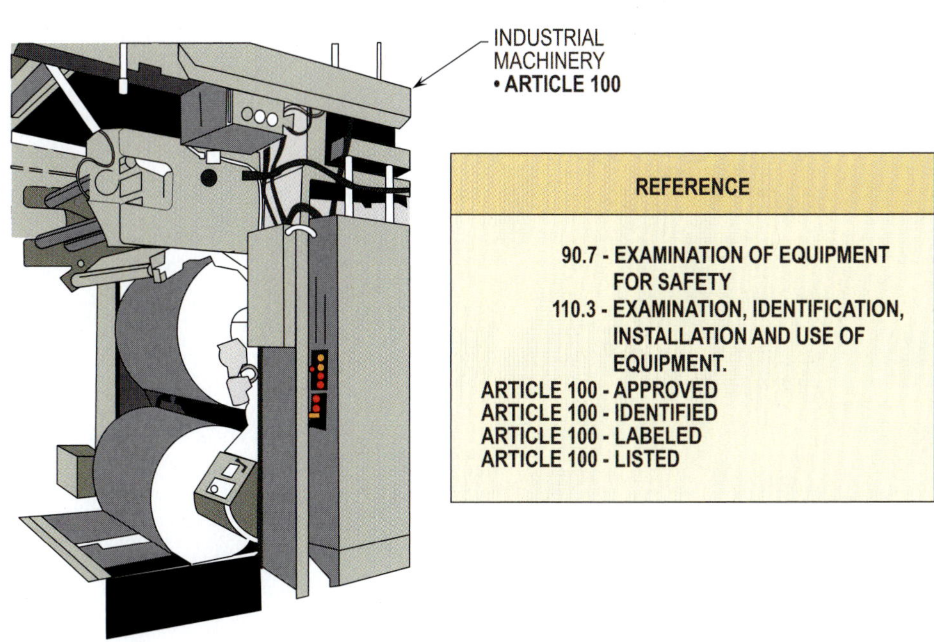

INDUSTRIAL
MACHINERY
• ARTICLE 100
REFERENCE
90.7 - EXAMINATION OF EQUIPMENT FOR SAFETY
110.3 - EXAMINATION, IDENTIFICATION, INSTALLATION AND USE OF EQUIPMENT.
ARTICLE 100 - APPROVED
ARTICLE 100 - IDENTIFIED
ARTICLE 100 - LABELED
ARTICLE 100 - LISTED

**EQUIPMENT
ARTICLE 100**

Purpose of Change: The term "machinery" has been added to ensure that electrical machinery is included in the definition for electrical equipment.

NEC Ch. 1 - Article 100
Part I - Grounding Conductor, Equipment

Type of Change	Panel Action	UL	UL 508	API 500 - 1997	API 505 - 1997	OSHA - 1994
Revision	Accept	-	-	-	-	1910.399(a)
ROP	**ROC**	**NFPA 70E - 2004**	**NFPA 70B - 2006**	**NFPA 79 - 2007**	**NFPA**	**NEMA**
pg. 15 # 5-6	pg. 11 # 5-3	Article 100	3.3.24	3.3.51	-	-
log: 1513 CMP: 1	log: 1671	Submitter: Technical Correlating Committee on NEC		2005 NEC: Article 100		IEC: -

2005 NEC - Article 100

Grounding Conductor, Equipment. The ~~conductor used~~ to connect ~~the~~ non-current-carrying metal parts of equipment, ~~raceways, and other enclosures~~ and to the system grounded conductor, ~~the~~ grounding electrode conductor, or both, ~~at the service equipment or at the source of a separately derived system.~~

2008 NEC - Article 100

Grounding Conductor, Equipment (EGC). The conductive path installed to connect normally non-current-carrying metal parts of equipment together and to the system grounded conductor, grounding electrode conductor, or both.

FPN No. 1: It is recognized that the equipment grounding conductor also performs bonding.

FPN No. 2: See 250.118 for a list of acceptable equipment grounding conductors.

Author's Comment: This revision has been written to apply generally throughout the NEC and describe the purpose and function of the equipment grounding conductor.

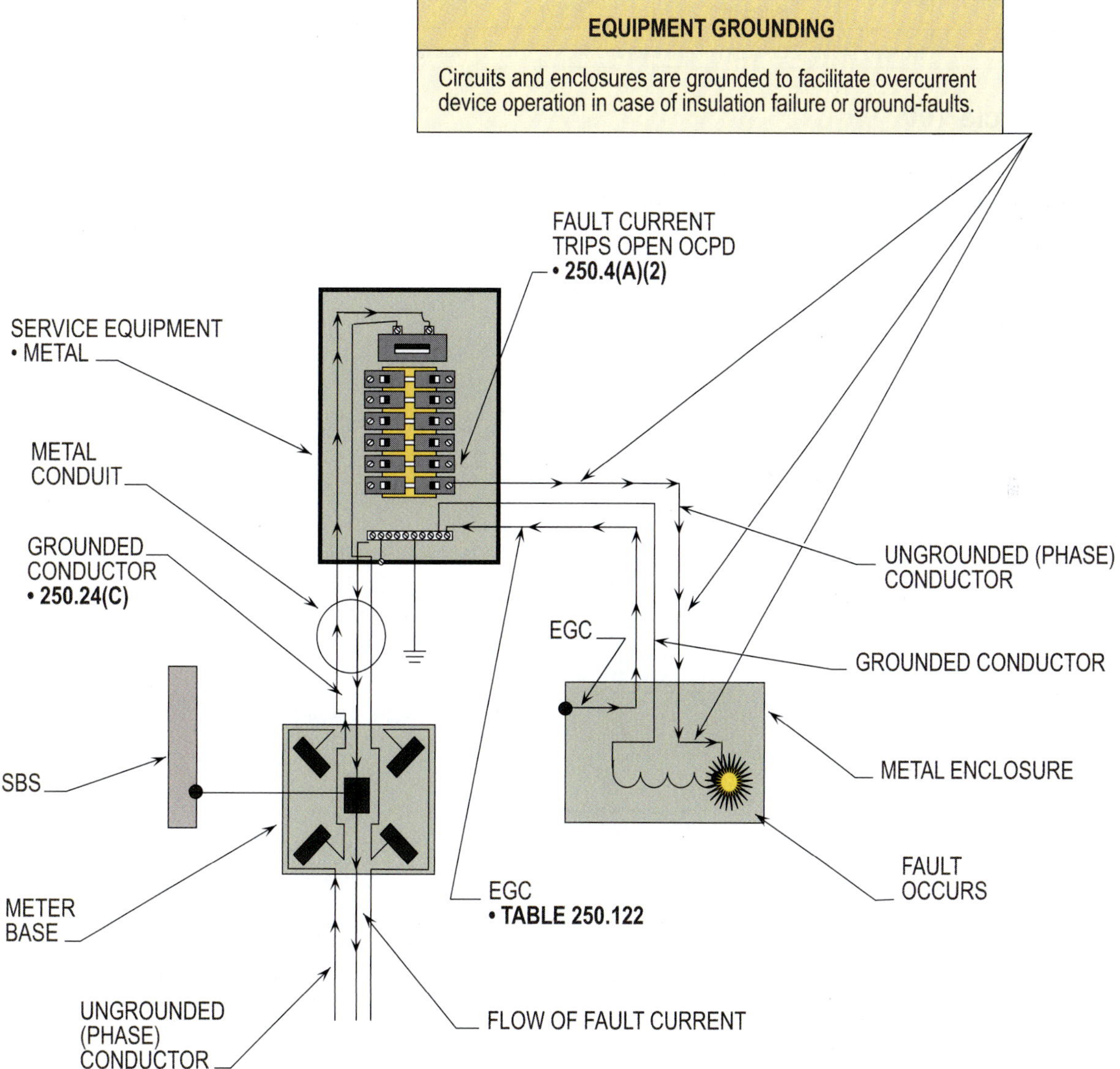

GROUNDING CONDUCTOR, EQUIPMENT
ARTICLE 100

Purpose of Change: To revise the definition of equipment grounding conductor (EGC) to apply generally throughout the NEC and describe the purpose and function of the equipment grounding conductor.

NEC Ch. 1 - Article 100
Part I - Ground

Type of Change	Panel Action	UL	UL 508	API 500 - 1997	API 505 - 1997	OSHA - 1994
Revision	Accept	-	-	-	-	1910.399(a)
ROP	**ROC**	**NFPA 70E - 2004**	**NFPA 70B - 2006**	**NFPA 79 - 2007**	**NFPA**	**NEMA**
pg. 16 # 5-8	pg. - # -	Article 100	3.3.27	3.3.47	-	-
log: 1515 CMP: 1	log: -	Submitter: Technical Correlating Committee on NEC		2005 NEC: Article 100		IEC: -

2005 NEC - Article 100

Ground. ~~A conducting connection, whether intentional or accidental, between an electrical circuit or equipment and the earth or to some conducting body that serves in place of~~ the earth.

2008 NEC - Article 100

Ground. The earth.

Author's Comment: The definition of ground has been simplified by deleting the intentional or accidental connection to earth or to some body serving in place of the earth.

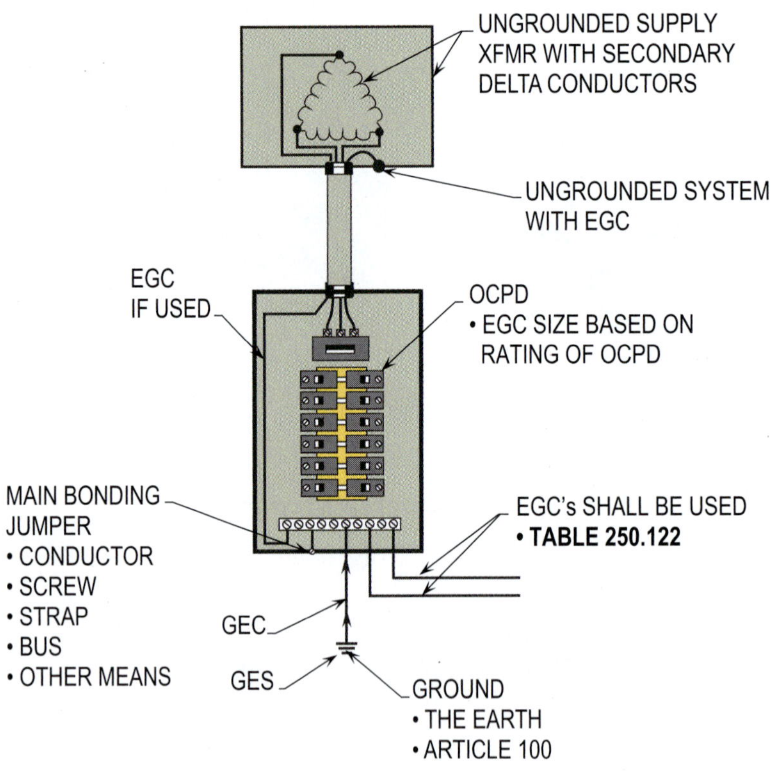

GROUND
ARTICLE 100

Purpose of Change: This revision clarifies that the ground is the earth whether it is intentional or accidental.

NEC Ch. 1 - Article 100
Part I - Grounded (Grounding)

Type of Change	Panel Action	UL	UL 508	API 500 - 1997	API 505 - 1997	OSHA - 1994
Revision	Accept in Principle	-	-	-	-	1910.399(a)
ROP	ROC	NFPA 70E - 2004	NFPA 70B - 2006	NFPA 79 - 2007	NFPA	NEMA
pg. 17 # 5-9	pg. - # -	Article 100	3.3.33	3.3.48	-	-
log: 1517 CMP: 1	log: -	Submitter: Technical Correlating Commitee on NEC		2005 NEC: Article 100		IEC: -

2005 NEC - Article 100

Grounded. Connected to ~~earth~~ ground or to ~~some conducting~~ body ~~that serves in place of the earth~~.

2008 NEC - Article 100

Grounded (Grounding). Connected (connecting) to ground or to a conductive body that extends the ground connection.

Author's Comment: The change to this definition is to more accurately describe the function of a grounded conductor.

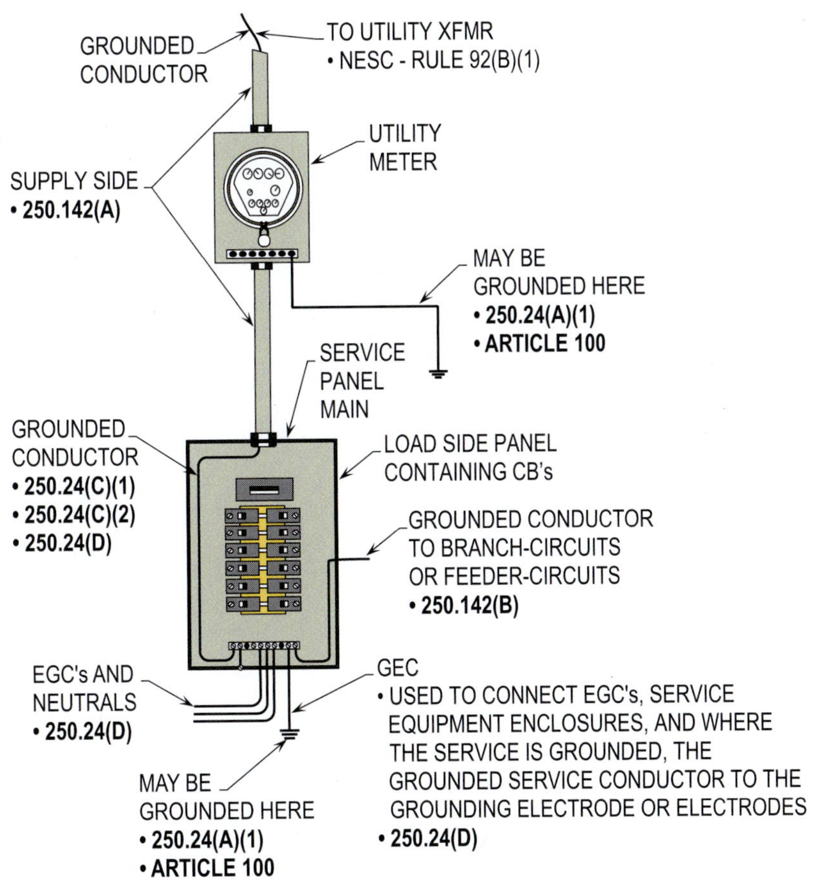

GROUNDED (GROUNDING)
ARTICLE 100

Purpose of Change: To revise the definition of grounded to apply generally throughout the NEC and describe the purpose and function of grounded.

NEC Ch. 1 - Article 100
Part I - Grounded, Effectively

Type of Change	Panel Action	UL	UL 508	API 500 - 1997	API 505 - 1997	OSHA - 1994
Deletion	Accept	-	-	-	-	1910.399(a)
ROP	**ROC**	**NFPA 70E - 2004**	**NFPA 70B - 2006**	**NFPA 79 - 2007**	**NFPA**	**NEMA**
pg. 18 # 5-12	pg. - # -	Article 100	3.3.17	-	-	-
log: 1516 CMP: 1	log: -	Submitter: Technical Correlating Committee on NEC		2005 NEC: Article 100		IEC: -

2008 NEC - Article 100

~~**Grounded, Effectively.** Intentionally connected to earth through a ground connection or connections of sufficiently low impedance and having sufficient current-carrying capacity to prevent the buildup of voltages that may result in undue hazards to connected equipment or to persons.~~

Author's Comment: The definition "Effectively Grounded" is very subjective and without any defined values or parameters for one to judge grounding as either "effective" or "ineffective." "Effective" is described in Section **250.4(A)** and **(B)**, but it relates to the effective ground-fault current path as a performance criteria. Deleting the term through-out the NEC and the definition is logical because there are no definitive parameters for Code users to determe what constitutes "effectively grounded."

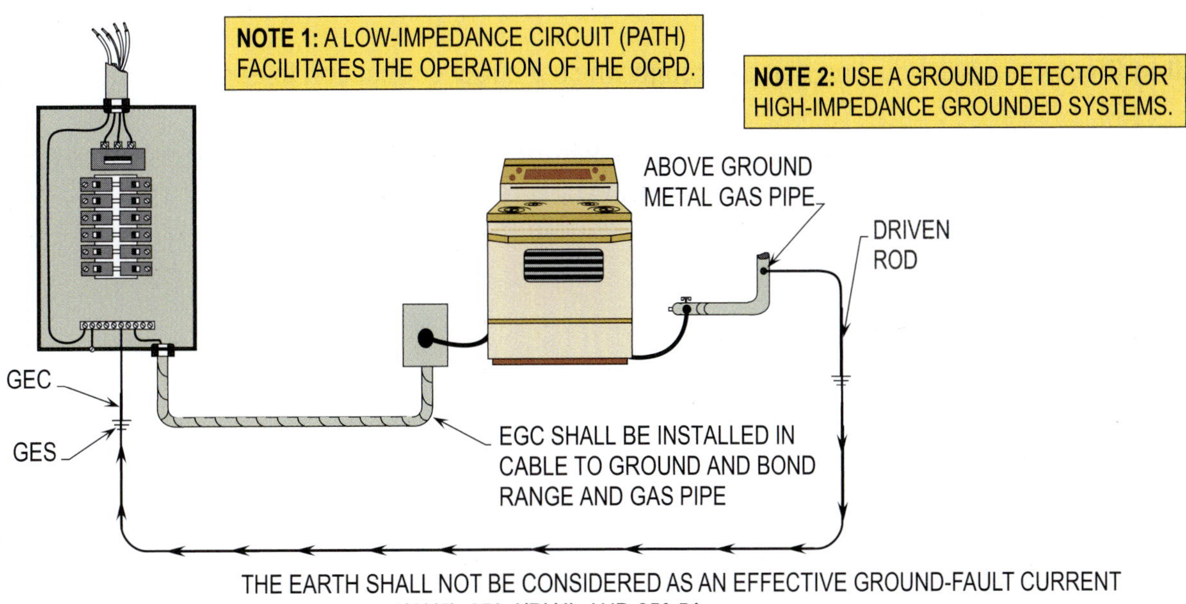
NOTE 1: A LOW-IMPEDANCE CIRCUIT (PATH) FACILITATES THE OPERATION OF THE OCPD.
NOTE 2: USE A GROUND DETECTOR FOR HIGH-IMPEDANCE GROUNDED SYSTEMS.
ABOVE GROUND METAL GAS PIPE
DRIVEN ROD
GEC
GES
EGC SHALL BE INSTALLED IN CABLE TO GROUND AND BOND RANGE AND GAS PIPE
THE EARTH SHALL NOT BE CONSIDERED AS AN EFFECTIVE GROUND-FAULT CURRENT PATH PER 250.4(A)(5), 250.4(B)(4), AND 250.54.

GROUNDED, EFFECTIVELY
ARTICLE 100

Purpose of Change: The definition "grounded, effectively" has been deleted because it is subjective and without any defined values or parameters for one to judge grounding as either "effective" or "ineffective."

NEC Ch. 1 - Article 100
Part I - Grounding Electrode

Type of Change	Panel Action	UL	UL 508	API 500 - 1997	API 505 - 1997	OSHA - 1994
Revision	Accept in Principle	-	-	-	-	-
ROP	**ROC**	**NFPA 70E - 2004**	**NFPA 70B - 2006**	**NFPA 79 - 2007**	**NFPA**	**NEMA**
pg. 18 # 5-14	pg. - # -	-	3.3.38	-	-	-
log: 1514 CMP: 1	log: -	Submitter: Technical Correlating Committee on NEC		2005 NEC: Article 100		IEC: -

2005 NEC - Article 100

Grounding Electrode. ~~A device that establishes an electrical connection to the earth.~~

2008 NEC - Article 100

Grounding Electrode. A conducting object through which a direct connection to earth is established.

Author's Comment: The definition for grounding electrode was rewritten to better describe its function. The grounding electrode establishes and maintains a direct connection to earth.

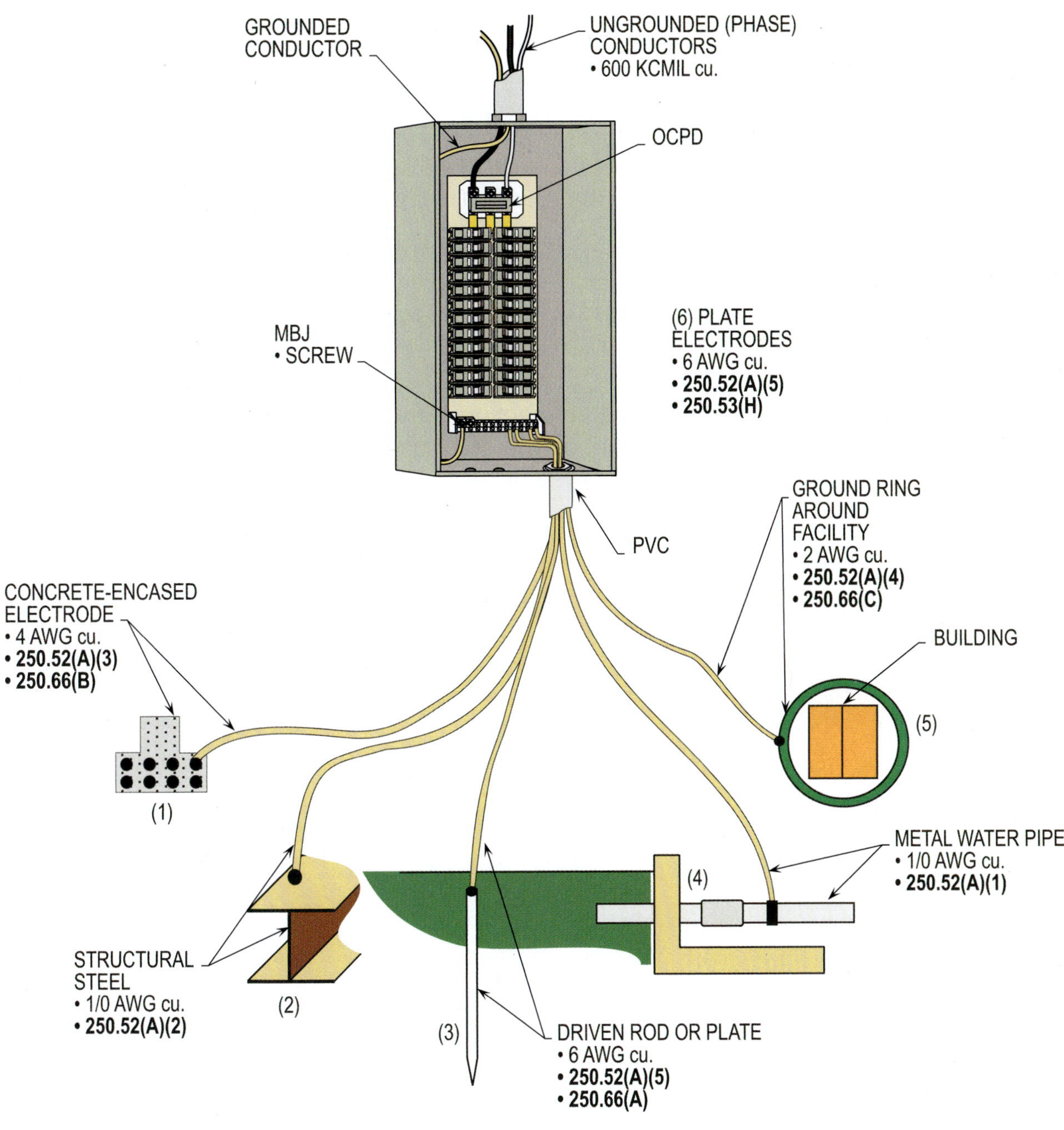

GROUNDING ELECTRODE
ARTICLE 100

Purpose of Change: To revise the definition of grounding electrode to apply generally throughout the NEC and describe the purpose and function of the grounding electrode.

NEC Ch. 1 - Article 100
Part I - Grounding Electrode Conductor

Type of Change	Panel Action	UL	UL 508	API 500 - 1997	API 505 - 1997	OSHA - 1994
Revision	Accept	-	-	-	-	1910.399(a)
ROP	**ROC**	**NFPA 70E - 2004**	**NFPA 70B - 2006**	**NFPA 79 - 2007**	**NFPA**	**NEMA**
pg. 19 / # 5-18	pg. 13 / # 5-6	Article 100	3.3.39	-	-	-
log: 3554 / CMP: 1	log: 2079	Submitter: Paul Dobrowsky		2005 NEC: Article 100		IEC: -

2005 NEC - Article 100

Grounding Electrode Conductor. ~~The~~ conductor used to connect ~~the grounding electrode(s) to the equipment grounding conductor, to the grounded conductor, or to both, at the service, at each building or structure where supplied by a feeder(s) or branch circuit(s), or at the source of a separately derived system.~~

2008 NEC - Article 100

Grounding Electrode Conductor. A conductor used to connect the system grounded conductor or the equipment to a grounding electrode or to a point on the grounding electrode system.

Author's Comment: This revision covers the connection of the grounding electrode conductor to the grounded conductor, the equipment, or a point on the grounding electrode system.

There are times where a grounding electrode conductor is connected to equipment, such as permitted where an auxiliary (supplementary) electrode is installed in accordance with **250.54**. An example would be a driven ground rod at a metal lighting pole in a parking lot where the ground rod is used as an auxiliary (supplementary) grounding electrode.

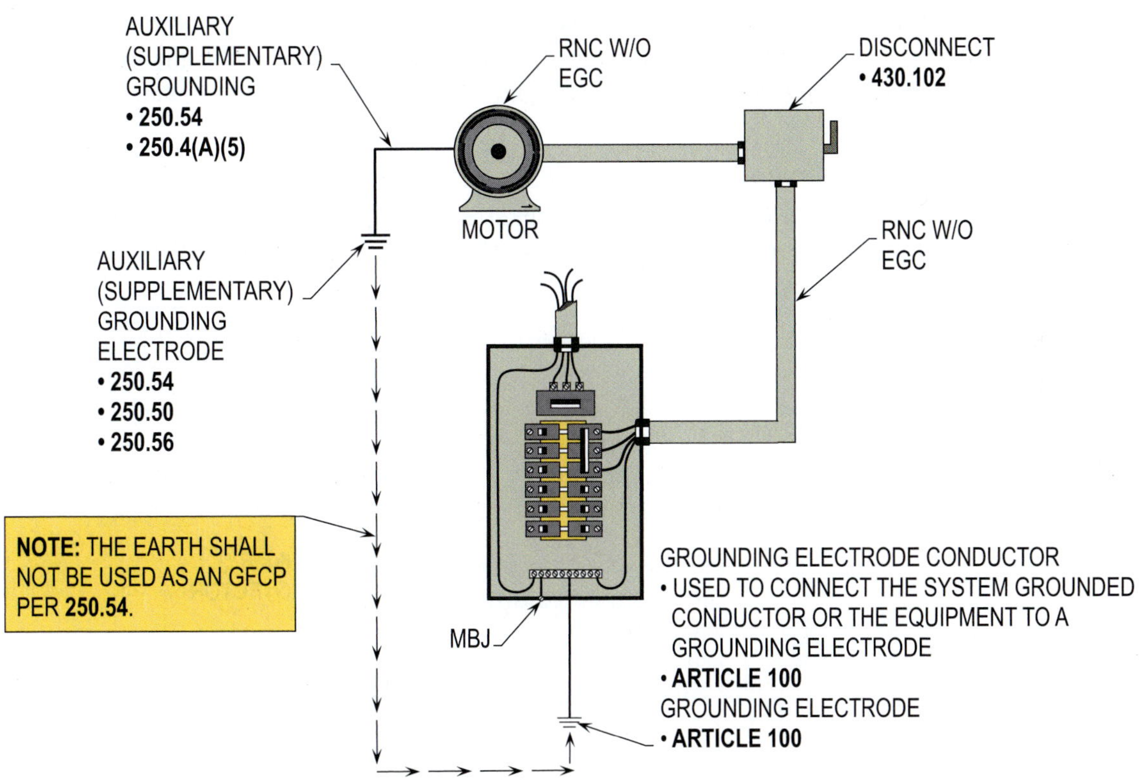

GROUNDING ELECTRODE CONDUCTOR
ARTICLE 100

Purpose of Change: To revise the definition of grounding electrode conductor to apply generally throughout the NEC and describe the purpose and function of the grounding electrode conductor.

NEC Ch. 1 - Article 100
Part I - Intersystem Bonding Termination

Type of Change		Panel Action		UL	UL 508	API 500 - 1997	API 505 - 1997	OSHA - 1994
New Definition		Accept in Principle		-	-	-	-	-
ROP		ROC		NFPA 70E - 2004	NFPA 70B - 2006	NFPA 79 - 2007	NFPA	NEMA
pg. 20	# 5-20	pg. 13	# 5-11	-	-	-	-	-
log: 1885	CMP: 1	log: 1670	Submitter: Jeffrey Boksiner			2005 NEC: -		IEC: -

2008 NEC - Article 100

Intersystem Bonding Termination. A device that provides a means for connecting communications system(s) grounding conductor(s) and bonding conductor(s) at the service equipment or at the disconnecting means for buildings or structures supplied by a feeder or branch circuit.

Author's Comment: A new definition has been added to apply to connections for communications grounding and bonding at the service disconnecting means or at the disconnecting means for buildings or structures supplied by a feeder or branch circuit, such as those located in **225.30** and **225.32** for outside branch circuits and feeders.

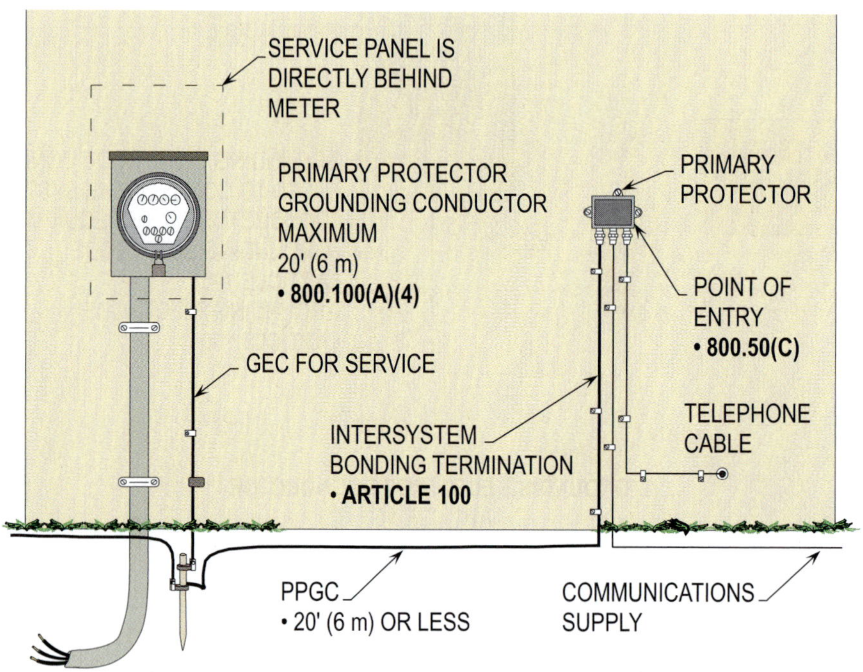

INTERSYSTEM BONDING TERMINATION
ARTICLE 100

Purpose of Change: A new definition, "intersystem bonding termination", has been added to apply to connections for grounding and bonding for communications systems at the service disconnecting means or at the feeder or branch circuit disconnecting means for buildings or structures.

NEC Ch. 1 - Article 100
Part I - Luminaire

Type of Change	Panel Action	UL	UL 508	API 500 - 1997	API 505 - 1997	OSHA - 1994
Revision	Accept	1598	-	-	-	-
ROP	**ROC**	**NFPA 70E - 2004**	**NFPA 70B - 2006**	**NFPA 79 - 2007**	**NFPA**	**NEMA**
pg. 22 # 18-4b	pg. - # -	-	-	-	-	-
log: CP 1800 CMP: 1	log: -	Submitter: Code Making Panel 18		2005 NEC: Article 100		IEC: -

2005 NEC - Article 100

Luminaire. A complete lighting unit consisting of a lamp or lamps together with the parts designed ~~to distribute the light,~~ to position and ~~protect the lamps and ballast (where applicable), and to~~ connect ~~the lamps~~ to the power supply.

2008 NEC - Article 100

Luminaire. A complete lighting unit consisting of a light source, such as a lamp or lamps, together with the parts designed to position the light source and connect it to the power supply. It may also include parts to protect the light source, ballast, or to distribute the light. A lampholder itself is not a luminaire.

Author's Comment: The rewording of the definition clarifies the fact that a lampholder is not a luminaire. The remainder of the text was changed for clarification.

NOTE: A LAMPHOLDER ITSELF IS NOT A LUMINAIRE.

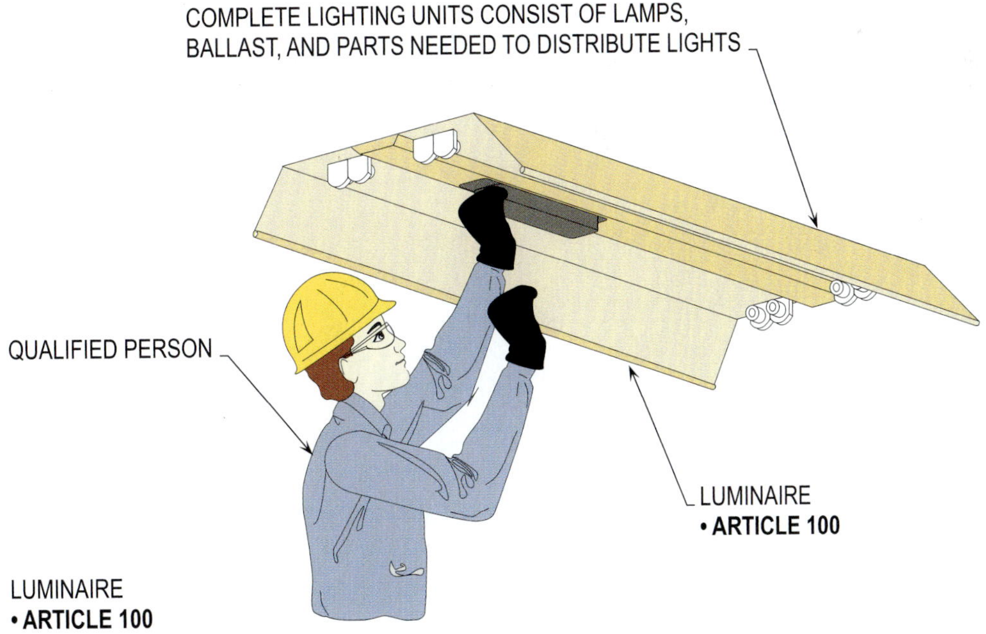

LUMINAIRE
ARTICLE 100

Purpose of Change: This revision clarifies that a lampholder is not a luminaire.

NEC Ch. 1 - Article 100
Part I - Neutral Conductor and Neutral Point

Type of Change		Panel Action		UL	UL 508	API 500 - 1997	API 505 - 1997	OSHA - 1994
New Definitions		Accept in Principle		-	-	-	-	-
ROP		ROC		NFPA 70E - 2004	NFPA 70B - 2006	NFPA 79 - 2007	NFPA	NEMA
pg. 26	# 5-36	pg. -	# -	-	-	-	-	-
log: 1554	CMP: 1	log: -	Submitter: Technical Correlating Committee on NEC			2005 NEC: -		IEC: -

2008 NEC - Article 100

Neutral Conductor. The conductor connected to the neutral point of a system that is intended to carry current under normal conditions.

Neutral point. The common point on a wye-connection in a polyphase system or midpoint on a single-phase, 3-wire system, or midpoint of a single-phase portion of a 3-phase delta system, or a midpoint of a 3-wire, direct current system.

FPN: At the neutral point of the system, the vectorial sum of the nominal voltages from all other phases within the system that utilize the neutral, with respect to the neutral point, is zero potential.

Author's Comment: Added two new definitions to provide information on what constitutes a neutral conductor and a neutral point.

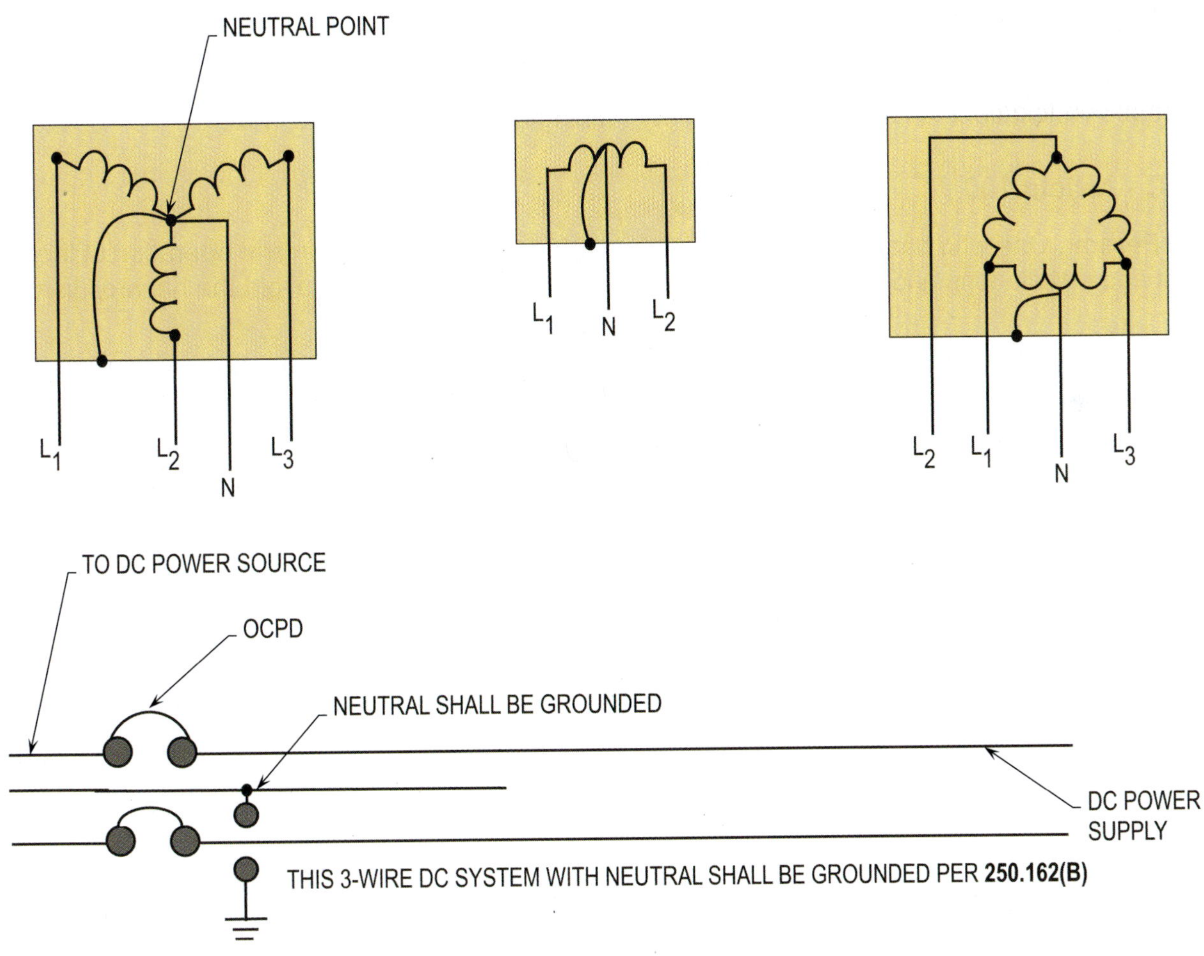

NEUTRAL CONDUCTOR AND NEUTRAL POINT ARTICLE 100

Purpose of Change: Two new definitions have been added to clarify what constitutes a neutral conductor and a neutral point.

NEC Ch. 1 - Article 100
Part I - Qualified Person

Type of Change	Panel Action	UL	UL 508	API 500 - 1997	API 505 - 1997	OSHA - 1994
Revision	Accept	-	-	-	-	1910.399(a)
ROP	**ROC**	**NFPA 70E - 2004**	**NFPA 70B - 2006**	**NFPA 79 - 2007**	**NFPA**	**NEMA**
pg. 27 # 1-45	pg. - # -	Article 100	3.3.58	3.3.79	-	-
log: 2589 CMP: 1	log: -	Submitter: Jebediah Novak		2005 NEC: Article 100		IEC: -

2005 NEC - Article 100

Qualified Person. One who has skills and knowledge related to the construction and operation of the electrical equipment and installations and has received safety training ~~on~~ the hazards involved.

2008 NEC - Article 100

Qualified Person. One who has skills and knowledge related to the construction and operation of the electrical equipment and installations and has received safety training to recognize and avoid the hazards involved.

Author's Comment: Adding the phrase "to recognize and avoid" to the definition of "qualified person" provides a reason for the safety training.

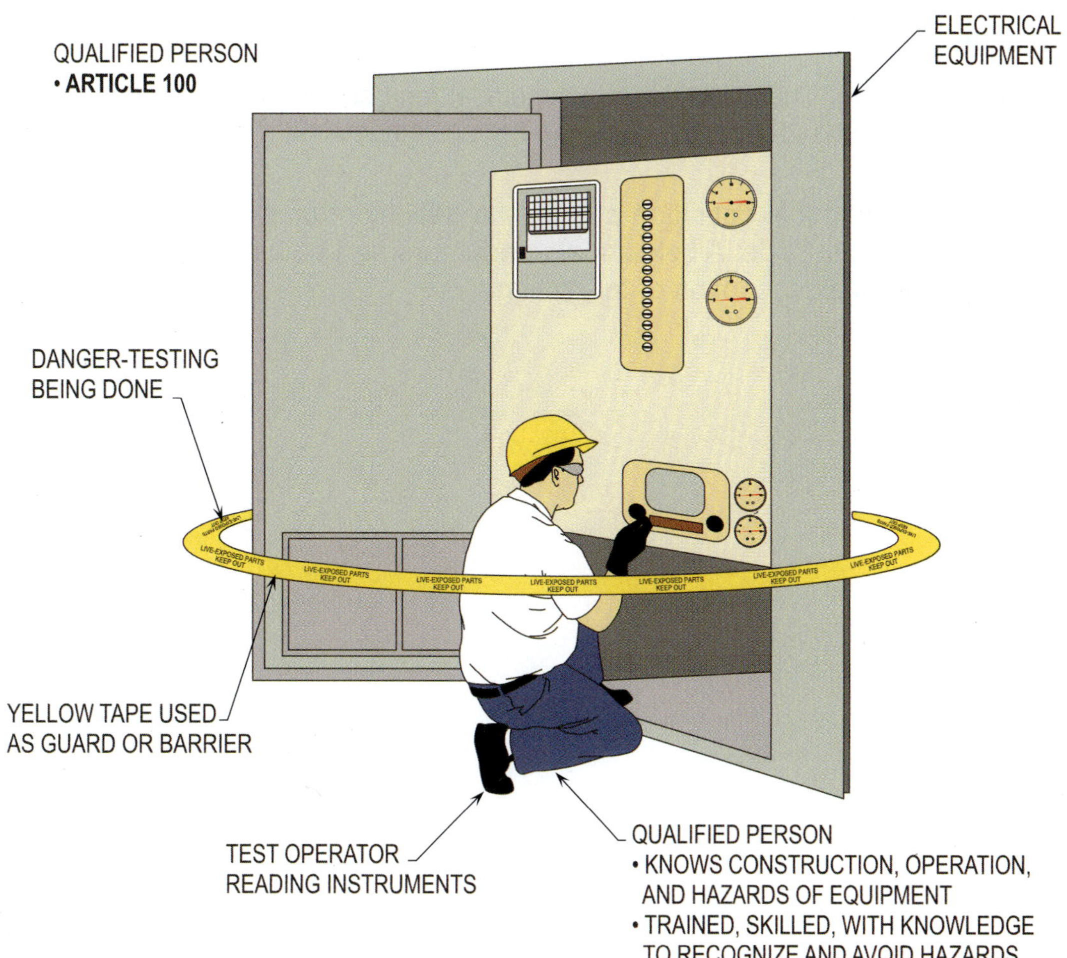

QUALIFIED PERSON
ARTICLE 100

Purpose of Change: This revision clarifies that qualified persons that have received safety training and knows how to recognize and avoid the hazards involved.

NEC Ch. 1 - Article 100
Part I - Short-Circuit Current Rating

Type of Change		Panel Action		UL	UL 508	API 500 - 1997	API 505 - 1997	OSHA - 1994
New Definition		Accept in Principle		-	-	-	-	-
ROP		**ROC**		**NFPA 70E - 2004**	**NFPA 70B - 2006**	**NFPA 79 - 2007**	**NFPA**	**NEMA**
pg. 28	# 10-2	pg. -	# -	-	-	-	-	-
log: 1743	CMP: 1	log: -	Submitter: David Sroka			2005 NEC: -		IEC: -

2008 NEC - Article 100

Short-Circuit Current Rating. The prospective symmetrical fault current at a nominal voltage to which an apparatus or system is able to be connected without sustaining damage exceeding defined acceptance criteria.

Author's Comment: Adding this definition provides a definition for the phrase used in 110.10 and various other locations in the Code. All electrical conductors and equipment have a withstand rating and this definition provides a description of this rating.

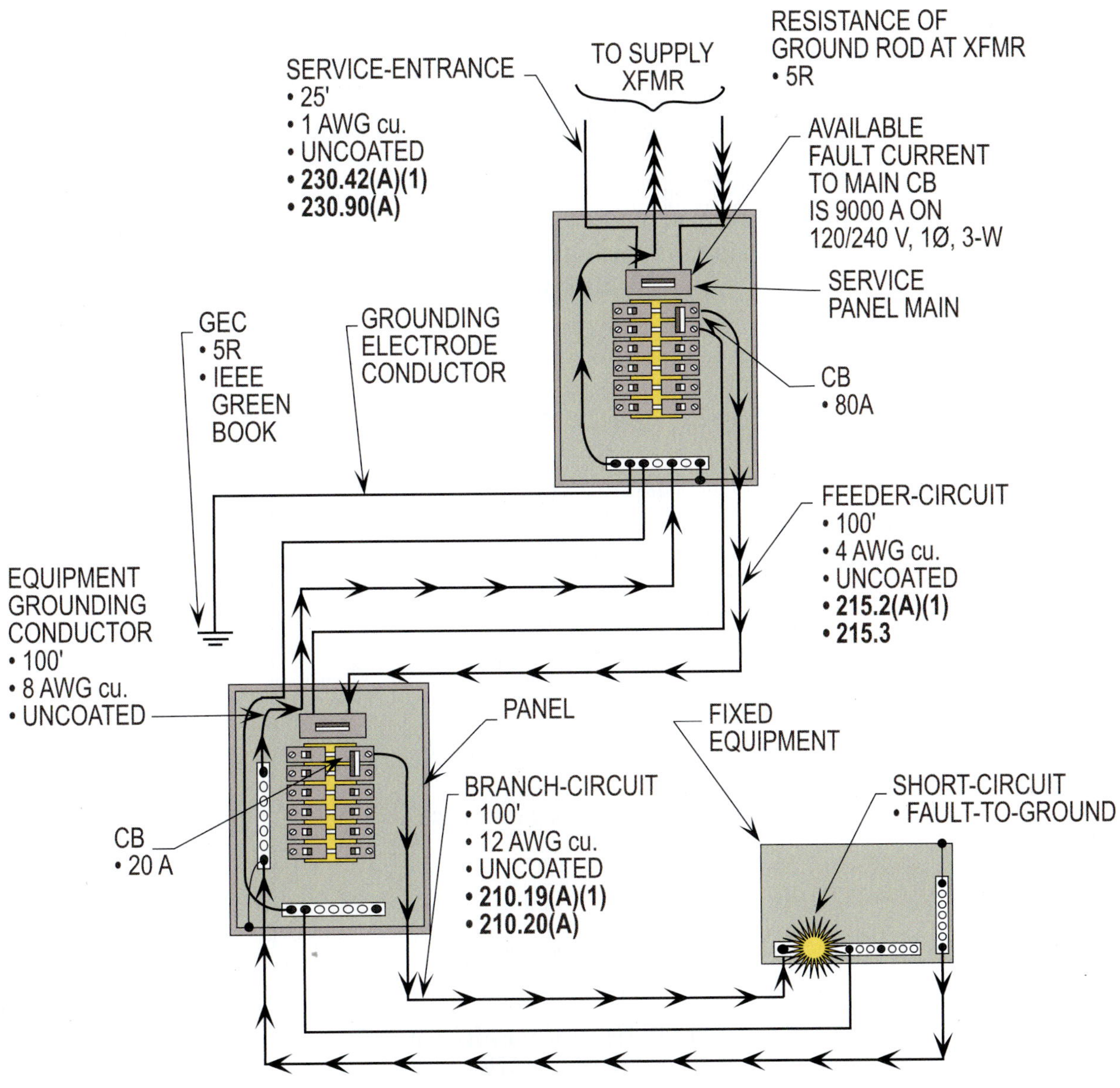

SHORT-CIRCUIT CURRENT RATING
ARTICLE 100

Purpose of Change: A new definition has been added for the phrase used in **110.10** and various other locations in the NEC.

FINDING TOTAL RESISTANCE
TABLE 8, CH. 9

Feeder-circuit conductors (HOT)

Step 1: Finding known values
Table 8, Ch. 9
100'
4 AWG cu. = 0.308 R

Step 2: Calculating resistance

$$R = \frac{100'}{1000'} \times 0.308\ R$$

R = 0.0308 (√)

Branch-circuit conductors (HOT)

Step 1: Finding known values
Table 8, Ch. 9
100'
12 AWG cu. = 1.93 R

Step 2: Calculating resistance

$$R = \frac{100'}{1000'} \times 1.93\ R$$

R = 0.193 R (√)

THE CURRENT FLOW THROUGH EARTH, IF EARTH WAS USED AS SOLE GROUND PATH

I = E ÷ R
I = 120 V ÷ 10R = 12 A

POINT OF FAULT AT EQUIPMENT

Branch-circuit conductor (EGC)

Step 1: Finding known values
Table 8, Ch. 9
100'
12 AWG cu. = 1.93 R

Step 2: Calculating resistance
$R = \frac{100'}{1000'} \times 1.93\ R$
R = 0.193 R (√)

Feeder-circuit conductors (EGC)

Step 1: Finding known values
Table 8, Ch. 9
100'
8 AWG cu. = 0.778 R

Step 2: Calculating resistance
$R = \frac{100'}{1000'} \times 0.778\ R$
R = 0.0778 R (√)

Total resistance

Step 1: Finding total resistance (√)
Branch-circuit = 0.193 R (HOT) + 0.193 R (EGC)
Branch-circuit = 0.386 R
Feeder-circuit = 0.0308 (HOT) + 0.0778 (EGC)
Feeder-circuit = 0.1086 R

Step 2: Calculating total resistance
Branch-circuit = 0.386 R
Feeder-circuit = 0.1086 R
Total = 0.4946

Calculating fault-current

I = E ÷ R

I = 120 ÷ 0.50 R

I = 240 A

Fault-current is equal to about 240 amps when 0.4946 R is rounded up to 0.50 R.

NOTE: *ALL WIRES ARE STRANDED*

NEC Ch. 1 - Article 100
Part - Surge Arrester and Surge Protective Devices (SPDs)

Type of Change	Panel Action	UL	UL 508	API 500 - 1997	API 505 - 1997	OSHA - 1994
New Definitions	Accept in Principle	1449	-	-	-	-
ROP	**ROC**	**NFPA 70E - 2004**	**NFPA 70B - 2006**	**NFPA 79 - 2007**	**IEEE**	**NEMA - 1999**
pg. 250 # 5-340	pg. - # -	-	-	-	C 62.1, 2, 11, and 22	LA 1
log: 2601 CMP: 5	log: -	Submitter: Joseph P. DeGregoria		2005 NEC: -		IEC: -

2008 NEC - Article 100

Surge Arrester. A protective device for limiting surge voltages by discharging or bypassing surge current, and it also prevents continued flow of follow current while remaining capable of repeating these functions.

Surge Protective Devices (SPDs). A protective device for limiting transient voltages by diverting or limiting surge current; it also prevents continued flow of follow current while remaining capable of repeating these functions and designated as follows:

Type 1- Permanently connected SPDs intended for installation between the secondary of the service transformer and the line side of the service disconnect overcurrent device.

Type 2- Permanently connected SPDs intended for installation on the load side of the service disconnect overcurrent device; including SPDs located at the branch panel.

Type 3 – Point of utilization SPDs.

Type 4 – Component SPDs, including discrete components, as well as assemblies.

FPN: For further information on Type 1, Type 2, Type 3, and Type 4 SPDs, see UL 1449, Standard for Surge Protective Devices.

Author's Comment: A new definition has been added to replace the term "transient voltage surge suppressor" used in **285.2**. The definition of surge arrester has been moved from **280.2** to **Article 100**. The new definition for surge protective device corresponds to a rewrite of UL 1449, **Article 280** for surge arresters, over 1 kV, and **Article 285** for transient voltage surge suppressors.

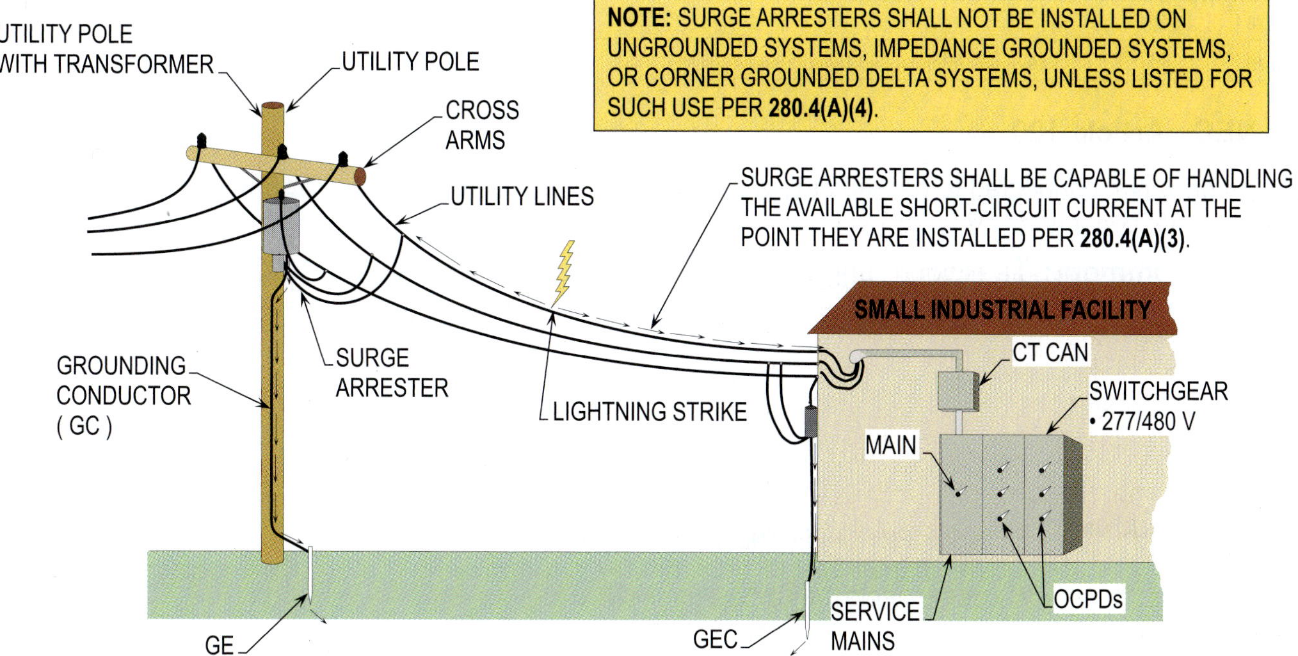
UTILITY POLE
WITH TRANSFORMER
UTILITY POLE
CROSS
ARMS
NOTE: SURGE ARRESTERS SHALL NOT BE INSTALLED ON UNGROUNDED SYSTEMS, IMPEDANCE GROUNDED SYSTEMS, OR CORNER GROUNDED DELTA SYSTEMS, UNLESS LISTED FOR SUCH USE PER 280.4(A)(4).
UTILITY LINES
SURGE ARRESTERS SHALL BE CAPABLE OF HANDLING THE AVAILABLE SHORT-CIRCUIT CURRENT AT THE POINT THEY ARE INSTALLED PER 280.4(A)(3).
SMALL INDUSTRIAL FACILITY
GROUNDING
CONDUCTOR
(GC)
SURGE
ARRESTER
LIGHTNING STRIKE
CT CAN
SWITCHGEAR
• 277/480 V
MAIN
SERVICE
MAINS
OCPDs
GE
GEC

SURGE PROTECTIVE DEVICES (SPDS)
TYPE 1: INSTALLED BETWEEN THE SECONDARY OF THE SERVICE TRANSFORMER AND THE LINE SIDE OF THE SERVICE DISCONNECT OVERCURRENT DEVICE.
TYPE 2: INSTALLED ON THE LOAD SIDE OF THE SERVICE DISCONNECT OVERCURRENT DEVICE; INCLUDED SPD'S LOCATED AT THE BRANCH PANEL.
TYPE 3: POINT OF UTILIZATION SPD'S.
TYPE 4: COMPONENT SPD'S, INCLUDING DISCRETE COMPONENTS, AS WELL AS ASSEMBLIES.

SURGE ARRESTER AND
SURGE PROTECTION DEVICES (SPDS)
ARTICLE 100

Purpose of Change: A new definition has been added to replace the term "transient voltage surge suppressor" used in **285.2**.

NEC Ch. 1 - Article 100
Part I - Ungrounded

Type of Change		Panel Action		UL	UL 508	API 500 - 1997	API 505 - 1997	OSHA - 1994
New Definition		Accept		-	-	-	-	-
ROP		ROC		NFPA 70E - 2004	NFPA 70B - 2006	NFPA 79 - 2007	NFPA	NEMA
pg. 28	# 5-38	pg. -	# -	-	-	-	-	-
log: 1518	CMP: 1	log: -	Submitter: Technical Correlating Committee on NEC			2005 NEC: -		IEC: -

2008 NEC - Article 100

Ungrounded. Not connected to ground or to a conductive body that extends the ground connection.

Author's Comment: The word "ungrounded" is used extensively in the NEC, and was without a definition previously.

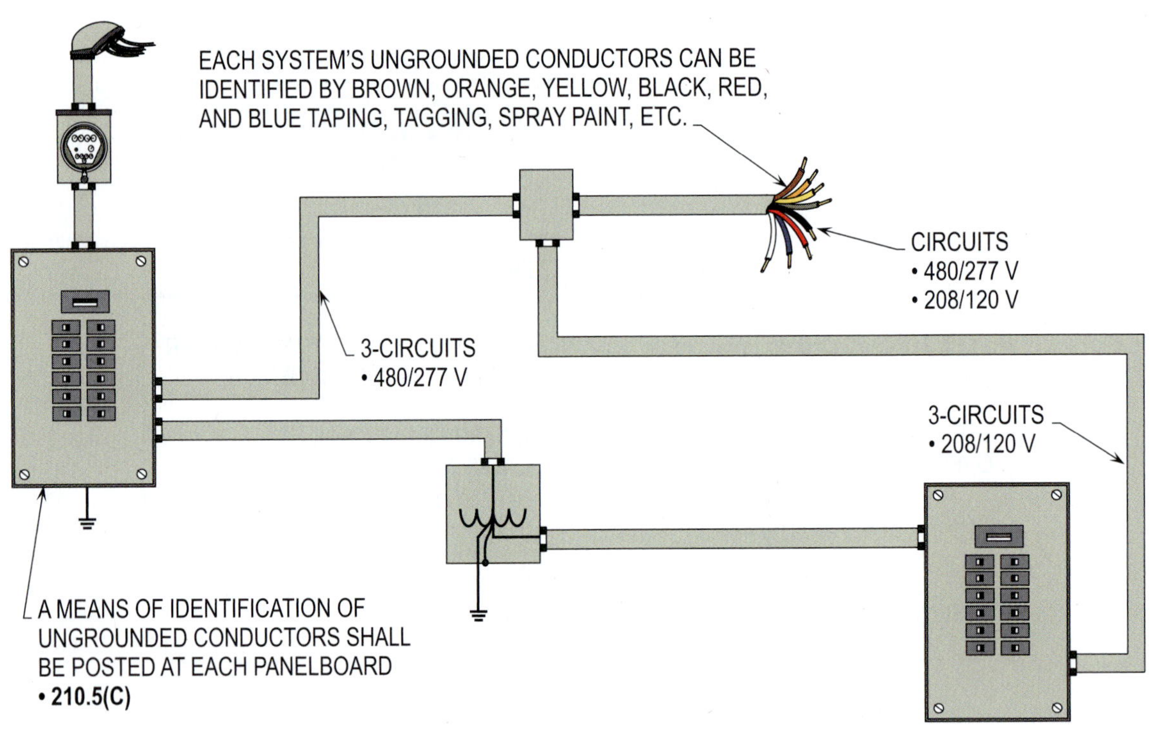
EACH SYSTEM'S UNGROUNDED CONDUCTORS CAN BE IDENTIFIED BY BROWN, ORANGE, YELLOW, BLACK, RED, AND BLUE TAPING, TAGGING, SPRAY PAINT, ETC.
CIRCUITS
• 480/277 V
• 208/120 V
3-CIRCUITS
• 480/277 V
3-CIRCUITS
• 208/120 V
A MEANS OF IDENTIFICATION OF UNGROUNDED CONDUCTORS SHALL BE POSTED AT EACH PANELBOARD
• 210.5(C)

UNGROUNDED
ARTICLE 100

Purpose of Change: A new definition has been added to apply generally throughout the NEC and describe the purpose and function of the term "ungrounded."

NEC Ch. 1 - Article 100
Part I - Utility-Interactive Inverter

Type of Change		Panel Action		UL	UL 508	API 500 - 1997	API 505 - 1997	OSHA - 1994
New Definition		Accept in Principle		1741	-	-	-	-
ROP		ROC		NFPA 70E - 2004	NFPA 70B - 2006	NFPA 79 - 2007	NFPA	NEMA
pg. 29	# 13-3	pg. -	# -	-	-	-	-	-
log: 2580	CMP: 1	log: -	Submitter: Timothy M. Croushore			2005 NEC: -		IEC: -

2008 NEC - Article 100

Utility-Interactive Inverter. An inverter intended for use in parallel with an electric utility to supply common loads that may deliver power to the utility.

Author's Comment: The purpose of this change is to add a new definition of Utility-Interactive Inverter that will be used in **Articles 690**, **692**, and **705**. Code Making Panel 13 will be given authority for this definition to appear in Article 100. The definition is the same as the definition in Section 2.41 of Underwriters Laboratory Standard 1741 – Inverters, Converters and Controllers for Use in Independent Power Systems. This new term used in the phrase of "Interactive System" in **Articles 100**, **690**, **692** and **705**. This change is as part of a re-write of **Articles 690**, **692** and **705** with respect to the interconnection of systems and equipment for use with distributed energy resources.

UL 1741 is currently under revision with a title change from "Inverters, Converters, and Controllers for Use in Independent Power Systems" to "Inverters, Converters, Controllers and Interconnection Systems Equipment Use with Distributed Energy Resources." The definition of Utility-Interactive Inverter remains unchanged in the UL standard.

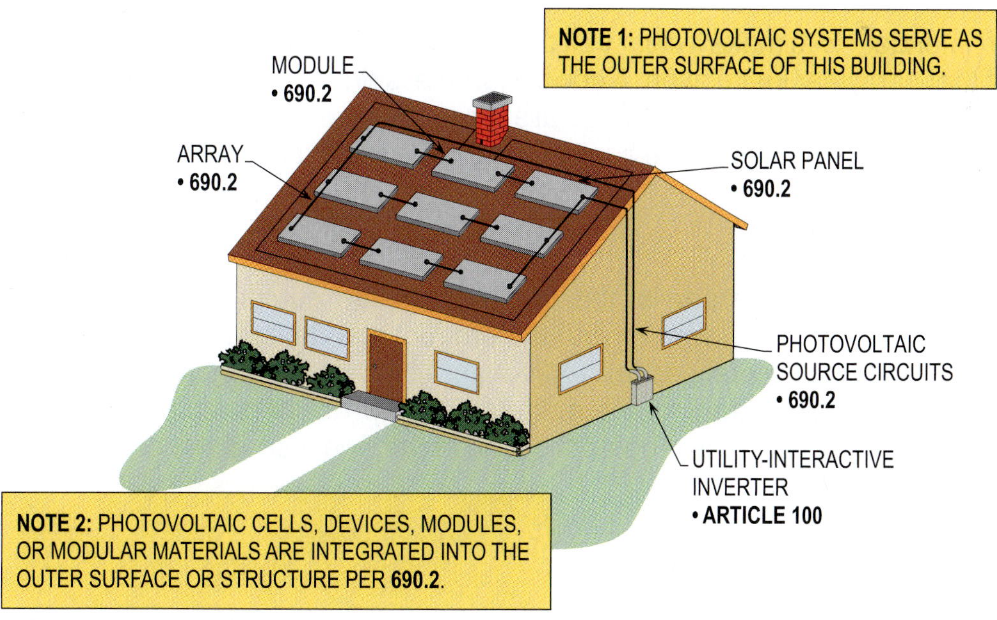

UTILITY-INTERACTIVE INVERTER
ARTICLE 100

Purpose of Change: A new definition has been added as part of a re-write of **Articles 690**, **692** and **705** with respect to the interconnection of systems and equipment for use with distributed energy resources.

NEC Ch. 1 - Article 100
Part I - 110.12(A)

Type of Change	Panel Action	UL	UL 508	API 500 - 1997	API 505 - 1997	OSHA - 1994
Revision	Accept	-	-	-	-	-
ROP	**ROC**	**NFPA 70E - 2004**	**NFPA 70B - 2006**	**NFPA 79 - 2007**	**NFPA**	**NEMA**
pg. 34 # 1-71	pg. 21 # 1-50	400.8(A)	-	11.4	-	-
log: 2677 CMP: 1	log: 834 Submitter: Dorothy Kellogg			2005 NEC: 110.12(A)		IEC: -

2005 NEC - 110.12

(A) Unused Openings. Unused ~~cable or raceway~~ openings ~~in boxes, raceways, auxiliary gutters, cabinets, cutout boxes, meter socket enclosures, equipment cases, or housings~~ shall be ~~effectively~~ closed to afford protection substantially equivalent to the wall of the equipment. Where metallic plugs or plates are used with nonmetallic enclosures, they shall be recessed at least 6 mm (1/ 4 in.) from the outer surface of the enclosure.

2008 NEC - 110.12

(A) Unused Openings. Unused openings, other than those intended for the operation of equipment, those intended for mounting purposes, or those permitted as part of the design for listed equipment, shall be closed to afford protection substantially equivalent to the wall of the equipment. Where metallic plugs or plates are used with nonmetallic enclosures, they shall be recessed at least 6 mm (1/ 4 in.) from the outer surface of the enclosure.

Author's Comment: The change indicates all unused openings should be closed except for those required for the functional operation of the equipment or enclosure or for the proper mounting of such equipment or enclosure.

The addition of "intended" is to make it clear that there may be openings that are there, but not necessarily used as part of that particular installation. A good example is drainage openings in a Type 3R enclosure that is installed indoors. Clearly the openings are there, but are not used in that application. Another example is mounting holes that are in the back of the enclosure, but are not used because the enclosure is mounted from the side.

The new words "or permitted as part of the design for listed equipment" is intended to address a conflict that is created by the new wording and the product standards. For example, UL 50 - Standard for Enclosures has specific allowances for additional (albeit small) openings in an enclosure. These openings may have been necessary for the manufacturing process (such as drain openings for paint during the painting process) and have no application in the final use of the product.

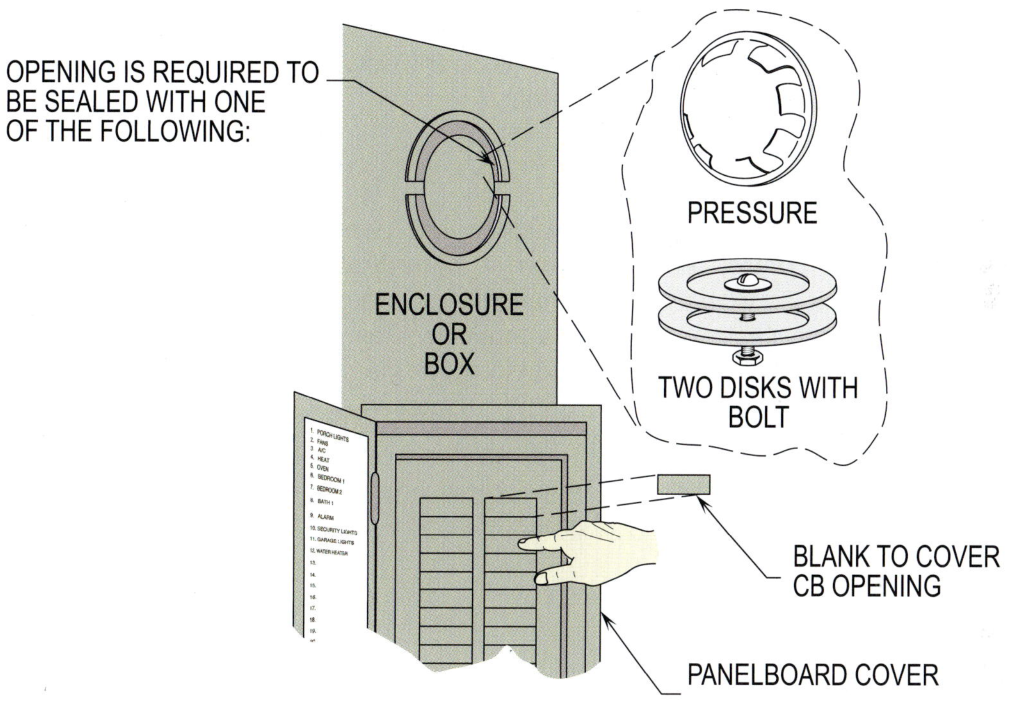

UNUSED OPENINGS
110.12(A)

Purpose of Change: This revision indicates all unused openings should be closed, except for those required for the proper mounting of such equipment or enclosure.

NEC Ch. 1 - Article 110
Part I - 110.16

Type of Change	Panel Action	UL	UL 508	API 500 - 1997	API 505 - 1997	OSHA - 1994
Revision	Accept	-	-	-	-	-
ROP	**ROC**	**NFPA 70E - 2004**	**NFPA 70B - 2006**	**NFPA 79 - 2007**	**ANSI - 2006**	**NEMA**
pg. 37 # 1-84	pg. - # -	400.11	7.6	-	Z535.6	-
log: 2338 CMP: 1	log: -	Submitter: Larry G. McManhill		2005 NEC: 110.16		IEC: -

2005 NEC - 110.16

Flash Protection. Switchboards, panelboards, industrial control panels, meter socket enclosures, and motor control centers that are in other than dwelling occupancies and are likely to require examination, adjustment, servicing, or maintenance while energized shall be field marked to warn qualified persons of potential electric arc flash hazards. The marking shall be located so as to be clearly visible to qualified persons before examination, adjustment, servicing, or maintenance of the equipment.

2008 NEC - 110.16

Flash Protection. Electrical equipment, such as switchboards, panelboards, industrial control panels, meter socket enclosures, and motor control centers that are in other than dwelling occupancies and are likely to require examination, adjustment, servicing, or maintenance while energized shall be field marked to warn qualified persons of potential electric arc flash hazards. The marking shall be located so as to be clearly visible to qualified persons before examination, adjustment, servicing, or maintenance of the equipment.

Author's Comment: Adding the words "electrical equipment, such as" now requires flash protection marking for large fusible switches and similar electrical equipment where a flash hazard may exist, such as a 1600 amp fusible wall-mounted switch.

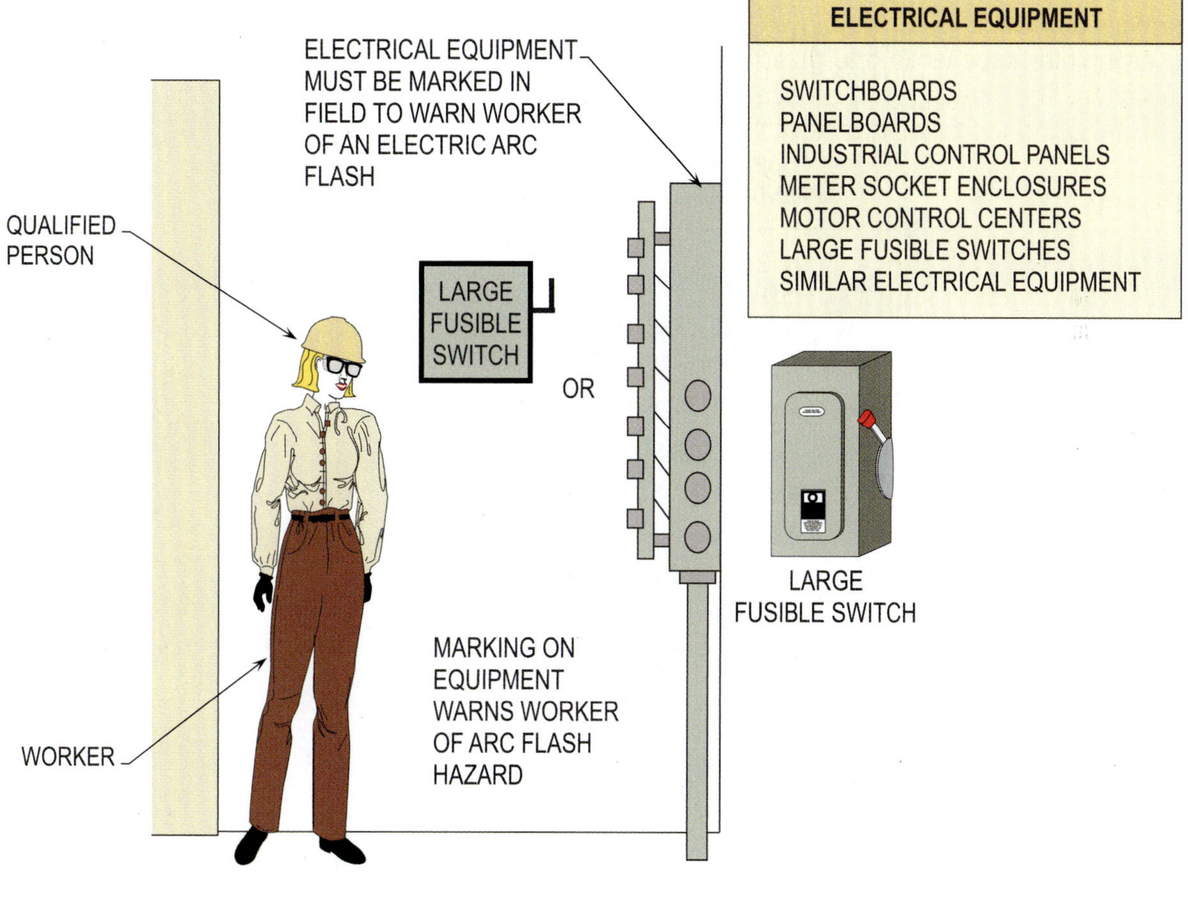

FLASH PROTECTION
110.16

Purpose of Change: This revision now clarifies that flash protection markings are required for large fusible switches and similar equipment where a flash hazard may exist.

GIVEN:
APPENDIX, B-5.2 IN NFPA 70E

- E_{MB} = $1038.7\ D_B^{-1.4738}\ t_A\ (.0093\ F^2 - .3453F + .59675)$
- D_B = 18 INCHES
- t_A = 1/2 CYCLE OR .0084 SEC.
- F = 35 kA
- MVA = 11

APPLICATION OF FORMULA:
APPENDIX B-5.2 IN PART II OF NFPA 70E

Step 1: $E_{MB} = (1038.7)(0.014125)(.0084)(11.3925 - 12.0855 + 5.9675)$

Step 2: $E_{MB} = (14.671638)(.0084)(5.2745)$

Step 3: $E_{MB} = (14.671638)(.0443)$

Step 4: $E_{MB} = 0.649954\ cal/cm^2$

Solution: From Table 3-3.9.3 in Part II of NFPA 70E, 0.649954 cal/cm^2 requires a hazard risk category of 0 with one layer of untreated cotton (total weight oz/yd^2 is 4.5-7).

Note: This procedure is used only for 600 volts or less systems.

(Review NFPA 70E carefully)

Step 5: Finding flash protection boundary

$D_C = \sqrt{(53 \times MVA \times t)}$

$D_C = \sqrt{(53 \times 11 \times .0084)}$

$D_C = 2.21$ ft.

Solution: Based on the given parameters, the authors suggest an arc-flash boundary of 2.21 ft and an incident energy value of 0.649954 cal/cm^2.

Note: Authors recommend that only engineers or qualified personnel calculate and determine the boundary.

NEC Ch. 1 - Article 110
Part I - 110.20

Type of Change	Panel Action	UL	UL 508	API 500 - 1997	API 505 - 1997	OSHA - 1994
Relocated - Renumbered	Accept in Principle in Part	-	-	-	-	-
ROP	**ROC**	**NFPA 70E - 2004**	**NFPA 70B - 2006**	**NFPA 79 - 2007**	**NFPA**	**NEMA - 2003**
pg. 41 # 1-95	pg. 24 # 1-64	-	-	-	-	250
log: 1980 CMP: 1	log: 976 Submitter: Vince Baclawski			2005 NEC: -		IEC: -

2008 NEC - 110.20

Enclosure Types. Enclosures (other than surrounding fences or walls) of switchboards, panelboards, industrial control panels, motor control centers, meter sockets, and motor controllers, rated not over 600 volts nominal and intended for such locations, shall be marked with an enclosure type number as shown in Table 110.20.

Table 110.20 shall be used for selecting enclosures for use in specific locations, other than hazardous (classified) locations. The enclosures are not intended to protect against conditions, such as condensation, icing, corrosion, or contamination that may occur within the enclosure or enter via the conduit or unsealed openings.

Author's Comment: Table 430.91, covering motor controller enclosure types, has been used by many NEC users as enclosure types for numerous kinds of equipment, even though it stated in **430.91** that it only applied to motor controller enclosures. Moving the requirements of **430.91** and **Table 430.91** into a general application area of the NEC and specifically stating the kinds of equipment to which they apply will add clarity.

FOR OUTDOOR USE										
PROVIDES A DEGREE OF PROTECTION AGAINST THE FOLLOWING ENVIRONMENTAL CONDITIONS	ENCLOSURE TYPE NUMBER									
	3	3R	3S	3X	3RX	3SX	4	4X	6	6P
INCIDENTAL CONTACT WITH THE ENCLOSED EQUIPMENT	X	X	X	X	X	X	X	X	X	X
RAIN, SNOW, AND SLEET	X	X	X	X	X	X	X	X	X	X
SLEET[1]	_	_	X	_	_	X	_	_	_	_
WINDBLOWN DUST	X	_	X	X	_	X	X	X	X	X
HOSEDOWN	_	_	_	_	_	_	X	X	X	X
CORROSIVE AGENTS	_	_	_	X	X	X	_	X	_	X
TEMPORARY SUBMERSION	_	_	_	_	_	_	_	_	X	X
PROLONGED SUBMERSION	_	_	_	_	_	_	_	_	_	X

FOR INDOOR USE										
PROVIDES A DEGREE OF PROTECTION AGAINST THE FOLLOWING ENVIRONMENTAL CONDITIONS	**ENCLOSURE TYPE NUMBER**									
	1	**2**	**4**	**4X**	**5**	**6**	**6P**	**12**	**12K**	**13**
INCIDENTAL CONTACT WITH THE ENCLOSED EQUIPMENT	X	X	X	X	X	X	X	X	X	X
FALLING DIRT	X	X	X	X	X	X	X	X	X	X
FALLING LIQUIDS AND LIGHT SPLASHING	_	X	X	X	X	X	X	X	X	X
CIRCULATING DUST, LINT, FIBERS, AND FLYINGS	_	_	X	X	_	X	X	X	X	X
SETTLING AIRBORNE DUST, LINT, FIBERS, AND FLYINGS	_	_	X	X	X	X	X	X	X	X
HOSEDOWN AND SPLASHING WATER	_	_	X	X	_	X	X	_	_	_
OIL AND COOLANT SEEPAGE	_	_	_	_	_	_	_	X	X	X
OIL OR COOLANT SPRAYING AND SPLASHING	_	_	_	_	_	_	_	_	_	X
CORROSIVE AGENTS	_	_	_	X	_	_	X	_	_	_
TEMPORARY SUBMERSION	_	_	_	_	_	X	X	_	_	_
PROLONGED SUBMERSION	_	_	_	_	_	_	X	_	_	_

[1]Mechanism shall be operable when ice-covered.

FPN: The term *raintight* is typically used in conjunction with Enclosure Types 3, 3S, 3SX, 3X, 4, 4X, 6, 6P. The term *rainproof* is typically used in conjunction with Enclosure Types 3R, 3RX. The term *driptight* is typically used in conjunction with Enclosure Types 2, 5, 12, 12K, 13. The term *dusttight* is typically used in conjunction with Enclosure Types 3, 3S, 3SX, 5, 12, 12K, 13.

ENCLOSURE TYPES
110.20

Purpose of Change: The requirements of **430.91** and **Table 430.91** have been relocated to **Article 110** in the general application area of the NEC and specifically stating the kinds of equipment to which it applies.

NEC Ch. 1 - Article 110
Part I - 110.22(B)

Type of Change	Panel Action	UL	UL 508	API 500 - 1997	API 505 - 1997	OSHA - 1994
Revision	Accept	-	-	-	-	1910.303(f)
ROP	**ROC**	**NFPA 70E - 2004**	**NFPA 70B - 2006**	**NFPA 79 - 2007**	**NFPA**	**NEMA**
pg. 43 # 1-98	pg. 25 # 1-70	400.14(A); (B)	-	-	-	-
log: 1978 CMP: 1	log: 350	Submitter: Vince Baclawski		2005 NEC: 110.22		IEC: -

2005 NEC - 110.22

Identification of Disconnecting Means. Each disconnecting means shall be legibly marked to indicate its purpose unless located and arranged so the purpose is evident. The marking shall be of sufficient durability to withstand the environment involved.

Where circuit breakers or fuses are applied in compliance with the series combination ratings marked on the equipment by the manufacturer, the equipment enclosure(s) shall be legibly marked in the field to indicate the equipment has been applied with a series combination rating. The marking shall be readily visible and state the following:

CAUTION __ SERIES COMBINATION SYSTEM

RATED__ AMPERES. IDENTIFIED

REPLACEMENT COMPONENTS REQUIRED.

FPN: See 240.86(B) ~~for interrupting rating marking for end-use equipment~~

2008 NEC - 110.22

Identification of Disconnecting Means.

(A) General. Each disconnecting means shall be legibly marked to indicate its purpose unless located and arranged so the purpose is evident. The marking shall be of sufficient durability to withstand the environment involved.

(B) Engineered Series Combination Systems. Where circuit breakers or fuses are applied in compliance with the series combination ratings selected under engineer supervision and marked on the equipment as directed by the engineer, the equipment enclosure(s) shall be legibly marked in the field to indicate the equipment has been applied with a series combination rating. The marking shall be readily visible and state the following:

CAUTION - ENGINEERED SERIES COMBINATION

SYSTEM RATED__ AMPERES.

IDENTIFIED REPLACEMENT COMPONENTS REQUIRED.

FPN: See 240.86(A) for Engineered Series Combination Systems.

(C) Tested Series Combination Systems. Where circuit breakers or fuses are applied in compliance with the series combination ratings marked on the equipment by the manufacturer, the equipment enclosure(s) shall be legibly marked in the field to indicate the equipment has been applied with a series combination rating. The marking shall be readily visible and state the following:

CAUTION - SERIES COMBINATION SYSTEM

RATED__ AMPERES. IDENTIFIED

REPLACEMENT COMPONENTS REQUIRED.

FPN: See 240.86(B) Tested Series Combination Systems.

Author's Comment: The addition of the text in **110.22** for series-combination ratings selected under engineering supervision in existing installations provides coordination between Sections **110.22** and **240.86(A)**, where permission for series rated systems selected by calculation was added in the 2005 NEC.

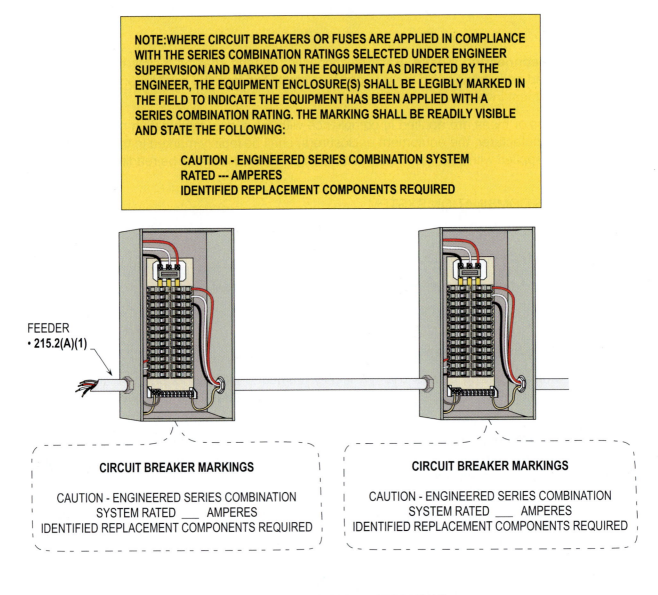

NOTE:WHERE CIRCUIT BREAKERS OR FUSES ARE APPLIED IN COMPLIANCE WITH THE SERIES COMBINATION RATINGS SELECTED UNDER ENGINEER SUPERVISION AND MARKED ON THE EQUIPMENT AS DIRECTED BY THE ENGINEER, THE EQUIPMENT ENCLOSURE(S) SHALL BE LEGIBLY MARKED IN THE FIELD TO INDICATE THE EQUIPMENT HAS BEEN APPLIED WITH A SERIES COMBINATION RATING. THE MARKING SHALL BE READILY VISIBLE AND STATE THE FOLLOWING:

CAUTION - ENGINEERED SERIES COMBINATION SYSTEM
RATED --- AMPERES
IDENTIFIED REPLACEMENT COMPONENTS REQUIRED

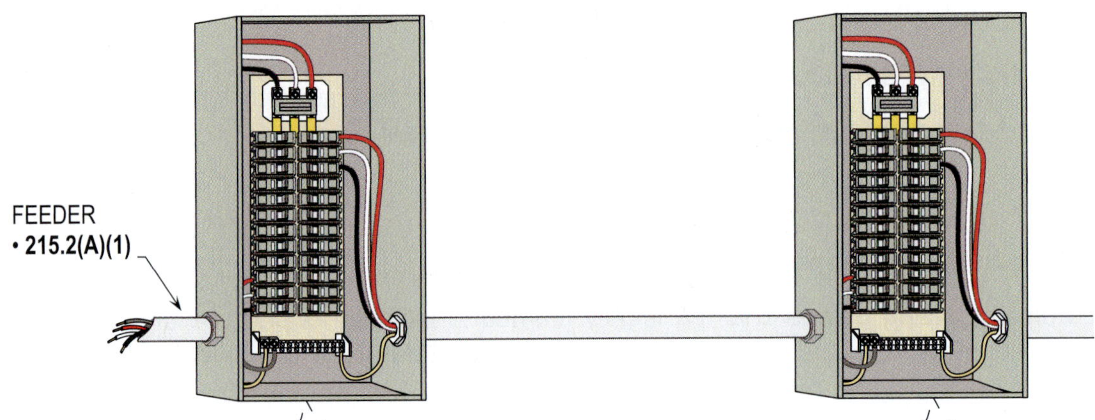

CIRCUIT BREAKER MARKINGS

CAUTION - ENGINEERED SERIES COMBINATION
SYSTEM RATED ___ AMPERES
IDENTIFIED REPLACEMENT COMPONENTS REQUIRED

CIRCUIT BREAKER MARKINGS

CAUTION - ENGINEERED SERIES COMBINATION
SYSTEM RATED ___ AMPERES
IDENTIFIED REPLACEMENT COMPONENTS REQUIRED

IDENTIFICATION OF DISCONNECTING MEANS
110.22(B)

Purpose of Change: This revision clarifies that series-combination ratings selected under engineering supervision in accordance with **240.86** in existing installations shall be identified.

NEC Ch. 1 - Article 110
Part II - 110.26(C)(1) and (C)(2)

Type of Change	Panel Action	UL	UL 508	API 500 - 1997	API 505 - 1997	OSHA - 1994
Revision	Accept in Part	-	-	-	-	1910.303(g)(1)(iii)
ROP	**ROC**	**NFPA 70E - 2004**	**NFPA 70B - 2006**	**NFPA 79 - 2007**	**NFPA**	**NEMA**
pg. 48 # 1-119	pg. 29 # 1-84	400.15(C)(1); (C)(2)	-	-	-	-
log: 2169 CMP: 1	log: 389	Submitter: Noel Williams		2005 NEC: 110.26(C)(1); (C)(2)		IEC: -

2005 NEC - 110.26

(C) Entrance to Working Space.

(1) Minimum Required. At least one entrance of sufficient area shall be provided to give access to working space about electrical equipment.

(2) Large Equipment. For equipment rated 1200 amperes or more that contains overcurrent devices, switching devices, or control devices, there shall be one entrance to the required working space not less than 610 mm (24 in.) wide and 2.0 m (6 1/ 2 ft) high at each end of the working space. Where the entrance has a personnel door(s), the door(s) shall open in the direction of egress and be equipped with panic bars, pressure plates, or other devices that are normally latched but open under simple pressure.

A single entrance to the required working space shall be permitted where either of the conditions in 110.26(C)(2)(a) or (C)(2)(b) is met.

(a) Unobstructed ~~Exit~~. Where the location permits a continuous and unobstructed way of ~~exit~~ travel, a single entrance to the working space shall be permitted.

(b) Extra Working Space. Where the depth of the working space is twice that required by 110.26(A)(1), a single entrance shall be permitted. It shall be located so that the distance from the equipment to the nearest edge of the entrance is not less than the minimum clear distance specified in Table 110.26(A)(1) for equipment operating at that voltage and in that condition.

2008 NEC - 110.26

(C) Entrance to and Egress From Working Space.

(1) Minimum Required. At least one entrance of sufficient area shall be provided to give access to and egress from working space about electrical equipment.

(2) Large Equipment. For equipment rated 1200 amperes or more and over 1.8 m or (6 ft) wide that contains overcurrent devices, switching devices, or control devices, there shall be one entrance to and egress from the required working space not less than 610 mm (24 in.) wide and 2.0 m (6 1/ 2 ft) high at each end of the working space.

A single entrance to and egress from the required working space shall be permitted where either of the conditions in 110.26(C)(2)(a) or (C)(2)(b) is met.

(a) Unobstructed Egress. Where the location permits a continuous and unobstructed way of egress travel, a single entrance to the working space shall be permitted.

(b) Extra Working Space. Where the depth of the working space is twice that required by 110.26(A)(1), a single entrance shall be permitted. It shall be located so that the distance from the equipment to the nearest edge of the entrance is not less than the minimum clear distance specified in Table 110.26(A)(1) for equipment operating at that voltage and in that condition.

Author's Comment: The previous Code text appeared to only address the entrance into a working space. By adding "egress from the working space" there will be equal emphasis on requiring sufficient room leaving the electrical equipment work space.

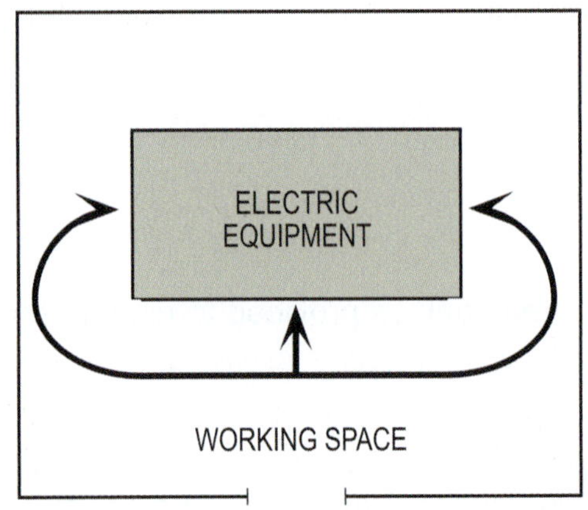

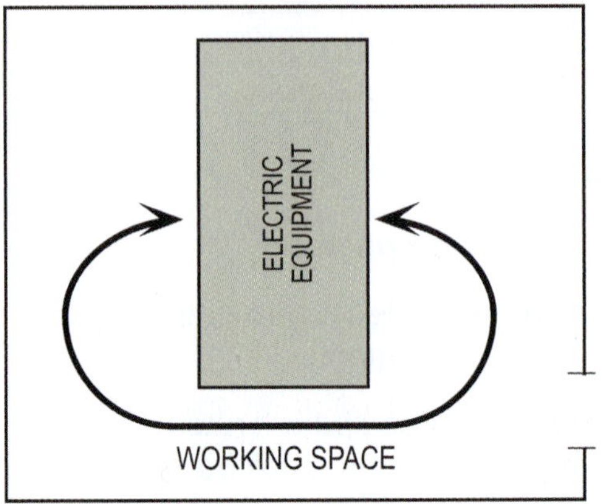

AT LEAST ONE ENTRANCE TO AND EGRESS FROM THE WORK SPACE SHALL BE REQUIRED TO PROVIDE ACCESS TO AND EGRESS FROM THE WORKING SPACE AROUND ELECTRIC EQUIPMENT.

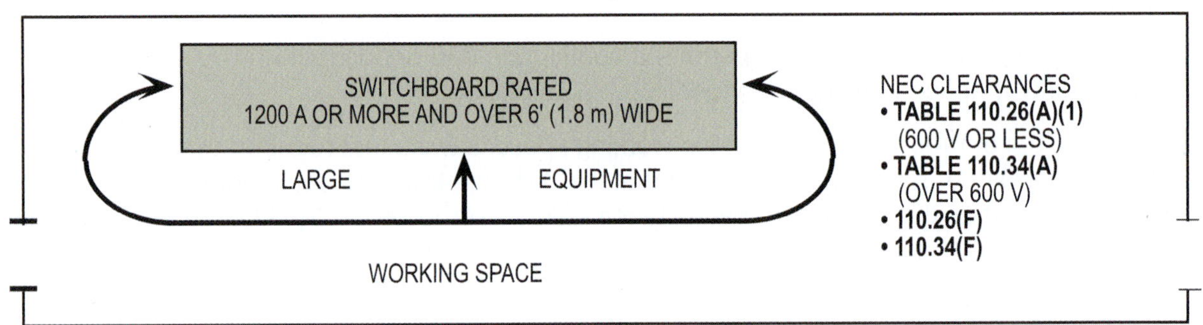

FOR EQUIPMENT RATED 1200 A OR MORE AND OVER 6' WIDE (1.8 m) THERE SHALL BE ONE ENTRANCE TO AND EGRESS FROM AND NOT LESS THAN 24" (610 mm) WIDE AND 6' 6" (2 m) HIGH AT EACH END OR WORKSPACE SHALL BE DOUBLED.

ENTRANCE TO AND EGRESS FROM WORKING SPACE
110.26(C)(1) and (C)(2)

Purpose of Change: This revision clarifies that by adding "egress from the working space" there will be equal emphasis on requiring sufficient room leaving the electrical equipment work space.

NEC Ch. 1 - Article 110
Part II - 110.26(C)(3)

Type of Change	Panel Action	UL	UL 508	API 500 - 1997	API 505 - 1997	OSHA - 1994
Revision	Accept in Principle in Part	-	-	-	-	1910.303(g)(1)(iii)
ROP	**ROC**	**NFPA 70E - 2004**	**NFPA 70B - 2006**	**NFPA 79 - 2007**	**NFPA**	**NEMA**
pg. 51 # 1-127	pg. 32 # 1-92	400.15(C)(1); (C)(2)	-	-	-	-
log: 3487 CMP: 1	log: 2330	Submitter: Alan Manche		2005 NEC: 110.26(C)(2)		IEC: -

2005 NEC - 110.26

(C) Entrance to Working Space

(2) Large Equipment. For equipment rated 1200 amperes or more that contains overcurrent devices, switching devices, or control devices, there shall be one entrance to the required working space not less than 610 mm (24 in.) wide and 2.0 m (6 1/ 2 ft) high at each end of the working space. Where the entrance has a personnel door(s), the door(s) shall open in the direction of egress and be equipped with panic bars, pressure plates, or other devices that are normally latched but open under simple pressure.

A single entrance to the required working space shall be permitted where either of the conditions in 110.26(C)(2)(a) or (C)(2)(b) is met.

(a) Unobstructed ~~Exit~~. Where the location permits a continuous and unobstructed way of ~~exit~~ travel, a single entrance to the working space shall be permitted.

(b) Extra Working Space. Where the depth of the working space is twice that required by 110.26(A)(1), a single entrance shall be permitted. It shall be located so that the distance from the equipment to the nearest edge of the entrance is not less than the minimum clear distance specified in Table 110.26(A)(1) for equipment operating at that voltage and in that condition.

2008 NEC - 110.26

(C) Entrance to Working Space

(2) Large Equipment. For equipment rated 1200 amperes or more and over 1.8 m (6 ft) wide that contains overcurrent devices, switching devices, or control devices, there shall be one entrance to the required working space not less than 610 mm (24 in.) wide and 2.0 m (6 1/ 2 ft) high at each end of the working space.

A single entrance to and egress from the required working space shall be permitted where either of the conditions in 110.26(C)(2)(a) or (C)(2)(b) is met.

(a) Unobstructed Egress. Where the location permits a continuous and unobstructed way of egress travel, a single entrance to the working space shall be permitted.

(b) Extra Working Space. Where the depth of the working space is twice that required by 110.26(A)(1), a single entrance shall be permitted. It shall be located so that the distance from the equipment to the nearest edge of the entrance is not less than the minimum clear distance specified in Table 110.26(A)(1) for equipment operating at that voltage and in that condition.

(3) Personnel Doors. Where equipment rated 1200 A or more that contains overcurrent devices, switching devices, or control devices is installed and there is a personnel door(s) intended for entrance to and egress from the working space less than 7.6 m (25 ft) from the nearest edge of the working space, the door(s) shall open in the direction of egress and be equipped with panic bars, pressure plates, or other devices that are normally latched but open under simple pressure.

Author's Comment: This change effectively returns the requirement for large equipment to be considered over 6 ft in width, as well as being 1200 amperes or more. The over 6 ft requirement had been deleted for the 2005 NEC. The personnel door requirement applies to all equipment 1200 amp or more, not just over 6 feet in width.

The requirements covering personnel doors intended for entrance and egress from large equipment has been relocated into new **110.26(C)(3)** and the related text in **(2)** deleted. New **(3)** will now require any door within 25 feet of large equipment to have panic hardware and open in the direction of egress.

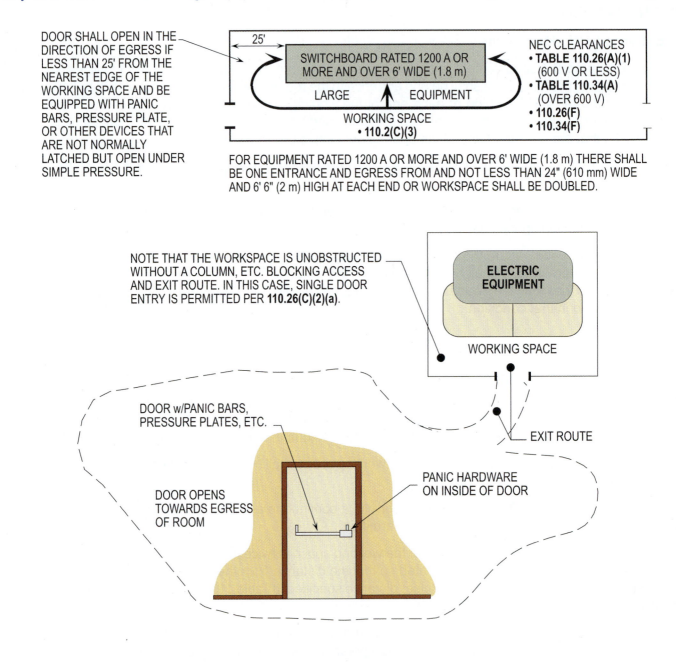

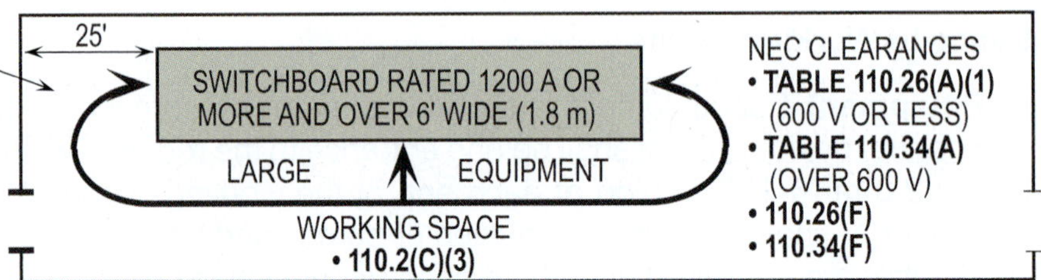

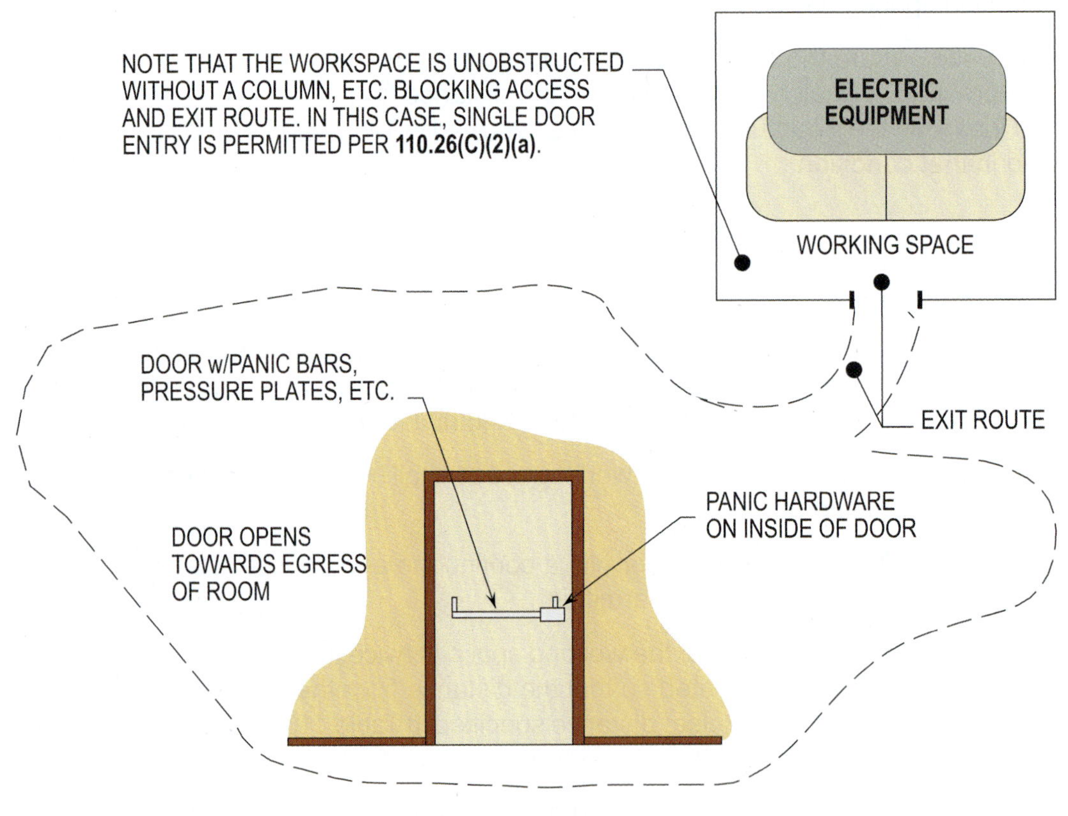

PERSONNEL DOORS
110.26(C)(3)

Purpose of Change: This revision clarifies that the personnel door(s) shall have panic bars and similar equipment and open in the direction of egress if located less than 25 ft from the nearest edge of the working space.

NEC Ch. 1 - Article 110
Part III - 110.33(A)

Type of Change	Panel Action	UL	UL 508	API 500 - 1997	API 505 - 1997	OSHA - 1994
Revision	Accept in Principle	-	-	-	-	1910.303(h)(4)(i)
ROP	**ROC**	**NFPA 70E - 2004**	**NFPA 70B - 2006**	**NFPA 79 - 2007**	**NFPA**	**NEMA**
pg. 55 # 1-147	pg. - # -	404.20(A)	-	-	-	-
log: 3223 CMP: 1	log: -	Submitter: Robert Carbone		2005 NEC: 110.33(A)		IEC: -

2005 NEC - 110.33 Entrance and Access to ~~Work~~ Space.

(A) Entrance. At least one entrance not less than 610 mm (24 in.) wide and 2.0 m (6 1/ 2 ft) high shall be provided to give access to the working space about electric equipment. Where the entrance has a personnel door(s), the door(s) shall open in the direction of egress and be equipped with panic bars, pressure plates, or other devices that are normally latched but open under simple pressure.

2008 NEC - 110.33 Entrance to Enclosures and Access to Working Space.

(A) Entrance. At least one entrance to enclosures for electrical installations as described in 110.31 not less than 610 mm (24 in.) wide and 2.0 m (6 1/ 2 ft) high shall be provided to give access to the working space about electric equipment.

Author's Comment: The change in this section requires at least one entrance to provide access to the working space for a high voltage enclosure as described in Section **110.31**.

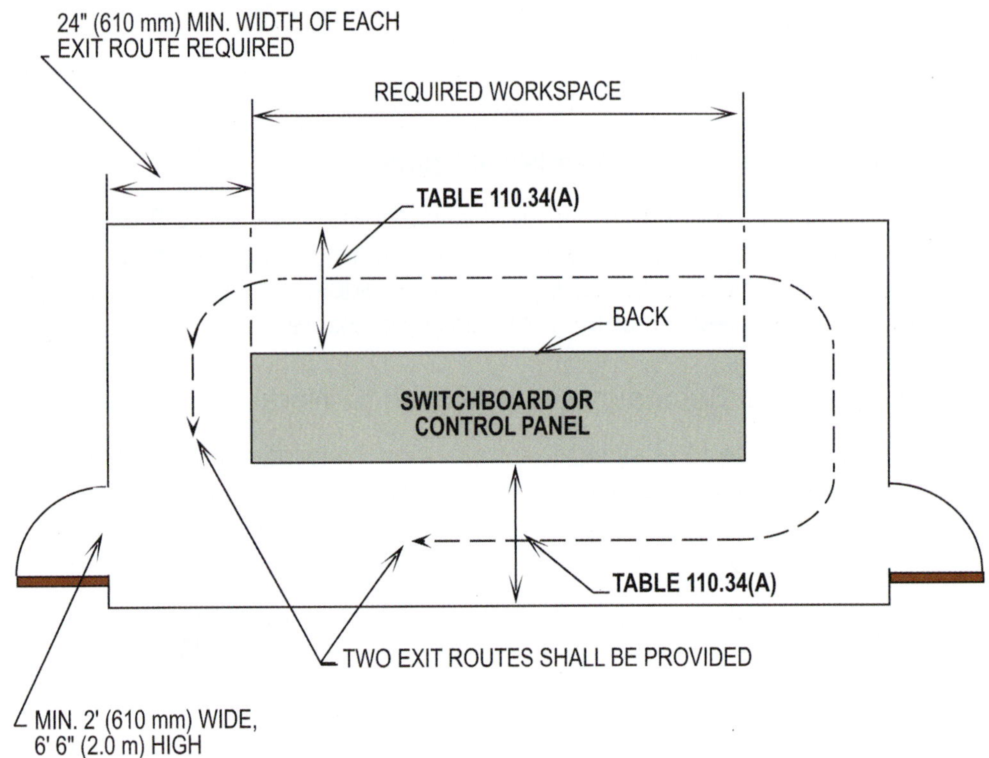

ENCLOSURES FOR ELECTRICAL INSTALLATIONS

- ELECTRICAL INSTALLATIONS IN VAULTS
- ELECTRICAL INSTALLATIONS IN ROOMS
- ELECTRICAL INSTALLATIONS IN CLOSETS
- LOCATIONS SURROUNDED BY A:

(A) WALL
(B) SCREEN
(C) FENCE

ENTRANCE
110.33(A)

Purpose of Change: This revision clarifies that at least one entrance is required to provide access to the working space for a high-voltage enclosure as described in **110.31**.

NEC Ch. 1 - Article 110
Part III - 110.33(A)(3)

Type of Change	Panel Action	UL	UL 508	API 500 - 1997	API 505 - 1997	OSHA - 1994
Revision	Accept in Principle in Part	-	-	-	-	1910.303(h)(4)(i)
ROP	**ROC**	**NFPA 70E - 2004**	**NFPA 70B - 2006**	**NFPA 79 - 2007**	**NFPA**	**NEMA**
pg. 56 # 1-148	pg. 34 # 1-99	404.20(A)	-	-	-	-
log: 1197 CMP: 1	log: 2332 Submitter: Lanny G. McMahill			2005 NEC: 110.33(A)		IEC: -

2005 NEC - 110.33 Entrance and Access to Work Space.

(A) Entrance. At least one entrance not less than 610 mm (24 in.) wide and 2.0 m (6 1/ 2 ft) high shall be provided to give access to the working space about electric equipment. ~~Where the entrance has a personnel door(s), the door(s) shall open in the direction of egress and be equipped with panic bars, pressure plates, or other devices that are normally latched but open under simple pressure.~~

2008 NEC - 110.33 Entrance and Access to Work Space.

(A) Entrance. At least one entrance not less than 610 mm (24 in.) wide and 2.0 m (6 1/ 2 ft) high shall be provided to give access to the working space about electric equipment.

<u>**(3) Personnel Doors.** Where there is a personnel door(s) intended for entrance to and egress from the working space less than 7.6 m (25 ft) from the nearest edge of the working space, the door(s) shall open in the direction of egress and be equipped with panic bars, pressure plates, or other devices that are normally latched but open under simple pressure.</u>

Author's Comment: The new text clarifies that the personnel door(s) requirement for panic bars and similar hardware and the opening in the direction of egress applies to personnel doors that are less than 25 feet from the equipment working space. This requirement would not apply to doors located 25 feet or more from the working space.

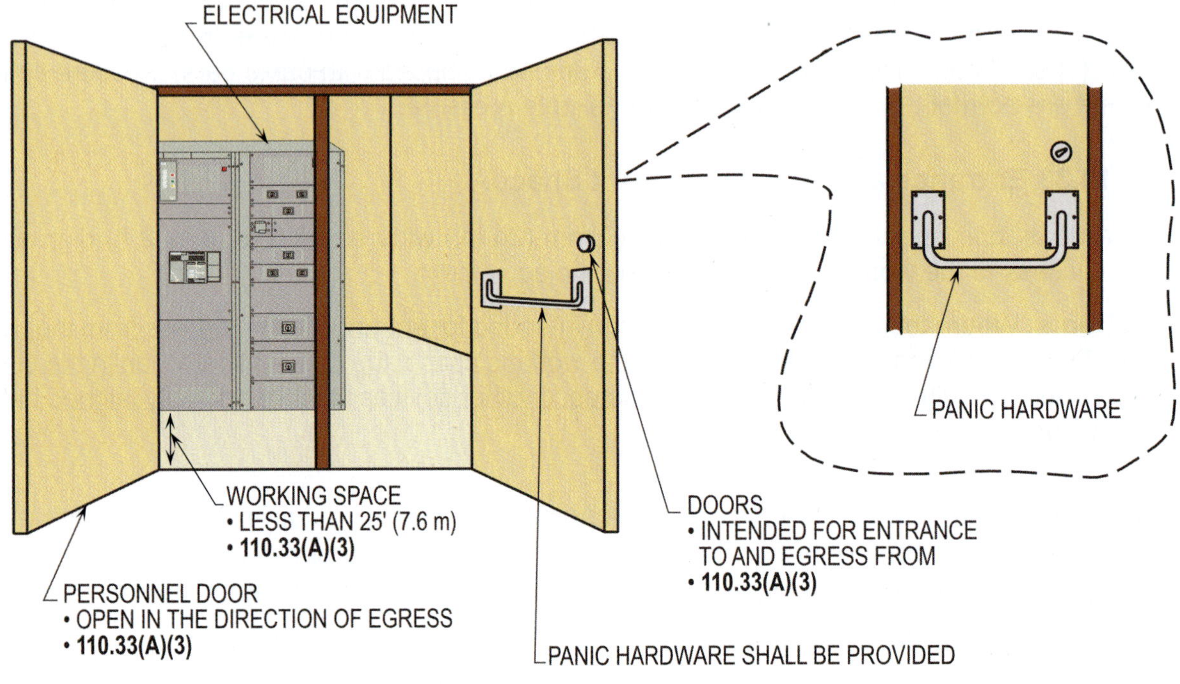

PERSONNEL DOORS
110.33(A)(3)

Purpose of Change: This revision clarifies that the personnel door(s) shall have panic bars and similar equipment and open in the direction of egress if located less than 25 ft (7.6 m) from the nearest edge of the working space.

2

Wiring and Protection

Chapter 2 of the NEC has always been referred to as the "Designing Chapter" and used by engineers, electrical contractors and electricians who have the responsibility of calculating loads and sizing the elements of the electrical system.

Chapter 2 is the starting point to begin calculating ampacities for branch-circuits and feeder-circuits. Even the ampacity for sizing service-entrance conductors is calculated by applying the rules of **Article 220**, which are found in **Parts I, II,** and **III**, as well as **Part IV** of **Article 230**.

The key number for finding the requirements necessary for calculating loads in **Chapter 2** is 200. In other words, all articles and sections will be identified by using a 200 series number. When designing electrical systems, **Chapter 2** and the 200 series is utilized with other pertinent articles and sections. It is nearly impossible for a designer to calculate loads and determine the size of various elements of the electrical system, if he or she is not properly acquainted with the calculation requirements of **Chapter 2**.

For example, if the user wanted to calculate the load in amps for a motor feeder-circuit, he or she would refer to **220.14(C)** and this section references **430.24** since **Article 220** does not list the rules for sizing loads for motor circuits. If **Article 220** does not contain the rules for calculating the load, it will refer the user to the required section in other articles of the NEC.

NEC Ch. 2 - Article 200
Part I - 200.2(B)

Type of Change		Panel Action		UL	UL 508	API 500 - 1997	API 505 - 1997	OSHA - 1994
New Sections		Accept		-	-	-	-	-
ROP		ROC		NFPA 70E - 2004	NFPA 70B - 2006	NFPA 79 - 2007	NFPA	NEMA
pg. 187	# 5-90	pg. 125	# 5-47	-	-	-	-	-
log: 3389	CMP: 5	log: 1906	Submitter: Frederic P. Hartwell			2005 NEC: -		IEC: -

2005 NEC - 200.2 General.

All premises wiring systems, other than circuits and systems exempted or prohibited by 210.10, 215.7, 250.21, 250.22, 250.162, 503.155, 517.63, 668.11, 668.21, and 690.41 Exception, shall have a grounded conductor that is identified in accordance with 200.6.

The grounded conductor, where insulated shall have insulation that is (1) suitable, other than color, for any ungrounded conductor of the same circuit on circuits of less than 1000 volts or impedance grounded neutral systems of 1 kV and over, or (2) rated not less than 600 volts for solidly grounded neutral systems of 1 kV and over as described in 250.184(A).

2008 NEC - 200.2 General.

All premises wiring systems, other than circuits and systems exempted or prohibited by 210.10, 215.7, 250.21, 250.22, 250.162, 503.155, 517.63, 668.11, 668.21, and 690.41 Exception, shall have a grounded conductor that is identified in accordance with 200.6. The grounded conductor shall comply with (A) and (B).

(A) Insulation. The grounded conductor, where insulated shall have insulation that is (1) suitable, other than color, for any ungrounded conductor of the same circuit on circuits of less than 1000 volts or impedance grounded neutral systems of 1 kV and over, or (2) rated not less than 600 volts for solidly grounded neutral systems of 1 kV and over as described in 250.184(A).

(B) Continuity. The continuity of a grounded conductor shall not depend on a connection to a metallic enclosure, raceway, or cable armor.

Author's Comment: The new subsection is intended to ensure that the continuity of the connection of a grounded conductor does not depend upon a connection to the metal enclosure, raceway, or cable armor. A busbar or conductor must be used to make the connection between the neutral bar and the equipment ground bar, not the metal of the enclosure.

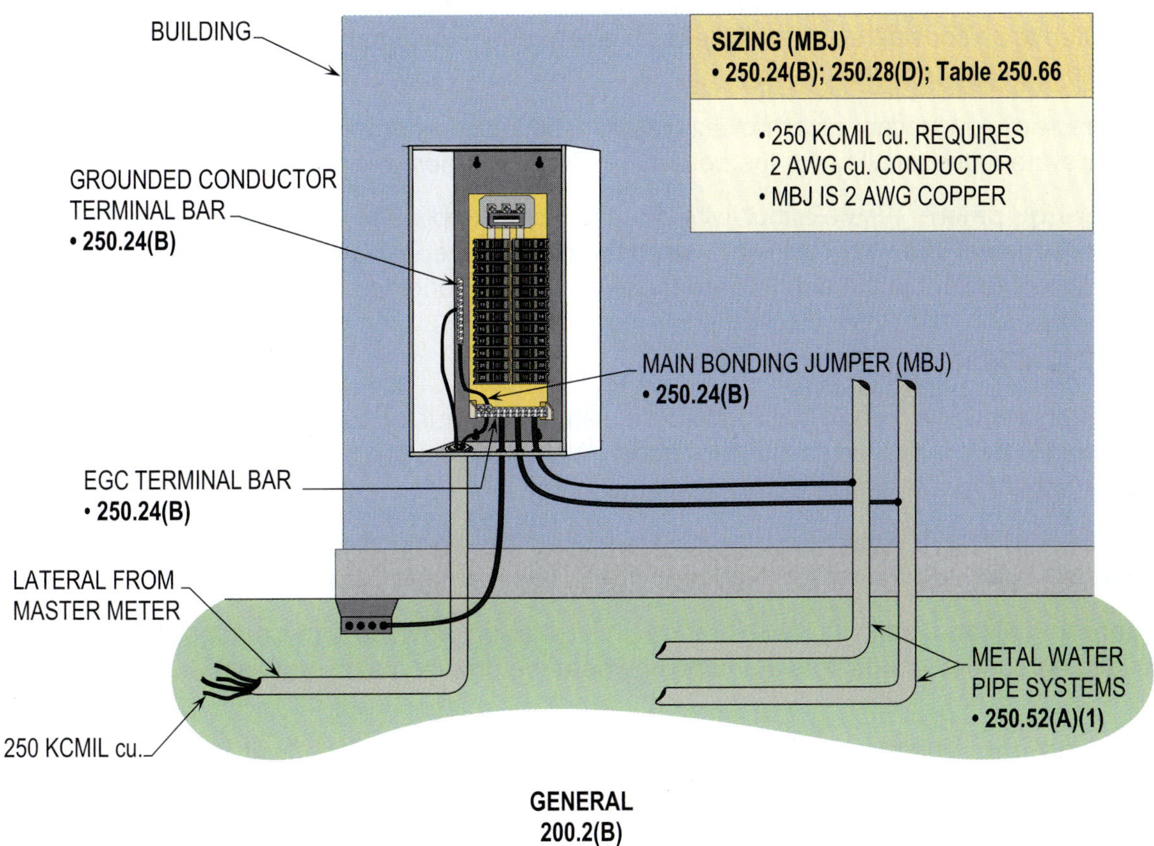

GENERAL
200.2(B)

Purpose of Change: This new subsection clarifies that a busbar or conductor shall be used to make the connection between the neutral bar and the equipment grounding terminal bar, not the metal of the enclosure.

NEC Ch. 2 - Article 210
Part I - 210.4(A) and (B)

Type of Change	Panel Action	UL	UL 508	API 500 - 1997	API 505 - 1997	OSHA - 1994
Revision	Accept in Principle	-	-	-	-	-
ROP	ROC	NFPA 70E - 2004	NFPA 70B - 2006	NFPA 79 - 2007	NFPA	NEMA
pg. 61 # 2-10	pg. - # -	-	-	-	-	-
log: 2679 CMP: 2	log: -	Submitter: Dorothy Kellogg		2005 NEC: 210.4(A) and (B)		IEC: -

2005 NEC - 210.4 Multiwire Branch Circuits.

(A) General. Branch circuits recognized by this article shall be permitted as multiwire circuits. A multiwire circuit shall be permitted to be considered as multiple circuits. All conductors shall originate from the same panelboard or similar distribution equipment.

FPN: A 3-phase, 4-wire, wye-connected power system used to supply power to nonlinear loads may necessitate that the power system design allow for the possibility of high harmonic neutral currents.

(B) ~~Devices or Equipment.~~ ~~Where a~~ multiwire branch circuit ~~supplies more than one device or equipment on the same yoke, a means shall be provided to disconnect~~ simultaneously all ungrounded conductors ~~supplying those devices or equipment~~ at the point where the branch circuit originates.

2008 NEC - 210.4 Multiwire Branch Circuits.

(A) General. Branch circuits recognized by this article shall be permitted as multiwire circuits. A multiwire circuit shall be permitted to be considered as multiple circuits. All conductors of a multiwire branch circuit shall originate from the same panelboard or similar distribution equipment.

FPN: A 3-phase, 4-wire, wye-connected power system used to supply power to nonlinear loads may necessitate that the power system design allow for the possibility of high harmonic neutral currents.

(B) Disconnecting Means. Each multiwire branch circuit shall be provided with a means that will simultaneously disconnect all ungrounded conductors at the point where the branch circuit originates.

Author's Comment: There were two reasons for the changes to the text in this section. The first reason was to clarify that all conductors of a multiwire branch circuit must originate from the same panelboard or distribution equipment. The change in **(B)** was done to emphasize the safety concerns associated with unintentional voltage being present on multiwire branch circuits, especially during maintenance. Deleting the text that applied simultaneous disconnecting requirements for a multiwire branch circuit to a single device or equipment on the same yoke will require simultaneous disconnect of all ungrounded conductors on any multiwire branch circuit at its origin.

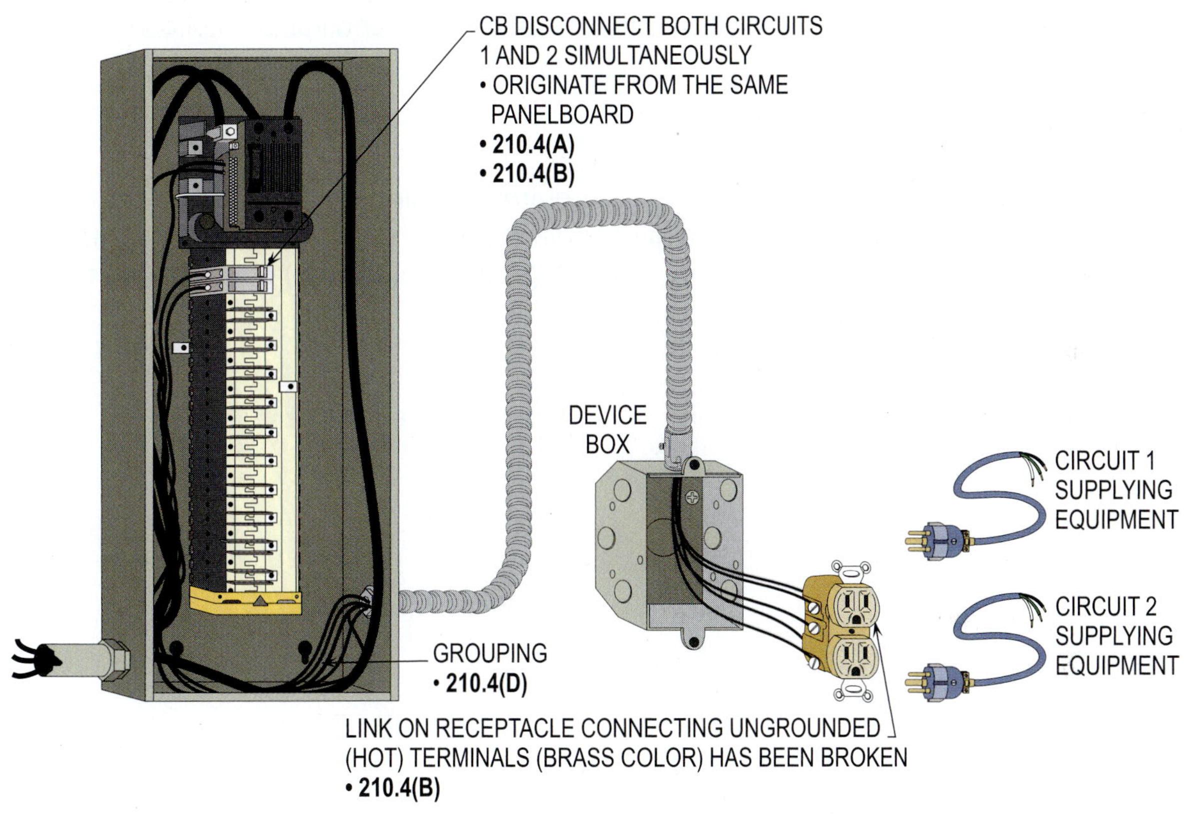

MULTIWIRE BRANCH CIRCUITS
210.4(A) and (B)

Purpose of Change: This revision clarifies that multiwire branch circuits shall originate from the same panelboard or similar distribution equipment.

NEC Ch. 2 - Article 210
Part I - 210.4(D)

Type of Change		Panel Action		UL	UL 508	API 500 - 1997	API 505 - 1997	OSHA - 1994
New Subsection		Accept in Principle		-	-	-	-	-
ROP		ROC		NFPA 70E - 2004	NFPA 70B - 2006	NFPA 79 - 2007	NFPA	NEMA
pg. 63	# 2-17	pg. -	# -	-	-	-	-	-
log: 3378	CMP: 2	log: -	Submitter: Frederic P. Hartwell			2005 NEC: -		IEC: -

2008 NEC - 210.4 Multiwire Branch Circuits.

(D) Grouping. The ungrounded and grounded conductors of each multiwire branch circuit shall be grouped by wire ties or similar means in at least one location within the panelboard or other point of origination.

Exception: The requirement for grouping shall not apply if the circuit enters from a cable or raceway unique to the circuit that makes the grouping obvious.

Author's Comment: This proposed change will require grouping all the conductors of a multiwire branch circuit within a panelboard or other point of origin to help ensure that all ungrounded conductors in a multiwire branch circuit are disconnected and the grounded (neutral) conductor is de-energized, especially during service work.

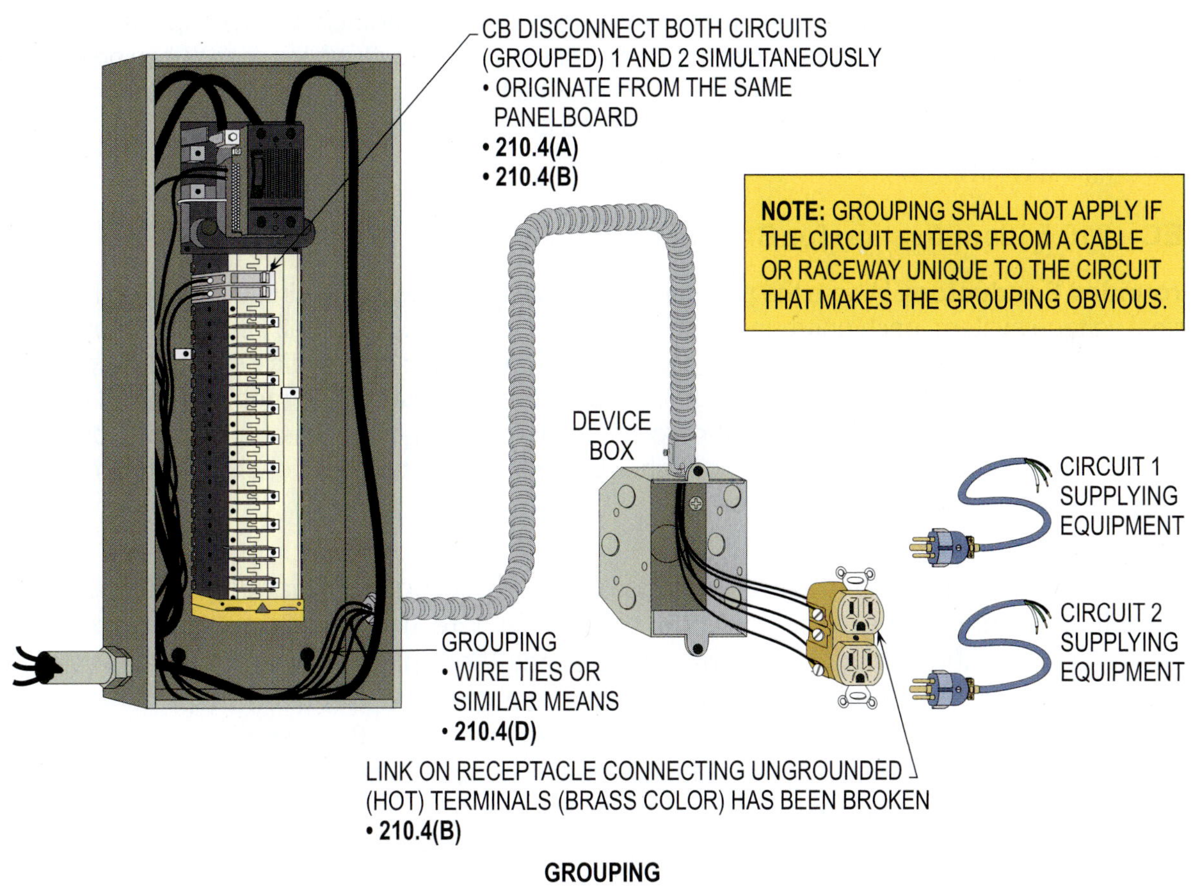
CB DISCONNECT BOTH CIRCUITS
(GROUPED) 1 AND 2 SIMULTANEOUSLY
• ORIGINATE FROM THE SAME PANELBOARD
• 210.4(A)
• 210.4(B)
NOTE: GROUPING SHALL NOT APPLY IF THE CIRCUIT ENTERS FROM A CABLE OR RACEWAY UNIQUE TO THE CIRCUIT THAT MAKES THE GROUPING OBVIOUS.
DEVICE BOX
CIRCUIT 1 SUPPLYING EQUIPMENT
CIRCUIT 2 SUPPLYING EQUIPMENT
GROUPING
• WIRE TIES OR SIMILAR MEANS
• 210.4(D)
LINK ON RECEPTACLE CONNECTING UNGROUNDED (HOT) TERMINALS (BRASS COLOR) HAS BEEN BROKEN
• 210.4(B)

GROUPING
210.4(D)

Purpose of Change: This new subsection requires grouping all the conductors of a multiwire branch circuit to help ensure that all ungrounded conductors are disconnected and grounded (neutral) conductor is de-energized, especially during service work.

NEC Ch. 2 - Article 210
Part I - 210.5(C)

Type of Change	Panel Action	UL	UL 508	API 500 - 1997	API 505 - 1997	OSHA - 1994
Revision	Accept in Part	-	-	-	-	1910.304(b)(1)
ROP	**ROC**	**NFPA 70E - 2004**	**NFPA 70B - 2006**	**NFPA 79 - 2007**	**NFPA**	**NEMA**
pg. 64 # 2-23	pg. 39 # 2-18	410.2(A)	-	13.2.4	-	-
log: 2681 CMP: 2	log: 1897	Submitter: Dorothy Kellogg		2005 NEC: 210.5(C)		IEC: -

2005 NEC - 210.5 Identification for Branch Circuits.

(C) Ungrounded Conductors. Where the premises wiring system has branch circuits supplied from more than one nominal voltage system, each ungrounded conductor of a branch circuit~~, where accessible,~~ shall be identified by system. The means of identification shall be permitted to be by separate color coding, marking tape, tagging, or other approved means ~~and~~ shall be permanently posted at each branch-circuit panelboard or similar branch-circuit distribution equipment.

2008 NEC - 210.5 Identification for Branch Circuits.

(C) Ungrounded Conductors. Where the premises wiring system has branch circuits supplied from more than one nominal voltage system, each ungrounded conductor of a branch circuit shall be identified by phase or line and system at all termination, connection, and splice points. The means of identification shall be permitted to be by separate color coding, marking tape, tagging, or other approved means. The method utilized for conductors originating within each branch-circuit panelboard or similar branch-circuit distribution equipment shall be documented in a manner that is readily available or shall be permanently posted at each branch-circuit panelboard or similar branch-circuit distribution equipment.

Author's Comment: The current wording in the 2005 NEC requires marking of the conductors at every conduit fitting and pull box or any other location where the branch circuit is accessible. The locations where the branch circuit is terminated, connected or spliced are the critical locations where the marking is needed. Now each ungrounded conductor must be identified by phase or line and by system. This will affect all conductor color coding, even at fluorescent fixture whips. The revised wording would account for branch circuits installed as cables and branch circuits installed as single conductors in raceways. The change also expands the methods that can be used for identification to permit documentation that is readily available or permanent posting at the panelboard or distribution equipment.

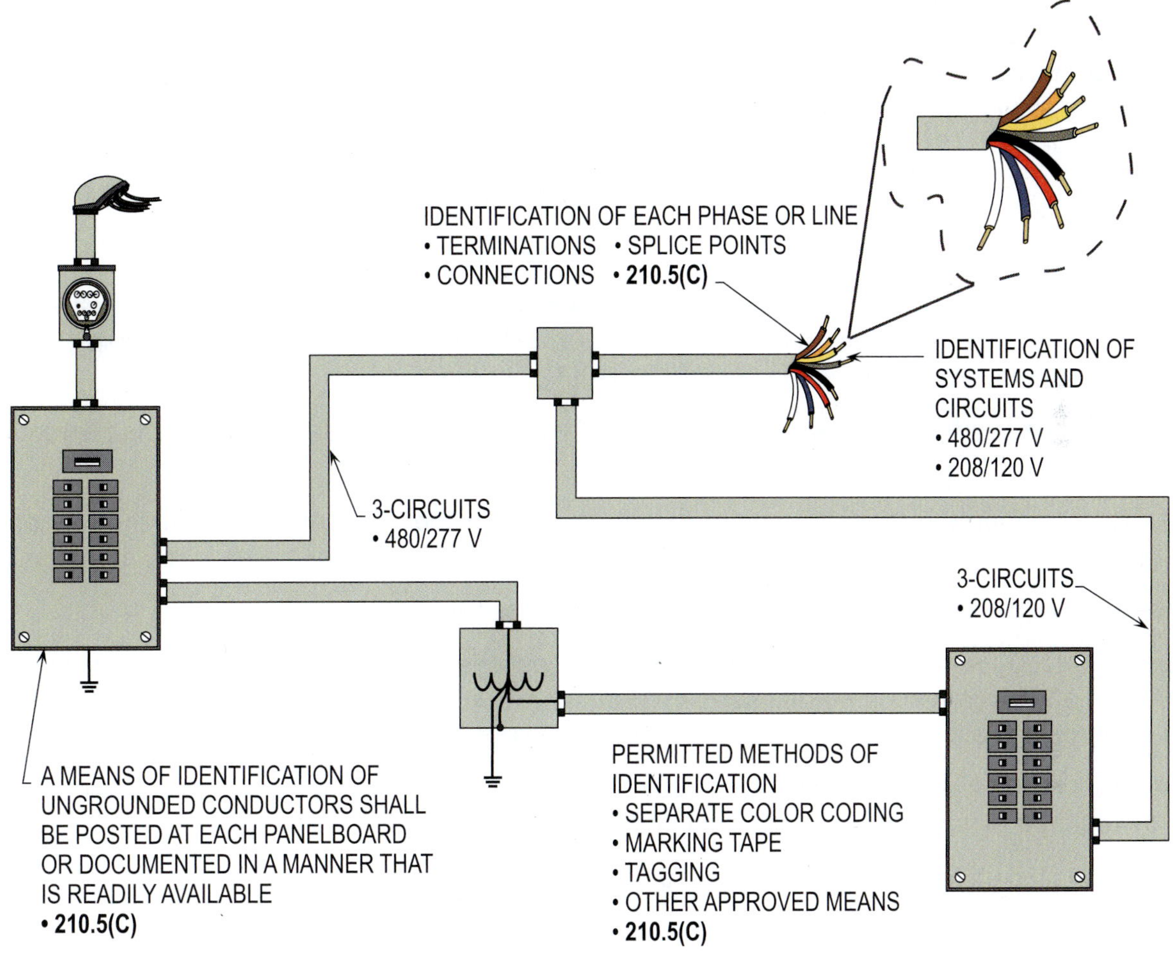

UNGROUNDED CONDUCTORS
210.5(C)

Purpose of Change: This revision clarifies that each ungrounded conductor of a branch circuit shall be identified by phase or line and by system at all termination, connection, and splice points.

NEC Ch. 2 - Article 210
Part I - 210.8(A)(2) and (A)(5)

Type of Change	Panel Action	UL	UL 508	API 500 - 1997	API 505 - 1997	OSHA - 1994
Deletion	Accept in Principle	-	-	-	-	-
ROP	**ROC**	**NFPA 70E - 2004**	**NFPA 70B - 2006**	**NFPA 79 - 2007**	**NFPA**	**NEMA - 2002**
pg. 67 # 2-40	pg. 42 # 2-35	-	-	-	-	WD 6
log: 3601 CMP: 2	log: 809	Submitter: Douglas Hansen		2005 NEC: 210.8(A)(2); (A)(5); Exceptions		IEC: -

2005 NEC - 210.8 Ground-Fault Circuit-Interrupter Protection for Personnel.

(A) Dwelling Units. All 125-volt, single-phase, 15- and 20-ampere receptacles installed in the locations specified in (1) through (8) shall have ground-fault circuit-interrupter protection for personnel.

(2) Garages, and also accessory buildings that have a floor located at or below grade level not intended as habitable rooms and limited to storage areas, work areas, and areas of similar use

~~Exception No. 1 to (2): Receptacles that are not readily accessible.~~

~~Exception No. 2 to (2): A single receptacle or a duplex receptacle for two appliances located within dedicated space for each appliance that, in normal use, is not easily moved from one place to another and that is cord-and-plug connected in accordance with 400.7(A)(6), (A)(7), or (A)(8).~~

~~Receptacles installed under the exceptions to 210.8(A)(2) shall not be considered as meeting the requirements of 210.52(G).~~

(5) Unfinished basements — for purposes of this section, unfinished basements are defined as portions or areas of the basement not intended as habitable rooms and limited to storage areas, work areas, and the like

~~Exception No. 1 to (5): Receptacles that are not readily accessible.~~

~~Exception No. 2 to (5): A single receptacle or a duplex receptacle for two appliances located within dedicated space for each appliance that, in normal use, is not easily moved from one place to another and that is cord-and-plug connected in accordance with 400.7(A)(6), (A)(7), or (A)(8).~~

Exception ~~No. 3~~ to (5): A receptacle supplying only a permanently installed fire alarm or burglar alarm system shall not be required to have ground-fault circuit-interrupter protection.

Receptacles installed under the exception~~s~~ to 210.8(A)(5) shall not be considered as meeting the requirements of 210.52(G).

2008 NEC - 210.8 Ground-Fault Circuit-Interrupter Protection for Personnel.

(A) Dwelling Units. All 125-volt, single-phase, 15- and 20-ampere receptacles installed in the locations specified in (1) through (8) shall have ground-fault circuit-interrupter protection for personnel.

(2) Garages, and also accessory buildings that have a floor located at or below grade level not intended as habitable rooms and limited to storage areas, work areas, and areas of similar use

(5) Unfinished basements — for purposes of this section, unfinished basements are defined as portions or areas of the basement not intended as habitable rooms and limited to storage areas, work areas, and the like

Exception to (5): A receptacle supplying only a permanently installed fire alarm or burglar alarm system shall not be required to have ground-fault circuit-interrupter protection.

Receptacles installed under the exception to 210.8(A)(5) shall not be considered as meeting the requirements of 210.52(G).

Author's Comment: The two exceptions permitting receptacles that were not readily accessible and single or duplex receptacles for two appliances within a dedicated space for each appliance to not have GFCI protection for personnel in garages and unfinished basements were deleted. The substantiation in the proposal and the discussion during the Panel deliberation indicated that the present generation of GFCI devices do not have the problems of nuisance tripping that plagued earlier devices.

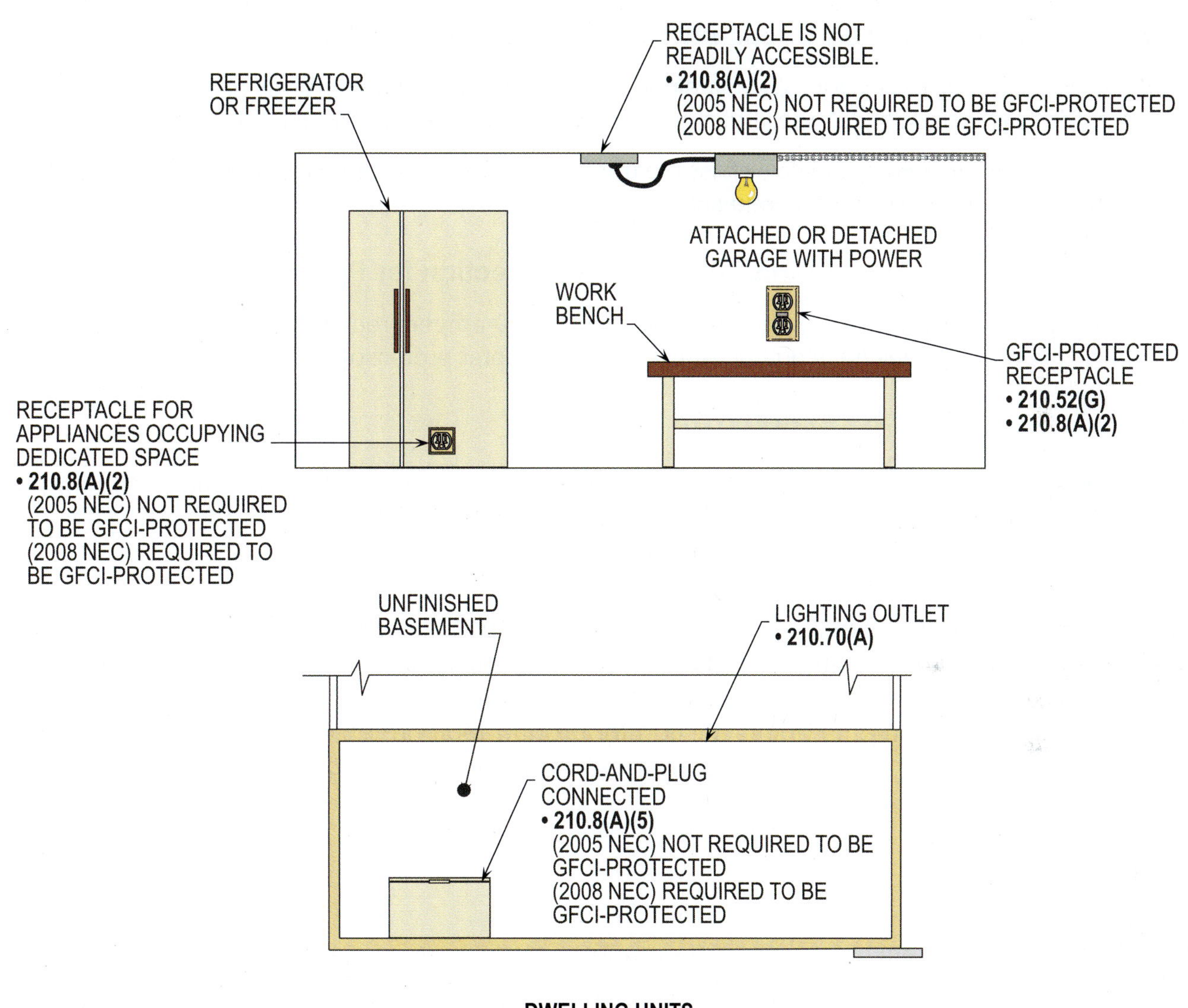

DWELLING UNITS
210.8(A)(2) and (A)(5)

Purpose of Change: The exceptions permitting receptacles that were not readily accessible and single or duplex receptacles for two appliances within a dedicated space for each appliance to not have GFCI-protection for personnel in garages and unfinished basements were deleted.

NEC Ch. 2 - Article 210
Part I - 210.8(B)(2)

Type of Change	Panel Action	UL	UL 508	API 500 - 1997	API 505 - 1997	OSHA - 1994
Revision	Accept in Principle	-	-	-	-	-
ROP	**ROC**	**NFPA 70E - 2004**	**NFPA 70B - 2006**	**NFPA 79 - 2007**	**NFPA**	**NEMA - 2002**
pg. 73 # 2-73	pg. 47 # 2-50	410.4(A)	-	-	-	WD 6
log: 1724 CMP: 2	log: 1357	Submitter: Richard P. Owen		2005 NEC: 210.8(B)(2)		IEC: -

2005 NEC - 210.8 Ground-Fault Circuit-Interrupter Protection for Personnel.

(B) Other Than Dwelling ~~Units~~. All 125-volt, single-phase, 15- and 20-ampere receptacles installed in the locations specified in (1) through (5) shall have ground-fault circuit-interrupter protection for personnel:

(2) ~~Commercial and institutional~~ kitchens ~~— for the purposes of this section, a kitchen is an area with a sink and permanent facilities for food preparation and cooking~~

2008 NEC - 210.8 Ground-Fault Circuit-Interrupter Protection for Personnel.

(B) Other Than Dwellings. All 125-volt, single-phase, 15- and 20-ampere receptacles installed in the locations specified in (1) through (5) shall have ground-fault circuit-interrupter protection for personnel:

(2) kitchens

Author's Comment: The definition of "kitchen" has been moved to Article 100 since it is used in so many different locations within the NEC so the definition has been deleted out of 210.8(B)(2)

An office "break area" with a sink and permanent facilities for food preparation and cooking does not really qualify as either commercial or institutional, but the break area or room with a sink and permanent cooking facilities has the same shock hazard potential. This change clarifies that these break rooms or areas fall under the requirements of this section for ground-fault protection. By deleting "commercial and institutional" this section now applies to all non-dwelling kitchens, regardless of the type of facility.

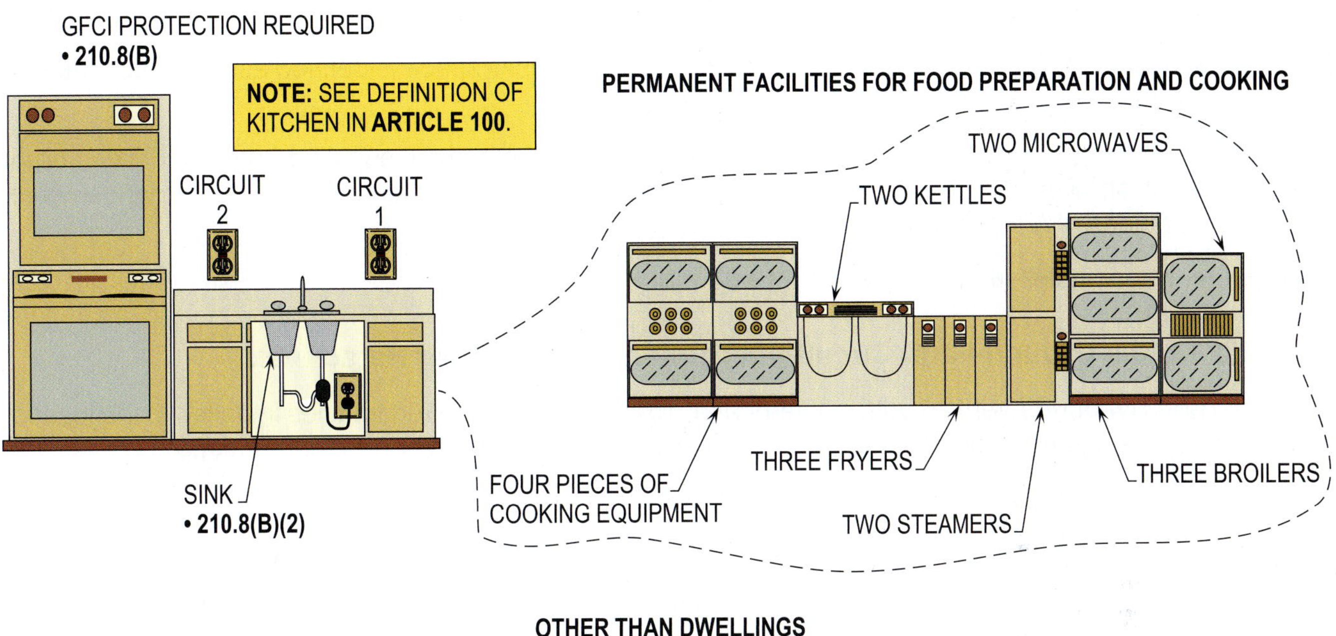

OTHER THAN DWELLINGS
210.8(B)(2)

Purpose of Change: This revision clarifies that GFCI protection shall be provided for receptacles in kitchens. An office "break room" with a sink and permanent cooking facilities falls under the definition of a kitchen.

NEC Ch. 2 - Article 210
Part I - 210.8(B)(4)

Type of Change	Panel Action	UL	UL 508	API 500 - 1997	API 505 - 1997	OSHA - 1994
Revision	Accept in Principle in Part	-	-	-	-	-
ROP	**ROC**	**NFPA 70E - 2004**	**NFPA 70B - 2006**	**NFPA 79 - 2007**	**NFPA**	**NEMA - 2002**
pg. 71 # 2-70	pg. 45 # 2-41	410.4(A)	-	-	-	WD 6
log: 1443 CMP: 2	log: 1358 Submitter: Ryan Jackson			2005 NEC: 210.8(B)(4)		IEC: -

2005 NEC - 210.8 Ground-Fault Circuit-Interrupter Protection for Personnel.

(B) Other Than Dwelling ~~Units~~. All 125-volt, single-phase, 15- and 20-ampere receptacles installed in the locations specified in (1) through (5) shall have ground-fault circuit-interrupter protection for personnel:

(3) Rooftops

(4) Outdoors ~~in public spaces—for the purpose of this section a public space is defined as any space that is for use by, or is accessible to, the public~~

Exception to (3) and (4): Receptacles that are not readily accessible and are supplied from a dedicated branch circuit for electric snow-melting or deicing equipment shall be permitted to be installed ~~in accordance with the applicable provisions of Article 426~~.

~~(5) Outdoors, where installed to comply with 210.63~~

2008 NEC - 210.8 Ground-Fault Circuit-Interrupter Protection for Personnel.

(B) Other Than Dwellings. All 125-volt, single-phase, 15- and 20-ampere receptacles installed in the locations specified in (1) through (5) shall have ground-fault circuit-interrupter protection for personnel:

(3) Rooftops

(4) Outdoors

Exception No. 1 to (3) and (4): Receptacles that are not readily accessible and are supplied from a dedicated branch circuit for electric snow-melting or deicing equipment shall be permitted to be installed without GFCI protection.

Exception No. 2 to (4): In industrial establishments only, where the conditions of maintenance and supervision ensure that only qualified personnel are involved, An assured equipment grounding conductor program as specified in 590.6(B)(2) shall be permitted for only those receptacle outlets used to supply equipment that would create a greater hazard if power is interrupted or having a design that is not compatible with GFCI protection.

Author's Comment: This revision requires GFCI protection for all 15- and 20-ampere, 125-volt single-phase receptacles installed outdoors, rather than just those in public spaces. An exception was added to exempt those receptacles in industrial establishments only, where the conditions of maintenance and supervision ensure that only qualified personnel are involved, that are limited to use with equipment qualified under an assured equipment grounding conductor program as specified in **590.6(B)(2)**.

The words "in accordance with the applicable provisions of **Article 426**" were replaced with the words "without GFCI protection" to comply with Section 4.1.1 of the NEC Style Manual to not reference an entire article.

A PUBLIC SPACE IS ANY SPACE THAT "IS FOR USE BY, OR IS ACCESSIBLE TO, THE PUBLIC" HAS BEEN DELETED. NOW APPLIES TO ALL 15- AND 20-AMPERE, 125 VOLT SINGLE-PHASE RECEPTACLES
• 210.8(B)(4)
OUTDOORS
• 210.8(B)(4)
ALL NON-DWELLING
RECEPTACLE OUTLETS SHALL BE GFCI PROTECTED
• 210.8(B)(4)

NOTE: WHERE RECEPTACLES ARE USED IN INDUSTRIAL ESTABLISHMENTS ONLY, GFCI PROTECTION ON RECEPTACLES SHALL NOT BE REQUIRED WHERE THE CONDITIONS OF MAINTENANCE AND SUPERVISION ENSURE THAT ONLY QUALIFIED PERSONNEL ARE INVOLVED, THAT ARE LIMITED TO USE WITH EQUIPMENT QUALIFIED UNDER AN ASSURED EQUIPMENT GROUNDING CONDUCTOR PROGRAM AS SPECIFIED IN **590.6(B)(2)**.

OTHER THAN DWELLINGS
210.8(B)(4)

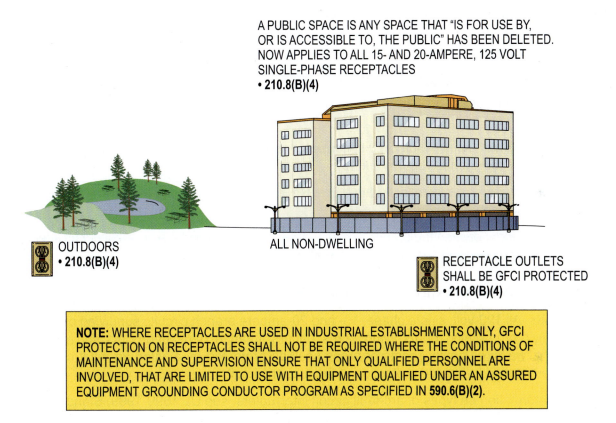

Purpose of Change: This revision requires GFCI protection for all 15- and 20-ampere, 125-volt single-phase receptacles installed outdoors.

NEC Ch. 2 - Article 210
Part I - 210.12(B)

Type of Change		Panel Action		UL	UL 508	API 500 - 1997	API 505 - 1997	OSHA - 1994
Revision		Accept		-	-	-	-	-
ROP		ROC		NFPA 70E - 2004	NFPA 70B - 2006	NFPA 79 - 2007	NFPA	NEMA - 2002
pg. 89	# 2-142	pg. 73	# 2-137	-	-	-	-	WD 6
log: 3488	CMP: 2	log: 840	Submitter: Alan Manche			2005 NEC: 210.12(B)		IEC: -

2005 NEC - 210.12 Arc-Fault Circuit-Interrupter Protection.

(B) Dwelling Unit ~~Bedrooms~~. All 120-volt, single phase, 15- and 20-ampere branch circuits supplying outlets installed in dwelling unit bedrooms shall be protected by a listed arc-fault circuit interrupter, combination type installed to provide protection of the branch circuit.

~~Branch/feeder AFCIs shall be permitted to be used to meet the requirements of 210.12(B) until January 1, 2008.~~

~~**FPN:**~~ For information on types of arc-fault circuit interrupters, see UL 1699-1999, Standard for Arc-Fault Circuit Interrupters.

~~**Exception:** The location of the arc-fault circuit interrupter shall be permitted to be at other than the origination of the branch circuit in compliance with (a) and (b):~~

~~(a) The arc-fault circuit interrupter installed within 1.8 m (6 ft) of the branch circuit overcurrent device as measured along the branch circuit conductors.~~

~~(b) The circuit conductors between the branch circuit overcurrent device and the arc-fault circuit interrupter shall be installed in a metal raceway or a cable with a metallic sheath.~~

2008 NEC - 210.12 Arc-Fault Circuit-Interrupter Protection.

(B) Dwelling Units. All 120-volt, single phase, 15- and 20-ampere branch circuits supplying outlets installed in dwelling unit family rooms, dining rooms, living rooms, parlors, libraries, dens, bedrooms, sun rooms, recreation rooms, closets, hallways, or similar rooms or areas shall be protected by a listed arc-fault circuit interrupter, combination type installed to provide protection of the branch circuit.

FPN No. 1: For information on types of arc-fault circuit interrupters, see UL 1699-1999, Standard for Arc-Fault Circuit Interrupters.

FPN No. 2: See 11.6.3(5) of NFPA 72®-2007, National Fire Alarm Code®, for information related to secondary power supply requirements for smoke alarms installed in dwelling units.

FPN No. 3: See 760.41(B) and 760.121(B) for power-supply requirements for fire alarm systems.

Author's Comment: The action in these proposals and comments require AFCI protection for all 120-volt, single-phase, 15- and 20-ampere branch circuits in family rooms, dining rooms, living rooms, parlors, libraries, dens, bedrooms, sun rooms, recreation rooms, closets, hallways, or similar rooms or areas. In addition, the last sentence, permitting branch/feeder AFCI devices until January 1, 2008, has been deleted and the effect is to now require only listed combination AFCI branch circuit protection.

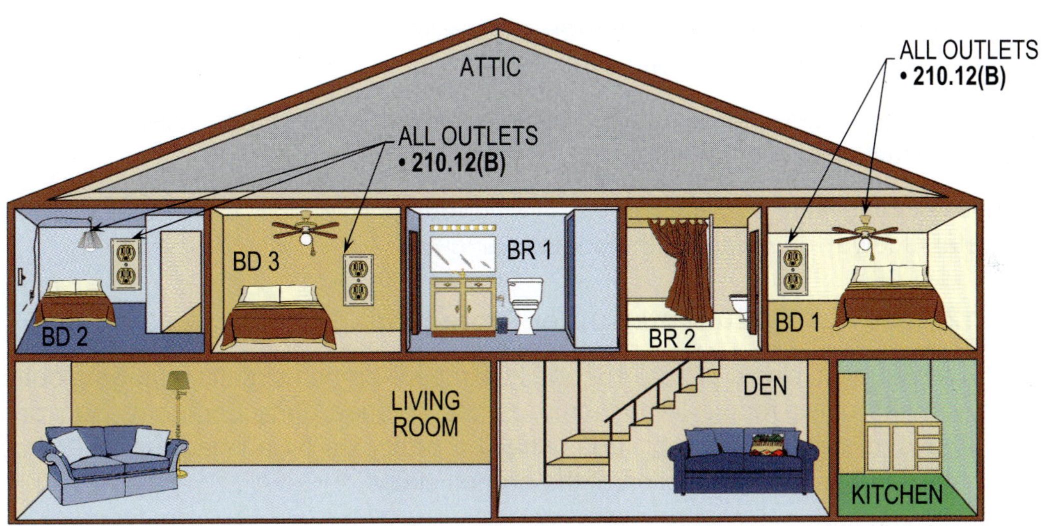

DWELLING UNITS
210.12(B)

Purpose of Change: This revision now requires AFCI protection for all 120-volt, single-phase, 15- and 20-ampere branch circuits supplying outlets installed in dwelling units in family rooms, dining rooms, living rooms, parlors, libraries, dens, bedrooms, sun rooms, recreation rooms, closets, hallways, or similar rooms or areas.

NEC Ch. 2 - Article 210
Part I - 210.12(B), Exception

11

Type of Change	Panel Action	UL	UL 508	API 500 - 1997	API 505 - 1997	OSHA - 1994
Revision	Accept in Principle	-	-	-	-	-
ROP	**ROC**	**NFPA 70E - 2004**	**NFPA 70B - 2006**	**NFPA 79 - 2007**	**NFPA**	**NEMA - 2002**
pg. 91 # 2-147	pg. - # -	-	-	-	-	WD 6
log: 3360 CMP: 2	log: -	Submitter: Aaron B. Chase		2005 NEC: 210.12(B)		IEC: -

2005 NEC - 210.12 Arc-Fault Circuit-Interrupter Protection.

(B) Dwelling Unit ~~Bedrooms~~. All 120-volt, single phase, 15- and 20-ampere branch circuits supplying outlets installed in dwelling unit ~~bedrooms~~ shall be protected by a listed arc-fault circuit interrupter, combination type installed to provide protection of the branch circuit.

~~**Exception:** The location of the arc-fault circuit interrupter shall be permitted to be at other than the origination of the branch circuit in compliance with (a) and (b):~~

~~(a) Where the arc-fault circuit interrupter is installed within 1.8 m (6 ft) of the branch circuit overcurrent device as measured along the branch circuit conductors.~~

~~(b) The circuit conductors between the branch circuit overcurrent device and the arc-fault circuit interrupter shall be installed in a metal raceway or a cable with a metallic sheath.~~

2008 NEC - 210.12 Arc-Fault Circuit-Interrupter Protection.

(B) Dwelling Units. All 120-volt, single phase, 15- and 20-ampere branch circuits supplying outlets installed in dwelling unit in family rooms, dining rooms, living rooms, parlors, libraries, dens, bedrooms, sun rooms, recreation rooms, closets, hallways, or similar rooms or areas shall be protected by a listed arc-fault circuit interrupter, combination type installed to provide protection of the branch circuit.

Author's Comment: The requirement in the exception to enclose the six feet of circuit conductors in a metal raceway or metallic-sheathed cable has been deleted. The effect of the change is to permit circuit conductors to originate in a panel, extend out of the panel using any appropriate wiring method, and locate the AFCI device within 6 feet of the branch circuit overcurrent protective device as measured along the branch circuit conductors.

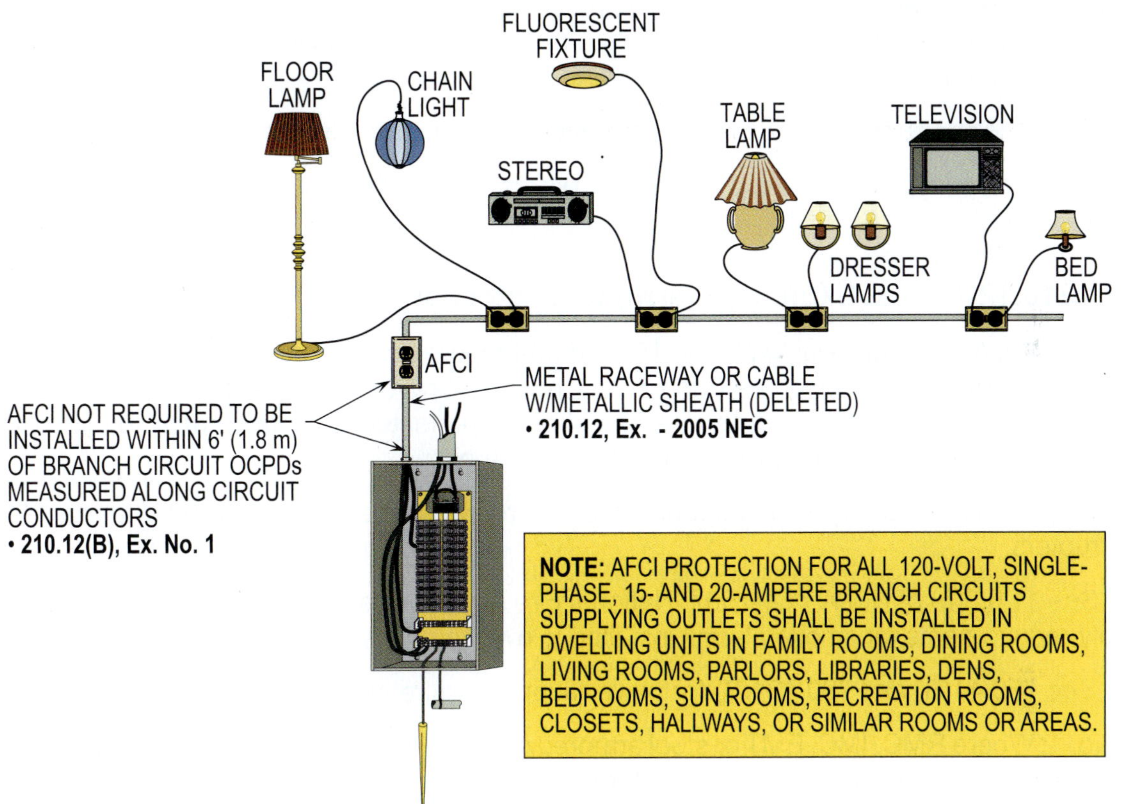
FLUORESCENT FIXTURE
FLOOR LAMP
CHAIN LIGHT
TABLE LAMP
TELEVISION
STEREO
DRESSER LAMPS
BED LAMP
AFCI
METAL RACEWAY OR CABLE W/METALLIC SHEATH (DELETED)
• 210.12, Ex. - 2005 NEC
AFCI NOT REQUIRED TO BE INSTALLED WITHIN 6' (1.8 m) OF BRANCH CIRCUIT OCPDs MEASURED ALONG CIRCUIT CONDUCTORS
• 210.12(B), Ex. No. 1
NOTE: AFCI PROTECTION FOR ALL 120-VOLT, SINGLE-PHASE, 15- AND 20-AMPERE BRANCH CIRCUITS SUPPLYING OUTLETS SHALL BE INSTALLED IN DWELLING UNITS IN FAMILY ROOMS, DINING ROOMS, LIVING ROOMS, PARLORS, LIBRARIES, DENS, BEDROOMS, SUN ROOMS, RECREATION ROOMS, CLOSETS, HALLWAYS, OR SIMILAR ROOMS OR AREAS.

DWELLING UNITS
210.12(B), Exception

Purpose of Change: This revision clarified that enclosing the 6 ft (1.8 m) of circuit conductors in a metal raceway or metallic-sheathed cable is not required.

NEC Ch. 2 - Article 210
Part I - 210.12(B), Exception No. 2

Type of Change	Panel Action	UL	UL 508	API 500 - 1997	API 505 - 1997	OSHA - 1994
Revision	Accept	-	-	-	-	-
ROP	**ROC**	**NFPA 70E - 2004**	**NFPA 70B - 2006**	**NFPA 79 - 2007**	**NFPA**	**NEMA - 2002**
pg. 89 # 2-142	pg. 73 # 2-137	-	-	-	-	WD 6
log: 3488 CMP: 2	log: 840	Submitter: Alan Manche		2005 NEC: 210.12(B)		IEC: -

2005 NEC - 210.12 Arc-Fault Circuit-Interrupter Protection.

(B) Dwelling Unit ~~Bedrooms~~. All 120-volt, single phase, 15- and 20-ampere branch circuits supplying outlets installed in dwelling unit ~~bedrooms~~ shall be protected by a listed arc-fault circuit interrupter, combination type installed to provide protection of the branch circuit.

~~Branch/feeder AFCIs shall be permitted to be used to meet the requirements of 210.12(B) until January 1, 2008.~~

~~FPN:~~ For information on types of arc-fault circuit interrupters, see UL 1699-1999, Standard for Arc-Fault Circuit Interrupters.

2008 NEC - 210.12 Arc-Fault Circuit-Interrupter Protection.

(B) Dwelling Units. All 120-volt, single phase, 15- and 20-ampere branch circuits supplying outlets installed in dwelling unit in family rooms, dining rooms, living rooms, parlors, libraries, dens, bedrooms, sun rooms, recreation rooms, closets, hallways, or similar rooms or areas shall be protected by a listed arc-fault circuit interrupter, combination type installed to provide protection of the branch circuit.

FPN No. 1: For information on types of arc-fault circuit interrupters, see UL 1699-1999, Standard for Arc-Fault Circuit Interrupters.

FPN No. 2: See 11.6.3(5) of NFPA 72®-2007, National Fire Alarm Code®, for information related to secondary power supply requirements for smoke alarms installed in dwelling units.

FPN No. 3: See 760.41(B) and 760.121(B) for power-supply requirements for fire alarm systems.

Exception No. 1: Where RMC, IMC, EMT or steel armored cable, Type AC, meeting the requirements of 250.118 using metal outlet and junction boxes is installed for the portion of the branch circuit between the branch-circuit overcurrent device and the first outlet, it shall be permitted to install a combination AFCI at the first outlet to provide protection for the remaining portion of the branch circuit.

Exception No. 2: Where a branch circuit to a fire alarm system installed in accordance with 760.41(B) and 760.121(B) is installed in in RMC, IMC, EMT, or steel armored cable, Type AC, meeting the requirements of 250.118, with metal outlet and junction boxes, AFCI protection shall be permitted to be omitted.

Author's Comment: A new Exception No. 1 has been added to permit a combination AFCI to be installed at the first outlet to provide protection for the remaining portion of the branch circuit where RMC, IMC, EMT or steel armored cable, Type AC, meeting the requirements of 250.118 using metal outlet and junction boxes is installed for the portion of the branch circuit between the branch-circuit overcurrent device and the first outlet. A new Exception No. 2 has been added permitting rigid metal conduit, intermediate metal conduit, or electrical metallic tubing to protect the branch circuit supplying a fire alarm system branch ciruit in lieu of an AFCI device.

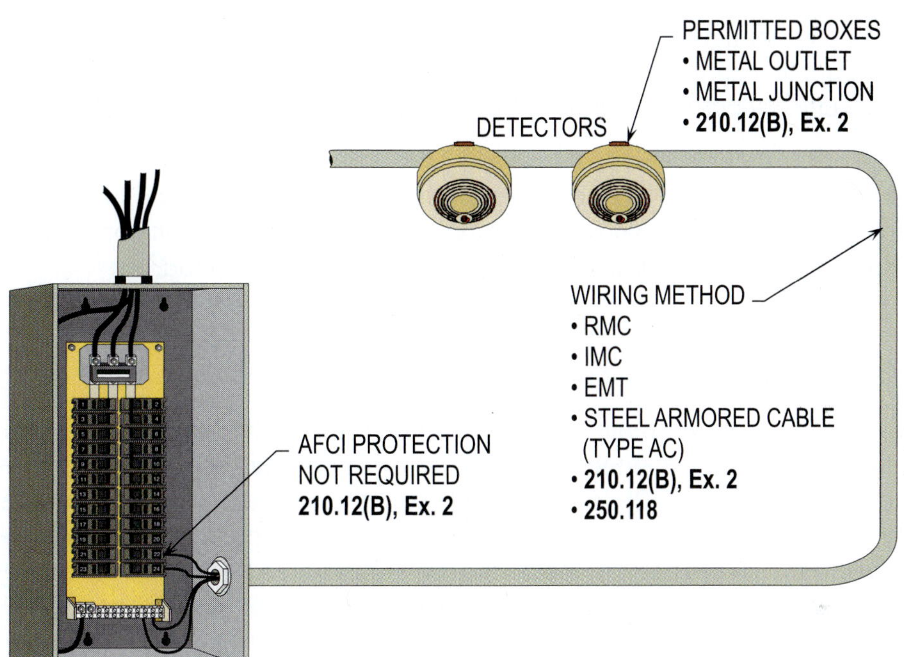

DWELLING UNITS
210.12(B), Exception No. 2

Purpose of Change: A new **Exception No. 2** has been added to permit AFCI protection to be omitted if the branch circuit to a fire alarm system is installed in RMC, IMC, EMT, or steel armored cable (Type AC).

NEC Ch. 2 - Article 210
Part II - 210.19(A)(1), Exception No. 2

Type of Change	Panel Action	UL	UL 508	API 500 - 1997	API 505 - 1997	OSHA - 1994
New Exception	Accept in Principle	-	-	-	-	-
ROP	**ROC**	**NFPA 70E - 2004**	**NFPA 70B - 2006**	**NFPA 79 - 2007**	**NFPA**	**NEMA**
pg. 95 # 2-166	pg. 84 # 2-188	-	-	1.4	-	-
log: 1319 CMP: 2	log: 1900 Submitter: Mike Holt			2005 NEC: -		IEC: -

2005 NEC - 210.19 Conductors — Minimum Ampacity and Size.

(A) Branch Circuits Not More Than 600 Volts.

(1) General. Branch-circuit conductors shall have an ampacity not less than the maximum load to be served. Where a branch circuit supplies continuous loads or any combination of continuous and noncontinuous loads, the minimum branch-circuit conductor size, before the application of any adjustment or correction factors, shall have an allowable ampacity not less than the noncontinuous load plus 125 percent of the continuous load.

Exception: Where the assembly, including the overcurrent devices protecting the branch circuit(s), is listed for operation at 100 percent of its rating, the allowable ampacity of the branch circuit conductors shall be permitted to be not less than the sum of the continuous load plus the noncontinuous load.

Annex D

Example D3a Industial Feeders in a Common Raceway

~~Although~~ the neutral runs between the main switchboard and the building panelboard, likely terminating on a busbar at both locations, ~~the busbar connections are part of listed devices and are not "separately installed pressure devices." Therefore 110.14(C)(2) does not apply, and the normal termination temperature limits apply. In addition, the listing requirements to gain exemption from the additional sizing allowance under continuous loading (*see 215.3 Exception*) covers not just the overcurrent protective device, but its entire assembly as well. Therefore, sincle the lighting load is continuous, the minimum conductor size is based on 1.25 x~~ (11,600 VA / 277V) = ~~52~~ amperes, to be evaluated under the 75°C column of Table 310.16. The minimum size of the neutral ~~is 6~~ AWG. ~~The size is also the minimum size required by 215.2(A)(1), because~~ the minimum size equipment grounding conductor for a 150 ampere circuit, as covered in Table 250.122, is 6 AWG.

2008 NEC - 210.19 Conductors — Minimum Ampacity and Size.

(A) Branch Circuits Not More Than 600 Volts.

(1) General. Branch-circuit conductors shall have an ampacity not less than the maximum load to be served. Where a branch circuit supplies continuous loads or any combination of continuous and noncontinuous loads, the minimum branch-circuit conductor size, before the application of any adjustment or correction factors, shall have an allowable ampacity not less than the noncontinuous load plus 125 percent of the continuous load.

Exception No. 1: Where the assembly, including the overcurrent devices protecting the branch circuit(s), is listed for operation at 100 percent of its rating, the allowable ampacity of the branch circuit conductors shall be permitted to be not less than the sum of the continuous load plus the noncontinuous load.

Exception No. 2: Grounded conductors that are not connected to an overcurrent device shall be permitted to be sized at 100 percent of the continuous and non-continuous load.

Annex D

Example D3a Industrial Feeders in a Common Raceway

Feeder Neutral Conductor (see 220.61)

Because 210.11(B) does not apply to these buildings, the load cannot be assumed to be evenly distributed across phases. Therefore the maximum imbalance must be assumed to be the full lighting load in this case, or 11,600 VA. (11,600 VA / 277V = 42 amperes.) The ability of the neutral to return fault current [see 250.32(B)(2)(2)] is not a factor in this calculation.

Because the neutral runs between the main switchboard and the building panelboard, likely terminating on a busbar at both locations, and not on overcurrent devices, the effects of continuous loading can be disregarded in evaluating its terminations (see 215.2(A)(1), Exception No. 2. That calculation is (11,600 VA / 277V) = 42 amperes, to be evaluated under the 75°C column of Table 310.16. The minimum size of the neutral might seem to be 8 AWG, but that size would not be sufficient to be depended upon in the event of a line-to-neutral short circuit (see 215.2(A)(1), second paragraph). Therefore, since the minimum size equipment grounding conductor for a 150 ampere circuit, as covered in Table 250.122, is 6 AWG, that is the minimum neutral size required for this feeder.

Author's Comment: A new **Exception No. 2** has been added to permit grounded conductors that are not connected to an overcurrent protective device to be sized at 100 percent of continuous and non-continuous loads.

The text in the last paragraph of **Annex D3a** was revised to agree with the addition of new **Exception No. 2** permitting the grounded conductor not connected to an overcurrent protective device to be sized at 100 percent of continuous and non-continuous loads, not 125 percent as previously required.

WHAT SIZE NEUTRAL IS REQUIRED, BASED ON LOAD?

Ungrounded Conductors	**Grounded Conductors**
Step 1: Calculating load **210.19(A)(1); 210.20(A)** Continuous 12 A x 125% = 15 A Noncontinuous 5 A x 100% = 5 A **Total** = 20 A	**Step 1:** Calculating load **210.19(A)(1); 210.20(A)** Continuous 12 A x 100% = 12 A Noncontinuous 5 A x 100% = 5 A **Total** = 17 A
Step 2: Finding conductors **Table 310.16,** Asterisk 20 A requires 12 AWG cu.	**Step 2:** Finding conductors **Table 310.16** 17 A requires 12 AWG cu.
Solution: The size of the ungrounded conductors are 12 AWG cu.	**Solution: The size of the grounded conductor is a 12 AWG cu.**

GENERAL
210.19(A)(1), Exception No. 2

Purpose of Change: A new exception has been added to clarify that the grounded conductor shall be permitted to be sized at 100 percent of continuous and noncontinuous loads where not connected to an overcurrent protective device.

NEC Ch. 2 - Article 210
Part III - 210.52(C)

Type of Change	Panel Action	UL	UL 508	API 500 - 1997	API 505 - 1997	OSHA - 1994
Revision	Accept in Principle	-	-	-	-	-
ROP	**ROC**	**NFPA 70E - 2004**	**NFPA 70B - 2006**	**NFPA 79 - 2007**	**NFPA**	**NEMA - 2002**
pg. 104 # 2-207	pg. 90 # 2-218	-	-	-	-	WD 6
log: 3382 CMP: 2	log: 973 Submitter: Frederic P. Hartwell			2005 NEC: 210.52(C)		IEC: -

2005 NEC - 210.52 Dwelling Unit Receptacle Outlets.

(C) Countertops. In kitchens and dining rooms of dwelling units, receptacle outlets for ~~counter~~ spaces shall be installed in accordance with 210.52(C)(1) through (C)(5).

(2) Island ~~Counter~~ Spaces. At least one receptacle shall be installed at each island ~~counter~~ space with a long dimension of 600 mm (24 in.) or greater and a short dimension of 300 mm (12 in.) or greater. ~~Where a rangetop or sink is installed in an island counter and the width of the counter behind the rangetop or sink is less than 300 mm (12 in.), the rangetop or sink is considered to divide the island into two separate countertop spaces as defined in 210.52(C)(4).~~

Figure 210.52(C)(1) Determination of Area Behind ~~Sink or~~ Range

2008 NEC - 210.52 Dwelling Unit Receptacle Outlets.

(C) Countertops. In kitchens and dining rooms, breakfast rooms, pantries, and similar areas of dwelling units, receptacle outlets for countertop spaces shall be installed in accordance with 210.52(C)(1) through (C)(5). Where a range, counter-mounted cooking unit, or sink is installed in an island or peninsular countertop and the width of the counter behind the range, counter-mounted cooking unit, or sink is less than 300 mm (12 in.), the range, counter-mounted cooking unit, or sink is considered to divide the countertop into two separate countertop spaces as defined in 210.52(C)(4). Each separate countertop space shall comply with the applicable requirements in 210.52(C).

(2) Island Countertop Spaces. At least one receptacle shall be installed at each island countertop space with a long dimension of 600 mm (24 in.) or greater and a short dimension of 300 mm (12 in.) or greater.

Figure 210.52(C)(1) Determination of Area Behind Range, Counter-Mounted Cooking Unit, or Sink

Author's Comment: Where a range, a counter-mounted cooking unit, or sink is installed in either an island or a peninsular countertop and if the width of the space behind these units are 12 inches or less, the countertop is considered to be divided into two separate countertop spaces. Both island and peninsular countertop spaces must comply with **210.52(C)** for placement of receptacles. Moving this text from **210.52(C)(2)** into the base rule of **210.52** now provides consistency for both islands and peninsulas.

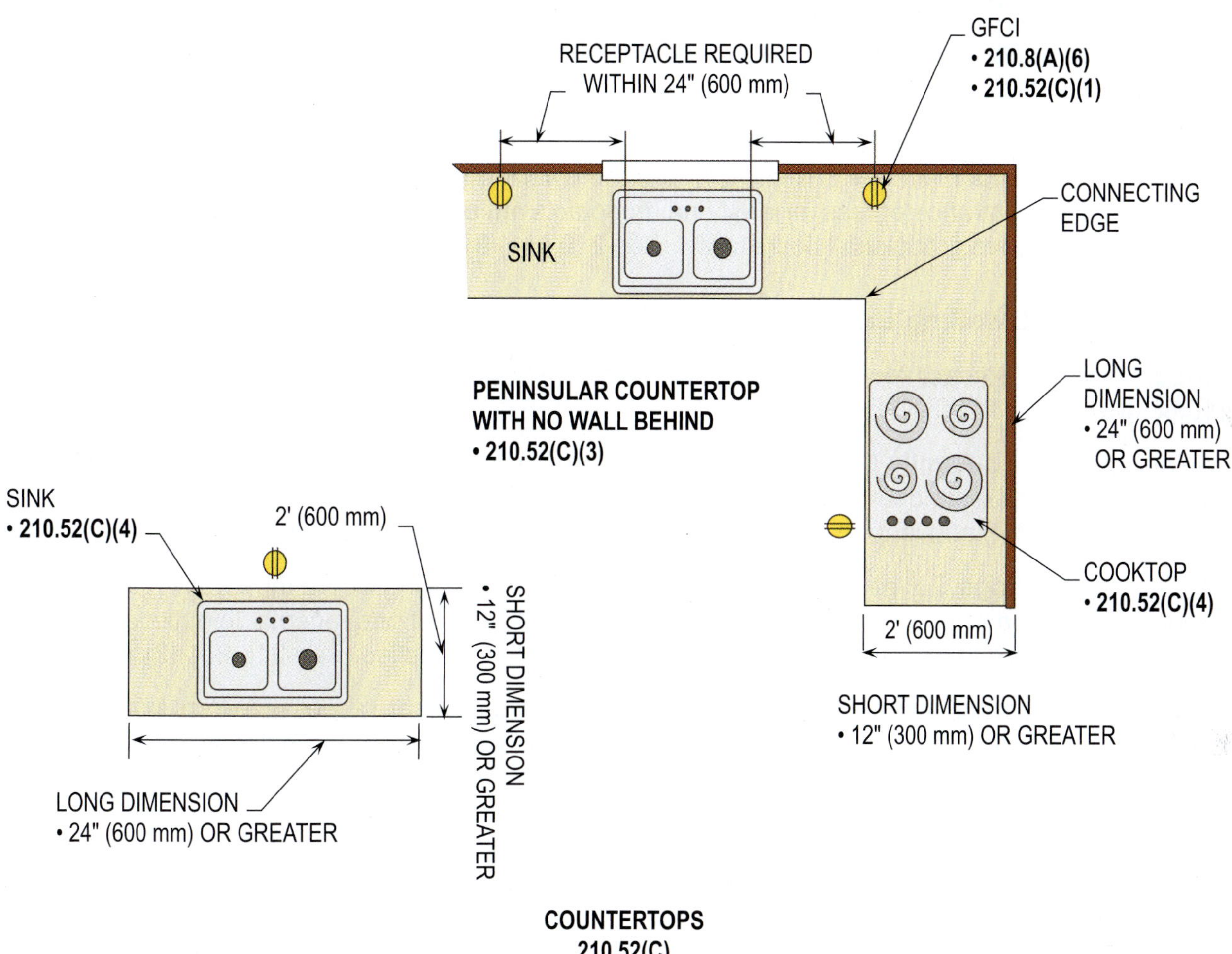

COUNTERTOPS
210.52(C)

Purpose of Change: This revision clarifies that where a range, counter-mounted cooking unit, or sink is installed in either an island or a peninsular countertop, such that the width of the space behind these units is 12 in. (300 mm) or less, the countertop is considered to be divided into two separate countertop spaces.

NEC Ch. 2 - Article 210
Part III - 210.52(E)(1) through (E)(3)

13

Type of Change	Panel Action	UL	UL 508	API 500 - 1997	API 505 - 1997	OSHA - 1994
New Subsections	Accept in Principle	-	-	-	-	-
ROP	**ROC**	**NFPA 70E - 2004**	**NFPA 70B - 2006**	**NFPA 79 - 2007**	**NFPA**	**NEMA - 2002**
pg. 108 # 2-229	pg. 93 # 2-230	-	-	-	-	WD 6
log: 637 CMP: 2	log: 2023 Submitter: Vince Baclawski			2005 NEC: 210.52(E)		IEC: -

2005 NEC - 210.52 Dwelling Unit Receptacle Outlets.

(E) Outdoor Outlets. For a one-family dwelling and each unit of a two-family dwelling that is at grade level, at least one receptacle outlet accessible at grade level and not more than 2.0 m (6 1/2 ft) above grade shall be installed at the front and back of the dwelling.

For each dwelling unit of a multifamily dwelling where the dwelling unit is located at grade level and provided with individual exterior entrance/egress, at least one receptacle outlet accessible from grade level and not more than 2.0 m (6 1/2 ft) above grade shall be installed. See 210.8(A)(3).

2008 NEC - 210.52 Dwelling Unit Receptacle Outlets.

(E) Outdoor Outlets. Outdoor receptacle outlets shall be installed in accordance with (E)(1) through (E)(3). [See 210.8(A)(3)]

(1) One-Family and Two-Family Dwellings. For a one-family dwelling and each unit of a two-family dwelling that is at grade level, at least one receptacle outlet accessible while standing at grade level and located not more than 2.0 m (6 1/2 ft) above grade shall be installed at the front and back of the dwelling.

(2) Multi-Family Dwellings. For each dwelling unit of a multifamily dwelling where the dwelling unit is located at grade level and provided with individual exterior entrance/egress, at least one receptacle outlet accessible from grade level and not more than 2.0 m (6 1/2 ft) above grade shall be installed. See 210.8(A)(3).

(3) Balconies, Decks, and Porches. Balconies, decks, and porches that are accessible from inside the dwelling unit shall have at least one receptacle outlet installed within the perimeter of the balcony, deck, or porch. The receptacle shall not be located more than 2.0 m (6 1/2 ft) above the balcony, deck, or porch surface.

Exception to (3): Balconies, decks, or porches with a useable area of less than 1.86m^2 (20 sq ft) are not required to have a receptacle installed.

Author's Comment: The new item **(3)** added by the Panel requires a receptacle be installed on any porch, deck or balcony where the balcony, deck, or porch is accessible from inside the dwelling unit. This receptacle was intended to be in addition to those that are installed to meet **(1)** or **(2)**.

Comment 2-225 added text to clarify that at least one receptacle must be accessible while standing at grade level. The word "located" was added to make the language more technically correct.

Comment 2-227 clarifies that balconies, decks, and porches accessible from inside the dwelling units must have at least one receptacle outlet installed within the perimeter of the balcony, deck, or porch. The receptacle must also be located at not more than 6 1/2 feet above the balcony, deck, or porch.

Comment 2-230 has added an exception to not require a receptacle to be installed for a balcony, deck, or porch that has a useable area of less than 20 square feet, especially if these areas are used for decorative or architectural purposes.

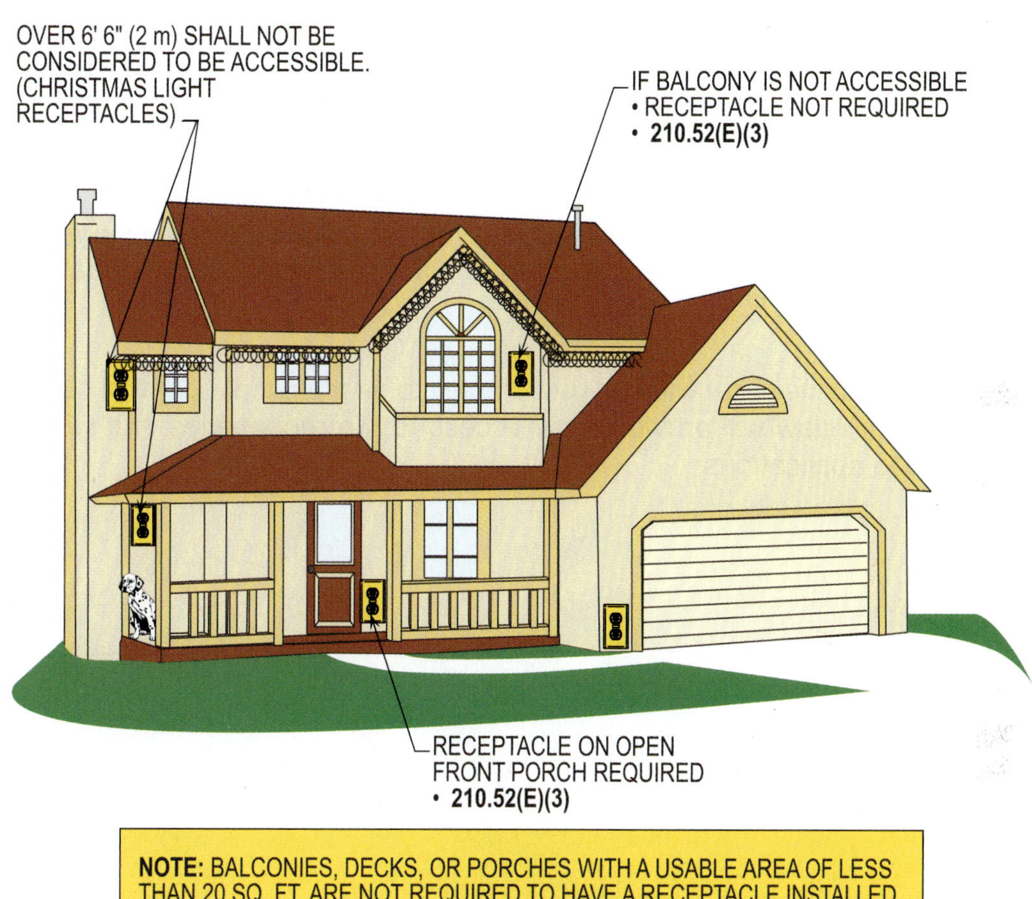
OVER 6' 6" (2 m) SHALL NOT BE CONSIDERED TO BE ACCESSIBLE. (CHRISTMAS LIGHT RECEPTACLES)
IF BALCONY IS NOT ACCESSIBLE
• RECEPTACLE NOT REQUIRED
• 210.52(E)(3)
RECEPTACLE ON OPEN FRONT PORCH REQUIRED
• 210.52(E)(3)
NOTE: BALCONIES, DECKS, OR PORCHES WITH A USABLE AREA OF LESS THAN 20 SQ. FT. ARE NOT REQUIRED TO HAVE A RECEPTACLE INSTALLED.

OUTDOOR OUTLETS
210.52(E)(1) through (E)(3)

Purpose of Change: New subsection titles have been added for usability. A new subsection **(3)** has been added to require a receptacle be installed on any porch, deck, or balcony where accessible from inside the dwelling unit. A new **Exception** to **(3)** does not require a receptacle on a balcony, deck, or porch with a usable area of less than 20 sq. ft.

NEC Ch. 2 - Article 210
Part III - 210.62

Type of Change	Panel Action	UL	UL 508	API 500 - 1997	API 505 - 1997	OSHA - 1994
Revision	Accept	-	-	-	-	-
ROP	**ROC**	**NFPA 70E - 2004**	**NFPA 70B - 2006**	**NFPA 79 - 2007**	**NFPA**	**NEMA - 2002**
pg. 111 # 2-244	pg. - # -	-	-	-	-	WD 6
log: 1895 CMP: 2	log: -	Submitter: James W. Carpenter		2005 NEC: 210.62		IEC: -

2005 NEC - 210.62 Show Windows.

At least one receptacle outlet shall be installed ~~directly above~~ a show window for each 3.7 linear m (12 linear ft) or major fraction thereof of show window area measured horizontally at its maximum width.

2008 NEC - 210.62 Show Windows.

At least one receptacle outlet shall be installed within 450 mm (18 in) of the top of a show window for each 3.7 linear m (12 linear ft) or major fraction thereof of show window area measured horizontally at its maximum width.

Author's Comment: Requiring show window receptacles to be placed at height of not greater than 18 inches above the show window will provide easy access to the receptacles and will limit the use of extension cords in these applications.

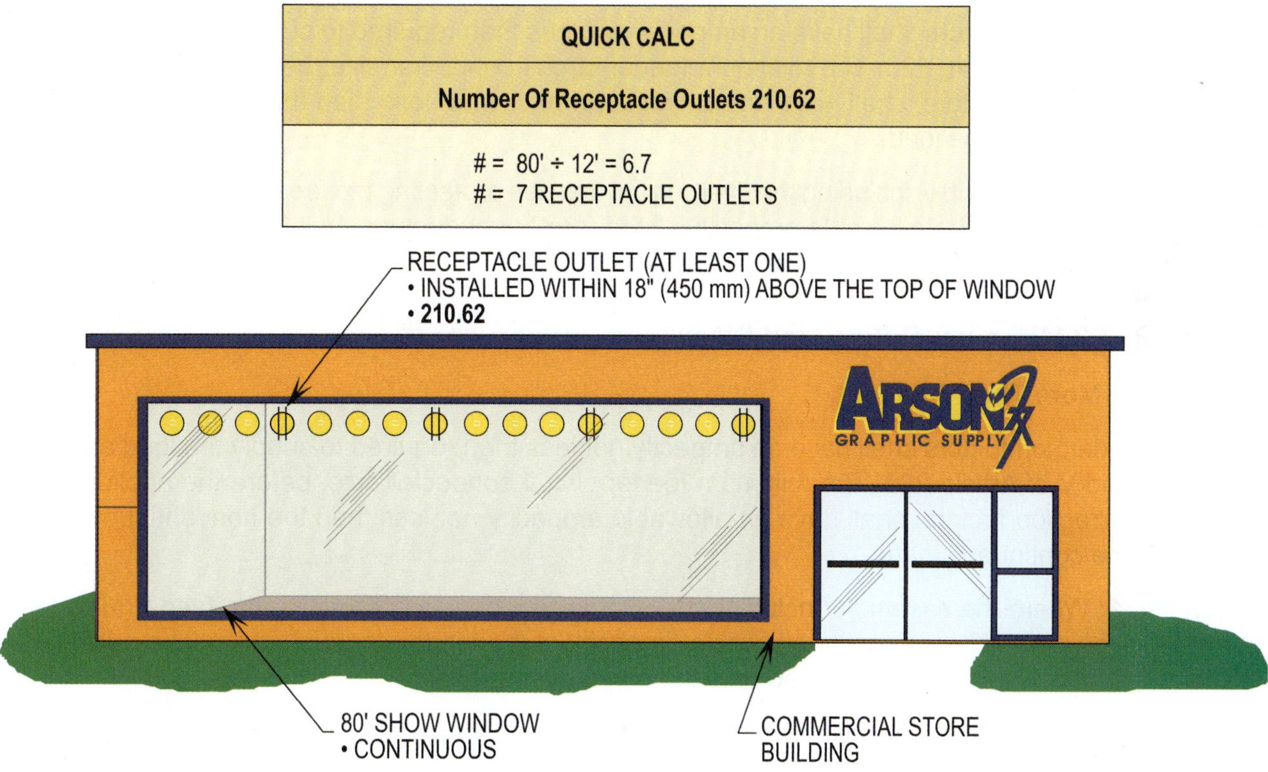

SHOW WINDOWS
210.62

Purpose of Change: This revision clarifies that at least one receptacle outlet shall be installed within 18 in. (450 mm) above the top of a show window.

NEC Ch. 2 - Article 215
Part - 215.2(A)(1), Exception No. 2

Type of Change		Panel Action		UL	UL 508	API 500 - 1997	API 505 - 1997	OSHA - 1994
New Exception		Accept in Principle		-	-	-	-	-
ROP		ROC		NFPA 70E - 2004	NFPA 70B - 2006	NFPA 79 - 2007	NFPA	NEMA
pg. 116	# 2-275	pg. -	# -	-	-	1.4	-	-
log: 1322	CMP: 2	log: -	Submitter: Mike Holt			2005 NEC: -		IEC: -

2005 NEC - 215.2 Minimum Rating and Size.

(A) Feeders Not More Than 600 Volts.

(1) General. Feeder conductors shall have an ampacity not less than required to supply the load as calculated in Parts III, IV, and V of Article 220. The minimum feeder-circuit conductor size, before the application of any adjustment or correction factors, shall have an allowable ampacity not less than the noncontinuous load plus 125 percent of the continuous load.

Exception: Where the assembly, including the overcurrent devices protecting the feeder(s), is listed for operation at 100 percent of its rating, the allowable ampacity of the feeder conductors shall be permitted to be not less than the sum of the continuous load plus the noncontinuous load.

2008 NEC - 215.2 Minimum Rating and Size.

(A) Feeders Not More Than 600 Volts.

(1) General. Feeder conductors shall have an ampacity not less than required to supply the load as calculated in Parts III, IV, and V of Article 220. The minimum feeder-circuit conductor size, before the application of any adjustment or correction factors, shall have an allowable ampacity not less than the noncontinuous load plus 125 percent of the continuous load.

Exception No. 1: Where the assembly, including the overcurrent devices protecting the feeder(s), is listed for operation at 100 percent of its rating, the allowable ampacity of the feeder conductors shall be permitted to be not less than the sum of the continuous load plus the noncontinuous load.

Exception No. 2: Grounded conductors that are not connected to an overcurrent device shall be permitted to be sized at 100 percent of the continuous and noncontinuous load.

Author's Comment: Conductors connected to an overcurrent protective device are required to be sized based on 125 percent of a continuous load plus 100 percent of any non-continuous loads. The conductors act as a heat sink for any heat generated in the overcurrent protective device and the combination of the conductor and the overcurrent device are tested together with consideration of this heat transfer. Since a grounded conductor is not normally connected to an overcurrent protective device, the grounded conductor can be sized based on 100 percent of continuous and non-continuous loads.

SIZING OCPD AND THWN COPPER CONDUCTORS

Sizing OCPD based on loads

Step 1: Calculating loads
215.3; 215.2(A)(1)
Table 430.248; 220.14

- Lighting load
  - 40 A x 125% = 50 A
- Receptacle load
  - 37.5 A x 100% = 37.5 A
- Appliance load
  - 30 A x 125% = 37.5 A
- Heat or A/C load
  - 10 kVA x 1,000 x 100% / 240 V = 42 A
- Motor load
  - 5 HP = 28 A x 100% = 28 A
  - 1 HP = 16 A x 100% = 16 A
  - 2 HP = 24 A x 100% = 24 A
- Largest motor load
  - 28 A x 25% = 7 A

Total load = 242 A

Step 2: Selecting OCPD based on load
215.3; 240.4(G); 430.63
242 A allows 250 A OCPD

Solution: **The size OCPD based upon calculated load is 250 amps.**

Sizing OCPD based on motor load

Step 1: Calculating OCPD
430.52(A)(1); 430.62(A); 240.4(G)
Table 430.52
Motor loads
- 5 HP = 28 A x 250% = 70 A
- 1 HP = 16 A x 100% = 16 A
- 2 HP = 24 A x 100% = 24 A

Step 2: Other loads
- Lighting load = 50 A
- Receptacle load = 37.5 A
- Appliance load = 37.5 A
- Heating load = 42 A

Total load = 277 A

Step 3: Selecting OCPD
430.62(A); 240.6(A)
250 A is the next size below 277 A

Solution: **The size OCPD based upon motor loads is 250 amps.**

Note: In most cases, the calculated load produces the largest or same size OCPD, unless there is an unusually large motor involved.

Sizing conductors

Step 1: Calculating loads
215.2(A)(1)

• Lighting load		
40 A x 125%	=	50 A
• Receptacle load		
37.5 A x 100%	=	37.5 A
• Appliance load		
30 A x 125%	=	37.5 A
• Heat or A/C load		
10 kVA x 1,000 x 100% ÷ 240 V	=	42 A
• Motor load		
5 HP = 28 A x 100%	=	28 A
1 HP = 16 A x 100%	=	16 A
2 HP = 24 A x 100%	=	24 A
• Largest motor load		
28 A x 25%	=	7 A
Total load	=	242 A

Step 2: Selecting conductors
310.10, FPN (2); Table 310.16
242 A requires 250 KCMIL

Solution: **The size conductors are 250 KCMIL THWN cu.**

Sizing neutral

Step 1: Calculating load
430.24; 220.61

• Lighting load		
40 A x 100%	=	40 A
• Receptacle load		
37.5 A x 100%	=	37.5 A
• Appliance load		
30 A x 100%	=	30 A
• Motor load		
1 HP = 16 A x 100%	=	16 A
2 HP = 24 A x 100%	=	24 A
• Largest motor load		
24 A x 25%	=	6 A
Total load	=	153.5 A

Step 2: Selecting conductor
310.10, FPN (2); Table 310.16
153.5 A requires 2/0 AWG cu.

Solution: **The size of the conductor based upon 100% are 2/0 AWG THWN copper.**

Note: **The AHJ may require the continuous loads to be multiplied by 125% when calculating the neutral load**

GENERAL
215.2(A)(1), Exception No. 2

Purpose of Change: A new exception has been added to clarify that the grounded conductor shall be sized at 100 percent of continuous and noncontinuous loads when not connected to an overcurrent protective device.

NEC Ch. 2 - Article 215
Part - 215.6

Type of Change	Panel Action	UL	UL 508	API 500 - 1997	API 505 - 1997	OSHA - 1994
Revision	Accept	-	-	-	-	-
ROP	**ROC**	**NFPA 70E - 2004**	**NFPA 70B - 2006**	**NFPA 79 - 2007**	**NFPA**	**NEMA**
pg. 118 # 2-283	pg. 98 # 2-250	-	-	-	-	-
log: 524 CMP: 2	log: 24	Submitter: Michael J. Johnston		2005 NEC: 215.6		IEC: -

2005 NEC - 215.6 Feeder ~~Conductor~~ Grounding ~~Means~~.

Where a feeder supplies branch circuits in which equipment grounding conductors are required, the feeder shall include or provide ~~a grounding means~~, in accordance with the provisions of 250.134, to which the equipment grounding conductors of the branch circuits shall be connected.

2008 NEC - 215.6 Feeder Equipment Grounding Conductor.

Where a feeder supplies branch circuits in which equipment grounding conductors are required, the feeder shall include or provide an equipment grounding conductor, in accordance with the provisions of 250.134, to which the equipment grounding conductors of the branch circuits shall be connected. Where the feeder supplies a separate building or structure, the requirements of 250.32(B) shall apply.

Author's Comment: The last sentence was added for installations of feeders where a separate building or structure is not supplied with an equipment grounding conductor but utilizes the grounded conductor for the fault return path, as well as any neutral current. Section **250.32(B)** has been revised to only permit the grounded (neutral) conductor to provide the fault return path as well as the neutral current for existing premises wiring systems only.

WHAT SIZE NEUTRAL IS REQUIRED, BASED ON LOAD?

Step 1: Calculating neutral load
220.61(A); 225.3(B); 310.15(B)(4)(c)
50 A x 100% = 50 A

Step 2: Sizing neutral
250.32(B), Ex.; Table 310.16
50 A requires 8 AWG cu.

Step 3: Sizing neutral to be used as EGC
250.32(B), Ex.; Table 250.122
225 A OCPD requires 4 AWG cu.

Solution: **The grounded neutral conductor must be 4 AWG THNN copper to serve as an EGC.**

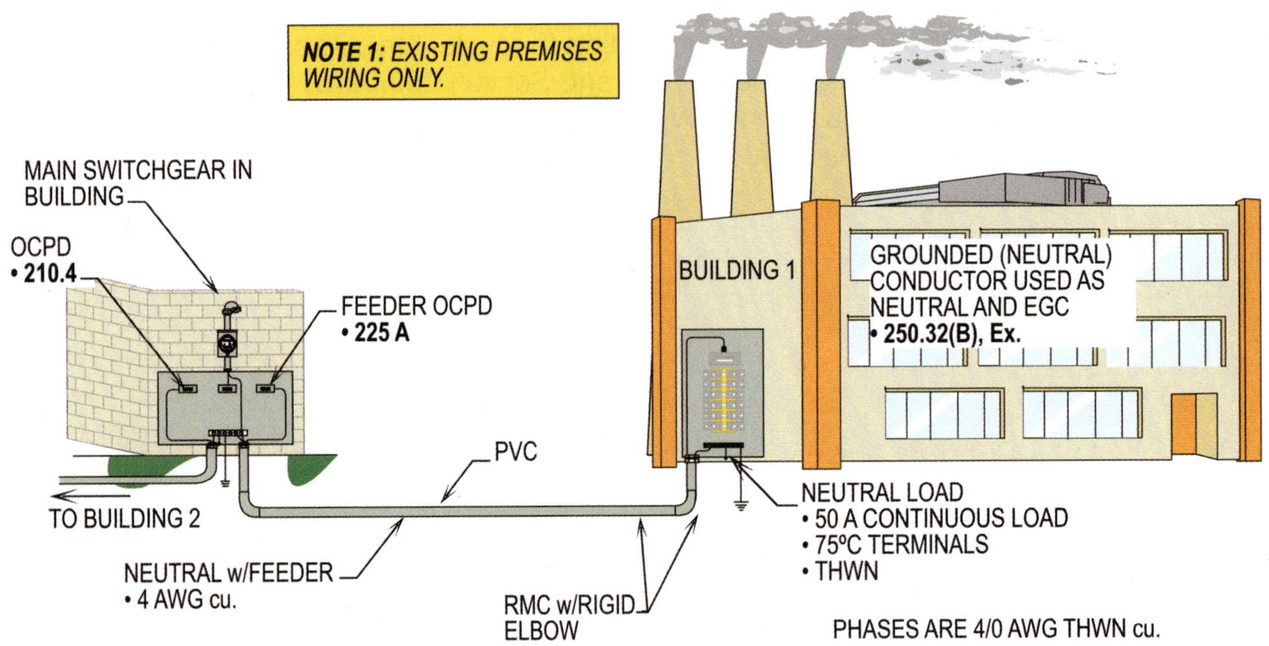

FEEDER EQUIPMENT GROUNDING CONDUCTOR
215.6

Purpose of Change: This revision correlates with installations where a separate building or structure is not supplied with an equipment grounding conductor but utilizes the grounded conductor for the fault return path but for existing premises wiring systems only.

NEC Ch. 2 - Article 215
Part - 215.12(C)

Type of Change	Panel Action	UL	UL 508	API 500 - 1997	API 505 - 1997	OSHA - 1994
Revision	Accept in Principle	-	-	-	-	1910.304(a)(1)
ROP	**ROC**	**NFPA 70E - 2004**	**NFPA 70B - 2006**	**NFPA 79 - 2007**	**NFPA**	**NEMA**
pg. 120 # 2-292	pg. 98 # 2-254	410.3	-	13.2.4	-	-
log: 2736 CMP: 2	log: 1898	Submitter: Jim Pauley		2005 NEC: 215.12(C)		IEC: -

2005 NEC - 215.12 Identification for Feeders.

(C) Ungrounded Conductors. Where the premises wiring system has feeders supplied from more than one nominal voltage system, each ungrounded conductor of a feeder~~, where accessible,~~ shall be identified by system. The ~~means of identification shall be permitted to be by separate color coding, marking tape, tagging, or other approved means and~~ shall be permanently posted at each feeder panelboard or similar feeder distribution equipment.

2008 NEC - 215.12 Identification for Feeders.

(C) Ungrounded Conductors. Where the premises wiring system has feeders supplied from more than one nominal voltage system, each ungrounded conductor of a feeder shall be identified by phase or line and system at all termination, connection, and splice points. The method utilized for conductors originating within each feeder panelboard or similar feeder distribution equipment shall be documented in a manner that is readily available or shall be permanently posted at each feeder panelboard or similar feeder distribution equipment.

Author's Comment: Requiring identification of each ungrounded conductor of the feeder by phase or by line in a single phase system and identification by system will help electrical installation and maintenance personnel identify each phase of the system. The Comment has revised the last line in this section to require the identification posted at the panelboard or be documented in a manner that is readily available to identify the conductors that originate at each panelboard.

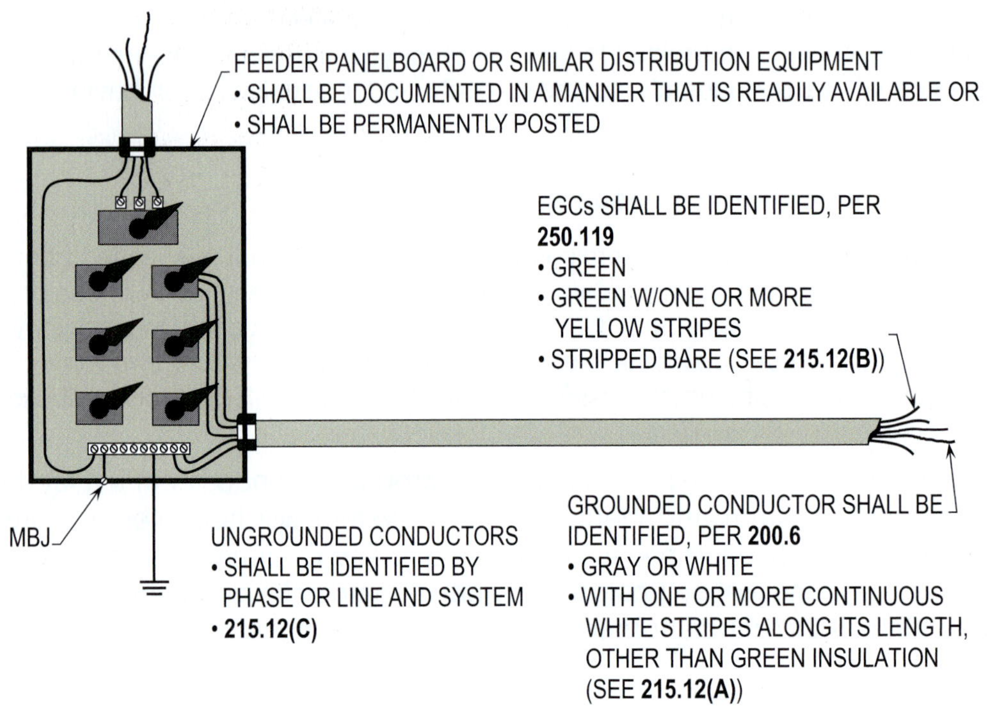
FEEDER PANELBOARD OR SIMILAR DISTRIBUTION EQUIPMENT
• SHALL BE DOCUMENTED IN A MANNER THAT IS READILY AVAILABLE OR
• SHALL BE PERMANENTLY POSTED
EGCs SHALL BE IDENTIFIED, PER 250.119
• GREEN
• GREEN W/ONE OR MORE YELLOW STRIPES
• STRIPPED BARE (SEE 215.12(B))
MBJ
UNGROUNDED CONDUCTORS
• SHALL BE IDENTIFIED BY PHASE OR LINE AND SYSTEM
• 215.12(C)
GROUNDED CONDUCTOR SHALL BE IDENTIFIED, PER 200.6
• GRAY OR WHITE
• WITH ONE OR MORE CONTINUOUS WHITE STRIPES ALONG ITS LENGTH, OTHER THAN GREEN INSULATION (SEE 215.12(A))

UNGROUNDED CONDUCTORS
215.12(C)

Purpose of Change: This revision clarifies the identification of each ungrounded conductor of the feeder by phase or line and system to help maintenance personnel identify each phase of the system. Identification shall be posted or documented at the feeder panelboard or similar distribution equipment or be documented in a manner that is readily available.

NEC Ch. 2 - Article 225
Part - 225.18

Type of Change	Panel Action	UL	UL 508	API 500 - 1997	API 505 - 1997	OSHA - 1994
Revision	Accept in Principle	-	-	-	-	1910.304(c)(2)
ROP	**ROC**	**NFPA 70E - 2004**	**NFPA 70B - 2006**	**NFPA 79 - 2007**	**NFPA**	**NEMA**
pg. 137 # 4-12	pg. - # -	410.7(B)	-	-	-	-
log: 620 CMP: 4	log: - Submitter: Michael J. Johnston			2005 NEC: 225.18		IEC: -

2005 NEC - 225.18 Clearance ~~from Ground.~~

Overhead spans of open conductors and open multiconductor cables of not over 600 volts, nominal, shall have a clearance of not less than the following:

2008 NEC - 225.18 Clearance for Overhead Conductors and Cables.

Overhead spans of open conductors and open multiconductor cables of not over 600 volts, nominal, shall have a clearance of not less than the following:

Author's Comment: The term "grade" was too subjective to replace "from Ground", as suggested in the proposal, since there is initial grade at time of construction, there is intermediate grade during construction, and then final grade at the end of construction. "Grade" could ultimately be changed at any future time. The title uses the actual text of "Clearance for Overhead Conductors and Cables" within the section itself.

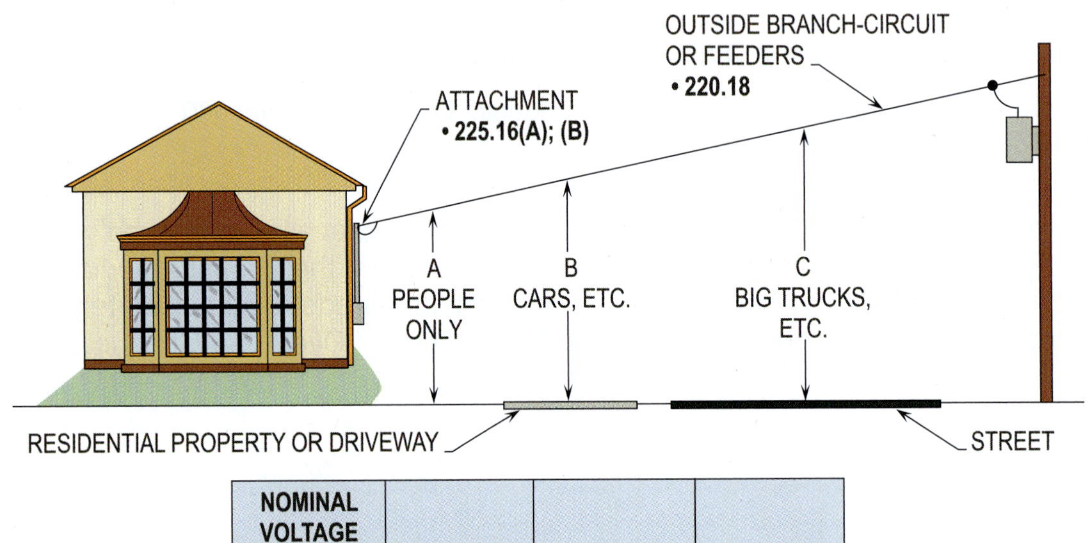

NOMINAL VOLTAGE TO GROUND			
150 V	10' (3.0 m)	12' (3.7 m)	18' (5.5 m)
300 V	12' (3.7 m)	12' (3.7 m)	18' (5.5 m)
600 V	15' (4.5 m)	15' (4.5 m)	18' (5.5 m)

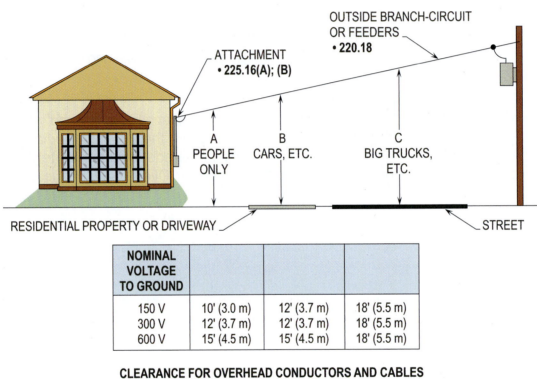

CLEARANCE FOR OVERHEAD CONDUCTORS AND CABLES
225.18

Purpose of Change: This revision to the title clarifies that the clearance for overhead conductors and cables is subject to change at any given time period and shall be maintained.

NEC Ch. 2 - Article 225
Part - 225.39

14

Type of Change	Panel Action	UL	UL 508	API 500 - 1997	API 505 - 1997	OSHA - 1994
Revision	Accept in Principle	-	-	-	-	-
ROP	**ROC**	**NFPA 70E - 2004**	**NFPA 70B - 2006**	**NFPA 79 - 2007**	**NFPA**	**NEMA**
pg. 140 # 4-25	pg. - # -	-	-	-	-	-
log: 2156 CMP: 4	log: -	Submitter: James Grant		2005 NEC: 225.39		IEC: -

2005 NEC - 225.39 Rating of Disconnect.

The feeder or branch-circuit disconnecting means shall have a rating of not less than the load to be supplied, determined in accordance with Parts I and II of Article 220 for branch circuits, Parts III or IV of Article 220 for feeders, or Part V of Article 220 for farm loads. In no case shall the rating be lower than specified in 225.39(A), (B), (C), or (D).

2008 NEC - 225.39 Rating of Disconnect.

The feeder or branch-circuit disconnecting means shall have a rating of not less than the calculated load to be supplied, determined in accordance with Parts I and II of Article 220 for branch circuits, Parts III or IV of Article 220 for feeders, or Part V of Article 220 for farm loads. Where the branch circuit or feeder disconnecting means consists of more than one switch or circuit breaker, as permitted by 225.33, combining the ratings of all the switches or circuit breakers for determining the rating of the disconnecting means shall be permitted. In no case shall the rating be lower than specified in 225.39(A), (B), (C), or (D).

Author's Comment: Adding the word "calculated" agrees with changes throughout the NEC in 2005. The second change involves an added sentence for dealing with multiple disconnects. Since a disconnect is a device or group of devices, permission is inherent to add each device to reach a total rating in compliance with this section. The new revised text will eliminate any confusion that there is inherent permission for breakers to be additive in calculating the rating of a disconnect means. For example, where the feeder or branch circuit disconnecting means is required to be not less than 60 amps, the ratings of each of the multiple disconnects can be combined. Two 30 amp disconnects would satisfy the minimum 60 amp requirement.

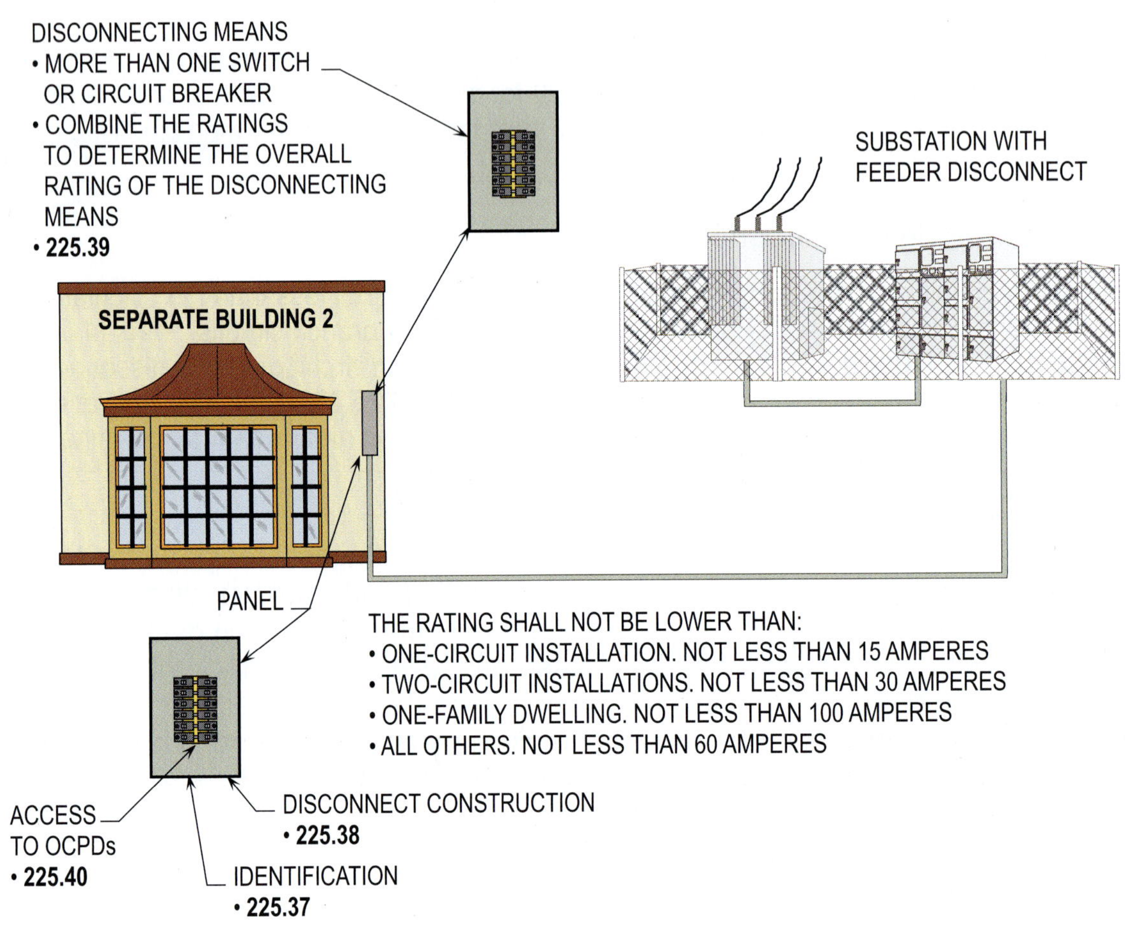
DISCONNECTING MEANS
• MORE THAN ONE SWITCH OR CIRCUIT BREAKER
• COMBINE THE RATINGS TO DETERMINE THE OVERALL RATING OF THE DISCONNECTING MEANS
• 225.39
SUBSTATION WITH FEEDER DISCONNECT
SEPARATE BUILDING 2
PANEL
THE RATING SHALL NOT BE LOWER THAN:
• ONE-CIRCUIT INSTALLATION. NOT LESS THAN 15 AMPERES
• TWO-CIRCUIT INSTALLATIONS. NOT LESS THAN 30 AMPERES
• ONE-FAMILY DWELLING. NOT LESS THAN 100 AMPERES
• ALL OTHERS. NOT LESS THAN 60 AMPERES
ACCESS TO OCPDs
• 225.40
DISCONNECT CONSTRUCTION
• 225.38
IDENTIFICATION
• 225.37

RATING OF DISCONNECT
225.39

Purpose of Change: This revision clarifies that the ratings of all switches or circuit breakers shall be permitted to be combined to determine the overall rating of the disconnecting means.

NEC Ch. 2 - Article 230
Part II - 230.24(B)

Type of Change	Panel Action	UL	UL 508	API 500 - 1997	API 505 - 1997	OSHA - 1994
Revision	Accept in Principle	-	-	-	-	1910.304(c)(2)
ROP	**ROC**	**NFPA 70E - 2004**	**NFPA 70B - 2006**	**NFPA 79 - 2007**	**NFPA**	**NEMA**
pg. 142 # 4-40	pg. - # -	410.7(B)	-	-	-	-
log: 619 CMP: 4	log: -	Submitter: Michael J. Johnston		2005 NEC: 230.24(B)		IEC: -

2005 NEC - 230.24 Clearances.

(B) Vertical Clearance ~~from Ground~~. Service-drop conductors, where not in excess of 600 volts, nominal, shall have the following minimum clearance from final grade:

(1) 3.0 m (10 ft) – at the electric service entrance to buildings, also at the lowest point of the drip loop of the building electric entrance, and above areas or sidewalks accessible only to pedestrians, measured from final grade or other accessible surface only for service-drop cables supported on and cabled together with a grounded bare messenger where the voltage does not exceed 150 volts to ground

(2) 3.7 m (12 ft) – over residential property and driveways, and those commercial areas not subject to truck traffic where the voltage does not exceed 300 volts to ground

(3) 4.5 m (15 ft) – for those areas listed in the 3.7-m (12 ft) classification where the voltage exceeds 300 volts to ground

(4) 5.5 m (18 ft) – over public streets, alleys, roads, parking areas subject to truck traffic, driveways on other than residential property, and other land such as cultivated, grazing, forest, and orchard

2008 NEC - 230.24 Clearances.

(B) Vertical Clearance for Service-drop Conductors. Service-drop conductors, where not in excess of 600 volts, nominal, shall have the following minimum clearance from final grade:

(1) 3.0 m (10 ft) – at the electric service entrance to buildings, also at the lowest point of the drip loop of the building electric entrance, and above areas or sidewalks accessible only to pedestrians, measured from final grade or other accessible surface only for service-drop cables supported on and cabled together with a grounded bare messenger where the voltage does not exceed 150 volts to ground

(2) 3.7 m (12 ft) – over residential property and driveways, and those commercial areas not subject to truck traffic where the voltage does not exceed 300 volts to ground

(3) 4.5 m (15 ft) – for those areas listed in the 3.7-m (12 ft) classification where the voltage exceeds 300 volts to ground

(4) 5.5 m (18 ft) – over public streets, alleys, roads, parking areas subject to truck traffic, driveways on other than residential property, and other land such as cultivated, grazing, forest, and orchard

Author's Comment: The term "grade" was not specific enough to replace "from Ground", as proposed, since there is initial grade at time of construction, there is intermediate grade during construction, and then final grade at the end of construction. "Grade" could ultimately be changed at any future time. The title uses the actual text "for Service-drop Conductors" within the section itself.

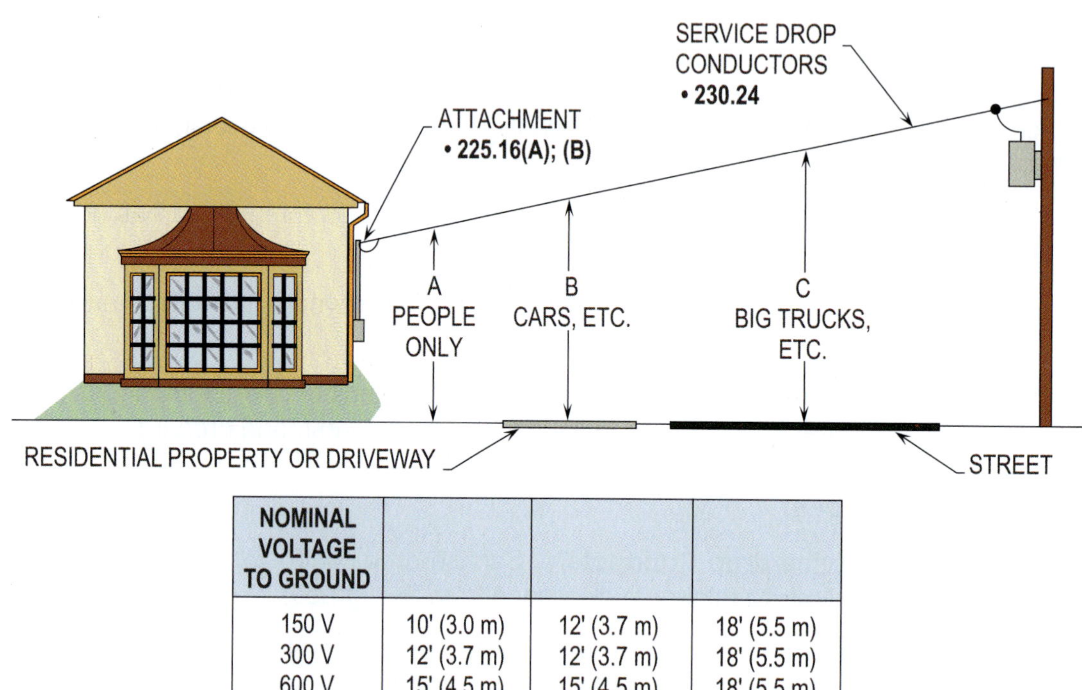

NOMINAL VOLTAGE TO GROUND			
150 V	10' (3.0 m)	12' (3.7 m)	18' (5.5 m)
300 V	12' (3.7 m)	12' (3.7 m)	18' (5.5 m)
600 V	15' (4.5 m)	15' (4.5 m)	18' (5.5 m)

VERTICAL CLEARANCE FOR SERVICE-DROP CONDUCTORS
230.24(B)

Purpose of Change: This revision clarifies that clearances for service-drop conductors shall be based on the minimum clearances from final grade.

NEC Ch. 2 - Article 230
Part IV - 230.40, Exception No. 1

Type of Change	Panel Action	UL	UL 508	API 500 - 1997	API 505 - 1997	OSHA - 1994
Revision	Accept in Principle	-	-	-	-	-
ROP	**ROC**	**NFPA 70E - 2004**	**NFPA 70B - 2006**	**NFPA 79 - 2007**	**NFPA**	**NEMA**
pg. 143 # 4-42	pg. - # -	-	-	-	-	-
log: 2188 CMP: 4	log: - Submitter: Dann Strube			2005 NEC: 230.40, Ex. 1		IEC: -

2005 NEC - 230.40 Number of Service-Entrance Conductor Sets.

Each service drop or lateral shall supply only one set of service-entrance conductors.

Exception No. 1: A building shall be permitted to have one set of service-entrance conductors for each service, as defined in 230.2, run to each occupancy or group of occupancies.

2008 NEC - 230.40 Number of Service-Entrance Conductor Sets.

Each service drop or lateral shall supply only one set of service-entrance conductors.

Exception No. 1: A building <u>with more than one occupancy</u> shall be permitted to have one set of service-entrance conductors for each service, as defined in 230.2, run to each occupancy or group of occupancies.

Author's Comment: Section **230.2** and **230.40** main rules are clear that a building with one occupancy can have more than one service. It is also clear that a set of entrance conductors is allowed for each service to a one-occupant building. The only application for **Exception No. 1** is where a multi-occupancy situation is involved. To reduce confusionin the use of this exception, that condition has been clarified.

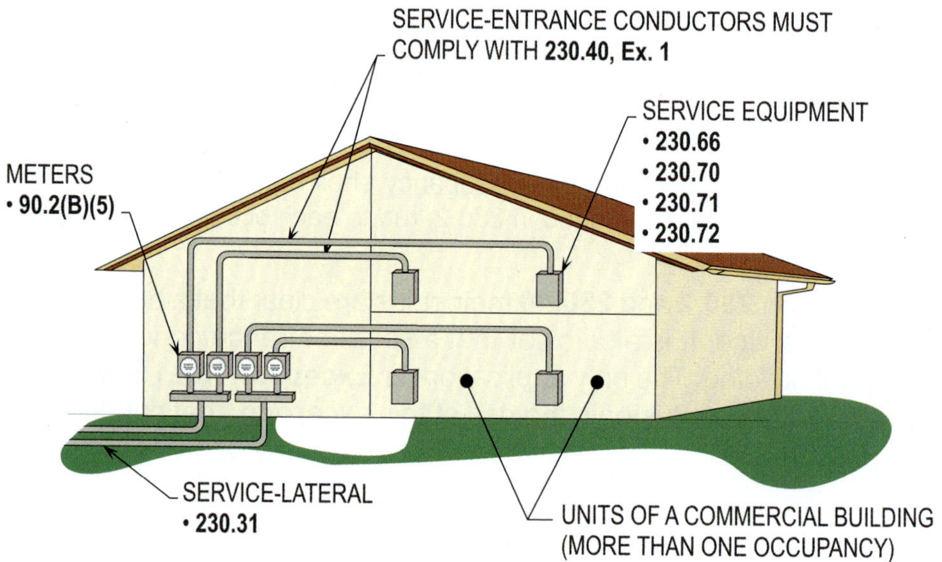

NUMBER OF SERVICE-ENTRANCE CONDUCTOR SETS
230.40, Exception No. 1

Purpose of Change: This revision clarifies that a building with more than one occupancy shall be permitted to have one set of service-entrance conductors run to each occupancy or group of occupancies.

NEC Ch. 2 - Article 230
Part IV - 230.44, Exception

15

Type of Change	Panel Action	UL	UL 508	API 500 - 1997	API 505 - 1997	OSHA - 1994
Revision	Accept in Principle	-	-	-	-	-
ROP	**ROC**	**NFPA 70E - 2004**	**NFPA 70B - 2006**	**NFPA 79 - 2007**	**NFPA**	**NEMA - 2002**
pg. 146 # 4-52	pg. - # -	-	-	13.5.10	-	VE 1
log: 1898 CMP: 4	log: -	Submitter: James W. Carpenter		2005 NEC: 230.44, Ex.		IEC: -

2005 NEC - 230.44 Cable Trays.

Cable tray systems shall be permitted to support service-entrance conductors. Cable trays used to support service-entrance conductors shall contain only service-entrance conductors.

Exception: Conductors, other than service-entrance conductors, shall be permitted to be installed in a cable tray with service-entrance conductors, provided a solid fixed barrier of a material compatible with the cable tray is installed to separate the service-entrance conductors from other conductors installed in the cable tray.

2008 NEC - 230.44 Cable Trays.

Cable tray systems shall be permitted to support service-entrance conductors. Cable trays used to support service-entrance conductors shall contain only service-entrance conductors.

Exception: Conductors, other than service-entrance conductors, shall be permitted to be installed in a cable tray with service-entrance conductors, provided a solid fixed barrier of a material compatible with the cable tray is installed to separate the service-entrance conductors from other conductors installed in the cable tray. Cable trays shall be identified with permanently affixed labels with the wording "Service-Entrance Conductors." The labels shall be located so as to be visible after installation and placed so that the service-entrance conductors may be readily traced through the entire length of the cable tray.

Author's Comment: Text has been added to the exception where conductors, other than service entrance conductors, are installed in cable trays with service entrance conductors. Where these other conductors are added, they must be separated by a fixed solid barrier but, now, in addition, labels must be installed with the wording "service-entrance conductor" so as to be readily visible after installation and placed so the service entrance conductors can be readily traced through the entire length to the cable tray.

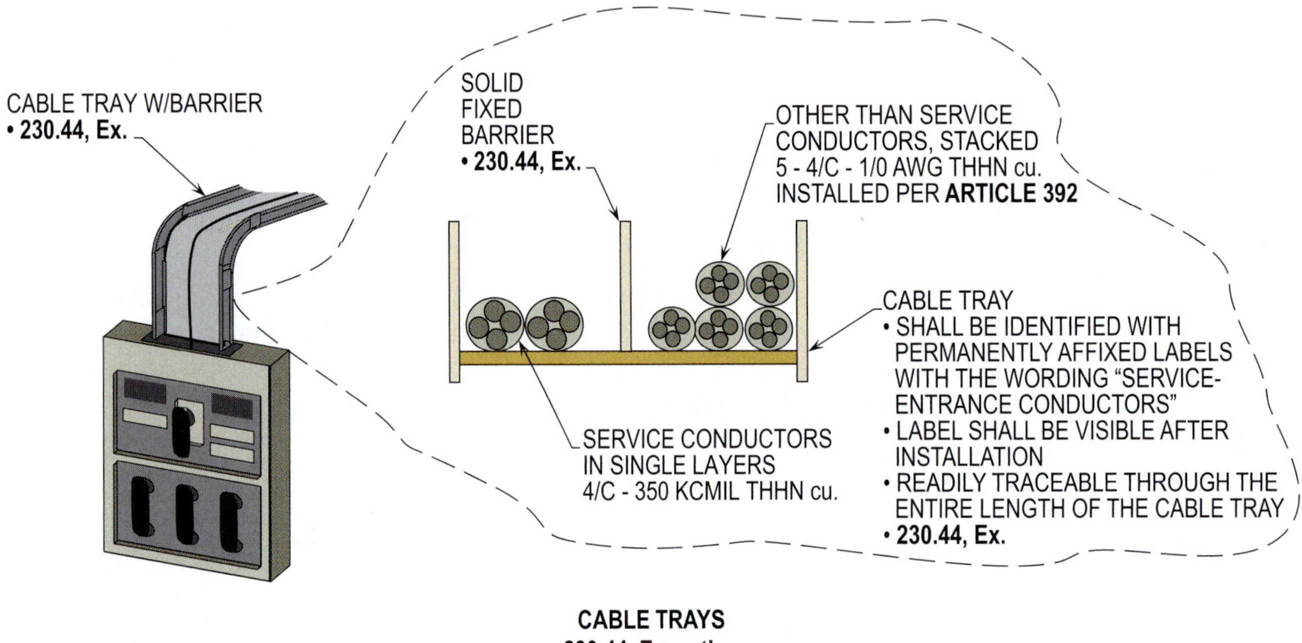

CABLE TRAYS
230.44, Exception

Purpose of Change: This revision clarifies that service-entrance conductors installed in cable trays with other than service-entrance conductor shall be separated with a barrier and shall be identified with permanently affixed labels on the cable tray that are readily traced through the entire length.

NEC Ch. 2 - Article 230
Part V - 230.71(B)

Type of Change		Panel Action		UL	UL 508	API 500 - 1997	API 505 - 1997	OSHA - 1994
Revision		Accept		-	-	-	-	-
ROP		ROC		NFPA 70E - 2004	NFPA 70B - 2006	NFPA 79 - 2007	NFPA	NEMA - 2002
pg. 148	# 4-65	pg. -	# -	-	-	-	-	AB 1
log: 1327	CMP: 4	log: -	Submitter: Mike Holt			2005 NEC: 230.71(B)		IEC: -

2005 NEC - 230.71 Maximum Number of Disconnects.

(B) Single-Pole Units. Two or three single-pole switches or breakers, capable of individual operation, shall be permitted on multiwire circuits, one pole for each ungrounded conductor, as one multipole disconnect, provided they are equipped with handle ties or a master handle to disconnect all conductors of the service with no more than six operations of the hand.

2008 NEC - 230.71 Maximum Number of Disconnects.

(B) Single-Pole Units. Two or three single-pole switches or breakers, capable of individual operation, shall be permitted on multiwire circuits, one pole for each ungrounded conductor, as one multipole disconnect, provided they are equipped with identified handle ties or a master handle to disconnect all conductors of the service with no more than six operations of the hand.

Author's Comment: The word "identified" has been inserted to match the text accepted in **240.20(B)** for handle ties in the 2005 NEC.

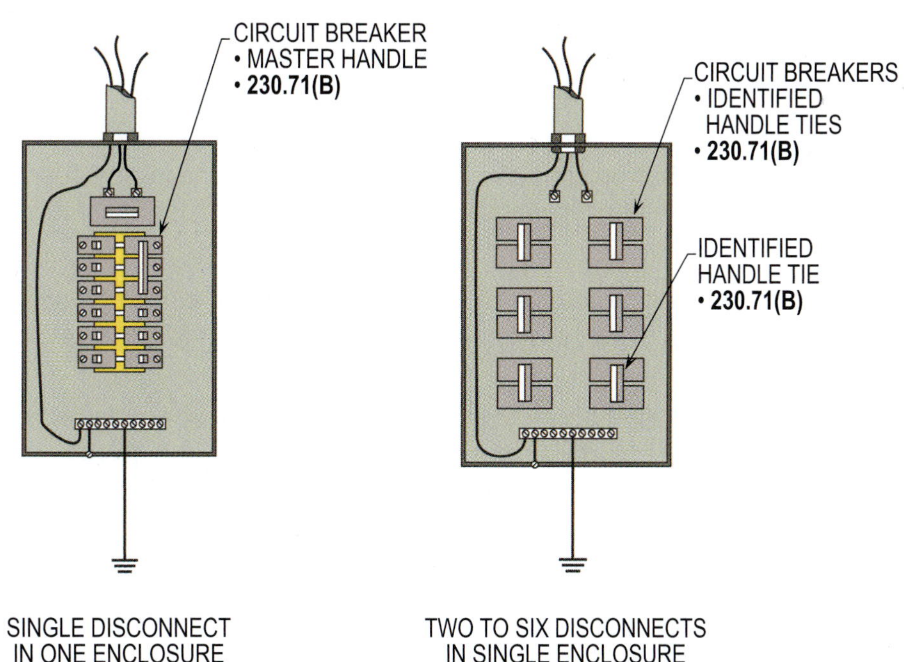

SINGLE-POLE UNITS
230.71(B)

Purpose of Change: This revision clarifies that identified handle ties or a master handle shall be used to disconnect all conductors of the service with no more than six operations of the hand.

NEC Ch. 2 - Article 230
Part VII - 230.95, ExceptionNo. 2

Type of Change	Panel Action	UL	UL 508	API 500 - 1997	API 505 - 1997	OSHA - 1994
Revision	Accept	-	-	-	-	-
ROP	**ROC**	**NFPA 70E - 2004**	**NFPA 70B - 2006**	**NFPA 79 - 2007**	**NFPA**	**NEMA - 2004**
pg. 153 # 4-84	pg. - # -	-	-	-	-	PB 2.2
log: 3500 CMP: 4	log: -	Submitter: Jim Pauley		2005 NEC: 230.95, Ex. 2		IEC: -

2005 NEC - 230.95 Ground-Fault Protection of Equipment.

Ground-fault protection of equipment shall be provided for solidly grounded wye electrical services of more than 150 volts to ground but not exceeding 600 volts phase-to-phase for each service disconnect rated 1000 amperes or more. The grounded conductor for the solidly grounded wye system shall be connected directly to ground without inserting any resistor or impedance device.

The rating of the service disconnect shall be considered to be the rating of the largest fuse that can be installed or the highest continuous current trip setting for which the actual overcurrent device installed in a circuit breaker is rated or can be adjusted.

Exception ~~No. 1~~: The ground-fault protection provisions of this section shall not apply to a service disconnect for a continuous industrial process where a nonorderly shutdown will introduce additional or increased hazards.

~~Exception No. 2: The ground-fault protection provisions of this section shall not apply to fire pumps.~~

2008 NEC- 230.95 Ground-Fault Protection of Equipment.

Ground-fault protection of equipment shall be provided for solidly grounded wye electrical services of more than 150 volts to ground but not exceeding 600 volts phase-to-phase for each service disconnect rated 1000 amperes or more. The grounded conductor for the solidly grounded wye system shall be connected directly to ground, through a grounding electrode system, as specified in 250.50, without inserting any resistor or impedance device.

The rating of the service disconnect shall be considered to be the rating of the largest fuse that can be installed or the highest continuous current trip setting for which the actual overcurrent device installed in a circuit breaker is rated or can be adjusted.

Exception: The ground-fault protection provisions of this section shall not apply to a service disconnect for a continuous industrial process where a nonorderly shutdown will introduce additional or increased hazards.

Author's Comment: During the 2005 NEC Cycle, CMP 13 added the provision to prohibit GFPE on fire pumps in **695.6(H)**. The section reads as follows: **(H) Ground Fault Protection of Equipment.** Ground fault protection of equipment shall not be permitted for fire pumps.

Since **Chapter 6** can supplement or modify requirements in **Chapter 2**, there is no need for the exception in **230.95**. In fact, the exception adds confusion because code users misinterpret that an exception needs to be in **Chapter 2** for **Chapter 6** to exempt it.

The same change has occurred in **215.10** with the deletion of **Exception No. 2**, ground-fault protection of equipment is no longer required on a fire pump feeder.

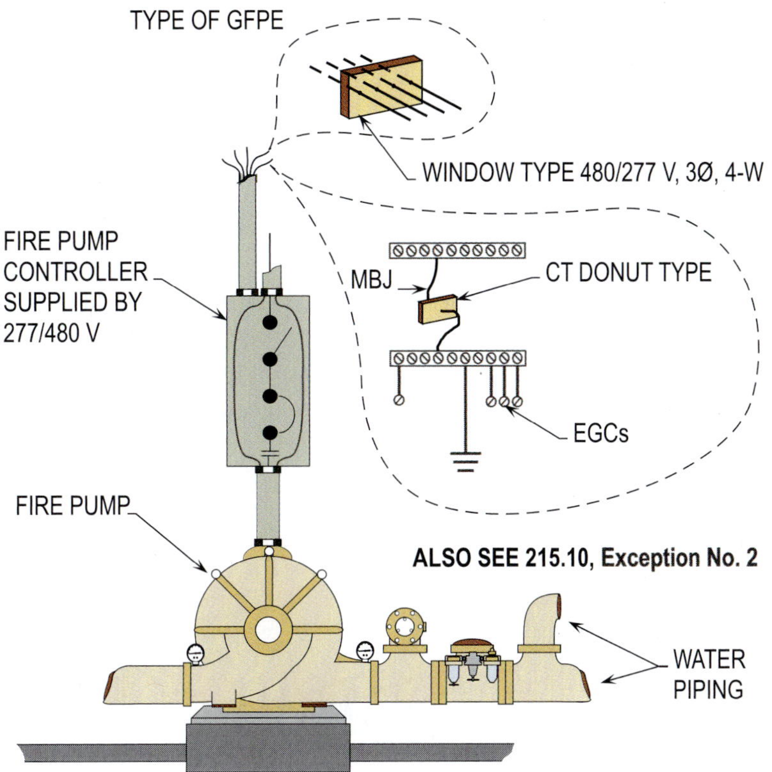

GROUND-FAULT PROTECTION OF EQUIPMENT
230.95, Exception No. 2

Purpose of Change: Exception No. 2 has been deleted.

NEC Ch. 2 - Article 230
Part VIII - 230.205(A)

Type of Change		Panel Action		UL	UL 508	API 500 - 1997	API 505 - 1997	OSHA - 1994
Revision		Accept		-	-	-	-	-
ROP		ROC		NFPA 70E - 2004	NFPA 70B - 2006	NFPA 79 - 2007	NFPA	NEMA
pg. 154	# 4-87	pg. -	# -	-	-	-	-	-
log: 2952	CMP: 4	log: -	Submitter: James J. Rogers			2005 NEC: 230.205(A)		IEC: -

2005 NEC - 230.205 Disconnecting Means.

(A) Location. The service disconnecting means shall be located in accordance with 230.70.

2008 NEC - 230.205 Disconnecting Means.

(A) Location. The service disconnecting means shall be located in accordance with 230.70.

For either overhead or underground primary distribution systems on private property, the service disconnect shall be permitted to be located in a location that is not readily accessible.

Author's Comment: There are many installations where the service point is at the edge of the property and a high voltage switch at the top of a pole is the actual service disconnect for the distribution system which then becomes a feeder for multiple buildings on the property, the disconnect requirements in **Article 225** would apply to the buildings. The change in this section permits the service disconnect to be not readily accessible.

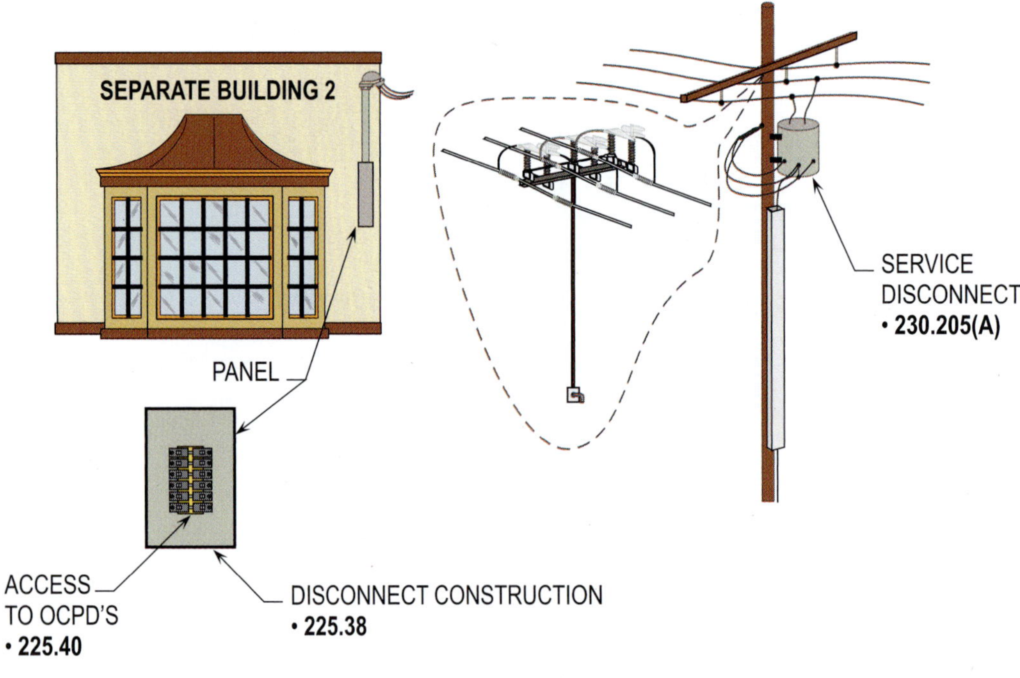

LOCATION
230.205(A)

Purpose of Change: This revision clarifies that on private property, the service disconnect shall be permitted to be in a location that is not readily accessible, such as the top of a pole.

NEC Ch. 2 - Article 240
Part I - 240.4(D)(1) through (D)(7)

Type of Change		Panel Action		UL	UL 508	API 500 - 1997	API 505 - 1997	OSHA - 1994
Revision		Accept in Principle		-	-	-	-	-
ROP		ROC		NFPA 70E - 2004	NFPA 70B - 2006	NFPA 79 - 2007	NFPA	NEMA
pg. 156	# 10-10	pg. -	# -	-	-	-	-	-
log: 2931	CMP: 10	log: -	Submitter: Todd Lottmann			2005 NEC: 240.4(D)		IEC: -

2005 NEC - 240.4 Protection of Conductors.

(D) Small Conductors. Unless specifically permitted in 240.4(E) or ~~240.4~~(G), the overcurrent protection shall not exceed 15 amperes for 14 AWG, 20 amperes for 12 AWG, and 30 amperes for 10 AWG copper; or 15 amperes for 12 AWG, and 25 amperes for 10 AWG aluminum and copper-clad aluminum after any correction factors for ambient temperature and number of conductors have been applied.

2008 NEC - 240.4 Protection of Conductors.

(D) Small Conductors. Unless specifically permitted in 240.4(E) or (G), the overcurrent protection shall not exceed that required by (D)(1) through (D)(7) after any correction factors for ambient temperature and number of conductors have been applied.

(1) 18 AWG Copper. 7 amperes provided all the following conditions are met:

(1) Continuous loads do not exceed 5.6 amperes

(2) Overcurrent protection is provided by one of the following:

a. Branch-circuit rated circuit breakers listed and marked for use with 18 AWG copper wire

b. Branch-circuit rated fuses listed and marked for use with 18 AWG copper wire

c. Class CC, Class J, or Class T fuses

(2) 16 AWG Copper. 10 amperes provided all the following conditions are met:

(1) Continuous loads do not exceed 8 amperes

(2) Overcurrent protection is provided by one of the following:

a. Branch-circuit rated circuit breakers listed and marked for use with 16 AWG copper wire

b. Branch-circuit rated fuses listed and marked for use with 16 AWG copper wire

c. Class CC, Class J, or Class T fuses

(3) 14 AWG Copper. 15 amperes.

(4) 12 AWG Aluminum and Copper-Clad Aluminum. 15 amperes.

(5) 12 AWG Copper. 20 amperes.

(6) 10 AWG Aluminum and Copper-clad Aluminum. 25 amperes.

(7) 10 AWG copper. 30 amperes.

Author's Comment: Small conductors, such as 18 AWG and 16 AWG sizes, were added to correspond to the use of these smaller conductors based on NFPA 79, UL 508, UL 508A, and various small conductors specified and permitted in Article 400 for flexible cords and cables, as well as control circuits in **Articles 725**, **760**, and communications circuits in **Chapter 8**.

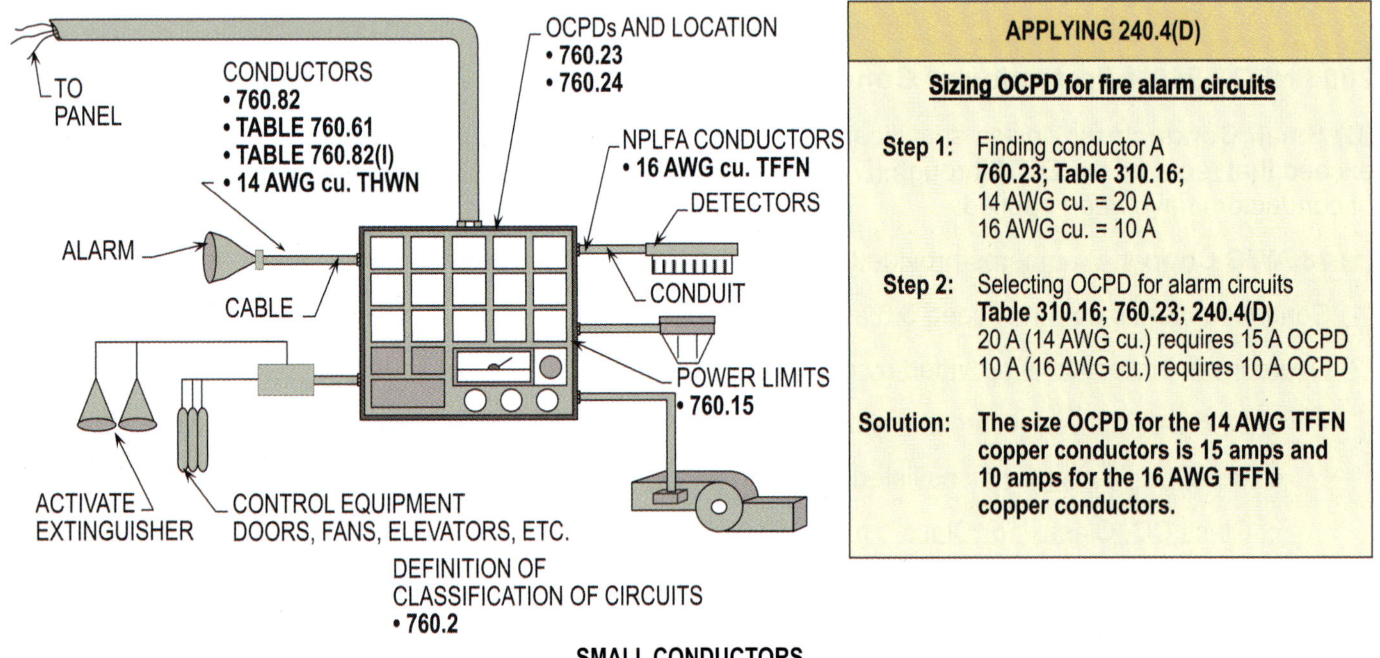

TO PANEL
CONDUCTORS
• 760.82
• TABLE 760.61
• TABLE 760.82(I)
• 14 AWG cu. THWN
OCPDs AND LOCATION
• 760.23
• 760.24
NPLFA CONDUCTORS
• 16 AWG cu. TFFN
DETECTORS
ALARM
CABLE
CONDUIT
POWER LIMITS
• 760.15
ACTIVATE EXTINGUISHER
CONTROL EQUIPMENT
DOORS, FANS, ELEVATORS, ETC.
DEFINITION OF
CLASSIFICATION OF CIRCUITS
• 760.2
APPLYING 240.4(D)
Sizing OCPD for fire alarm circuits
Step 1: Finding conductor A
760.23; Table 310.16;
14 AWG cu. = 20 A
16 AWG cu. = 10 A
Step 2: Selecting OCPD for alarm circuits
Table 310.16; 760.23; 240.4(D)
20 A (14 AWG cu.) requires 15 A OCPD
10 A (16 AWG cu.) requires 10 A OCPD
Solution: The size OCPD for the 14 AWG TFFN copper conductors is 15 amps and 10 amps for the 16 AWG TFFN copper conductors.

SMALL CONDUCTORS
240.4(D)(1) through (D)(7)

Purpose of Change: This revision adds small conductors in 18 AWG and 16 AWG sizes to correspond to the permission to use these smaller conductors.

NEC Ch. 2 - Article 240
Part II - 240.21(C)(2)(4)

Type of Change	Panel Action	UL	UL 508	API 500 - 1997	API 505 - 1997	OSHA - 1994
New Subdivision	Accept	-	-	-	-	-
ROP	ROC	NFPA 70E - 2004	NFPA 70B - 2006	NFPA 79 - 2007	NFPA	NEMA - 1997
pg. 163 # 10-27	pg. - # -	-	-	7.2.7	-	ST 20
log: 3387 CMP: 10	log: -	Submitter: Frederic P. Hartwell		2005 NEC: -		IEC: -

2005 NEC - 240.21 Location in Circuit.

(C) Transformer Secondary Conductors.

(2) Transformer Secondary Conductors Not Over 3 m (10 ft) Long. Where the length of secondary conductor does not exceed 3 m (10 ft) and complies with all of the following:

(1) The ampacity of the secondary conductors is

a. Not less than the combined calculated loads on the circuits supplied by the secondary conductors, and

b. Not less than the rating of the device supplied by the secondary conductors or not less than the rating of the overcurrent-protective device at the termination of the secondary conductors, and

~~c. Not less than one-tenth of the rating of the overcurrent device protecting the primary of the transformer, multiplied by the primary to secondary transformer voltage ratio~~

2008 NEC - 240.21 Location in Circuit.

(C) Transformer Secondary Conductors.

(2) Transformer Secondary Conductors Not Over 3 m (10 ft) Long. Where the length of secondary conductor does not exceed 3 m (10 ft) and complies with all of the following:

(4) For field installations where the secondary conductors leave the enclosure or vault in which the supply connection is made, the rating of the overcurrent device protecting the primary of the transformer, multiplied by the primary to secondary transformer voltage ratio, shall not exceed 10 times the ampacity of the secondary conductor.

Author's Comment: This language correlates this section with the conventional 10-foot tap rule in 240.21(B)(1).

Suppose, for example, a unit substation with a 2000 ampere secondary is installed, with five sets of 600 kcmil secondary conductors between the transformer and the disconnecting means for the separately derived system. Suppose the ampacity (2100 amperes) reflects the transformer-winding ratio. According to the literal text of the 2005 NEC, conductors on the secondary side of this transformer must not be smaller than 3/0 copper. This could be a problem for instrumentation supplied with the gear. In addition, if the installation is located within a vault, there is little likelihood of a problem in short conductors affecting the building as a whole. These allowances were inserted in the rules for conventional 10-foot taps (in the 1993 NEC), and the added text now applies equally for transformer secondaries.

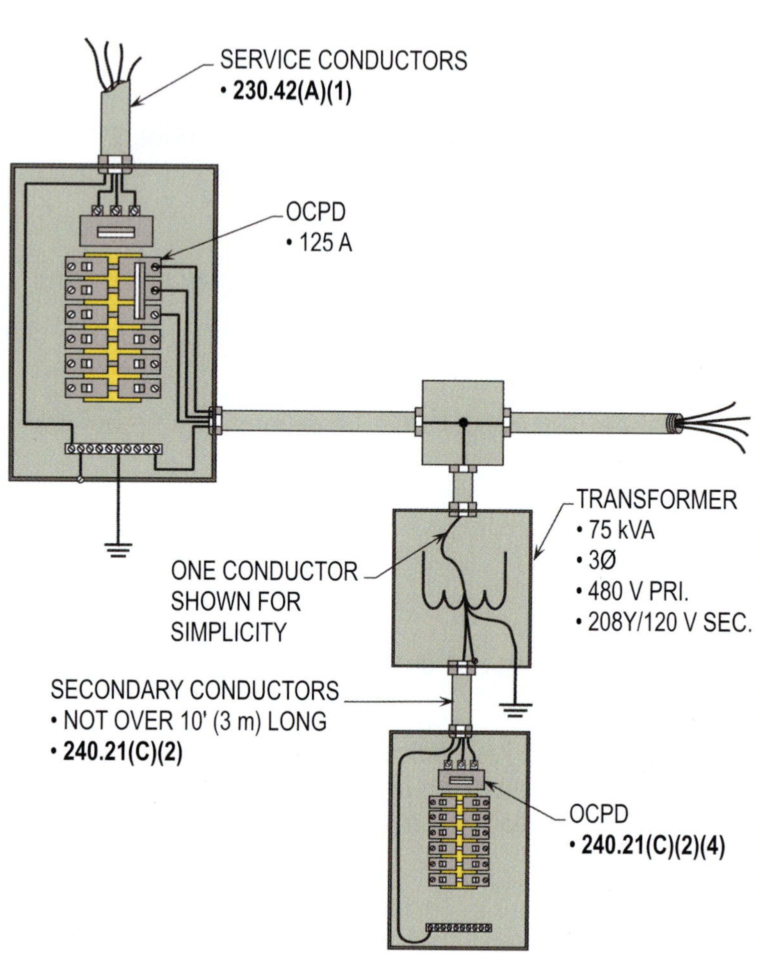

APPLYING 240.21(C)(2)(4)

Sizing primary OCPD

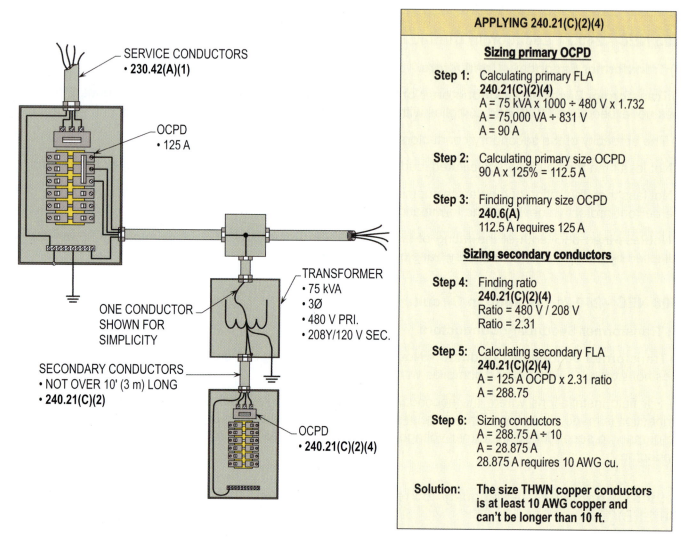

Step 1: Calculating primary FLA
240.21(C)(2)(4)
A = 75 kVA x 1000 ÷ 480 V x 1.732
A = 75,000 VA ÷ 831 V
A = 90 A

Step 2: Calculating primary size OCPD
90 A x 125% = 112.5 A

Step 3: Finding primary size OCPD
240.6(A)
112.5 A requires 125 A

Sizing secondary conductors

Step 4: Finding ratio
240.21(C)(2)(4)
Ratio = 480 V / 208 V
Ratio = 2.31

Step 5: Calculating secondary FLA
240.21(C)(2)(4)
A = 125 A OCPD x 2.31 ratio
A = 288.75

Step 6: Sizing conductors
A = 288.75 A ÷ 10
A = 28.875 A
28.875 A requires 10 AWG cu.

Solution: **The size THWN copper conductors is at least 10 AWG copper and can't be longer than 10 ft.**

TRANSFORMER SECONDARY CONDUCTORS NOT OVER 3 m (10 FT) LONG 240.21(C)(2)(4)

Purpose of Change: This revision clarifies that this section correlates with the conventional 10-foot tap rule in **240.21(B)(1)**.

NEC Ch. 2 - Article 240
Part II - 240.21(C)(3)(1)

Type of Change	Panel Action	UL	UL 508	API 500 - 1997	API 505 - 1997	OSHA - 1994
Revision	Accept	-	-	-	-	-
ROP	**ROC**	**NFPA 70E - 2004**	**NFPA 70B - 2006**	**NFPA 79 - 2007**	**NFPA**	**NEMA - 1997**
pg. 163 # 10-3a	pg. - # -	-	-	7.2.7	-	ST 20
log: CP 1000 CMP: 10	log: - Submitter: Code Making Panel 10			2005 NEC: 240.21(C)(3)		IEC: -

2005 NEC - 240.21 Location in Circuit.

(C) Transformer Secondary Conductors.

(3) Industrial Installation Secondary Conductors Not Over 7.5 m (25 ft) Long. For industrial installations only, where the length of the secondary conductors does not exceed 7.5 m (25 ft) and complies with all of the following:

~~(1)~~ The ampacity of the secondary conductors is not less than the secondary current rating of the transformer, and the sum of the ratings of the overcurrent devices does not exceed the ampacity of the secondary conductors.

~~(2)~~ All overcurrent devices are grouped.

~~(3)~~ The secondary conductors are protected from physical damage by being enclosed in an approved raceway or by other approved means.

2008 NEC - 240.21 Location in Circuit.

(C) Transformer Secondary Conductors.

(3) Industrial Installation Secondary Conductors Not Over 7.5 m (25 ft) Long. For industrial installations only, where the length of the secondary conductors does not exceed 7.5 m (25 ft) and complies with all of the following:

(1) Conditions of maintenance and supervision ensure that only qualified persons service the systems.

(2) The ampacity of the secondary conductors is not less than the secondary current rating of the transformer, and the sum of the ratings of the overcurrent devices does not exceed the ampacity of the secondary conductors.

(3) All overcurrent devices are grouped.

(4) The secondary conductors are protected from physical damage by being enclosed in an approved raceway or by other approved means.

Author's Comment: This new text provides requirements with conditions of maintenance for the equipment and proper supervision of qualified persons to use this maximum 25-foot extension of the transformer secondary conductors.

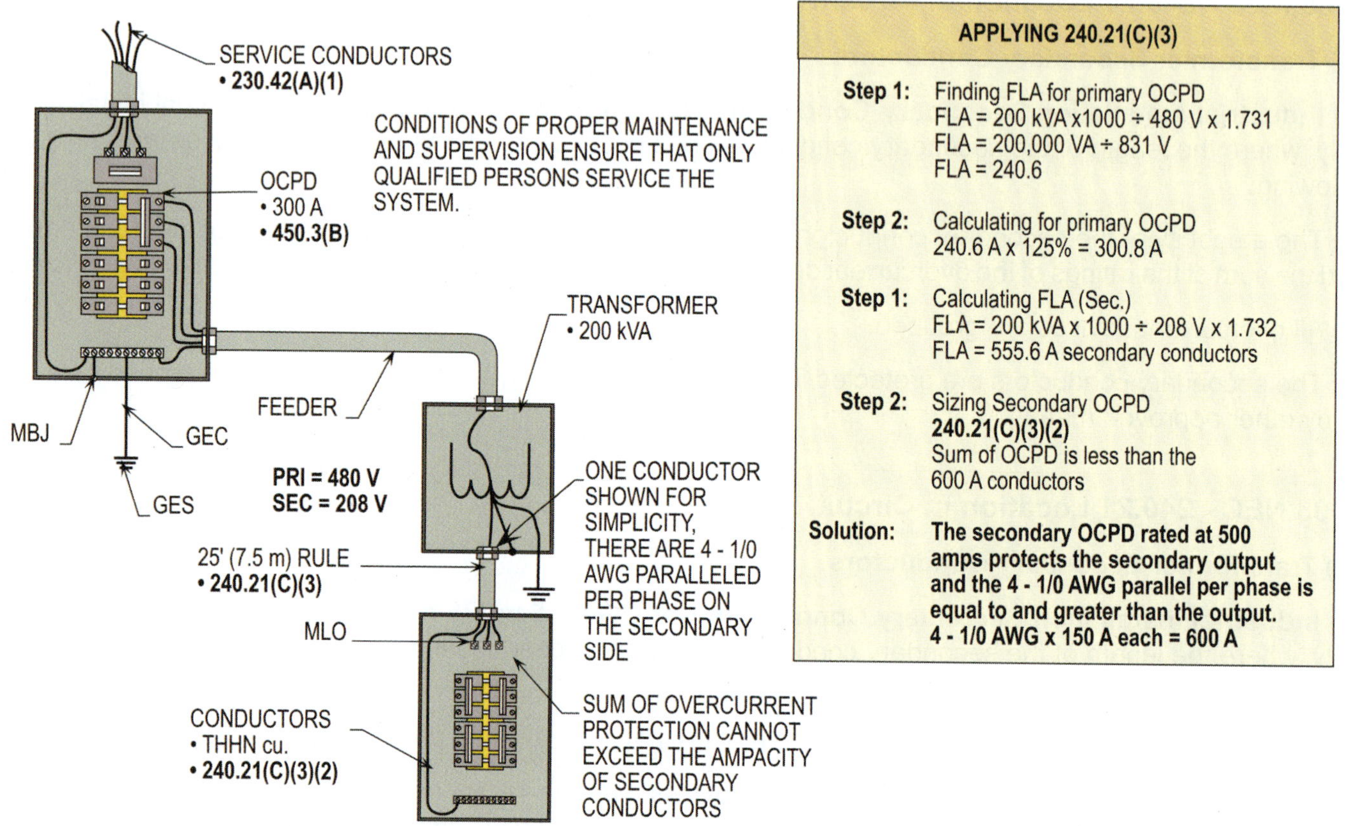

SERVICE CONDUCTORS
• 230.42(A)(1)
CONDITIONS OF PROPER MAINTENANCE AND SUPERVISION ENSURE THAT ONLY QUALIFIED PERSONS SERVICE THE SYSTEM.
OCPD
• 300 A
• 450.3(B)
TRANSFORMER
• 200 kVA
FEEDER
MBJ
GEC
GES
PRI = 480 V
SEC = 208 V
ONE CONDUCTOR SHOWN FOR SIMPLICITY, THERE ARE 4 - 1/0 AWG PARALLELED PER PHASE ON THE SECONDARY SIDE
25' (7.5 m) RULE
• 240.21(C)(3)
MLO
CONDUCTORS
• THHN cu.
• 240.21(C)(3)(2)
SUM OF OVERCURRENT PROTECTION CANNOT EXCEED THE AMPACITY OF SECONDARY CONDUCTORS
APPLYING 240.21(C)(3)
Step 1: Finding FLA for primary OCPD
FLA = 200 kVA x1000 ÷ 480 V x 1.731
FLA = 200,000 VA ÷ 831 V
FLA = 240.6
Step 2: Calculating for primary OCPD
240.6 A x 125% = 300.8 A
Step 1: Calculating FLA (Sec.)
FLA = 200 kVA x 1000 ÷ 208 V x 1.732
FLA = 555.6 A secondary conductors
Step 2: Sizing Secondary OCPD
240.21(C)(3)(2)
Sum of OCPD is less than the 600 A conductors
Solution: The secondary OCPD rated at 500 amps protects the secondary output and the 4 - 1/0 AWG parallel per phase is equal to and greater than the output.
4 - 1/0 AWG x 150 A each = 600 A

INDUSTRIAL INSTALLATION SECONDARY CONDUCTORS NOT OVER 7.5 m (25 FT) LONG 240.21(C)(3)(1)

Purpose of Change: This revision clarifies that proper supervision by qualified persons and proper conditions of maintenance for the equipment is required to use this 25 ft (7.5 m) extension of the transformer secondary conductors without terminating in a single overcurrent device.

NEC Ch. 2 - Article 240
Part II - 240.21(H)

Type of Change		Panel Action		UL	UL 508	API 500 - 1997	API 505 - 1997	OSHA - 1994
New Subsection		Accept in Principle		-	-	-	-	-
ROP		ROC		NFPA 70E - 2004	NFPA 70B - 2006	NFPA 79 - 2007	NFPA	NEMA - 2003
pg. 165	# 10-33	pg. -	# 1013	-	-	-	-	PE 1
log: 1672	CMP: 10	log: 1846	Submitter: Paul E. Guidry		2005 NEC: -			IEC: -

2008 NEC - 240.21 Location in Circuit.

Overcurrent protection shall be provided in each ungrounded circuit conductor and shall be located at the point where the conductors receive their supply exept as specified in 240.21(A) through (H). No conductor supplied under the provisions of 240.21(A) through (G) shall supply another conductor under those provisions, except through an overcurrent protective device meeting the requirements of 240.4

(H) Battery Conductors. Overcurrent protection shall be permitted to be installed as close as practicable to the storage battery terminals in a non-hazardous location. Installation of the overcurrent protection within a hazardous location shall also be permitted.

Author's Comment: Overcurrent protection for battery conductors in UPS systems and similar applications is now covered in new **240.21(H)** and permits the overcurrent protection device to be installed as close as practicable to the battery terminals. (See new **480.5** requiring a disconnecting means be installed to disconnect all ungrounded conductors)

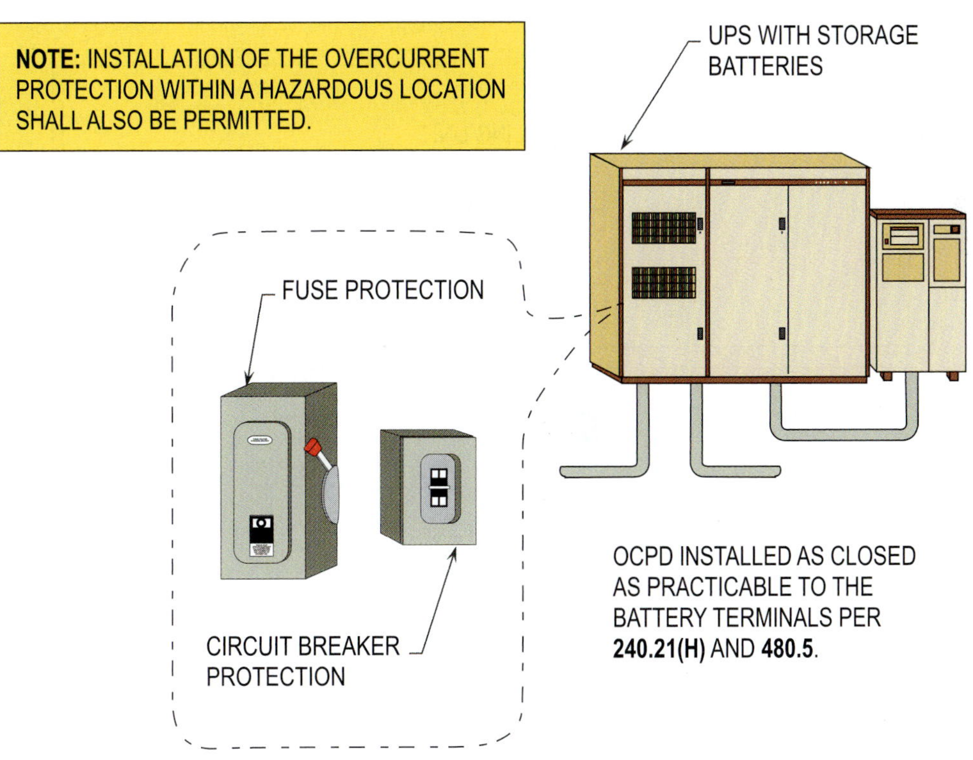

NOTE: INSTALLATION OF THE OVERCURRENT PROTECTION WITHIN A HAZARDOUS LOCATION SHALL ALSO BE PERMITTED.
UPS WITH STORAGE BATTERIES
FUSE PROTECTION
CIRCUIT BREAKER PROTECTION
OCPD INSTALLED AS CLOSED AS PRACTICABLE TO THE BATTERY TERMINALS PER 240.21(H) AND 480.5.

BATTERY CONDUCTORS
240.21(H)

Purpose of Change: A new subsection has been added to permit overcurrent protection to be installed as close as practicable to the storage battery terminals in a non-hazardous location.

NEC Ch. 2 - Article 240
Part II - 240.24(F)

Type of Change		Panel Action		UL	UL 508	API 500 - 1997	API 505 - 1997	OSHA - 1994
New Subsection		Accept in Principle		-	-	-	-	-
ROP		ROC		NFPA 70E - 2004	NFPA 70B - 2006	NFPA 79 - 2007	NFPA	NEMA
pg. 167	# 10-40	pg. 113	# 10-15	-	-	-	-	-
log: 1633	CMP: 10	log: 267	Submitter: L. Keith Lofland			2005 NEC: -		IEC: -

2008 NEC - 240.24 Location in or on Premises.

(F) Not located over Steps. Overcurrent devices shall not be located over steps of a stairway.

Author's Comment: This new subsection does not permit overcurrent protection devices to be located over the riser steps of a stairway since anyone trying to work on the devices would not be able to have a level workplace and it may be dangerous. However, many stairways have horizontal landings that could prove suitable for installations where appropriate working space exists so this new section just applies to the riser part of the stairs.

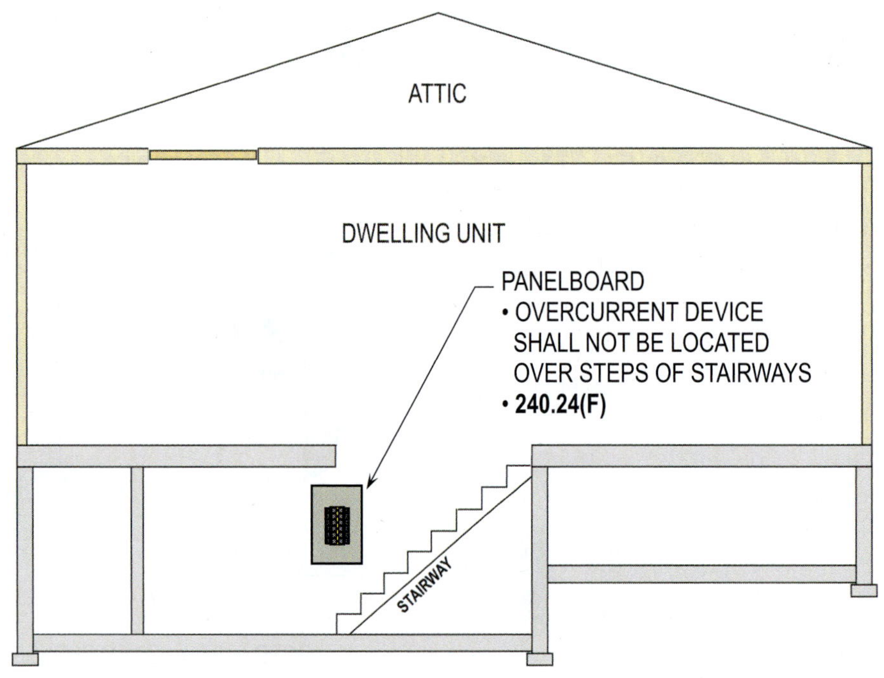

NOT LOCATED OVER STEPS
240.24(F)

Purpose of Change: A new subsection has been added to clarify that overcurrent devices shall not be located over steps of stairways.

NEC Ch. 2 - Article 240
Part VII - 240.86(A)

20

Type of Change	Panel Action	UL	UL 508	API 500 - 1997	API 505 - 1997	OSHA - 1994
Revision	Accept	-	-	-	-	-
ROP	**ROC**	**NFPA 70E - 2004**	**NFPA 70B - 2006**	**NFPA 79 - 2007**	**NFPA**	**NEMA**
pg. 170 # 10-50a	pg. - # -	-	-	-	-	-
log: CP 1001 CMP: 10	log: -	Submitter: Code Making Panel 10		2005 NEC: 240.86(A)		IEC: -

2005 NEC - 240.86 Series Ratings.

Where a circuit breaker is used on a circuit having an available fault current higher than the marked interrupting rating by being connected on the load side of an acceptable overcurrent protective device having a higher rating, the circuit breaker shall meet the requirements specified in (A) or (B), and (C).

(A) Selected Under Engineering Supervision in Existing Installations. The series rated combination devices shall be selected by a licensed professional engineer engaged primarily in the design or maintenance of electrical installations. The selection shall be documented and stamped by the professional engineer. This documentation shall be available to those authorized to design, install, inspect, maintain, and operate the system. This series combination rating, including identification of the upstream device, shall be field marked on the end use equipment.

2008 NEC - 240.86 Series Ratings.

Where a circuit breaker is used on a circuit having an available fault current higher than the marked interrupting rating by being connected on the load side of an acceptable overcurrent protective device having a higher rating, the circuit breaker shall meet the requirements specified in (A) or (B), and (C).

FPN: See 110.22 for marking of Series Combination Systems.

(A) Selected Under Engineering Supervision in Existing Installations. The series rated combination devices shall be selected by a licensed professional engineer engaged primarily in the design or maintenance of electrical installations. The selection shall be documented and stamped by the professional engineer. This documentation shall be available to those authorized to design, install, inspect, maintain, and operate the system. This series combination rating, including identification of the upstream device, shall be field marked on the end use equipment. For calculated applications, the engineer shall ensure that the downstream circuit breaker(s) that are part of the series combination remain passive during the interruption period of the line side fully rated, current-limiting device.

Author's Comment: The new sentence that was added provides some clarification to the overall application of calculations for existing installations. Devices that are part of the series combination system must be passive downstream during the reaction time of the upstream device and the engineer must be able to ensure that the downstream devices are passive devices as part of the overall calculation.

The passive downstream device ensures an increased impedance will not occur due to arcing between the contacts of the downstream device.

The added text provides specific marking requirements that must be readily visible for both the upstream and the down stream device in this series combination system.

A new Fine Print Note was added to clarify that marking for series Combination systems is located in 110.22.

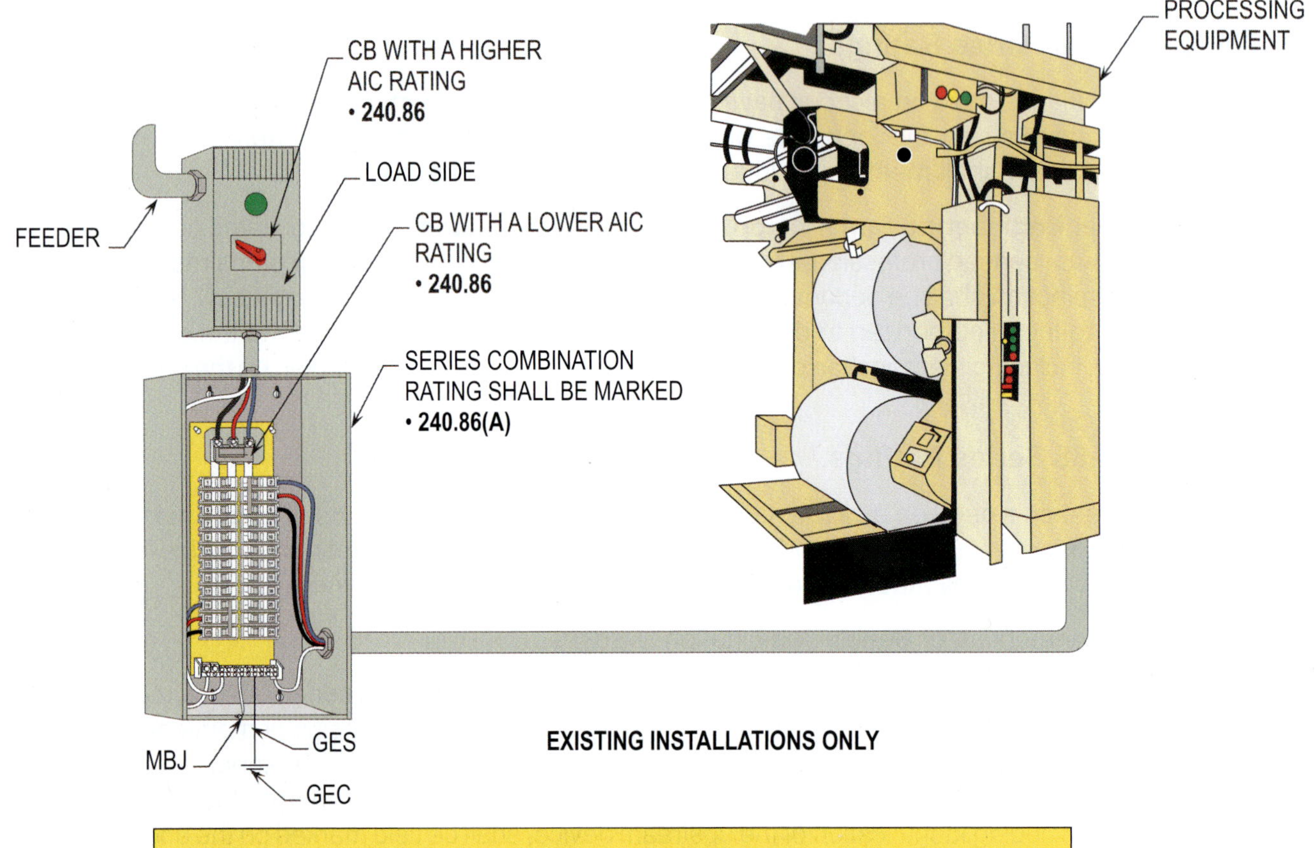
CB WITH A HIGHER
AIC RATING
• 240.86
LOAD SIDE
CB WITH A LOWER AIC
RATING
• 240.86
FEEDER
SERIES COMBINATION
RATING SHALL BE MARKED
• 240.86(A)
PROCESSING
EQUIPMENT
MBJ
GES
GEC
EXISTING INSTALLATIONS ONLY
NOTE: FOR CALCULATED APPLICATIONS, THE ENGINEER SHALL ENSURE THAT THE DOWNSTREAM CIRCUIT BREAKER(S) THAT ARE PART OF THE SERIES COMBINATION REMAIN PASSIVE DURING THE INTERRUPTION PERIOD OF THE LINE SIDE FULLY RATED, CURRENT-LIMITING DEVICE.

SELECTED UNDER ENGINEERING SUPERVISION IN EXISTING INSTALLATIONS 240.86(A)

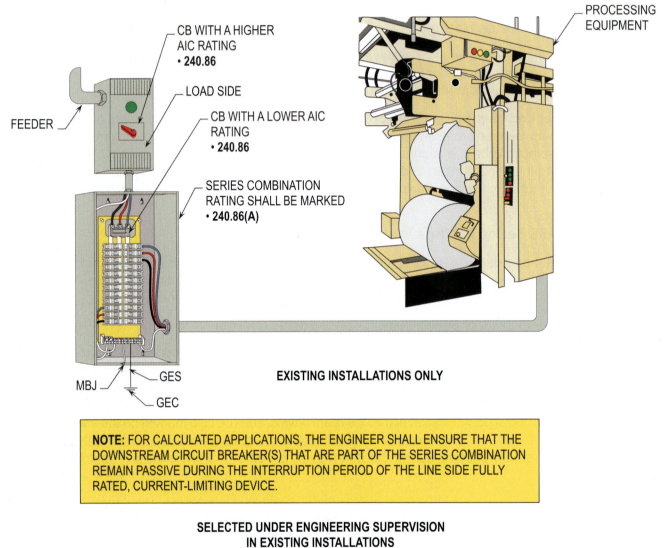

Purpose of Change: This revision clarifies that for calculated applications, the engineer shall ensure that the downstream circuit breaker(s) that are part of the series combination remain passive during the interruption period of the line side fully rated, current-limiting device.

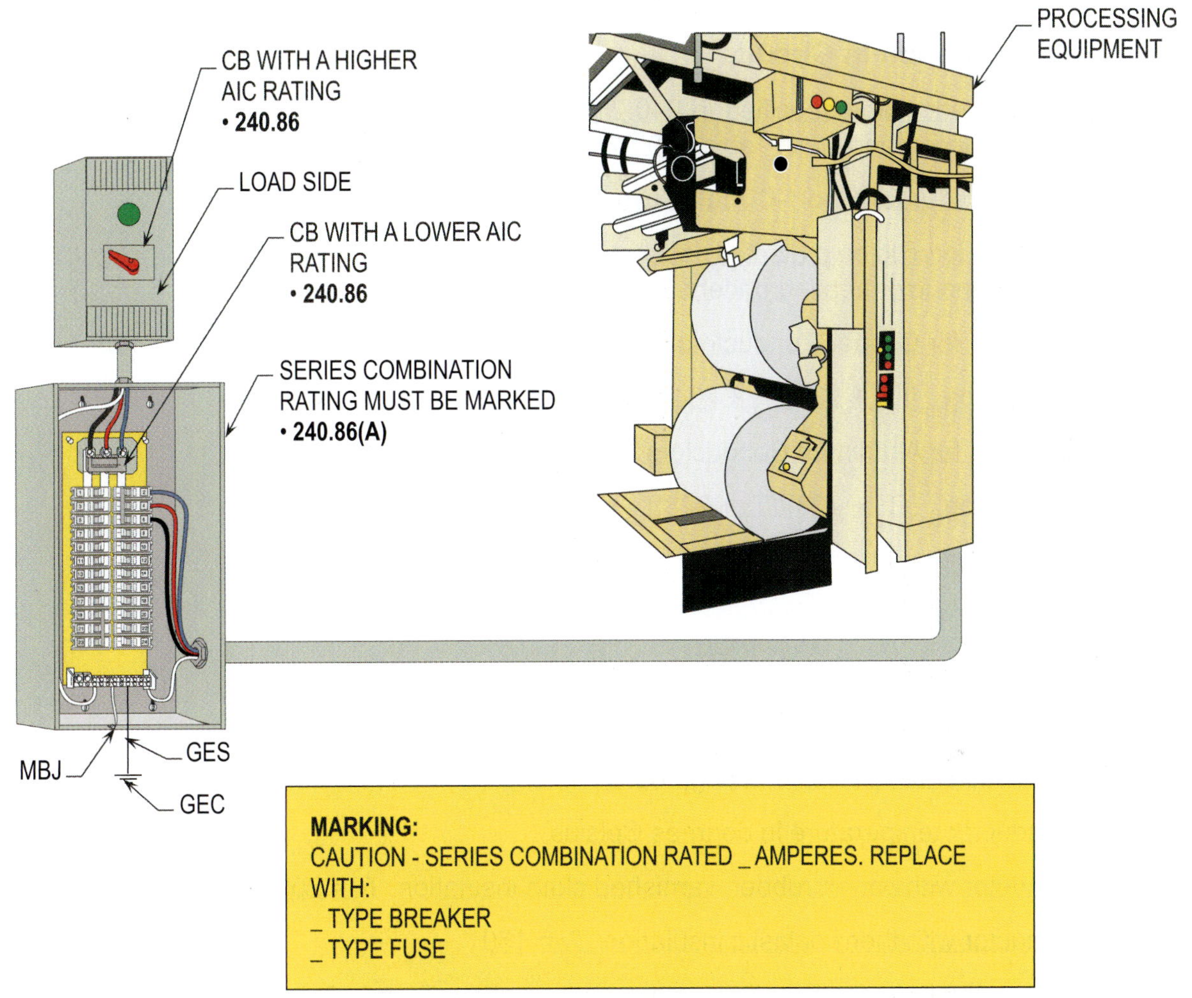

SELECTED UNDER ENGINEERING SUPERVISION IN EXISTING INSTALLATIONS 240.86(A)

Purpose of Change: This revision clarifies that specific markings must be readily visible for both the upstream and the downstream devices in series combination systems.

NEC Ch. 2 - Article 240
Part VIII - 240.92(B)

Type of Change	Panel Action	UL	UL 508	API 500 - 1997	API 505 - 1997	OSHA - 1994
New Subsection	Accept in Principle	-	-	-	-	-
ROP	**ROC**	**NFPA 70E - 2004**	**NFPA 70B - 2006**	**NFPA 79 - 2007**	**NFPA**	**NEMA**
pg. 160 / # 10-21	pg. - / # -	-	-	-	-	-
log: 2877 / CMP: 10	log: -	Submitter: Robert Padgham		2005 NEC: 240.92(A); (B)		IEC: -

2008 NEC - 240.92 Location in Circuit.

(B) Feeder Taps. For feeder taps specified in 240.21(B)(2), (B)(3), and (B)(4), the tap conductors shall be permitted to be sized in accordance with Table 240.92(B).

Table 240.92(B) Tap Conductor Short-Circuit Current Ratings.

Tap conductors are considered to be protected under short-circuit conditions when their short-circuit temperature limit is not exceeded. Conductor heating under short-circuit conditions is determined by (1) or (2):

(1) Short-Circuit Formula for Copper Conductors

$$(I^2/A^2)t = 0.0297 \log_{10} [(T_2 + 234)(T_1 + 234)]$$

(2) Short-Circuit Formula for Aluminum Conductors

$$(I^2/A^2)t = 0.0125 \log_{10} [(T_2 + 228)/(T_1 + 228)]$$

where:

I = short-circuit current in amperes

A = conductor area in circular mils

t = time of short-circuit in seconds (for times less than or equal to 10 seconds)

T_1 = initial conductor temperature in degrees Celsius.

T_2 = final conductor temperature in degrees Celsius.

Copper conductor with paper, rubber, varnished cloth insulation, T_2 = 200

Copper conductor with thermoplastic insulation, T_2 = 150

Copper conductor with cross-linked polyethylene insulation, T_2 = 250

Copper conductor with ethylene propylene rubber insulation, T_2 = 250

Aluminum conductor with paper, rubber, varnished cloth insulation, T_2 = 200

Aluminum conductor with thermoplastic insulation, T_2 = 150

Aluminum conductor with cross-linked polyethylene insulation, T_2 = 250

Aluminum conductor with ethylene propylene rubber insulation, T_2 = 250

Author's Comment: This new subsection will increase the enforceability necessary by the inspection community and limit the application of expanding feeder tap conductors for supervised industrial installations. This action recognizes the performance of the overcurrent device as a factor in determining the tap conductor size by considering the short circuit temperature limit. There is a formula for copper and one for aluminum.

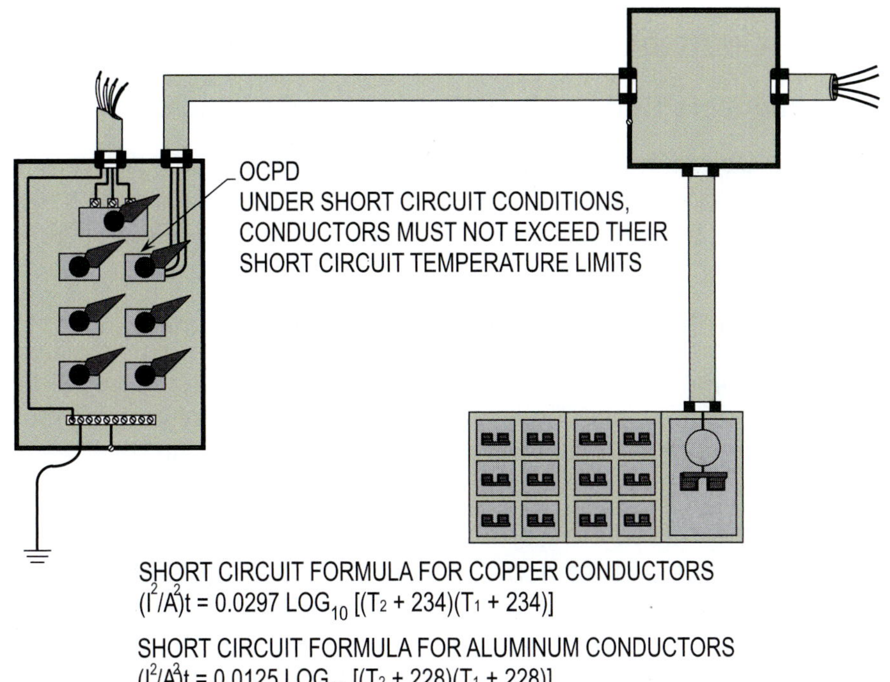
OCPD
UNDER SHORT CIRCUIT CONDITIONS, CONDUCTORS MUST NOT EXCEED THEIR SHORT CIRCUIT TEMPERATURE LIMITS
SHORT CIRCUIT FORMULA FOR COPPER CONDUCTORS
$(I^2/A^2)t = 0.0297\ LOG_{10}\ [(T_2 + 234)(T_1 + 234)]$
SHORT CIRCUIT FORMULA FOR ALUMINUM CONDUCTORS
$(I^2/A^2)t = 0.0125\ LOG_{10}\ [(T_2 + 228)(T_1 + 228)]$

SHORT-CIRCUIT FORMULA VALUES

- I = SHORT-CIRCUIT CURRENT IN AMPERES
- A = CONDUCTOR ARE IN CIRCULAR MILS
- t = TIME OF SHORT-CIRCUIT IN SECONDS (FOR TIMES LESS THAN OR EQUAL TO 10 SECONDS)
- T_1 = INITIAL CONDUCTOR TEMPERATURE IN DEGREES CELSIUS
- T_2 = FINAL CONDUCTOR TEMPERATURE IN DEGREES CELSIUS
- COPPER CONDUCTOR WITH PAPER, RUBBER, VARNISHED CLOTH INSULATION, $T_2 = 200$
- COPPER CONDUCTOR WITH TERMOPLASTIC INSULATION, $T_2 = 250$
- COPPER CONDUCTOR WITH CROSS-LINKED POLYETHYLENE INSULATION, $T_2 = 250$
- COPPER CONDUCTOR WITH ETHYLENE PROPYLENE RUBBER INSUALTION, $T_2 = 250$
- ALUMINUM CONDUCTOR WITH PAPER, RUBBER, VARNISHED CLOTH INSULATION, $T_2 = 200$
- ALUMINUM CONDUCTOR WITH THERMOPLASTIC INSULATION, $T_2 = 150$
- ALUMINUM CONDUCTOR WITH CROSS-LINKED POLYETHYLENE INSULATION, $T_2 = 250$
- ALUMINUM CONDUCTOR WITH ETHYLENE PROPYLENE RUBBER INSULATION, $T_2 = 250$

FEEDER TAPS
240.92(B)

Purpose of Change: This new subsection will increase the enforceability necessary by the inspection community and will limit the application of expanding feeder tap conductors for supervised industrial installations.

NEC Ch. 2 - Article 250
Part I - 250.6(A)

Type of Change	Panel Action	UL	UL 508	API 500 - 1997	API 505 - 1997	OSHA - 1994
Revision	Accept	-	-	-	-	-
ROP	**ROC**	**NFPA 70E - 2004**	**NFPA 70B - 2006**	**NFPA 79 - 2007**	**NFPA**	**NEMA - 2000**
pg. 179 # 5-75a	pg. - # -	-	-	-	-	LS 1
log: CP 500 CMP: 5	log: -	Submitter: Code Making Panel 5		2005 NEC: 250.6(A)		IEC: -

2005 NEC - 250.6 Objectionable Current Over Grounding Conductors.

(A) Arrangement to Prevent Objectionable Current. The grounding of electrical systems, circuit conductors, surge arresters, and conductive non–current-carrying ~~materials and~~ equipment shall be installed and arranged in a manner that will prevent objectionable current ~~over the grounding conductors or grounding paths~~.

2008 NEC - 250.6 Objectionable Current Over Grounding Conductors.

(A) Arrangement to Prevent Objectionable Current. The grounding of electrical systems, circuit conductors, surge arresters, <u>surge protective devices,</u> and conductive <u>normally</u> non–current-carrying <u>metal parts</u> of equipment shall be installed and arranged in a manner that will prevent objectionable current.

Author's Comment: The panel added the words "normally" and changed the words "materials and" to "metal parts of" since the grounding is for non-current carrying metal parts of equipment. The words "equipment grounding conductors or equipment grounding conductor paths" were deleted to reflect that objectionable current could be in all paths, including the equipment grounding conductor. The words "surge protective devices " were added as a result of the action on Proposal 5-74 to revise the name "transient voltage surge suppressor" to "surge protective devices" based on a change in UL 1449.

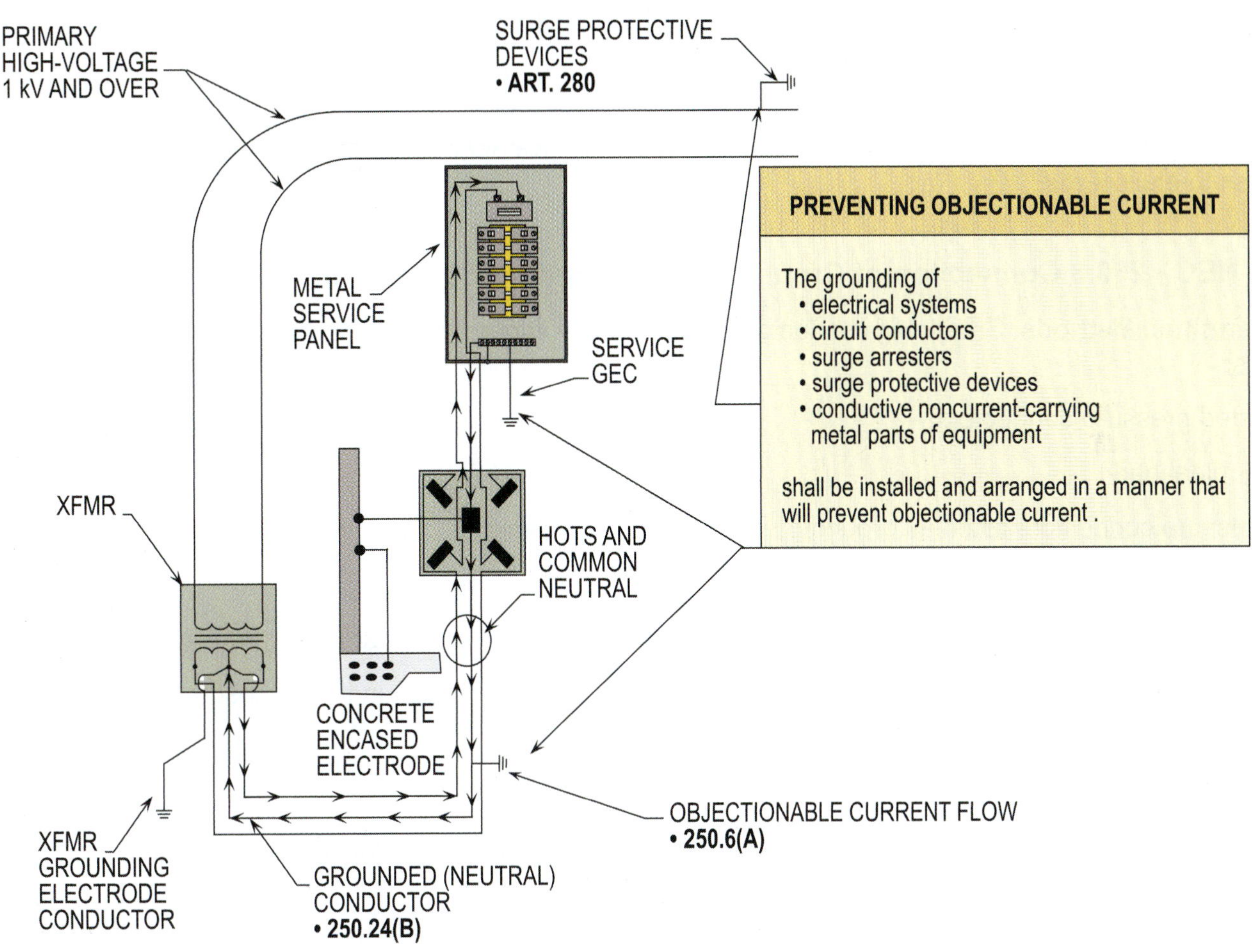

ARRANGEMENT TO PREVENT OBJECTIONABLE CURRENT
250.6(A)

Purpose of Change: This revision clarifies that objectionable current could be in all paths, including the equipment grounding conductor.

NEC Ch. 2 - Article 250
Part I - 250.8(A) and (B)

22

Type of Change	Panel Action	UL	UL 508	API 500 - 1997	API 505 - 1997	OSHA - 1994
Revision	Accept in Principle	-	-	-	-	-
ROP	**ROC**	**NFPA 70E - 2004**	**NFPA 70B - 2006**	**NFPA 79 - 2007**	**NFPA**	**NEMA - 2005**
pg. 186 # 5-84	pg. - # -	-	-	8.2.1.3	-	WD 1
log: 3365 CMP: 5	log: -	Submitter: Paul Dobrowsky		2005 NEC: 250.8		IEC: -

2005 NEC - 250.8 Connection of Grounding and Bonding Equipment.

Grounding conductors and bonding jumpers shall be connected by exothermic welding, listed pressure connectors, ~~listed clamps,~~ or other listed means. Connection devices or fittings that depend solely on solder shall not be used. ~~Sheet metal screws shall not be used to connect grounding conductors or connection devices to enclosures.~~

2008 NEC - 250.8 Connection of Grounding and Bonding Equipment.

(A) Permitted Methods. Grounding conductors and bonding jumpers shall be connected by one of the following means:

(1) Listed pressure connectors

(2) Terminal bars

(3) Pressure connectors listed as grounding and bonding equipment

(4) The exothermic welding process

(5) Machine screw-type fasteners that engage not less than two threads or are secured with a nut

(6) Thread-forming machine screws that engage not less than two threads in the enclosure

(7) Connections that are part of a listed assembly

(8) Other listed means

(B) Methods Not Permitted. Connection devices or fittings that depend solely on solder shall not be used.

Author's Comment: Section **250.8** has been revised by deleting the specific prohibition of sheet metal screws. This was done because there are many other types of screws that are also prohibited, such as drywall screws, self-tapping tech screws with less than two threads into an enclosure, and similar screws that may not provide an acceptable ground return path. Connections that are part of a listed assembly where part of the listing process are acceptable because the grounding connections have been tested. Machine screws having at least two threads or secured with a nut are acceptable. Also, any self-tapping screw where at least two threads are formed are acceptable. Section **250.8** has been formed into a list format for ease of use.

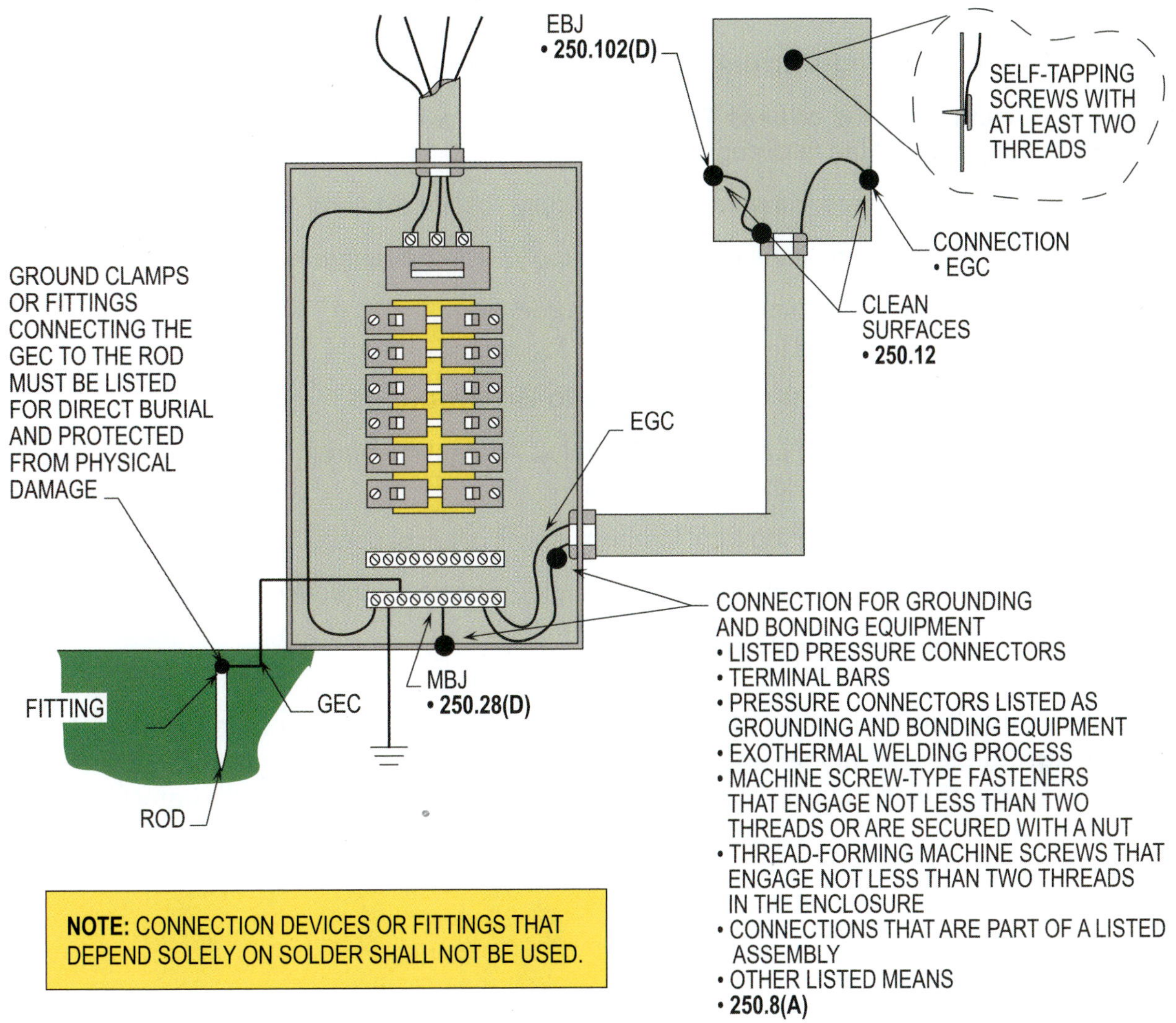
EBJ
• 250.102(D)
SELF-TAPPING SCREWS WITH AT LEAST TWO THREADS
CONNECTION
• EGC
CLEAN SURFACES
• 250.12
GROUND CLAMPS OR FITTINGS CONNECTING THE GEC TO THE ROD MUST BE LISTED FOR DIRECT BURIAL AND PROTECTED FROM PHYSICAL DAMAGE
EGC
FITTING
GEC
MBJ
• 250.28(D)
ROD
CONNECTION FOR GROUNDING AND BONDING EQUIPMENT
• LISTED PRESSURE CONNECTORS
• TERMINAL BARS
• PRESSURE CONNECTORS LISTED AS GROUNDING AND BONDING EQUIPMENT
• EXOTHERMAL WELDING PROCESS
• MACHINE SCREW-TYPE FASTENERS THAT ENGAGE NOT LESS THAN TWO THREADS OR ARE SECURED WITH A NUT
• THREAD-FORMING MACHINE SCREWS THAT ENGAGE NOT LESS THAN TWO THREADS IN THE ENCLOSURE
• CONNECTIONS THAT ARE PART OF A LISTED ASSEMBLY
• OTHER LISTED MEANS
• 250.8(A)
NOTE: CONNECTION DEVICES OR FITTINGS THAT DEPEND SOLELY ON SOLDER SHALL NOT BE USED.

CONNECTION OF GROUNDING AND BONDING EQUIPMENT
250.8(A) and (B)

Purpose of Change: This revision clarifies the permitted methods for connections of grounding and bonding equipment.

NEC Ch. 2 - Article 250
Part II - 250.20(A)(3)

Type of Change	Panel Action	UL	UL 508	API 500 - 1997	API 505 - 1997	OSHA - 1994
Revision	Accept in Principle	-	-	-	-	1910.304(f)(1)(iii)
ROP	**ROC**	**NFPA 70E - 2004**	**NFPA 70B - 2006**	**NFPA 79 - 2007**	**NFPA**	**NEMA - 2002**
pg. 187 # 5-92	pg. - # -	410.10(C)(3)	-	4.3.2	-	PB 1.1
log: 1107 CMP: 5	log: -	Submitter: Daniel Leaf		2005 NEC: 250.20(A)		IEC: -

2005 NEC - 250.20 Alternating-Current Systems to Be Grounded.

(A) Alternating-Current Systems of Less Than 50 Volts. Alternating-current systems of less than 50 volts shall be grounded under any of the following conditions:

(1) Where supplied by transformers, if the transformer supply system exceeds 150 volts to ground

(2) Where supplied by transformers, if the transformer supply system is ungrounded

(3) Where installed as overhead conductors ~~outside of buildings~~.

2008 NEC - 250.20 Alternating-Current Systems to Be Grounded.

(A) Alternating-Current Systems of Less Than 50 Volts. Alternating-current systems of less than 50 volts shall be grounded under any of the following conditions:

(1) Where supplied by transformers, if the transformer supply system exceeds 150 volts to ground

(2) Where supplied by transformers, if the transformer supply system is ungrounded

(3) Where installed outside as overhead conductors.

Author's Comment: This change expands the requirement for ac circuits of less than 50 volts to be grounded. Where these conductors are installed outside, lightning and high voltage crossover can occur so AC systems of less than 50 volts must be grounded, wherever installed anywhere outside.

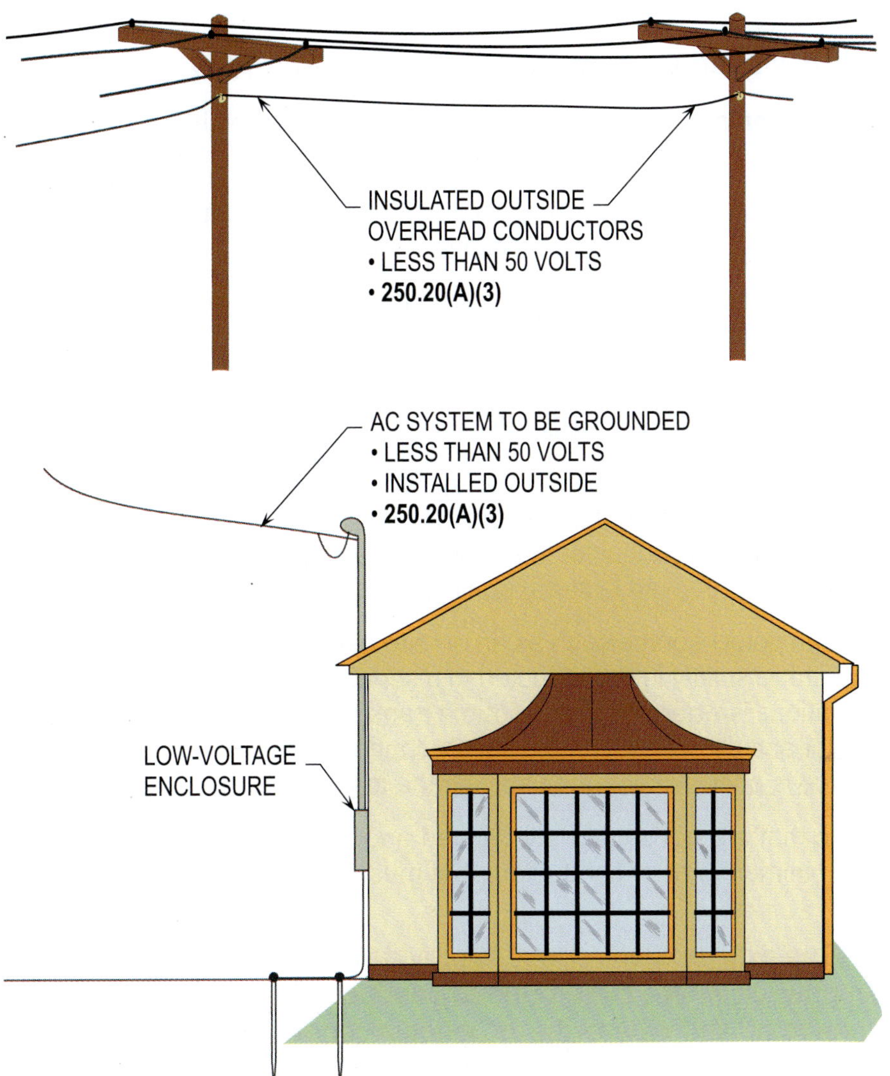

ALTERNATING-CURRENT SYSTEM OF LESS THAN 50 VOLTS
250.20(A)(3)

Purpose of Change: This revision clarifies that AC systems of less than 50 volts shall be grounded where installed outside as overhead conductors.

NEC Ch. 2 - Article 250
Part II - 250.20(D)

23

Type of Change	Panel Action	UL	UL 508	API 500 - 1997	API 505 - 1997	OSHA - 1994
Revision	Accept in Principle	-	-	-	-	-
ROP	**ROC**	**NFPA 70E - 2004**	**NFPA 70B - 2006**	**NFPA 79 - 2007**	**NFPA**	**NEMA - 2002**
pg. 188 # 5-95	pg. - # -	-	-	4.3.2	-	PB 1.1
log: 341 CMP: 5	log: -	Submitter: Michael J. Johnston		2005 NEC: 250.20(D)		IEC: -

2005 NEC - 250.20 Alternating-Current Systems to Be Grounded.

(D) Separately Derived Systems. Separately derived systems, as covered in 250.20(A) or (B), shall be grounded as specified in 250.30.

FPN No. 1: An alternate ac power source such as an on-site generator is not a separately derived system if the neutral is solidly interconnected to a service-supplied system neutral.

FPN No. 2: For systems that are not separately derived and are not required to be grounded as specified in 250.30, see 445.13 for minimum size of conductors that must carry fault current.

2008 NEC - 250.20 Alternating-Current Systems to Be Grounded.

(D) Separately Derived Systems. Separately derived systems, as covered in 250.20(A) or (B), shall be grounded as specified in 250.30(A). Where an alternate source such as an on site generator is provided with transfer equipment that includes a grounded conductor that is not solidly interconnected to the service supplied grounded conductor, the alternate source (derived system) shall be grounded in accordance with 250.30(A).

FPN No. 1: An alternate ac power source such as an on-site generator is not a separately derived system if the grounded conductor neutral is solidly interconnected to a service-supplied system grounded conductor neutral. An example of such situations is where alternate source transfer equipment does not include a switching action in the grounded conductor and allows it to remain solidly connected to the service supplied grounded conductor when the alternate source is operational and supplying the load served.

FPN No. 2: For systems that are not separately derived and are not required to be grounded as specified in 250.30, see 445.13 for minimum size of conductors that must carry fault current.

Author's Comment: The extra text was added to provide clear direction in positive text that an alternate source with a transfer switch where the grounded conductor is not solidly connected with the normal power grounded conductor and where the grounded conductor is switched with the ungrounded conductors. Where the grounded conductor is switched with the ungrounded conductors, the system is a separately derived system and must be grounded based on **250.30**.

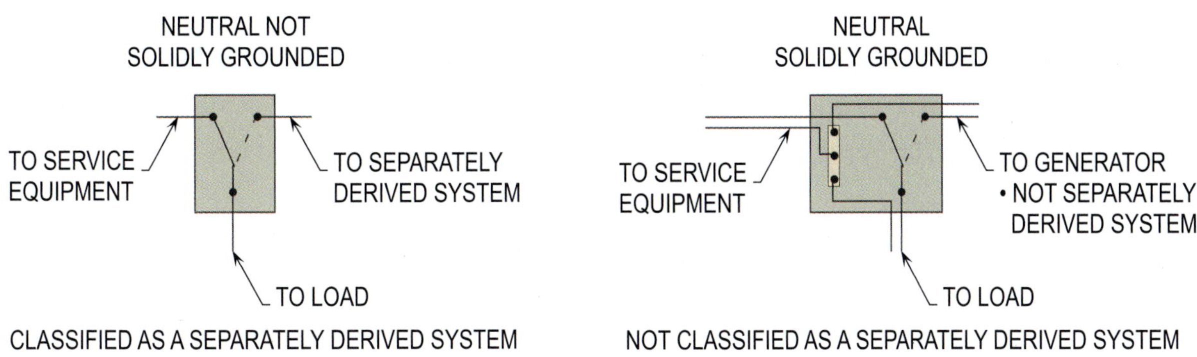

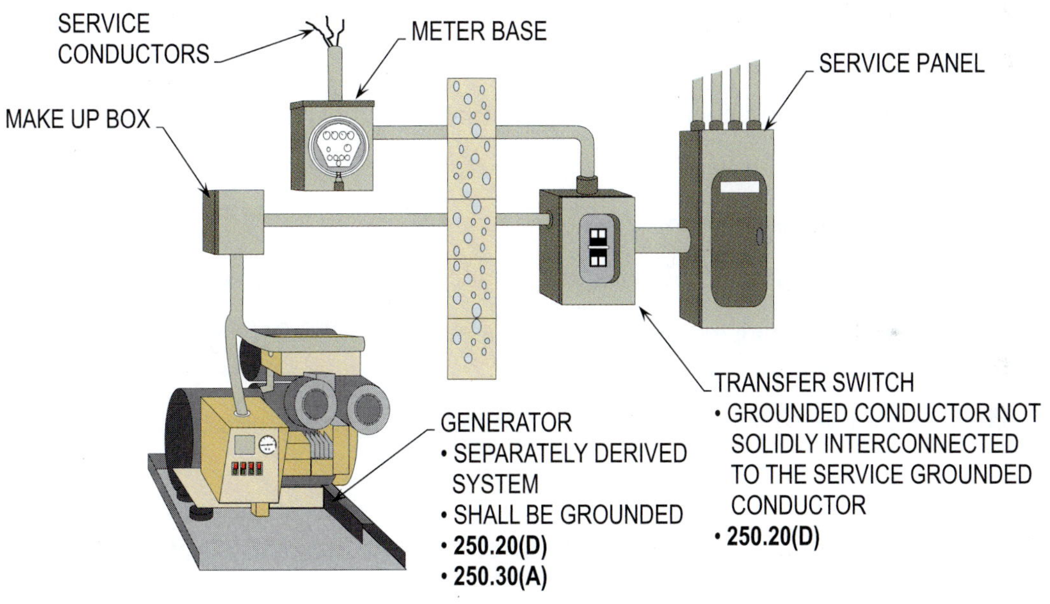

SEPARATELY DERIVED SYSTEMS
250.20(D)

Purpose of Change: This revision clarifies that a separately derived systems shall be grounded where transfer equipment is provided that includes switching the grounded conductor and where the grounded conductor is not solidly interconnected to the service supplied grounded conductor.

NEC Ch. 2 - Article 250
Part II - 250.21(B)

Type of Change	Panel Action	UL	UL 508	API 500 - 1997	API 505 - 1997	OSHA - 1994
Revision	Accept in Principle	-	-	-	-	1910.304(f)(1)(v)[C]
ROP	**ROC**	**NFPA 70E - 2004**	**NFPA 70B - 2006**	**NFPA 79 - 2007**	**NFPA**	**NEMA - 2002**
pg. 188 # 5-98	pg. 126 # 5-49	410.10(C)(5)(3)	-	4.3.2	-	PB 1.1
log: 3341 CMP: 5	log: 604	Submitter: Jamie mCNamara		2005 NEC: 250.21(B)		IEC: -

2005 NEC - 250.21 Alternating-Current Systems of 50 Volts to 1000 Volts Not Required to Be Grounded.

The following ac systems of 50 volts to 1000 volts shall be permitted to be grounded but shall not be required to be grounded:

(1) Electric systems used exclusively to supply industrial electric furnaces for melting, refining, tempering, and the like

(2) Separately derived systems used exclusively for rectifiers that supply only adjustable-speed industrial drives

(3) Separately derived systems supplied by transformers that have a primary voltage rating less than 1000 volts, provided that all the following conditions are met:

a. The system is used exclusively for control circuits.

b. The conditions of maintenance and supervision ensure that only qualified persons service the installation.

c. Continuity of control power is required.

~~d. Ground detectors are installed on the control system.~~

(4) Other systems that are not required to be grounded in accordance with the requirements of 250.20(B).

~~Where an alternating-current system is not grounded as permitted in 250.21(1) through (4), ground detectors shall be installed on the system.~~

~~Exception: Systems of less than 120 volts to ground as permitted by this Code shall not be required to have ground detectors.~~

2008 NEC - 250.21 Alternating-Current Systems of 50 Volts to 1000 Volts Not Required to Be Grounded.

(A) General. The following ac systems of 50 volts to 1000 volts shall be permitted to be grounded but shall not be required to be grounded:

(1) Electric systems used exclusively to supply industrial electric furnaces for melting, refining, tempering, and the like

(2) Separately derived systems used exclusively for rectifiers that supply only adjustable-speed industrial drives

(3) Separately derived systems supplied by transformers that have a primary voltage rating less than 1000 volts, provided that all the following conditions are met:

a. The system is used exclusively for control circuits.

b. The conditions of maintenance and supervision ensure that only qualified persons service the installation.

c. Continuity of control power is required.

(4) Other systems that are not required to be grounded in accordance with the requirements of 250.20(B).

(B) Ground Detectors. Ungrounded alternating current systems as permitted in 250.21(A)(1) through (A)(4) operating at not less than 120 volts and not exceeding 1000 volts shall have ground detectors installed on the system.

Author's Comment: The NEC TCC has directed that the text in **(B)** be changed to the text rewritten by the TCC to be in compliance with the NEC Style Manual. The change has been done below.

The exception requiring ground detectors was deleted and turned into positive text in a new subsection **(B)** with a new **(A)** covering general requirements for ungrounded systems.

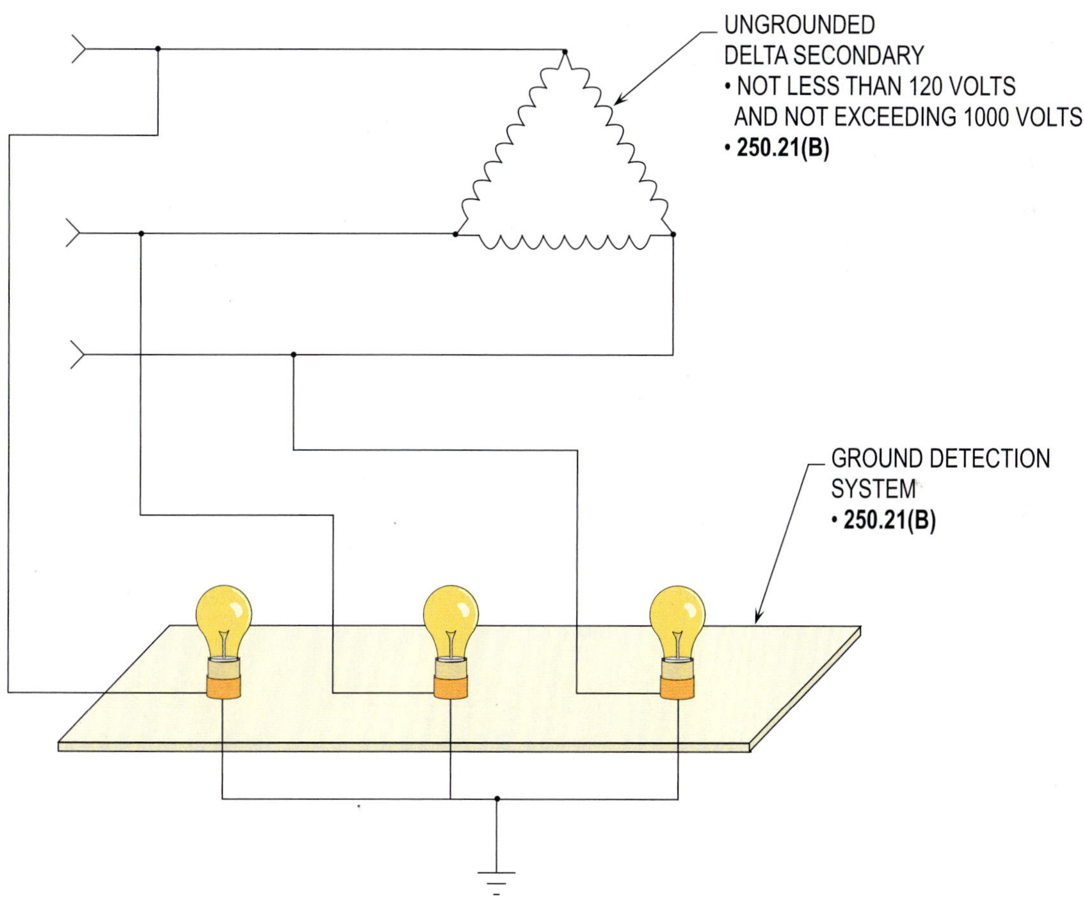

GROUND DETECTORS
250.21(B)

Purpose of Change: This revision for ground detectors has been placed into positive text in a new subsection (B).

NEC Ch. 2 - Article 250
Part II - 250.22(5)

Type of Change	Panel Action	UL	UL 508	API 500 - 1997	API 505 - 1997	OSHA - 1994
Revision	Accept	-	-	-	-	-
ROP	**ROC**	**NFPA 70E - 2004**	**NFPA 70B - 2006**	**NFPA 79 - 2007**	**NFPA**	**NEMA - 2002**
pg. 189 # 5-100	pg. - # -	-	-	4.3.2	-	PB 1.1
log: 1205 CMP: 5	log: - Submitter: James Tente			2005 NEC: 250.22		IEC: -

2005 NEC - 250.22 Circuits Not to Be Grounded.

The following circuits shall not be grounded:

(1) Circuits for electric cranes operating over combustible fibers in Class III locations, as provided in 503.155

(2) Circuits in health care facilities as provided in 517.61 and 517.160

(3) Circuits for equipment within electrolytic cell working zone as provided in Article 668

(4) Secondary circuits of lighting systems as provided in 411.5(A)

2008 NEC - 250.22 Circuits Not to Be Grounded.

The following circuits shall not be grounded:

(1) Circuits for electric cranes operating over combustible fibers in Class III locations, as provided in 503.155.

(2) Circuits in health care facilities as provided in 517.61 and 517.160.

(3) Circuits for equipment within electrolytic cell working zone as provided in Article 668.

(4) Secondary circuits of lighting systems as provided in 411.5(A)

(5) Secondary circuits of lighting systems as provided in 680.23(A)(2).

Author's Comment: The new (5) provides a reference to low voltage (15-volt or less) transformers for underwater lighting in permanent swimming pools that are not permitted to have a grounded secondary for the transformer.

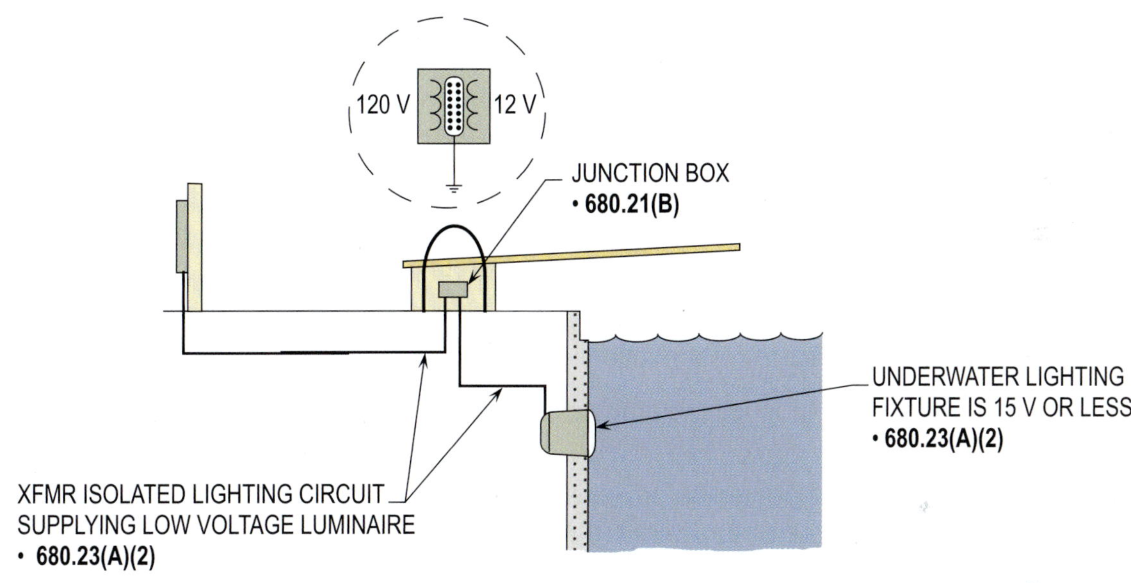

CIRCUITS NOT TO BE GROUNDED
250.22(5)

Purpose of Change: To clarify that these low voltage lighting systems are not to be grounded.

NEC Ch. 2 - Article 250
Part II - 250.24(D)

Type of Change	Panel Action	UL	UL 508	API 500 - 1997	API 505 - 1997	OSHA - 1994
Revision	Accept	-	-	-	-	1910.304(f)(3)(i)
ROP	**ROC**	**NFPA 70E - 2004**	**NFPA 70B - 2006**	**NFPA 79 - 2007**	**NFPA**	**NEMA - 2005**
pg. 189 # 5-103	pg. - # -	410.10(D)(1)	-	-	-	GR 1
log: 1707 CMP: 5	log: - Submitter: Danny Thomas			2005 NEC: 250.24(D)		IEC: -

2005 NEC - 250.24 Grounding Service-Supplied Alternating-Current Systems.

(D) Grounding Electrode Conductor. A grounding electrode conductor shall be used to connect the equipment grounding conductors, the service-equipment enclosures, and, where the system is grounded, the grounded service conductor to the grounding electrode(s) required by Part III of this article.

High-impedance grounded neutral system connections shall be made as covered in 250.36.

FPN: See 250.24(A) for ac system grounding connections.

2008 NEC - 250.24 Grounding Service-Supplied Alternating-Current Systems.

(D) Grounding Electrode Conductor. A grounding electrode conductor shall be used to connect the equipment grounding conductors, the service-equipment enclosures, and, where the system is grounded, the grounded service conductor to the grounding electrode(s) required by Part III of this article. This conductor shall be sized in accordance with 250.66.

High-impedance grounded neutral system connections shall be made as covered in 250.36.

FPN: See 250.24(A) for ac system grounding connections.

Author's Comment: This added sentence provides direction on how to size the grounding electrode conductor for service-supplied alternating current systems.

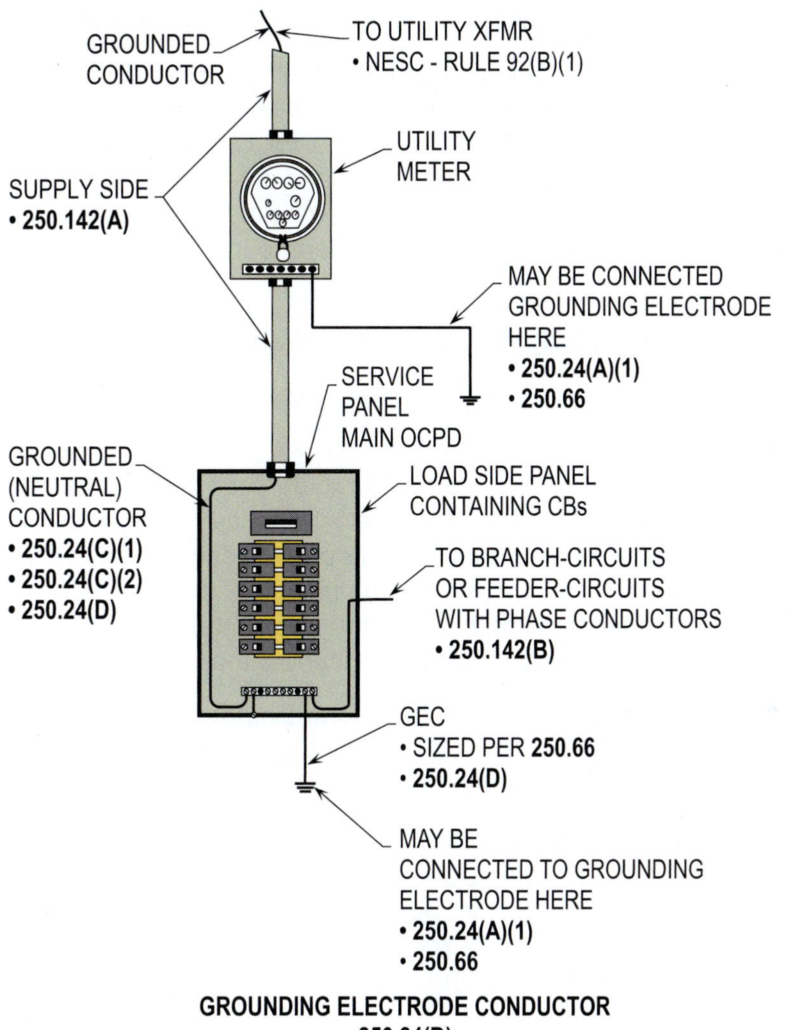

GROUNDING ELECTRODE CONDUCTOR
250.24(D)

Purpose of Change: This revision clarifies that the grounding electrode conductors shall be sized in accordance with **250.66**.

NEC Ch. 2 - Article 250
Part II - 250.28(D)(1) through (D)(3)

24

Type of Change		Panel Action		UL	UL 508	API 500 - 1997	API 505 - 1997	OSHA - 1994
Revision		Accept in Principle		-	-	-	-	-
ROP		ROC		NFPA 70E - 2004	NFPA 70B - 2006	NFPA 79 - 2007	NFPA	NEMA
pg. 190	# 5-107	pg. 126	# 5-53	-	-	-	-	-
log: 2012	CMP: 5	log: 1758	Submitter: Michael J. Johnston			2005 NEC: 250.28(D)		IEC: -

2005 NEC - 250.28 Main Bonding Jumper and System Bonding Jumper.

(D) Size. Main bonding jumpers and system bonding jumpers shall not be smaller than the sizes shown in Table 250.66. Where the supply conductors are larger than 1100 kcmil copper or 1750 kcmil aluminum, the bonding jumper shall have an area that is not less than 12 1/2 percent of the area of the largest phase conductor except that, where the phase conductors and the bonding jumper are of different materials (copper or aluminum), the minimum size of the bonding jumper shall be based on the assumed use of phase conductors of the same material as the bonding jumper and with an ampacity equivalent to that of the installed phase conductors.

2008 NEC - 250.28 Main Bonding Jumper and System Bonding Jumper.

(D) Size. Main bonding jumpers and system bonding jumpers shall be sized in accordance with 250.28(D)(1) through (D)(3).

(1) General. Main bonding jumpers and system bonding jumpers shall not be smaller than the sizes shown in Table 250.66. Where the supply conductors are larger than 1100 kcmil copper or 1750 kcmil aluminum, the bonding jumper shall have an area that is not less than 12 1/2 percent of the area so the largest phase conductor except that, where the phase conductors and the bonding jumper are of different materials (copper or aluminum), the minimum size of the bonding jumper shall be based on the assumed use of phase conductors of the same material as the bonding jumper and with an ampacity equivalent to that of the installed phase conductors.

(2) Main Bonding Jumper for Service With More Than One Enclosure. Where a service consists of more than a single enclosure as permitted in 230.71(A), the main bonding jumper for each enclosure shall be sized in accordance with 250.28(D)(1) based on the largest ungrounded service conductor serving that enclosure.

(3) Separately Derived System With More Than One Enclosure. Where a separately derived system supplies more than a single enclosure, the system bonding jumper for each enclosure shall be sized in accordance with 250.28(D)(1) based on the largest ungrounded feeder conductor serving that enclosure or a single system bonding jumper shall be installed at the source and sized in accordance with 250.28(D)(1) based on the equivalent size of the largest supply conductor determined by the largest sum of the areas of the corresponding conductors of each set.

Author's Comment: New **250.28(D)(2)** and **(D)(3)** provide direction sizing main bonding jumpers for services with more than one enclosure and sizing system bonding jumpers for separately derived systems with more than one enclosure. Section **250.28(D)(1)** applies to either services or separately derived systems with only one enclosure.

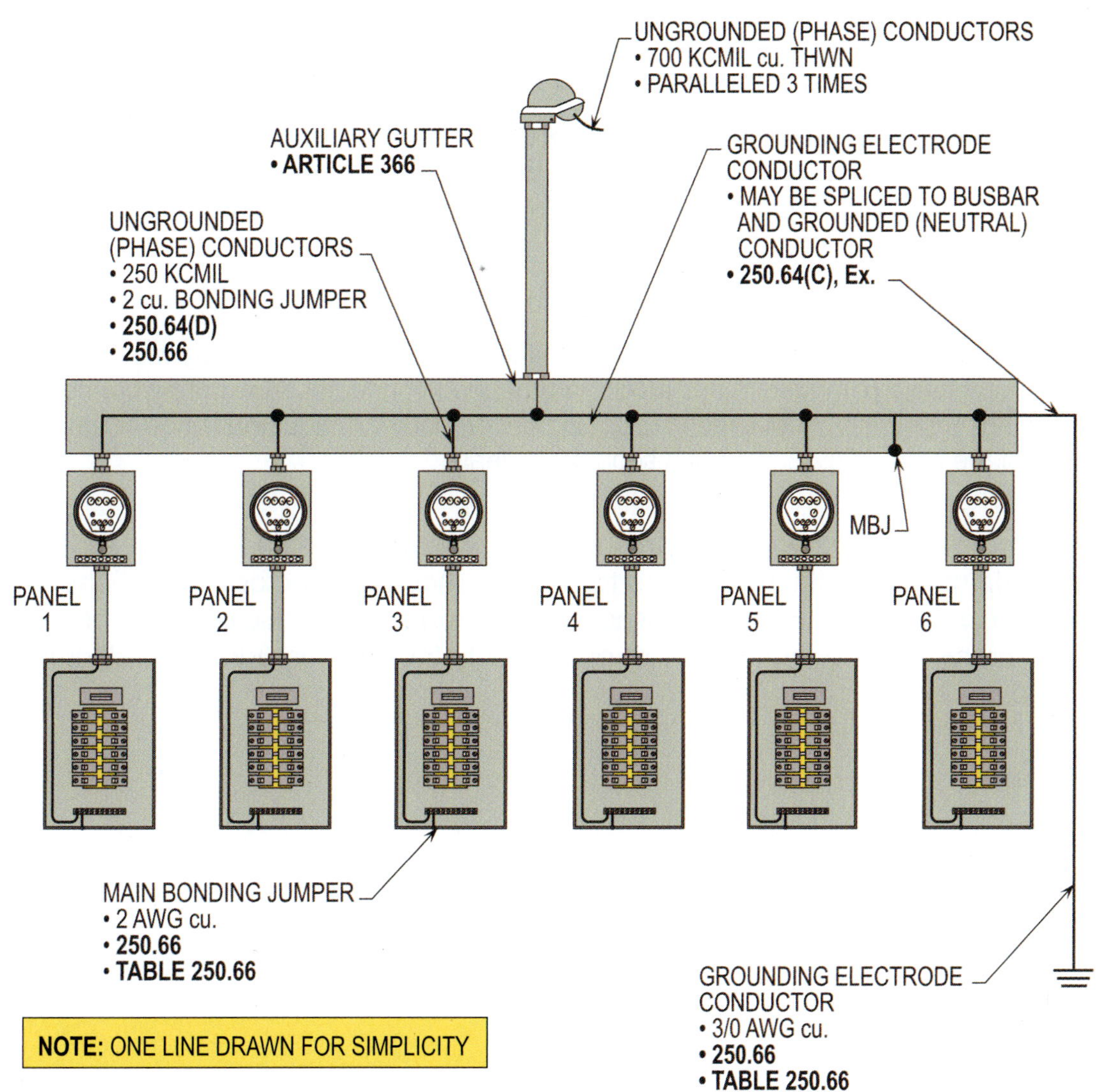

SIZE
250.28(D)(1) through (D)(3)

Purpose of Change: This revision clarifies how to size main bonding jumpers for services with more than one enclosure and to size system bonding jumpers for separately derived systems with more than one enlcosure.

NEC Ch. 2 - Article 250
Part II - 250.30(A)

Type of Change	Panel Action	UL	UL 508	API 500 - 1997	API 505 - 1997	OSHA - 1994
Revision	Accept	-	-	-	-	-
ROP	**ROC**	**NFPA 70E - 2004**	**NFPA 70B - 2006**	**NFPA 79 - 2007**	**NFPA**	**NEMA**
pg. 190 # 5-107a	pg. - # -	-	-	-	-	-
log: CP 502 CMP: 5	log: -	Submitter: Code Making Panel 5		2005 NEC: 250.30(A)		IEC: -

2005 NEC - 250.30 Grounding Separately Derived Alternating-Current Systems.

(A) Grounded Systems. A separately derived ac system that is grounded shall comply with 250.30(A)(1) through (A)(8). A ~~grounding connection~~ shall not be ~~made~~ to ~~any grounded circuit conductor~~ on the load side of the point of grounding of the separately derived system ~~except as otherwise permitted in this article~~.

2008 NEC - 250.30 Grounding Separately Derived Alternating-Current Systems.

(A) Grounded Systems. A separately derived ac system that is grounded shall comply with 250.30(A)(1) through (A)(8). Except as otherwise permitted in this article, a grounded conductor shall not be connected to normally non-current carrying metal parts of equipment, to equipment grounding conductors, or be reconnected to ground on the load side of the point of grounding of a separately derived system.

Author's Comment: These changes clarify the present requirement in more prescriptive language. This revision adds more specific restrictions for the grounded conductor connections to any ground connection on the load side of the service disconnect. A grounded conductor must not be connected to normally non-current-carrying metal parts of equipment since the grounded conductor is a current-carrying conductor.

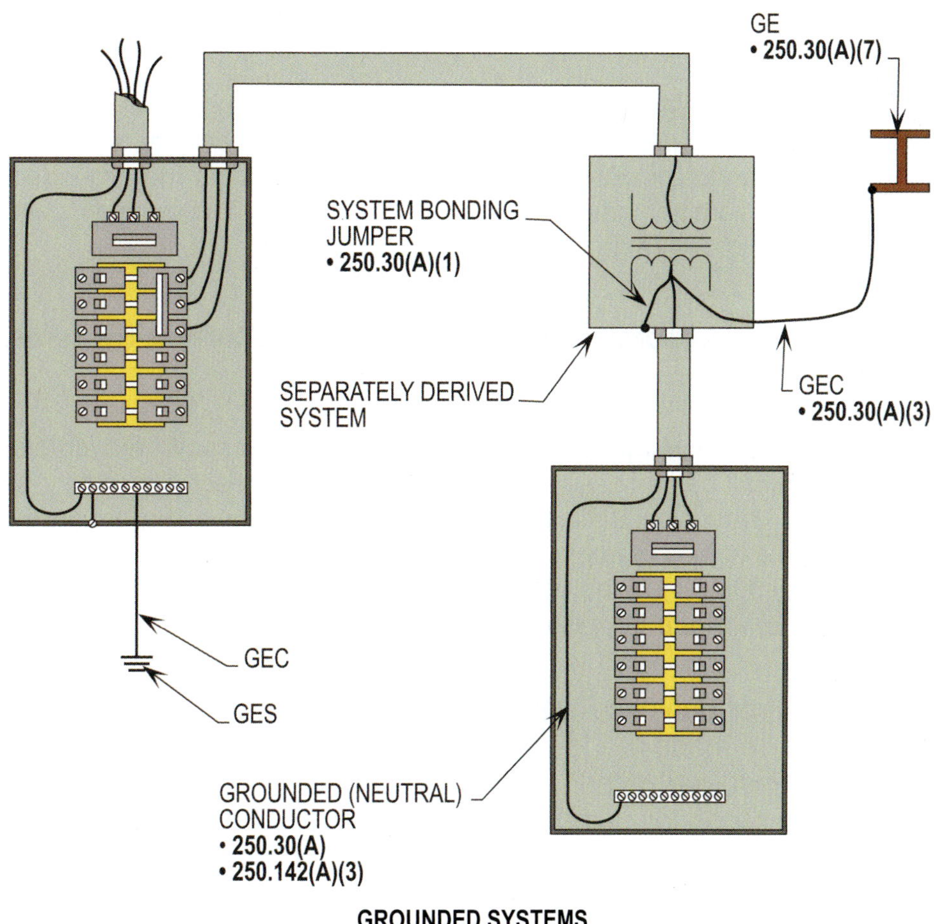

GROUNDED SYSTEMS
250.30(A)

Purpose of Change: This revision clarifies that the grounded conductor shall not be connected to normally non-current carrying metal parts of equipment, to EGC, or re-connected to ground on the load side of the point of grounding of a separately derived system.

NEC Ch. 2 - Article 250
Part II - 250.30(A)(4)

Type of Change		Panel Action		UL	UL 508	API 500 - 1997	API 505 - 1997	OSHA - 1994
Revision		Accept		-	-	-	-	-
ROP		ROC		NFPA 70E - 2004	NFPA 70B - 2006	NFPA 79 - 2007	NFPA	NEMA
pg. 191	# 5-110	pg. -	# -	-	-	-	-	-
log: 1649	CMP: 5	log: -	Submitter: Jim Davis			2005 NEC: 250.30(A)(4)		IEC: -

2005 NEC - 250.30 Grounding Separately Derived Alternating-Current Systems.

(A) Grounded Systems.

(4) Grounding Electrode Conductor, Multiple Separately Derived Systems. Where more than one separately derived system is installed, it shall be permissible to connect a tap from each separately derived system to a common grounding electrode conductor. Each tap conductor shall connect the grounded conductor of the separately derived system to the common grounding electrode conductor. The grounding electrode conductors and taps shall comply with 250.30(A)(4)(a) through (A)(4)(c).

2008 NEC - 250.30 Grounding Separately Derived Alternating-Current Systems.

(A) Grounded Systems.

(4) Grounding Electrode Conductor, Multiple Separately Derived Systems. Where more than one separately derived system is installed, it shall be permissible to connect a tap from each separately derived system to a common grounding electrode conductor. Each tap conductor shall connect the grounded conductor of the separately derived system to the common grounding electrode conductor. The grounding electrode conductors and taps shall comply with 250.30(A)(4)(a) through (A)(4)(c). This connection shall be made at the same point on the separately derived system where the system bonding jumper is installed.

Author's Comment: This final sentence was added to make it clear that the grounding electrode conductor connection must be made at the same point on the separately derived system where the system bonding jumper is installed, either within the separately derived system or within the first disconnecting means on the secondary side of the separately derived system, where provided, but not both. This sentence is similar to the one in **250.30(A)(3)** for a single separately derived system.

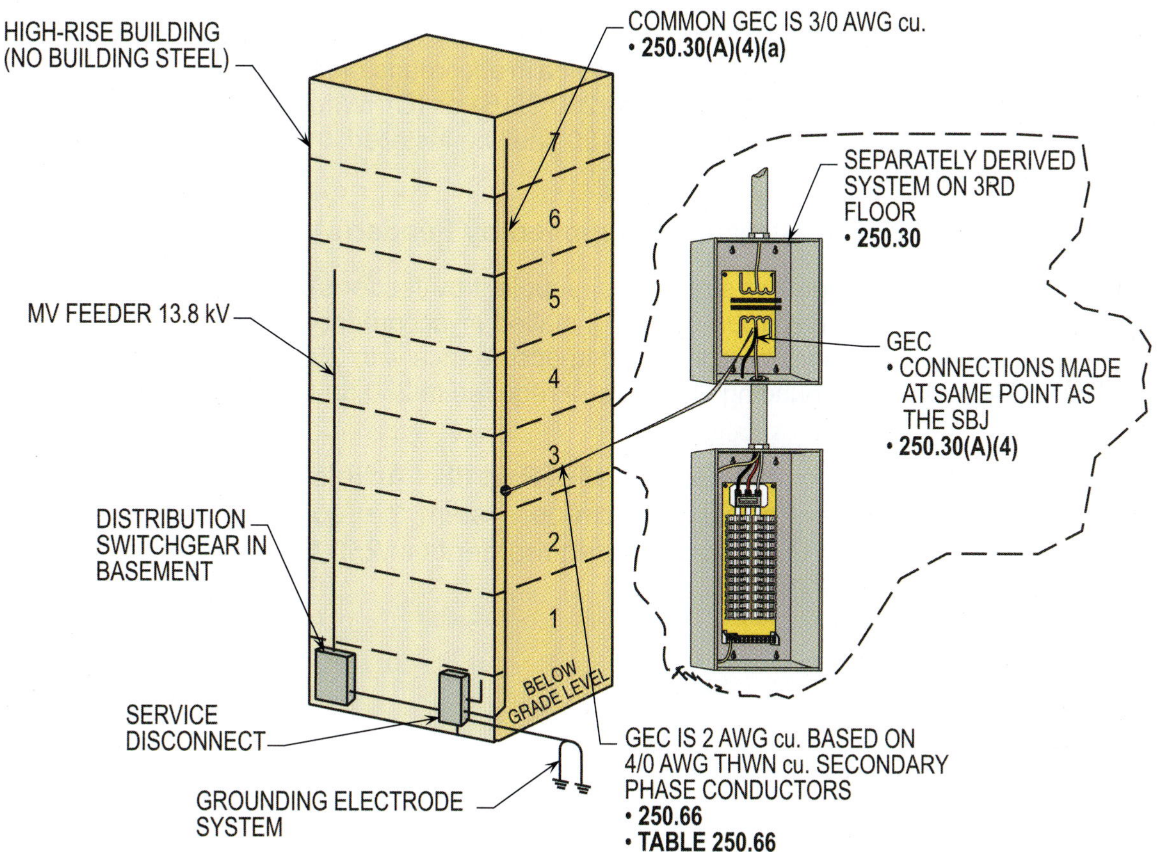

GROUNDING ELECTRODE CONDUCTOR, MULTIPLE SEPARATELY DERIVED SYSTEMS
250.30(A)(4)

Purpose of Change: This revision clarifies that for multiple separately derived systems, the grounding electrode conductor (GEC) connection shall be made at the same point on the separately derived system as where the system bonding jumper (SBJ) is installed, either in the source or the first disconnecting mean.

NEC Ch. 2 - Article 250
Part II - 250.32(A)

Type of Change	Panel Action	UL	UL 508	API 500 - 1997	API 505 - 1997	OSHA - 1994
Revision	Accept	-	-	-	-	1910.304(f)(3)(i)
ROP	**ROC**	**NFPA 70E - 2004**	**NFPA 70B - 2006**	**NFPA 79 - 2007**	**NFPA**	**NEMA - 2005**
pg. 194 # 5-120	pg. - # -	410.10(D)(1)	-	-	-	GR 1
log: 1298 CMP: 5	log: -	Submitter: Joseph Whitt		2005 NEC: 250.32(A)		IEC: -

2005 NEC - 250.32 Buildings or Structures Supplied by Feeder(s) or Branch Circuit(s).

(A) Grounding Electrode. Building(s) or structure(s) supplied by feeder(s) or branch circuit(s) shall have a grounding electrode or grounding electrode system installed in accordance with ~~250.50~~ . The grounding electrode conductor(s) shall be connected in accordance with 250.32(B) or (C). Where there is no existing grounding electrode, the grounding electrode(s) required in 250.50 shall be installed.

2008 NEC - 250.32 Buildings or Structures Supplied by Feeder(s) or Branch Circuit(s).

(A) Grounding Electrode. Building(s) or structure(s) supplied by feeder(s) or branch circuit(s) shall have a grounding electrode or grounding electrode system installed in accordance with Part III of Article 250. The grounding electrode conductor(s) shall be connected in accordance with 250.32(B) or (C). Where there is no existing grounding electrode, the grounding electrode(s) required in 250.50 shall be installed.

Author's Comment: Requiring compliance with all of **Part III** of **Article 250** makes it clear that feeders and branch circuits must have a grounding electrode system that complies with all the appropriate electrodes in **250.50**, **250.52** and the installation requirements in **250.53**, as well as other appropriate requirements in the remainder of **Part III**.

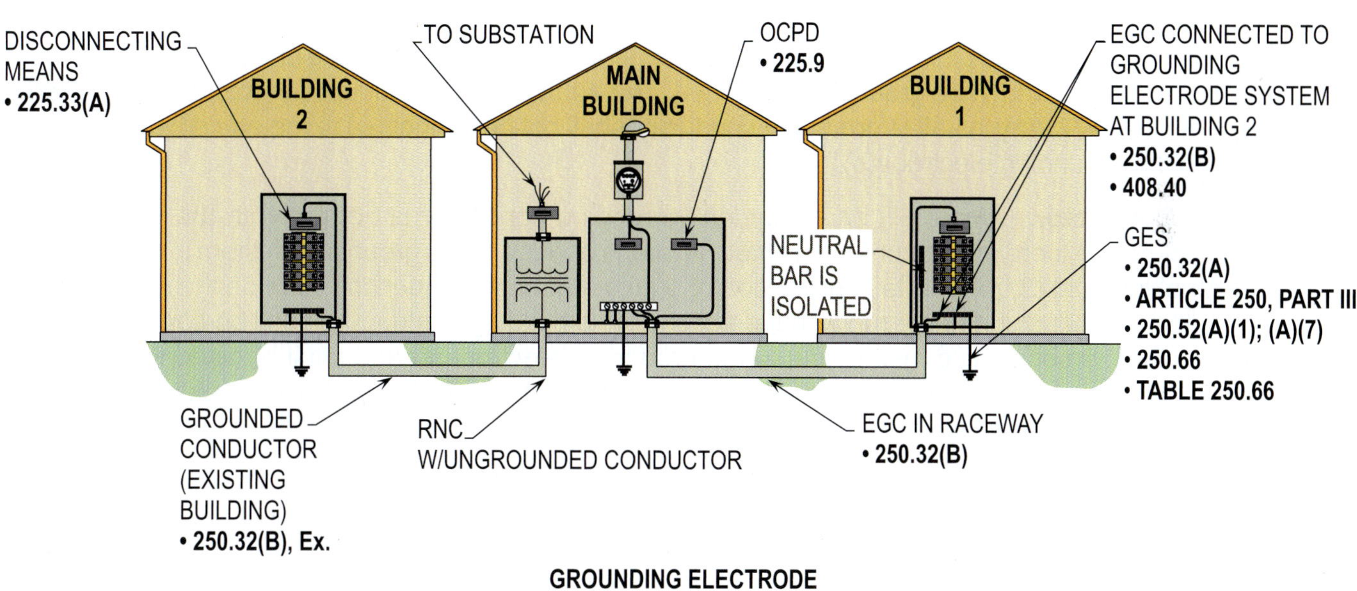

GROUNDING ELECTRODE
250.32(A)

Purpose of Change: This revision requires buildings or structures supplied by branch circuits or feeders to have a grounding electrode system that complies with all of the requirements in **Part III** of **Article 250**, not only **250.50**.

NEC Ch. 2 - Article 250
Part II - 250.32(B), Exception

Type of Change	Panel Action	UL	UL 508	API 500 - 1997	API 505 - 1997	OSHA - 1994
Revision	Accept in Principle	-	-	-	-	1910.304(f)(3)(i)
ROP	**ROC**	**NFPA 70E - 2004**	**NFPA 70B - 2006**	**NFPA 79 - 2007**	**NFPA**	**NEMA - 2005**
pg. 193 # 5-119	pg. 127 # 5-58	410.10(D)(1)	-	-	-	GR 1
log: 2395 CMP: 5	log: 1518 Submitter: Mike Holt			2005 NEC: 250.30(B)(1); (B)(2)		IEC: -

2005 NEC - 250.32 Buildings or Structures Supplied by Feeder(s) or Branch Circuit(s).

(B) Grounded Systems. For a grounded system at the separate building or structure, ~~the connection to the grounding electrode and grounding or bonding of equipment, structures, or frames required to be grounded or bonded shall comply with either 250.32(B)(1) or (B)(2).~~

~~**(1) Equipment Grounding Conductor.**~~ An equipment grounding conductor as described in 250.118 shall be run with the supply conductors and connected to the building or structure disconnecting means and to the grounding electrode(s). The equipment grounding conductor shall be used for grounding or bonding of equipment, structures, or frames required to be grounded or bonded. The equipment grounding conductor shall be sized in accordance with 250.122. Any installed grounded conductor shall not be connected to the equipment grounding conductor or to the grounding electrode(s).

~~**(2) Grounded Conductor.** Where (1)~~ an equipment grounding conductor is not run with the supply to the building or structure, (2) there are no continuous metallic paths bonded to the grounding system in each building or structure involved, and (3) ground-fault protection of equipment has not been installed on the supply side of the feeder(s), the grounded conductor run with the supply to the building or structure shall be connected to the building or structure disconnecting means and to the grounding electrode(s) and shall be used for grounding and bonding of equipment, structures, or frames required to be grounded or bonded. The size of the grounded conductor shall not be smaller than the larger of either of the following:

(1) That required by 220.61

(2) That required by 250.122

2008 NEC - 250.32 Buildings or Structures Supplied by Feeder(s) or Branch Circuit(s).

(B) Grounded Systems. For a grounded system at the separate building or structure, an equipment grounding conductor as described in 250.118 shall be run with the supply conductors and be connected to the building or structure disconnecting means and to the grounding electrode(s). The equipment grounding conductor shall be used for grounding or bonding of equipment, structures, or frames required to be grounded or bonded. The equipment grounding conductor shall be sized in accordance with 250.122. Any installed grounded conductor shall not be connected to the equipment grounding conductor or to the grounding electrode(s).

Exception: For existing premises wiring systems only, the grounded conductor run with the supply to the building or structure shall be permitted to be connected to the building or structure disconnecting means and to the grounding electrode(s) and shall be used for grounding or bonding of equipment, structures, or frames required to be grounded or bonded where all the requirements of (1), (2), and (3) are met:

(1) An equipment grounding conductor is not run with the supply to the building or structure,

(2) There are no continuous metallic paths bonded to the grounding system in each building or structure involved, and

(3) Ground-fault protection of equipment has not been installed on the supply side of the feeder(s),

Where the grounded conductor is used for grounding in accordance with the provision of this exception, the size of the grounded conductor shall not be smaller than the larger of either of the following:

(1) That required by 220.61

(2) That required by 250.122

Author's Comment: Section **250.32(B)(2)** permitted the equipment grounding conductor to be deleted in an installation where there was no continuous metallic path bonded to the grounding system from building or structure to building or structure. The grounded conductor was then used for a path back to the original building or structure for any fault current. There was no way to effectively ensure that the two buildings or structures would remain isolated in the future. This change, applying the exception only for existing premises wiring systems, helps reduce the number of designs or installations where inappropriate neutral-to-ground connections can and often do happen at a later date where the installation of a grounding path between the buildings or structures occurs at that later date.

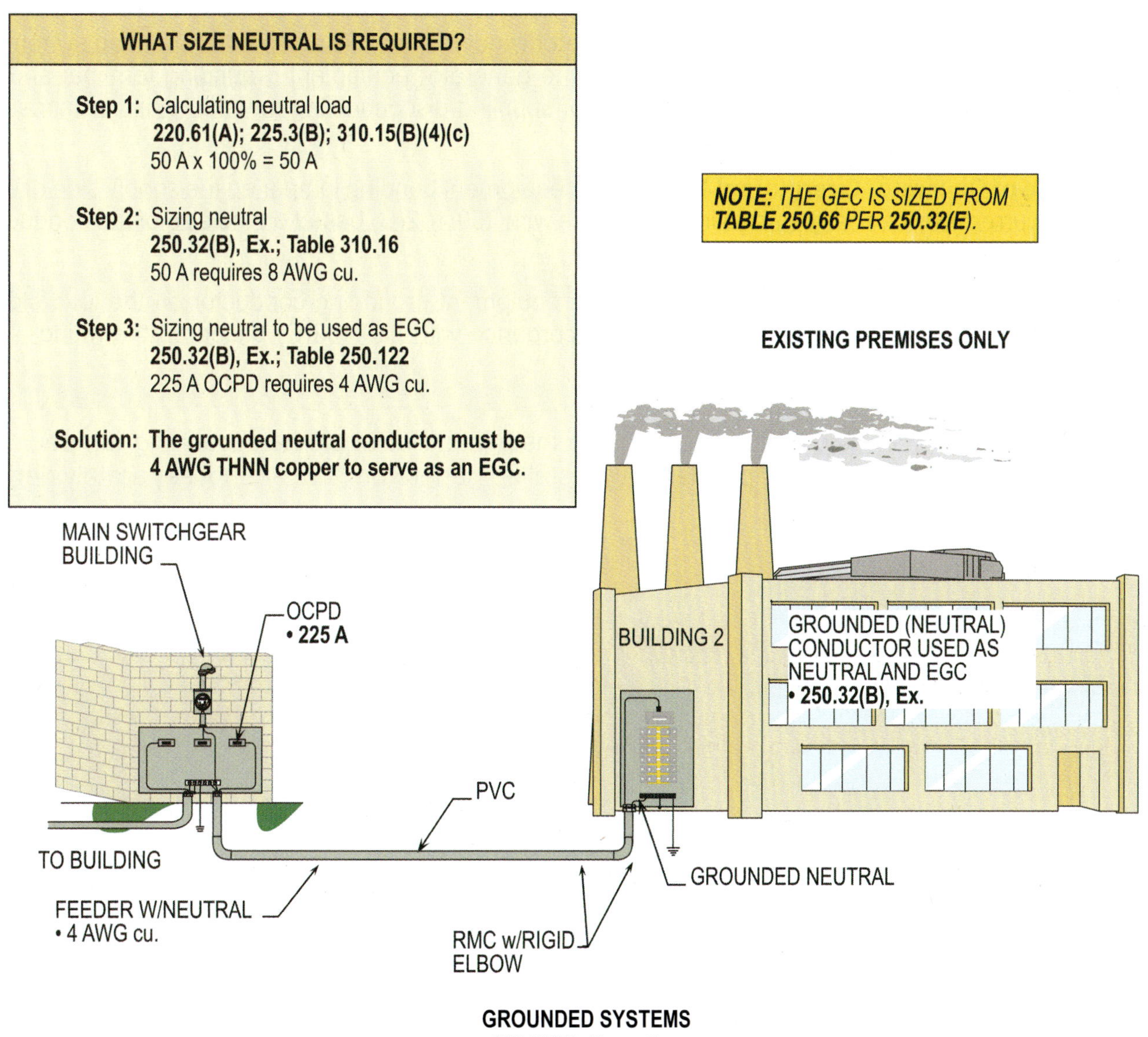

WHAT SIZE NEUTRAL IS REQUIRED?
Step 1: Calculating neutral load
220.61(A); 225.3(B); 310.15(B)(4)(c)
50 A x 100% = 50 A
Step 2: Sizing neutral
250.32(B), Ex.; Table 310.16
50 A requires 8 AWG cu.
Step 3: Sizing neutral to be used as EGC
250.32(B), Ex.; Table 250.122
225 A OCPD requires 4 AWG cu.
Solution: The grounded neutral conductor must be 4 AWG THHN copper to serve as an EGC.
NOTE: THE GEC IS SIZED FROM TABLE 250.66 PER 250.32(E).
EXISTING PREMISES ONLY
MAIN SWITCHGEAR BUILDING
OCPD
• 225 A
BUILDING 2
GROUNDED (NEUTRAL) CONDUCTOR USED AS NEUTRAL AND EGC
• 250.32(B), Ex.
PVC
TO BUILDING
GROUNDED NEUTRAL
FEEDER W/NEUTRAL
• 4 AWG cu.
RMC w/RIGID ELBOW

GROUNDED SYSTEMS
250.32(B), Exception

Purpose of Change: This revision clarifies that the grounded (neutral) conductor for other buildings or structures supplied by feeders or branch circuits shall be permitted to be used as the equipment grounding conductor for existing premises wiring system only.

NEC Ch. 2 - Article 250
Part II - 250.35

Type of Change	Panel Action	UL	UL 508	API 500 - 1997	API 505 - 1997	OSHA - 1994
New Section	Accept in Principle	-	-	-	-	-
ROP	**ROC**	**NFPA 70E - 2004**	**NFPA 70B - 2006**	**NFPA 79 - 2007**	**NFPA**	**NEMA - 2003**
pg. 195 # 5-128	pg. 129 # 5-67	-	-	-	-	MG 1
log: 3503 CMP: 5	log: 1787	Submitter: Paul Dobrowsky		2005 NEC: -		IEC: -

2008 NEC - 250.35 Permanently Installed Generators.

A conductor that provides an effective ground-fault current path shall be installed with the supply conductors from a permanently installed generator(s) to the first disconnecting mean(s) in accordance with (A) or (B).

(A) Separately Derived System. Where the generator is installed as a separately derived system, the requirements in 250.30 shall apply.

(B) Non-separately Derived System. Where the generator is not installed as a separately derived system, an equipment bonding jumper shall be installed between the generator equipment grounding terminal and the equipment grounding terminal or bus of the enclosure of supplied disconnecting mean(s) in accordance with (B)(1) or (B)(2).

(1) Supply Side of Generator Overcurrent Device. The equipment bonding jumper on the supply side of each generator overcurrent device shall be sized in accordance with 250.102(C) based on the size of the conductors supplied by the generator.

(2) Load Side of Generator Overcurrent Device. The equipment grounding conductor on the load side of each generator overcurrent device shall be sized in accordance with 250.102(D) based on the rating of the overcurrent device supplied.

Author's Comment: This new text deals with the installation of equipment bonding jumpers and equipment grounding conductors for permanently installed generators that are not separately derived systems.

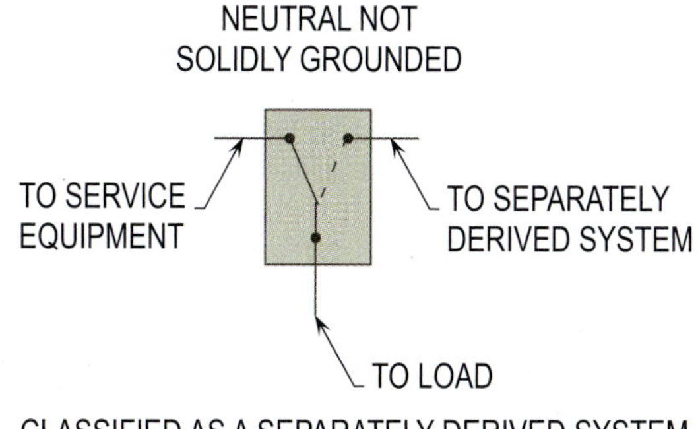

NOTE 1: THE EQUIPMENT BONDING JUMPER ON THE SUPPLY SIDE OF EACH GENERATOR OVERCURRENT DEVICE SHALL BE SIZED IN ACCORDANCE WITH **250.102(C)** BASED ON THE SIZE OF THE CONDUCTORS SUPPLIED BY THE GENERATOR.

NOTE 2: THE EQUIPMENT GROUNDING CONDUCTOR ON THE LOAD SIDE OF EACH GENERATOR OVERCURRENT DEVICE SHALL BE SIZED IN ACCORDANCE WITH **250.102(D)** BASED ON THE RATING OF THE OVERCURRENT DEVICE SUPPLIED.

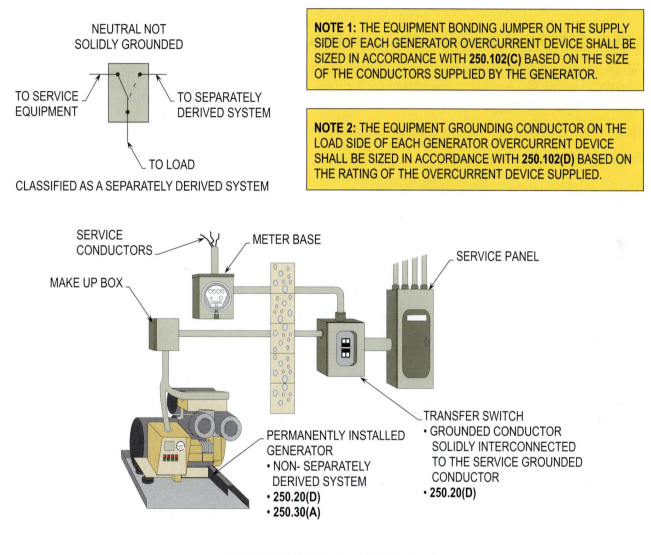

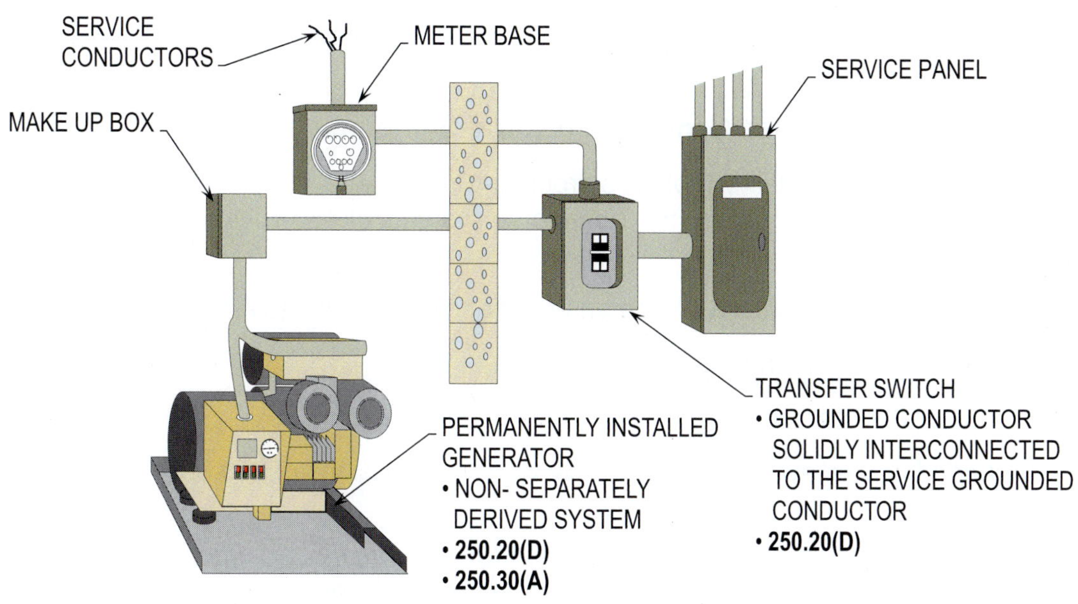

PERMANENTLY INSTALLED GENERATORS
250.35

Purpose of Change: A new section had been added to address the installation of equipment bonding jumpers and equipment grounding conductors for permanently installed generators that are not separately derived systems.

NEC Ch. 2 - Article 250
Part II - 250.36

Type of Change	Panel Action	UL	UL 508	API 500 - 1997	API 505 - 1997	OSHA - 1994
Revision	Accept	-	-	-	-	-
ROP	**ROC**	**NFPA 70E - 2004**	**NFPA 70B - 2006**	**NFPA 79 - 2007**	**NFPA**	**NEMA**
pg. 196 # 5-129	pg. - # -	410.10(C)(5)(4)	-	-	-	-
log: 1457 CMP: 5	log: -	Submitter: Ryan Jackson		2005 NEC: 250.36		IEC: -

2005 NEC - 250.36 High-Impedance Grounded Neutral Systems.

High-impedance grounded neutral systems in which a grounding impedance, usually a resistor, limits the ground-fault current to a low value shall be permitted for 3-phase ac systems of 480 volts to 1000 volts where all the following conditions are met.

(1) The conditions of maintenance and supervision ensure that only qualified persons service the installation.

(2) ~~Continuity of power is required.~~

(3) Ground detectors are installed on the system.

~~(4)~~ Line-to-neutral loads are not served.

High-impedance grounded neutral systems shall comply with the provisions of 250.36(A) through (G).

2008 NEC - 250.36 High-Impedance Grounded Neutral Systems.

High-impedance grounded neutral systems in which a grounding impedance, usually a resistor, limits the ground-fault current to a low value shall be permitted for 3-phase ac systems of 480 volts to 1000 volts where all the following conditions are met.

(1) The conditions of maintenance and supervision ensure that only qualified persons service the installation.

(2) Ground detectors are installed on the system.

(3) Line-to-neutral loads are not served.

High-impedance grounded neutral systems shall comply with the provisions of 250.36(A) through (G).

Author's Comment: Continuity of power is not required for high impedance grounded neutral systems. However, it is a design issue where loss of power may be critical to the operation of an installation. The purpose of a high impedance grounded neutral system is to ensure continuity of power but, with the deletion of this item as a requirement, continuity is not a required reason to qualify for this type of grounding.

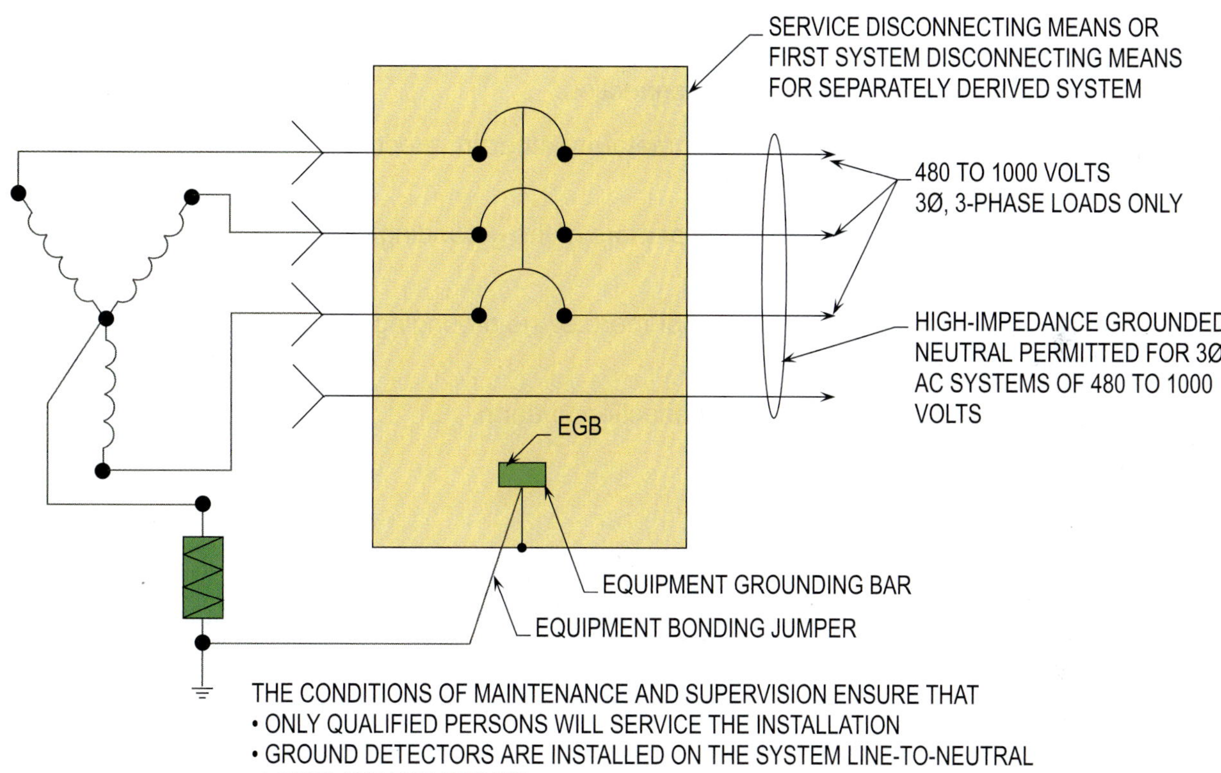
SERVICE DISCONNECTING MEANS OR FIRST SYSTEM DISCONNECTING MEANS FOR SEPARATELY DERIVED SYSTEM
480 TO 1000 VOLTS 3Ø, 3-PHASE LOADS ONLY
HIGH-IMPEDANCE GROUNDED NEUTRAL PERMITTED FOR 3Ø AC SYSTEMS OF 480 TO 1000 VOLTS
EGB
EQUIPMENT GROUNDING BAR
EQUIPMENT BONDING JUMPER
THE CONDITIONS OF MAINTENANCE AND SUPERVISION ENSURE THAT
• ONLY QUALIFIED PERSONS WILL SERVICE THE INSTALLATION
• GROUND DETECTORS ARE INSTALLED ON THE SYSTEM LINE-TO-NEUTRAL LOADS ARE NOT SERVED

HIGH-IMPEDANCE GROUNDED NEUTRAL SYSTEMS
250.36

Purpose of Change: This revision clarifies that continuity of power is not required but is a design consideration where loss of power could be critical to the operation of an installation.

NEC Ch. 2 - Article 250
Part III - 250.52(A)(2)

27

Type of Change	Panel Action	UL	UL 508	API 500 - 1997	API 505 - 1997	OSHA - 1994
Revision	Accept	-	-	-	-	1910.304(f)(3)(i)
ROP	**ROC**	**NFPA 70E - 2004**	**NFPA 70B - 2006**	**NFPA 79 - 2007**	**NFPA**	**NEMA**
pg. 201 # 5-148	pg. 134 # 5-85	410.10(D)(1)	-	-	-	-
log: 342 CMP: 5	log: 1521	Submitter: Michael J. Johnston		2005 NEC: 250.52(A)(3)		IEC: -

2005 NEC - 250.52 Grounding Electrodes.

(A) Electrodes Permitted for Grounding.

(2) Metal Frame of the Building or Structure. The metal frame of the building or structure~~, where~~ any of the following methods ~~are used to make an earth connection~~:

(1) 3.0 m (10 ft) or more of a single structural metal member in direct contact with the earth or encased in concrete that is in direct contact with the earth

(2) The structural metal frame ~~is bonded to one or more of the grounding electrodes as defined in 250.52(A)(1), (A)(3), or (A)(4)~~

(3) The structural metal frame ~~is bonded~~ to one or more of the grounding electrodes as defined in 250.52(A)(5) or (A)~~(6)~~ that comply with 250.56, or

(4) Other approved means of establishing a connection to earth.

2008 NEC - 250.52 Grounding Electrodes.

(A) Electrodes Permitted for Grounding.

(2) Metal Frame of the Building or Structure. The metal frame of the building or structure that is connected to the earth by any of the following methods:

(1) 3.0 m (10 ft) or more of a single structural metal member in direct contact with the earth or encased in concrete that is in direct contact with the earth

(2) Connecting the structural metal frame to the reinforcing bars of a concrete-encased electrode as provided in 250.52(A)(3) or ground ring as provided in 250.52(A)(4)

(3) Bonding the structural metal frame to one or more of the grounding electrodes as defined in 250.52(A)(5) or (A)(7) that comply with 250.56 or

(4) Other approved means of establishing a connection to earth.

Author's Comment: This section in the 2005 NEC created a possible problem by using the metal water pipe as a method of making the metal frame of the building or structure into a grounding electrode. Where only the metal water pipe and the metal frame of the building or structure are the electrodes, 250.53(D)(2) requires an additional supplemental electrode, of which one can be the metal of a building or structure. If, in the future, the building metal water piping system is replaced with nonmetallic water pipe, the building metal would no longer be an electrode and total loss of any electrode for the building electrical system would be the result. The text was revised to correct this problem by requiring the building metal frame to be in direct contact with the earth for 10 feet or more, use the concrete-encased electrode, the ground ring, rods, pipe or plate electrodes, or other approved means of connecting to earth.

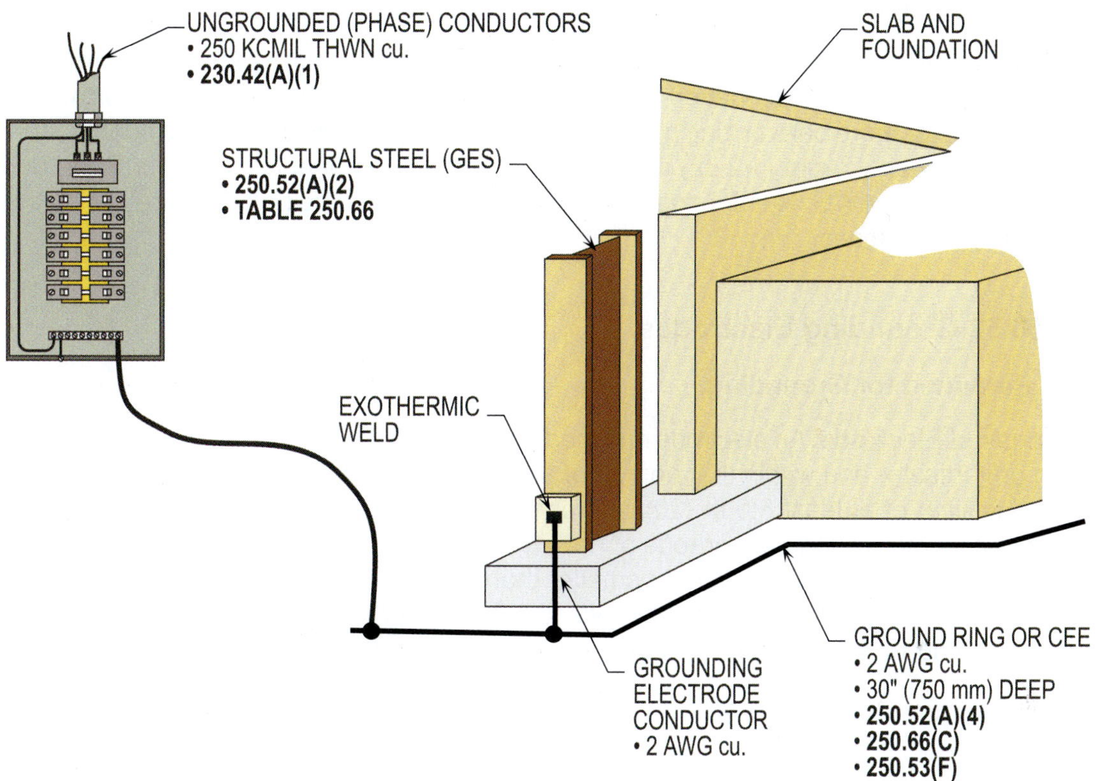

METAL FRAME OF THE BUILDING OR STRUCTURE
250.52(A)(2)

Purpose of Change: This revision clarifies that the structural steel shall be permitted to serve as a grounding electrode if connected to the reinforcing bars of a concrete encased electrode or ground ring.

NEC Ch. 2 - Article 250
Part III - 250.52(A)(3)

Type of Change	Panel Action	UL	UL 508	API 500 - 1997	API 505 - 1997	OSHA - 1994
Revision	Accept in Principle	-	-	-	-	1910.304(f)(3)(i)
ROP	**ROC**	**NFPA 70E - 2004**	**NFPA 70B - 2006**	**NFPA 79 - 2007**	**NFPA**	**NEMA**
pg. 197 # 5-137	pg. 134 # 5-86	410.10(D)(1)	-	-	-	-
log: 2642 CMP: 5	log: 381	Submitter: Robert A. Jones		2005 NEC: 250.52(A)(3)		IEC: -

2005 NEC - 250.52 Grounding Electrodes.

(A) Electrodes Permitted for Grounding.

(3) Concrete-Encased Electrode. An electrode encased by at least 50 mm (2 in.) of concrete, located ~~within and~~ near the bottom of a concrete foundation or footing that is in direct contact with the earth, consisting of at least 6.0 m (20 ft) of one or more bare or zinc galvanized or other electrically conductive coated steel reinforcing bars or rods of not less than 13 mm (1/2 in.) in diameter, or consisting of at least 6.0 m (20 ft) of bare copper conductor not smaller than 4 AWG. Reinforcing bars shall be permitted to be bonded together by the usual steel tie wires or other effective means.

2008 NEC - 250.52 Grounding Electrodes.

(A) Electrodes Permitted for Grounding.

(3) Concrete-Encased Electrode. An electrode encased by at least 50 mm (2 in.) of concrete, located horizontally near the bottom or vertically, and within that portion of a concrete foundation or footing that is in direct contact with the earth, consisting of at least 6.0 m (20 ft) of one or more bare or zinc galvanized or other electrically conductive coated steel reinforcing bars or rods of not less than 13 mm (1/2 in.) in diameter, or consisting of at least 6.0 m (20 ft) of bare copper conductor not smaller than 4 AWG. Reinforcing bars shall be permitted to be bonded together by the usual steel tie wires or other effective means. Where multiple concrete-encased electrodes are present at a building or structure, it shall be permissible to bond only one into the grounding electrode system.

Author's Comment: This change permits the concrete-encased electrode to be located horizontally near the bottom of the footing or vertically within the portion of the footing that is in direct contact with the earth, rather than just horizontally at the bottom of the footing. An additional sentence has been added recognizing that there may be more than one concrete-encased electrode and thus requiring only one of those to be used in the grounding electrode system.

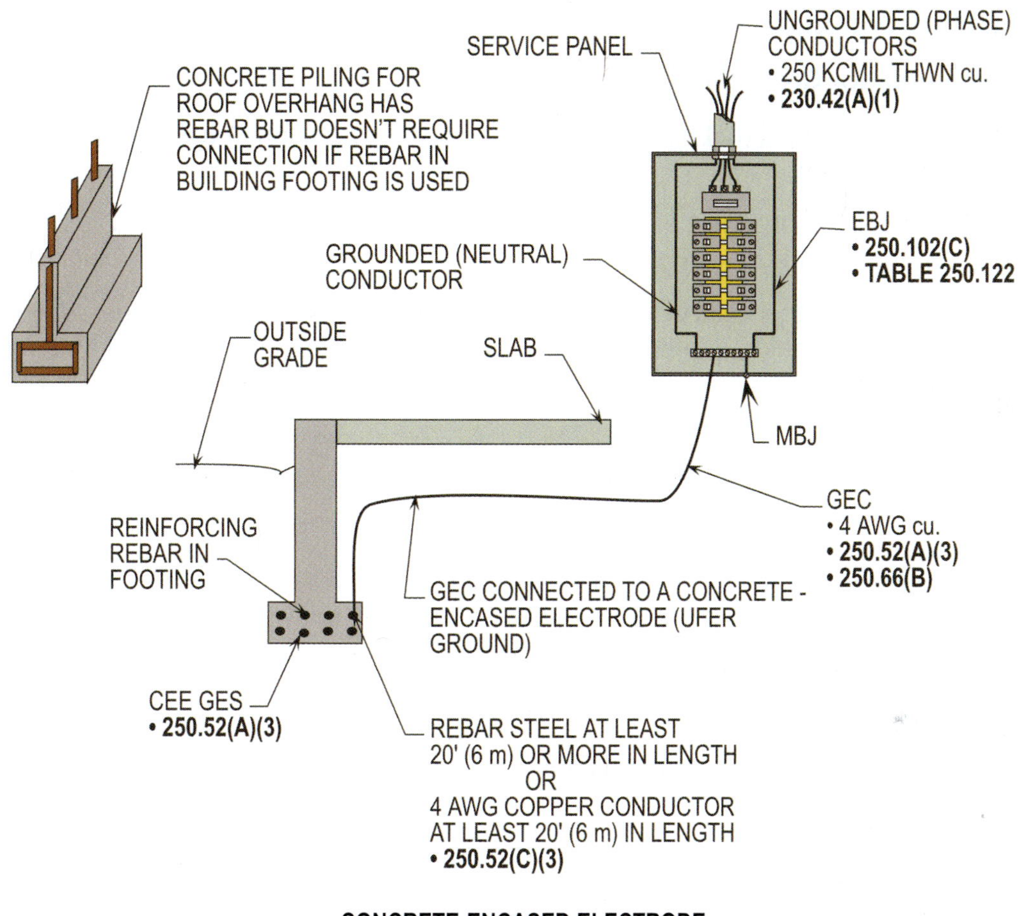

CONCRETE-ENCASED ELECTRODE
250.52(A)(3)

Purpose of Change: This revision clarifies that it shall be permissible to bond only one concrete-encased electrode (CEE) into the grounding electrode system (GES) where multiple concrete electrodes are present. This change also permits the concrete-encased electrode to be located horizontally near the bottom of the footing or vertically where foundation or footing is in direct contact with the earth.

NEC Ch. 2 - Article 250
Part III - 250.52(A)(5)

Type of Change	Panel Action	UL	UL 508	API 500 - 1997	API 505 - 1997	OSHA - 1994
Revision	Accept in Principle	-	-	-	-	1910.304(f)(3)(i)
ROP	**ROC**	**NFPA 70E - 2004**	**NFPA 70B - 2006**	**NFPA 79 - 2007**	**NFPA**	**NEMA - 2005**
pg. 204 # 5-160	pg. - # -	410.10(D)(1)	-	-	-	GR 1
log: 1985 CMP: 5	log: -	Submitter: Roger J. Montambo		2005 NEC: 250.52(A)(5)		IEC: -

2005 NEC - 250.52 Grounding Electrodes.

(A) Electrodes Permitted for Grounding.

(5) Rod and Pipe Electrodes. Rod and pipe electrodes shall not be less than ~~2.5 m~~ (8 ft) in length and shall consist of the following materials.

(a) Electrodes of pipe or conduit shall not be smaller than metric designator 21 (trade size 3/4) and, where of iron or steel, shall have the outer surface galvanized or otherwise metal-coated for corrosion protection.

(b) Electrodes ~~of rods~~ of ~~iron or~~ steel shall be at least 15.87 mm (5/8 in.) in diameter, unless ~~Stainless steel rods less than 16 mm (5/8 in.) in diameter, nonferrous rods, or their equivalent shall be~~ listed and ~~shall~~ not ~~be~~ less than ~~13~~ mm (1/2 in.) in diameter.

2008 NEC - 250.52 Grounding Electrodes.

(A) Electrodes Permitted for Grounding.

(5) Rod and Pipe Electrodes. Rod and pipe electrodes shall not be less than <u>2.44 m</u> (8 ft) in length and shall consist of the following materials.

(a) <u>Grounding</u> electrodes of pipe or conduit shall not be smaller than metric designator 21 (trade size 3/4) and, where of iron or steel, shall have the outer surface galvanized or otherwise metal-coated for corrosion protection.

(b) <u>Grounding</u> electrodes of <u>stainless steel, and copper or zinc coated</u> steel shall be at least 15.87 mm (5/8 in.) in diameter, unless listed and not less than <u>12.70</u> mm (1/2 in.) in diameter.

Author's Comment: Copper- and zinc-coated rods are both produced from "ferrous" steel cores and coated with appropriate "non-ferrous" coating (copper or zinc). The new text recognizes listed stainless steel, copper- and zinc-coated ground rods that are not smaller than 1/2 inch in diameter as acceptable ground electrodes.

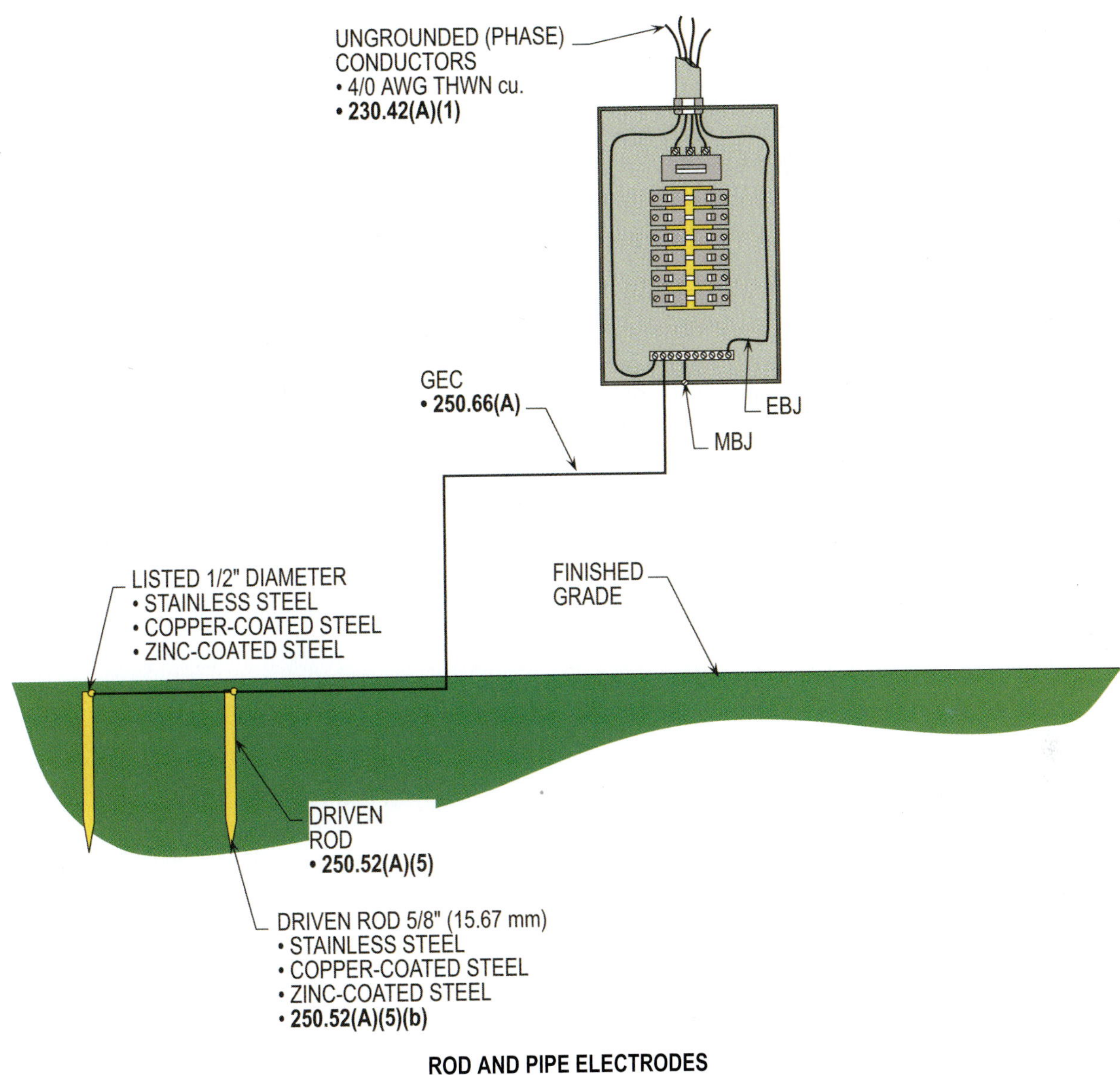

ROD AND PIPE ELECTRODES
250.52(A)(5)

Purpose of Change: This revision clarifies that rod and pipe electrodes can be stainless steel, copper, or zinc-coated steel if at least 5/8 inch in diameter or not smaller than 1/2 inch in diameter if listed.

NEC Ch. 2 - Article 250
Part III - 250.54

Type of Change	Panel Action	UL	UL 508	API 500 - 1997	API 505 - 1997	OSHA - 1994
Revision	Accept in Principle	-	-	-	-	-
ROP	**ROC**	**NFPA 70E - 2004**	**NFPA 70B - 2006**	**NFPA 79 - 2007**	**NFPA**	**NEMA - 2005**
pg. 207 / # 5-170	pg. - / # -	-	-	8.2.1.3.5	-	GR 1
log: 1398 / CMP: 5	log: - / Submitter: George Stolz II			2005 NEC: 250.54		IEC: -

2005 NEC - 250.54

~~**Supplementary**~~ **Grounding Electrodes.** ~~Supplementary~~ grounding electrodes shall be permitted to be connected to the equipment grounding conductors specified in 250.118 and shall not be required to comply with the electrode bonding requirements of 250.50 or 250.53(C) or the resistance requirements of 250.56, but the earth shall not be used as an effective ground-fault current path as specified in 250.4(A)(5) and 250.4(B)(4).

2008 NEC - 250.54

Auxiliary Grounding Electrodes. One or more grounding electrodes shall be permitted to be connected to the equipment grounding conductors specified in 250.118 and shall not be required to comply with the electrode bonding requirements of 250.50 or 250.53(C) or the resistance requirements of 250.56, but the earth shall not be used as an effective ground-fault current path as specified in 250.4(A)(5) and 250.4(B)(4).

Author's Comment: The two terms, "supplemental" and "supplementary" are synonymous in the English language. It is easier to distinguish between the two concepts of supplemental grounding electrode as used in **250.53(D)(2)** and auxiliary grounding electrode as will now be used in **250.54**.

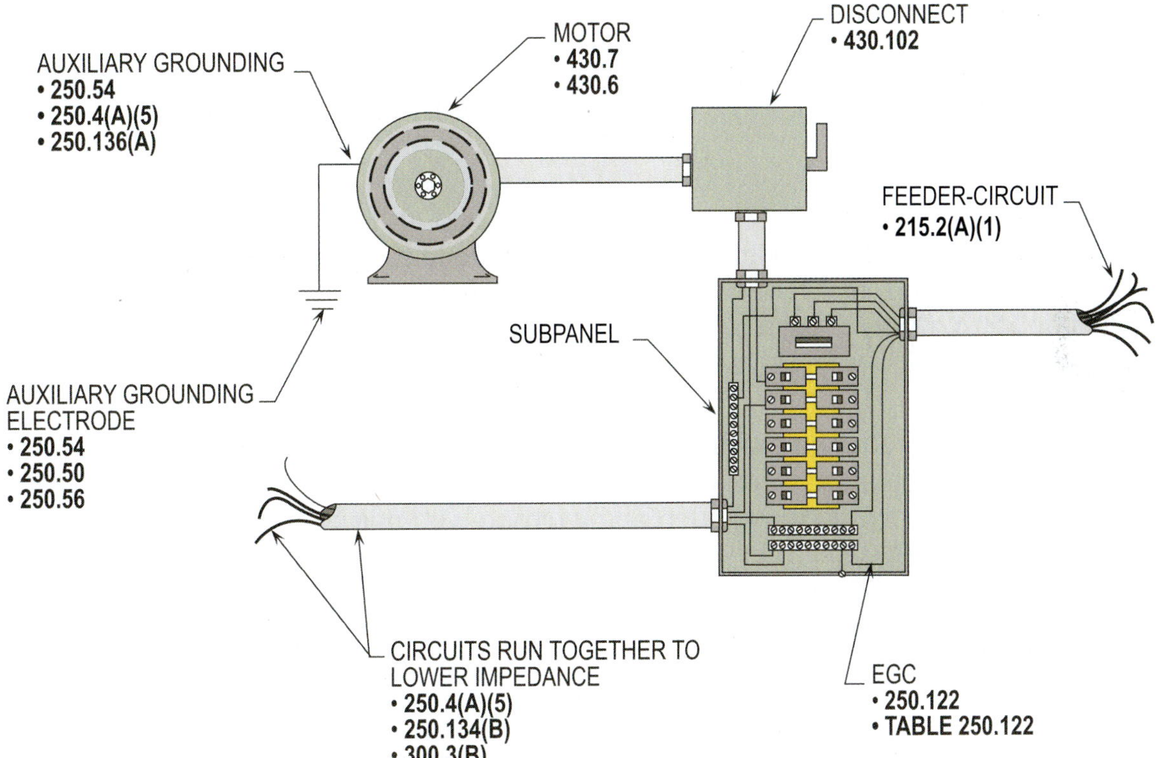

AUXILIARY GROUNDING ELECTRODES
250.54

Purpose of Change: This revision clarifies the term "auxiliary" will now be used instead of "supplementary" **250.54**.

NEC Ch. 2 - Article 250
Part III - 250.56

Type of Change	Panel Action	UL	UL 508	API 500 - 1997	API 505 - 1997	OSHA - 1994
Revision	Accept in Principle	-	-	-	-	1910.304(f)(3)(i)
ROP	**ROC**	**NFPA 70E - 2004**	**NFPA 70B - 2006**	**NFPA 79 - 2007**	**NFPA**	**NEMA - 2005**
pg. 207 # 5-174	pg. 138 # 5-100	410.10(D)(1)	-	-	-	GR 1
log: 2967 CMP: 5	log: 1122	Submitter: Andre R. Cartal		2005 NEC: 250.56		IEC: -

2005 NEC - 250.56

Resistance of Rod, Pipe, and Plate Electrodes. A single electrode consisting of a rod, pipe, or plate that does not have a resistance to ground of 25 ohms or less shall be augmented by one additional electrode of any of the types specified by 250.52~~(A)(2)~~ through ~~(A)(7)~~. Where multiple rod, pipe, or plate electrodes are installed to meet the requirements of this section, they shall not be less than 1.8 m (6 ft) apart.

2008 NEC - 250.56

Resistance of Rod, Pipe, and Plate Electrodes. A single electrode consisting of a rod, pipe, or plate that does not have a resistance to ground of 25 ohms or less shall be augmented by one additional electrode of any of the types specified by 250.52(A)(4) through (A)(8). Where multiple rod, pipe, or plate electrodes are installed to meet the requirements of this section, they shall not be less than 1.8 m (6 ft) apart.

Author's Comment: The requirement for a single electrode that does not have a resistance 25 ohms or less to be augmented by an additional electrode does not apply to any electrodes other than rod, pipe, or plate. It is required to augment such an electrode with one additional electrode of any of the types specified by 250.52(A)(4) through (A)(8) so (A)(1) for underground metal water pipes, (A)(2) for metal frame of a building or structure, and (A)(3) for concrete-encased electrodes were deleted from the list of permissible electrodes to be used for augmenting the rod, pipe, or plate electrodes.

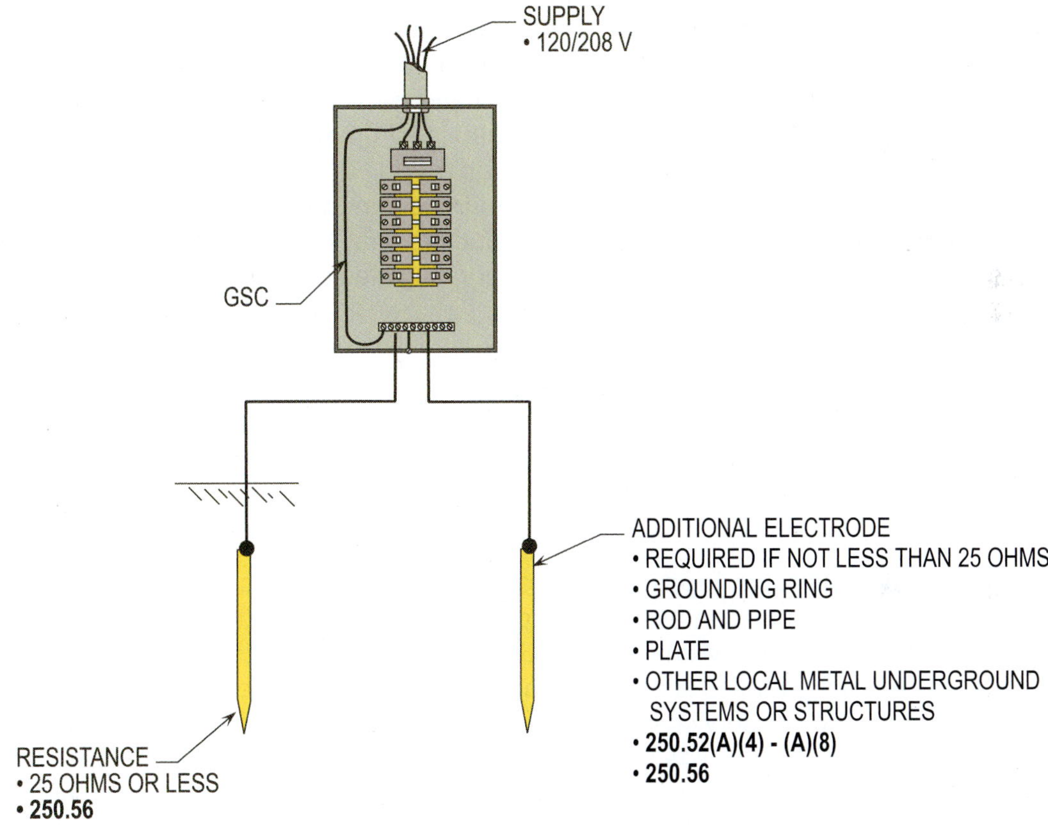

RESISTANCE OF ROD, PIPE, AND PLATE ELECTRODES
250.56

Purpose of Change: This revision clarifies that only the ground ring, rod and pipe, or other metal underground systems or structures shall be permitted to augment a rod, pipe, or plate electrodes where the resistance is more than 25 ohms.

NEC Ch. 2 - Article 250
Part III - 250.64(D)(1) through (D)(3)

30

Type of Change	Panel Action	UL	UL 508	API 500 - 1997	API 505 - 1997	OSHA - 1994
Revision	Accept in Principle	-	-	-	-	1910.304(f)(3)(i)
ROP	**ROC**	**NFPA 70E - 2004**	**NFPA 70B - 2006**	**NFPA 79 - 2007**	**NFPA**	**NEMA - 2005**
pg. 210 # 5-192	pg. - # -	410.10(D)(1)	-	-	-	GR 1
log: 2396 CMP: 5	log: -	Submitter: Mike Holt		2005 NEC: 250.64(D)		IEC: -

2005 NEC - 250.64 Grounding Electrode Conductor Installation.

(D) Grounding Electrode Conductor Taps. Where a service consists of more than a single enclosure as permitted in 230.71(A), ~~it shall be permitted to connect taps to the common grounding electrode conductor. Each such~~ tap conductor shall extend to the inside of each ~~such~~ enclosure. The common grounding electrode conductor shall be sized in accordance with 250.66, based on the sum of the circular mil area of the largest ungrounded service entrance conductor(s). Where ~~more than one set of~~ the service entrance conductors ~~as permitted by 230.40, Exception No. 2~~ connect directly to a service drop or service lateral, the common grounding electrode conductor shall be sized in accordance with Table 250.66, Note 1. ~~The tap conductors shall be permitted to be sized in accordance with the grounding electrode conductors specified in 250.66 for the largest conductor serving the respective enclosures.~~ The tap conductors shall be connected to the common grounding electrode conductor in such a manner that the common grounding electrode conductor remains without a splice or joint.

2008 NEC - 250.64 Grounding Electrode Conductor Installation.

(D) Service with Multiple Disconnecting Means Enclosures. Where a service consists of more than a single enclosure as permitted in 230.71(A), grounding electrode connections shall be made in accordance with (D)(1), (D)(2) or (D)(3).

(1) Grounding Electrode Conductor Taps. Where the service is installed as permitted by 230.40, Exception No. 2, a common grounding electrode conductor and grounding electrode conductor taps shall be installed. The common grounding electrode conductor shall be sized in accordance with 250.66, based on the sum of the circular mil area of the largest ungrounded service entrance conductor(s). Where the service entrance conductors connect directly to a service drop or service lateral, the common grounding electrode conductor shall be sized in accordance with Table 250.66, Note 1. A tap conductor shall extend to the inside of each service disconnecting means enclosure. The grounding electrode conductor taps shall be sized in accordance with 250.66 for the largest conductor serving the individual enclosure. The tap conductors shall be connected to the common grounding electrode conductor by exothermic welding or with connectors listed as grounding and bonding equipment in such a manner that the common grounding electrode conductor remains without a splice or joint.

(2) Individual Grounding Electrode Conductors. A grounding electrode conductor shall be connected between the grounded conductor in each service equipment disconnecting means enclosure and the grounding electrode system. Each grounding electrode conductor shall be sized in accordance with 250.66 based on the service-entrance conductor(s) supplying the individual service disconnecting means.

(3) Common Location. A grounding electrode conductor shall be connected to the grounded service conductor(s) in a wireway or other accessible enclosure on the supply side of the service disconnecting means. The connection shall be made with exothermic welding or a connector listed as grounding and bonding equipment. The grounding electrode conductor shall be sized in accordance with 250.66 based on the service-entrance conductor(s) at the common location where the connection is made.

Author's Comment: Section **250.64(D)** has been expanded to deal with grounding electrode conductor installation requirements for multiple disconnect enclosures. Reworded **(1)** contains installation requirements for two to six individual disconnects where the main grounding electrode conductor is installed under the disconnecting means with individual grounding electrode taps connected into each enclosure. New **(2)** was added to permit individual grounding electrode conductors to be installed to each individual disconnecting means enclosure. New **(3)** was added to include the provision in **250.24** for making the grounding electrode connection at an accessible location in a wireway or other accessible enclosure on the supply side of the service disconnecting means.

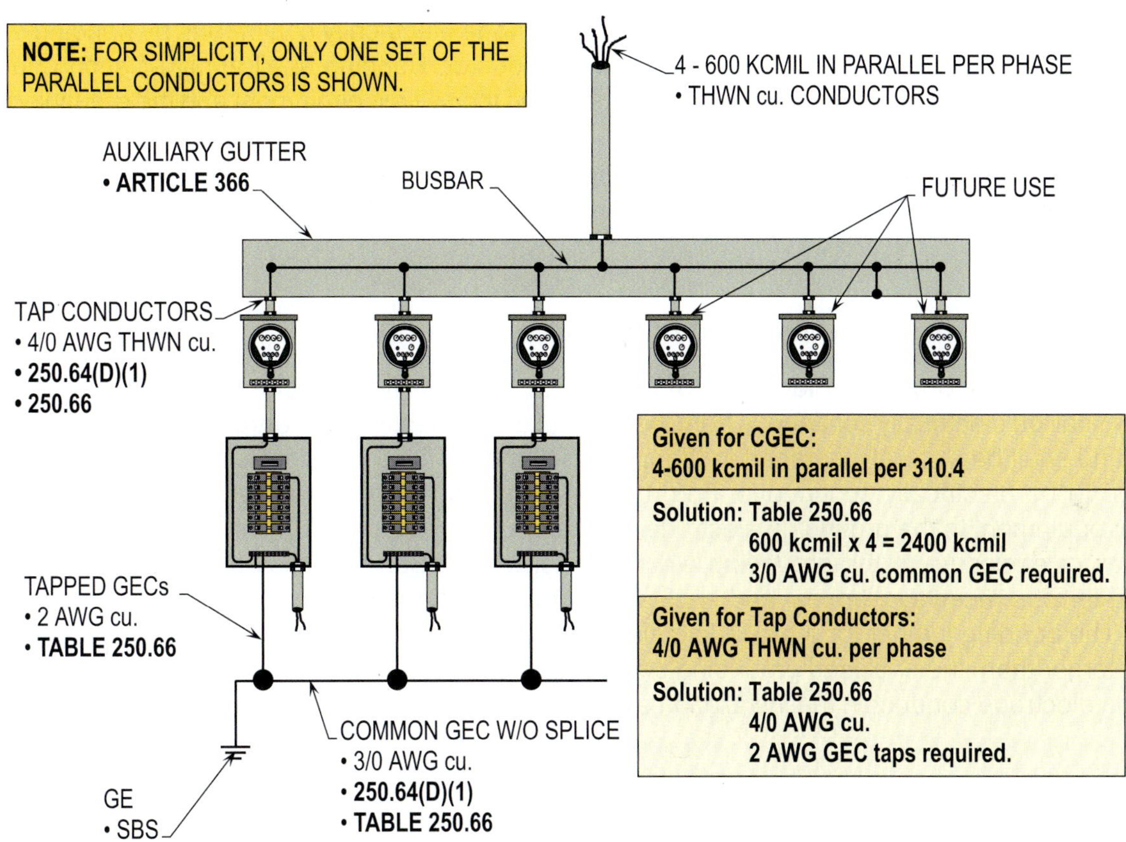

SERVICE WITH MULTIPLE DISCONNECTING MEANS ENCLOSURES
250.64(D)(1) through (D)(3)

Purpose of Change: This revision clarifies the sizing and installation requirements for the grounding electrode conductor where multiple disconnecting enclosures are installed.

NEC Ch. 2 - Article 250
Part III - 250.64(E)

Type of Change	Panel Action	UL	UL 508	API 500 - 1997	API 505 - 1997	OSHA - 1994
Revision	Accept in Principle	-	-	-	-	-
ROP	**ROC**	**NFPA 70E - 2004**	**NFPA 70B - 2006**	**NFPA 79 - 2007**	**NFPA**	**NEMA**
pg. 211 # 5-195	pg. 139 # 5-111	-	-	-	-	-
log: 2905 CMP: 5	log: 922	Submitter: Peter D. Noval, Jr.		2005 NEC: 250.64(E)		IEC: -

2005 NEC - 250.64 Grounding Electrode Conductor Installation.

(E) Enclosures for Grounding Electrode Conductors. Ferrous metal enclosures for grounding electrode conductors shall be electrically continuous from the point of attachment to cabinets or equipment to the grounding electrode and shall be securely fastened to the ground clamp or fitting. Nonferrous metal enclosures shall not be required to be electrically continuous. Ferrous metal enclosures that are not physically continuous from cabinets or equipment to the grounding electrode shall be made electrically continuous by bonding each end of the raceway or enclosure to the grounding electrode conductor. Bonding shall apply at each end and to all intervening ferrous raceways, boxes, and enclosures between the ~~service~~ equipment and the grounding electrode. The bonding jumper for a grounding electrode conductor raceway or cable armor shall be the same size as, or larger than, the ~~required~~ enclosed grounding electrode conductor. Where a raceway is used as protection for a grounding electrode conductor, the installation shall comply with the requirements of the appropriate raceway article.

2008 NEC - 250.64 Grounding Electrode Conductor Installation.

(E) Enclosures for Grounding Electrode Conductors. Ferrous metal enclosures for grounding electrode conductors shall be electrically continuous from the point of attachment to cabinets or equipment to the grounding electrode and shall be securely fastened to the ground clamp or fitting. Nonferrous metal enclosures shall not be required to be electrically continuous. Ferrous metal enclosures that are not physically continuous from cabinets or equipment to the grounding electrode shall be made electrically continuous by bonding each end of the raceway or enclosure to the grounding electrode conductor. Bonding shall apply at each end and to all intervening ferrous raceways, boxes, and enclosures between the cabinets or equipment and the grounding electrode. The bonding jumper for a grounding electrode conductor raceway or cable armor shall be the same size as, or larger than, the enclosed grounding electrode conductor. Where a raceway is used as protection for a grounding electrode conductor, the installation shall comply with the requirements of the appropriate raceway article.

Author's Comment: Changing "service equipment" to "cabinets or equipment" has expanded this section on installation of grounding electrode conductors to more than just services. This application now appropriately applies to service equipment, as well as outside feeders and branch circuits, in compliance with **Part II** of **Article 225**, where feeders and branch circuits are supplied from one building or structure to another building or structure. The word "service equipment" was deleted to ensure someone did not apply the bonding requirement to cabinets and equipment related to services.

The word "required" was deleted since field experience has shown that grounding electrode conductors are sometimes installed in sizes larger than the minimum required by the NEC or the design professional. Frequently, this "oversizing" occurs because the larger size is readily available on the job site and the smaller "required" size is not. Subsequently, bonding jumpers are then installed on the enclosing raceway to provide electrical continuity, but they are the same size as the "required" grounding electrode conductor. In these instances, the bonding jumpers are smaller than the enclosed grounding electrode conductor.

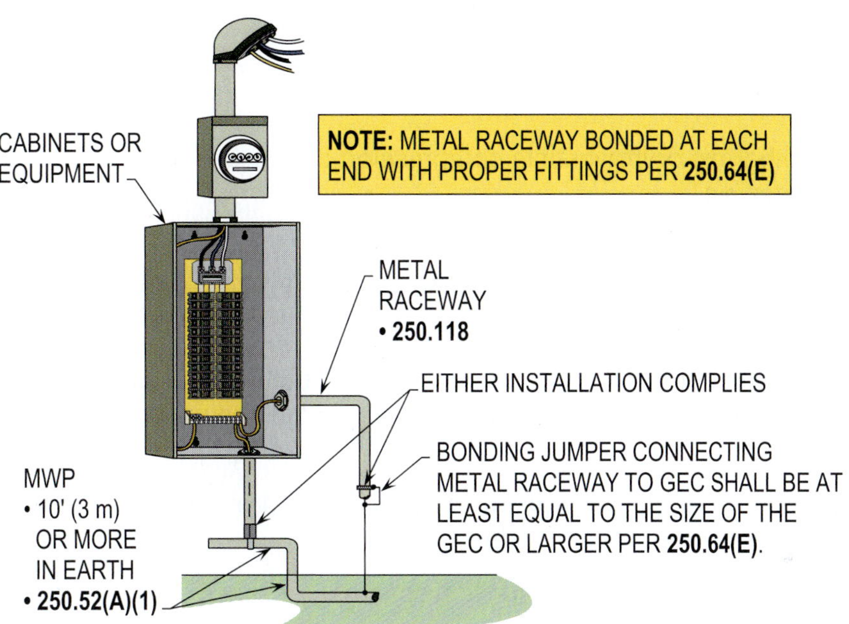
CABINETS OR
EQUIPMENT
NOTE: METAL RACEWAY BONDED AT EACH END WITH PROPER FITTINGS PER 250.64(E)
METAL
RACEWAY
• 250.118
EITHER INSTALLATION COMPLIES
BONDING JUMPER CONNECTING
METAL RACEWAY TO GEC SHALL BE AT
LEAST EQUAL TO THE SIZE OF THE
GEC OR LARGER PER 250.64(E).
MWP
• 10' (3 m)
OR MORE
IN EARTH
• 250.52(A)(1)

ENCLOSURES FOR GROUNDING ELECTRODE CONDUCTORS
250.64(E)

Purpose of Change: This revision clarifies that bonding applies to service equipment, as well as other applications, such as outside feeders and branch circuits.

NEC Ch. 2 - Article 250
Part III - 250.68(A), Exception No. 2

Type of Change	Panel Action	UL	UL 508	API 500 - 1997	API 505 - 1997	OSHA - 1994
Revision	Accept in Principle	-	-	-	-	1910.304(f)(3)(i)
ROP	ROC	NFPA 70E - 2004	NFPA 70B - 2006	NFPA 79 - 2007	NFPA	NEMA - 2005
pg. 215 # 5-213	pg. 142 # 5-119	410.10(D)(1)	-	-	-	GR 1
log: 3393 CMP: 5	log: 1524	Submitter: Frederic P. Hartwell		2005 NEC: 250.68(A)		IEC: -

2005 NEC - 250.68 Grounding Electrode Conductor and Bonding Jumper Connection to Grounding Electrodes.

(A) Accessibility. ~~The connection of~~ a grounding electrode conductor or bonding jumper to a grounding electrode shall be accessible.

Exception No 1: An encased or buried connection to a concrete encased, driven, or buried grounding electrode shall not be required to be accessible.

Exception No. 2: ~~An~~ exothermic or irreversible compression connection to fire-proofed structural metal shall not be required to be accessible.

2008 NEC - 250.68 Grounding Electrode Conductor and Bonding Jumper Connection to Grounding Electrodes.

The connection of a grounding electrode conductor at the service, at each building or structure where supplied by a feeder(s) or branch circuit(s), or at a separately derived system and associated bonding jumper(s) shall be made as specified in 250.68(A) and (B).

(A) Accessibility. All mechanical elements used to terminate a grounding electrode conductor or bonding jumper to a grounding electrode shall be accessible.

Exception No 1: An encased or buried connection to a concrete encased, driven, or buried grounding electrode shall not be required to be accessible.

Exception No. 2: Exothermic or irreversible compression connections used at terminations, together with the mechanical means used to attach such terminations to fireproofed structural metal whether or not the mechanical means is reversible, shall not be required to be accessible.

Author's Comment: The main rule was changed to indicate that all mechanical connections of a grounding electrode conductor must be accessible after installation. Exception No. 2 was revised to permit the mechanical means, such as a nut and bolt, used to connect irreversible compression connectors to the fireproofed metal structure to be inaccessible.

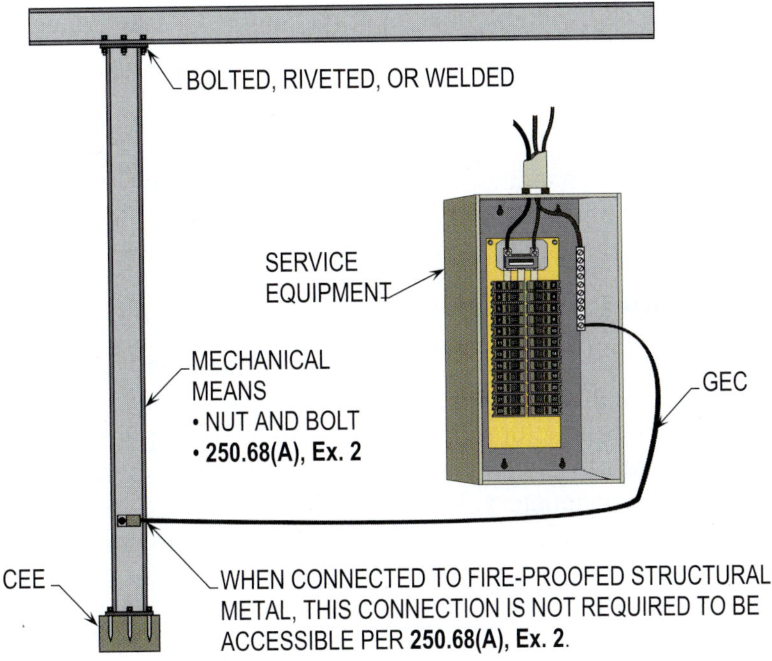

ACCESSIBILITY
250.68(A), Exception No. 2

Purpose of Change: This revision clarifies that the mechanical means, such as a nut and bolt, used to connect irreversible compression connectors to the metal structure is not required to be accessible.

NEC Ch. 2 - Article 250
Part V - 250.94 and Exception

Type of Change	Panel Action	UL	UL 508	API 500 - 1997	API 505 - 1997	OSHA - 1994
Revision	Accept in Principle	-	-	-	-	-
ROP	**ROC**	**NFPA 70E - 2004**	**NFPA 70B - 2006**	**NFPA 79 - 2007**	**NFPA**	**NEMA**
pg. 217 # 5-220	pg. 143 # 5-122	410.10(B)	-	8.2.3	-	-
log: 1886 CMP: 5	log: 1151	Submitter: Jeffrey Boksiner		2005 NEC: 250.94		IEC: -

2005 NEC - 250.94 Bonding for Other Systems.

An accessible means external to enclosures for connecting intersystem bonding and grounding electrode conductors shall be provided at the service equipment and at the disconnecting means for any additional buildings or structures by at least one of the following means:

(1) Exposed nonflexible metallic raceways

(2) Exposed grounding electrode conductor

(3) Approved means for the external connection of a copper or other corrosion-resistant bonding or grounding conductor to the grounded raceway or equipment

2008 NEC - 250.94 Bonding for Other Systems.

An intersystem bonding termination for connecting intersystem bonding and grounding conductors required for other systems shall be provided external to enclosures at the service equipment and at the disconnecting means for any additional buildings or structures. The intersystem bonding termination shall be accessible for connection and inspection. The intersystem bonding termination shall have the capacity for connection of not less than three intersystem bonding conductors. The intersystem bonding termination device shall not interfere with opening a service or metering equipment enclosure. The intersystem bonding termination shall be one of the following:

(1) A set of terminals securely mounted to the meter enclosure and electrically connected to the meter enclosure. The terminals shall be listed as grounding and bonding equipment.

(2) A bonding bar near the service equipment enclosure, meter enclosure or raceway for service conductors. The bonding bar shall be connected with a minimum 6 AWG copper conductor to an equipment grounding conductor(s) in the service equipment enclosure, meter enclosure, or exposed nonflexible metallic raceway.

(3) A bonding bar near the grounding electrode conductor. The bonding bar shall be connected to the grounding electrode conductor with a minimum 6 AWG copper conductor.

Exception: In existing buildings or structures where any of the intersystem bonding and grounding conductors required by 770.93, 800.100(B), 810.21(F), 820.100(B), 830.100(B) exist, installation of the intersystem bonding termination is not required. An accessible means external to enclosures for connecting intersystem bonding and grounding electrode conductors shall be permitted provided at the service equipment and at the disconnecting means for any additional buildings or structures by at least one of the following means:

(1) Exposed nonflexible metallic raceways

(2) Exposed grounding electrode conductor

(3) Approved means for the external connection of a copper or other corrosion-resistant bonding or grounding conductor to the grounded raceway or equipment

Author's Comment: Section **250.94**, covering intersystem bonding and grounding requirements has been totally rewritten to provide a more clear, concise, and consistent method of connecting the electrical power system grounding to telecommunications system grounding and bonding systems to ensure there is no difference of potential between the two systems. The comment addressed an issue where the device that is used to provide this intersystem bonding termination is often installed on the cover or door of service and/or meter enclosures thus preventing their routine opening. In these cases, the only option available to properly operate or maintain this equipment is to remove the intersystem bonding termination device from the service or meter enclosure. Text has been added to ensure the intersystem bonding will not prevent opening of covers or doors of equipment. This additional wording provides prescriptive requirements that mandate the proper installation of bonding terminations at these locations and will correlate with the actions on proposals of Code Making Panel 16 in Chapter 8.

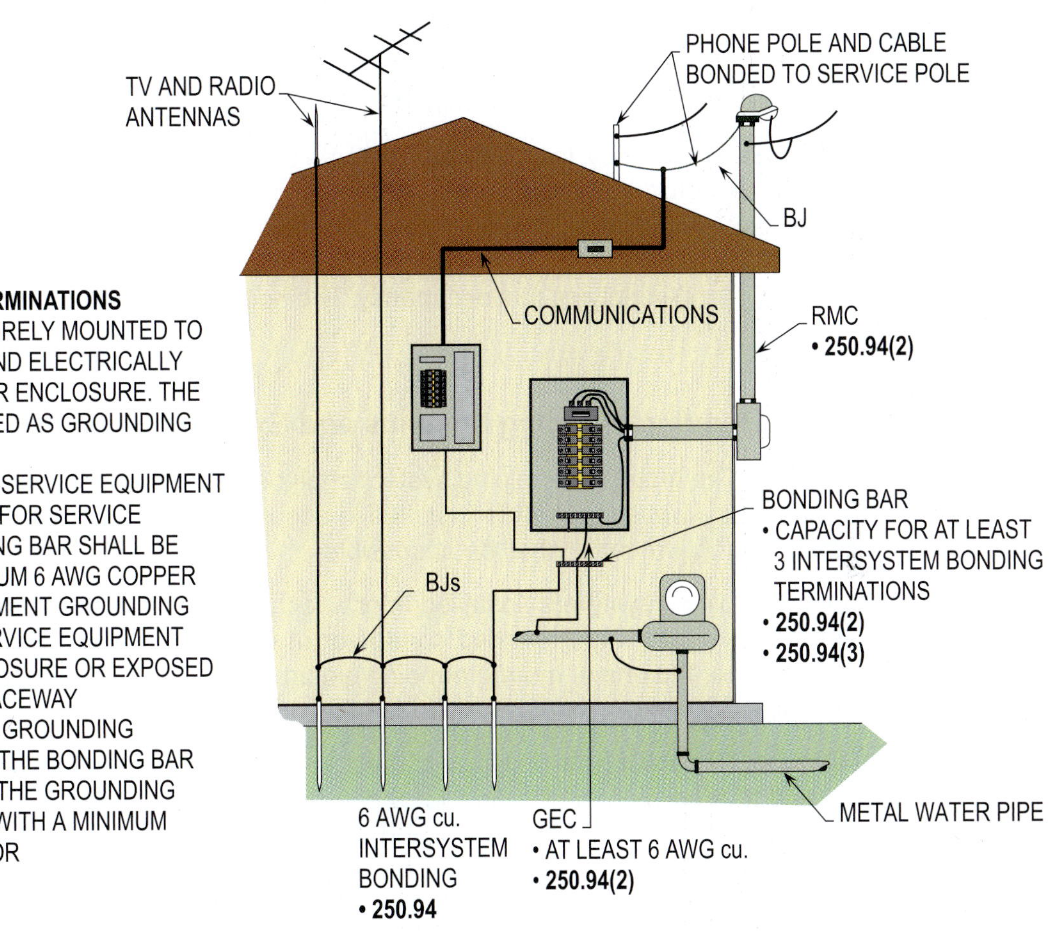

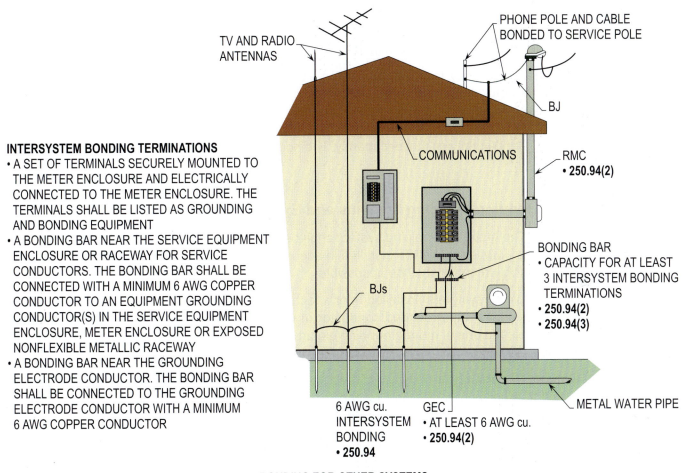

INTERSYSTEM BONDING TERMINATIONS

- A SET OF TERMINALS SECURELY MOUNTED TO THE METER ENCLOSURE AND ELECTRICALLY CONNECTED TO THE METER ENCLOSURE. THE TERMINALS SHALL BE LISTED AS GROUNDING AND BONDING EQUIPMENT
- A BONDING BAR NEAR THE SERVICE EQUIPMENT ENCLOSURE OR RACEWAY FOR SERVICE CONDUCTORS. THE BONDING BAR SHALL BE CONNECTED WITH A MINIMUM 6 AWG COPPER CONDUCTOR TO AN EQUIPMENT GROUNDING CONDUCTOR(S) IN THE SERVICE EQUIPMENT ENCLOSURE, METER ENCLOSURE OR EXPOSED NONFLEXIBLE METALLIC RACEWAY
- A BONDING BAR NEAR THE GROUNDING ELECTRODE CONDUCTOR. THE BONDING BAR SHALL BE CONNECTED TO THE GROUNDING ELECTRODE CONDUCTOR WITH A MINIMUM 6 AWG COPPER CONDUCTOR

BONDING FOR OTHER SYSTEMS
250.94 and Exception

Purpose of Change: This revision clarifies the methods for connecting the electrical power system grounding to telecommunications system grounding and bonding systems to ensure there is no difference of potential between the two systems.

NEC Ch. 2 - Article 250
Part V - 250.104(A)(2)

Type of Change		Panel Action		UL	UL 508	API 500 - 1997	API 505 - 1997	OSHA - 1994
Revision		Accept in Principle		-	-	-	-	-
ROP		ROC		NFPA 70E - 2004	NFPA 70B - 2006	NFPA 79 - 2007	NFPA	NEMA
pg. 219	# 5-229	pg. -	# -	410.10(B)	-	8.2.3	-	-
log: 1484	CMP: 5	log: -	Submitter: Ryan Jackson			2005 NEC: 250.104(A)		IEC: -

2005 NEC - 250.104 Bonding of Piping Systems and Exposed Structural Steel.

(A) Metal Water Piping. The metal water piping system shall be bonded as required in (A)(1), (A)(2), or (A)(3) of this section. The bonding jumper(s) shall be installed in accordance with 250.64(A), (B), and (E). The points of attachment of the bonding jumper(s) shall be accessible.

(1) General. Metal water piping system(s) installed in or attached to a building or structure shall be bonded to the service equipment enclosure, the grounded conductor at the service, the grounding electrode conductor where of sufficient size, or to the one or more grounding electrodes used. The bonding jumper(s) shall be sized in accordance with Table 250.66 except as permitted in 250.104(A)(2) and (A)(3).

(2) Buildings of Multiple Occupancy. In buildings of multiple occupancy where the metal water piping system(s) installed in or attached to a building or structure for the individual occupancies is metallically isolated from all other occupancies by use of nonmetallic water piping, the metal water piping system(s) for each occupancy shall be permitted to be bonded to the equipment grounding terminal of the panelboard or switchboard enclosure (other than service equipment) supplying that occupancy. The bonding jumper shall be sized in accordance with Table 250.122.

2008 NEC - 250.104 Bonding of Piping Systems and Exposed Structural Steel.

(A) Metal Water Piping. The metal water piping system shall be bonded as required in (A)(1), (A)(2), or (A)(3) of this section. The bonding jumper(s) shall be installed in accordance with 250.64(A), (B), and (E). The points of attachment of the bonding jumper(s) shall be accessible.

(1) General. Metal water piping system(s) installed in or attached to a building or structure shall be bonded to the service equipment enclosure, the grounded conductor at the service, the grounding electrode conductor where of sufficient size, or to the one or more grounding electrodes used. The bonding jumper(s) shall be sized in accordance with Table 250.66 except as permitted in 250.104(A)(2) and (A)(3).

(2) Buildings of Multiple Occupancy. In buildings of multiple occupancy where the metal water piping system(s) installed in or attached to a building or structure for the individual occupancies is metallically isolated from all other occupancies by use of nonmetallic water piping, the metal water piping system(s) for each occupancy shall be permitted to be bonded to the equipment grounding terminal of the panelboard or switchboard enclosure (other than service equipment) supplying that occupancy. The bonding jumper shall be sized in accordance with Table 250.122, based on the rating of the overcurrent protective device for the circuit supplying the occupancy.

Author's Comment: Simply referring the Code user to **Table 250.122** is not adequate. The user must know which overcurrent protection device is being used when referring to the table so the size of overcurrent protective device can be used to determine the bonding jumper size.

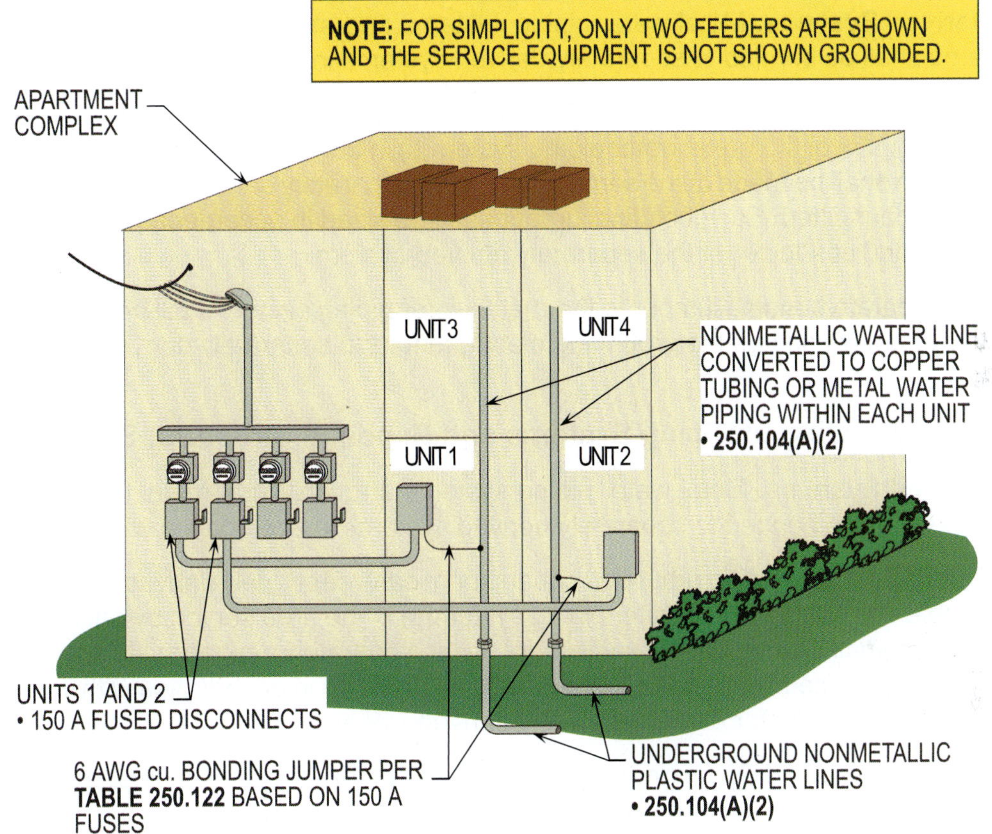

BUILDINGS OF MULTIPLE OCCUPANCY
250.104(A)(2)

Purpose of Change: This revision clarifies that the bonding jumper shall be sized in accordance with **Table 250.122** based on the rating of the overcurrent protective device for the circuit supplying the occupancy.

NEC Ch. 2 - Article 250
Part V - 250.104(D)(1), Exception No. 1

Type of Change	Panel Action	UL	UL 508	API 500 - 1997	API 505 - 1997	OSHA - 1994
Revision	Accept in Principle	-	-	-	-	-
ROP	**ROC**	**NFPA 70E - 2004**	**NFPA 70B - 2006**	**NFPA 79 - 2007**	**NFPA**	**NEMA**
pg. 222 # 5-245	pg. - # -	410.10(B)	-	8.2.3	-	-
log: 2785 CMP: 5	log: -	Submitter: Ted Smith		2005 NEC: 250.104(D), Ex. 1		IEC: -

2005 NEC - 250.104 Bonding of Piping Systems and Exposed Structural Steel.

(D) Separately Derived Systems. Metal water piping systems and structural metal that is interconnected to form a building frame shall be bonded to separately derived systems in accordance with (D)(1) through (D)(3).

(1) Metal Water Piping System(s). The grounded conductor of each separately derived system shall be bonded to the nearest available point of the metal water piping system(s) in the area served by each separately derived system. This connection shall be made at the same point on the separately derived system where the grounding electrode conductor is connected. Each bonding jumper shall be sized in accordance with Table 250.66 based on the largest ungrounded conductor of the separately derived system.

Exception No. 1: A separate bonding jumper to the metal water piping system shall not be required where the metal water piping system is used as the grounding electrode for the separately derived system.

2008 NEC - 250.104 Bonding of Piping Systems and Exposed Structural Steel.

(D) Separately Derived Systems. Metal water piping systems and structural metal that is interconnected to form a building frame shall be bonded to separately derived systems in accordance with (D)(1) through (D)(3).

(1) Metal Water Piping System(s). The grounded conductor of each separately derived system shall be bonded to the nearest available point of the metal water piping system(s) in the area served by each separately derived system. This connection shall be made at the same point on the separately derived system where the grounding electrode conductor is connected. Each bonding jumper shall be sized in accordance with Table 250.66 based on the largest ungrounded conductor of the separately derived system.

Exception No. 1: A separate bonding jumper to the metal water piping system shall not be required where the metal water piping system is used as the grounding electrode for the separately derived system and the water piping system is in the area served.

Author's Comment: The intent is to not require both a grounding electrode conductor under **250.30** and a bonding jumper under **250.104** from the separately derived system to the same point on the metallic water pipe. The purpose of this section is to ensure the water piping in the area served by the separately derived system is bonded back to the source (the separately derived system). This ensures that all electrical equipment and the water piping connected to the equipment is at the same zero potential.

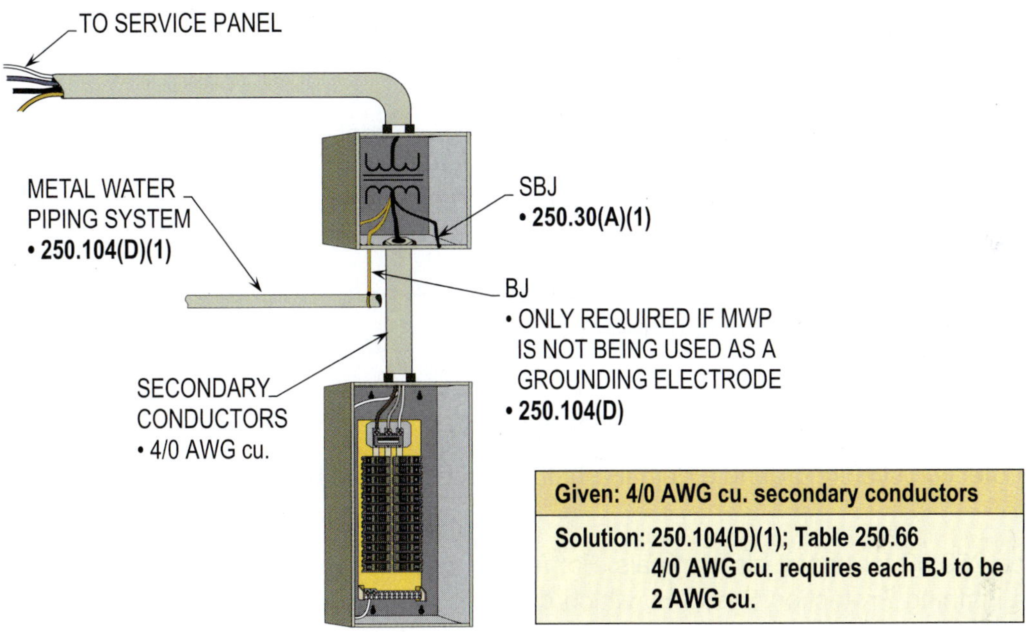

METAL WATER PIPING SYSTEM(S)
250.104(D)(1), Exception No. 1

Purpose of Change: To ensure that the section rules are specific on how to determine the correct size of the bonding jumper to bond to the metal water pipe and the water piping in the area served by the separately derived system are bonded together.

NEC Ch. 2 - Article 250
Part V - 250.106, FPN No. 1

Type of Change	Panel Action	UL	UL 508	API 500 - 1997	API 505 - 1997	OSHA - 1994
Revision	Accept	-	-	-	-	1910.308(a)(3)(i)[b]
ROP	**ROC**	**NFPA 70E - 2004**	**NFPA 70B - 2006**	**NFPA 79 - 2007**	**NFPA 780**	**NEMA**
pg. 222 # 5-246	pg. - # -	450.5(C)	-	7.8.1	4.19 - 4.21	-
log: 1414 CMP: 5	log: -	Submitter: Technical Committee on Lightning Protection		2005 NEC: 250.106, FPN 1		IEC: -

2005 NEC - 250.106 Lightning Protection Systems.

The lightning protection system ground terminals shall be bonded to the building or structure grounding electrode system.

FPN No. 1: See 250.60 for use of air terminals. For further information, see NFPA 780-2008, Standard for the Installation of Lightning Protection Systems, which contains detailed information on grounding, bonding, and ~~spacing~~ from lightning protection systems.

FPN No. 2: Metal raceways, enclosures, frames, and other non-current-carrying metal parts of electric equipment installed on a building equipped with lightning protection system may require bonding or spacing from the lightning protection conductors in accordance with NFPA 780-2004, Standard for the Installation of Lightning Protection Systems.

~~Separation from lightning protection conductors is typically 1.8 m (6 ft) through air or 900 mm (3 ft) through dense materials such as concrete, brick, or wood.~~

2008 NEC - 250.106 Lightning Protection Systems.

The lightning protection system ground terminals shall be bonded to the building or structure grounding electrode system.

FPN No. 1: See 250.60 for use of air terminals. For further information, see NFPA 780-2008, Standard for the Installation of Lightning Protection Systems, which contains detailed information on grounding, bonding, and sideflash distance from lightning protection systems.

FPN No. 2: Metal raceways, enclosures, frames, and other non-current carrying metal parts of electric equipment installed on a building equipped with lightning protection system may require bonding or spacing from the lightning protection conductors in accordance with NFPA 780-2008, Standard for the Installation of Lightning Protection Systems.

Author's Comment: NFPA 780 shows the method for calculating sideflash distance based on factors related to the design of the lightning protection system. Grounded metallic bodies within the sideflash distance are required to be interconnected with the system, and those outside the calculated distance are not. The term "spacing" misleads the user, who may believe that a grounded system could be separated from the lightning protection system with no connection, when in reality all grounded systems must be interconnected at grade level minimum. The evaluation of sideflash distance then occurs at roof level, and at intermediate height levels. Sections 4.19 to 4.21 of NFPA 780 cover this method of determination. (The sections referenced in the commentary may be different and the references in the diagram may be slightly different due to changes in NFPA 780)

Grounded metal bodies must be evaluated for additional interconnections at intermediate vertical heights. Ungrounded or floating metal bodies must be evaluated based on their ability to provide a short circuit path from the lightning protection system to another grounded building system. The indication of a "typical" distance is misleading, since evaluation of the factors involved can led to a sideflash potential distance as small as 1 foot or larger than 10 feet depending on the various system design factors.

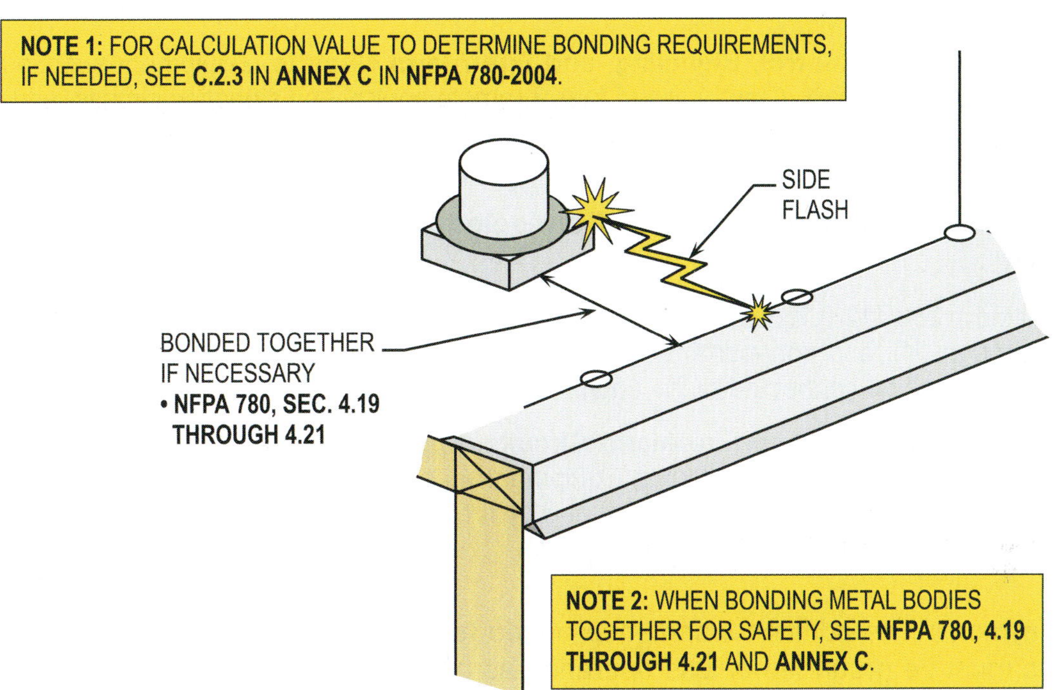

NOTE 1: FOR CALCULATION VALUE TO DETERMINE BONDING REQUIREMENTS, IF NEEDED, SEE C.2.3 IN ANNEX C IN NFPA 780-2004.
SIDE FLASH
BONDED TOGETHER IF NECESSARY
• NFPA 780, SEC. 4.19 THROUGH 4.21
NOTE 2: WHEN BONDING METAL BODIES TOGETHER FOR SAFETY, SEE NFPA 780, 4.19 THROUGH 4.21 AND ANNEX C.

LIGHTNING PROTECTION SYSTEMS
250.106, FPN No. 1

Purpose of Change: This revision to **250.106, FPN No. 1** clarifies that the method for calculating side flash distance based on factors related to the design of the lightning protection system should be in accordance with NFPA 780.

NEC Ch. 2 - Article 250
Part VI - 250.112(I)

Type of Change	Panel Action	UL	UL 508	API 500 - 1997	API 505 - 1997	OSHA - 1994
Revision	Accept	-	-	-	-	1910.304(f)(5)(iv)
ROP	**ROC**	**NFPA 70E - 2004**	**NFPA 70B - 2006**	**NFPA 79 - 2007**	**NFPA**	**NEMA**
pg. 224 # 5-252	pg. - # -	410.10(E)(4)	-	8.2.1.2	-	-
log: 3339 CMP: 5	log: -	Submitter: Frederic P. Hartwell		2005 NEC: 250.112(I)		IEC: -

2005 NEC - 250.112 Fastened in Place or Connected by Permanent Wiring Methods (Fixed) - Specific.

Exposed, non–current-carrying metal parts of the kinds of equipment described in 250.112(A) through (K), and non–current-carrying metal parts of equipment and enclosures described in 250.112(L) and (M), shall be ~~grounded~~ regardless of voltage.

(I) ~~Power-Limited~~ Remote Control, Signaling, and Fire Alarm Circuits. Equipment supplied by Class 1 power-limited circuits ~~and Class 1~~, Class 2, and Class 3 remote control and signaling circuits, and by fire alarm circuits, shall be grounded where system grounding is required by Part II or Part VIII of this article.

2008 NEC - 250.112 Fastened in Place or Connected by Permanent Wiring Methods (Fixed) - Specific.

<u>Except as permitted in 250.112(I),</u> exposed, non–current carrying metal parts of the kinds of equipment described in 250.112(A) through (K), and non–current-carrying metal parts of equipment and enclosures described in 250.112(L) and (M), shall be <u>connected to the equipment grounding conductor</u> regardless of voltage.

(I) Remote Control, Signaling, and Fire Alarm Circuits. <u>Equipment supplied by Class 1 circuits shall be grounded unless operating at less than 50 volts.</u> Equipment supplied by Class 1 power-limited circuits, Class 2, and Class 3 remote control and signaling circuits, and by fire alarm circuits, shall be grounded where system grounding is required by Part II or Part VIII of this article.

Author's Comment: The additional sentence added to **(I)** makes it clear that Class 1 non-power-limited circuits must be grounded unless operating as power limited Class 1 circuits at less than 50 volts.

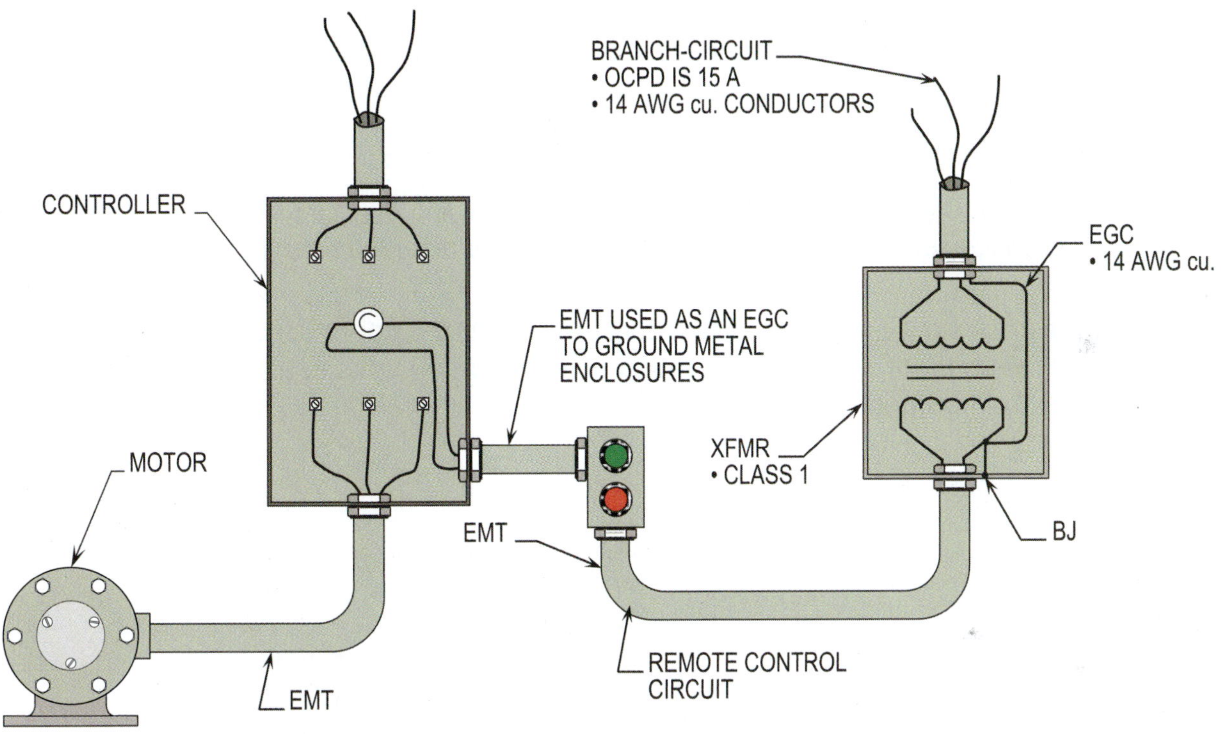

REMOTE-CONTROL, SIGNALING, AND FIRE ALARM CIRCUITS
250.112(I)

Purpose of Change: This revision clarifies that equipment supplied by Class I circuits shall be grounded unless operating at less than 50 volts.

NEC Ch. 2 - Article 250
Part VI - 250.119, Exception

Type of Change		Panel Action		UL	UL 508	API 500 - 1997	API 505 - 1997	OSHA - 1994
New Exception		Accept in Principle		-	-	-	-	1910.304(a)(1)
ROP		ROC		NFPA 70E - 2004	NFPA 70B - 2006	NFPA 79 - 2007	NFPA	NEMA
pg. 227	# 5-265	pg. 147	# 5-141	410.1(A)	-	13.2.2	-	-
log: 2553	CMP: 5	log: 603	Submitter: Tom Baker			2005 NEC: -		IEC: -

2008 NEC - 250.119 Identification of Equipment Grounding Conductor.

Unless required elsewhere in this Code, equipment grounding conductors shall be permitted to be bare, covered, or insulated. Individually covered or insulated equipment grounding conductors shall have a continuous outer finish that is either green or green with one or more yellow stripes except as permitted in this section. Conductors with insulation or individual covering that is green, green with one or more yellow stripes, or otherwise identified as permitted by this sectin shall not be used for ungrounded or grounded circuit conductors.

Exception. Power-limited, Class 2, or Class 3 circuit cables containing only circuits operating at less than 50 volts shall be permitted to use a conductor with green insulation for other than equipment grounding purposes.

Author's Comment: A new exception was added to **250.119** to permit power-limited, Class 2, or Class 3 circuit cables containing circuits operating at less than 50 volts to use a conductor with green insulation for other than grounding purposes.

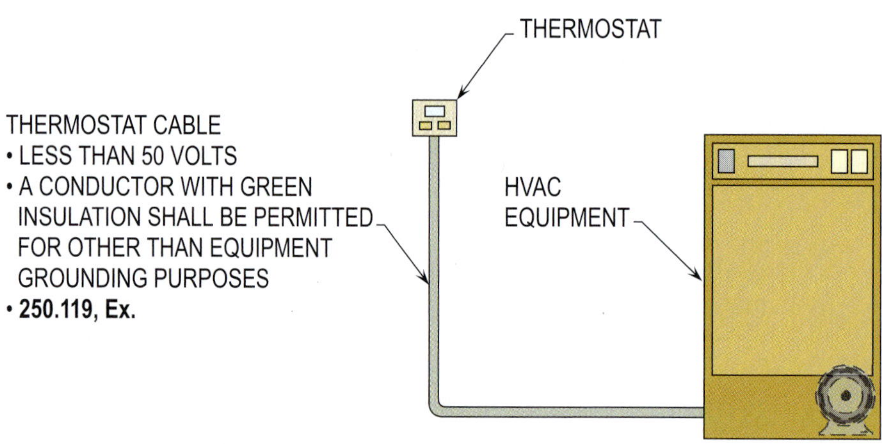
THERMOSTAT
THERMOSTAT CABLE
• LESS THAN 50 VOLTS
• A CONDUCTOR WITH GREEN INSULATION SHALL BE PERMITTED FOR OTHER THAN EQUIPMENT GROUNDING PURPOSES
• 250.119, Ex.
HVAC EQUIPMENT

IDENTIFICATION OF EQUIPMENT GROUNDING CONDUCTOR
250.119, Exception

Purpose of Change: A new exception has been added to permit power-limited, Class 2 or Class 3 circuit cables containing circuits operating at less than 50 volts to use a conductor with green insulation for other than grounding purposes.

NEC Ch. 2 - Article 250
Part VI - 250.119(A)

Type of Change	Panel Action	UL	UL 508	API 500 - 1997	API 505 - 1997	OSHA - 1994
Revision	Accept	-	-	-	-	1910.304(a)(1)
ROP	**ROC**	**NFPA 70E - 2004**	**NFPA 70B - 2006**	**NFPA 79 - 2007**	**NFPA**	**NEMA**
pg. 227 # 5-266	pg. - # -	410.1(A)	-	13.2.2	-	-
log: 454 CMP: 5	log: - Submitter: W. Creighton Schwan			2005 NEC: 250.119(A)(2)		IEC: -

2005 NEC - 250.119 Identification of Equipment Grounding Conductors.

(A) Conductors Larger Than 6 AWG. Equipment grounding conductors larger than 6 AWG shall comply with 250.119(A)(1) and (A)(2).

(1) An insulated or covered conductor larger than 6 AWG shall be permitted, at the time of installation, to be permanently identified as an equipment grounding conductor at each end and at every point where the conductor is accessible.

Exception: Conductors larger than 6 AWG shall not be required to be marked in conduit bodies that contain no splices or unused hubs.

(2) Identification shall encircle the conductor and shall be accomplished by one of the following:

a. Stripping the insulation or covering from the entire exposed length

b. Coloring the ~~exposed~~ insulation or covering green

c. Marking the ~~exposed~~ insulation or covering with green tape or green adhesive labels

2008 NEC - 250.119 Identification of Equipment Grounding Conductors.

(A) Conductors Larger Than 6 AWG. Equipment grounding conductors larger than 6 AWG shall comply with 250.119(A)(1) and (A)(2).

(1) An insulated or covered conductor larger than 6 AWG shall be permitted, at the time of installation, to be permanently identified as an equipment grounding conductor at each end and at every point where the conductor is accessible.

Exception: Conductors larger than 6 AWG shall not be required to be marked in conduit bodies that contain no splices or unused hubs.

(2) Identification shall encircle the conductor and shall be accomplished by one of the following:

a. Stripping the insulation or covering from the entire exposed length

b. Coloring the insulation or covering green at the termination

c. Marking the insulation or covering with green tape or green adhesive labels at the termination

Author's Comment: The new text clarifies that the identification shall encircle the conductor at the termination.

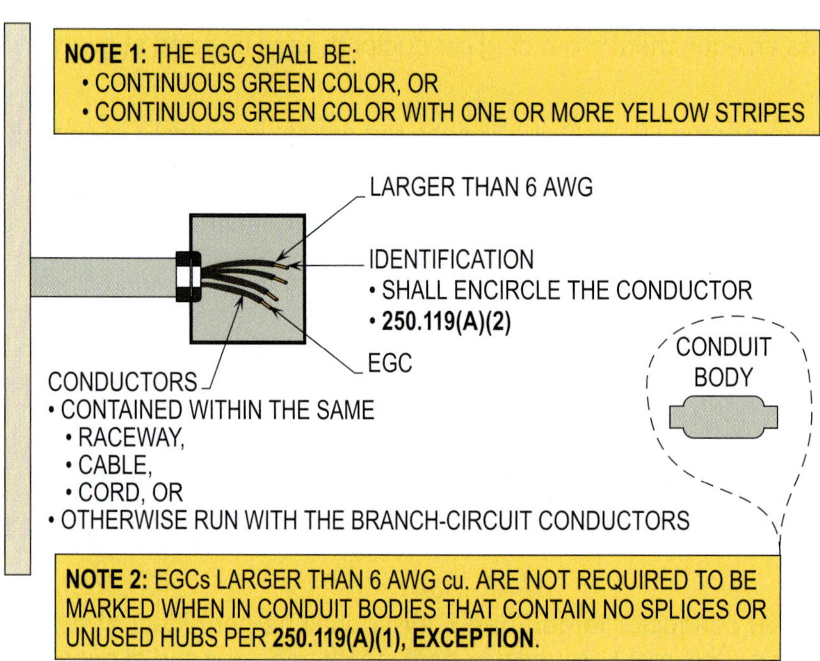
NOTE 1: THE EGC SHALL BE:
• CONTINUOUS GREEN COLOR, OR
• CONTINUOUS GREEN COLOR WITH ONE OR MORE YELLOW STRIPES
LARGER THAN 6 AWG
IDENTIFICATION
• SHALL ENCIRCLE THE CONDUCTOR
• 250.119(A)(2)
EGC
CONDUIT BODY
CONDUCTORS
• CONTAINED WITHIN THE SAME
• RACEWAY,
• CABLE,
• CORD, OR
• OTHERWISE RUN WITH THE BRANCH-CIRCUIT CONDUCTORS
NOTE 2: EGCs LARGER THAN 6 AWG cu. ARE NOT REQUIRED TO BE MARKED WHEN IN CONDUIT BODIES THAT CONTAIN NO SPLICES OR UNUSED HUBS PER 250.119(A)(1), EXCEPTION.

CONDUCTORS LARGER THAN 6 AWG
250.119(A)

Purpose of Change: This revision clarifies that the identification of the equipment grounding conductor (EGC), larger than the 6 AWG, shall encircle the conductor at the termination.

NEC Ch. 2 - Article 250
Part VI - 250.122(C)

nec
33

Type of Change		Panel Action		UL	UL 508	API 500 - 1997	API 505 - 1997	OSHA - 1994
Revision		Accept		-	-	-	-	-
ROP		ROC		NFPA 70E - 2004	NFPA 70B - 2006	NFPA 79 - 2007	NFPA	NEMA
pg. 231	# 5-282a	pg. -	# -	-	-	8.2.1.3	-	-
log: CP 513	CMP: -	log: -	Submitter: Code Making Panel 5			2005 NEC: 250.122(C)		IEC: -

2005 NEC - 250.122 Size of Equipment Grounding Conductors.

(C) Multiple Circuits. Where a single equipment grounding conductor is run with multiple circuits in the same raceway, ~~or~~ cable, it shall be sized for the largest overcurrent device protecting conductors in the raceway ~~or~~ cable.

2008 NEC - 250.122 Size of Equipment Grounding Conductors.

(C) Multiple Circuits. Where a single equipment grounding conductor is run with multiple circuits in the same raceway, cable, <u>or cable tray</u>, it shall be sized for the largest overcurrent device protecting conductors in the raceway, cable <u>or cable tray. Equipment grounding conductors installed in cable trays shall meet the minimum requirements of 392.3(B)(1)(c)</u>.

Author's Comment: Since cable trays are not a raceway but are permitted by **250.118(11), 392.3(B)(1)(c),** and **392.7(B)** as an equipment grounding conductor, adding cable trays into **250.122** provides the proper reference to tie these sections together.

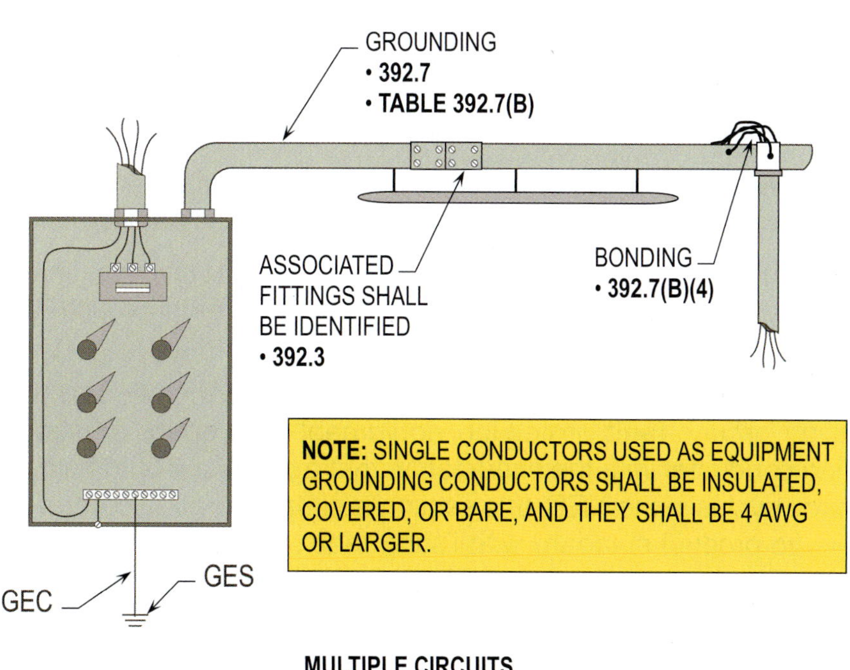
GROUNDING
• 392.7
• TABLE 392.7(B)
ASSOCIATED FITTINGS SHALL BE IDENTIFIED
• 392.3
BONDING
• 392.7(B)(4)
NOTE: SINGLE CONDUCTORS USED AS EQUIPMENT GROUNDING CONDUCTORS SHALL BE INSULATED, COVERED, OR BARE, AND THEY SHALL BE 4 AWG OR LARGER.
GEC
GES

**MULTIPLE CIRCUITS
250.122(C)**

Purpose of Change: This revision clarifies that where a single equipment grounding conductor is run with multiple circuits in the same cable tray, the equipment grounding conductor must be sized for the largest OCPD protecting conductors in the cable tray and must meet the minimum requirements in **392.3(B)(1)(c)**.

NEC Ch. 2 - Article 250
Part VI - 250.122(F)

Type of Change	Panel Action	UL	UL 508	API 500 - 1997	API 505 - 1997	OSHA - 1994
Revision	Accept	-	-	-	-	-
ROP	**ROC**	**NFPA 70E - 2004**	**NFPA 70B - 2006**	**NFPA 79 - 2007**	**NFPA**	**NEMA**
pg. 232 # 5-287	pg. - # -	-	-	8.2.1.3	-	-
log: 3491 CMP: 5	log: - Submitter: Alan Manche			2005 NEC: 250.122(F)(1); (F)(2)		IEC: -

2005 NEC - 250.122 Size of Equipment Grounding Conductors.

(F) Conductors in Parallel. Where conductors are run in parallel in multiple raceways or cables as permitted in 310.4, the equipment grounding conductors, where used, shall be run in parallel in each raceway or cable. ~~One of the methods in 250.122(F)(1) or (F)(2) shall be used to ensure the equipment grounding conductors are protected.~~

~~(1) Based on Rating of Overcurrent Protective Device.~~ Each parallel equipment grounding conductor shall be sized on the basis of the ampere rating of the overcurrent device protecting the circuit conductors in the raceway or cable in accordance with Table 250.122.

~~(2) Ground-Fault Protection of Equipment. Installed Where ground-fault protection of equipment is installed, each parallel equipment grounding conductor in a multiconductor cable shall be permitted to be sized in accordance with Table 250.122 on the basis of the trip rating of the ground-fault protection where the following conditions are met:~~

~~(1) Conditions of maintenance and supervision ensure that only qualified persons will service the installation.~~

~~(2) The ground-fault protection equipment is set to trip at not more than the ampacity of a single ungrounded conductor of one of the cables in parallel.~~

~~(3) The ground-fault protection is listed for the purpose of protecting the equipment grounding conductor.~~

2008 NEC - 250.122 Size of Equipment Grounding Conductors.

(F) Conductors in Parallel. Where conductors are run in parallel in multiple raceways or cables as permitted in 310.4, the equipment grounding conductors, where used, shall be run in parallel in each raceway or cable.

Each parallel equipment grounding conductor shall be sized on the basis of the ampere rating of the overcurrent device protecting the circuit conductors in the raceway or cable in accordance with Table 250.122.

Author's Comment: Section **250.122(F)(2)** was introduced into the NEC as the concept of placing a listing requirement on ground-fault protection equipment with specific listing for protection of the equipment grounding conductor with the expectation that a product would be listed for this application. After a number of code cycles in the NEC, this section is simply creating confusion among the electrical community. There is no product standard with listing requirements for this application, and there has been no product introduced to support this application so it has been deleted.

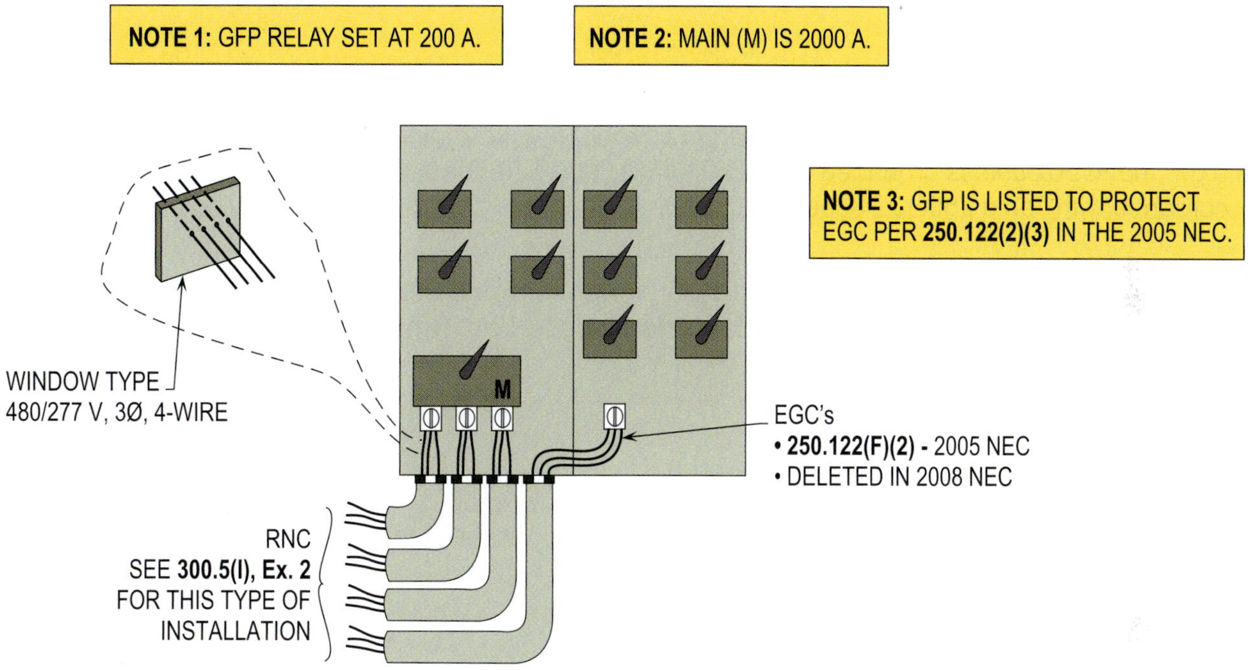

NOTE 1: GFP RELAY SET AT 200 A.

NOTE 2: MAIN (M) IS 2000 A.

NOTE 3: GFP IS LISTED TO PROTECT EGC PER **250.122(2)(3)** IN THE 2005 NEC.

WINDOW TYPE
480/277 V, 3Ø, 4-WIRE

M

EGC's
• **250.122(F)(2)** - 2005 NEC
• DELETED IN 2008 NEC

RNC
SEE **300.5(I), Ex. 2**
FOR THIS TYPE OF
INSTALLATION

CONDUCTORS IN PARALLEL
250.122(F)

Purpose of Change: This revision clarifies that there is no product standard with listing requirements for this application, and there has been no product introduced to support this application so it has been deleted.

NEC Ch. 2 - Article 250
Part VII - 250.146

34

Type of Change	Panel Action	UL	UL 508	API 500 - 1997	API 505 - 1997	OSHA - 1994
Revision	Accept	-	-	-	-	1910.304(a)(3)
ROP	**ROC**	**NFPA 70E - 2004**	**NFPA 70B - 2006**	**NFPA 79 - 2007**	**NFPA**	**NEMA**
pg. 234 # 5-297	pg. - # -	410.2(B)(2)	-	-	-	-
log: 628 CMP: 5	log: -	Submitter: Michael J. Johnston		2005 NEC: 250.146		IEC: -

2005 NEC - 250.146 Connecting Receptacle Grounding Terminal to Box.

An equipment bonding jumper shall be used to connect the grounding terminal of a grounding-type receptacle to a grounded box unless grounded as in 250.146(A) thru (D).

2008 NEC - 250.146 Connecting Receptacle Grounding Terminal to Box.

An equipment bonding jumper shall be used to connect the grounding terminal of a grounding-type receptacle to a grounded box unless grounded as in 250.146(A) thru (D). The equipment bonding jumper shall be sized in accordance with Table 250.122 based on the rating of the overcurrent device protecting the circuit conductors.

Author's Comment: The addition of the last sentence makes it clear that the sizing of the equipment bonding jumper from the box to the receptacle is based on **Table 250.122** for the rating of the branch circuit overcurrent protective device.

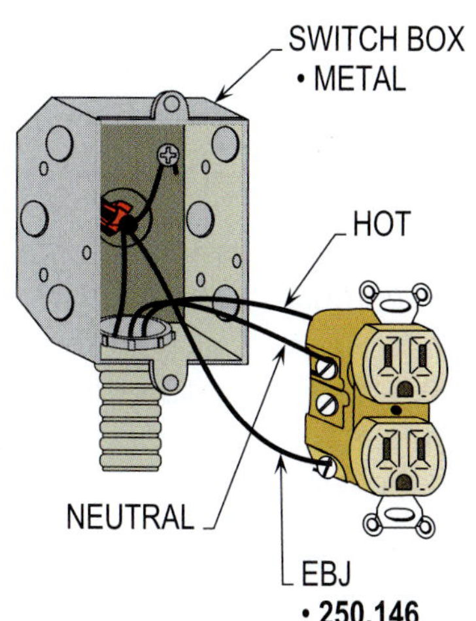

EXAMPLE

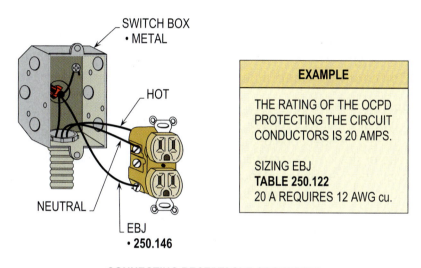

THE RATING OF THE OCPD PROTECTING THE CIRCUIT CONDUCTORS IS 20 AMPS.

SIZING EBJ
TABLE 250.122
20 A REQUIRES 12 AWG cu.

CONNECTING RECEPTACLE GROUNDING TERMINAL TO BOX
250.146

Purpose of Change: This revision clarifies that the sizing of the equipment bonding jumper (EBJ) from the box to the receptacle is based on **Table 250.122** for the rating of the branch circuit overcurrent protective device.

NEC Ch. 2 - Article 250
Part VII - 250.146(A)

Type of Change	Panel Action	UL	UL 508	API 500 - 1997	API 505 - 1997	OSHA - 1994
Revision	Accept	-	-	-	-	1910.304(a)(3)
ROP	**ROC**	**NFPA 70E - 2004**	**NFPA 70B - 2006**	**NFPA 79 - 2007**	**NFPA**	**NEMA**
pg. 234 # 5-300	pg. - # -	410.2(B)(2)	-	-	-	-
log: 2484 CMP: 5	log: -	Submitter: William Slater		2005 NEC: 250.146(A)		IEC: -

2005 NEC - 250.146 Connecting Receptacle Grounding Terminal to Box.

(A) Surface Mounted Box. Where the box is mounted on the surface, direct metal-to-metal contact between the device yoke and the box or a contact yoke or device that complies with 250.146(B) shall be permitted to ground the receptacle to the box. At least one of the insulating washers shall be removed from receptacles that do not have a contact yoke or device that complies with 250.146(B) to ensure direct metal-to-metal contact. This provision shall not apply to cover-mounted receptacles unless the box and cover combination are listed as providing satisfactory ground continuity between the box and the receptacle.

2008 NEC - 250.146 Connecting Receptacle Grounding Terminal to Box.

(A) Surface Mounted Box. Where the box is mounted on the surface, direct metal-to-metal contact between the device yoke and the box or a contact yoke or device that complies with 250.146(B) shall be permitted to ground the receptacle to the box. At least one of the insulating washers shall be removed from receptacles that do not have a contact yoke or device that complies with 250.146(B) to ensure direct metal-to-metal contact. This provision shall not apply to cover-mounted receptacles unless the box and cover combination are listed as providing satisfactory ground continuity between the box and the receptacle.

A listed exposed work cover shall be permitted to be the grounding and bonding means when (1) the device is attached to the cover with at least two fasteners that are permanent (such as a rivet) or have a thread locking or screw locking means and (2) when the cover mounting holes are located on a flat non-raised portion of the cover.

Author's Comment: The new paragraph provides permission to use a listed exposed work cover on a surface-mounted box where the device is attached to the cover with two permanent fasteners, such as rivets, or thread-locking or screw locking means when the cover mounting holes are on a flat part of the cover.

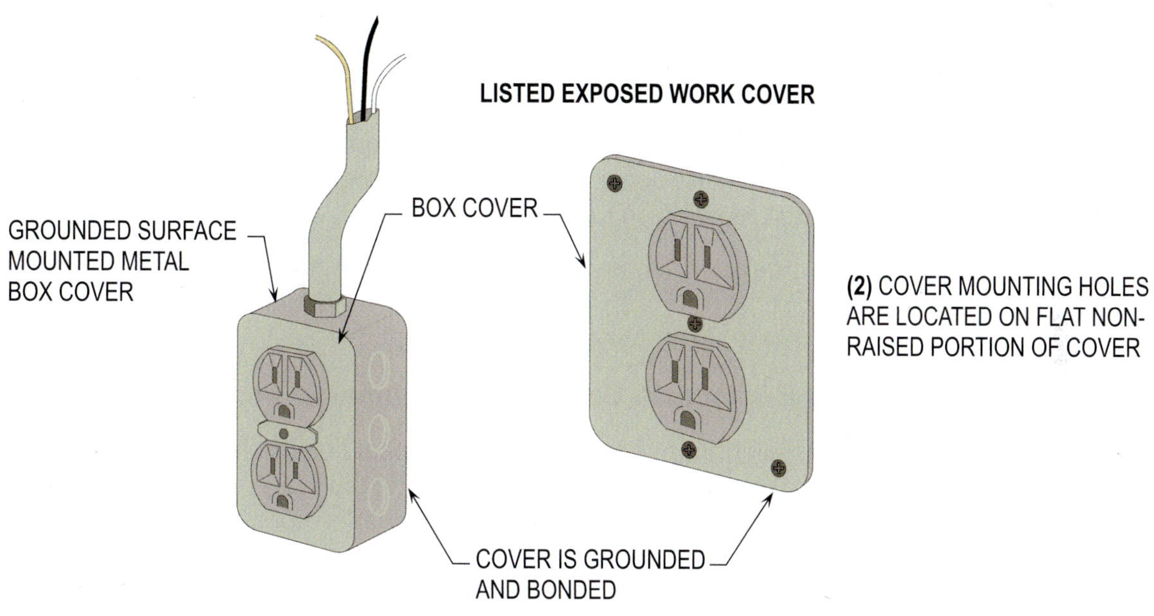
LISTED EXPOSED WORK COVER
GROUNDED SURFACE MOUNTED METAL BOX COVER
BOX COVER
(2) COVER MOUNTING HOLES ARE LOCATED ON FLAT NON-RAISED PORTION OF COVER
COVER IS GROUNDED AND BONDED
(1) AT LEAST TWO FASTENERS PERMANENTLY ATTACH DEVICE SUCH AS:
• TWO FASTENERS (SUCH AS A RIVET)
• THREAD LOCKING
• SCREW LOCKING MEANS

SURFACE MOUNTED BOX
250.146(A)

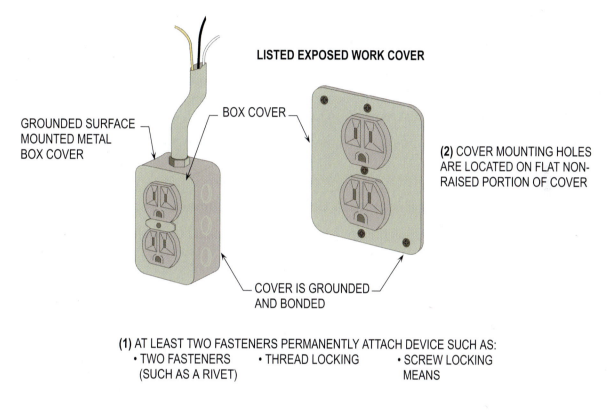

Purpose of Change: This revision clarifies that a listed exposed work cover shall be permitted as a grounding and bonding means on a surface mounted box where the device is attached to the cover with two permanent fasteners.

NEC Ch. 2 - Article 250
Part VII - 250.146(D)

Type of Change		Panel Action		UL	UL 508	API 500 - 1997	API 505 - 1997	OSHA - 1994
Revision		Accept		-	-	-	-	1910.304(a)(3)
ROP		ROC		NFPA 70E - 2004	NFPA 70B - 2006	NFPA 79 - 2007	NFPA	NEMA
pg. 235	# 5-301a	pg. 150	# 5-155	420.1(A)(2)	-	-	-	-
log: CP 514	CMP: 5	log: 337	Submitter: Code Making Panel 5			2005 NEC: 250.146(D)		IEC: -

2005 NEC - 250.146 Connecting Receptacle Grounding Terminal to Box.

(D) Isolated Receptacles. Where ~~required~~ for the reduction of electrical noise (electromagnetic interference) on the grounding circuit, a receptacle in which the grounding terminal is purposely insulated from the receptacle mounting means shall be permitted. The receptacle grounding terminal shall be ~~groundd by~~ an insulated equipment grounding conductor run with the circuit conductors. This grounding conductor shall be permitted to pass through one or more panelboards without connection to the panelboard grounding terminal as permitted in 408.40, Exception, so as to terminate within the same building or structure directly at an equipment grounding conductor terminal of the applicable derived system or service.

FPN: Use of an isolated equipment grounding conductor does not relieve the requirement for grounding the raceway system and outlet box.

2008 NEC - 250.146 Connecting Receptacle Grounding Terminal to Box.

(D) Isolated Receptacles. Where installed for the reduction of electrical noise (electromagnetic interference) on the grounding circuit, a receptacle in which the grounding terminal is purposely insulated from the receptacle mounting means shall be permitted. The receptacle grounding terminal shall be connect to an insulated equipment grounding conductor run with the circuit conductors. This equipment grounding conductor shall be permitted to pass through one or more panelboards without a connection to the panelboard grounding terminal bar as permitted in 408.40, Exception, so as to terminate within the same building or structure directly at an equipment grounding conductor terminal of the applicable derived system or service. Where installed in accordance with the provisions of this section, this equipment grounding conductor shall also be permitted to pass through boxes, wireways, or other enclosures without being connected to such enclosures.

FPN: Use of an isolated equipment grounding conductor does not relieve the requirement for grounding the raceway system and outlet box.

Author's Comment: The new paragraph provides permission to use a listed exposed work cover on a surface-mounted box where the device is attached to the cover with two permanent fasteners, such as rivets, or thread-locking or screw locking means when the cover mounting holes are on a flat part of the cover.

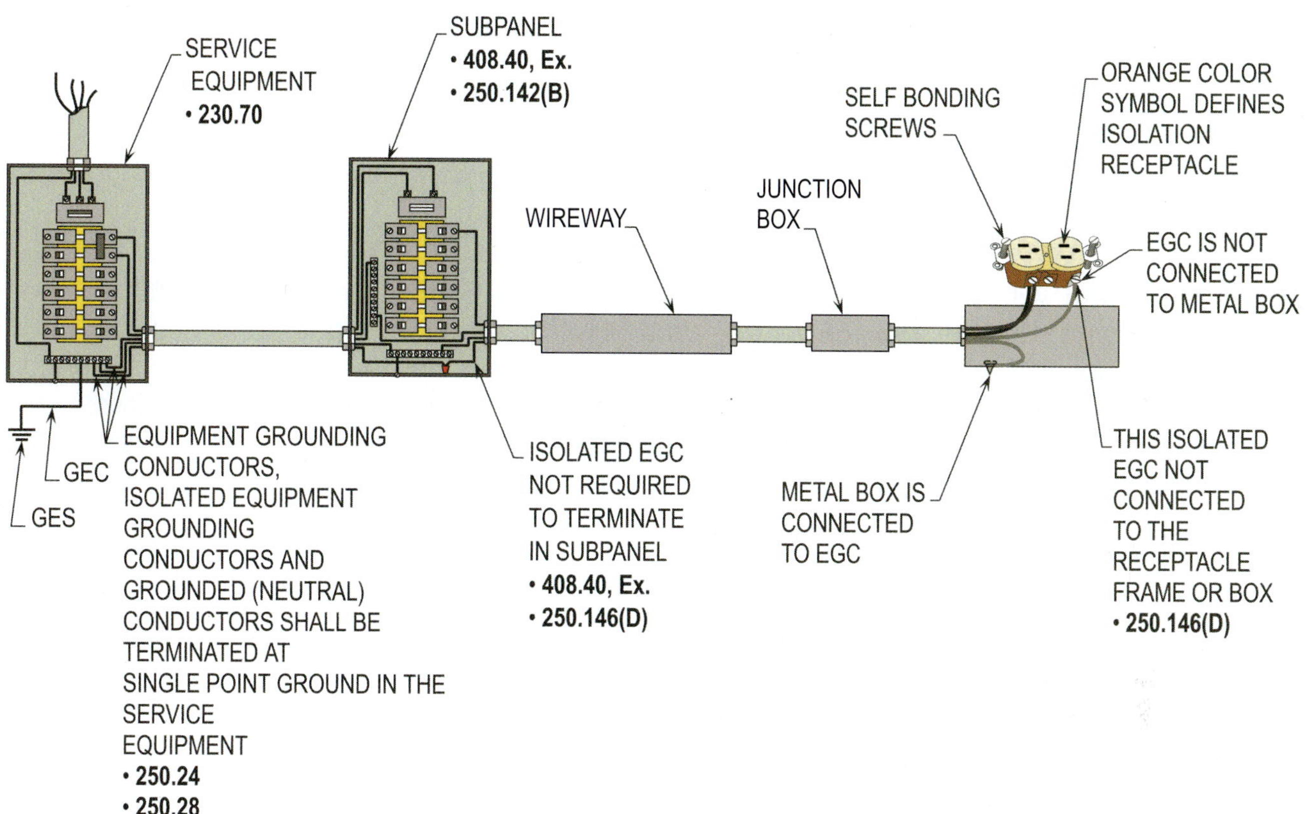

ISOLATED RECEPTACLES
250.146(D)

Purpose of Change: This revision permits the equipment grounding conductor to also pass through boxes, wireway or other enclosures without being connected to such enclosures.

NEC Ch. 2 - Article 250
Part VIII - 250.166

Type of Change	Panel Action	UL	UL 508	API 500 - 1997	API 505 - 1997	OSHA - 1994
Revision	Accept	-	-	-	-	-
ROP	ROC	NFPA 70E - 2004	NFPA 70B - 2006	NFPA 79 - 2007	NFPA	NEMA
pg. 236 # 5-310	pg. - # -	-	-	-	-	-
log: 2064 CMP: 5	log: -	Submitter: John C. Wiles		2005 NEC: 250.166		IEC: -

2005 NEC - 250.166 Size of Direct-Current Grounding Electrode Conductor.

The size of the grounding electrode conductor for a dc system shall be as specified in 250.166 (A) ~~through (E)~~.

2008 NEC - 250.166 Size of the Direct-Current Grounding Electrode Conductor.

The size of the grounding electrode conductor for a dc system shall be as specified in 250.166 (A) and (B), except as permitted by 250.166 (C) through (E).

Author's Comment: As Section **250.166** was written in the 2005 NEC, Sections **250.166 (A)** and **(B)** were in conflict with sections **250.166(C)**, **(D)**, and **(E)**. For example, many dc systems are not as described in Section **260.166 (A)**, so Section **250.166 (B)** applies. However, where using a ground rod electrode Section **250.166(C)** would apply. Sections **250.166(B)** and **(C)** dictate two different sizes of grounding electrode conductors. Revising the first sentence in this section as shown adds clarity and ensures these sections are no longer conflicting with each other.

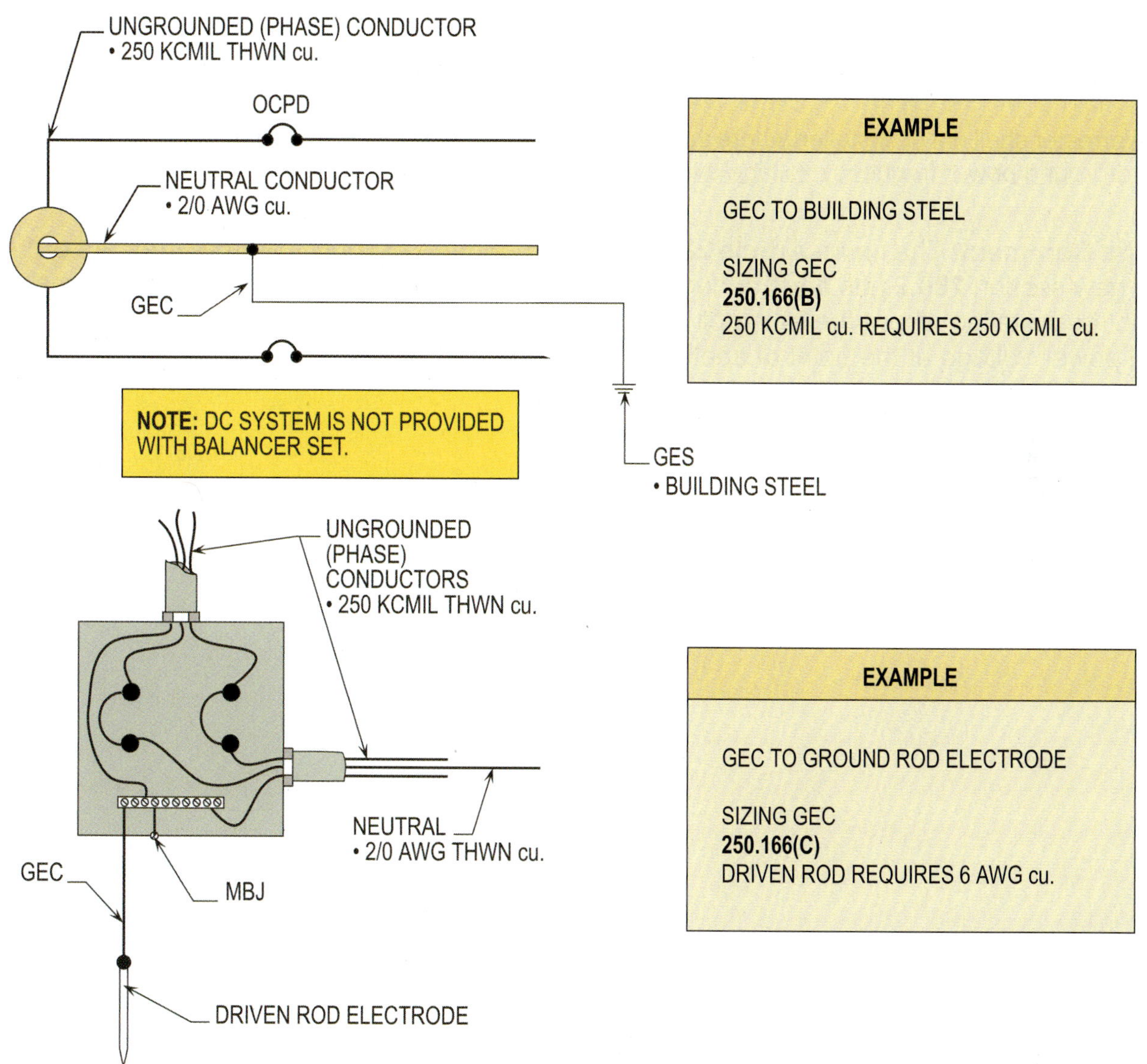

SIZE OF THE DIRECT-CURRENT GROUNDING ELECTRODE CONDUCTOR 250.166

Purpose of Change: The revision adds clarity for sizing the direct-current grounding electrode conductor.

NEC Ch. 2 - Article 280
Part I - 280.2

Type of Change		Panel Action		UL	UL 508	API 500 - 1997	API 505 - 1997	OSHA - 1994
New Section		Accept in Principle		-	-	-	-	-
ROP		**ROC**		**NFPA 70E - 2004**	**NFPA 70B - 2006**	**NFPA 79 - 2007**	**NFPA**	**NEMA - 1999**
pg. 246	# 5-335	pg. 151	# 5-162	-	-	-	-	LA 1
log: 2599	CMP: 5	log: 975	Submitter: Joseph P. Degregoria			2005 NEC: -		IEC: -

2008 NEC - 280.2 Uses Not Permitted.

A surge arrester shall not be installed where the rating of the surge arrester is less than the maximum continuous phase-to-ground power frequency voltage available at the point of application.

Author's Comment: The entire **Article 280** has been rewritten to apply to surge arresters of 1 kV and over. A new section **280.2** has been added to show uses not permitted. For example, a surge arrester must not be installed where the rating of the arrester is less than the maximum continuous phase to ground power voltage at the point of application.

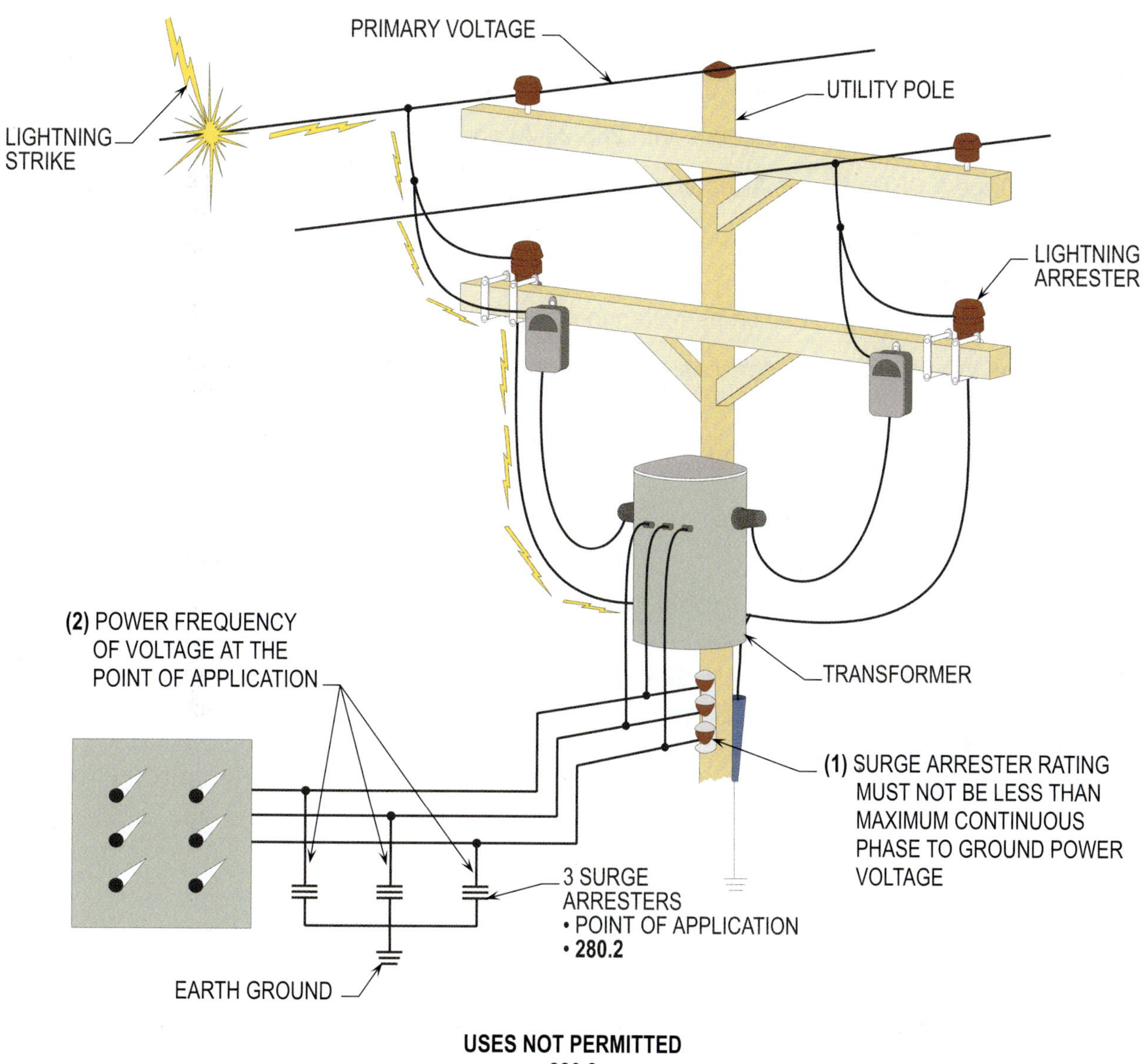

USES NOT PERMITTED
280.2

Purpose of Change: A new section has been added to clarify the uses not permitted for surge arresters and the article has been rewritten to apply to 1 kV and greater surge arresters.

NEC Ch. 2 - Article 285
Part III - 285.23(A) and (B)

Type of Change	Panel Action	UL	UL 508	API 500 - 1997	API 505 - 1997	OSHA - 1994
New Section	Accept in Principle	-	-	-	-	-
ROP	**ROC**	**NFPA 70E - 2004**	**NFPA 70B - 2006**	**NFPA 79 - 2007**	**NFPA**	**NEMA - 1999**
pg. 252 # 5-349	pg. 154 # 5-168	-	-	-	-	LA 1
log: 2604 CMP: 5	log: 2351	Submitter: Joseph P. DeGregoria		2005 NEC: -		IEC: -

2008 NEC - 285.23 Type 1 SPDs (Surge Arresters).

Type 1 SPDs shall be installed in accordance with 285.23(A) and (B).

(A) Installation. Type 1 SPDs (Surge Arrester) shall be installed as follows:

(1) Type 1 SPDs (surge arresters) shall be permitted to be connected to the supply side of the service disconnect as permitted in 230.82(4) or

(2) Type 1 SPDs (surge arrester) shall be permitted to be connected as specified in 285.24.

(B) At the Service. When installed at services, the grounding conductor of a Type 1 SPD shall be connected to one of the following:

(1) Grounded service conductor

(2) Grounding electrode conductor

(3) Grounding electrode for the service

(4) Equipment grounding terminal in the service equipment

Author's Comment: The entire **Article 285** has been rewritten and new title with the chagne from "Transient Voltage Surge Arrester" to "Surge Protective Devices (SPDs) 1 kV or Less". Surge arresters less than 1 kV are Type 1 SPDs and these are transient voltage surge suppressors that are Type 2 and Type 3 SPDs. A new section has been added to address the installation requirements for Type 1 SPDs (TVSS).

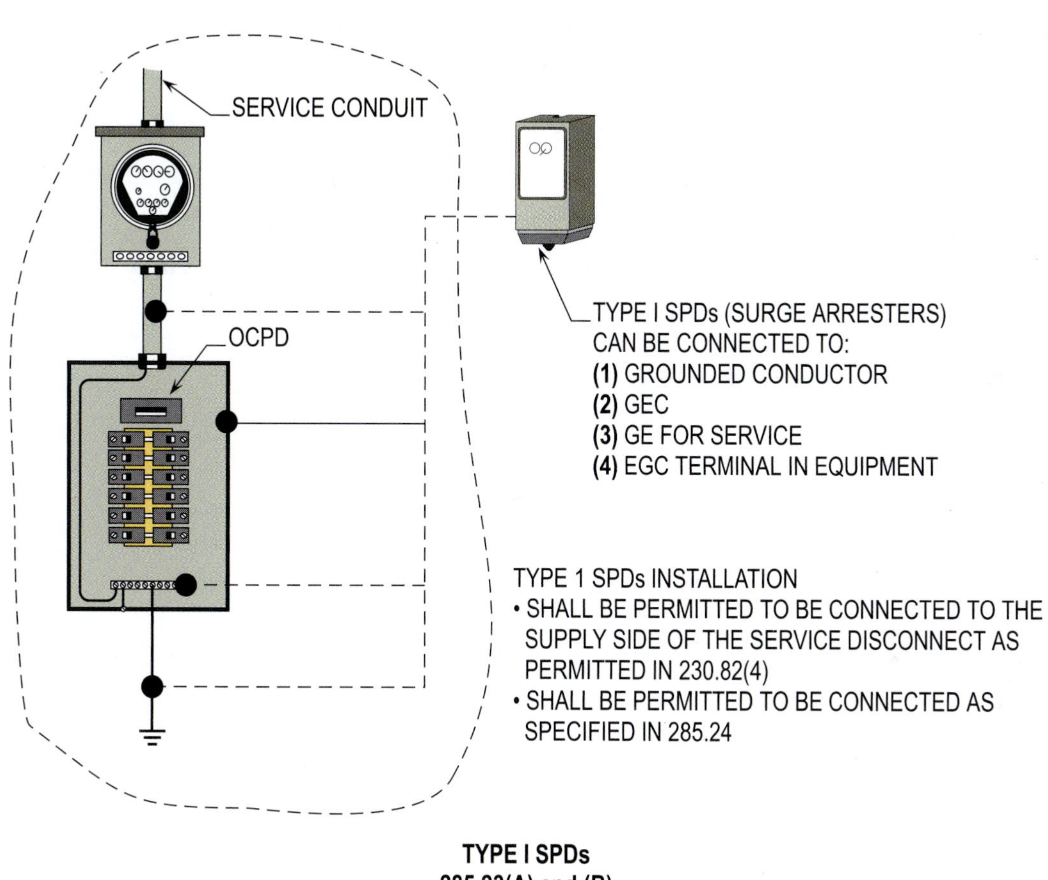

SERVICE CONDUIT
OCPD
TYPE I SPDs (SURGE ARRESTERS)
CAN BE CONNECTED TO:
(1) GROUNDED CONDUCTOR
(2) GEC
(3) GE FOR SERVICE
(4) EGC TERMINAL IN EQUIPMENT
TYPE 1 SPDs INSTALLATION
• SHALL BE PERMITTED TO BE CONNECTED TO THE SUPPLY SIDE OF THE SERVICE DISCONNECT AS PERMITTED IN 230.82(4)
• SHALL BE PERMITTED TO BE CONNECTED AS SPECIFIED IN 285.24

TYPE I SPDs
285.23(A) and (B)

Purpose of Change: The entire **Article 285** has been rewritten with a new title for the article. A new section has been added to clarify the installation requirements for Type I SPDs.

NEC Ch. 2 - Article 285
Part III - 285.24

Type of Change		Panel Action		UL	UL 508	API 500 - 1997	API 505 - 1997	OSHA - 1994
New Section		Accept in Principle		-	-	-	-	-
ROP		**ROC**		**NFPA 70E - 2004**	**NFPA 70B - 2006**	**NFPA 79 - 2007**	**NFPA**	**NEMA - 1999**
pg. 252	# 5-349	pg. -	# 5-168	-	-	-	-	LA 1
log: 2604	CMP: 5	log: 2351	Submitter: Joseph P. DeGregoria			2005 NEC: -		IEC: -

2008 NEC - 285.24 Type 2 SPDs (TVSSs).

Type 2 SPDs (TVSS) shall be installed in accordance with 285.24(A) through (C).

(A) Service Supplied Building or Structure. Type 2 SPDs (TVSSs) shall be connected anywhere on the load side of a service disconnect overcurrent device required in 230.91, unless installed in accordance with 230.82(8).

(B) Feeder Supplied Building or Structure. Type 2 SPDs (TVSSs) shall be connected at the building or structure anywhere on the load side of the first overcurrent device at the building structure.

(C) Separately Derived System. The SPD (TVSS) shall be connected on the load side of the first disconnecting overcurrent device in a separately derived system.

Author's Comment: A new section has been added to address the installation requirements for Type 2 SPDs (TVSS) at a service-supplied building or structure, feeder-supplied building or structure, or separately derived system.

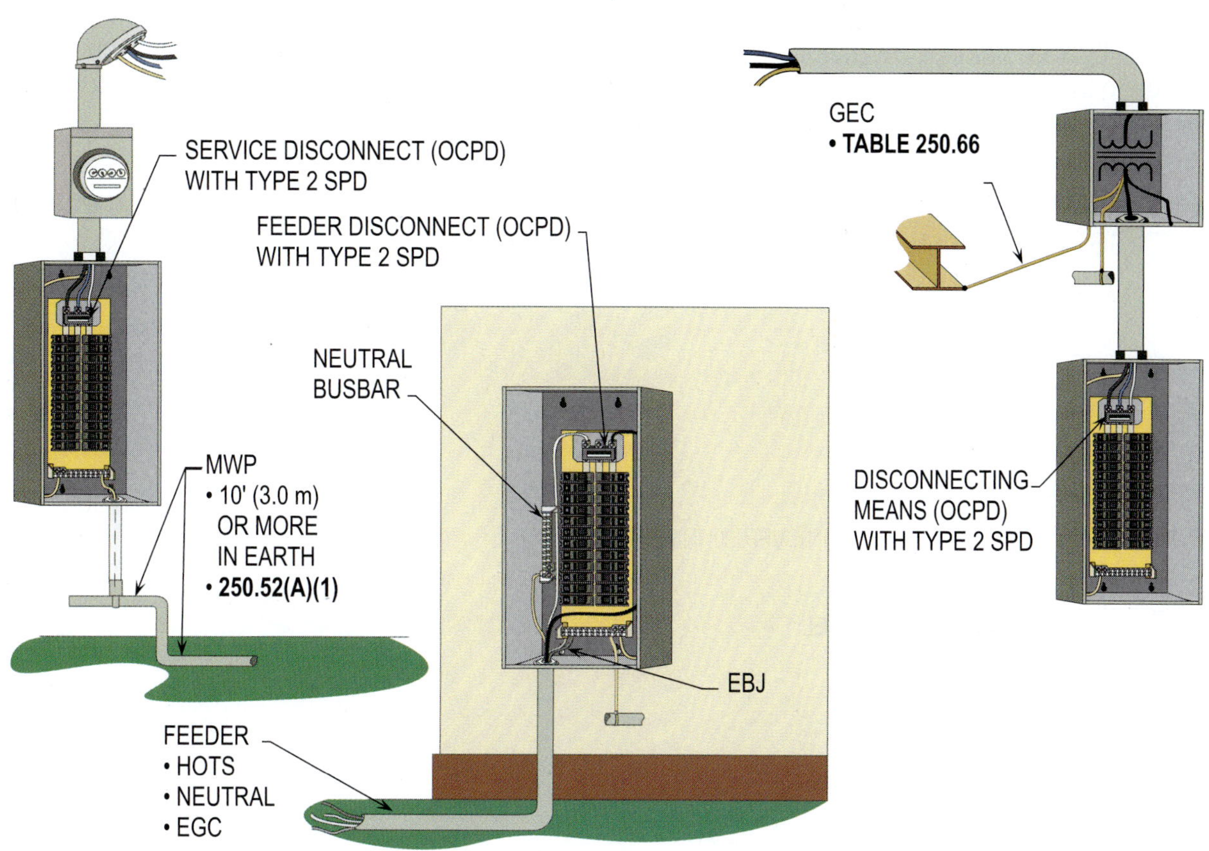

TYPE SDPs (TVSS) CAN BE CONNECTED ANYWHERE ON THE LOAD SIDE OF THE OCPD FOR SERVICE, FEEDER, OR SEPARATELY DERIVED SYSTEM

TYPE 2 SPDs (TVSSs)
285.24

Purpose of Change: A new section has been added to clarify the installation requirements for Type 2 SPDs.

NEC Ch. 2 - Article 285
Part III - 285.25

Type of Change	Panel Action	UL	UL 508	API 500 - 1997	API 505 - 1997	OSHA - 1994
New Section	Accept in Principle	-	-	-	-	-
ROP	**ROC**	**NFPA 70E - 2004**	**NFPA 70B - 2006**	**NFPA 79 - 2007**	**NFPA**	**NEMA - 1999**
pg. 252 # 5-349	pg. 154 # 5-168	-	-	-	-	LA 1
log: 2604 CMP: 5	log: 2351	Submitter: Joseph P. DeGregoria		2005 NEC: -		IEC: -

2008 NEC - 285.25 Type 3 SPDs.

Type 3 SPDs (TVSSs) shall be permitted to be installed anywhere on the load side of branch circuit overcurrent protection up to the equipment served, provided the connection is a minimum 10 m (30 ft.) of conductor distance from the service or separately derived system disconnect.

Author's Comment: A new section has been added to address the installation requirements for Type 3 SPDs (TVSS).

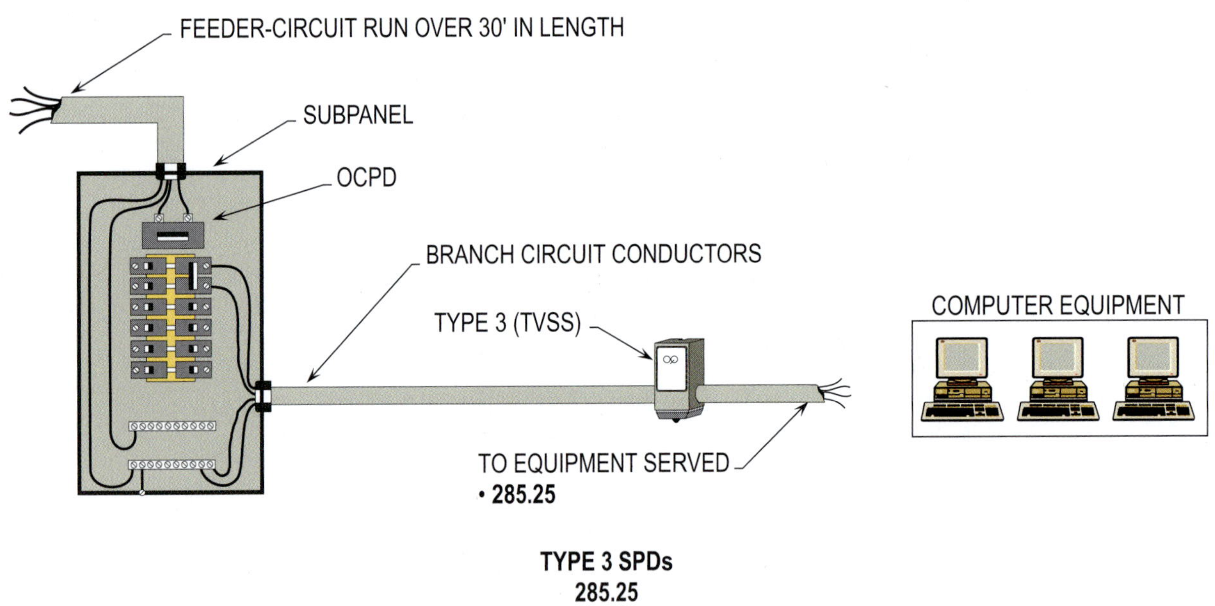

TYPE 3 SPDs
285.25

Purpose of Change: A new section has been added to clarify the installation requirements for Type 3 SPDs.

3

Wiring Methods and Materials

Chapter 3 of the NEC has always been referred to as the "Installation Chapter" and used by electricians to install electrical systems in a safe, dependable and reliable manner. **Chapter 3** covers the requirements needed by on-the-job electricians who are installing service, feeder, and branch circuits and thus providing electrical power to the end of branch-circuits and on to the point of use.

Chapter 3 is also called the "Rough-In Chapter" by electricians. All articles in **Chapter 3** are of the 300 series and each contains the rules that pertain to installing electrical wiring methods and accessories. On-the-job work procedures of the average electrician bring them into almost daily contact with an article in **Chapter 3** concerning these installation rules.

For example, when installing wiring in cable trays, electricians cannot install and fill cable trays with different wiring, cables, and systems without knowing the requirements of **Article 392**. The same is true for electricians installing rigid metal conduit. Installers would not know how many 90° bends are permitted or how often supports are required for rigid metal conduit, without first studying **Article 344**.

NEC Ch. 3 - Article 300
Part I - 300.4(E)

35

Type of Change		Panel Action		UL	UL 508	API 500 - 1997	API 505 - 1997	OSHA - 1994
New Subsection		-		514B - 651	-	-	-	-
ROP		ROC		NFPA 70E - 2004	NFPA 70B - 2006	NFPA 79 - 2007	NFPA	NEMA
pg. 260	# 3-31	pg. 157	# 3-10	-	-	-	-	-
log: 3310	CMP: 3	log: 2040	Submitter: William Benard			2005 NEC: -		IEC: -

2008 NEC - 300.4 Protection Against Physical Damage.

(E) Cables and Raceways Installed Under Roof Decking. A cable- or raceway-type wiring method, installed in exposed or concealed locations under metal-corrugated sheet roof decking, shall be installed and supported so the nearest outside surface of the cable or raceway is not less than 38 mm (1 1/2 in.) from the nearest surface of the roof decking.

FPN: Roof decking material is often repaired or replaced after the initial raceway or cabling and roofing installation and may be penetrated by the screws or other mechanical devices designed to provide "hold down" strength of the waterproof membrane or roof insulating material.

Exception. Rigid metal conduit and intermediate metal conduit shall not be required to comply with 300.4(E).

Author's Comment: A new subection has been added to address physical protection for cables and raceway methods installed below roof decks with insulating and waterproofing material secured above using screws intended to penetrate the decking by at least 1 1/2 in. (38 mm) to meet minimum manufacturer windsheer specifications.

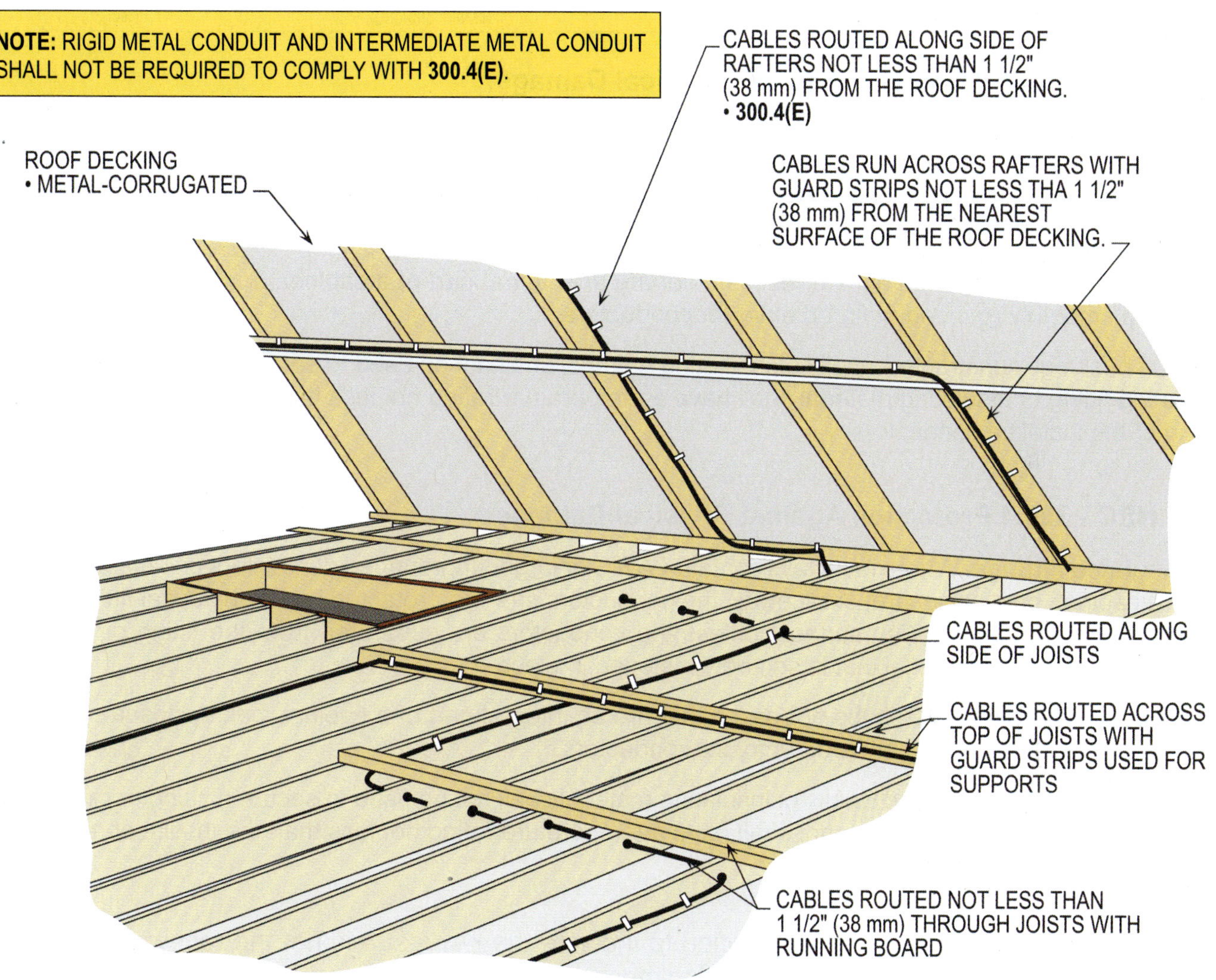
NOTE: RIGID METAL CONDUIT AND INTERMEDIATE METAL CONDUIT SHALL NOT BE REQUIRED TO COMPLY WITH **300.4(E)**.
CABLES ROUTED ALONG SIDE OF RAFTERS NOT LESS THAN 1 1/2" (38 mm) FROM THE ROOF DECKING. • **300.4(E)**
ROOF DECKING • METAL-CORRUGATED
CABLES RUN ACROSS RAFTERS WITH GUARD STRIPS NOT LESS THA 1 1/2" (38 mm) FROM THE NEAREST SURFACE OF THE ROOF DECKING.
CABLES ROUTED ALONG SIDE OF JOISTS
CABLES ROUTED ACROSS TOP OF JOISTS WITH GUARD STRIPS USED FOR SUPPORTS
CABLES ROUTED NOT LESS THAN 1 1/2" (38 mm) THROUGH JOISTS WITH RUNNING BOARD

CABLES AND RACEWAY INSTALLED UNDER ROOF DECKING 300.4(E)

Purpose of Change: A new subsection has been added to required at least 1 1/2 in. clearance for a cable- or raceway-type wiring method to be installed in exposed or concealed locations under metal-corrugated sheet roof decking.

NEC Ch. 3 - Article 300
Part I - 300.4(G)

Type of Change	Panel Action	UL	UL 508	API 500 - 1997	API 505 - 1997	OSHA - 1994
Revision	Accept in Principle	514B - 651	-	-	-	-
ROP	**ROC**	**NFPA 70E - 2004**	**NFPA 70B - 2006**	**NFPA 79 - 2007**	**NFPA**	**NEMA - 2003**
pg. 261 # 3-33	pg. 158 # 3-12	-	-	-	-	250
log: 1338 CMP: 3	log: 2255	Submitter: Mike Holt		2005 NEC: 300.4(F)		IEC: -

2005 NEC - 300.4 Protection Against Physical Damage.

~~(F)~~ **Insulated Fittings.** Where raceways containing ungrounded conductors 4 AWG or larger enter a cabinet, box, enclosure, or raceway, the conductors shall be protected by a substantial fitting providing a smoothly rounded insulating surface, unless the conductors are separated from the fitting or raceway by substantial insulating material that is securely fastened in place.

Exception: Where threaded hubs or bosses that are an integral part of a cabinet, box enclosure, or raceway provide a smoothly rounded or flared entry for conductors.

Conduit bushings constructed wholly of insulating material shall not be used to secure a fitting or raceway. The insulating fitting or insulating material shall have a temperature rating not less than the insulation temperature rating of the installed conductors.

2008 NEC - 300.4 Protection Against Physical Damage.

<u>(G)</u> **Insulated Fittings.** Where raceways contain 4 AWG or larger ungrounded <u>conductors and these conductors</u> enter a cabinet, box, enclosure, or raceway, the conductors shall be protected by a substantial fitting providing a smoothly rounded insulating surface, unless the conductors are separated from the fitting or raceway by substantial insulating material that is securely fastened in place.

Exception: Where threaded hubs or bosses that are an integral part of a cabinet, box enclosure, or raceway provide a smoothly rounded or flared entry for conductors.

Conduit bushings constructed wholly of insulating material shall not be used to secure a fitting or raceway. The insulating fitting or insulating material shall have a temperature rating not less than the insulation temperature rating of the installed conductors.

Author's Comment: The text was clarified to indicate the key issue is not the raceway entering the cabinet or box but that the conductors entering these enclosures are properly protected with a substantial fitting providing an insulated surface.

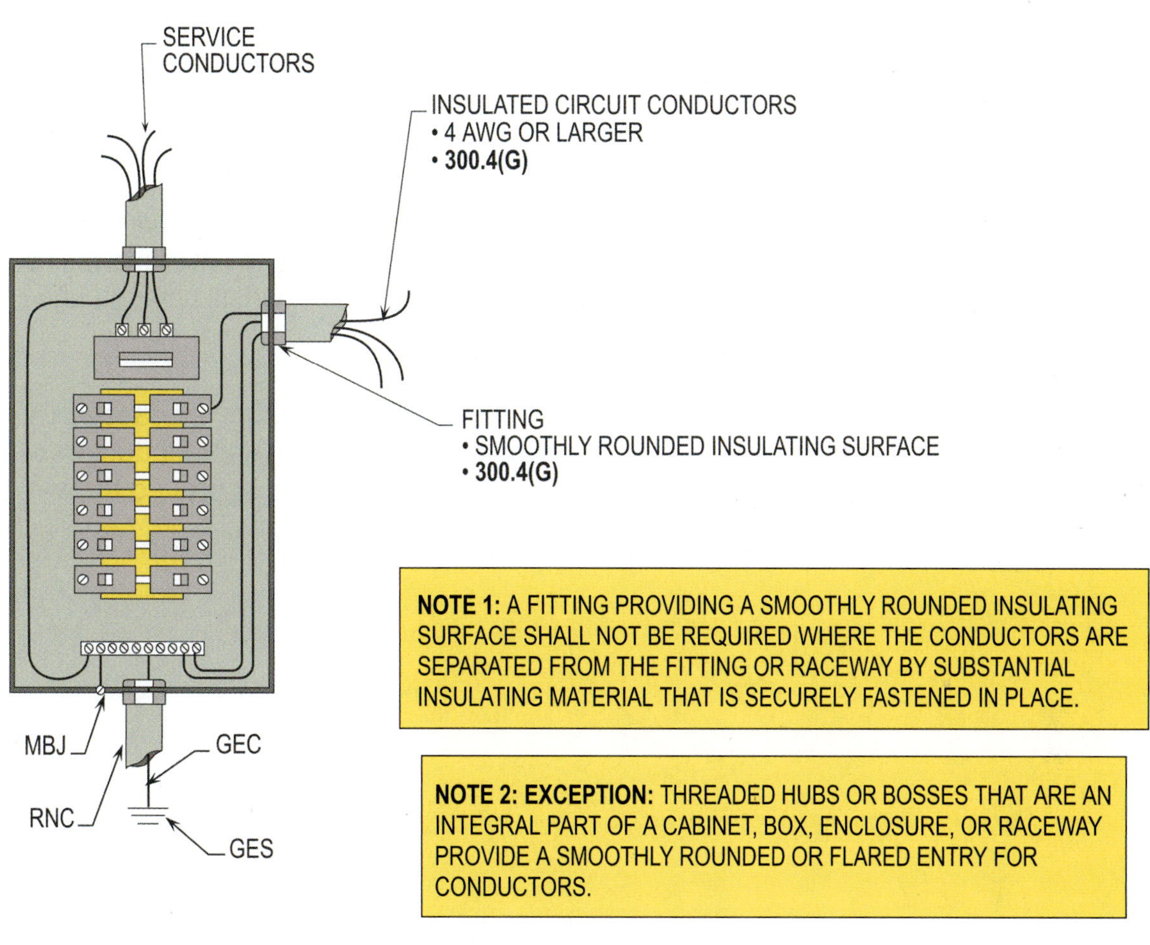
SERVICE CONDUCTORS
INSULATED CIRCUIT CONDUCTORS
• 4 AWG OR LARGER
• 300.4(G)
FITTING
• SMOOTHLY ROUNDED INSULATING SURFACE
• 300.4(G)
NOTE 1: A FITTING PROVIDING A SMOOTHLY ROUNDED INSULATING SURFACE SHALL NOT BE REQUIRED WHERE THE CONDUCTORS ARE SEPARATED FROM THE FITTING OR RACEWAY BY SUBSTANTIAL INSULATING MATERIAL THAT IS SECURELY FASTENED IN PLACE.
NOTE 2: EXCEPTION: THREADED HUBS OR BOSSES THAT ARE AN INTEGRAL PART OF A CABINET, BOX, ENCLOSURE, OR RACEWAY PROVIDE A SMOOTHLY ROUNDED OR FLARED ENTRY FOR CONDUCTORS.
MBJ
GEC
RNC
GES

INSULATED FITTINGS
300.4(G)

Purpose of Change: This revision clarifies that insulated grounded conductors shall be protected.

NEC Ch. 3 - Article 300
Part I - 300.5(B)

Type of Change	Panel Action	UL	UL 508	API 500 - 1997	API 505 - 1997	OSHA - 1994
Revision	Accept	-	-	-	-	-
ROP	**ROC**	**NFPA 70E - 2004**	**NFPA 70B - 2006**	**NFPA 79 - 2007**	**NFPA**	**NEMA - 2003**
pg. 263 # 3-42	pg. 160 # 3-20a	-	-	-	-	250
log: 1206 CMP: 3	log: CC 300	Submitter: Dennis Downer		2005 NEC: 300.5(B)		IEC: -

2005 NEC - 300.5(B) ~~Listing.~~

~~Cables and insulated conductors installed in enclosures or raceways in underground installations shall be listed for use in wet locations.~~

2008 NEC - 300.5(B) Wet Locations.

The interior of enclosures or raceways installed underground shall be considered to be a wet location. Insulated conductors and cables installed in these enclosures or raceways in underground installations shall be listed for use in wet locations and shall comply with 310.8(C). Any connections or splices in an underground installation shall be approved for wet locations.

Author's Comment: This revision clarifies that the interior of enclosure or raceways installed underground shall be considered to be classified as a wet location. All insulated conductors and cables shall be lised for use in wet locations and connections or splices shall be approved for wet locations when installed in these enclosures or raceways in underground installations.

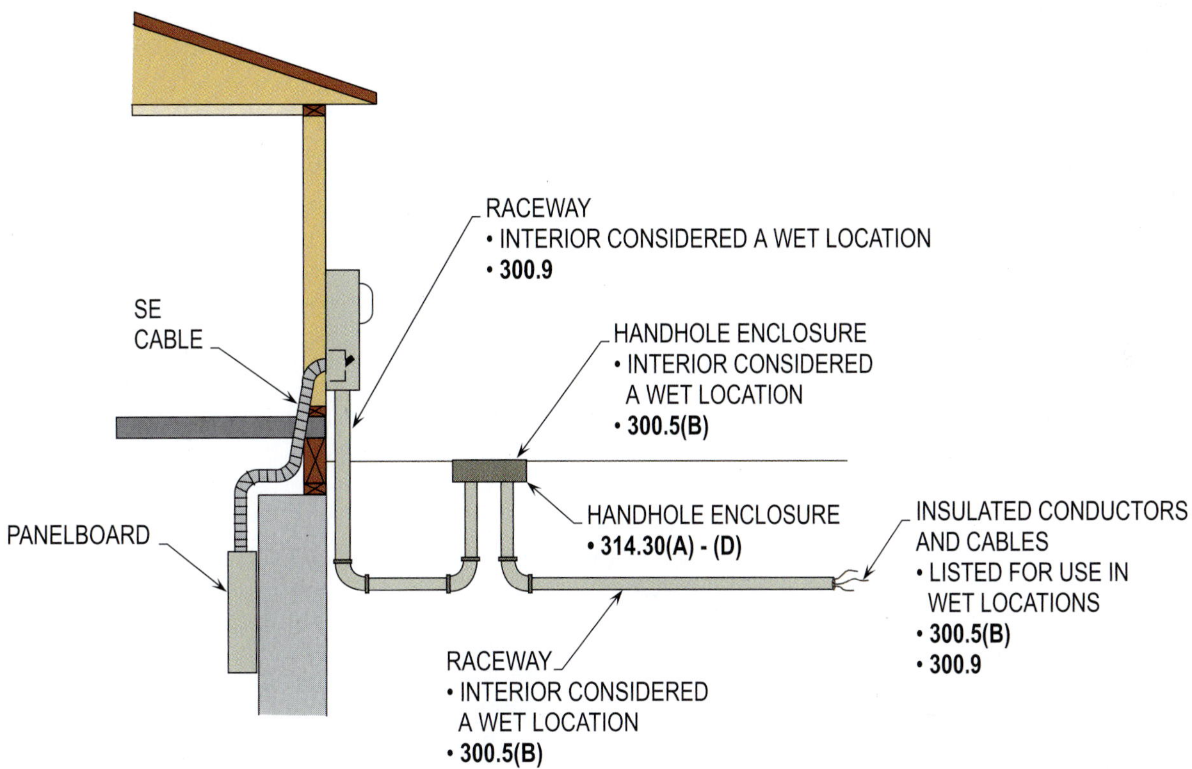

WET LOCATIONS
300.5(B)

Purpose of Change: A new subsection has been added to clarify that the interior of enclosures or raceways shall be considered as a wet location. All insulated conductors and cables shall be listed for use in wet locations and shall comply with **310.8(C)**.

NEC Ch. 3 - Article 300
Part I - 300.9

Type of Change		Panel Action		UL	UL 508	API 500 - 1997	API 505 - 1997	OSHA - 1994
New Section		Accept in Principle		-	-	-	-	-
ROP		ROC		NFPA 70E - 2004	NFPA 70B - 2006	NFPA 79 - 2007	NFPA	NEMA
pg. 267	# 3-63	pg. 166	# 3-52	-	-	-	-	-
log: 2234	CMP: 3	log: 2257	Submitter: Donald A. Ganiere			2005 NEC: -		IEC: -

2008 NEC - 300.9 Raceways in Wet Locations Above Grade.

Where raceways are installed in wet locations abovegrade, the interior of these raceways shall be considered to be a wet location. Insulated conductors and cables installed in raceways in wet locations abovegrade shall comply with 310.8(C).

Author's Comment: A new section has been added to clarify that the interior of raceways shall be considered be a wet location if installed in wet locations abovegrade.

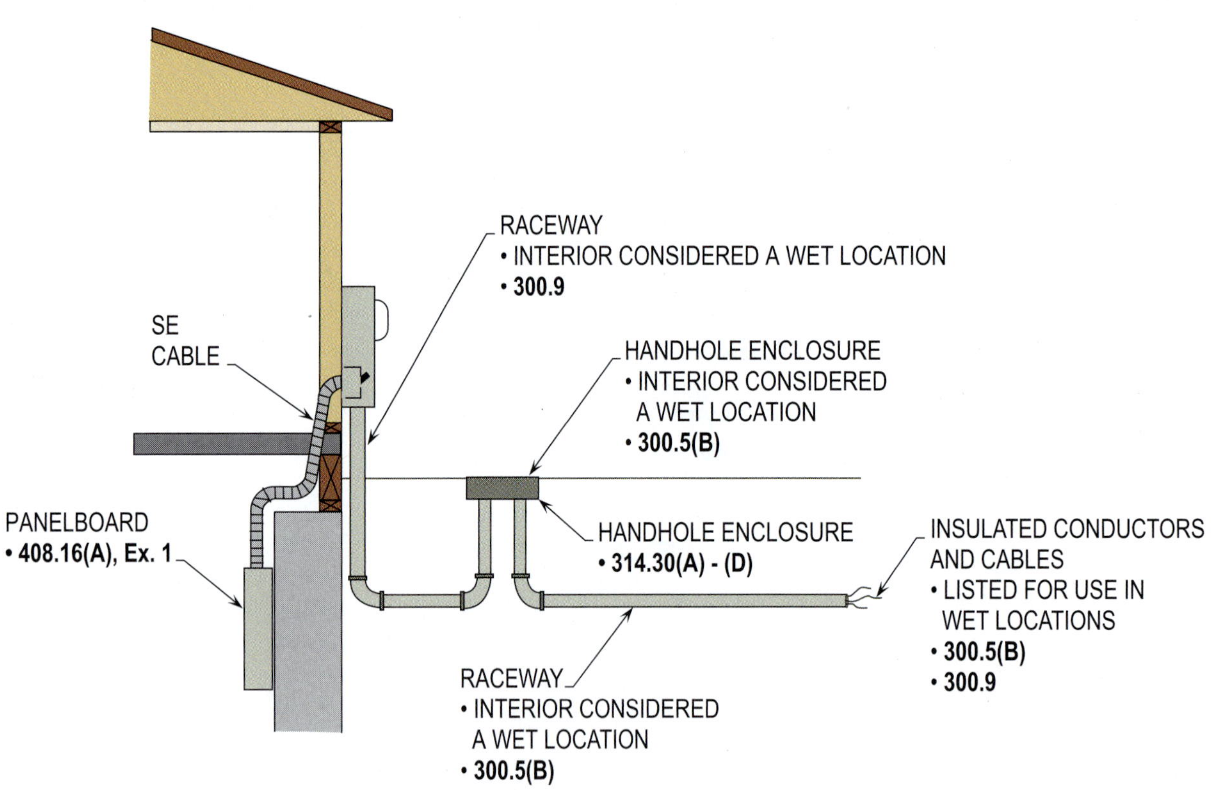

RACEWAYS IN WET LOCATIONS ABOVE GRADE
300.9

Purpose of Change: A new section has been added to clarify that the interior of enclosures or raceways shall be considered to be a wet location if installed in wet locations above grade. All insulated conductors and cables shall be listed for use in wet locations.

NEC Ch. 3 - Article 300
Part I - 300.12, Exception No. 2

Type of Change		Panel Action		UL	UL 508	API 500 - 1997	API 505 - 1997	OSHA - 1994
Revision		Accept in Principle		-	-	-	-	1910.305(a)(3)(i)
ROP		ROC		NFPA 70E - 2004	NFPA 70B - 2006	NFPA 79 - 2007	NFPA	NEMA
pg. 267	# 3-65	pg. 166	# 3-53	420.1(A)(1)	29.1.2	8.2.1.2	-	-
log: 2228	CMP: 3	log: 1841	Submitter: Donald A. Ganiere			2005 NEC: -		IEC: -

2008 NEC - 300.12 Mechanical Continuity - Raceways and Cables.

Metal or nonmetallic raceways, cable armors, and cable sheaths shall be continuous between cabinets, boxes, fittings, or other enclosures or outlets.

Exception No. 2: Raceways and cables installed into the bottom of open bottom equipment, such as switchboards, motor control centers, and floor or pad-mounted transformers, shall not be required to be mechanically secured to the equipment.

Author's Comment: A new exception has been added to permit raceways and cables installed into the bottom of open bottom equipment, such as switchboards, motor control centers, and transformers, to not be mechanically secured to the equipment. Metal raceways must still be bonded to the open bottom equipment in accordance with **250.102(D)**.

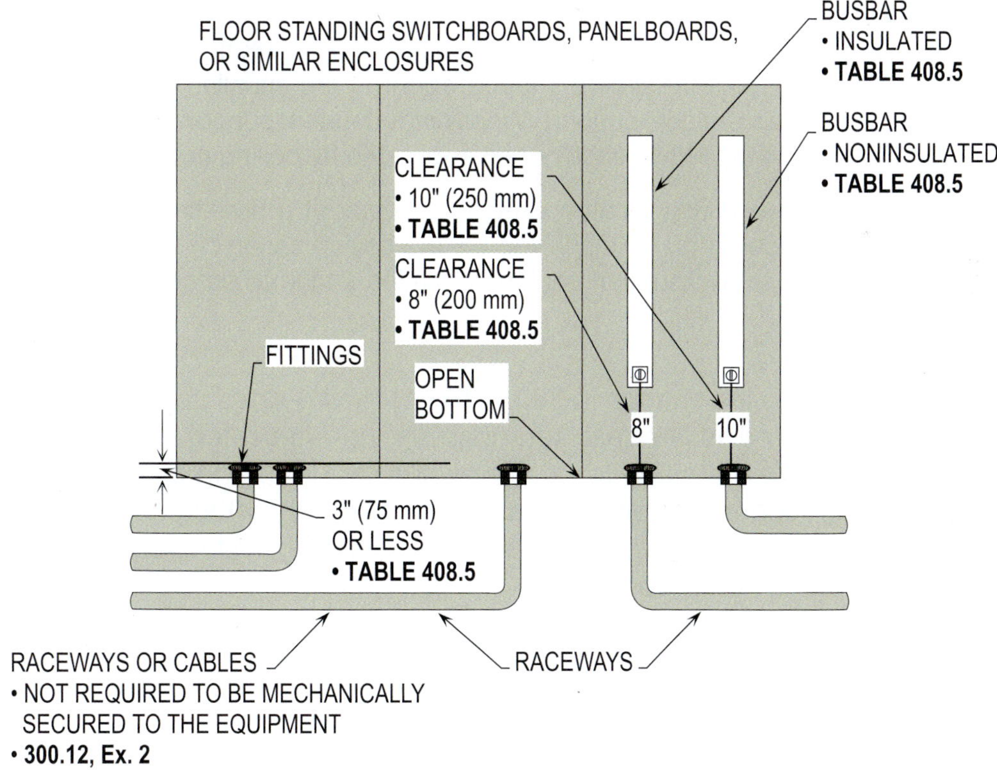

MECHANICAL CONTINUITY - RACEWAYS AND CABLES
300.12, Exception No. 2

Purpose of Change: A new exception has been added to clarify that raceways and cables shall not be required to be mechanically secured to the equipment where installed into the bottom of open bottom equipment.

NEC Ch. 3 - Article 300
Part I - 300.11(A)(2)

Type of Change	Panel Action	UL	UL 508	API 500 - 1997	API 505 - 1997	OSHA - 1994
Revision	Accept	-	-	-	-	-
ROP	**ROC**	**NFPA 70E - 2004**	**NFPA 70B - 2006**	**NFPA 79 - 2007**	**NFPA**	**NEMA**
pg. 268 # 3-67a	pg. - # -	-	-	-	-	-
log: CP 300 CMP: 3	log: -	Submitter: Code Making Panel 3		2005 NEC: 300.11(A)(2)		IEC: -

2005 NEC - 300.11 Securing and Supporting.

(A) Secured in Place. Raceways, cable assemblies, boxes, cabinets, and fittings shall be securely fastened in place. Support wires that do not provide secure support shall not be permitted as the sole support. Support wires and associated fittings that provide secure support and that are installed in addition to the ceiling grid support wires shall be permitted as the sole support. Where independent support wires are used, they shall be secured at both ends. Cables and raceways shall not be supported by ceiling grids.

(2) Non–Fire-Rated Assemblies. Wiring located within the cavity of a non–fire-rated floor–ceiling or roof–ceiling assembly shall not be secured to, or supported by, the ceiling assembly, including the ceiling support wires. An independent means of secure support shall be provided.

2008 NEC - 300.11 Securing and Supporting.

(A) Secured in Place. Raceways, cable assemblies, boxes, cabinets, and fittings shall be securely fastened in place. Support wires that do not provide secure support shall not be permitted as the sole support. Support wires and associated fittings that provide secure support and that are installed in addition to the ceiling grid support wires shall be permitted as the sole support. Where independent support wires are used, they shall be secured at both ends. Cables and raceways shall not be supported by ceiling grids.

(2) Non–Fire-Rated Assemblies. Wiring located within the cavity of a non–fire-rated floor–ceiling or roof–ceiling assembly shall not be secured to, or supported by, the ceiling assembly, including the ceiling support wires. An independent means of secure support shall be provided <u>and shall be permitted to be attached to the assembly</u>.

Author's Comment: Similar to the change in **300.11(A)(1)** for fire-rated assemblies in the 2005 NEC, additional text has been added to the last sentence for non-fire-rated assemblies clarifying that independent support wires are permitted to be connected to the ceiling assembly at the grid. This change ensures that independent support wires are permitted to be attached at both the top and the bottom of the ceiling cavity. By permitting the raceway support to be attached at the grid, the support wires will not be able to trapeze back and forth.

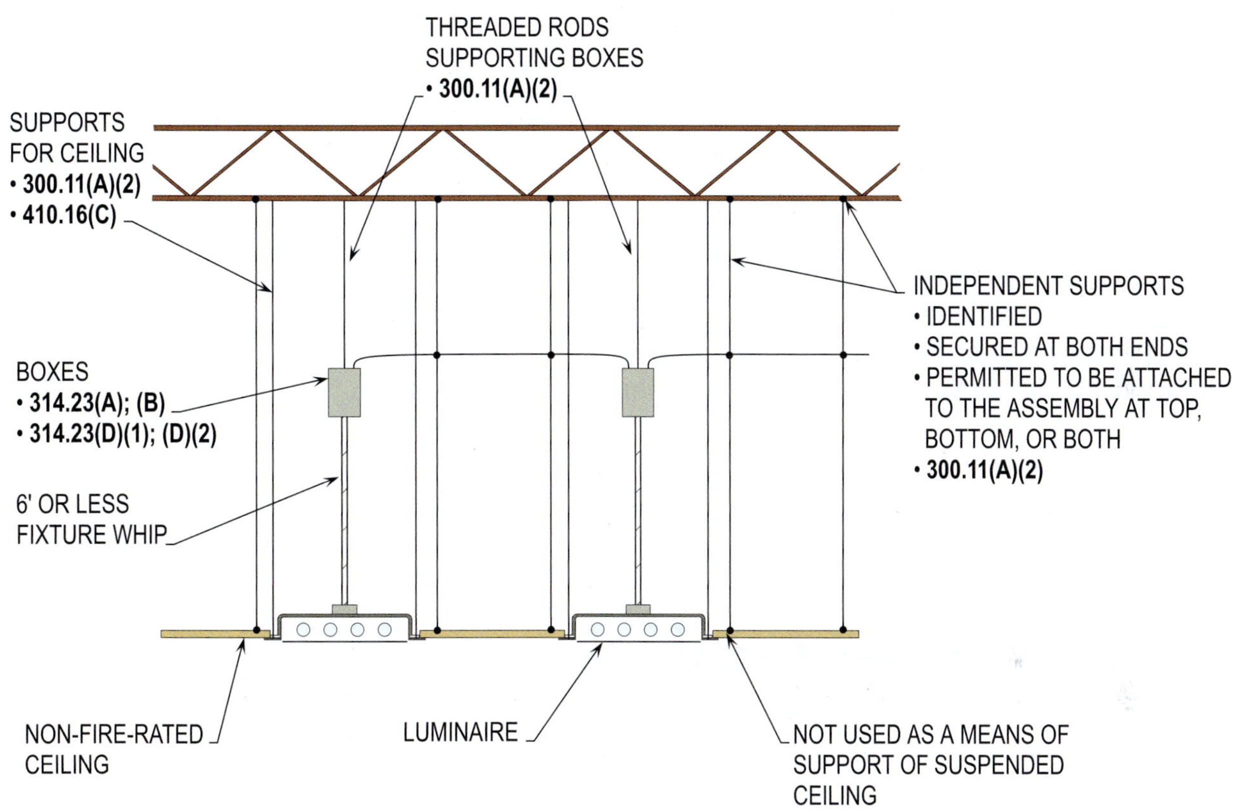

NON-FIRE-RATED ASSEMBLIES
300.11(A)(2)

Purpose of Change: This revision clarifies that independent supports for electrical wiring shall be permitted to be attached to the assembly.

NEC Ch. 3 - Article 300
Part II - Table 300.50, Note 3

Type of Change	Panel Action	UL	UL 508	API 500 - 1997	API 505 - 1997	OSHA - 1994
New Note	Accept in Principle	-	-	-	-	-
ROP	**ROC**	**NFPA 70E - 2004**	**NFPA 70B - 2006**	**NFPA 79 - 2007**	**NFPA**	**NEMA - 2005**
pg. 278 # 3-105	pg. - # -	-	-	-	-	RV 2
log: 3038 CMP: 3	log: -	Submitter: Melvin K. Sanders		2005 NEC: Table 300.50, Notes		IEC: -

2008 NEC - Table 300.50 Minimum Cover[a] Requirements

Notes:

3. In industrial establishments, where conditions of maintenance and supervision ensure that qualified persons will service the installation, the minimum cover requirements, for other than rigid metal conduit and intermediate metal conduit, shall be permitted to be reduced 150 mm (6 inches) for each 50 mm (2 inches) of concrete or equivalent placed entirely within the trench over the underground installation.

Author's Comment: New **Note 3** was added to **Table 300.50** to permit industrial establishments with qualified personnel to reduce the minimum cover requirements by 6 inches for each 2 inches of concrete or equivalent placed entirely within the trench over the underground installation for all underground wiring methods in the Table except rigid and intermediate metal conduit.

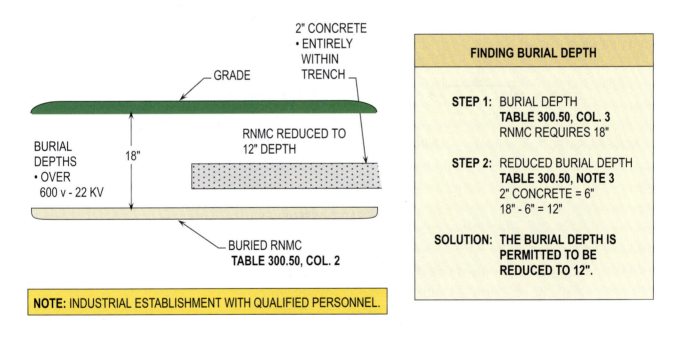

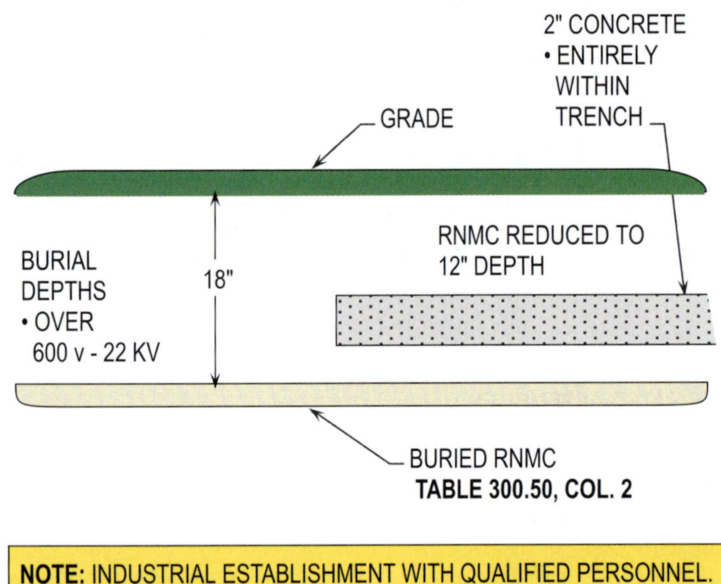

FINDING BURIAL DEPTH

STEP 1: BURIAL DEPTH
TABLE 300.50, COL. 3
RNMC REQUIRES 18"

STEP 2: REDUCED BURIAL DEPTH
TABLE 300.50, NOTE 3
2" CONCRETE = 6"
18" - 6" = 12"

SOLUTION: **THE BURIAL DEPTH IS PERMITTED TO BE REDUCED TO 12".**

MINIMUM COVER[a] REQUIREMENTS
TABLE 300.50, NOTE 3

Purpose of Change: A new Note 3 has been added to **Table 300.50** to permit industrial establishments with qualified persons to reduce the minimum cover requirements by 6 in. for each 2 in. of concrete. A similar rule was permitted for many NEC cycles all installations over 600 volts. This was deleted for the 2005 NEC but industrial facilities asked Panel 3 to reinstate the rule for very limited industrial facilities.

NEC Ch. 3 - Article 300
Part II - Table 300.50[d]

Type of Change		Panel Action		UL	UL 508	API 500 - 1997	API 505 - 1997	OSHA - 1994
New Note		Accept in Principle		-	-	-	-	-
ROP		ROC		NFPA 70E - 2004	NFPA 70B - 2006	NFPA 79 - 2007	NFPA	NEMA - 2005
pg. 278	# 3-108	pg. -	# -	-	-	-	-	RV 2
log: 679	CMP: 3	log: -	Submitter: Jamie McNamara			2005 NEC: Table 300.50, Notes		IEC: -

2008 NEC - Table 300.50 Minimum Cover[a] Requirements

Specific Footnotes:

[d] Underground direct-buried cables that are not encased or protected by concrete and are buried 750 mm (30 inches) or more below grade shall have their location identified by a warning ribbon that is placed in the trench at least 300 mm (12 inches) above the cables.

Author's Comment: Direct-buried cables at over 600 volts may be damaged where excavation is being done at existing installations and where the cables are damaged, people are subject to injury. High voltage cables installed in rigid nonmetallic conduit or raceways installed under concrete slabs at least 4 inches thick are less likely to be damaged during excavation. Placing a warning ribbon over direct buried cables should provide a warning for excavators.

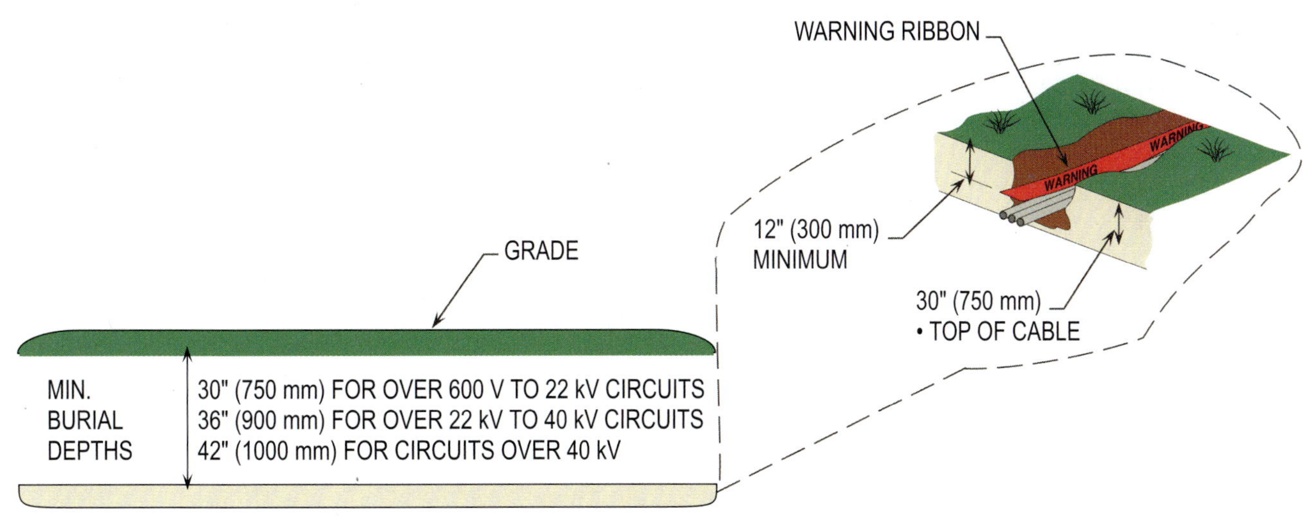

MINIMUM COVER[a] REQUIREMENTS
TABLE 300.50[d]

Purpose of Change: A new superscript [d] has been added to **Table 300.50** to require direct-buried cables that are not concrete-encased, protected by concrete or sleeved with proper protection and are buried at least 30 in. (750 mm) or more below grade must have location identified by warning ribbon that is installed in the trench at least 12 in. (300 mm) above the cables.

NEC Ch. 3 - Article 310
Part - 310.7, Exception No. 2

39

Type of Change	Panel Action	UL	UL 508	API 500 - 1997	API 505 - 1997	OSHA - 1994
New Exception	-	-	-	-	-	-
ROP	**ROC**	**NFPA 70E - 2004**	**NFPA 70B - 2006**	**NFPA 79 - 2007**	**NFPA**	**NEMA - 2005**
pg. 285 # 6-25	pg. 181 # 6-36	-	-	-	-	RV 2
log: 2744 CMP: 6	log: 38 Submitter: Daniel Baker			2005 NEC: -		IEC: -

2008 NEC - 310.7 Direct Burial Conductors.

Conductors used for direct burial applications shall be of a type identified for such use.

Cables rated above 2000 volts shall be shielded.

Exception No. 1: Nonshielded multiconductor cables rated 2001 - 2400 volts shall be permitted if the cable has an overall metallic sheath or armor.

The metallic shield, sheath, or armor shall be connected to a grounding electrode conductor, grounding busbar, or a grounding electrode.

Exception No. 2. Airfield lighting cable used in series circuits that are rated up to 5000 volts and are powered by regulators shall be permitted to be nonshielded.

FPN to Exception No. 2. Federal Aviation Adminstration (FAA) Advisory Circulars (ACs) provide additional practices and methods for airport lighting.

Author's Comment: A new Exception No. 2 has been added to 310.7 to clarify that ungrounded airfield lighting series circuits may be unshielded when rated up to 5000 volts and powered by regulators.

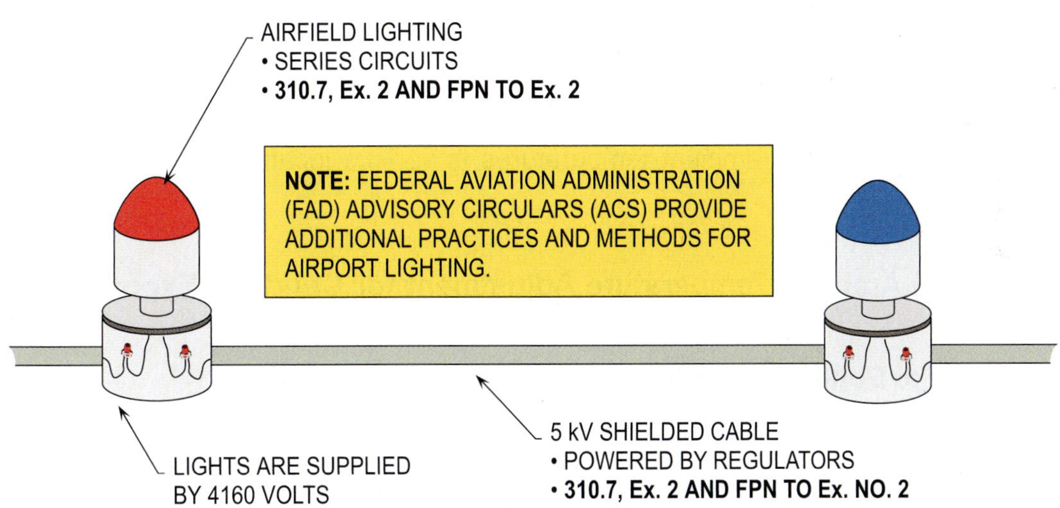
AIRFIELD LIGHTING
• SERIES CIRCUITS
• 310.7, Ex. 2 AND FPN TO Ex. 2
NOTE: FEDERAL AVIATION ADMINISTRATION (FAD) ADVISORY CIRCULARS (ACS) PROVIDE ADDITIONAL PRACTICES AND METHODS FOR AIRPORT LIGHTING.
LIGHTS ARE SUPPLIED BY 4160 VOLTS
5 kV SHIELDED CABLE
• POWERED BY REGULATORS
• 310.7, Ex. 2 AND FPN TO Ex. NO. 2

DIRECT BURIAL CONDUCTORS
310.7, Exception No. 2

Purpose of Change: A new exception and FPN has been added to clarify that ungrounded airfield lighting series circuits may be unshielded where rated up to 5000 volts and powered by regulators.

NEC Ch. 3 - Article 310
Part - 310.15(B)(2)(c)

Type of Change		Panel Action		UL	UL 508	API 500 - 1997	API 505 - 1997	OSHA - 1994
New Subdivision		Accept in Principle		-	-	-	-	-
ROP		ROC		NFPA 70E - 2004	NFPA 70B - 2006	NFPA 79 - 2007	NFPA	NEMA
pg. 295	# 6-51	pg. 183	# 6-45	-	-	-	-	-
log: 3150	CMP: 6	log: 1070	Submitter: Travis Lindsey			2005 NEC: 310.15(B)(2)		IEC: -

2008 NEC - 310.15 Ampacities for Conductors Rated 0–2000 Volts.

(B) Tables.

(2) Adjustment Factors.

(c) Conduits Exposed to Sunlight on Rooftops. Where conductors or cables are installed in conduits exposed to direct sunlight on or above rooftops, the adjustments shown in Table 310.15(B)(2)(c) shall be added to the outdoor temperature to determine the applicable ambient temperature for application of the correction factors in Table 310.16 and Table 310.18.

FPN: One source for the average ambient temperatures in various locations is the ASHRAE Handbook - Fundamentals.

Table 310.15(B)(2)(c) Ambient Temperature Adjustment for Conduits Exposed to Sunlight On or Above Rooftops

Distance Above Roof to Bottom of Conduit	Temperature Adder C°	Temperature Adder F°
0 - 13 mm (1/2 in.)	33	60
Above 13 mm (1/2 in.) - 90 mm (3 1/2 in.)	22	40
Above 90 mm (3 1/2 in.) - 300 mm (12 in.)	17	30
Above 300 mm (12 in.) - 900 mm (36 in.)	14	25

FPN to Table 310.15(B)(2)(c): The temperature adders in Table 310.15(B)(2)(c) are based on the results of averaging the ambient temperatures.

Author's Comment: Section **310.10** stipulates that no conductor shall be used in such a manner that its operating temperature exceeds that designated for the type of insulated conductor involved. The air inside conduit in direct sunlight is significantly hotter than the surrounding air, and appropriate ampacity corrections must be made to comply with **310.10**. This new subsection and table will provide the necessary ampacity derating requirements for conduits containing conductors installed on rooftops where exposed to the sun. Comment 6-45 adds a **Fine Print Note** that explains that the temperature adders in **Table 310.15(B)(2)(c)** are based on the averaging of ambient temperatures. Cables that were exposed to direct sunlight were not added to the new table since the text data only dealt with conduits exposed to sunlight on rooftops and not to multiconductor cables. Comment 6-54 added a **Fine Print Note** that provides a reference to the ASHRAE Fundamentals Handbook as a source for average ambient temperatures for various locations.

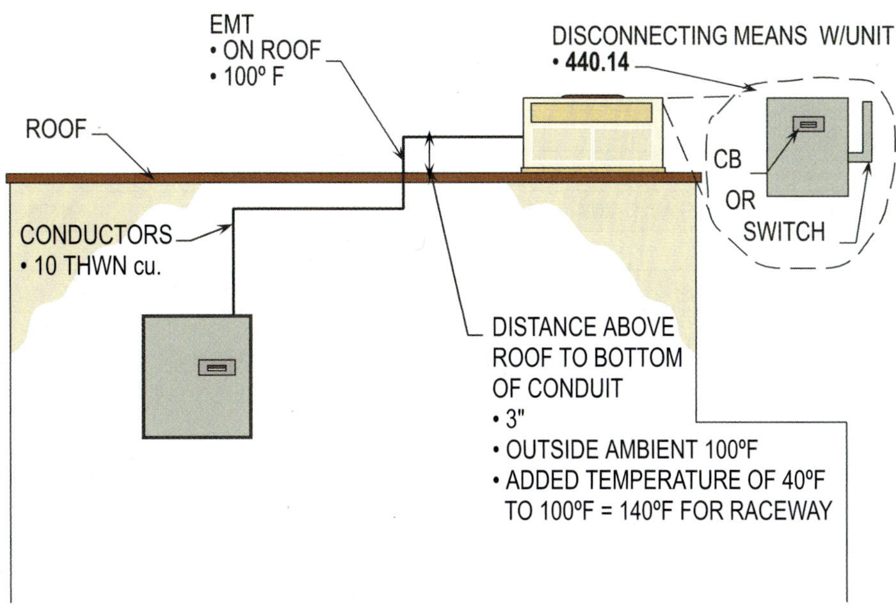

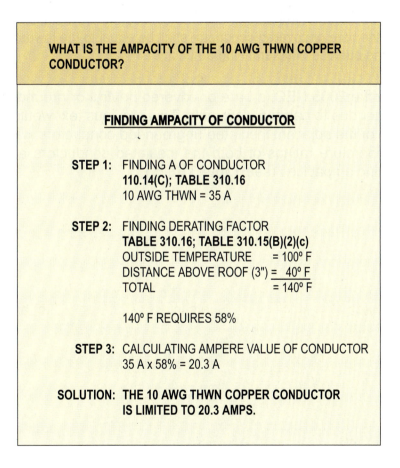

WHAT IS THE AMPACITY OF THE 10 AWG THWN COPPER CONDUCTOR?

FINDING AMPACITY OF CONDUCTOR

STEP 1: FINDING A OF CONDUCTOR
110.14(C); TABLE 310.16
10 AWG THWN = 35 A

STEP 2: FINDING DERATING FACTOR
TABLE 310.16; TABLE 310.15(B)(2)(c)
OUTSIDE TEMPERATURE = 100° F
DISTANCE ABOVE ROOF (3") = 40° F
TOTAL = 140° F

140° F REQUIRES 58%

STEP 3: CALCULATING AMPERE VALUE OF CONDUCTOR
35 A x 58% = 20.3 A

SOLUTION: **THE 10 AWG THWN COPPER CONDUCTOR IS LIMITED TO 20.3 AMPS.**

CONDUITS EXPOSED TO SUNLIGHT ON ROOFTOPS
310.15(B)(2)(c)

Purpose of Change: A new subdivision has been added to require an additional ambient temperature to be applied for conduits exposed to sunlight on rooftops.

NEC Ch. 3 - Article 310
Part - 310.15(B)(3)

Type of Change	Panel Action	UL	UL 508	API 500 - 1997	API 505 - 1997	OSHA - 1994
Revision	Accept in Principle in Part	-	-	-	-	-
ROP	**ROC**	**NFPA 70E - 2004**	**NFPA 70B - 2006**	**NFPA 79 - 2007**	**NFPA**	**NEMA**
pg. 296 # 6-57	pg. - # -	-	-	-	-	-
log: 1065 CMP: 6	log: - Submitter: Daniel Leaf			2005 NEC: 310.15(B)(3)		IEC: -

2005 NEC - 310.15 Ampacities for Conductors Rated 0–2000 Volts.

(B) Tables.

(3) Bare or Covered Conductors. Where bare or covered conductors are ~~used~~ with insulated conductors, ~~their allowable ampacities shall be limited to those permitted for the adjacent~~ insulated conductors.

2008 NEC - 310.15 Ampacities for Conductors Rated 0–2000 Volts.

(B) Tables.

(3) Bare or Covered Conductors. Where bare or covered conductors are <u>installed</u> with insulated conductors, <u>the temperature rating of the bare or covered conductor shall be equal to the lowest temperature rating of the</u> insulated conductors <u>for the purpose of determining ampacity</u>.

Author's Comment: In the 2005 NEC, a bare or covered neutral or grounded conductor was often required to be sized larger due to harmonic currents. This previous text would then have the effect of also increasing the size of the adjacent insulated ungrounded conductors. The text in the 2008 NEC will now match the temperature ratings of both the insulated conductors and the bare or covered conductors for determining ampacities (see **Table 310.21**).

BARE OR COVERED CONDUCTORS
310.15(B)(3)

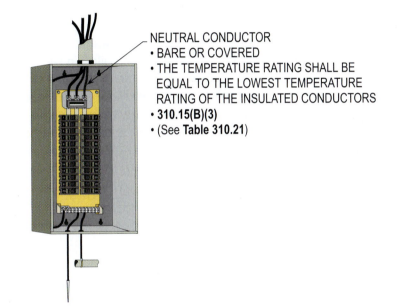

Purpose of Change: This revision clarifies that bare or covered conductors installed with insulated conductors shall have a temperature rating equal to the lowest temperature rating of the insulated conductors for the purpose of determining ampacity.

NEC Ch. 3 - Article 310
Part - 310.15(B)(6)

41

Type of Change	Panel Action	UL	UL 508	API 500 - 1997	API 505 - 1997	OSHA - 1994
Revision	Accept	-	-	-	-	-
ROP	**ROC**	**NFPA 70E - 2004**	**NFPA 70B - 2006**	**NFPA 79 - 2007**	**NFPA**	**NEMA**
pg. 298 # 6-61	pg. 188 # 6-63	-	-	-	-	-
log: 194 CMP: 6	log: 1915 Submitter: Frederic P. Hartwell			2005 NEC: 310.15(B)(6)		IEC: -

2005 NEC - 310.15 Ampacities for Conductors Rated 0–2000 Volts.

(B) Tables.

(6) 120/240-Volt, 3-Wire, Single-Phase Dwelling Services and Feeders. For individual dwelling units of one family, two-family, and multifamily dwellings, conductors, as listed in Table 310.15(B)(6), shall be permitted as 120/240-volt, 3-wire, single-phase service-entrance conductors, service lateral conductors, and feeder conductors that serve as the main power feeder to each dwelling unit and are installed in raceway or cable with or without an equipment grounding conductor. For application of this section, the main power feeder shall be the feeder(s) between the main disconnect and the ~~lighting and appliance branch-circuit~~ panelboard~~(s)~~. The feeder conductors to a dwelling unit shall not be required to have an allowable ampacity rating greater than their service-entrance conductors. The grounded conductor shall be permitted to be smaller than the ungrounded conductors, provided the requirements of 215.2, 220.61, and 230.42 are met.

2008 NEC - 310.15 Ampacities for Conductors Rated 0–2000 Volts.

(B) Tables.

(6) 120/240-Volt, 3-Wire, Single-Phase Dwelling Services and Feeders. For individual dwelling units of one family, two-family, and multifamily dwellings, conductors, as listed in Table 310.15(B)(6), shall be permitted as 120/240-volt, 3-wire, single-phase service-entrance conductors, service lateral conductors, and feeder conductors that serve as the main power feeder to each dwelling unit and are installed in raceway or cable with or without an equipment grounding conductor. For application of this section, the main power feeder shall be the feeder(s) between the main disconnect and the panelboard that supplies, either by branch circuits, or by feeders, or both, all loads that are part or associated with the dwelling unit. The feeder conductors to a dwelling unit shall not be required to have an allowable ampacity rating greater than their service-entrance conductors. The grounded conductor shall be permitted to be smaller than the ungrounded conductors, provided the requirements of 215.2, 220.61, and 230.42 are met.

Author's Comment. Dwelling unit subpanel loads do not present the same diversity as dwelling unit panels serving the entire dwelling unit, and thereby undercut one of the traditional supporting assumptions underlying these allowances. It is the panel's intent that this allowance apply only to conductors carrying 100 percent of the dwelling unit's diversified load. The comment clarifies that the dwelling could have a subpanel, as long as it is fed from the main panel so that the load diversity of the subpanel loads is included within the load of the main feeder. The comment wording also includes associated loads, as would be the case where there was a detached garage. The comment deletes the terminology " lighting and appliance branch-circuit" because CMP 9 removed this from **Article 408** by Proposal 9- 117.

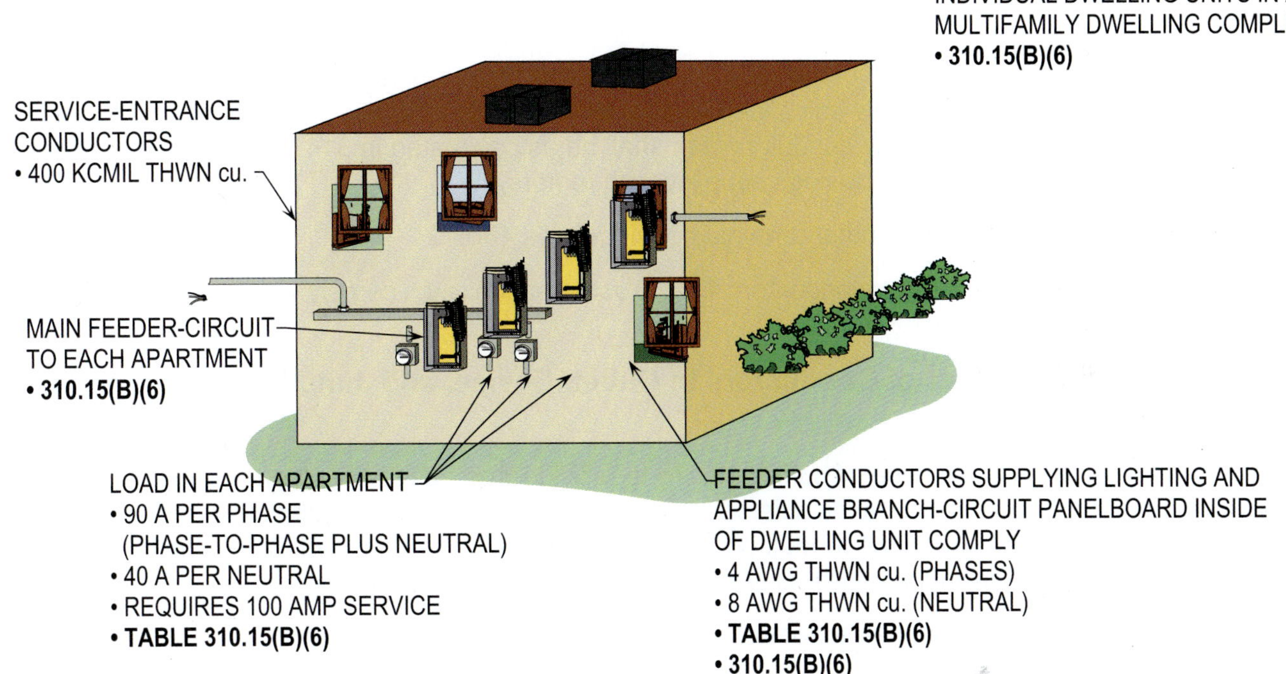

120/240 VOLTS, 3-WIRE, SINGLE-PHASE
DWELLING SERVICES AND FEEDERS
310.15(B)(6)

Purpose of Change: This revision clarifies that the main power feeder can be the feeder(s) between the main disconnect and the panelboards (Loads associated with the dwelling unit).

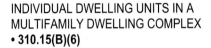

NEC Ch. 3 - Article 314
Part II - 314.16 and FPN

Type of Change		Panel Action		UL	UL 508	API 500 - 1997	API 505 - 1997	OSHA - 1994
Revision		Accept		-	-	-	-	-
ROP		**ROC**		**NFPA 70E - 2004**	**NFPA 70B - 2006**	**NFPA 79 - 2007**	**NFPA**	**NEMA - 2003**
pg. 306	# 9-26	pg. 191	# 9-12	-	19.7	13.3	-	250
log: 2923	CMP: 9	log: 963	Submitter: Peter D. Noval, Jr.			2005 NEC: 314.16		IEC: -

2005 NEC - 314.16 Number of Conductors in Outlet, Device, and Junction Boxes, and Conduit Bodies.

Boxes and conduit bodies, shall be of sufficient size to provide free space for all enclosed conductors. In no case shall the volume of the box, as calculated in 314.16(A), be less than the fill calculation as calculated in 314.16(B). The minimum volume for conduit bodies shall be as calculated in 314.16(C). The provisions of this section shall not apply to terminal housings supplied with motors.

FPN: For volume requirements of motor terminal housings, see 430.12.

Boxes and conduit bodies enclosing conductors 4 AWG or larger shall also comply with the provisions of 314.28.

2008 NEC - 314.16 Number of Conductors in Outlet, Device, and Junction Boxes, and Conduit Bodies.

Boxes and conduit bodies, shall be of sufficient size to provide free space for all enclosed conductors. In no case shall the volume of the box, as calculated in 314.16(A), be less than the fill calculation as calculated in 314.16(B). The minimum volume for conduit bodies shall be as calculated in 314.16(C). The provisions of this section shall not apply to terminal housings supplied with motors or generators.

FPN: For volume requirements of motor or generator terminal housings, see 430.12.

Boxes and conduit bodies enclosing conductors 4 AWG or larger shall also comply with the provisions of 314.28.

Author's Comment: Generators were added to the text in the last sentence to make it clear that Section **314.16** does not apply to generator terminal housings. The change in the **FPN** sends the user to **430.12** for both motor and generator terminal housing volume requirements. The comment assured that the final sentence of the 2005 NEC text in this section was to immediately follow the FPN and included with the modified text.

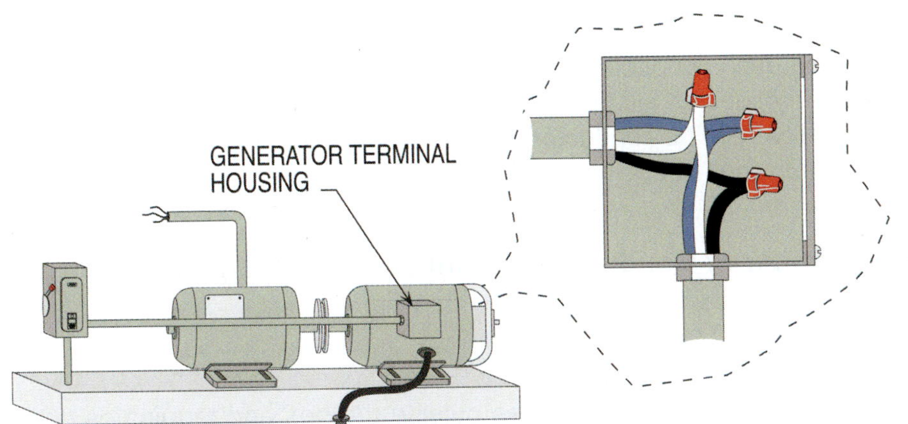
GENERATOR TERMINAL
HOUSING
TO DETERMINE MIN. SIZE OF GENERATOR TERMINAL HOUSING, COMPARE FLA OF GENERATOR TO MOTOR HP RATINGS IN **TABLES 430.247 THRU 430.250** AS PERMITTED IN **445.17.**

NUMBER OF CONDUCTORS IN OUTLET, DEVICE, AND JUNCTION BOXES, AND CONDUIT BODIES
314.16 and FPN

Purpose of Change: This revision clarifies that **314.16** does not apply to generator terminal housings.

NEC Ch. 3 - Article 314
Part II - 314.16(B)(1)

Type of Change	Panel Action	UL	UL 508	API 500 - 1997	API 505 - 1997	OSHA - 1994
Revision	Accept in Part	-	-	-	-	-
ROP	**ROC**	**NFPA 70E - 2004**	**NFPA 70B - 2006**	**NFPA 79 - 2007**	**NFPA**	**NEMA - 2003**
pg. 307 # 9-29	pg. - # -	-	19.7	13.3	-	250
log: 2488 CMP: 9	log: -	Submitter: Sukanta Sengupta		2005 NEC: 314.16(B)(1)		IEC: -

2005 NEC - 314.16 Number of Conductors in Outlet, Device, and Junction Boxes, and Conduit Bodies.

(B) Box Fill Calculations.

(1) Conductor Fill. Each conductor that originates outside the box and terminates or ~~is~~ spliced within the box shall be counted once, and each conductor that passes through the box without splice or termination shall be counted once. ~~A looped~~ unbroken conductor not less than twice the minimum length required for free conductors in 300.14 shall be counted twice. The conductor fill shall be calculated using Table 314.16(B). A conductor, no part of which leaves the box, shall not be counted.

2008 NEC - 314.16 Number of Conductors in Outlet, Device, and Junction Boxes, and Conduit Bodies.

(B) Box Fill Calculations.

(1) Conductor Fill. Each conductor that originates outside the box and terminates or spliced within the box shall be counted once, and each conductor that passes through the box without splice or termination shall be counted once. Each loop or coil of unbroken conductor not less than twice the minimum length required for free conductors in 300.14 shall be counted twice. The conductor fill shall be calculated using Table 314.16(B). A conductor, no part of which leaves the box, shall not be counted.

Author's Comment: "Each loop or coil" was substituted for "A loop" in the second sentence because it clarifies the need to account for all loops in the box. Electricians were installing a loop or a coil and not counting the conductor as two, then cutting it later for termination at a device. This corrects that problem.

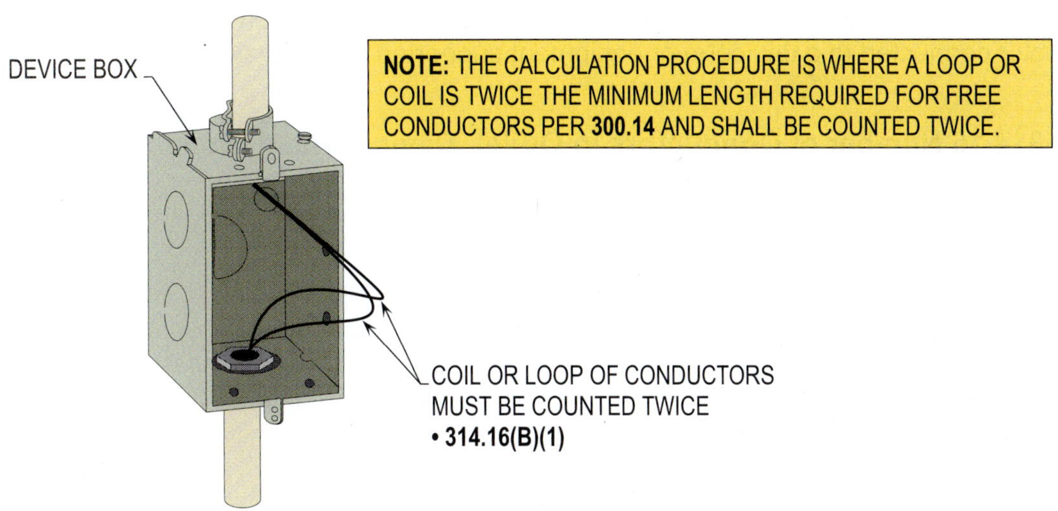

DEVICE BOX
NOTE: THE CALCULATION PROCEDURE IS WHERE A LOOP OR COIL IS TWICE THE MINIMUM LENGTH REQUIRED FOR FREE CONDUCTORS PER 300.14 AND SHALL BE COUNTED TWICE.
COIL OR LOOP OF CONDUCTORS MUST BE COUNTED TWICE
• 314.16(B)(1)

CONDUCTOR FILL
314.16(B)(1)

Purpose of Change: This revision clarifies that each loop or coil of conductors shall be counted twice.

NEC Ch. 3 - Article 314
Part II - 314.16(B)(4)

Type of Change	Panel Action	UL	UL 508	API 500 - 1997	API 505 - 1997	OSHA - 1994
Revision	Accept	-	-	-	-	-
ROP	**ROC**	**NFPA 70E - 2004**	**NFPA 70B - 2006**	**NFPA 79 - 2007**	**NFPA**	**NEMA - 2003**
pg. 307 # 9-31	pg. 191 # 9-16	-	19.7	13.3	-	250
log: 3398 CMP: 9	log: 1916	Submitter: Frederic P. Hartwell		2005 NEC: 310.16(B)(4)		IEC: -

2005 NEC - 314.16 Number of Conductors in Outlet, Device, and Junction Boxes, and Conduit Bodies.

(B) Box Fill Calculations.

(4) Device or Equipment Fill. For each yoke or strap containing one or more devices or equipment, a double volume allowance in accordance with Table 314.16(B) shall be made for each yoke or strap based on the largest conductor connected to a device(s) or equipment supported by that yoke or strap.

2008 NEC - 314.16 Number of Conductors in Outlet, Device, and Junction Boxes, and Conduit Bodies.

(B) Box Fill Calculations.

(4) Device or Equipment Fill. For each yoke or strap containing one or more devices or equipment, a double volume allowance in accordance with Table 314.16(B) shall be made for each yoke or strap based on the largest conductor connected to a device(s) or equipment supported by that yoke or strap. A device or utilization equipment wider than a single 50 mm (2 in.) device box as described in Table 314.16(A) shall have double volume allowances provided for each gang required for mounting.

Author's Comment: A large device that cannot be mounted in a conventional single-gang box and must be installed in a box with multiple gangs should carry the conductor allowances that multiple devices in adjacent gangs already carry. A 3-pole 3-wire non-grounding dryer receptacle, for example, installed prior to the 1996 NEC, could be installed in a single-gang box, though a decision to use a two-gang box could have been a design choice on the part of the installer. However, the current version of this receptacle, a 3-pole 4-wire grounding-type dryer receptacle, will not mount in a single-gang box, and would be subject to this new requirement.

The comment modified the proposed last sentence in **314.16(B)(4)** to require a device or utilization equipment that is larger than a single 2-inch-wide device box to have double volume allowance for each gang required to mount the device.

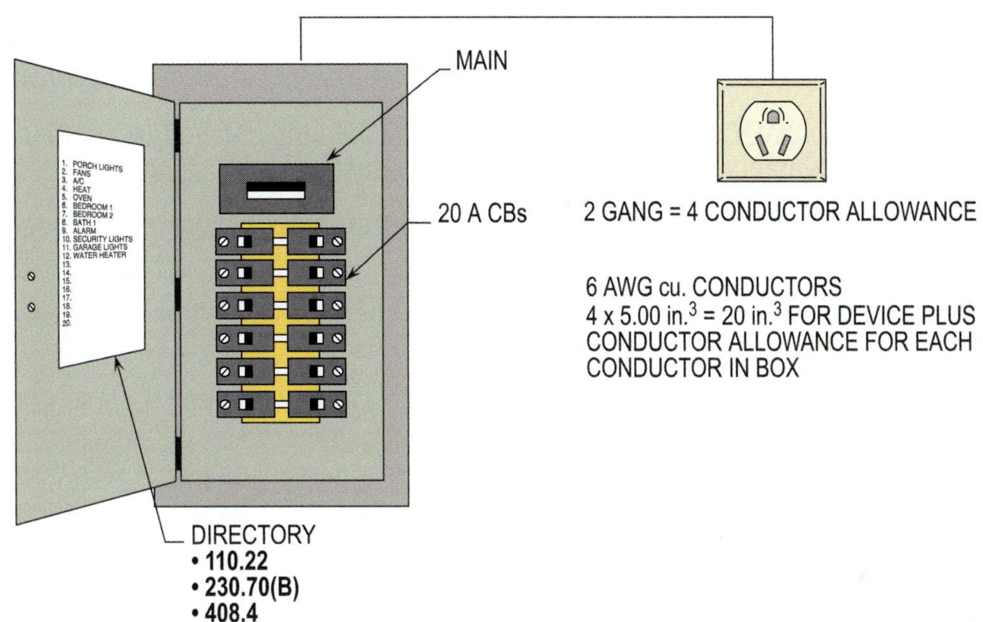

DEVICE OR EQUIPMENT FILL
314.16(B)(4)

Purpose of Change: This revision clarifies that a device or utilization equipment wider than a single (2 in.) device box shall have double volume allowance provided for each gang.

NEC Ch. 3 - Article 314
Part II - 314.24(A) through (C)

Type of Change	Panel Action	UL	UL 508	API 500 - 1997	API 505 - 1997	OSHA - 1994
Revision	Accept	514	-	-	-	-
ROP	**ROC**	**NFPA 70E - 2004**	**NFPA 70B - 2006**	**NFPA 79 - 2007**	**NFPA**	**NEMA - 2003**
pg. 312 # 9-52	pg. 192 # 9-20	-	19.7	13.3	-	250
log: 3322 CMP: 9	log: 1414	Submitter: Frederic P. Hartwell		2005 NEC: 314.24		IEC: -

2005 NEC - 314.24 Depth of ~~Outlet~~ Boxes.

No box shall have an internal depth of less than 12.7 mm (1/2 in.). Boxes intended to enclose flush devices shall have an internal depth of not less than 23.8 mm (15/16 in.).

2008 NEC - 314.24 Minimum Depth of Boxes for Outlets, Devices, and Utilization Equipment.

Outlet and device boxes shall have sufficient depth to allow equipment installed within them to be mounted properly and with sufficient clearance to prevent damage to conductors within the box.

(A) Outlet Boxes Without Enclosed Devices or Utilization Equipment. No box shall have an internal depth of less than 12.7 mm (1/2 in.).

(B) Outlet and Device Boxes With Enclosed Devices. Boxes intended to enclose flush devices shall have an internal depth of not less than 23.8 mm (15/16 in.).

(C) Utilization Equipment. Outlet and device boxes that enclose utilization equipment shall have a minimum internal depth that accomodates the rearward projection of the equipment and the size of the conductors that supply the equipment. The internal depth shall include, where used, that of any extension boxes, plaster rings, or raised covers, the internal depth shall comply with all applicable provisions of (C)(1) through (C)(5).

(1) Large Equipment. Boxes that enclose utilization equipment that projects more than 48 mm (1 7/8 in.) rearward from the mounting plane of the box shall have a depth that is not less than the depth of the equipment plus 6 mm (1/4 in.).

(2) Conductors Larger Than 4 AWG. Boxes that enclose utilization equipment supplied by conductors larger than 4 AWG shall be identified for their specific function.

(3) Conductors 8, 6, or 4 AWG. Boxes that enclose utilization equipment supplied by 8, 6, or 4 AWG conductors shall have an internal depth that is not less than 52.4 mm (2 1/16 in.).

(4) Conductors 12 or 10 AWG. Boxes that enclose utilization equipment supplied by 12 or 10 AWG conductors shall have an internal depth that is not less than 30.2 mm (1 3/16 in.). Where the equipment projects rearward from the mounting plane of the box by more than 25 mm (1 in.), the box shall have a depth not less than that of the equipment plus 6 mm (1/4 in.).

(5) Conductors 14 AWG and Smaller. Boxes that enclose equipment supplied by 14 AWG or smaller conductors shall have a depth that is not less than 23.8 mm (15/16 in.).

Exception to (C)(1) through (C)(5): Utilization equipment that is listed to be installed with specified boxes shall be permitted.

Author's Comment: This rewrite of **314.24** is meant to provide a solution to the problem of installing large switches, receptacles, and utilization equipment into outlet and device boxes with enough space to also accommodate the installation of conductors into the enclosures.

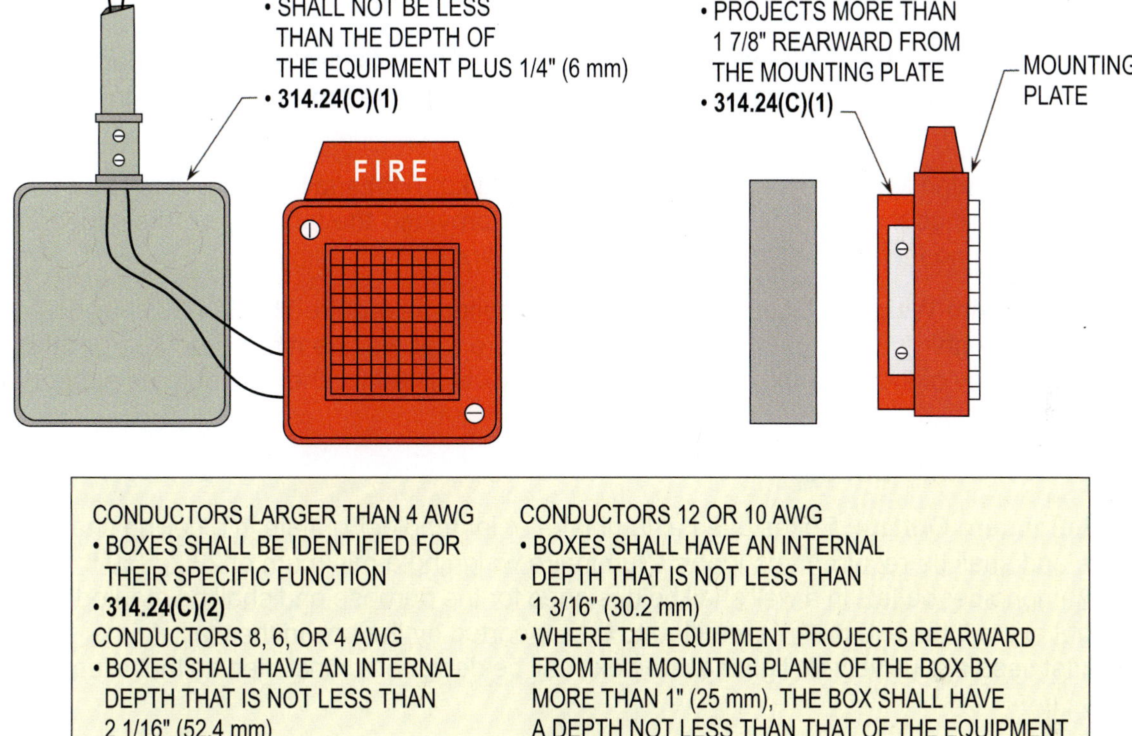
DEPTH OF BOX
• SHALL NOT BE LESS THAN THE DEPTH OF THE EQUIPMENT PLUS 1/4" (6 mm)
• 314.24(C)(1)
FIRE
DEPTH
• PROJECTS MORE THAN 1 7/8" REARWARD FROM THE MOUNTING PLATE
• 314.24(C)(1)
MOUNTING PLATE
CONDUCTORS LARGER THAN 4 AWG
• BOXES SHALL BE IDENTIFIED FOR THEIR SPECIFIC FUNCTION
• 314.24(C)(2)
CONDUCTORS 8, 6, OR 4 AWG
• BOXES SHALL HAVE AN INTERNAL DEPTH THAT IS NOT LESS THAN 2 1/16" (52.4 mm)
• 314.24(C)(3)
CONDUCTORS 12 OR 10 AWG
• BOXES SHALL HAVE AN INTERNAL DEPTH THAT IS NOT LESS THAN 1 3/16" (30.2 mm)
• WHERE THE EQUIPMENT PROJECTS REARWARD FROM THE MOUNTNG PLANE OF THE BOX BY MORE THAN 1" (25 mm), THE BOX SHALL HAVE A DEPTH NOT LESS THAN THAT OF THE EQUIPMENT PLUS 1/4" (6 mm)
• 314.24(C)(4)
CONDUCTORS 14 AWG AND SMALLER
• BOXES SHALL HAVE A DEPTH THAT IS NOT LESS THAN 15/16" (23.8 mm)
• 314.24(C)(5)

MINIMUM DEPTH OF BOXES FOR OUTLETS, DEVICES, AND UTILIZATION EQUIPMENT
314.24(A) through (C)

Purpose of Change: The rewrite provides requirements for the installation of large switches, receptacles, and utilization equipment into outlet and device boxes.

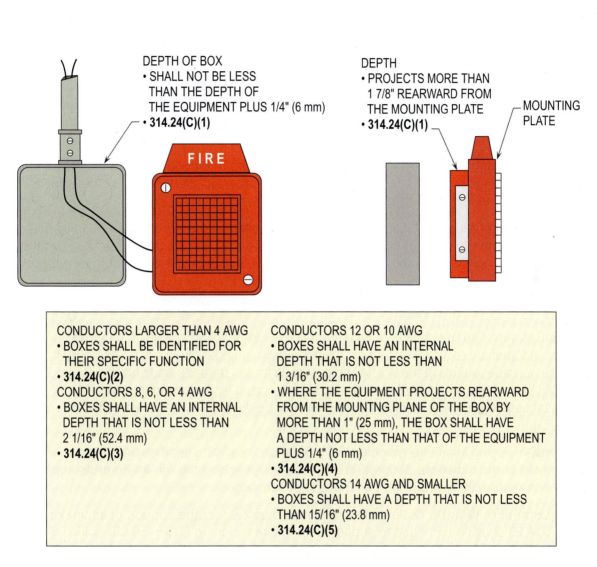

NEC Ch. 3 - Article 314
Part II - 314.27(A) and (B)

45

Type of Change	Panel Action	UL	UL 508	API 500 - 1997	API 505 - 1997	OSHA - 1994
Revision	Accept in Principle	514	-	-	-	-
ROP	**ROC**	**NFPA 70E - 2004**	**NFPA 70B - 2006**	**NFPA 79 - 2007**	**NFPA**	**NEMA**
pg. 314 # 9-56	pg. - # -	-	-	-	-	-
log: 2611 CMP: 9	log: -	Submitter: David H. Kendall		2005 NEC: 314.27(A); (B)		IEC: -

2005 NEC - 314.27 Outlet Boxes.

(A) Boxes at Luminaire Outlets. Boxes used at luminaire (lighting fixture) or lampholder outlets shall be designed for the purpose at every outlet used exclusively for lighting, the box shall be designed or installed so that a luminaire (lighting fixture) may be attached.

Exception: A wall-mounted luminaire weighing not more than 3 kg (6 lb) shall be permitted to be supported on other boxes or plaster rings that are secured to other boxes, provided the luminaire or its supporting yoke is secured to the box with no fewer than two No. 6 or larger screws.

(B) Maximum Luminaire Weight. Outlet boxes or fittings installed as required by 314.23 shall be permitted to support a luminaire weighing 23 kg (50 Lb) or less. A luminaire that weighs more than 23 kg (50 lb) shall be supported independently of the outlet box unless the outlet box is listed for the weight to be supported.

2008 NEC - 314.27 Outlet Boxes.

(A) Boxes at Luminaire Outlets. Boxes used at luminaire or lampholder outlets in a ceiling shall be designed for the purpose and shall be required to support a luminaire weighing a minimum of 23 kg (50 lb.). Boxes used at luminaire or lampholder outlets in a wall shall be designed for the purpose and shall be marked to indicate the maximum weight of the luminaire that is permitted to be supported by the box in the wall, if other than 23 kg (50 lb). At every outlet used exclusively for lighting, the box shall be designed or installed so that a luminaire may be attached.

Exception: A wall-mounted luminaire weighing not more than 3 kg (6 lb) shall be permitted to be supported on other boxes or plaster rings that are secured to other boxes, provided the luminaire or its supporting yoke is secured to the box with no fewer than two No. 6 or larger screws.

(B) Maximum Luminaire Weight. Outlet boxes or fittings designed for the support of luminaires and installed as required by 314.23 shall be permitted to support a luminaire weighing 23 kg (50 Lb) or less. A luminaire that weighs more than 23 kg (50 lb) shall be supported independently of the outlet box unless the outlet box is listed and marked for the maximum weight to be supported.

Author's Comment: This proposed change clarifies the requirements for listed boxes used for luminaire support. Currently, there are outlet boxes on the market that are listed and marked for luminaires weighing between 3 kg (6 lb) and 23 kg (50 lb) for ceiling applications. Panel 9 intends that a ceiling box for luminaire support shall be required to support a minimum of 23 kg (50 lb) or less. Markings for boxes between 3 kg (6 lb) and 23 kg (50 lb) were initially permitted for wall-mounted luminaires only and were not intended for ceiling-mounted luminaires. Homeowners may unknowingly change a ceiling luminaire to a heavier weight than the box is listed to handle. Section **314.27(B)** was revised to indicate that boxes listed for luminaires weighing more than 23 kg (50 lb) are required to be marked with the maximum weight.

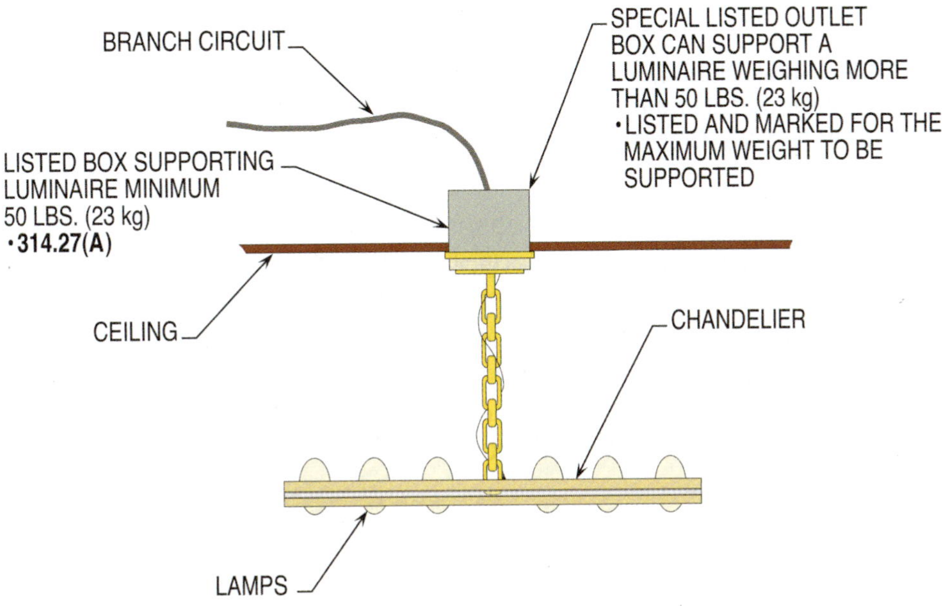
BRANCH CIRCUIT
SPECIAL LISTED OUTLET BOX CAN SUPPORT A LUMINAIRE WEIGHING MORE THAN 50 LBS. (23 kg)
•LISTED AND MARKED FOR THE MAXIMUM WEIGHT TO BE SUPPORTED
LISTED BOX SUPPORTING LUMINAIRE MINIMUM 50 LBS. (23 kg)
•314.27(A)
CEILING
CHANDELIER
LAMPS

BOXES AT LUMINAIRE OUTLETS
314.27(A) and (B)

Purpose of Change: This revision clarifies the requirements for listed boxes used for luminaire support.

NEC Ch. 3 - Article 314
Part II - 314.27(E) and Exception

46

Type of Change	Panel Action	UL	UL 508	API 500 - 1997	API 505 - 1997	OSHA - 1994
New Subsection and Ex.	Accept in Principle	514	-	-	-	-
ROP	**ROC**	**NFPA 70E - 2004**	**NFPA 70B - 2006**	**NFPA 79 - 2007**	**NFPA**	**NEMA**
pg. 317 # 9-63	pg. 195 # 9-35	-	-	-	-	-
log: 3399 CMP: 9	log: 1918 Submitter: Frederic P. Hartwell			2005 NEC: -		IEC: -

2008 NEC - 314.27 Outlet Boxes.

(E) Utilization Equipment. Boxes used for the support of utilization equipment other than ceiling-suspended (paddle) fans shall meet the requirements of 314.27(A) and (B) for the support of a luminaire that is the same size and weight.

Exception: Utilization equipment weighing not more than 3 kg (6 lb) shall be permitted to be supported on other boxes or plaster rings that are secured to other boxes, provided the equipment or its supporting yoke is secured to the box with no fewer than two No. 6 or larger screws.

Author's Comment: The purpose of this new subsection is to require heavier utilization equipment (up to 50 lb), other than luminaires and ceiling fans, be mounted to outlet boxes, rather than device boxes. Utilization equipment heavier than 50 lb are required to be supported independent of the outlet box. The exception permits smaller utilization equipment, such as smoke detectors, to be mounted to plaster rings on boxes or to device boxes. The comment deleted the proposed phrase "of comparable size and weight" which was not enforceable and inserted the phrase "that is the same size and weight" which is enforceable based on the NEC Style Manual.

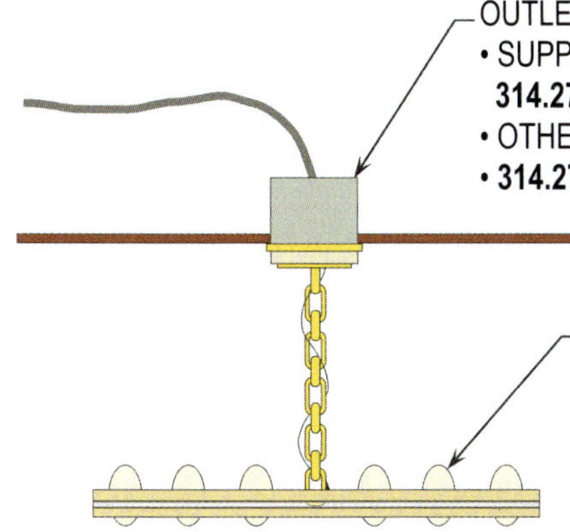
OUTLET BOX
• SUPPORT OF UTILIZATION EQUIPMENT MEETS
314.27(A) AND (B) REQUIREMENTS
• OTHER THAN CEILING-SUSPENDED (PADDLE) FANS
• 314.27(E)
UTILIZATION EQUIPMENT
• 50 LBS OR LESS
• MOUNTED TO OUTLET BOXES, NOT DEVICE BOXES
NOTE: UTILIZATION EQUIPMENT HEAVIER THAN 50 LBS. SHALL BE SUPPORTED INDEPENDENT OF THE OULET BOX, UNLESS LISTED FOR GREATER WEIGHT.

OULET BOX
• NOT WEIGHING
MORE THAN 6 LBS (3 kg)
• AT LEAST 2 - #6 OR
LARGER SCREWS
• 314.27(E), Ex.

UTILIZATION EQUIPMENT
314.27(E) and Exception

Purpose of Change: A new subsection and exception have been added to address the installation requirements for utilization equipment used with outlet boxes.

NEC Ch. 3 - Article 314
Part II - 314.28(A)

Type of Change	Panel Action	UL	UL 508	API 500 - 1997	API 505 - 1997	OSHA - 1994
Revision	Accept	514 - 514A	-	-	-	-
ROP	**ROC**	**NFPA 70E - 2004**	**NFPA 70B - 2006**	**NFPA 79 - 2007**	**NFPA**	**NEMA**
pg. 318 # 9-66	pg. - # -	-	-	-	-	-
log: 3400 CMP: 9	log: -	Submitter: Frederic P. Hartwell		2005 NEC: 314.28(A)		IEC: -

2005 NEC - 314.28 Pull and Junction Boxes and Conduit Bodies.

Boxes and conduit bodies used as pull or junction boxes shall comply with 314.28(A) through (D).

Exception: Terminal housings supplied with motors shall comply with the provisions of 430.12.

(A) Minimum Size. For raceways containing conductors of 4 AWG or larger, and for cables containing conductors of 4 AWG or larger, the minimum dimensions of pull or junction boxes installed in a raceway or cable run shall comply with (A)(1) through (A)(3). Where an enclosure dimension is to be calculated based on the diameter of entering raceways, the diameter shall be the metric designator (trade size) expressed in the units of measurement employed.

2008 NEC - 314.28 Pull and Junction Boxes and Conduit Bodies.

Boxes and conduit bodies used as pull or junction boxes shall comply with 314.28(A) through (D).

Exception: Terminal housings supplied with motors shall comply with the provisions of 430.12.

(A) Minimum Size. For raceways containing conductors of 4 AWG or larger that are required to be insulated, and for cables containing conductors of 4 AWG or larger, the minimum dimensions of pull or junction boxes installed in a raceway or cable run shall comply with (A)(1) through (A)(3). Where an enclosure dimension is to be calculated based on the diameter of entering raceways, the diameter shall be the metric designator (trade size) expressed in the units of measurement employed.

Author's Comment: Raceways are commonly used to route grounding electrode conductors that are not required to be insulated and may be installed bare. Such raceway runs often use conduit bodies to make changes of direction. The rules in **314.28** are related to the safety objective of making sure that conductor insulation is not degraded through excessively cramped bending radii in pull boxes and conduit bodies, whether during installation or through continuing stress after installation. The requirements in the previous Code are irrelevant to conductors that need not be insulated. This change corrects that oversight.

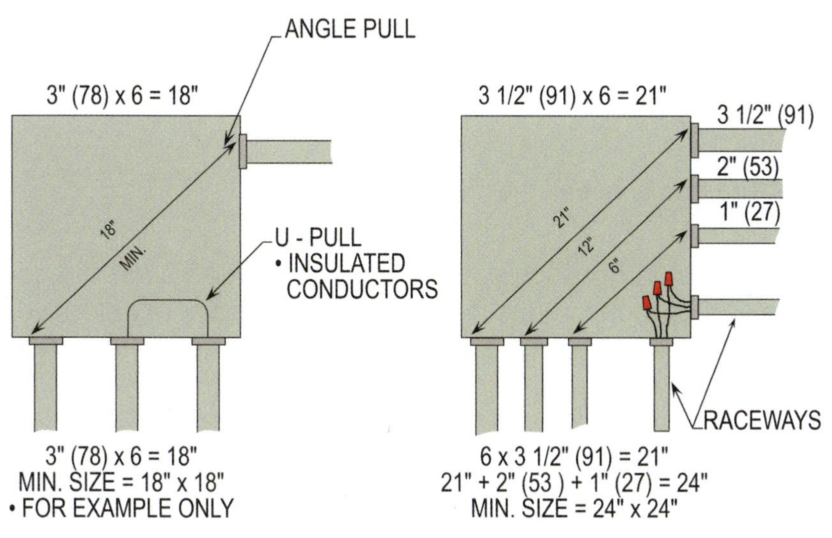

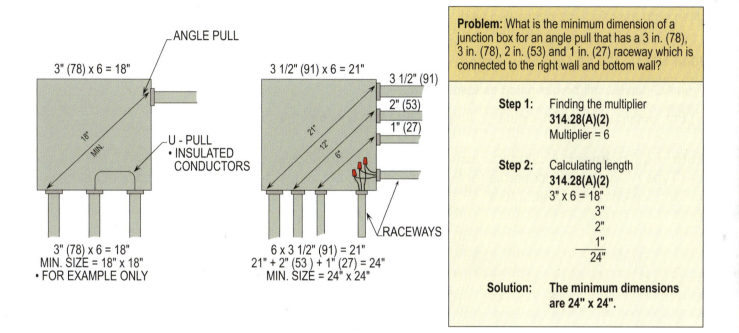

Problem: What is the minimum dimension of a junction box for an angle pull that has a 3 in. (78), 3 in. (78), 2 in. (53) and 1 in. (27) raceway which is connected to the right wall and bottom wall?

Step 1: Finding the multiplier
314.28(A)(2)
Multiplier = 6

Step 2: Calculating length
314.28(A)(2)
3" x 6 = 18"
3"
2"
1"
24"

Solution: **The minimum dimensions are 24" x 24".**

MINIMUM SIZE
314.28(A)

Purpose of Change: This revision clarifies that enclosure dimensions apply to raceways containing conductors 4 AWG or larger that are insulated and installed in pull, junction boxes, and conduit bodies.

NEC Ch. 3 - Article 320
Part II - 320.10(1)

Type of Change		Panel Action		UL	UL 508	API 500 - 1997	API 505 - 1997	OSHA - 1994
Revision		Accept		4	-	-	-	-
ROP		ROC		NFPA 70E - 2004	NFPA 70B - 2006	NFPA 79 - 2007	NFPA	NEMA - 2004
pg. 321	# 7-2	pg. -	# -	-	-	-	-	RV 1
log: 1495	CMP: 7	log: -	Submitter: Chris MacCreery			2005 NEC: 320.10(1)		IEC: -

2005 NEC - 320.10 Uses Permitted.

Type AC cable shall be permitted as follows:

(1) In both exposed and concealed work

2008 NEC - 320.10 Uses Permitted.

Type AC cable shall be permitted as follows:

(1) For feeders and branch circuits in both exposed and concealed work

Author's Comment: Adding the text "for feeders and branch circuits" in (1) provides clear direction that Type AC cable is acceptable for feeders and branch circuits in both exposed and concealed work, not service entrance conductors.

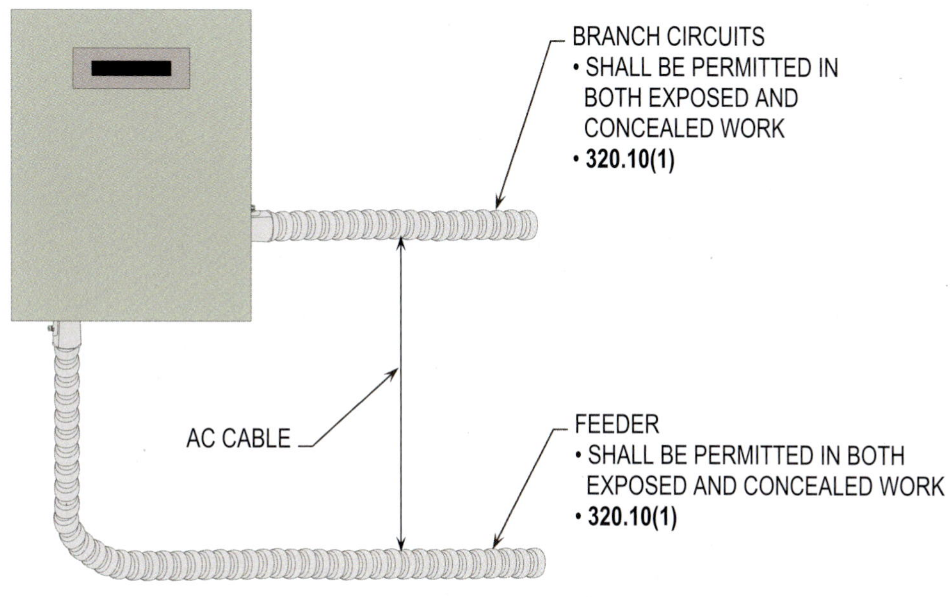

USES PERMITTED
320.10(1)

Purpose of Change: This revision clarifies that AC cable shall be permitted for feeder and branch circuits for both exposed and concealed work.

NEC Ch. 3 - Article 328
Part II - 328.10(3) and (6)

Type of Change	Panel Action	UL	UL 508	API 500 - 1997	API 505 - 1997	OSHA - 1994
Revision; New Item	Accept in Principle	-	-	-	-	-
ROP	**ROC**	**NFPA 70E - 2004**	**NFPA 70B - 2006**	**NFPA 79 - 2007**	**NFPA**	**NEMA**
pg. 328 # 7-22	pg. 198 # 7-9	-	-	-	-	-
log: 199 CMP: 7	log: 982	Submitter: Frederic P. Hartwell		2005 NEC: 328.10		IEC: -

2005 NEC - 328.10 Uses Permitted.

Type MV cable shall be permitted for use on power systems rated up to 35,000 volts nominal as follows:

(1) In wet or dry locations

(2) In raceways

(3) In cable trays ~~as specified in 392.3(B)(2)~~

(4) Direct buried in accordance with 300.50

(5) In messenger-supported wiring

2008 NEC - 328.10 Uses Permitted.

Type MV cable shall be permitted for use on power systems rated up to 35,000 volts nominal as follows:

(1) In wet or dry locations

(2) In raceways

(3) In cable trays, where identified for the use, in accordance with Article 392

Exception: Type MV cable that has an overall metallic sheath or armor, also complies with the requirements for Type MC cable and is identified as "MV or MC" shall be permitted to be installed in cable trays in accordance with 392.3(B)(2).

(4) Direct buried in accordance with 300.50

(5) In messenger-supported wiring in accordance with Part II of Article 396

(6) As exposed runs in accordance with 300.37

Exception: Type MV cable that has an overall metallic sheath or armor, also complies with the requirements for Type MC cable, and is identified as "MV or MC" shall be permitted to be installed as exposed runs of metal-clad cable in accordance with 300.37.

FPN: The "uses permitted" is not at all-inclusive text.

Author's Comment: Comment 7-9 replaced text in (3) so that Type MV cable is permitted to be used in a cable tray, where the MV cable is identified for cable tray use. The new (6) in **328.10** permits Type MV cable as exposed runs in an aboveground wiring method for voltages up to 35,000 volts. Comment 7-12 added two exceptions. The first exception was added to **328.10(3)** permitting Type MV cable with an overall metallic sheath or armor complying, with the requirements for Type MC cable to be installed in cable trays. The second exception was added to new **328.10(6)** to also permit metallic sheathed or armored Type MV cable to be installed as exposed runs of metal clad cable in accordance with 300.37.

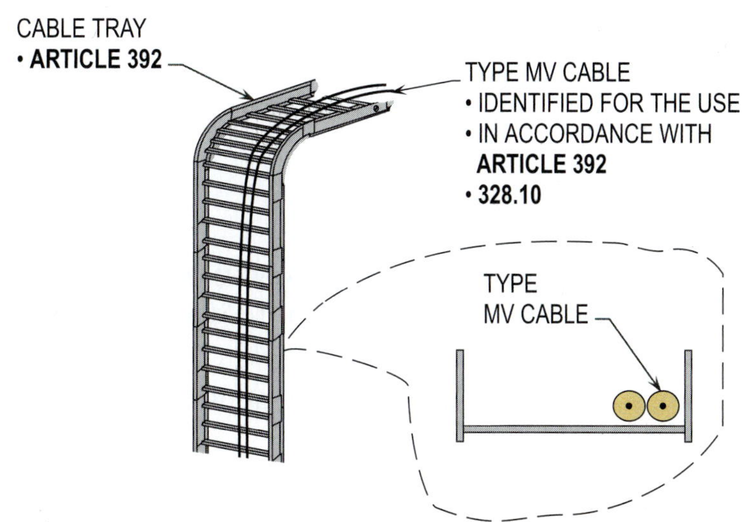

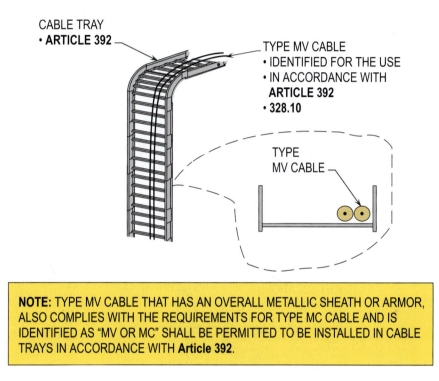

NOTE: TYPE MV CABLE THAT HAS AN OVERALL METALLIC SHEATH OR ARMOR, ALSO COMPLIES WITH THE REQUIREMENTS FOR TYPE MC CABLE AND IS IDENTIFIED AS "MV OR MC" SHALL BE PERMITTED TO BE INSTALLED IN CABLE TRAYS IN ACCORDANCE WITH **Article 392**.

USES PERMITTED
328.10(3) and (6)

Purpose of Change: This revision and new exception addresses the installation requirements for Type MV cable in cable trays.

NEC Ch. 3 - Article 330
Part II - 330.10(B)(4)

Type of Change		Panel Action		UL	UL 508	API 500 - 1997	API 505 - 1997	OSHA - 1994
Revision		Accept		1569	-	-	-	-
ROP		ROC		NFPA 70E - 2004	NFPA 70B - 2006	NFPA 79 - 2007	NFPA	NEMA - 2004
pg. 329	# 7-27	pg. -	# -	-	-	-	-	RV 1
log: 928	CMP: 7	log: -	Submitter: Daniel Leaf			2005 NEC: 330.10(B)(4)		IEC: -

2005 NEC - 330.10 Uses Permitted.

(B) Specific Uses.

(4) Installed Outside of Buildings or as Aerial Cable. Type MC cable installed outside of buildings or as aerial cable shall comply with 225.10, 396.10, and 396.12.

2008 NEC - 330.10 Uses Permitted.

(B) Specific Uses.

(4) Installed Outside of Buildings, or Structures, or as Aerial Cable. Type MC cable installed outside of buildings or structures or as aerial cable shall comply with 225.10, 396.10, and 396.12.

Author's Comment: Structures that are not buildings are added for specific uses of Type MC cable.

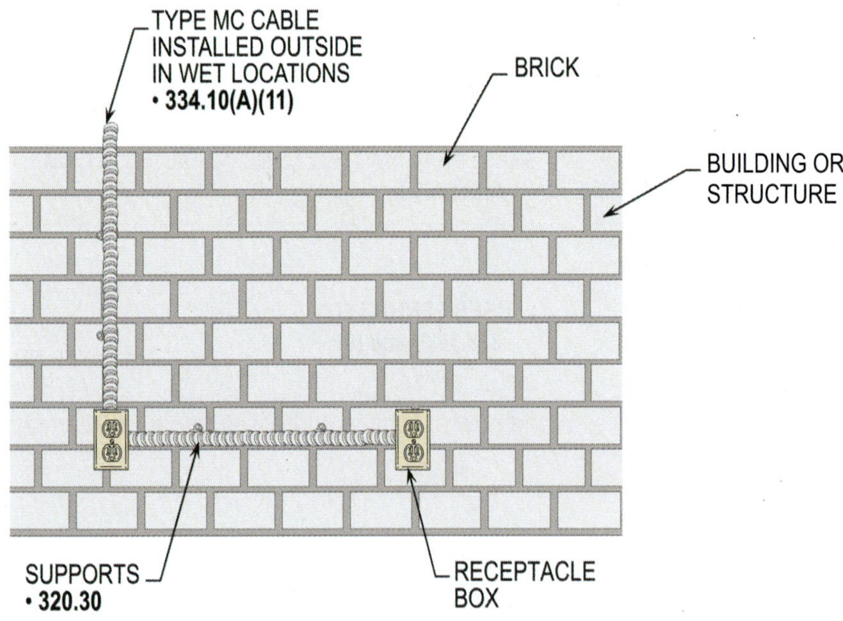

INSTALLED OUTSIDE OF BUILDINGS, OR STRUCTURES, OR AS AERIAL CABLE 330.10(B)(4)

Purpose of Change: This revision clarifies that type MC cable shall be permitted to be installed outside on structures, as well as buildings.

NEC Ch. 3 - Article 334
Part II - 334.12(A)(1), Exception

Type of Change		Panel Action		UL	UL 508	API 500 - 1997	API 505 - 1997	OSHA - 1994
New Exception		Accept in Principle		719	-	-	-	-
ROP		ROC		NFPA 70E - 2004	NFPA 70B - 2006	NFPA 79 - 2007	NFPA	NEMA
pg. 334	# 7-51	pg. -	# -	-	-	-	-	-
log: 202	CMP: 7	log: -	Submitter: Robert C. Duncan			2005 NEC: -		IEC: -

2005 NEC - 334.12 Uses Not Permitted.

(A) Types NM, NMC, and NMS. Types NM, NMC, and NMS cables shall not be permitted as follows:

(1) In any dwelling or structure not specifically permitted in 334.10(1), (2), and (3)

2008 NEC - 334.12 Uses Not Permitted.

(A) Types NM, NMC, and NMS. Types NM, NMC, and NMS cables shall not be permitted as follows:

(1) In any dwelling or structure not specifically permitted in 334.10(1), (2), and (3)

Exception: Type NM, NMC, and NMS cable shall be permitted in Type I and II construction when installed within raceways permitted to be installed in Type I and II construction.

Author's Comment: There is a new listed NM Hybrid Cable on the market consisting of power, communications and signaling conductors under a common overall jacket. This new exception will permit the use of this new cable installed in raceways recognized for use within Types I and II construction. The model building codes and **NFPA 220** recognize Type I as fire-resistive construction and Type II as noncombustible construction.

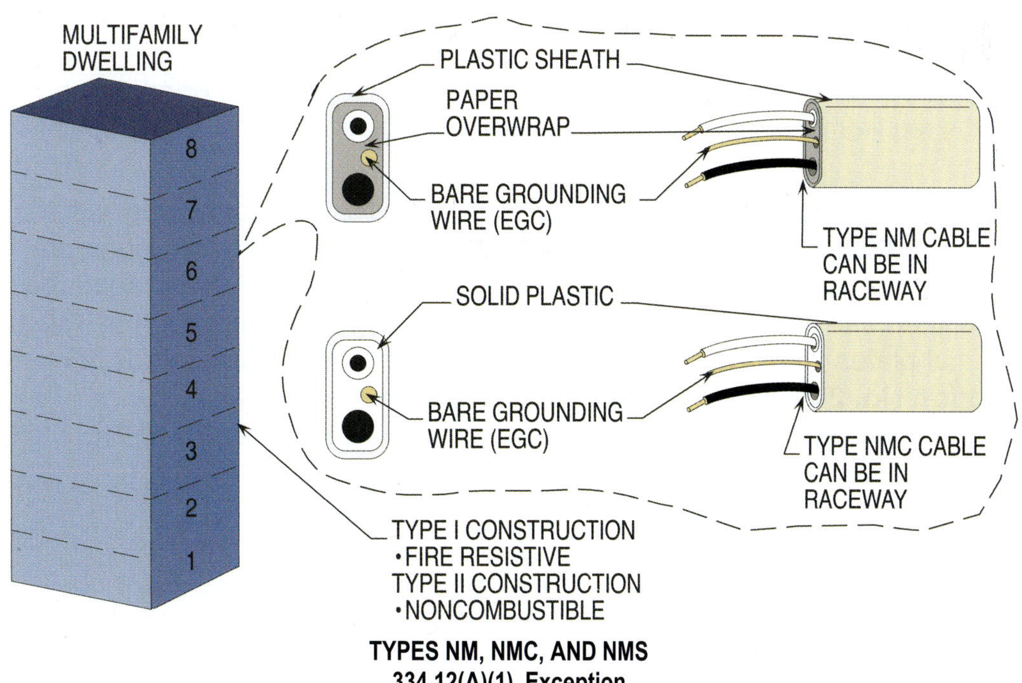

TYPES NM, NMC, AND NMS
334.12(A)(1), Exception

Purpose of Change: A new exception has been added to permit a new listed NM hybrid cable to be installed in Type I and II construction when installed within raceways permitted within that type of construction.

NEC Ch. 3 - Article 334
Part II - 334.15(B)

Type of Change	Panel Action	UL	UL 508	API 500 - 1997	API 505 - 1997	OSHA - 1994
Revision	Accept in Principle	719	-	-	-	-
ROP	**ROC**	**NFPA 70E - 2004**	**NFPA 70B - 2006**	**NFPA 79 - 2007**	**NFPA**	**NEMA**
pg. 336 # 7-61	pg. - # -	-	-	-	-	-
log: 1905 CMP: 7	log: -	Submitter: James W. Carpenter		2005 NEC: 334.15(B)		IEC: -

2005 NEC - 334.15 Exposed Work.

In exposed work, except as provided in 300.11(A), cable shall be installed as specified in 334.15(A) through (C).

(B) Protection from Physical Damage. Cable shall be protected from physical damage where necessary by rigid metal conduit, intermediate metal conduit, electrical metallic tubing, Schedule 80 PVC ~~rigid nonmetallic~~ conduit, or other approved means. Where passing through a floor, the cable shall be enclosed in rigid metal conduit, intermediate metal conduit, electrical metallic tubing, Schedule 80 PVC ~~rigid nonmetallic~~ conduit, or other approved means extending at least 150 mm (6 in.) above the floor.

~~Where~~ Type NMC cable ~~is~~ installed in shallow chases in masonry, concrete, or adobe~~, the cable~~ shall be protected ~~against nails or screws by a steel plate at least 1.59 mm (1/ 16 in.) thick~~ and covered with plaster, adobe, or similar finish.

2008 NEC - 334.15 Exposed Work.

In exposed work, except as provided in 300.11(A), cable shall be installed as specified in 334.15(A) through (C).

(B) Protection from Physical Damage. Cable shall be protected from physical damage where necessary by rigid metal conduit, intermediate metal conduit, electrical metallic tubing, Schedule 80 PVC conduit, or other approved means. Where passing through a floor, the cable shall be enclosed in rigid metal conduit, intermediate metal conduit, electrical metallic tubing, Schedule 80 PVC conduit, or other approved means extending at least 150 mm (6 in.) above the floor.

Type NMC cable installed in shallow chases <u>or groves</u> in masonry, concrete, or adobe shall be protected <u>in accordance with the requirements in 300.4(F)</u> and covered with plaster, adobe, or similar finish.

Author's Comment: Shallow is not defined anywhere in the NEC so complying with the requirements in **300.4(E)** is required where installing NMC in shallow chases in masonry, concrete, or adobe. For example, if the 1 1/4 in. spacing is maintained, then installing nail plates is not required. If the spacing is not maintained, then one of the other protection means in the first paragraph of **334.15(B)** would be permissible but nail plates would not be necessary as stated in **300.4(F), Exception No. 1**. Section **300.4(F), Exception No. 2** now permits listed, case-hardened plates less than a 1/16 in. thick so this text in **334.15(B)** now recognizes this method of protection where dealing with NM cable in a shallow chase in masonry, concrete, or adobe.

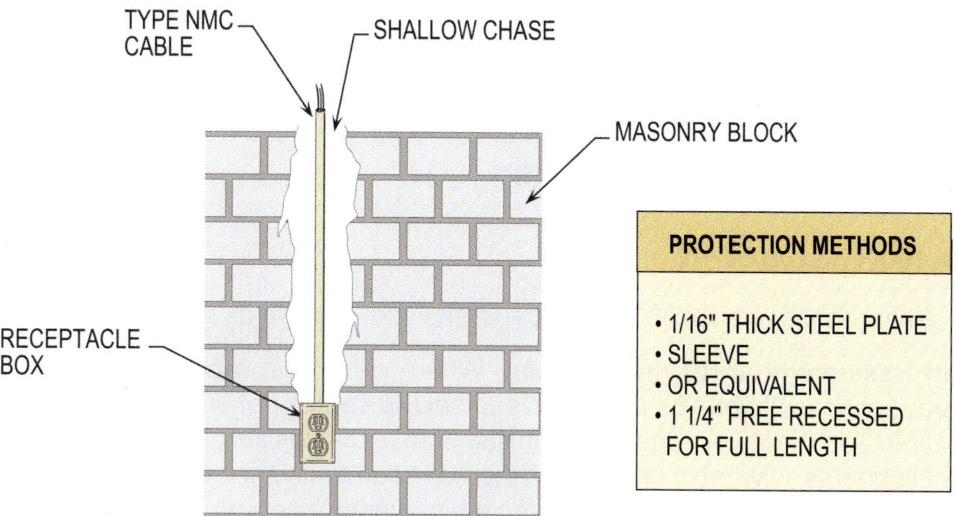

PROTECTION FROM PHYSICAL DAMAGE
334.15(B)

Purpose of Change: This revision clarifies that Type NMC cable installed in shallow chases shall be protected by a 1/16 in. thick steel plate, sleeve, or equivalent, or 1 1/4 in. recessed space for the full length.

NEC Ch. 3 - Article 334
Part II - 334.15(C)

47

Type of Change	Panel Action	UL	UL 508	API 500 - 1997	API 505 - 1997	OSHA - 1994
Revision	Accept	719	-	-	-	-
ROP	ROC	NFPA 70E - 2004	NFPA 70B - 2006	NFPA 79 - 2007	NFPA	NEMA
pg. 336 # 7-58	pg. - # -	-	-	-	-	-
log: 2399 CMP: 7	log: -	Submitter: Mike Holt		2005 NEC: 334.15(C)		IEC: -

2005 NEC - 334.15 Exposed Work.

In exposed work, except as provided in 300.11(A), cable shall be installed as specified in 334.15(A) through (C).

(C) In Unfinished Basements. Where cable is run at angles with joists in unfinished basements it shall be permissible to secure cables not smaller than two 6 AWG or three 8 AWG conductors directly to the lower edges of the joists. Smaller cables shall be run either through bored holes in joists or on running boards. NM cable ~~used~~ on ~~a~~ wall of an unfinished basement shall be permitted to be installed in a listed conduit ~~or~~ tubing. Conduit or tubing shall ~~utilize a nonmetallic~~ bushing or adapter at the point the cable enters the raceway. Metal conduit and tubings and metal outlet boxes shall be ~~grounded~~.

2008 NEC - 334.15 Exposed Work.

In exposed work, except as provided in 300.11(A), cable shall be installed as specified in 334.15(A) through (C).

(C) In Unfinished Basements and Crawl Spaces. Where cable is run at angles with joists in unfinished basements and crawl spaces, it shall be permissible to secure cables not smaller than two 6 AWG or three 8 AWG conductors directly to the lower edges of the joists. Smaller cables shall be run either through bored holes in joists or on running boards. NM cable installed on the wall of an unfinished basement shall be permitted to be installed in a listed conduit, tubing, or shall be protected in accordance with 300.4. Conduit or tubing shall be provided with a suitable insulating bushing or adapter at the point the cable enters the raceway. Metal conduit and tubings and metal outlet boxes shall be grounded. The NM cable sheath shall extend through the conduit or tubing and into the outlet or device box not less than 6 mm (1/4 in.). The cable shall be secured within 300 mm (12 in.) of the point where the cable enters the conduit or tubing. Metal conduit, tubing, and metal outlet boxes shall be connected to an equipment grounding conductor.

Author's Comment: Because the same dangers of damage exist for exposed NM cables in crawl spaces as in unfinished basements, "crawl spaces" has been added to both the title and the text for exposed NM cable.

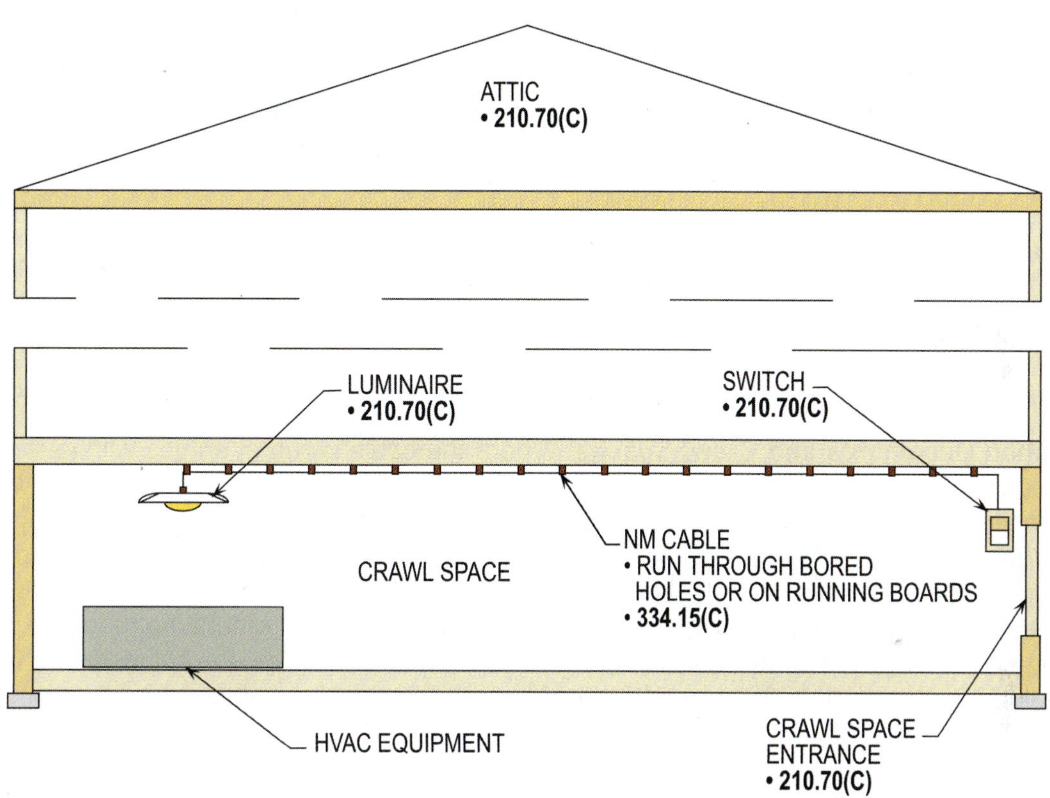

IN UNFINISHED BASEMENTS AND CRAWL SPACES
334.15(C)

Purpose of Change: This revision clarifies that NM cable shall be protected in crawl spaces.

NEC Ch. 3 - Article 334
Part II - 334.15(C)

 48

Type of Change	Panel Action	UL	UL 508	API 500 - 1997	API 505 - 1997	OSHA - 1994
Revision	Accept in Principle in Part	719	-	-	-	-
ROP	**ROC**	**NFPA 70E - 2004**	**NFPA 70B - 2006**	**NFPA 79 - 2007**	**NFPA**	**NEMA**
pg. 337 # 7-63	pg. - # -	-	-	-	-	-
log: 344 CMP: 7	log: - Submitter: Michael J. Johnston			2005 NEC: 334.15(C)		IEC: -

2005 NEC - 334.15 Exposed Work.

In exposed work, except as provided in 300.11(A), cable shall be installed as specified in 334.15(A) through (C).

(C) In Unfinished Basements. Where the cable is run at angles with joists in unfinished basements, it shall be permissible to secure cables not smaller than two 6 AWG or three 8 AWG conductors directly to the lower edges of the joists. Smaller cables shall be run either through bored holes in joists or on running boards. NM cable ~~used~~ on a wall of an unfinished basement shall be permitted to be installed in a listed conduit or tubing. Conduit or tubing shall ~~utilize~~ a ~~nonmetallic~~ bushing or adapter at the point the cable enters the raceway. Metal conduit ~~and~~ tubing and metal outlet boxes shall be ~~grounded~~.

2008 NEC - 334.15 Exposed Work.

In exposed work, except as provided in 300.11(A), cable shall be installed as specified in 334.15(A) through (C).

(C) In Unfinished Basements and Crawl Spaces. Where the cable is run at angles with joists in unfinished basements and crawl spaces, it shall be permissible to secure cables not smaller than two 6 AWG or three 8 AWG conductors directly to the lower edges of the joists. Smaller cables shall be run either through bored holes in joists or on running boards. NM cable installed on the wall of an unfinished basement shall be permitted to be installed in a listed conduit or tubing or shall be protected in accordance with 300.4. Conduit or tubing shall be provided with a suitable insulating bushing or adapter at the point the cable enters the raceway. The NM cable sheath shall extend through the conduit or tubing and into the outlet or device box not less than 6 mm (1/4 in.). The cable shall be secured within 300 mm (12 in.) of the point where the cable enters the conduit or tubing. Metal conduit, tubing, and metal outlet boxes shall be connected to an equipment grounding conductor.

Author's Comment: In the second sentence, the word "used" was replaced by the word "installed" because this word is more precise. Reference to **300.4** in the second sentence provides several methods of protection that can be used, such as steel plates, sleeves, and steel channels. The change to "insulating bushing" in the next sentence down was made to correlate with similar phrase in **300.16(B)**. The two sentences that were added provide requirements to extend the NM cable sheath to extend not less than 1/4 inch into the outlet or device box similar to the requirements in **314.17** and to require NM cable to be secured within 12 inches of where the cable enters into the conduit or tubing. These changes correspond and relate to the changes by Proposal 7-58 to add crawlspaces to both the title and the text.

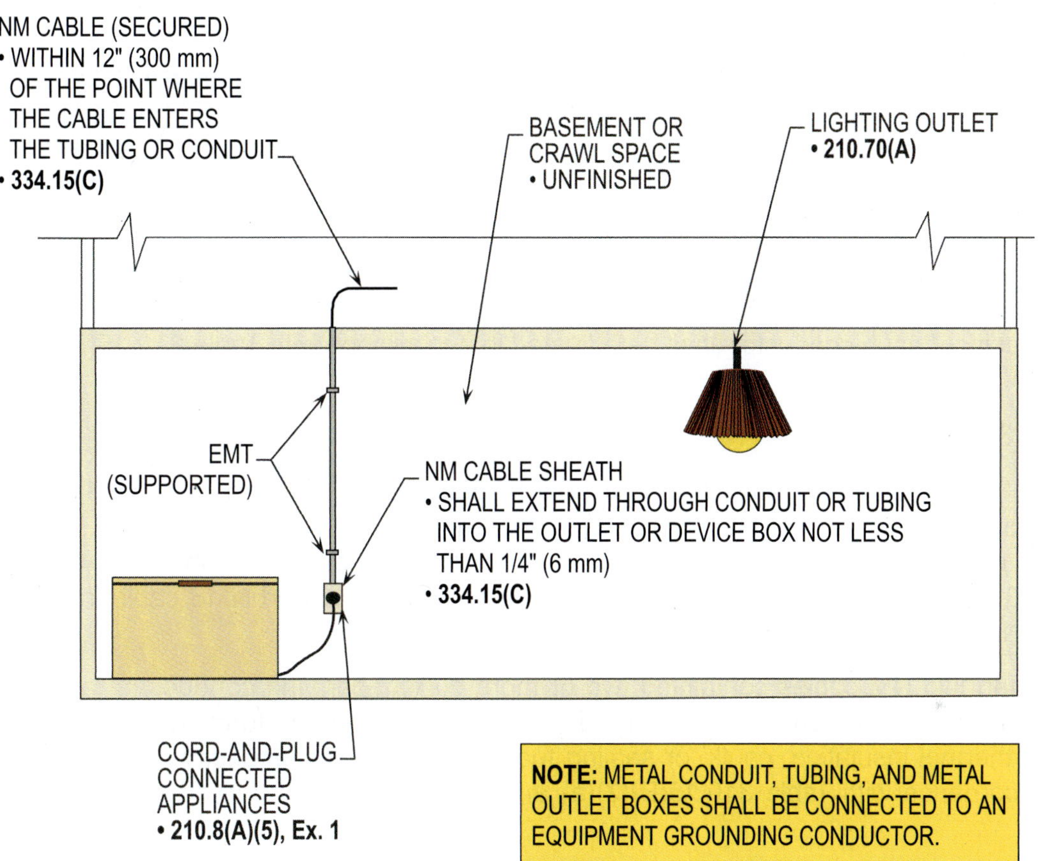

LIGHTING OUTLET
• 210.70(A)

EMT
(SUPPORTED)

NM CABLE SHEATH
• SHALL EXTEND THROUGH CONDUIT OR TUBING
 INTO THE OUTLET OR DEVICE BOX NOT LESS
 THAN 1/4" (6 mm)

CORD-AND-PLUG
CONNECTED
APPLIANCES
• 210.8(A)(5), Ex. 1

NOTE: METAL CONDUIT, TUBING, AND METAL
OUTLET BOXES SHALL BE CONNECTED TO AN
EQUIPMENT GROUNDING CONDUCTOR.

IN UNFINISHED BASEMENTS AND CRAWL SPACES
334.15(C)

Purpose of Change: This revision clarifies the installation requirements for NM cable in an unfinished basement or crawl space where the NM cable is installed in a conduit or tubing for protection.

NEC Ch. 3 - Article 334
Part II - 334.80

49

Type of Change	Panel Action	UL	UL 508	API 500 - 1997	API 505 - 1997	OSHA - 1994
Revision	Accept in Part	719	-	-	-	-
ROP	**ROC**	**NFPA 70E - 2004**	**NFPA 70B - 2006**	**NFPA 79 - 2007**	**NFPA**	**NEMA**
pg. 338 # 7-70	pg. 204 # 7-40	-	-	-	-	-
log: 2574 CMP: 7	log: 679	Submitter: Noel Williams		2005 NEC: 334.80		IEC: -

2005 NEC - 334.80 Ampacity.

The ampacity of Types NM, NMC, and NMS cable shall be determined in accordance with 310.15. The ampacity shall be in accordance with the 60°C (140°F) conductor temperature rating. The 90°C (194°F) rating shall be permitted to be used for ampacity derating purposes, provided the final derated ampacity does not exceed that for a 60°C (140°F) rated conductor. The ampacity of Types NM, NMC, and NMS cable installed in cable tray shall be determined in accordance with 392.11.

Where more than two NM cables containing two or more current-carrying conductors ~~are bundled together and pass~~ through wood framing that is to be fire- or draft-stopped using thermal insulation or sealing foam, the allowable ampacity of each conductor shall be adjusted in accordance with Table 310.15(B)(2)(a).

2008 NEC - 334.80 Ampacity.

The ampacity of Types NM, NMC, and NMS cable shall be determined in accordance with 310.15. The ampacity shall be in accordance with the 60°C (140°F) conductor temperature rating. The 90°C (194°F) rating shall be permitted to be used for ampacity derating purposes, provided the final derated ampacity does not exceed that for a 60°C (140°F) rated conductor. The ampacity of Types NM, NMC, and NMS cable installed in cable tray shall be determined in accordance with 392.11.

Where more than two NM cables containing two or more current-carrying conductors, <u>are installed, without maintaining spacing between cables,</u> through <u>the same opening in</u> wood framing that is to be fire- or draft-stopped using thermal insulation<u>, caulk,</u> or sealing foam, the allowable ampacity of each conductor shall be adjusted in accordance with Table 310.15(B)(2)(a) <u>and the provisions of 310.15(A)(2), Exception, shall not apply</u>.

(See next change for additional paragraph added for 2008 NEC)

Author's Comment: Section **310.15(A)(2), Exception** takes advantage of the heat-sinking capabilities of the wire by stating the following: "Where two different ampacities apply to adjacent portions of a circuit, the higher ampacity shall be permitted to be used beyond the point of transition, a distance equal to 3.0 m (10 ft) or 10 percent of the circuit length figured at the higher ampacity, whichever is less." The last sentence in **334.80** was amended to not permit the use of the heat-sinking characteristics of a circuit where more than two NM cables cabled together pass through wood framing.

Term "bundled" was deleted in the text since the use of bundled caused a lot of controversy as to what constituted bundled conductors. Many users wanted to argue that these cables are not bundled based on any code definition. This Panel Action eliminated the word "bundled" in **334.80** thus reducing the potential of confusion.

Caulking is a common method of fire or draft stoppage but an adjustment factor is not required by the 2005 Code if caulking is used. Adding the word "caulk" now ensures that the adjustment factors of **310.15(B)(2)(a)** and the accompanying **Table 310.15(B)(2)(a)** will apply to more than two NM cables containing two or more current-carrying conductors passed through wood framing.

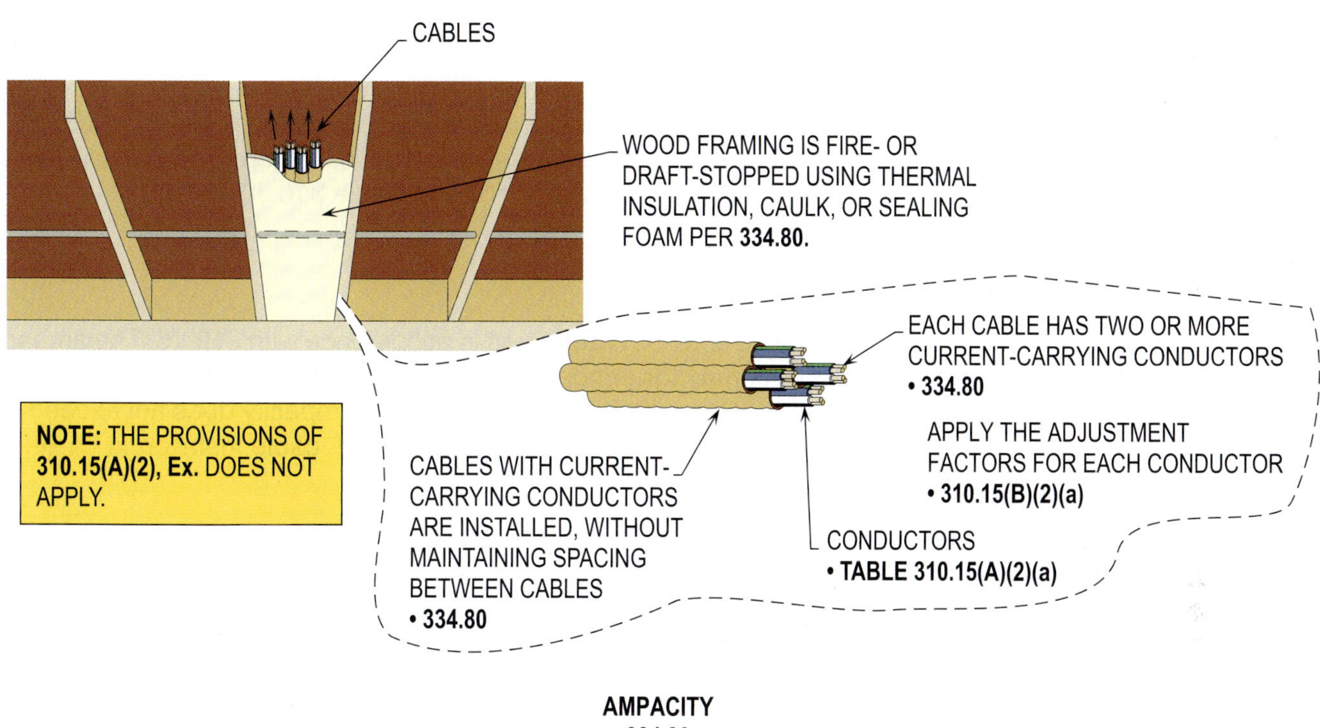
CABLES
WOOD FRAMING IS FIRE- OR DRAFT-STOPPED USING THERMAL INSULATION, CAULK, OR SEALING FOAM PER 334.80.
EACH CABLE HAS TWO OR MORE CURRENT-CARRYING CONDUCTORS • 334.80
APPLY THE ADJUSTMENT FACTORS FOR EACH CONDUCTOR • 310.15(B)(2)(a)
NOTE: THE PROVISIONS OF 310.15(A)(2), Ex. DOES NOT APPLY.
CABLES WITH CURRENT-CARRYING CONDUCTORS ARE INSTALLED, WITHOUT MAINTAINING SPACING BETWEEN CABLES • 334.80
CONDUCTORS • TABLE 310.15(A)(2)(a)

AMPACITY
334.80

Purpose of Change: This revision clarifies that where more than two NM cables containing two or more current-carrying conductors are installed, without maintaining spacing between cables, the provisions of **310.12(A)(2), Ex.** shall not apply. The term "bundled" has been deleted since the use of bundled caused a lot of controversy as to what constituted bundled conductors. The term "caulk" was added since it is a common method of fire or draft stopping.

NEC Ch. 3 - Article 334
Part II - 334.80

Type of Change	Panel Action	UL	UL 508	API 500 - 1997	API 505 - 1997	OSHA - 1994
Revision	Accept	719	-	-	-	-
ROP	**ROC**	**NFPA 70E - 2004**	**NFPA 70B - 2006**	**NFPA 79 - 2007**	**NFPA**	**NEMA**
pg. 339 # 7-74	pg. - # -	-	-	-	-	-
log: 3152 CMP: 7	log: -	Submitter: Travis Lindsey		2005 NEC: 334.80		IEC: -

2005 NEC - 334.80 Ampacity.

The ampacity of Types NM, NMC, and NMS cable shall be determined in accordance with 310.15. The ampacity shall be in accordance with the 60°C (140°F) conductor temperature rating. The 90°C (194°F) rating shall be permitted to be used for ampacity derating purposes, provided the final derated ampacity does not exceed that for a 60°C (140°F) rated conductor. The ampacity of Types NM, NMC, and NMS cable installed in cable tray shall be determined in accordance with 392.11.

Where more than two NM cables containing two or more current-carrying conductors ~~are bundled together and pass~~ through wood framing that is to be fire- or draft-stopped using thermal insulation or sealing foam, the allowable ampacity of each conductor shall be adjusted in accordance with Table 310.15(B)(2)(a).

2008 NEC - 334.80 Ampacity.

The ampacity of Types NM, NMC, and NMS cable shall be determined in accordance with 310.15. The ampacity shall be in accordance with the 60°C (140°F) conductor temperature rating. The 90°C (194°F) rating shall be permitted to be used for ampacity derating purposes, provided the final derated ampacity does not exceed that for a 60°C (140°F) rated conductor. The ampacity of Types NM, NMC, and NMS cable installed in cable tray shall be determined in accordance with 392.11.

Where more than two NM cables containing two or more current-carrying conductors<u>, are installed, without maintaining spacing between cables,</u> through <u>the same opening in</u> wood framing that is to be fire- or draft-stopped using thermal insulation<u>, caulk,</u> or sealing foam, the allowable ampacity of each conductor shall be adjusted in accordance with Table 310.15(B)(2)(a) <u>and the provisions of 310.15(A)(2), Exception, shall not apply</u>.

<u>Where more than two NM cables containing two or more current-carrying conductors are installed in contact with thermal insulation without maintaining spacing between cables, the allowable ampacity of each conductor shall be adjusted in accordance with Table 310.15(B)(2)(a).</u>

Author's Comment: The paragraph added by this proposal will now require the allowable ampacity to be adjusted for more than two NM cables containing two or more current-carrying conductors where the cables are installed in thermal insulation without maintaining spacing between the cables. An example would be three or four NM cables installed in insulation in an attic without maintaining spacing would require derating of the conductors. An alternative would be to maintain spacing between the cables and not derate.

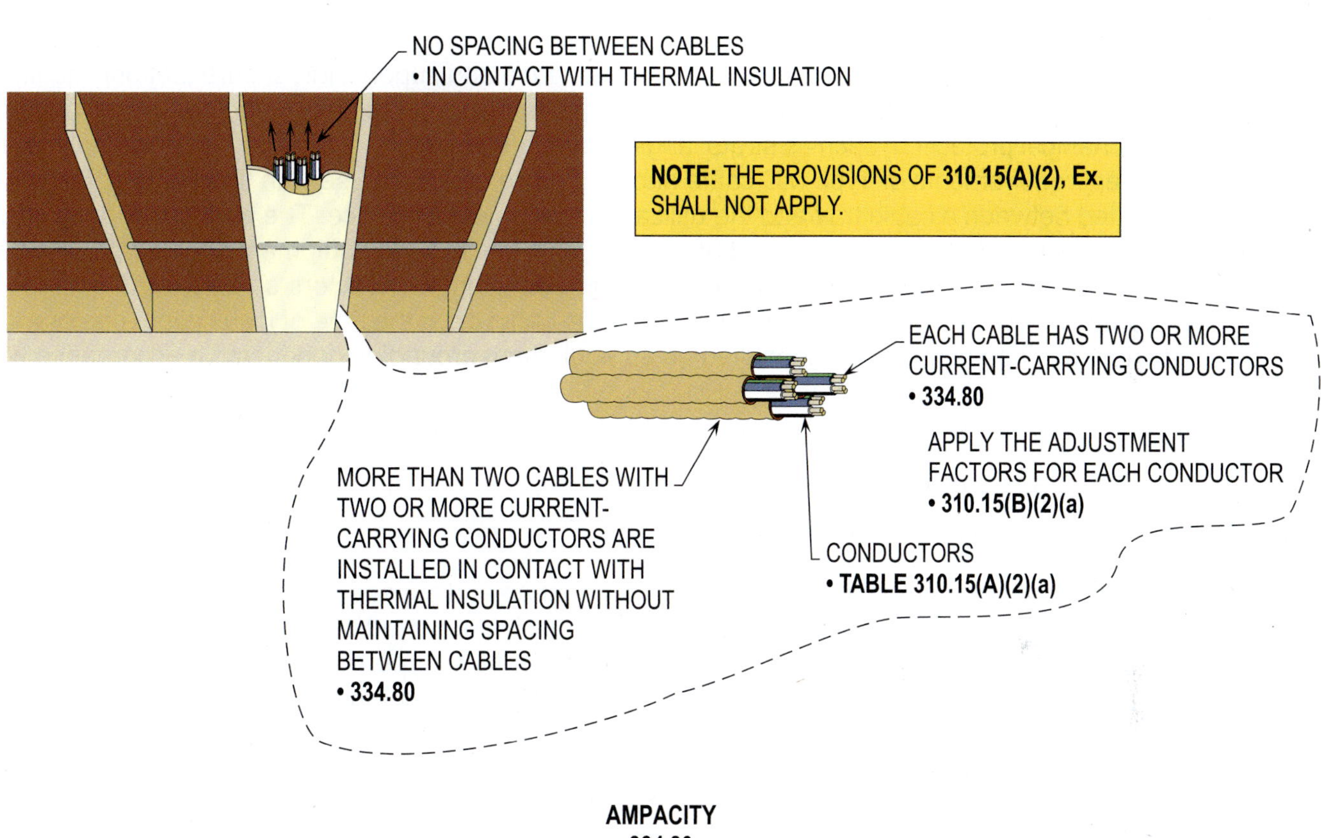
NO SPACING BETWEEN CABLES
• IN CONTACT WITH THERMAL INSULATION
NOTE: THE PROVISIONS OF 310.15(A)(2), Ex. SHALL NOT APPLY.
EACH CABLE HAS TWO OR MORE CURRENT-CARRYING CONDUCTORS
• 334.80
APPLY THE ADJUSTMENT FACTORS FOR EACH CONDUCTOR
• 310.15(B)(2)(a)
MORE THAN TWO CABLES WITH TWO OR MORE CURRENT-CARRYING CONDUCTORS ARE INSTALLED IN CONTACT WITH THERMAL INSULATION WITHOUT MAINTAINING SPACING BETWEEN CABLES
• 334.80
CONDUCTORS
• TABLE 310.15(A)(2)(a)

AMPACITY
334.80

Purpose of Change: This revision clarifies that where more than two NM cables containing two or more current-carrying conductors are installed in contact with thermal insulation without maintaining spacing between cables, the allowable ampacity of each conductor within the cables shall be adjusted.

NEC Ch. 3 - Article 336
Part II - 336.10(7), Exception

Type of Change		Panel Action		UL	UL 508	API 500 - 1997	API 505 - 1997	OSHA - 1994
New Exception		Accept in Principle		1277	-	-	-	-
ROP		ROC		NFPA 70E - 2004	NFPA 70B - 2006	NFPA 79 - 2007	NFPA	NEMA
pg. 341	# 7-80	pg. -	# -	-	-	-	-	-
log: 1720	CMP: 7	log: -	Submitter: Dennis A. Nielsen			2005 NEC: -		IEC: -

2005 NEC - 336.10 Uses Permitted.

Type TC cable shall be permitted to be used as follows:

(7) In industrial establishments where the conditions of maintenance and supervision ensure that only qualified persons service the installation, and where the cable is continuously supported and protected against physical damage using mechanical protection, such as struts, angles, or channels, Type TC tray cable that complies with the crush and impact requirements of Type MC cable and is identified for such use with the marking Type TC–ER shall be permitted between a cable tray and the utilization equipment or device. The cable shall be secured at intervals not exceeding 1.8 m (6 ft). Equipment grounding for the utilization equipment shall be provided by an equipment grounding conductor within the cable. In cables containing conductors sized 6 AWG or smaller, the equipment grounding conductor shall be provided within the cable or, at the time of installation, one or more insulated conductors shall be permanently identified as an equipment grounding conductor in accordance with 250.119(B).

2008 NEC - 336.10 Uses Permitted.

Type TC cable shall be permitted to be used as follows:

(7) In industrial establishments where the conditions of maintenance and supervision ensure that only qualified persons service the installation, and where the cable is continuously supported and protected against physical damage using mechanical protection, such as struts, angles, or channels, Type TC tray cable that complies with the crush and impact requirements of Type MC cable and is identified for such use with the marking Type TC–ER shall be permitted between a cable tray and the utilization equipment or device. The cable shall be secured at intervals not exceeding 1.8 m (6 ft). Equipment grounding for the utilization equipment shall be provided by an equipment grounding conductor within the cable. In cables containing conductors sized 6 AWG or smaller, the equipment grounding conductor shall be provided within the cable or, at the time of installation, one or more insulated conductors shall be permanently identified as an equipment grounding conductor in accordance with 250.119(B).

Exception: Where not subject to physical damage, Type TC-ER shall be permitted to transition between cable trays and between cable trays and utilization equipment or devices for a distance not to exceed 1.8 m (6 ft) without continuous support. The cable shall be mechanically supported where exiting the cable tray to ensure that the minimum bending radius is not exceeded.

Author's Comment: A new exception has been added to permit Type TC-ER cable (Type TC-ER cable complies with the crush and impact test requirements for Type MC cable with the ER designation used to identify this special type of TC cable) to be used as a transition between two cable trays or between cable tray and utilization equipment. While this cable can withstand possible damage, the new exception only permits its use where not subject to physical damage.

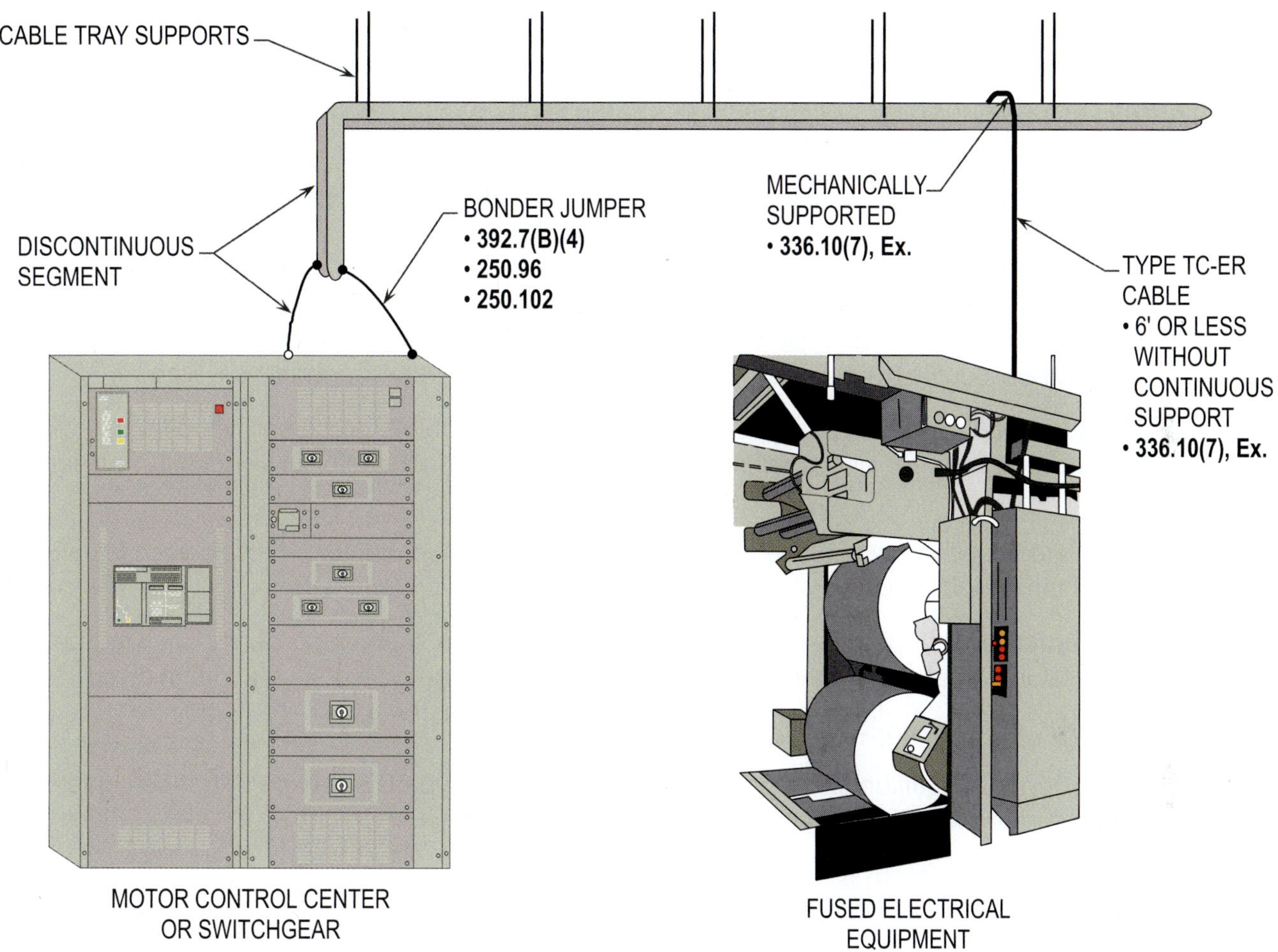

USES PERMITTED
336.10(7), Exception

Purpose of Change: A new exception has been added to permit 6 ft or less of Type TC-ER cable to transition between cable trays and between cable trays and utilization equipment or devices, where not subject to physical damage.

NEC Ch. 3 - Article 338
Part II - 338.12(A) and (B)

Type of Change	Panel Action	UL	UL 508	API 500 - 1997	API 505 - 1997	OSHA - 1994
Revision	Accept	854	-	-	-	-
ROP	**ROC**	**NFPA 70E - 2004**	**NFPA 70B - 2006**	**NFPA 79 - 2007**	**NFPA**	**NEMA**
pg. 342 # 7-84	pg. 206 # 7-49	-	-	-	-	-
log: 2635 CMP: 7	log: 45	Submitter: James M. Daly		2005 NEC: 338.10		IEC: -

2005 NEC - 338.10 Uses Permitted.

(A) Service-Entrance Conductors. Service-entrance cable shall be permitted to be used as service-entrance conductors and shall be installed in accordance with 230.6, 230.7, and Parts II, III, and IV of Article 230.

~~Type USE used for service laterals shall be permitted to emerge from the ground outside at terminations in meter bases or other enclosures where protected in accordance with 300.5(D).~~

(B) Branch Circuits or Feeders.

(1) Grounded Conductor Insulated. Type SE service-entrance cables shall be permitted in wiring systems where all of the circuit conductors of the cable are of the ~~rubber-covered~~ or thermoplastic type.

(2) Grounded Conductor Not Insulated. Type SE service-entrance cable shall be permitted for use where the insulated conductors are used for circuit wiring and the uninsulated conductor is used only for equipment grounding purposes.

Exception: Uninsulated conductors shall be permitted as a grounded conductor in accordance with ~~250.140, 250.32~~, and 225.30 through 225.40.

(3) Temperature Limitations. Type SE service-entrance cable used to supply appliances shall not be subject to conductor temperatures in excess of the temperature specified for the type of insulation involved.

(4) Installation Methods for Branch Circuits and Feeders.

(a) Interior Installations. In addition to the provisions of this article, Type SE service-entrance cable used for interior wiring shall comply with the installation requirements of ~~Parts I and~~ Part II of Article 334~~, excluding 334.80~~.

FPN: See 310.10 for temperature limitation of conductors.

(b) Exterior Installations. In addition to the provisions of this article, service-entrance cable used for feeders or branch circuits, where installed as exterior wiring, shall be installed in accordance with Part I of Article 225. The cable shall be supported in accordance with 334.30. ~~, unless used as messenger-supported wiring as permitted in Part II of Article 396~~. Type USE cable installed as underground feeder and branch circuit cable shall comply with Part II of Article 340. ~~Where Type USE cable emerges from the ground at terminations, it shall be protected in accordance with 300.5(D). Multiconductor service-entrance cable shall be permitted to be installed as messenger-supported wiring in accordance with 225.10 and Part II of Article 396.~~

2008 NEC - 338.10 Uses Permitted.

(A) Service-Entrance Conductors. Service-entrance cable shall be permitted to be used as service-entrance conductors and shall be installed in accordance with 230.6, 230.7, and Parts II, III, and IV of Article 230.

(B) Branch Circuits or Feeders.

(1) Grounded Conductor Insulated. Type SE service-entrance cables shall be permitted in wiring systems where all of the circuit conductors of the cable are of the thermoset or thermoplastic type.

(2) Grounded Conductor Not Insulated. Type SE service-entrance cable shall be permitted for use where the insulated conductors are used for circuit wiring and the uninsulated conductor is used only for equipment grounding purposes.

Exception: Uninsulated conductors shall be permitted as a grounded conductor in accordance with 250.32 and 250.140 where the uninsulated grounded conductor of the cable originates in service equipment, and 225.30 through 225.40.

(3) Temperature Limitations. Type SE service-entrance cable used to supply appliances shall not be subject to conductor temperatures in excess of the temperature specified for the type of insulation involved.

(4) Installation Methods for Branch Circuits and Feeders.

(a) Interior Installations. In addition to the provisions of this article, Type SE service-entrance cable used for interior wiring shall comply with the installation requirements of Part II of Article 334.

FPN: See 310.10 for temperature limitation of conductors.

(b) Exterior Installations. In addition to the provisions of this article, service-entrance cable used for feeders or branch circuits, where installed as exterior wiring, shall be installed in accordance with Part I of Article 225. The cable shall be supported in accordance with 334.30. Type USE cable installed as underground feeder and branch circuit cable shall comply with Part II of Article 340.

338.12 Uses Not Permitted.

(A) Service-Entrance Cable. Service-Entrance Cable (SE) shall not be used under the following conditions or in the following locations.

(1) Where subject to physical damage unless protected in accordance with 230.50(A)

(2) Underground with or without a raceway

(3) For exterior branch circuits and feeder wiring unless the installation complies with the provisions of Part I of Article 225 and is supported in accordance with 334.30 or is used as messenger supported wiring as permitted in Part II of Article 396

(B) Underground Service-Entrance Cable. Underground Service-Entrance Cable (USE) shall not be used under the following conditions or in the following locations

(1) For interior wiring

(2) For above ground installations except where USE cable emerges from the ground and is terminated in an enclosure at an outdoor location and the cable is protected in accordance with 300.5(D)

(3) As aerial cable unless it is a multiconductor cable identified for use above ground and installed as messenger supported wiring in accordance with 225.10 and Part II of Article 396.

Author's Comment: A Panel 7 Task Group added Section **338.12** to cover where service entrance cables are not permitted to provide uniformity to all of the other cable articles in **Chapter 3**. Some of the text from existing **338.10** dealt with uses not permitted and was moved down into **338.12**. Compliance with **Part I** of **Article 334** was deleted from **338.10(B)(4)(a)** since compliance with the general requirements for NM cable does not apply to service entrance cables (**Article 338**), where installing the SE cable as branch circuits or feeders. Proposal 7-86 and Comment 7-49 add text to **338.10(B)(2), Exception** to clarify that compliance with **250.24(A)(5)** must be maintained. The text in the **Exception** in the 2005 NEC does not restrict the use of uninsulated grounded conuctors downstream from the service. For 2008 NEC, this would only be permissible for services and **250.32(B)** installations.

Proposals 7-88 and 7-90 have deleted "excluding **334.80**." Deleting this phrase will require SE cables used for interior wiring to comply with the ampacity requirements for NM cable.

Proposal 7-93 deleted text under uses not permitted for SE cable as follows: "For underground use unless identified for the purpose" since SE is never identified for underground use, only USE is identified for underground use. The new text states cable is "not permitted underground with or without a raceway."

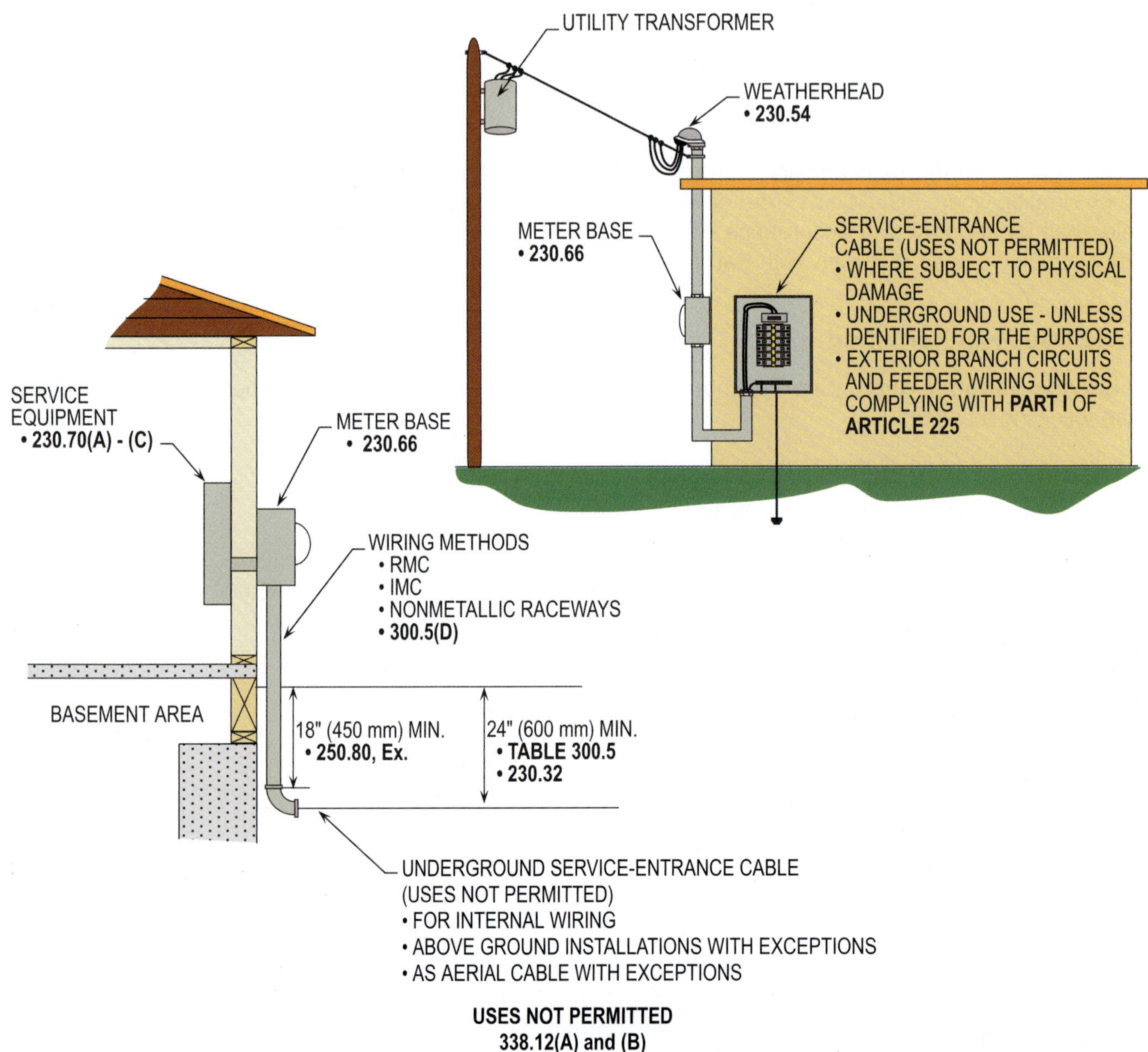

USES NOT PERMITTED
338.12(A) and (B)

Purpose of Change: This revision clarifies the uses permitted and not permitted for service-entrance and underground service-entrance cable.

NEC Ch. 3 - Article 342
Part II - 342.30(C)

Type of Change	Panel Action	UL	UL 508	API 500 - 1997	API 505 - 1997	OSHA - 1994
New Subsection	Accept	1242	-	-	-	-
ROP	**ROC**	**NFPA 70E - 2004**	**NFPA 70B - 2006**	**NFPA 79 - 2007**	**ANSI - 2005**	**NEMA - 2003**
pg. 345 # 8-9	pg. 209 # 8-7	-	11.4	13.5.3.2.3	C80.6	FB 2.10
log: 1345 CMP: 8	log: 781	Submitter: Mike Holt		2005 NEC: -		IEC: -

2005 NEC - 342.30 Securing and Supporting.

IMC shall be installed as a complete system in accordance with 300.18 and shall be securely fastened in place and supported in accordance with 342.30(A) and (B).

2008 NEC - 342.30 Securing and Supporting.

IMC shall be installed as a complete system in accordance with 300.18 and shall be securely fastened in place and supported in accordance with 342.30(A) and (B) or permitted to be unsupported in accordance with 342.30(C).

(C) Unsupported raceways. Where oversized, concentric or eccentric knockouts are not encountered, Type IMC shall be permitted to be unsupported where the raceway is not more than 450 mm (18 in.) and remains in unbroken lengths (without coupling). Such raceway shall terminate in an outlet box, junction box, device box, cabinet, or other termination at each end of the raceway.

Author's Comment: An 18-inch unbroken length of IMC does not have to be supported where an outlet box, junction box, device box, or cabinet is installed on either side of the IMC. The boxes or cabinet must be supported in accordance with other parts of the Code, such as **314.23** for boxes and **110.13(A)** for electrical equipment, and also provides support for the IMC.

Comment 8-7 provides the following information to justify changing from 3 feet in proposal to 18 inches in comment: The affects of weight and vibration of the raceway, especially with larger raceway sizes, may cause loosening at the raceway termination points. This loosening could certainly hamper the raceway's ability to safely carry the maximum fault current likely to be imposed on the raceway. Reducing the length from 900 mm (3 feet) to 450 mm (18 in.) would half the weight of the raceway. There is usually ample room to install normal supporting and securing hardware on longer lengths of IMC. Prohibiting unsupported raceways where oversized, concentric, or eccentric knockouts are encountered would serve to maintain the integrity of the equipment grounding function of the raceway.

Comment 8-8 adds an introduction to new (C) to the base text in 342.80.

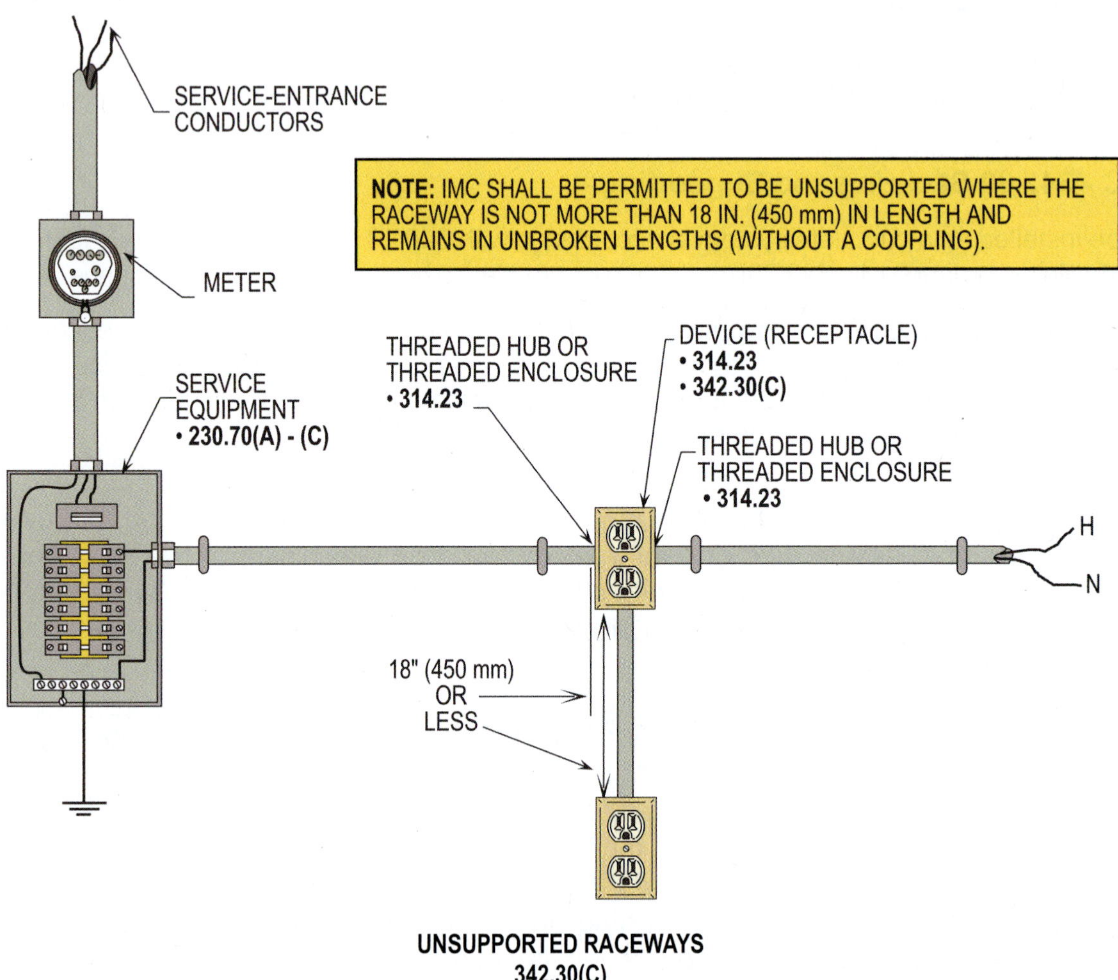

UNSUPPORTED RACEWAYS
342.30(C)

Purpose of Change: A new subsection has been added to permit IMC to be unsupported where the raceway is not more than 18 in. (450 mm) in length and remains in unbroken lengths (without a coupling), where oversized, concentric or eccentric knockouts are not encountered.

NEC Ch. 3 - Article 344
Part II - 344.10(A) through (C)

51

Type of Change	Panel Action	UL	UL 508	API 500 - 1997	API 505 - 1997	OSHA - 1994
Revision	Accept in Principle in Part	6 - 6A	-	-	-	-
ROP	**ROC**	**NFPA 70E - 2004**	**NFPA 70B - 2006**	**NFPA 79 - 2007**	**ANSI - 2005**	**NEMA - 2003**
pg. 346 # 8-15	pg. 210 # 8-14	-	11.4	13.5.3.2.3	C80.1	FB 2.10
log: 3456 CMP: 8	log: 984 Submitter: William A. Wolfe			2005 NEC: 344.10(A); (B); (C)		IEC: -

2005 NEC - 344.10 Uses Permitted.

(A) ~~All~~ Atmospheric Conditions and Occupancies. ~~Use of~~ RMC shall be permitted under all atmospheric conditions and occupancies. ~~Ferrous raceways and fittings protected from corrosion solely by enamel shall be permitted only indoors and in occupancies not subject to severe corrosive influences.~~

(B) ~~Corrosion~~ Environments. RMC, elbows, couplings, and fittings shall be permitted to be installed in concrete, in direct contact with the earth, or in areas subject to severe corrosive influences where protected by corrosion protection and judged suitable for the condition.

(C) Cinder Fill. RMC shall be permitted to be installed in or under cinder fill where subject to permanent moisture where protected on all sides by a layer of noncinder concrete not less than 50 mm (2 in.) thick; where the conduit is not less than 450mm (18 in.) under the fill; or where protected by corrosion protection and judged suitable for the condition.

2008 NEC - 344.10 Uses Permitted.

(A) Atmospheric Conditions and Occupancies.

(1) Galvanized Steel and Stainless Steel RMC. Galvanized steel and stainless steel RMC shall be permitted under all atmospheric conditions and occupancies.

(2) Red brass RMC. Red brass RMC shall be permitted to be installed for direct burial and swimming pool applications.

(3) Aluminum RMC. Aluminum RMC shall be permitted to be installed where judged suitable for the environment. Rigid aluminum conduit encased in concrete or in direct contact with the earth shall be provided with approved supplementary corrosion protection.

(4) Ferrous Raceways and Fittings. Ferrous raceways and fittings protected from corrosion solely by enamel shall be permitted only indoors and in occupancies not subject to severe corrosive influences.

(B) Corrosive Environments.

(1) Galvanized Steel, Stainless Steel, and Red Brass RMC, Elbows, Couplings, and Fittings. Galvanized steel, stainless steel and red brass RMC, elbows, couplings, and fittings shall be permitted to be installed in concrete, in direct contact with the earth, or in areas subject to severe corrosive influences where protected by corrosion protection and judged suitable for the condition.

(2) Supplementary Protection of Aluminum RMC. Aluminum RMC shall be provided with approved supplementary corrosion protection where encased in concrete or in direct contact with the earth.

(C) Cinder Fill. Galvanized steel, stainless steel and red brass RMC shall be permitted to be installed in or under cinder fill where subject to permanent moisture where protected on all sides by a layer of noncinder concrete not less than 50 mm (2 in.) thick; where the conduit is not less than 450mm (18 in.) under the fill; or where protected by corrosion protection and judged suitable for the condition.

Author's Comment: In UL's Electrical Construction Equipment Directory, the listing of galvanized rigid steel conduit installed in concrete or in contact with soil does not require supplementary corrosion protection. Aluminum conduit used in concrete or in contact with soil does require supplementary corrosion protection.

Article 344 has not differentiated between the corrosion protection required for ferrous vs. non-ferrous metal conduit, leading to confusion among installers/users. There have also been questions concerning whether or not galvanizing provides corrosion protection. This new text provides specific guidance for all types of rigid metal conduit and is consistent with listing requirements.

The comment added titles to each of the subsections in compliance with the NEC Style Manual. The proposed Fine Print Note providing information on corrosion protection was deleted because it was in conflict with the accepted text about corrosion protection for RMC.

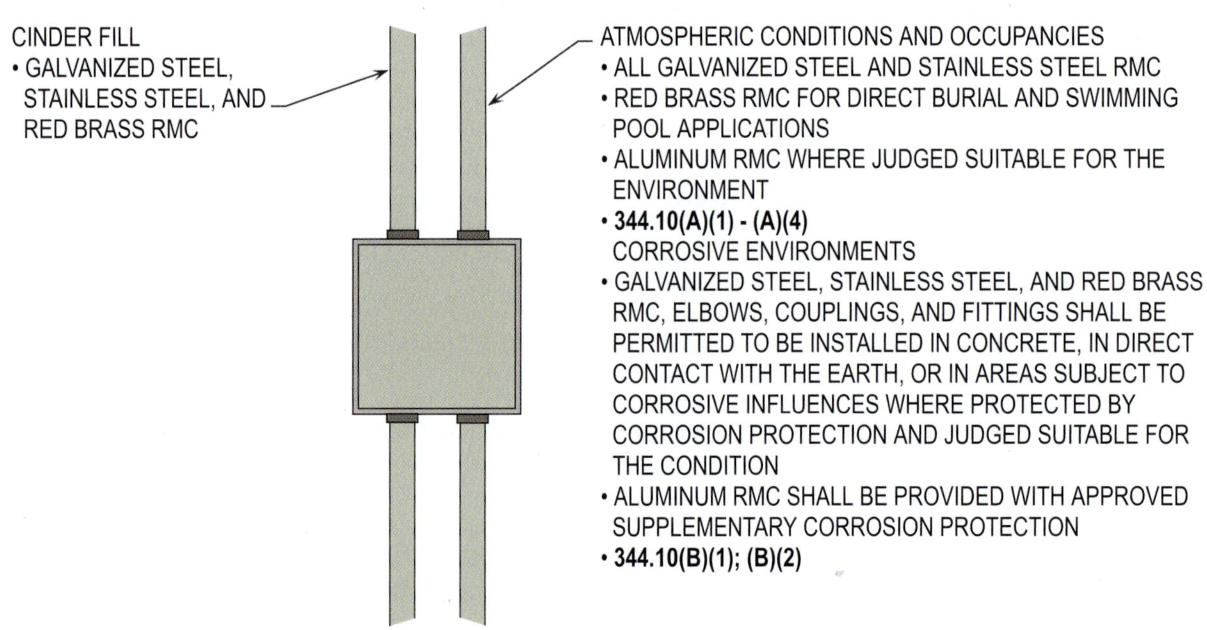
CINDER FILL
• GALVANIZED STEEL, STAINLESS STEEL, AND RED BRASS RMC
ATMOSPHERIC CONDITIONS AND OCCUPANCIES
• ALL GALVANIZED STEEL AND STAINLESS STEEL RMC
• RED BRASS RMC FOR DIRECT BURIAL AND SWIMMING POOL APPLICATIONS
• ALUMINUM RMC WHERE JUDGED SUITABLE FOR THE ENVIRONMENT
• 344.10(A)(1) - (A)(4)
CORROSIVE ENVIRONMENTS
• GALVANIZED STEEL, STAINLESS STEEL, AND RED BRASS RMC, ELBOWS, COUPLINGS, AND FITTINGS SHALL BE PERMITTED TO BE INSTALLED IN CONCRETE, IN DIRECT CONTACT WITH THE EARTH, OR IN AREAS SUBJECT TO CORROSIVE INFLUENCES WHERE PROTECTED BY CORROSION PROTECTION AND JUDGED SUITABLE FOR THE CONDITION
• ALUMINUM RMC SHALL BE PROVIDED WITH APPROVED SUPPLEMENTARY CORROSION PROTECTION
• 344.10(B)(1); (B)(2)

USES PERMITTED
344.10(A) through (C)

Purpose of Change: This revision clarifies the uses permitted for galvanized, stainless, red brass, and aluminum RMC and the requirements for additional corrosion protection where necessary.

NEC Ch. 3 - Article 344
Part II - 344.30(C)

Type of Change		Panel Action		UL	UL 508	API 500 - 1997	API 505 - 1997	OSHA - 1994
New Subsection		Accept		6	-	-	-	-
ROP		ROC		NFPA 70E - 2004	NFPA 70B - 2006	NFPA 79 - 2007	ANSI - 2005	NEMA - 2003
pg. 348	# 8-23	pg. 211	# 8-20	-	11.4	13.5.3.2.3	C80.1	FB 2.10
log: 1346	CMP: 8	log: 782	Submitter: Mike Holt			2005 NEC: -		IEC: -

2005 NEC - 344.30 Securing and Supporting.

RMC shall be installed as a complete system in accordance with 300.18 and shall be securely fastened in place and supported in accordance with 344.30(A) and (B).

2008 NEC - 344.30 Securing and Supporting.

RMC shall be installed as a complete system in accordance with 300.18 and shall be securely fastened in place and supported in accordance with 344.30(A) and (B) or permitted to be unsupported in accordance with 344.30(C).

(C) Unsupported Raceways: Where oversized, concentric or eccentric knockouts are not encountered, Type RMC shall be permitted to be unsupported where the raceway is not more than 450 mm (18 in.) and remains in unbroken lengths (without coupling). Such raceway shall terminate in an outlet box, junction box, device box, cabinet, or other termination at each end of the raceway.

Author's Comment: An 18-inch unbroken length of RMC does not have to be supported where an outlet box, junction box, device box, or cabinet is installed on either side of the RMC. The boxes or cabinet must be supported in accordance with other parts of the Code, such as **314.23** for boxes and **110.13(A)** for electrical equipment, and also provides the support for the RMC.

Comment 8-20 provides the following information to justify changing from 3 feet in proposal to 18 inches in comment: The affects of weight and vibration of the raceway, especially with larger raceway sizes, may cause loosening at the raceway termination points. This loosening could certainly hamper the raceway's ability to safely carry the maximum fault current likely to be imposed on the raceway. Reducing the length from 900 mm (3 feet) to 450 mm (18 in.) would half the weight of the raceway. There is usually ample room to install normal supporting and securing hardware on longer lengths of RMC. Prohibiting unsupported raceways where oversized, concentric, or eccentric knockouts are encountered would serve to maintain the integrity of the equipment grounding function of the raceway.

Comment 8-21 adds an introduction to new (C) to the base text in 344.30.

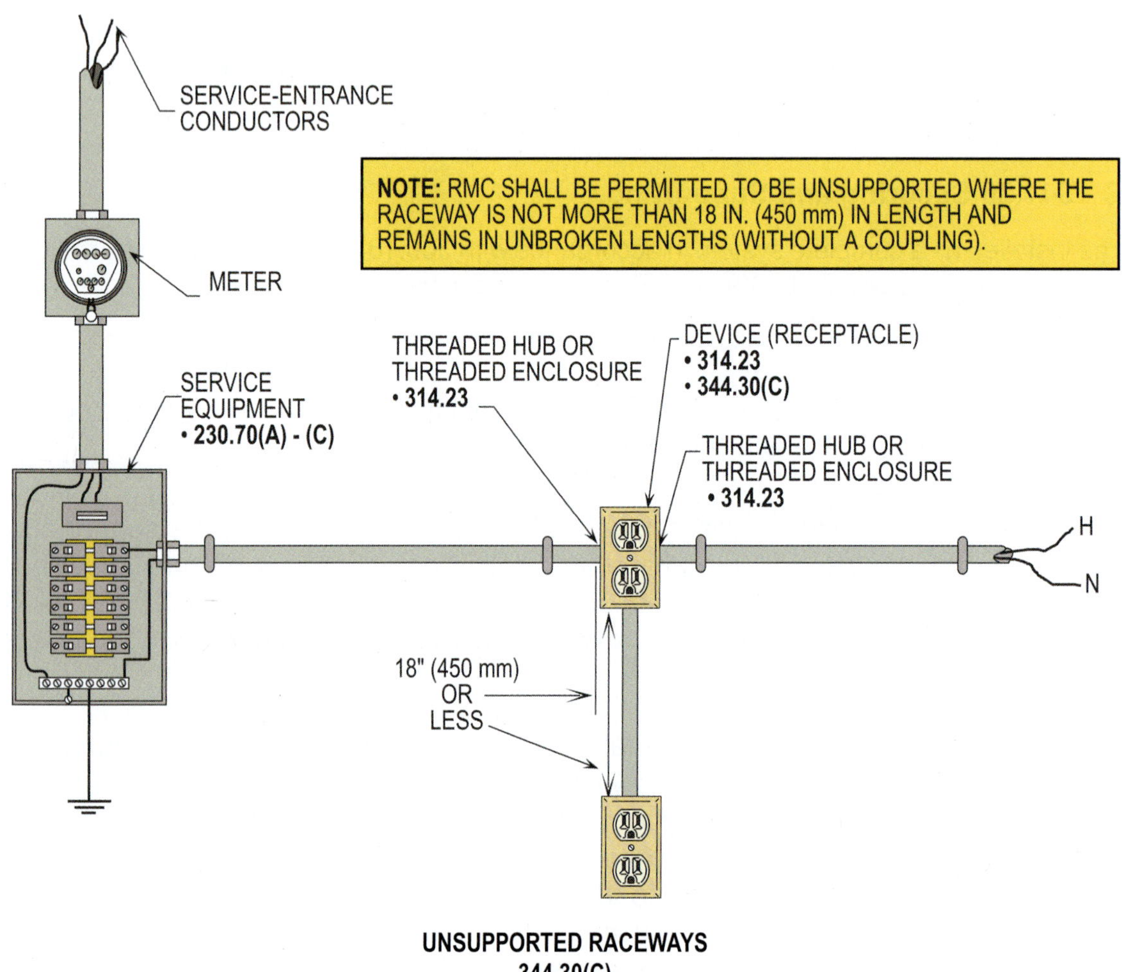

UNSUPPORTED RACEWAYS
344.30(C)

Purpose of Change: A new subsection has been added to permit RMC to be unsupported where the raceway is not more than 18 in. (450 mm) in length and remains in unbroken lengths (without a coupling), where oversized, concentric or eccentric knockouts are not encountered.

NEC Ch. 3 - Article 348
Part II - 348.12(1)

52

Type of Change	Panel Action	UL	UL 508	API 500 - 1997	API 505 - 1997	OSHA - 1994
Revision	Accept	1	-	-	-	-
ROP	**ROC**	**NFPA 70E - 2004**	**NFPA 70B - 2006**	**NFPA 79 - 2007**	**NFPA**	**NEMA - 2003**
pg. 349 # 8-26	pg. - # -	-	-	13.5.4	-	FB 2.20
log: 641 CMP: 8	log: -	Submitter: Vince Baclawski		2005 NEC: 348.12(1)(1)		IEC: -

2005 NEC - 348.12 Uses Not Permitted.

FMC shall not be used in the following:

(1) In wet locations ~~unless the conductors are approved for the specific conditions and the installation is such that liquid is not likely to enter raceways or enclosures to which the conduit is connected~~.

2008 NEC - 348.12 Uses Not Permitted.

FMC shall not be used in the following:

(1) In wet locations.

Author's Comment: The NEC is clear that liquid shall not enter the raceways or enclosures to which flexible metal conduit is connected. Flexible metal conduit does not have a continuous outer surface. It has an interlocking metal construction that along with its listed connectors readily permits the entrance of liquids. The present language "the installation is such that liquid is not likely to enter raceways or enclosures" does not provide the assurance necessary that the requirement to not permit liquid to enter raceways or enclosures will be met. Both liquidtight flexible metallic conduit and liquidtight flexible nonmetallic conduit products and associated liquidtight connectors are common, readily available, and assure compliance with the requirement that the entrance of liquid is not permitted.

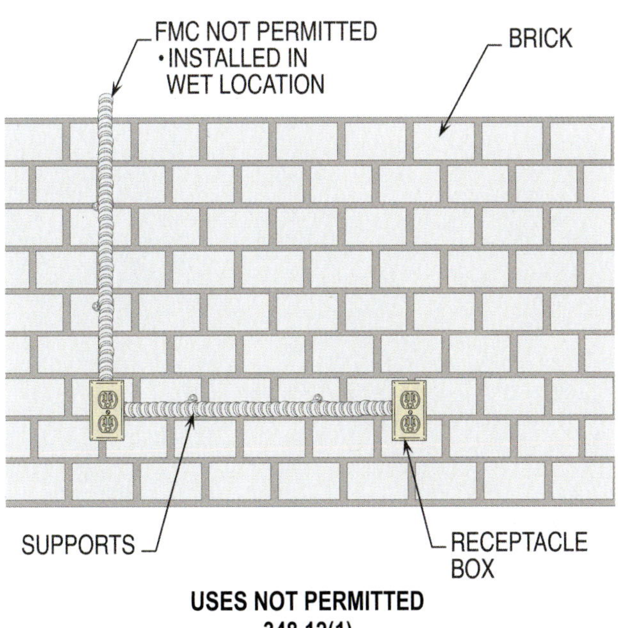

USES NOT PERMITTED
348.12(1)

Purpose of Change: This revision clarifies that FMC shall not be permitted to be installed in any wet locations.

NEC Ch. 3 - Article 348
Part II - 348.30(A), Exception No. 1

Type of Change		Panel Action		UL	UL 508	API 500 - 1997	API 505 - 1997	OSHA - 1994
Revision		Accept		1	-	-	-	-
ROP		**ROC**		**NFPA 70E - 2004**	**NFPA 70B - 2006**	**NFPA 79 - 2007**	**NFPA**	**NEMA - 2003**
pg. 350	# 8-33	pg. -	# -	-	-	13.5.4	-	FB 2.20
log: 3331	CMP: 8	log: -	Submitter: Daniel Leaf			2005 NEC: 348.30(A), Exception No,. 1		IEC: -

2005 NEC - 348.30 Securing and Supporting.

FMC shall be securely fastened in place and supported in accordance with 348.30(A) and (B).

(A) Securely Fastened. FMC shall be securely fastened in place by an approved means within 300 mm (12 in.) of each box, cabinet, conduit body, or other conduit termination and shall be supported and secured at intervals not to exceed 1.4 m (4 1/2 ft).

Exception No. 1: Where FMC is fished.

2008 NEC - 348.30 Securing and Supporting.

FMC shall be securely fastened in place and supported in accordance with 348.30(A) and (B).

(A) Securely Fastened. FMC shall be securely fastened in place by an approved means within 300 mm (12 in.) of each box, cabinet, conduit body, or other conduit termination and shall be supported and secured at intervals not to exceed 1.4 m (4 1/2 ft).

Exception No. 1: Where FMC is fished between access points through concealed spaces in finished buildings or structures and supporting is impractical.

Author's Comment: The additional text in the exception provides information that FMC can be fished between access points through concealed spaces in finished buildings without supporting the FMC.

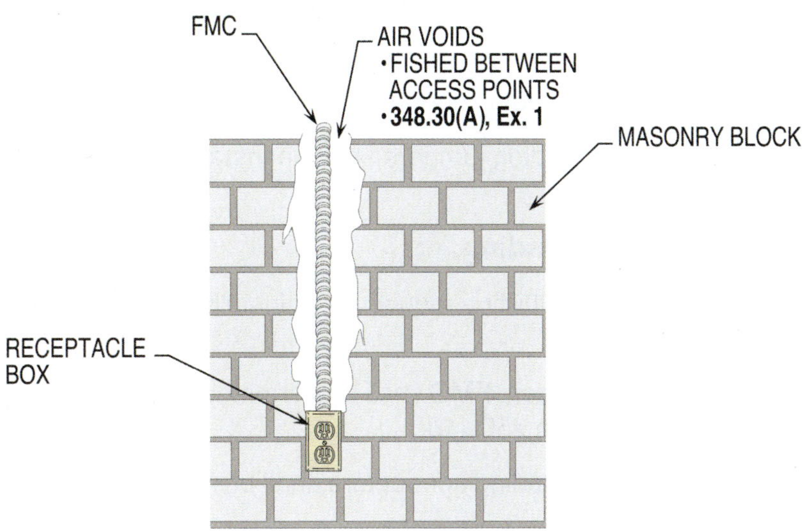

SECURELY FASTENED
348.30(A), Exception No. 1

Purpose of Change: This revision clarifies that FMC shall be permitted to be fished between access points through concealed spaces in finished buildings.

NEC Ch. 3 - Article 348
Part II - 348.60

Type of Change		Panel Action		UL	UL 508	API 500 - 1997	API 505 - 1997	OSHA - 1994
Revision		Accept		-	-	-	-	-
ROP		ROC		NFPA 70E - 2004	NFPA 70B - 2006	NFPA 79 - 2007	NFPA	NEMA - 2003
pg. 351	# 8-40	pg. -	# -	-	-	13.5.4	-	FB 2.20
log: 882	CMP: 8	log: -	Submitter: Noel Williams			2005 NEC: 348.60		IEC: -

2005 NEC - 348.60 Grounding and Bonding.

Where used to connect equipment where flexibility is required, an equipment grounding conductor shall be installed.

Where flexibility is not required, FMC shall be permitted to be used as an equipment grounding conductor when installed in accordance with 250.118(5).

Where required or installed, equipment grounding conductors shall be installed in accordance with 250.134(B).

Where required or installed, equipment bonding jumpers shall be installed in accordance with 250.102.

2008 NEC - 348.60 Grounding and Bonding.

Where used to connect equipment where flexibility is required after installation, an equipment grounding conductor shall be installed.

Where flexibility is not required after installation, FMC shall be permitted to be used as an equipment grounding conductor when installed in accordance with 250.118(5).

Where required or installed, equipment grounding conductors shall be installed in accordance with 250.134(B).

Where required or installed, equipment bonding jumpers shall be installed in accordance with 250.102.

Author's Comment: This change is intended to make this section agree with the requirements referenced in **250.118(5)** that describe installations where FMC may (and indirectly, where it may not) be used for grounding.

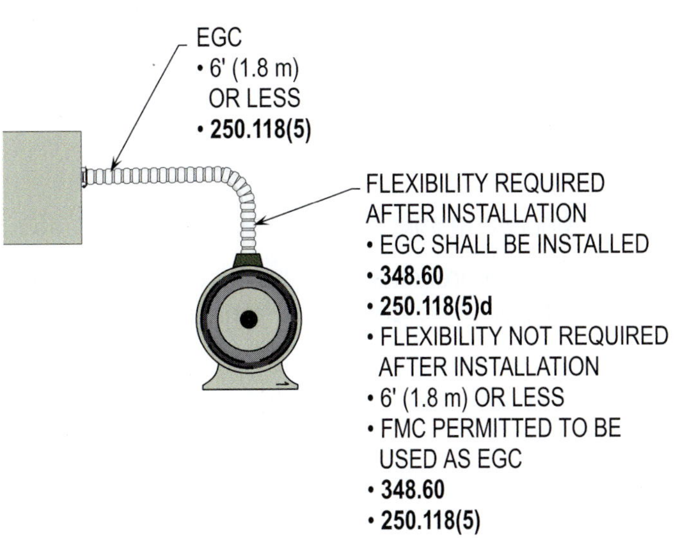

GROUNDING AND BONDING
348.60

Purpose of Change: This revision clarifies that FMC is not permitted to serve as EGC if more than 6 ft. (1.8 m) in length or 6 feet or less where flexibility is required after installation.

NEC Ch. 3 - Article 350
Part II - 350.30(A), Exception No. 1

Type of Change	Panel Action	UL	UL 508	API 500 - 1997	API 505 - 1997	OSHA - 1994
Revision	Accept	360	-	-	-	-
ROP	**ROC**	**NFPA 70E - 2004**	**NFPA 70B - 2006**	**NFPA 79 - 2007**	**NFPA**	**NEMA - 2003**
pg. 352, # 8-46	pg. -, # -	-	-	13.5.5	-	FB 2.20
log: 3330, CMP: 8	log: -	Submitter: Daniel Leaf		2005 NEC: 350.30(A), Exception No. 1		IEC: -

2005 NEC - 350.30 Securing and Supporting.

LFMC shall be securely fastened in place and supported in accordance with 350.30(A) and (B).

(A) Securely Fastened. LFMC shall be securely fastened in place by an approved means within 300 mm (12 in.) of each box, cabinet, conduit body, or other conduit termination and shall be supported and secured at intervals not to exceed 1.4 m (4 1/ 2 ft).

Exception No. 1: Where LFMC is fished.

2008 NEC - 350.30 Securing and Supporting.

LFMC shall be securely fastened in place and supported in accordance with 350.30(A) and (B).

(A) Securely Fastened. LFMC shall be securely fastened in place by an approved means within 300 mm (12 in.) of each box, cabinet, conduit body, or other conduit termination and shall be supported and secured at intervals not to exceed 1.4 m (4 1/ 2 ft).

Exception No. 1: Where LFMC is fished between access points through concealed spaces in finished buildings or structures and supporting is impractical.

Author's Comment: The additional text in the exception provides information that LFMC can be fished between access points through concealed spaces in finished buildings without supporting the LFMC.

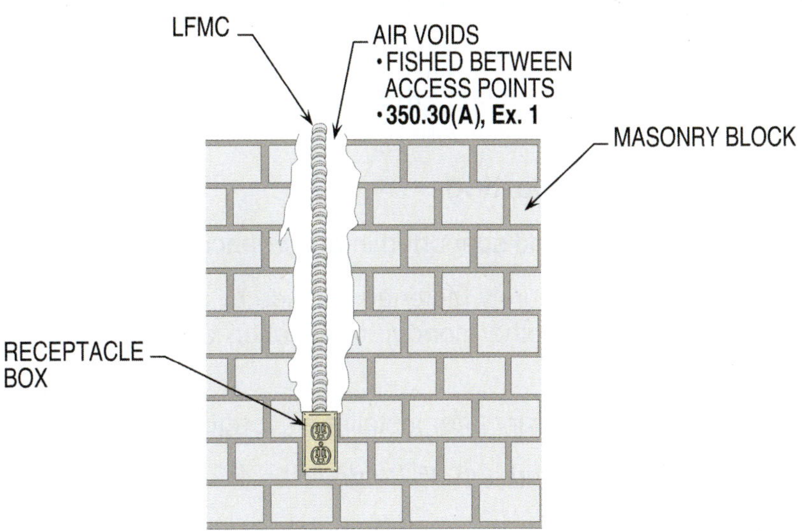

SECURELY FASTENED
350.30(A), Exception No. 1

Purpose of Change: This revision clarifies that LFMC shall be permitted to be fished between access points through concealed spaces in finished buildings.

NEC Ch. 3 - Article 350
Part II - 350.30(A), Exception No. 2

Type of Change	Panel Action	UL	UL 508	API 500 - 1997	API 505 - 1997	OSHA - 1994
Revision	Accept	360	-	-	-	-
ROP	**ROC**	**NFPA 70E - 2004**	**NFPA 70B - 2006**	**NFPA 79 - 2007**	**NFPA**	**NEMA - 2003**
pg. 352 # 8-44	pg. 213 # 8-29	-	-	13.5.5	-	FB 2.20
log: 1347 CMP: 8	log: 2044 Submitter: Mike Holt			2005 NEC: 350.30(A), Exception No. 2		IEC: -

2005 NEC - 350.30 Securing and Supporting.

LFMC shall be securely fastened in place and supported in accordance with 350.30(A) and (B).

(A) Securely Fastened. LFMC shall be securely fastened in place by an approved means within 300 mm (12 in.) of each box, cabinet, conduit body, or other conduit termination and shall be supported and secured at intervals not to exceed 1.4 m (4 1/2 ft).

Exception No. 2: ~~Lengths not exceeding 900 mm (3 ft) at terminals~~ where flexibility is necessary.

2008 NEC - 350.30 Securing and Supporting.

LFMC shall be securely fastened in place and supported in accordance with 350.30(A) and (B).

(A) Securely Fastened. LFMC shall be securely fastened in place by an approved means within 300 mm (12 in.) of each box, cabinet, conduit body, or other conduit termination and shall be supported and secured at intervals not to exceed 1.4 m (4 1/2 ft).

Exception No. 2: Where flexibility is necessary after installation, lengths shall not exceed the following:

(1) 900 mm (3 ft) for metric designators 16 through 35 (trade sizes 1/2 through 1 1/4)

(2) 1200 mm (4 ft) for metric designators 41 through 53 (trade sizes 1 1/2 through 2)

(3) 1500 mm (5 ft) for metric designators 63 (trade size 2 1/2) and larger

Author's Comment: Where flexibility of liquidtight flexible metal conduit is required, Exception No. 2 limits the length to 3 feet for 1/2-inch through 1 1/4-inch sizes, 4 feet for 1 1/2-inch and 2-inch, and 5 feet for 2 1/2-inch and larger. This change matches the change that occurred in the 2005 NEC for flexible metal conduit (FMC). Proposal 8-47 and Comment 8-29 add the words "after installation" to clarify that the flexibility concern is not during installation but is intended to apply after installation during operation of the equipment. Flexibility is not necessary for equipment permanently attached to the building. It is necessary for motor connections with motor belt tension adjustments, for example.

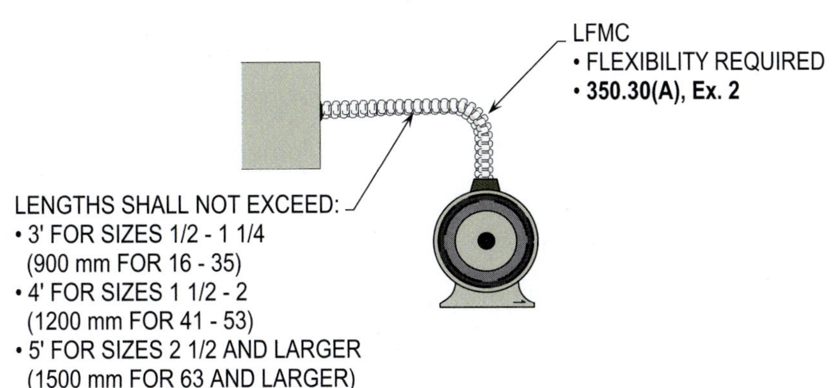

SECURELY FASTENED
350.30(A), Exception No. 2

Purpose of Change: This revision clarifies the lengths for LFMC where flexibility is required and provides limitation on lengths of LFMC where flexibility is required.

NEC Ch. 3 - Article 350
Part II - 350.60

Type of Change		Panel Action		UL	UL 508	API 500 - 1997	API 505 - 1997	OSHA - 1994
Revision		Accept		360	-	-	-	-
ROP		ROC		NFPA 70E - 2004	NFPA 70B - 2006	NFPA 79 - 2007	NFPA	NEMA - 2003
pg. 353	# 8-49	pg. -	# -	-	-	13.5.5	-	FB 2.20
log: 883	CMP: 8	log: -	Submitter: Noel Williams			2005 NEC: 350.60		IEC: -

2005 NEC - 350.60 Grounding and Bonding.

Where used to connect equipment where flexibility is required, an equipment grounding conductor shall be installed.

Where flexibility is not required, LFMC shall be permitted to be used as an equipment grounding conductor when installed in accordance with 250.118(6).

Where required or installed, equipment grounding conductors shall be installed in accordance with 250.134(B).

Where required or installed, equipment bonding jumpers shall be installed in accordance with 250.102.

FPN: See 501.30(B), 502.30(B), and 503.30(B) for types of equipment grounding conductors.

2008 NEC - 350.60 Grounding and Bonding.

Where used to connect equipment where flexibility is required after installation, an equipment grounding conductor shall be installed.

Where flexibility is not required after installation, LFMC shall be permitted to be used as an equipment grounding conductor when installed in accordance with 250.118(6).

Where required or installed, equipment grounding conductors shall be installed in accordance with 250.134(B).

Where required or installed, equipment bonding jumpers shall be installed in accordance with 250.102.

FPN: See 501.30(B), 502.30(B), 503.30(B), 505.25(B), and 506.25(B) for types of equipment grounding conductors.

Author's Comment: This change is intended to make this section agree with the requirements referenced in **250.118(6)** that describe installations where LFMC may (and indirectly, where it may not) be used for grounding. Proposal 8-52 added references in the FPN to cover equipment grounding conductors for Zones 0, 1, and 2 as well as Zones 20, 21, and 22.

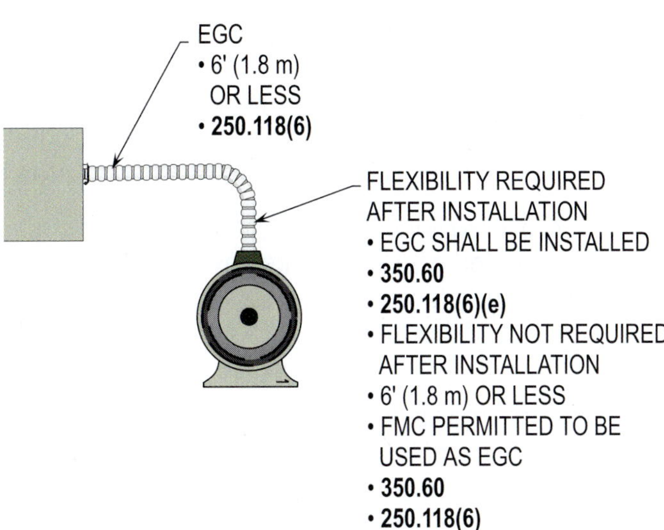

GROUNDING AND BONDING
350.60

Purpose of Change: This revision clarifies that LFMC is not permitted to serve as EGC if more than 6 ft. (1.8 m) in length or 6 feet or less where flexibility is required after installation.

NEC Ch. 3 - Article 352
Part II - 352.30(C)

Type of Change		Panel Action		UL	UL 508	API 500 - 1997	API 505 - 1997	OSHA - 1994
New Subsection		Accept in Principle		651	-	-	-	-
ROP		**ROC**		**NFPA 70E - 2004**	**NFPA 70B - 2006**	**NFPA 79 - 2007**	**NFPA**	**NEMA - 2003**
pg. 361	# 8-65	pg. 215	# 8-38	-	-	13.5.3.3	-	TC 2
log: 1348	CMP: 8	log: 785	Submitter: Mike Holt			2005 NEC: -		IEC: -

2005 NEC - 352.30 Securing and Supporting.

~~RNC~~ shall be installed as a complete system as provided in 300.18 and shall be fastened so that movement from thermal expansion or contraction is permitted. ~~RNC~~ shall be securely fastened and supported in accordance with 352.30(A) and (B).

2008 NEC - 352.30 Securing and Supporting.

PVC conduit shall be installed as a complete system as provided in 300.18 and shall be fastened so that movement from thermal expansion or contraction is permitted. PVC conduit shall be securely fastened and supported in accordance with 352.30(A) and (B) or permitted to be unsupported in accordance with 352.30(C).

(C) Unsupported raceways. Where oversized, concentric or eccentric knockouts are not encountered, PVC conduit shall be permitted to be unsupported where the raceway is not more than 450 mm (18 in.) and remains in unbroken lengths (without coupling). Such raceway shall terminate in an outlet box, junction box, device box, cabinet, or other termination at each end of the raceway.

Author's Comment: An 18-inch unbroken length of PVC conduit does not have to be supported where an outlet box, junction box, device box, or cabinet is installed on either side of the PVC conduit. The boxes or cabinet must be supported in accordance with other parts of the Code, such as **314.23** for boxes and **110.13(A)** for electrical equipment, and also provides the support for the PVC conduit.

Comment 8-65 provides the following information to justify changing from 3 feet in the proposal to 18 inches in the comment: The affects of weight and vibration of the raceway, especially with larger raceway sizes, may cause loosening at the raceway termination points. This loosening could certainly hamper the raceway's ability to safely carry the maximum fault current likely to be imposed on the raceway. Reducing the length from 900 mm (3 feet) to 450 mm (18 in.) would half the weight of the raceway. There is usually ample room to install normal supporting and securing hardware on longer lengths of PVC conduit. Prohibiting unsupported raceways where oversized, concentric, or eccentric knockouts are encountered would serve to maintain the integrity of the equipment grounding function of the raceway.

Comment 8-39 adds an introduction to new (C) dealing with PVC conduit to the base text in 352.30.

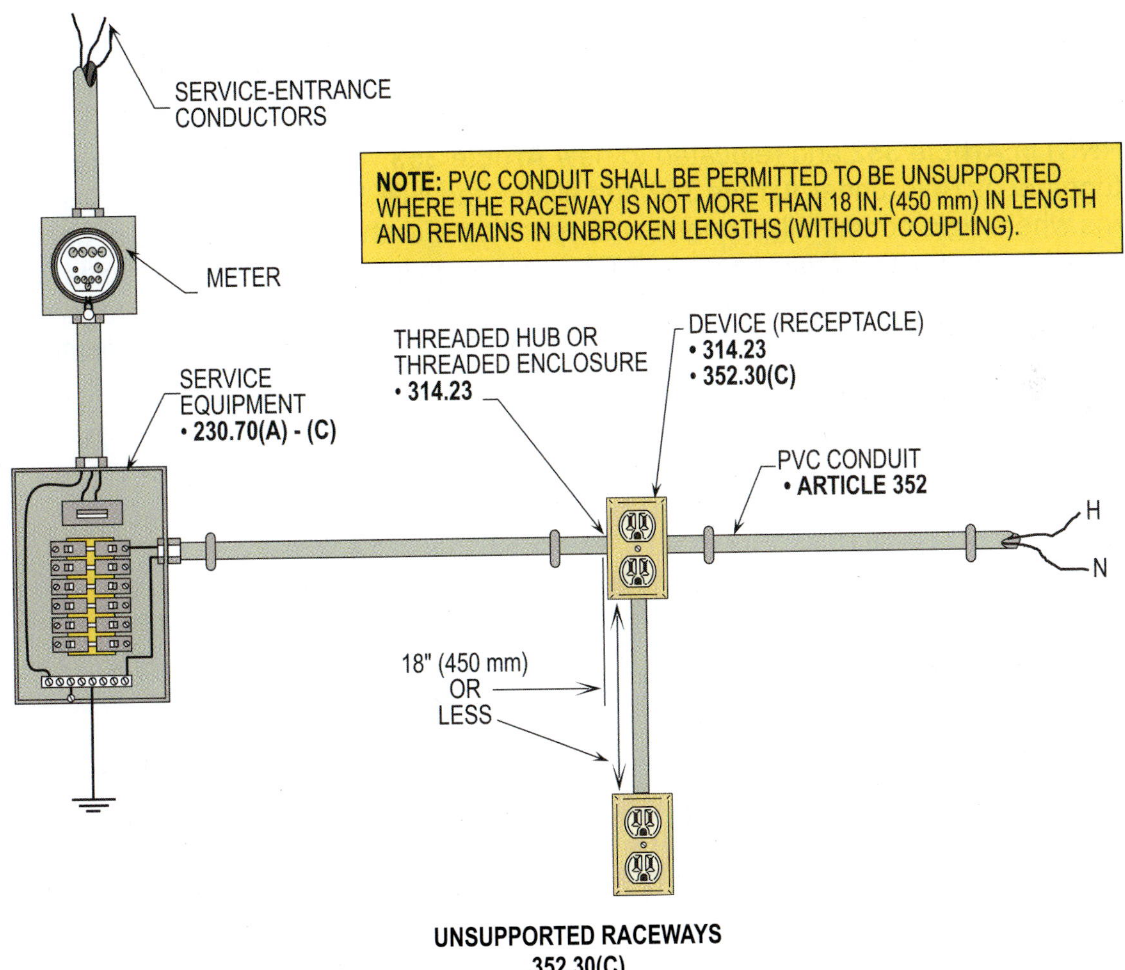

UNSUPPORTED RACEWAYS
352.30(C)

Purpose of Change: A new subsection has been added to permit PVC conduit to be unsupported where the raceway is not more than 18 in. (450 mm) in length and remains in unbroken lengths (without coupling), where oversized, concentric or eccentric knockouts are not encountered.

NEC Ch. 3 - Article 355
Parts I through III

53

Type of Change		Panel Action		UL	UL 508	API 500 - 1997	API 505 - 1997	OSHA - 1994
New Article		Accept in Principle		1684	-	-	-	-
ROP		ROC		NFPA 70E - 2004	NFPA 70B - 2006	NFPA 79 - 2007	NFPA	NEMA - 2002
pg. 365	# 8-78	pg. 219	# 8-49b	-	-	13.5.3.3	-	TC 14
log: 1920	CMP: 8	log: CC 800	Submitter: William Wagner			2005 NEC: -		IEC: -

2008 NEC - Article 355
Reinforced Thermosetting Resin Conduit: Type RTRC

Author's Comment: For the 2005 edition of the NEC, HDPE was separated from rigid nonmetallic conduit (RNC) in **Article 352** and relocated to new **Article 353**. This left two very dissimilar products grouped together as RNC under **Article 352** and eliminated HDPE as an acceptable wiring method in applications where rigid nonmetallic conduit was specified. For the 2008 NEC, the separation of PVC and reinforced thermosetting resin conduit (RTRC) products into two separate articles will complete the task begun during the 2005 NEC to provide three separate articles for nonmetallic conduits and will correct this situation by better defining the installation and construction specifications for each conduit type.

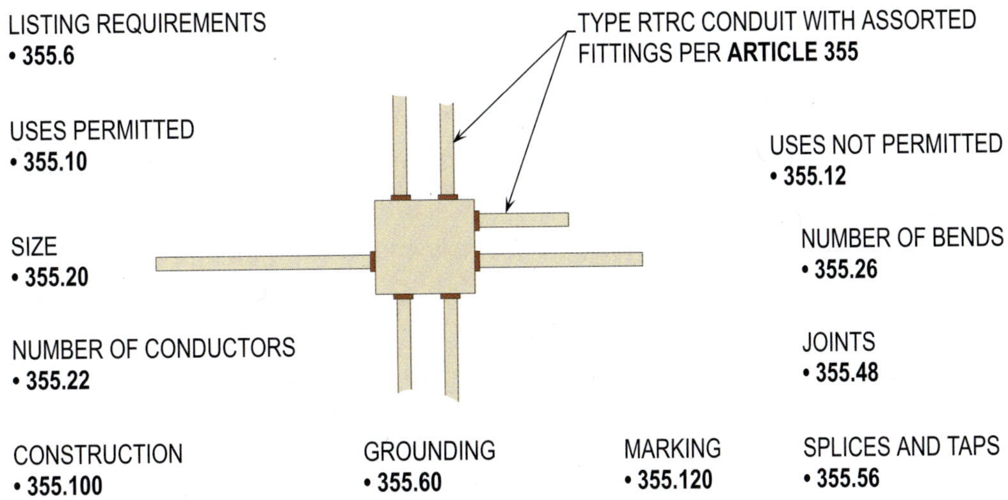

REINFORCED THERMOSETTING RESIN CONDUIT: TYPE RTRC ARTICLE 355

Purpose of Change: A new article has been added for reinforced thermosetting resin conduit (Type RTRC).

NEC Ch. 3 - Article 356
Part II - 356.10(7)

Type of Change		Panel Action		UL	UL 508	API 500 - 1997	API 505 - 1997	OSHA - 1994
New Item		Accept in Principle		1660	-	-	-	-
ROP		ROC		NFPA 70E - 2004	NFPA 70B - 2006	NFPA 79 - 2007	NFPA	NEMA - 2003
pg. 368	# 8-82	pg. -	# -	-	-	-	-	FB 2.20
log: 2622	CMP: 8	log: -	Submitter: David H. Kendall			2005 NEC: -		IEC: -

2005 NEC - 356.10 Uses Permitted.

LFNC shall be permitted to be used in exposed or concealed locations for the following purposes:

FPN: Extreme cold may cause some types of nonmetallic conduits to become brittle and therefore more susceptible to damage from physical contact.

(1) Where flexibility is required for installation, operation, or maintenance

(2) Where protection of the contained conductors is required from vapors, liquids, or solids

(3) For outdoor locations where listed and marked as suitable for the purpose

(4) For direct burial where listed and marked for the purpose

(5) Type LFNC-B shall be permitted to be installed in lengths longer than 1.8 m (6 ft) where secured in accordance with 356.30

(6) Type LFNC-B as a listed manufactured prewired assembly, metric designator 16 through 27 (trade size 1/ 2 through 1) conduit

2008 NEC - 356.10 Uses Permitted.

LFNC shall be permitted to be used in exposed or concealed locations for the following purposes:

FPN: Extreme cold may cause some types of nonmetallic conduits to become brittle and therefore more susceptible to damage from physical contact.

(1) Where flexibility is required for installation, operation, or maintenance

(2) Where protection of the contained conductors is required from vapors, liquids, or solids

(3) For outdoor locations where listed and marked as suitable for the purpose

(4) For direct burial where listed and marked for the purpose

(5) Type LFNC-B shall be permitted to be installed in lengths longer than 1.8 m (6 ft) where secured in accordance with 356.30

(6) Type LFNC-B as a listed manufactured prewired assembly, metric designator 16 through 27 (trade size 1/ 2 through 1) conduit

(7) For encasement in concrete where listed for direct burial and installed in compliance with 356.42

Author's Comment: This revision provides clarification to **356.10** Uses Permitted, making it clear that LFNC can be encased in concrete in accordance with the listing of liquidtight flexible nonmetallic conduit where straight fittings are used (angle fittings are not permitted to be encased in concrete).

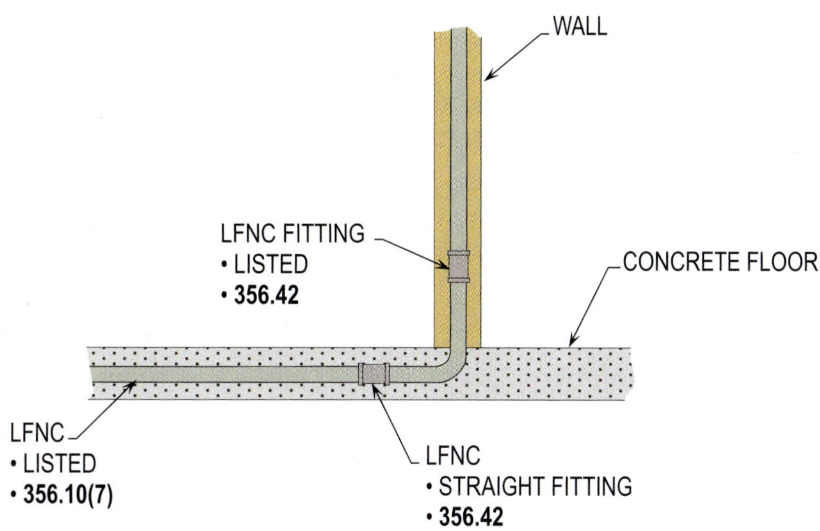

USES PERMITTED
356.10(7)

Purpose of Change: A new item has been added to permit liquidtight flexible nonmetallic conduit (LFNC) to be installed for encasement in concrete where listed for direct burial and concrete-encasement.

NEC Ch. 3 - Article 358
Part II - 358.30(C)

Type of Change		Panel Action		UL	UL 508	API 500 - 1997	API 505 - 1997	OSHA - 1994
New Subsection		Accept		797	-	-	-	-
ROP		ROC		NFPA 70E - 2004	NFPA 70B - 2006	NFPA 79 - 2007	ANSI	NEMA - 2003
pg. 372	# 8-104	pg. 220	# 8-57	-	-	13.5.3.2.4	C80.3	FB 1
log: 1349	CMP: 8	log: 784	Submitter: Mike Holt			2005 NEC: -		IEC: -

2005 NEC - 358.30 Securing and Supporting.

EMT shall be installed as a complete system in accordance with 300.18 and shall be securely fastened in place and supported in accordance with 358.30(A) and (B).

2008 NEC - 358.30 Securing and Supporting.

EMT shall be installed as a complete system in accordance with 300.18 and shall be securely fastened in place and supported in accordance with 358.30(A) and (B) or permitted to be unsupported in accordance with 358.30(C).

(C) Unsupported raceways. Where oversized, concentric, or eccentric knockouts are not encountered, Type EMT shall be permitted to be unsupported where the raceway is not more than 450 mm (18 in.) and remains in unbroken lengths (without coupling). Such raceways shall terminate in an outlet box, device box, cabinet, or other termination at each end of the raceway.

Author's Comment: An 18-inch unbroken length of EMT does not have to be supported where an outlet box, device box, or cabinet is installed on either side of the EMT. The box or cabinet must be supported in accordance with other parts of the Code, such as **314.23** for boxes and **110.13(A)** for electrical equipment, and also provides support for the EMT.

Comment 8-57 provides the following information to justify changing from 3 feet in the proposal to 18 inches in the comment: The affects of weight and vibration of the raceway, especially with larger raceway sizes, may cause loosening at the raceway termination points. This loosening could certainly hamper the raceway's ability to safely carry the maximum fault current likely to be imposed on the raceway. Reducing the length from 900 mm (3 feet) to 450 mm (18 in.) would half the weight of the raceway. There is usually ample room to install normal supporting and securing hardware on longer lengths of EMT. Prohibiting unsupported raceways where oversized, concentric, or eccentric knockouts are encountered would serve to maintain the integrity of the equipment grounding function of the raceway.

Comment 8-58 adds an introduction to new (C) to the base text in 358.30.

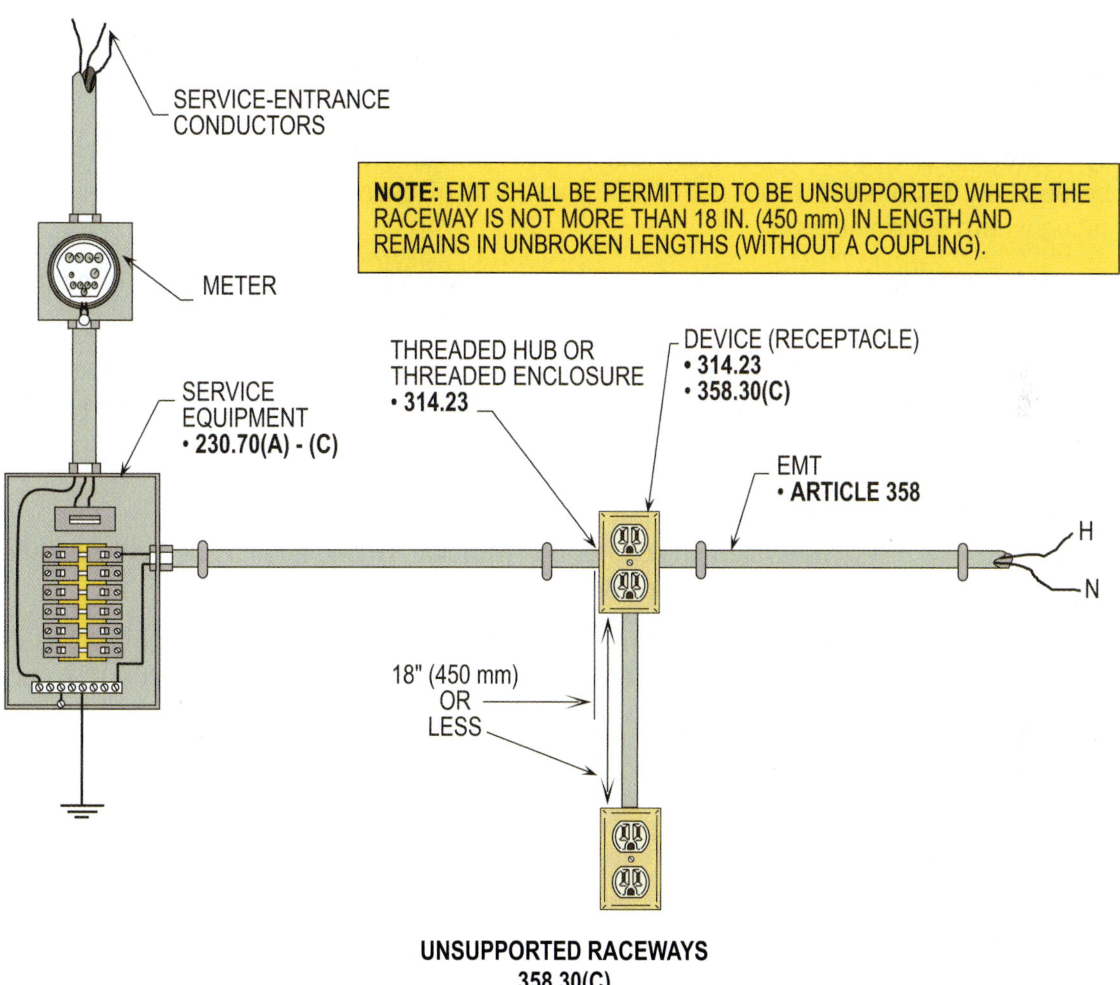

UNSUPPORTED RACEWAYS
358.30(C)

Purpose of Change: A new subsection has been added to permit EMT to be unsupported where the raceway is not more than 18 in. (450 mm) in length and remains in unbroken lengths (without a coupling), where oversized, concentric or eccentric knockouts are not encountered.

NEC Ch. 3 - Article 362
Part II - 362.30(A), Exception No. 3

Type of Change	Panel Action	UL	UL 508	API 500 - 1997	API 505 - 1997	OSHA - 1994
New Exception	Accept	1653	-	-	-	-
ROP	**ROC**	**NFPA 70E - 2004**	**NFPA 70B - 2006**	**NFPA 79 - 2007**	**NFPA**	**NEMA - 2005**
pg. 375 # 8-119	pg. - # -	-	-	-	-	TC 13
log: 2623 CMP: 8	log: - Submitter: David H. Kendall			2005 NEC: -		IEC: -

2005 NEC - 362.30 Securing and Supporting.

ENT shall be installed as a complete system in accordance with 300.18 and shall be securely fastened in place and supported in accordance with 362.30(A) and (B).

(A) Securely Fastened. ENT shall be securely fastened at intervals not exceeding 900 mm (3 ft). In addition, ENT shall be securely fastened in place within 900 mm (3 ft) of each outlet box, device box, junction box, cabinet, or fitting where it terminates.

Exception No. 1: Lengths not exceeding a distance of 1.8 m (6 ft) from a luminaire ~~(fixture)~~ terminal connection for tap connections to lighting luminaires ~~(fixtures)~~ shall be permitted without being secured.

Exception No. 2: Lengths not exceeding 1.8 m (6 ft) from the last point where the raceway is securely fastened for connections within an accessible ceiling to luminaire(s) ~~[lighting fixture(s)]~~ or other equipment.

2008 NEC - 362.30 Securing and Supporting.

ENT shall be installed as a complete system in accordance with 300.18 and shall be securely fastened in place and supported in accordance with 362.30(A) and (B).

(A) Securely Fastened. ENT shall be securely fastened at intervals not exceeding 900 mm (3 ft). In addition, ENT shall be securely fastened in place within 900 mm (3 ft) of each outlet box, device box, junction box, cabinet, or fitting where it terminates.

Exception No. 1: Lengths not exceeding a distance of 1.8 m (6 ft) from a luminaire terminal connection for tap connections to lighting luminaires shall be permitted without being secured.

Exception No. 2: Lengths not exceeding 1.8 m (6 ft) from the last point where the raceway is securely fastened for connections within an accessible ceiling to luminaire(s) or other equipment.

Exception No. 3: For concealed work in finished buildings or prefinished wall panels where such securing is impracticable, unbroken lengths (without coupling) of ENT shall be permitted to be fished.

Author's Comment: The flexibility of electrical nonmetallic tubing (ENT) permits it to be easily fished between access points in a finished wall in existing buildings, a common practice for the protection of conductors and cabling in these applications. This new exception will now permit this application.

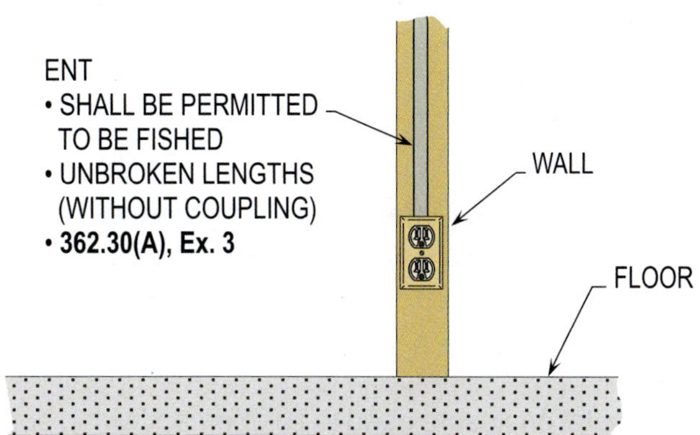

SECURELY FASTENED
362.30(A), Exception No. 3

Purpose of Change: A new exception has been added to permit ENT to be fished as concealed work in finished buildings or prefinished wall panels where installed in unbroken lengths (without coupling).

NEC Ch. 3 - Article 366
Part I - 366.2

Type of Change	Panel Action	UL	UL 508	API 500 - 1997	API 505 - 1997	OSHA - 1994
Revision	Accept	870	-	-	-	-
ROP	**ROC**	**NFPA 70E - 2004**	**NFPA 70B - 2006**	**NFPA 79 - 2007**	**NFPA**	**NEMA**
pg. 376 # 8-124	pg. - # -	-	-	-	-	-
log: 3403 CMP: 8	log: -	Submitter: Frederic P. Hartwell		2005 NEC: 366.2		IEC: -

2005 NEC - 366.2 Definitions.

Metallic Auxiliary Gutter. Sheet metal enclosure~~s with hinged or removable covers for housing and protecting electric wires, cable, and busbars in which conductors are laid in place after the wireway has~~ been installed as a complete system.

Nonmetallic Auxiliary Gutter. Flame retardant, nonmetallic ~~enclosures with removable covers for housing and protecting electric wires, cable, and busbars in which conductors are laid in place after the wireway has~~ been installed as a complete system.

366.10 Uses Permitted.

~~Auxiliary gutters shall be permitted to supplement wiring spaces at meter centers, distribution centers, switchboards, and similar points of wiring systems and may enclose conductors or busbars.~~

2008 NEC - 366.2 Definitions.

Metallic Auxiliary Gutter. A sheet metal enclosure used to supplement wiring spaces at meter centers, distribution centers, switchboards, and similar points of wiring systems. The enclosure has hinged or removable covers for housing and protecting electric wires, cable and busbars. The enclosure is designed for conductors to be laid or set in place after the enclosures have been installed as a complete system.

Nonmetallic Auxiliary Gutter. A flame retardant, nonmetallic enclosure used to supplement wiring spaces at meter centers, distribution centers, switchboards, and similar points of wiring systems. The enclosure has hinged or removable covers for housing and protecting electric wires, cable and busbars. The enclosure is designed for conductors to be laid or set in place after the enclosures have been installed as a complete system.

Author's Comment: The definitions for both metallic and nonmetallic auxiliary gutters have been expanded to indicate the use of auxiliary gutters as supplemental wiring spaces for meter centers, distribution centers, switchboards, and similar points to facilitate extra wiring spaces. The new definitions also state that the enclosures can have hinged or removable covers and are designed for installation of conductors or busbars after the gutters have been installed as a complete system. The wires or busbars are field-installed, not factory-installed. The text at the beginning of **366.10** has been deleted since the text has been included in the definitions in **366.2**.

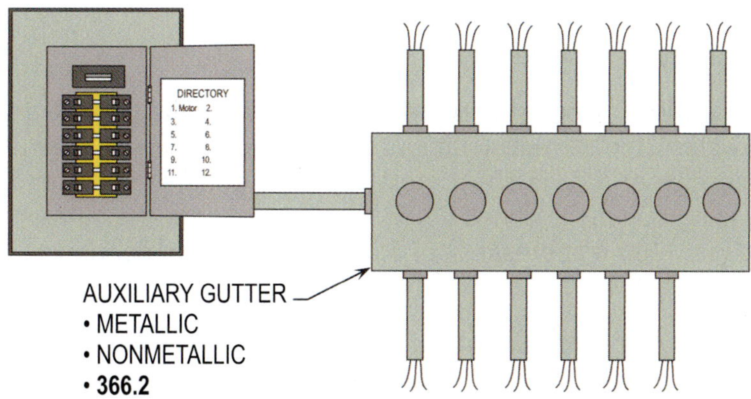

METALLIC AUXILIARY GUTTER
A SHEET METAL ENCLOSURE USED TO SUPPLEMENT WIRING SPACES AT METER CENTERS, DISTRIBUTION CENTERS, SWITCHBOARDS, AND SIMILAR POINTS OF WIRING SYSTEMS. THE ENCLOSURE HAS HINGED OR REMOVABLE COVERS FOR HOUSING AND PROTECTING ELECTRIC WIRES, CABLE, AND BUSBARS. THE ENCLOSURE IS DESIGNED FOR CONDUCTORS TO BE LAID OR SET IN PLACE AFTER THE ENCLOSURES HAVE BEEN INSTALLED AS A COMPLETE SYSTEM.

NONMETALLIC AUXILIARY GUTTER
A FLAME RETARDANT NONMETALLIC ENCLOSURE USED TO SUPPLEMENT WIRING SPACES AT METER CENTERS, DISTRIBUTION CENTERS, SWITCHBOARDS, AND SIMILAR POINTS OF WIRING SYSTEMS. THE ENCLOSURE HAS HINGED OR REMOVABLE COVERS FOR HOUSING AND PROTECTING ELECTRIC WIRES, CABLE AND BUSBARS. THE ENCLOSURE IS DESIGNED FOR CONDUCTORS TO BE LAID OR SET IN PLACE AFTER THE ENCLOSURES HAVE BEEN INSTALLED AS A COMPLETE SYSTEM.

DEFINITIONS
366.2

Purpose of Change: A revision has been made to the definitions of metallic and nonmetallic auxiliary gutters.

NEC Ch. 3 - Article 366
Part II - 366.23(A)

Type of Change	Panel Action	UL	UL 508	API 500 - 1997	API 505 - 1997	OSHA - 1994
Revision	Accept	870	-	-	-	-
ROP	**ROC**	**NFPA 70E - 2004**	**NFPA 70B - 2006**	**NFPA 79 - 2007**	**NFPA**	**NEMA**
pg. 376 # 8-127	pg. - # -	-	-	-	-	-
log: 2243 CMP: 8	log: -	Submitter: Donald A. Ganiere		2005 NEC: 366.23(A)		IEC: -

2005 NEC - 366.23 Ampacity of Conductors.

(A) Sheet Metal Auxiliary Gutters. Where the number of current-carrying conductors contained in the sheet metal auxiliary gutter is 30 or less, the ~~correction~~ factors specified in 310.15(B)(2)(a) shall not apply. The current carried continuously in bare copper bars in sheet metal auxiliary gutters shall not exceed 1.55 amperes/mm^2 (1000 amperes/in.^{2}) of cross section of the conductor. For aluminum bars, the current carried continuously shall not exceed 1.09 amperes/mm^2 (700 amperes/in.^{2}) of cross section of the conductor.

2008 NEC - 366.23 Ampacity of Conductors.

(A) Sheet Metal Auxiliary Gutters. Where the number of current-carrying conductors contained in the sheet metal auxiliary gutter is 30 or less, the adjustment factors specified in 310.15(B)(2)(a) shall not apply. The current carried continuously in bare copper bars in sheet metal auxiliary gutters shall not exceed 1.55 amperes/mm^2 (1000 amperes/in.^{2}) of cross section of the conductor. For aluminum bars, the current carried continuously shall not exceed 1.09 amperes/mm^2 (700 amperes/in.^{2}) of cross section of the conductor.

Author's Comment: The term "adjustment factors" is used for adjusting the conductor ampacity for more than three conductors in a raceway or cable. The term "correction factor" is used to correct the conductor ampacity for ambient temperature. Since the text is meant to deal with the permission to not apply **310.15(B)(2)(a)**, the text was modified to use "adjustment factors." This only applies to sheet metal auxiliary gutters since the metal of the gutter acts as a heat-sink for the conductors installed in the gutter.

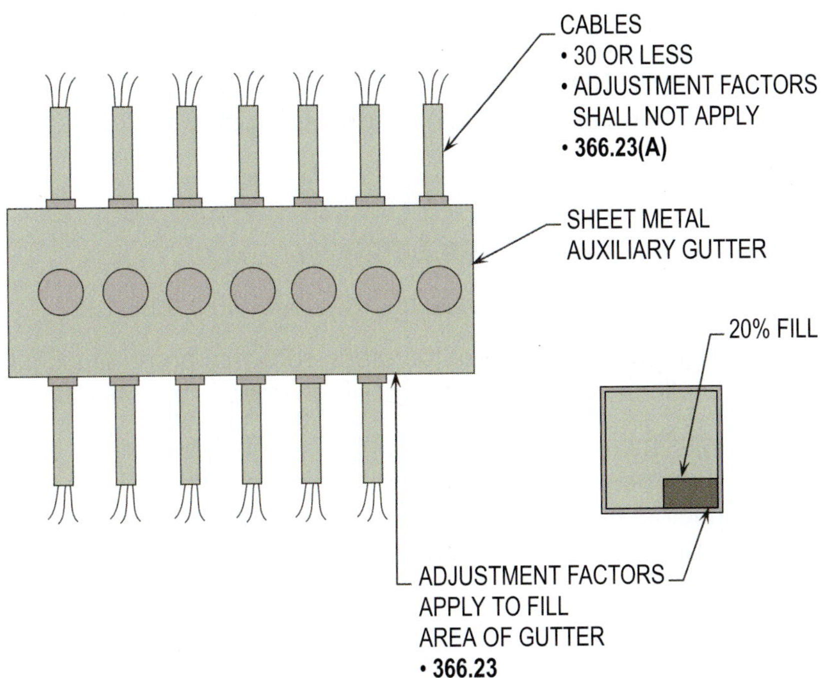

SHEET METAL AUXILIARY GUTTERS
366.23(A)

Purpose of Change: A revision has been made to clarify that adjustment factors shall not be required for current-carrying conductors that do not exceed 30 in sheet metal auxiliary gutters.

NEC Ch. 3 - Article 376
Part II - 376.22(A) and (B)

Type of Change	Panel Action	UL	UL 508	API 500 - 1997	API 505 - 1997	OSHA - 1994
Revision	Accept in Principle	870	-	-	-	-
ROP	**ROC**	**NFPA 70E - 2004**	**NFPA 70B - 2006**	**NFPA 79 - 2007**	**NFPA**	**NEMA**
pg. 381 # 8-157	pg. 223 # 8-73	-	-	13.5.6	-	-
log: 2754 CMP: 8	log: 969	Submitter: Jonathan R. Althouse		2005 NEC: 376.22		IEC: -

2005 NEC - 376.22 Number of Conductors.

The sum of the cross-sectional areas of all contained conductors at any cross section of a wireway shall not exceed 20 percent of the interior cross-sectional area of the wireway. The ~~derating~~ factors in 310.15(B)(2)(a) shall be applied only where the number of current-carrying conductors, including neutral conductors classified as current-carrying under the provisions of 310.15(B)(4), exceeds 30. Conductors for signaling circuits or controller conductors between a motor and its starter and used only for starting duty shall not be considered as current-carrying conductors.

2008 NEC - 376.22 Number of Conductors and Ampacity.

The number of conductors and their ampacity shall comply with 376.22(A) and (B).

(A) Cross-Sectional Areas of Wireway. The sum of the cross-sectional areas of all contained conductors at any cross section of a wireway shall not exceed 20 percent of the interior cross-sectional area of the wireway.

(B) Adjustment Factors. The adjustment factors in 310.15(B)(2)(a) shall be applied only where the number of current-carrying conductors, including neutral conductors classified as current-carrying under the provisions of 310.15(B)(4), exceeds 30. Conductors for signaling circuits or controller conductors between a motor and its starter and used only for starting duty shall not be considered as current-carrying conductors.

Author's Comment: The title of **376.22** has been changed to include ampacity to make it easier to located the ampacity requirements for conductors installed in metal wireways. The text in the section has been subdivided into **(A)** and **(B)** for ease of use. The comment added titles for the subdivided text.

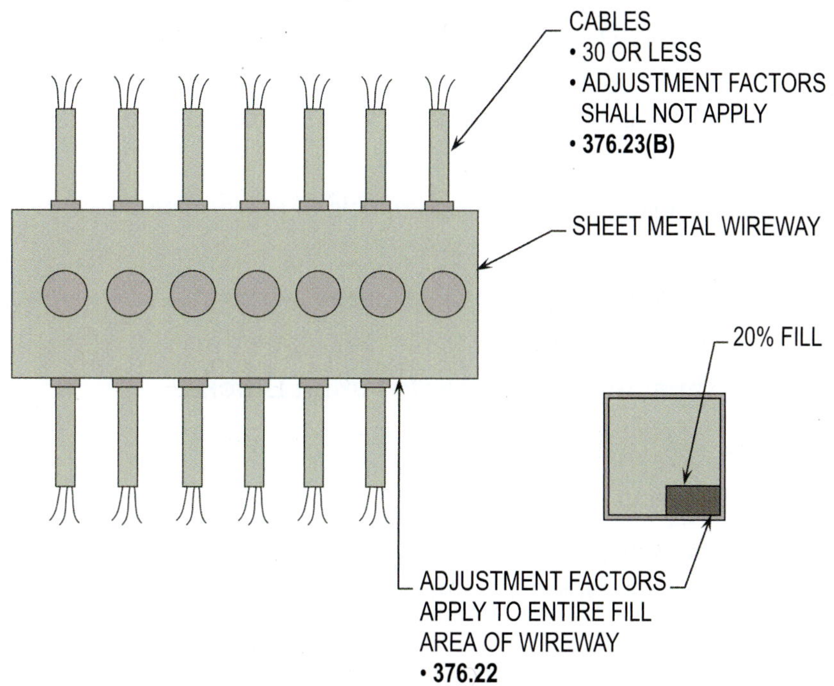

NUMBER OF CONDUCTORS AND AMPACITY
376.22(A) and (B)

Purpose of Change: A revision has been made to clarify that adjustment factors (not derating) shall be applied for current-carrying conductors that exceed 30 in sheet metal wireways.

NEC Ch. 3 - Article 376
Part II - 376.56(B)(4)

Type of Change	Panel Action	UL	UL 508	API 500 - 1997	API 505 - 1997	OSHA - 1994
Revision	Accept	870	-	-	-	-
ROP	**ROC**	**NFPA 70E - 2004**	**NFPA 70B - 2006**	**NFPA 79 - 2007**	**NFPA**	**NEMA**
pg. 381 # 8-155	pg. - # -	-	-	13.5.6	-	-
log: 3405 CMP: 8	log: - Submitter: Frederic P. Hartwell			2005 NEC: 376.56(B)(4)		IEC: -

2005 NEC - 376.56 Splices, Taps, and Power Distribution Blocks.

(B) Power Distribution Blocks

(1) Installation. Power distribution blocks installed in metal wireways shall be listed.

(2) Size of Enclosure. In addition to the wiring space requirement in 376.56(A), the power distribution block shall be installed in a wireway with dimensions not smaller than specified in the installation instructions of the power distribution block.

(3) Wire Bending Space. Wire bending space at the terminals of power distribution blocks shall comply with 312.6(B).

(4) Live Parts. Power distribution blocks shall not have ~~exposed~~ live part in a ~~the~~ wireway.

2008 NEC - 376.56 Splices, Taps, and Power Distribution Blocks.

(B) Power Distribution Blocks

(1) Installation. Power distribution blocks installed in metal wireways shall be listed.

(2) Size of Enclosure. In addition to the wiring space requirement in 376.56(A), the power distribution block shall be installed in a wireway with dimensions not smaller than specified in the installation instructions of the power distribution block.

(3) Wire Bending Space. Wire bending space at the terminals of power distribution blocks shall comply with 312.6(B).

(4) Live Parts. Power distribution blocks shall not have uninsulated live parts exposed within a wireway, whether or not the wireway cover is installed.

Author's Comment: With the cover in place, uninsulated power distribution blocks are "in the wireway" but no longer "exposed" and therefore comply with the literal text in the 2005 NEC. In addition, live parts include energized "insulated" conductors, so the word "uninsulated" has been added to this requirement so it will apply whether or not the cover is installed and to make it clear it does not apply to insulated live parts.

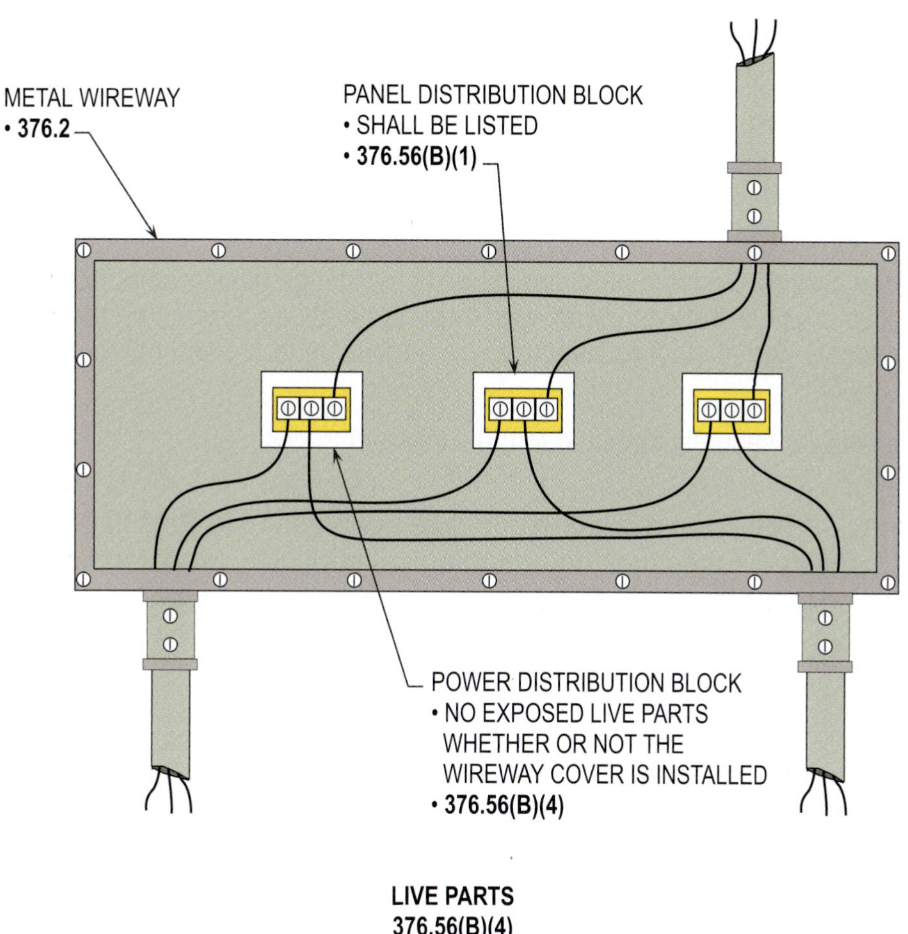

LIVE PARTS
376.56(B)(4)

Purpose of Change: This revision clarifies that power distribution blocks shall have no exposed live parts, whether or not the wireway cover is installed.

NEC Ch. 3 - Article 376
Part III - 376.100(A) through (D)

54

Type of Change	Panel Action	UL	UL 508	API 500 - 1997	API 505 - 1997	OSHA - 1994
New Section	Accept	870	-	-	-	-
ROP	**ROC**	**NFPA 70E - 2004**	**NFPA 70B - 2006**	**NFPA 79 - 2007**	**NFPA**	**NEMA**
pg. 381 / # 8-157a	pg. - / # -	-	-	13.5.6	-	-
log: CP 802 / CMP: 8	log: -	Submitter: Code Making Panel 8		2005 NEC: -		IEC: -

2008 NEC - 376.100 Construction.

(A) Electrical and Mechanical Continuity. Wireways shall be constructed and installed so that adequate electrical and mechanical continuity of the complete system is secured.

(B) Substantial Construction. Wireways shall be of substantial construction and shall provide a complete enclosure for the contained conductors. All surfaces, both interior and exterior, shall be suitably protected from corrosion. Corner joints shall be made tight, and where the assembly is held together by rivets, bolts, or screws, such fasteners shall be spaced not more than 300 mm (12 in.) apart.

(C) Smooth Rounded Edges. Suitable bushings, shields, or fittings having smooth, rounded edges shall be provided where conductors pass between wireways, through partitions, around bends, between wireways and cabinets or junction boxes, and at other locations where necessary to prevent abrasion of the insulation of the conductors.

(D) Covers. Covers shall be securely fastened to the wireway.

Author's Comment: Construction requirements have been added to **Part III** of **Article 376** to provide guidance on what constitutes electrical and mechanical continuity, substantial construction, smooth rounded edges, and covers being able to be securely fastened to the wireways.

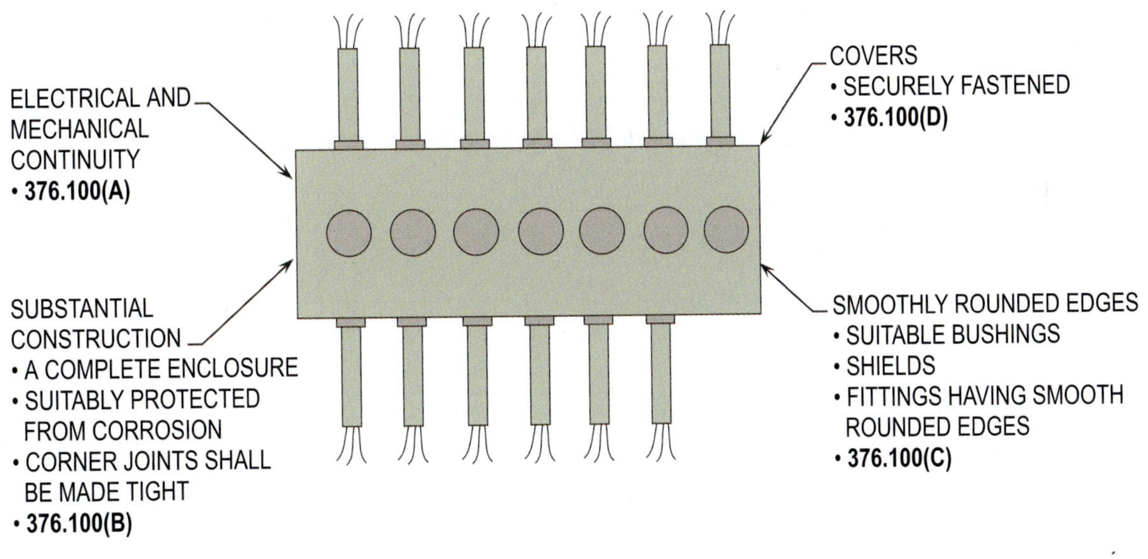

CONSTRUCTION
376.100(A) through (D)

Purpose of Change: A new section has been added for the construction requirements of a wireway.

NEC Ch. 3 - Article 382
Part I - 382.6

Type of Change	Panel Action	UL	UL 508	API 500 - 1997	API 505 - 1997	OSHA - 1994
New Section	Accept in Principle	-	-	-	-	-
ROP	**ROC**	**NFPA 70E - 2004**	**NFPA 70B - 2006**	**NFPA 79 - 2007**	**NFPA**	**NEMA**
pg. 383 # 7-98	pg. 224 # 7-56	-	-	-	-	-
log: 3450 CMP: 7	log: 2231	Submitter: Richard Temblador		2005 NEC: -		IEC: -

2005 NEC - 382.2 Definition.

Nonmetallic Extension. An assembly of two insulated conductors within a nonmetallic jacket or an extruded thermoplastic covering. The classification includes surface extensions intended for mounting directly on the surface of walls or ceilings.

2008 NEC - 382.6 Listing Requirements.

Concealable nonmetallic extensions and associated fittings and devices shall be listed. The starting/source tap device for the extension shall contain and provide the following protection for all loadside extensions and devices:

(1) Supplementary overcurrent protection

(2) Level of protection equivalent to a Class A GFCI

(3) Level of protection equivalent to a portable GFCI

(4) Line and load-side miswire protection

(5) Provide protection from the effects of arc faults

Author's Comment: A new section has been added to address the listing requirements for concealable nonmetallic extensions and associated fittings and devices.

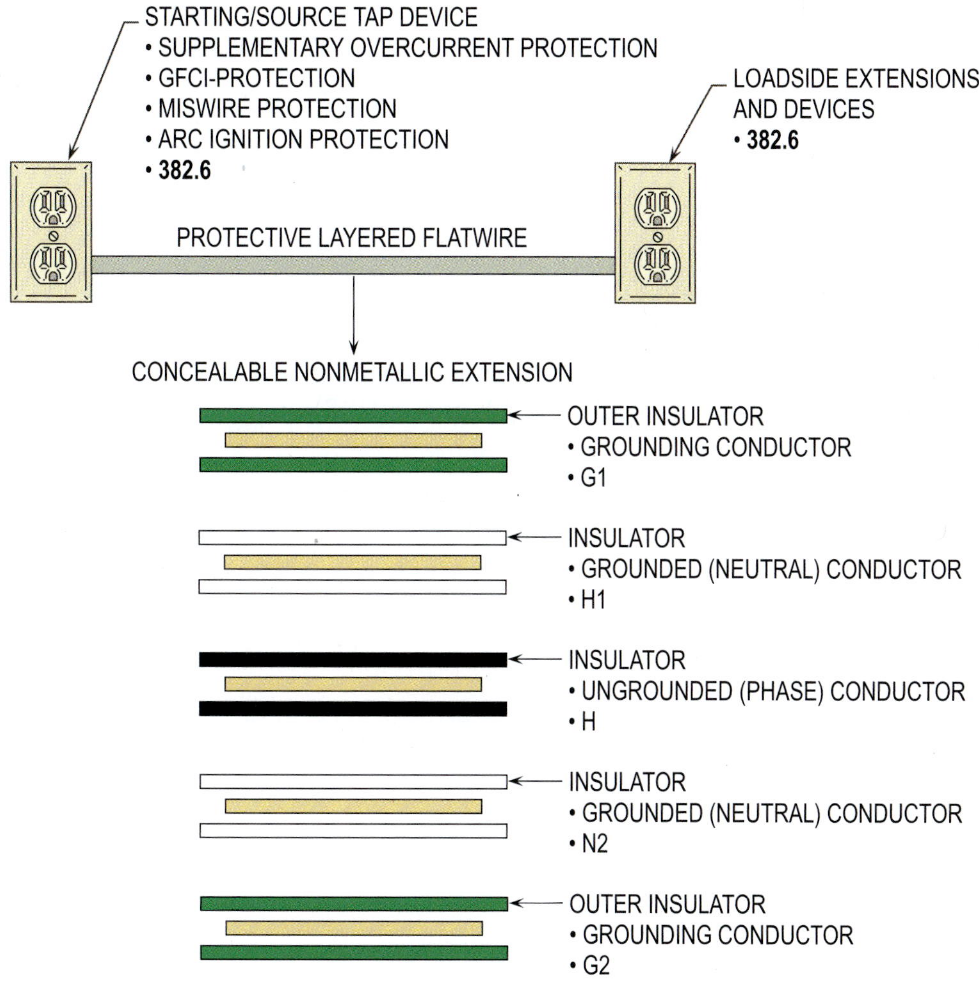

LISTING REQUIREMENTS
382.6

Purpose of Change: New sections have been added to address the listing requirements for concealable nonmetallic extensions and associated fittings and devices.

NEC Ch. 3 - Article 382
Part II - 382.10(A)

Type of Change		Panel Action		UL	UL 508	API 500 - 1997	API 505 - 1997	OSHA - 1994
Revision		Accept in Principle		-	-	-	-	-
ROP		**ROC**		**NFPA 70E - 2004**	**NFPA 70B - 2006**	**NFPA 79 - 2007**	**NFPA**	**NEMA**
pg. 383	# 7-98	pg. -	# -	-	-	-	-	-
log: 3450	CMP: 7	log: -	Submitter: Richard Temblador			2005 NEC: 382.10(A)		IEC: -

2005 NEC - 382.10 Uses Permitted.

Nonmetallic extensions shall be permitted only in accordance with 382.10(A), (B), and (C).

(A) From an Existing Outlet. The extension shall be from an existing outlet on a 15- or 20-ampere branch circuit.

2008 NEC - 382.10 Uses Permitted.

Nonmetallic extensions shall be permitted only in accordance with 382.10(A), (B), and (C).

(A) From an Existing Outlet. The extension shall be from an existing outlet on a 15- or 20-ampere branch circuit. Where a concealable nonmetallic extension originates from a non-grounding-type receptacle, the installation shall comply with 250.130(C), 406.3(D)(3)(b), or 406(D)(3)(c).

Author's Comment: This revision clarifies the installation requirements if a concealable nonmetallic extension originates from nongrounding type receptacles.

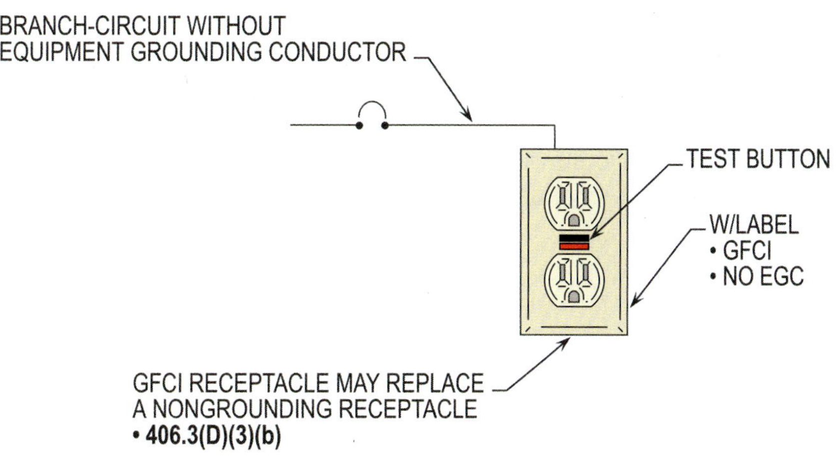
BRANCH-CIRCUIT WITHOUT
EQUIPMENT GROUNDING CONDUCTOR
TEST BUTTON
W/LABEL
• GFCI
• NO EGC
GFCI RECEPTACLE MAY REPLACE
A NONGROUNDING RECEPTACLE
• 406.3(D)(3)(b)

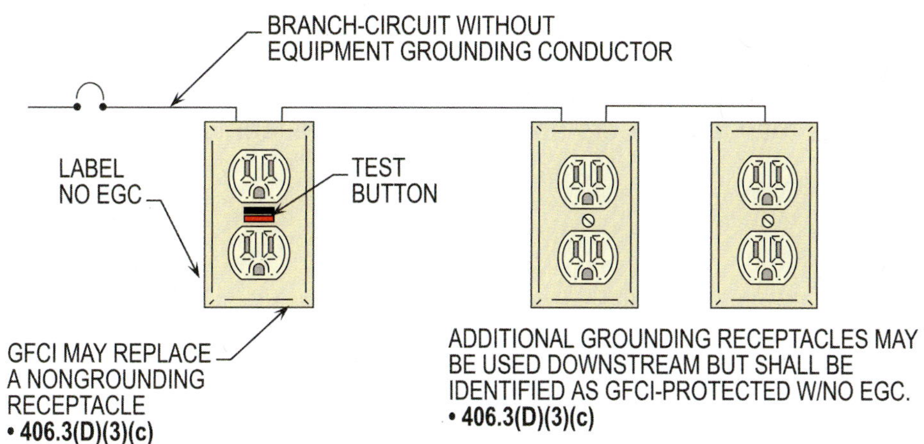
BRANCH-CIRCUIT WITHOUT
EQUIPMENT GROUNDING CONDUCTOR
LABEL
NO EGC
TEST
BUTTON
GFCI MAY REPLACE
A NONGROUNDING
RECEPTACLE
• 406.3(D)(3)(c)
ADDITIONAL GROUNDING RECEPTACLES MAY
BE USED DOWNSTREAM BUT SHALL BE
IDENTIFIED AS GFCI-PROTECTED W/NO EGC.
• 406.3(D)(3)(c)

FROM AN EXISTING OUTLET
382.10(A)

Purpose of Change: This revision clarifies the installation requirements if a concealable nonmetallic extension originates from nongrounding type receptacles.

NEC Ch. 3 - Article 382
Part II - 382.10(C)

Type of Change		Panel Action		UL	UL 508	API 500 - 1997	API 505 - 1997	OSHA - 1994
Revision		Accept in Principle		-	-	-	-	-
ROP		ROC		NFPA 70E - 2004	NFPA 70B - 2006	NFPA 79 - 2007	NFPA	NEMA
pg. 383	# 7-98	pg. -	# -	-	-	-	-	-
log: 3450	CMP: 7	log: -	Submitter: Richard Temblador			2005 NEC: 382.10(C)		IEC: -

2005 NEC - 382.10 Uses Permitted.

Nonmetallic extensions shall be permitted only in accordance with 382.10(A), (B), and (C).

(C) Residential or Offices. For nonmetallic surface extensions mounted directly on the surface of walls or ceilings, the building shall be occupied for residential or office purposes and shall not exceed three floors above grade.

2008 NEC - 382.10 Uses Permitted.

Nonmetallic extensions shall be permitted only in accordance with 382.10(A), (B), and (C).

(C) Residential or Offices. For nonmetallic surface extensions mounted directly on the surface of walls or ceilings, the building shall be occupied for residential or office purposes and shall not exceed three floors above grade. <u>Where identified for the use, concealable nonmetallic extensions shall be permitted more than three floors abovegrade.</u>

Author's Comment: This revision clarifies that concealable nonmetallic extensions shall be permitted to be installed more than than three floors above grade, where identified for the use.

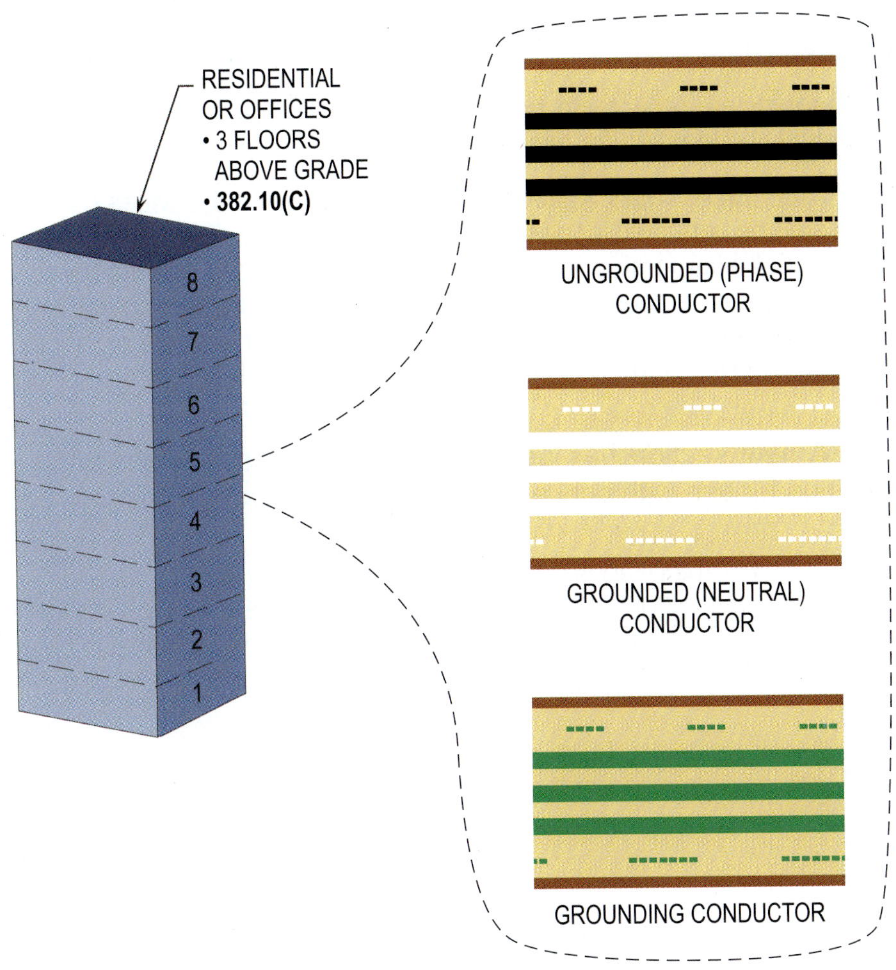

RESIDENTIAL OR OFFICES
382.10(C)

Purpose of Change: This revision clarifies that concealable nonmetallic extensions shall be permitted to be installed more than three floors above grade.

NEC Ch. 3 - Article 382
Part II - 382.42(B)

Type of Change		Panel Action		UL	UL 508	API 500 - 1997	API 505 - 1997	OSHA - 1994
New Subsection		Accept in Principle		-	-	-	-	-
ROP		ROC		NFPA 70E - 2004	NFPA 70B - 2006	NFPA 79 - 2007	NFPA	NEMA
pg. 383	# 7-98	pg. -	# -	-	-	-	-	-
log: 3450	CMP: 7	log: -	Submitter: Richard Temblador			2005 NEC: -		IEC: -

2008 NEC - 382.42 Devices.

(A) Receptacles. All receptacles, receptacle housings, and self-contained devices used with concealable nonmetallic extensions shall be identified for this use.

(B) Receptacles and Housings. Receptacle housings and self-contained devices designed either for surface or for recessed mounting shall be permitted for use with concealable nonmetallic extensions. Receptacle housings and self-contained devices shall incorporate means for facilitating entry and termination of concealable nonmetallic extensions and for electrically connecting the housing or device. Receptacle and self-contained devices shall comply with 406.3. Power and communications outlets installed together in common housing shall be permitted in accordance with 800.133(A)(1)(c), Exception No. 2.

Author's Comment: A new section has been added to address the installation requirements of receptacles and housings for use with concealable nonmetallic extensions.

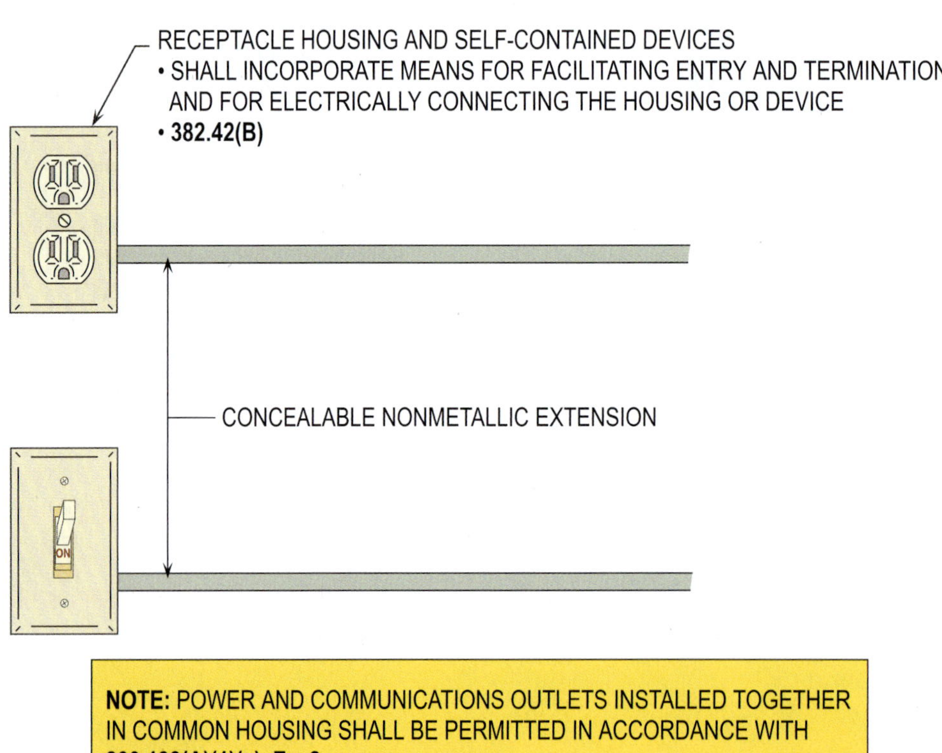

DEVICES
382.42(B)

Purpose of Change: A new section has been added to address the installation requirements of receptacles and housings for use with concealable nonmetallic extensions.

NEC Ch. 3 - Article 392
Part II - 392.8(A)

Type of Change	Panel Action	UL	UL 508	API 500 - 1997	API 505 - 1997	OSHA - 1994
Revision	Accept	-	-	-	-	1910.305(a)(3)(i)
ROP	**ROC**	**NFPA 70E - 2004**	**NFPA 70B - 2006**	**NFPA 79 - 2007**	**NFPA**	**NEMA - 2002**
pg. 391 # 8-192a	pg. 239 # 8-86	420.1(C)(4)	Chapter 24	13.5.10	-	VE 1
log: CP 800 CMP: 8	log: 2158	Submitter: Code Making Panel 8		2005 NEC: 392.8(A)		IEC: -

2005 NEC - 392.8 Cable Installation.

(A) Cable Splices. Cable splices made and insulated by approved methods shall be permitted to be located within a cable tray, provided they are accessible ~~and do not~~ project above the side rails.

2008 NEC - 392.8 Cable Installation.

(A) Cable Splices. Cable splices made and insulated by approved methods shall be permitted to be located within a cable tray, provided they are accessible. Splices shall be permitted to project above the side rails where not subject to physical damage.

Author's Comment: This revision is meant to emphasize that a cable splice must not extend above the side rails of the cable tray where the splice may be subject to physical damage. The comment provided text that is not negative and divided the existing text into two sentences for clarity.

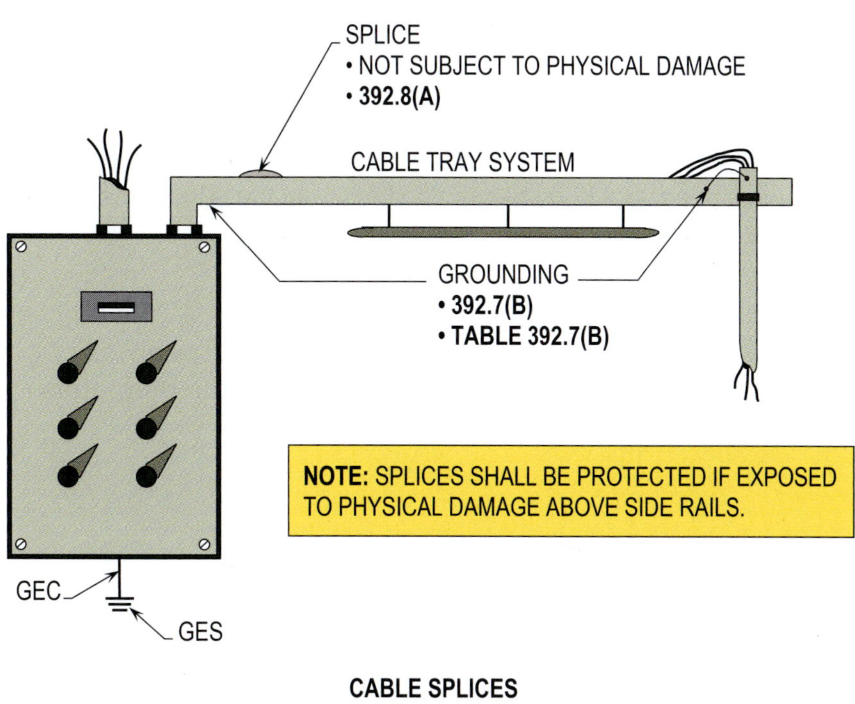

CABLE SPLICES
392.8(A)

Purpose of Change: This revision clarifies that cable splices shall be protected from physical damage if splices project above the side rails or make splices within cable tray without projecting above the side rails.

NEC Ch. 3 - Article 392
Part II - 392.9(A)(1)

Type of Change	Panel Action	UL	UL 508	API 500 - 1997	API 505 - 1997	OSHA - 1994
Revision	Accept	-	-	-	-	1910.305(a)(3)
ROP	**ROC**	**NFPA 70E - 2004**	**NFPA 70B - 2006**	**NFPA 79 - 2007**	**NFPA**	**NEMA - 2002**
pg. 391 # 8-194	pg. - # -	420.1(C)	Chapter 24	13.5.10	-	VE 1
log: 3129 CMP: 8	log: -	Submitter: Jonathan R. Althouse		2005 NEC: 392.9(A)(1)		IEC: -

2005 NEC - 392.9 Number of Multiconductor Cables, Rated 2000 Volts or Less, in Cable Trays.

The number of multiconductor cables, rated 2000 volts or less, permitted in a single cable tray shall not exceed the requirements of this section. The conductor sizes herein apply to both aluminum and copper conductors.

(A) Any Mixture of Cables. Where ladder or ventilated trough cable trays contain multiconductor power or lighting cables, or any mixture of multiconductor power, lighting, control, and signal cables, the maximum number of cables shall conform to the following:

(1) Where all of the cables are 4/0 AWG or larger, the sum of the diameters of all cables shall not exceed the cable tray width, and the cables shall be installed in a single layer.

2008 NEC - 392.9 Number of Multiconductor Cables, Rated 2000 Volts or Less, in Cable Trays.

The number of multiconductor cables, rated 2000 volts or less, permitted in a single cable tray shall not exceed the requirements of this section. The conductor sizes herein apply to both aluminum and copper conductors.

(A) Any Mixture of Cables. Where ladder or ventilated trough cable trays contain multiconductor power or lighting cables, or any mixture of multiconductor power, lighting, control, and signal cables, the maximum number of cables shall conform to the following:

(1) Where all of the cables are 4/0 AWG or larger, the sum of the diameters of all cables shall not exceed the cable tray width, and the cables shall be installed in a single layer. Where the cable ampacity is determined according to 392.11(A)(3), the cable tray width shall not be less than the sum of the diameters of the cables and the sum of the required spacing widths between the cables.

Author's Comment: Where multiconductor cables are installed in a single layer in uncovered trays with maintained spacing of not less than one cable diameter between cables in accordance with **392.11(A)(3)**, the cable tray width can not be less than the sum of the diameters of the cables and the sum of the required spacing widths between the cables.

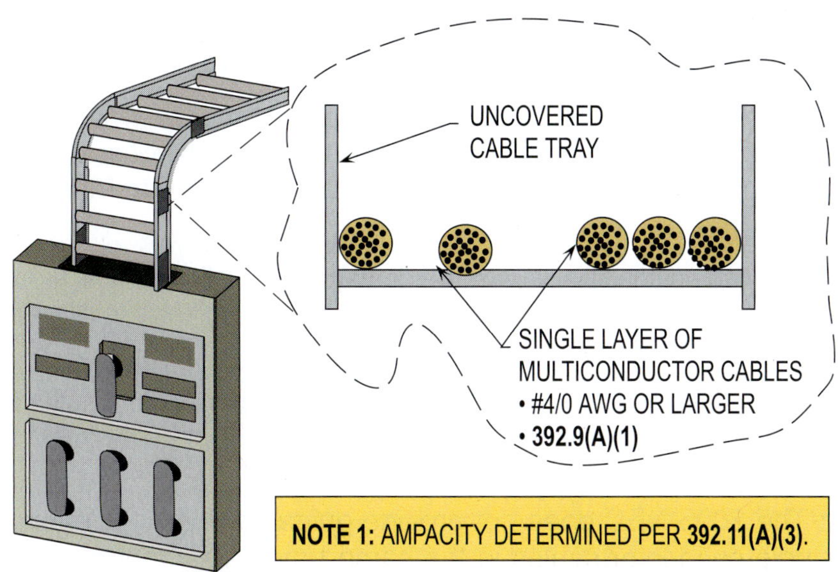

NOTE 2: WIDTH OF CABLE TRAY SHALL NOT BE LESS THAN THE SUM OF THE DIAMETERS OF THE CABLES PLUS THE SUM OF REQUIRED SPACING WIDTHS BETWEEN CABLES.

ANY MIXTURE OF CABLES
392.9(A)(1)

Purpose of Change: This revision clarifies that where multiconductor cables are installed in a single layer, the free-air ampacity calculation shall be permitted to be used in **392.11(A)(3)**, however, the cable tray width cannot be less than the sum of the diameters of the cables plus the sum of the required spacing between the cables.

NEC Ch. 3 - Article 392
Part II - 392.11(C)

Type of Change	Panel Action	UL	UL 508	API 500 - 1997	API 505 - 1997	OSHA - 1994
New Subsection	Accept in Principle	-	-	-	-	1910.305(a)(3)
ROP	**ROC**	**NFPA 70E - 2004**	**NFPA 70B - 2006**	**NFPA 79 - 2007**	**NFPA**	**NEMA - 2002**
pg. 391 # 8-197	pg. - # -	420.1(C)	Chapter 24	13.5.10	-	VE 1
log: 2703 CMP: 8	log: -	Submitter: Dorothy Kellogg		2005 NEC: -		IEC: -

2005 NEC - 392.11 Ampacity of Cables, Rated 2000 Volts or Less, in Cable Trays.

2008 NEC - 392.11 Ampacity of Cables, Rated 2000 Volts or Less, in Cable Trays.

(C) Combinations of Multiconductor and Single-Conductor Cables. Where a cable tray contains a combination of multiconductor and single-conductor cables, the allowable ampacities shall be as given in 392.11(A) for multiconductor cables and 392.11(B) for single-conductor cables, provided that the following conditions apply:

(1) The sum of the multiconductor cable fill area as a percentage of the allowable fill area for the tray calculated per 392.9, and the single-conductor cable fill area as a percentage of the allowable fill area for the tray calculated per 392.10, totals not more than 100%.

(2) Multiconductor cables are installed according to 392.9 and single conductor cables are installed according to 392.10 and 392.8(D) and (E).

Author's Comment: The 2005 NEC does not provide guidance for installation for a combination of multiconductor and single-conductor cables in the same tray. This change provides text stating that multiconductor and single-conductor cables installed in the same cable tray must not exceed the fill requirements for each type of cable and that all installation requirements and ampacity limits for each cable type would still apply.

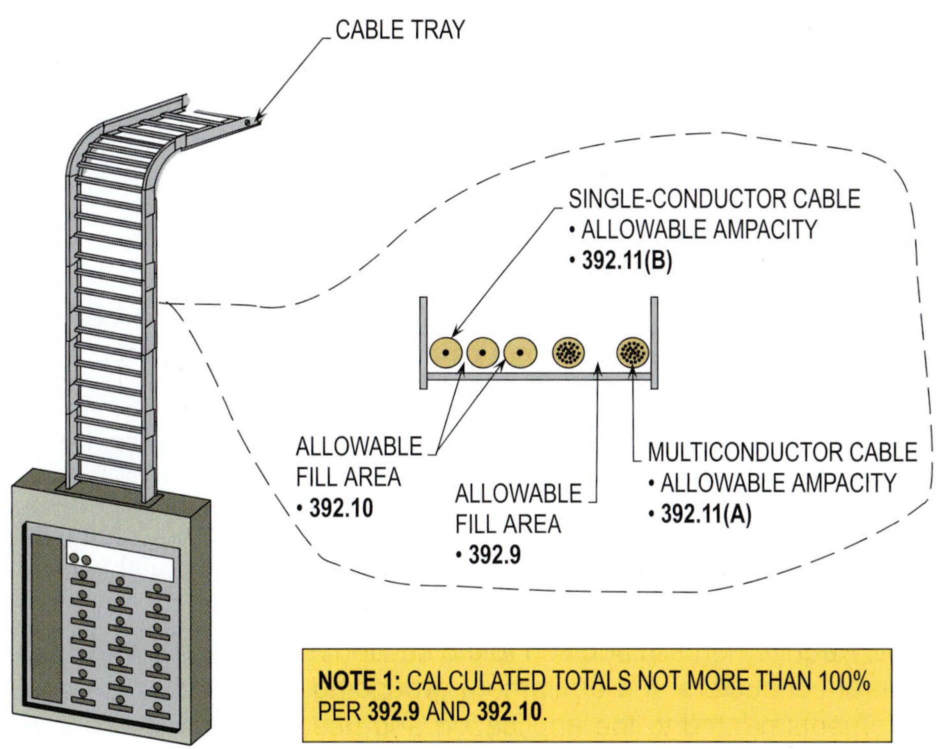

COMBINATIONS OF MULTICONDUCTOR AND SINGLE-CONDUCTOR CABLES 392.11(C)

Purpose of Change: A new subsection has been added to provide requirements for combinations of multiconductor and single-conductor cables installed in a cable tray.

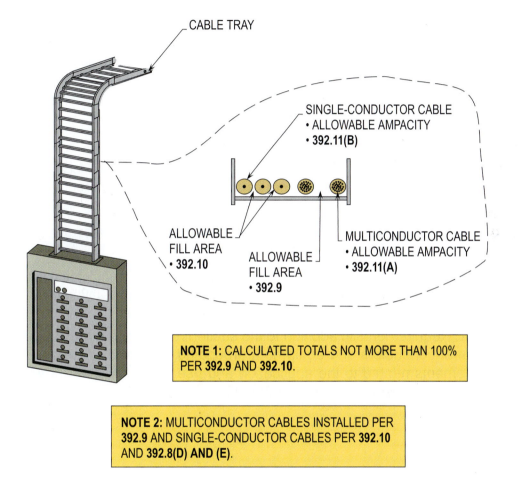

NEC Ch. 3 - Article 396
Part II - 396.30(A) through (C)

Type of Change	Panel Action	UL	UL 508	API 500 - 1997	API 505 - 1997	OSHA - 1994
Revision	Accept	-	-	-	-	-
ROP	**ROC**	**NFPA 70E - 2004**	**NFPA 70B - 2006**	**NFPA 79 - 2007**	**NFPA**	**NEMA**
pg. 393 # 7-106	pg. 243 # 7-76	-	-	-	-	-
log: 3361 CMP: 7	log: 1517	Submitter: Frederic P. Hartwell		2005 NEC: 396.60		IEC: -

2005 NEC - 396.30 Messenger ~~Support~~.

The messenger shall be supported at dead ends and at intermediate locations so as to eliminate tension on the conductors. The conductors shall not be permitted to come into contact with the messenger supports or any structural members, walls, or pipes.

2008 NEC - 396.30 Messenger.

(A) Support. The messenger shall be supported at dead ends and at intermediate locations so as to eliminate tension on the conductors. The conductors shall not be permitted to come into contact with the messenger supports or any structural members, walls, or pipes.

(B) Neutral Conductor. Where the messenger is used as a neutral conductor, it shall comply with the requirements of 225.4, 250.184(A), 250.184(B)(7), and 250.186(B).

(C) Equipment Grounding Conductor. Where the messenger is used as an equipment grounding conductor, it shall comply with the requirements of 250.32(B), 250.118, 250.184(B)(8) and 250.186(D).

Author's Comment: Requirements in addition to the section referenced in the Proposal already exist in the Code for neutral conductors and equipment grounding conductors. Expansion of **396.30** will locate all the requirements related to the application and use of the messenger within one section, rather than having some in **396.90** and some in **396.30**.

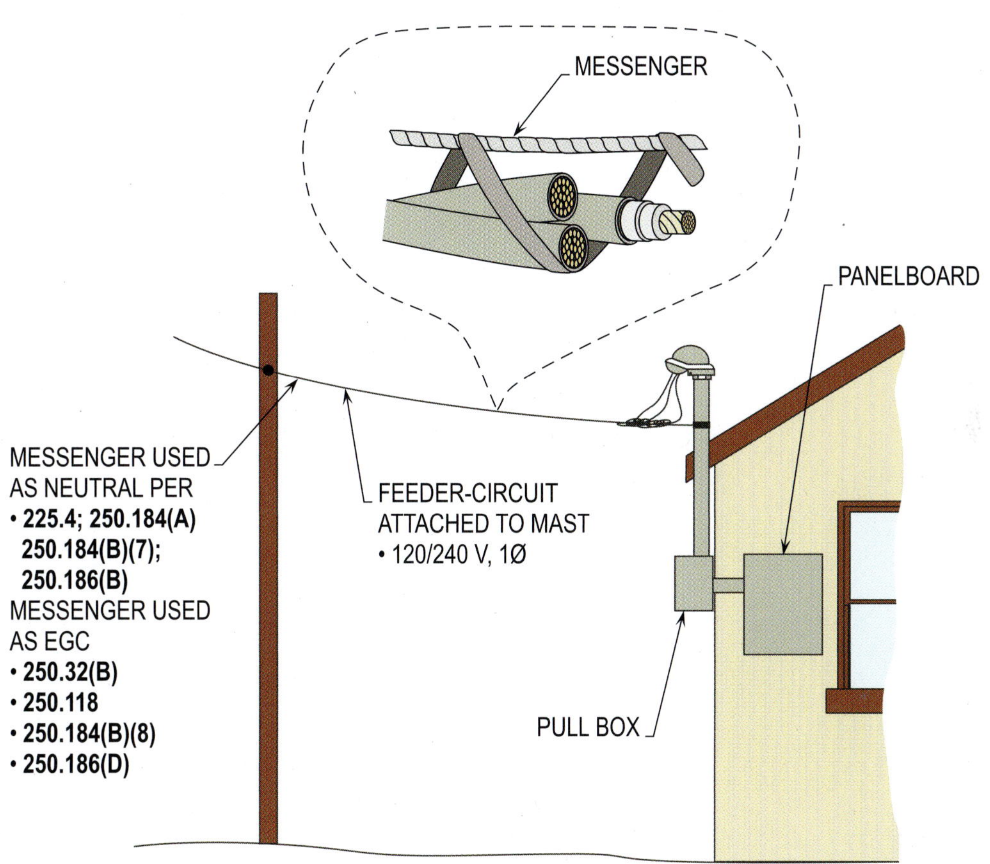

MESSENGER
396.30(A) through (C)

Purpose of Change: This revision clarifies the neutral conductor and equipment grounding conductor (EGC) being used as a messenger and locates the application and use.

4

Equipment for General Use

Chapter 4 has always been utilized by users and maintainers, who have the responsibility of installing general electrical equipment, luminaires, motors, and similar equipment.

If an electrician or user is replacing a ballast in a fluorescent luminaire, **410.73** of the NEC must be reviewed to ensure that the correct ballast and installation procedure are used. Consider a U frame motor that has been replaced with a T frame motor. In order to verify that the overloads are the proper size, **430.32(A)(1)** must be used. Switches are covered in **Article 404**, receptacles in **Article 406**, and panelboards or switchboards are covered in **Article 408**.

Electricians must not get confused when installing specialized equipment not included in **Chapter 4** of the NEC. **Chapter 4** applies to general electrical equipment and **Chapter 6** applies to special electrical equipment.

For example, where installing a crane or a hoist, electricians must reference **Article 610** of **Chapter 6**, which covers special equipment. For installing a elevator, **Article 620** must be referenced, not **Chapter 4**.

Article 445 applies to the installation of generators. If generators greater than 600 volt are being installed, **Article 490** covers the general equipment operating at over 600 volts.

NEC Ch. 4 - Article 402
Part - 402.11

Type of Change		Panel Action		UL	UL 508	API 500 - 1997	API 505 - 1997	OSHA - 1994
Revision		Accept in Principle		66	-	-	-	1910.305(i)(3)
ROP		ROC		NFPA 70E - 2004	NFPA 70B - 2006	NFPA 79 - 2007	NFPA	NEMA
pg. 401	# 6-92	pg. -	# -	420.9(C)	-	-	-	-
log: 997	CMP: 6	log: -	Submitter: Daniel Leaf			2005 NEC: 402.11		IEC: -

2005 NEC - 402.11 Uses Not Permitted.

Fixture wires shall not be used as branch-circuit conductors.

2008 NEC - 402.11 Uses Not Permitted.

Fixture wires shall not be used as branch-circuit conductors, except as permitted elsewhere in the Code.

Author's Comment: There is permission in other parts of the NEC to use fixture wire as a branch circuit conductor, so a phrase was added to **402.11** indicating that fact. A Class I motor control circuit installed in accordance with **430.72** must comply with **725.49(B)** [formerly **725.27(B)** in the 2005 NEC] if it is not tapped from the load side of a motor circuit protective device. This Class 1 circuit, with its overcurrent protection, constitutes a branch circuit based on the definition of a branch circuit in **Article 100**.

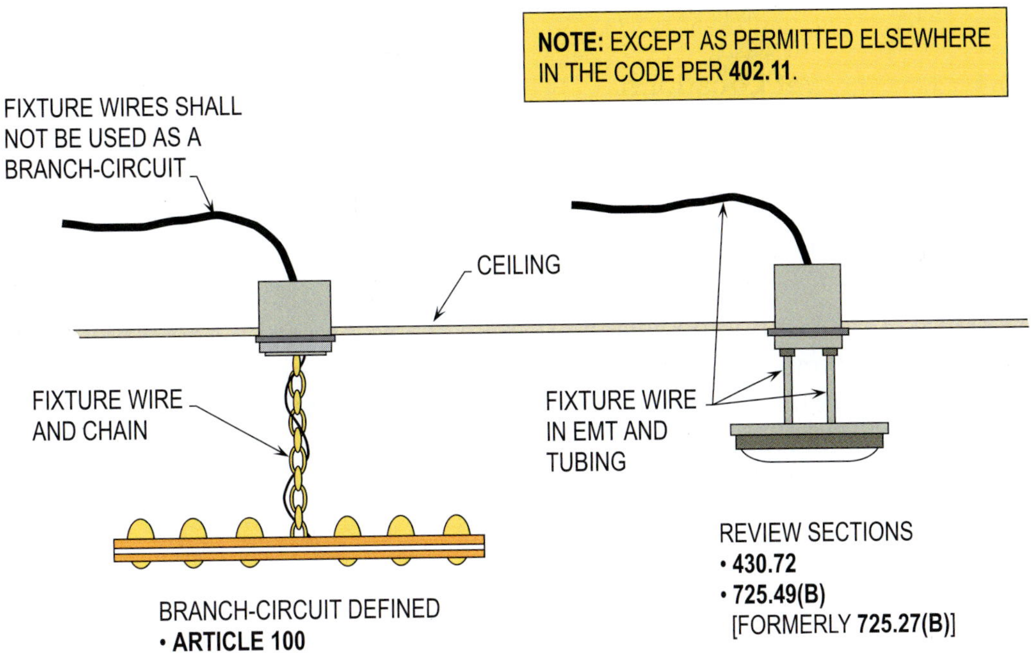

USES NOT PERMITTED
402.11

Purpose of Change: This revision clarifies that there is permission in other parts of the NEC to use fixture wire as a branch-circuit conductor.

NEC Ch. 4 - Article 404
Part I - 404.4

Type of Change	Panel Action	UL	UL 508	API 500 - 1997	API 505 - 1997	OSHA - 1994
Revision	Accept in Principle	20 - 98	-	-	-	1910.305(E)(1)
ROP	**ROC**	**NFPA 70E - 2004**	**NFPA 70B - 2006**	**NFPA 79 - 2007**	**NFPA**	**NEMA - 2003**
pg. 403 # 9-88	pg. - # -	420.5(A)	19.7	-	-	250
log: 974 CMP: 9	log: -	Submitter: Daniel Leaf		2005 NEC: 404.4		IEC: 60529 - 2004

2005 NEC - 404.4 Wet Locations.

A switch or circuit breaker in a wet location ~~or outside of a building~~ shall be enclosed in a weatherproof enclosure or cabinet that shall comply with 312.2(A). Switches shall not be installed within wet locations in tub or shower spaces unless installed as part of a listed tub or shower assembly.

2008 NEC - 404.4 Damp or Wet Locations.

A surface-mounted switch or circuit breaker in a damp or wet location shall be enclosed in a weatherproof enclosure or cabinet that shall comply with 312.2. A flush-mounted switch or circuit breaker in a damp or wet location shall be equipped with a weatherproof cover. Switches shall not be installed within wet locations in tub or shower spaces unless installed as part of a listed tub or shower assembly.

Author's Comment: A surface-mounted switch or circuit breaker must be enclosed in a weatherproof enclosure or cabinet in any damp or wet location because deterioration can occur and a serious shock hazard could develop. This could be a problem in *any* damp location or wet location, not just outside a building. A new second sentence was added to require a weatherproof cover to be installed on flush-mounted switches and circuit breakers.

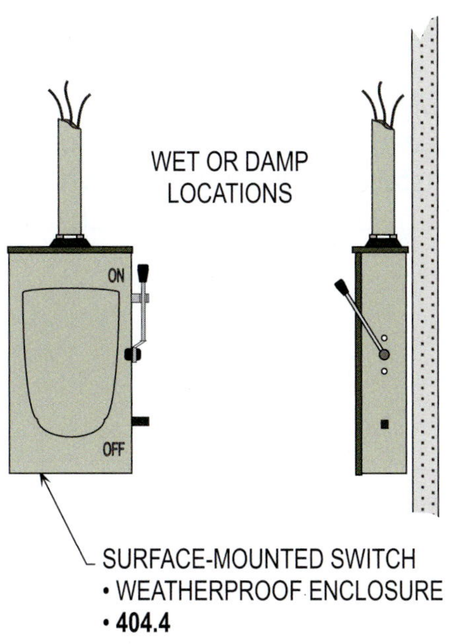

DAMP OR WET LOCATIONS
404.4

Purpose of Change: This revision clarifies that a flush-mounted switch or circuit breaker in either a wet or damp location shall be enclosed in a weatherproof enclosure or cabinet and equipped with a weatherproof cover. A surface-mounted switch or circuit breaker in a damp or wet location must have a weatherproof cover installed for protection.

NEC Ch. 4 - Article 404
Part I - 404.8(C)

Type of Change	Panel Action	UL	UL 508	API 500 - 1997	API 505 - 1997	OSHA - 1994
New Subsection	Accept	20	-	-	-	-
ROP	**ROC**	**NFPA 70E - 2004**	**NFPA 70B - 2006**	**NFPA 79 - 2007**	**NFPA**	**NEMA - 2004**
pg. 404 # 9-92	pg. - # -	-	19.5	-	-	410
log: 3406 CMP: 9	log: -	Submitter: Frederic P. Hartwell		2005 NEC: -		IEC: -

2008 NEC - 404.8 Accessibility and Grouping.

(C) Multipole Snap Switches. A multipole, general use snap switch shall not be permitted to be fed from more than a single circuit unless it is listed and marked as a two-circuit or three-circuit switch, or unless its voltage rating is not less than the nominal line-to-line voltage of the system supplying the circuits.

Author's Comment: The addition of this new subsection to **404.8** provides clear direction to installers that a multi-pole general-use switch can only be connected to one circuit, unless listed and marked for multiple circuits. In the past, some installers had connected a 120-volt circuit for an exhaust fan and a 277-volt circuit for luminaires to a two-pole, single-throw switch, even though the switch was only listed for a single circuit.

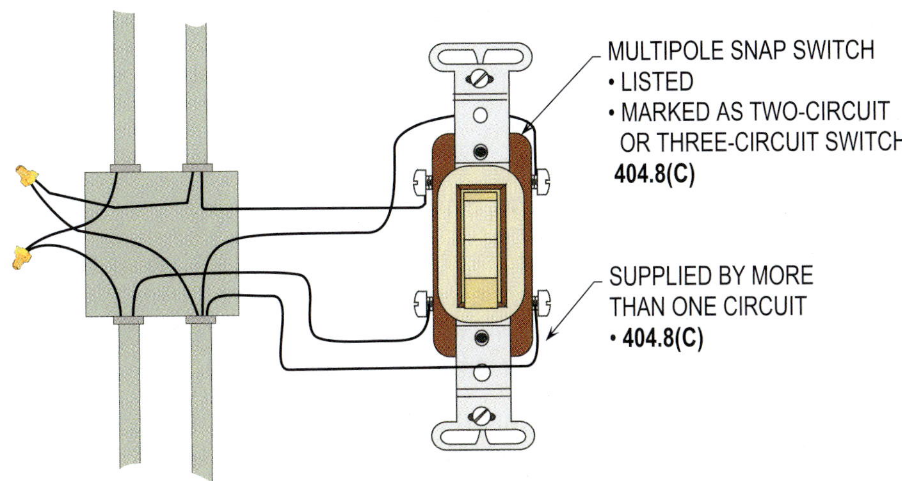

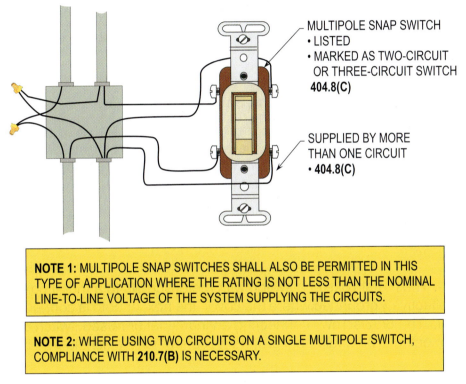

NOTE 1: MULTIPOLE SNAP SWITCHES SHALL ALSO BE PERMITTED IN THIS TYPE OF APPLICATION WHERE THE RATING IS NOT LESS THAN THE NOMINAL LINE-TO-LINE VOLTAGE OF THE SYSTEM SUPPLYING THE CIRCUITS.

NOTE 2: WHERE USING TWO CIRCUITS ON A SINGLE MULTIPOLE SWITCH, COMPLIANCE WITH **210.7(B)** IS NECESSARY.

MULTIPOLE SNAP SWITCHES
404.8(C)

Purpose of Change: A new subsection has been added to address the installation requirements for multipole snap switches.

NEC Ch. 4 - Article 404
Part I - 404.9(B)(1)

Type of Change	Panel Action	UL	UL 508	API 500 - 1997	API 505 - 1997	OSHA - 1994
Revision	Accept in Principle	20	-	-	-	-
ROP	**ROC**	**NFPA 70E - 2004**	**NFPA 70B - 2006**	**NFPA 79 - 2007**	**NFPA**	**NEMA - 2004**
pg. 405 # 9-96	pg. - # -	-	19.5	-	-	410
log: 2248 CMP: 9	log: - Submitter: Joseph Penachio			2005 NEC: 404.9(B)(1)		IEC: -

2005 NEC - 404.9 Provisions for General-Use Snap Switches.

(B) Grounding. Snap switches, including dimmer and similar control switches, shall be ~~effectively grounded~~ and shall provide a means to ~~ground~~ metal faceplates, whether or not a metal faceplate is installed. Snap switches shall be considered ~~effectively grounded~~ if either of the following conditions is met:

(1) The switch is mounted with metal screws to a metal box or to a nonmetallic box with integral means for ~~grounding devices~~.

2008 NEC - 404.9 Provisions for General-Use Snap Switches.

(B) Grounding. Snap switches, including dimmer and similar control switches, shall be connected to an equipment grounding conductor and shall provide a means to connect metal faceplates to the equipment grounding conductor, whether or not a metal faceplate is installed. Snap switches shall be considered to be part of an effective ground fault current path if either of the following conditions is met:

(1) The switch is mounted with metal screws to a metal box or metal cover that is connected to an equipment grounding conductor or to a nonmetallic box with integral means for connecting to an equipment grounding conductor.

Author's Comment: The metal yoke of a switch with metal screws to a metal box or to a metal cover is considered to be effectively grounded where the cover, the box, or both is connected to an equipment grounding conductor. The 8/32 screws connecting an industrial or raised switch cover to a metal box where the box is connected to an equipment grounding conductor are considered to sufficiently ground the switch.

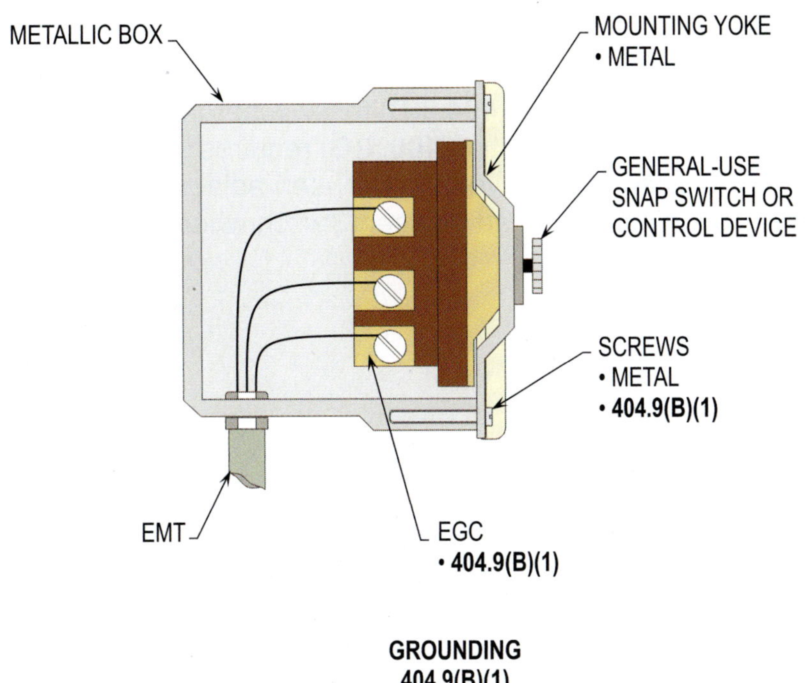

GROUNDING
404.9(B)(1)

Purpose of Change: This revision clarifies that the metal yoke of a switch in direct contact with a metal cover shall be considered effectively grounded where the cover and/or the box is connected to an equipment grounding conductor (EGC).

NEC Ch. 4 - Article 406
Part - 406.4(G)

57

Type of Change	Panel Action	UL	UL 508	API 500 - 1997	API 505 - 1997	OSHA - 1994
New Subsection	Accept	498	-	-	-	-
ROP	**ROC**	**NFPA 70E - 2004**	**NFPA 70B - 2006**	**NFPA 79 - 2007**	**NFPA**	**NEMA - 2002**
pg. 415 # 18-24	pg. 251 # 18-13	-	-	19.3	-	WD 6
log: 2129 CMP: 18	log: 216	Submitter: Russell LeBlanc		2005 NEC: -		IEC: -

2008 NEC - 406.4(G) Voltage Between Adjacent Devices.

A receptacle shall not be grouped or ganged in enclosures with other receptacles, snap switches, or similar devices, unless they are arranged so that the voltage between adjacent devices does not exceed 300 volts, or unless they are installed in enclosures equipped with identified, securely installed barriers between adjacent devices.

Author's Comment: The hazard addressed by **404.8(B)** for switches and other devices is also present for receptacles and other devices. New Subsection **406.4(G)** requires barriers where two receptacles, switches, or any devices are installed and the voltage between adjacent devices exceeds 300 volts. Therefore, a receptacle installed in a box with an adjacent switch where the voltage between devices exceeds 300 volts would require a barrier.

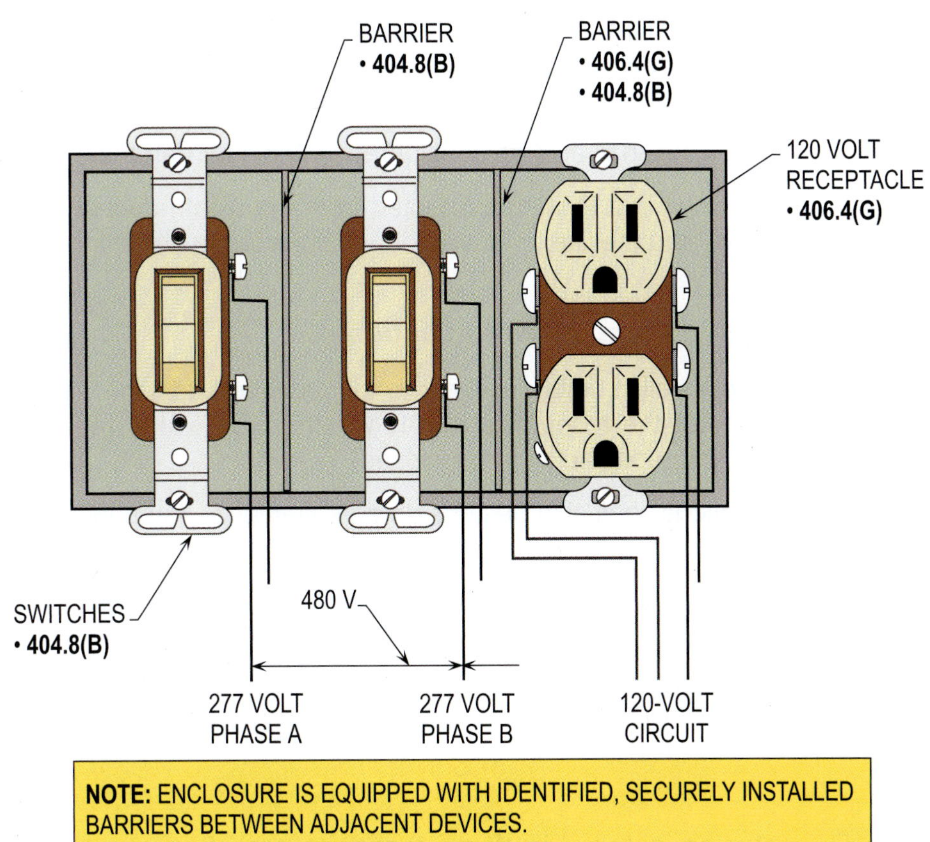

VOLTAGE BETWEEN ADJACENT DEVICES
406.4(G)

Purpose of Change: A new subsection has been added to address the hazard for receptacles, switches, and other devices where the voltage exceeds 300 volts between adjacent devices.

NEC Ch. 4 - Article 406
Part - 406.8(A) and FPN

Type of Change		Panel Action		UL	UL 508	API 500 - 1997	API 505 - 1997	OSHA - 1994
Revision		Accept in Principle		498	-	-	-	1910.305(j)(2)(ii)
ROP		ROC		NFPA 70E - 2004	NFPA 70B - 2006	NFPA 79 - 2007	NFPA	NEMA - 2002
pg. 415	# 18-28	pg. 252	# 18-16	420.10(C)(3)	19.3	-	-	WD 6
log: 3639	CMP: 18	log: 1424	Submitter: Aaron B. Chase			2005 NEC: 406.8(A)		IEC: -

2005 NEC - 406.8 Receptacles in Damp or Wet Locations.

(A) Damp Locations. A receptacle installed outdoors in a location protected from the weather or in other damp locations shall have an enclosure for the receptacle that is weatherproof when the receptacle is covered (attachment plug cap not inserted and receptacle covers closed).

An installation suitable for wet locations shall also be considered suitable for damp locations.

A receptacle shall be considered to be in a location protected from the weather where located under roofed open porches, canopies, marquees, and the like, and will not be subjected to a beating rain or water runoff.

2008 NEC - 406.8 Receptacles in Damp or Wet Locations.

(A) Damp Locations. A receptacle installed outdoors in a location protected from the weather or in other damp locations shall have an enclosure for the receptacle that is weatherproof when the receptacle is covered (attachment plug cap not inserted and receptacle covers closed).

An installation suitable for wet locations shall also be considered suitable for damp locations.

A receptacle shall be considered to be in a location protected from the weather where located under roofed open porches, canopies, marquees, and the like, and will not be subjected to a beating rain or water runoff. All 15- and 20-ampere, 125- and 250-volt nonlocking receptacles shall be a listed weather-resistant type.

FPN: The types of receptacles covered by this requirement are identified as 5-15, 5-20, 6-15, and 6-20 in ANSI/ NEMA WD 6-2002, National Electrical Manufacturers Association Standard for Dimensions of Attachment Plugs and Receptacles.

Author's Comment: Receptacles installed in damp locations are often exposed to water, sunlight (where installed outdoors with the resulting effects of ultraviolet light), and possible damage by impact of machinery or other objects. Even when receptacles are protected by weatherproof covers, some protective weatherproof covers are broken or have equipment inserted into the receptacle, thereby defeating the weatherproof aspects of the cover. In other cases, vertical cover plates may be installed in horizontal installations. The addition of the last sentence in the section will now require the actual receptacles be listed as weather-resistant. Even though the receptacle will be weather-resistant, the weatherproof cover will still be required.

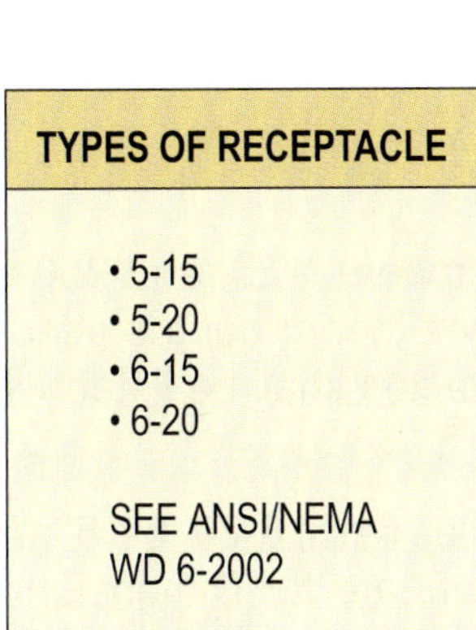

TYPES OF RECEPTACLE

- 5-15
- 5-20
- 6-15
- 6-20

SEE ANSI/NEMA WD 6-2002

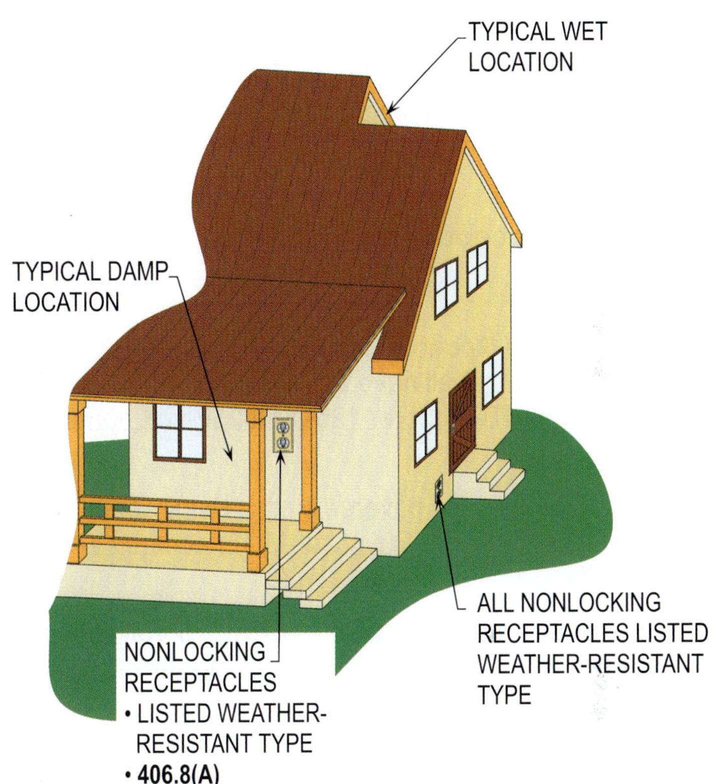

DAMP LOCATIONS
406.8(A) and FPN

Purpose of Change: This revision provides requirements that all nonlocking 15- and 20-ampere, 125- and 250-volt receptacles must be a listed weather-resistant type.

NEC Ch. 4 - Article 406
Part - 406.8(B)(1), Exception

59

Type of Change	Panel Action	UL	UL 508	API 500 - 1997	API 505 - 1997	OSHA - 1994
New Exception	Accept in Principle	498	-	-	-	1910.305(j)(2)(ii)
ROP	ROC	NFPA 70E - 2004	NFPA 70B - 2006	NFPA 79 - 2007	NFPA	NEMA - 2002
pg. 417 # 18-34	pg. - # -	420.10(C)(3)	19.3	-	-	WD 6
log: 648 CMP: 18	log: -	Submitter: Larry T. Smith		2005 NEC: -		IEC: -

2005 NEC - 406.8 Receptacles in Damp or Wet Locations.

(B) Wet Locations.

(1) 15- and 20-Ampere Receptacles in a Wet Location. 15- and 20-ampere, 125- and 250-volt receptacles installed in a wet location shall have an enclosure that is weatherproof whether or not the attachment plug cap is inserted.

2008 NEC - 406.8 Receptacles in Damp or Wet Locations.

(B) Wet Locations.

(1) 15- and 20-Ampere Receptacles in a Wet Location. 15- and 20-ampere, 125- and 250-volt receptacles installed in a wet location shall have an enclosure that is weatherproof whether or not the attachment plug cap is inserted. <u>All 15- and 20-ampere, 125- and 250-volt nonlocking receptacles shall be listed weather-resistant type.</u>

<u>FPN: The types of receptacles covered by this requirement are identified as 5-15, 5-20, 6-15, and 6-20 in ANSI/ NEMA WD 6 - 2002, National Electrical Manufacturers Association Standard for Dimensions of Attachment Plugs and Receptacles.</u>

<u>Exception: 15- and 20-ampere, 125- through 250-volt receptacles installed in a wet location and subject to routine high-pressure spray washing shall be permitted to have an enclosure that is weatherproof when the attachment plug is removed.</u>

Author's Comment: High-pressure spray wash cleaning is done on a regular basis in many establishments and at many buildings. There is the distinct possibility that spray will intrude into the enclosure through the cable openings in the in-use cover, causing corrosion of the receptacle. There is also an added danger of foreign materials being forced into the in-use enclosure by high-pressure spray. The new exception permits receptacles installed in wet locations with high-pressure spray washing to have an enclosure that is weatherproof when the attachment plug is removed, rather than weatherproof whether or not the attachment plug cap is installed.

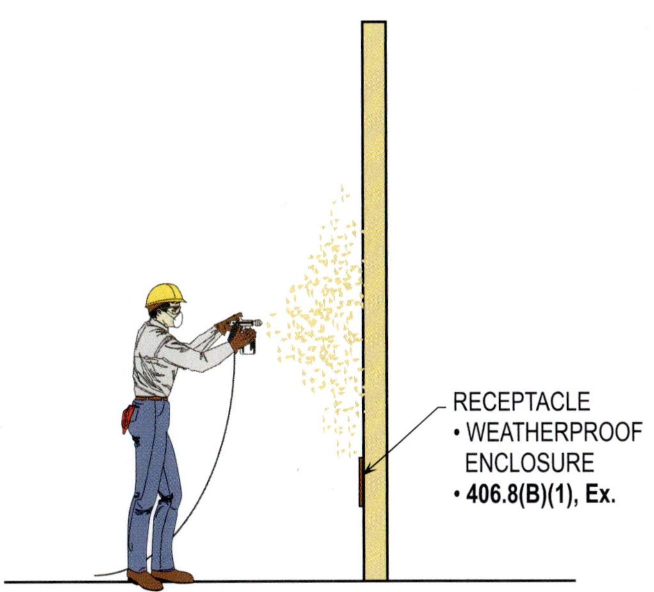

15- AND 20-AMPERE, 125- THROUGH 250-VOLT RECEPTACLES IN A WET LOCATION 406.8(B)(1), Exception

Purpose of Change: A new exception has been added to permit receptacles installed in wet locations subject to high pressure spray washing to have an enclosure that is weatherproof when the attachment plug is removed.

NEC Ch. 4 - Article 406
Part - 406.11

Type of Change		Panel Action		UL	UL 508	API 500 - 1997	API 505 - 1997	OSHA - 1994
New Section		Accept		498	-	-	-	-
ROP		**ROC**		**NFPA 70E - 2004**	**NFPA 70B - 2006**	**NFPA 79 - 2007**	**NFPA**	**NEMA - 2002**
pg. 418	# 18-40	pg. -	# -	-	-	-	-	WD 6
log: 1944	CMP: 18	log: -	Submitter: Vince Baclawski			2005 NEC: -		IEC: -

2008 NEC - <u>406.11 Tamper-Resistant Receptacles in Dwelling Units.</u>

<u>In all areas specified in 210.52, all 125-volt, 15- and 20-ampere receptacles shall be listed tamper resistant receptacles.</u>

Author's Comment: This new section requires 15- and 20-ampere, 125-volt receptacles installed in dwelling units to be tamper-resistant. The Panel was concerned about the possible increased insertion force of the cord cap into the tamper-resistant receptacle and asked for data on the amount of force necessary. There was substantial data submitted by the manufacturers indicating the number of child related electrical accidents involving children inserting metal objects into receptacles.

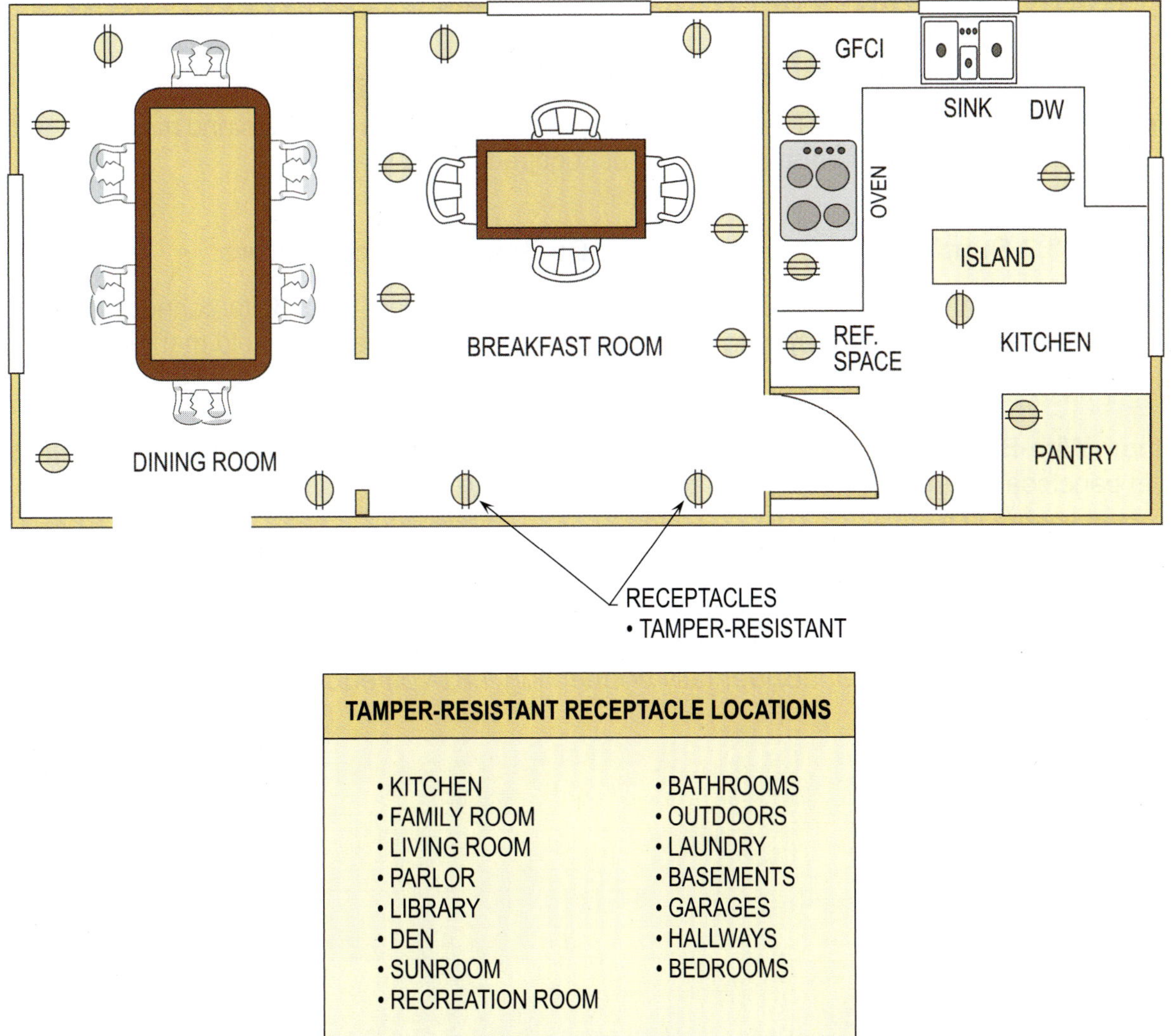

TAMPER-RESISTANT RECEPTACLES IN DWELLING UNITS
406.11

Purpose of Change: A new section has been added to require tamper-resistant receptacles in dwelling units for locations specified in **210.52**.

NEC Ch. 4 - Article 408
Part I - 408.3(F)

Type of Change	Panel Action	UL	UL 508	API 500 - 1997	API 505 - 1997	OSHA - 1994
New Subsection	Accept in Principle	891 - 67	-	-	-	-
ROP	**ROC**	**NFPA 70E - 2004**	**NFPA 70B - 2006**	**NFPA 79 - 2007**	**NFPA**	**NEMA - 2006**
pg. 421 # 9-109	pg. - # -	-	-	-	-	PB 1
log: 674 CMP: 9	log: -	Submitter: Jamie McNamara		2005 NEC: -		IEC: -

2005 NEC - 408.3 Support and Arrangement of Busbars and Conductors.

~~(F)~~ **Minimum Wire-Bending Space.** The minimum wire-bending space at terminals and minimum gutter space provided in panelboards and switchboards shall be as required in 312.6.

2008 NEC - 408.3 Support and Arrangement of Busbars and Conductors.

(F) High-Leg Identification. A switchboard or panelboard containing a 4-wire, delta-connected system where the midpoint of one phase winding is grounded shall be legibly and permanently field marked as follows:

"Caution __ Phase Has __ Volts to Ground"

(G) Minimum Wire-Bending Space. The minimum wire-bending space at terminals and minimum gutter space provided in panelboards and switchboards shall be as required in 312.6.

Author's Comment: The field-marking of panelboards and switchboards containing a delta high leg system will help installation and maintenance personnel recognize a system where the conductor and bus bar has a voltage to ground of 208-volts. This requirement will provide visible marking to help eliminate some of the hazards of accidentally connecting 120-volt equipment to the 208-volt high leg.

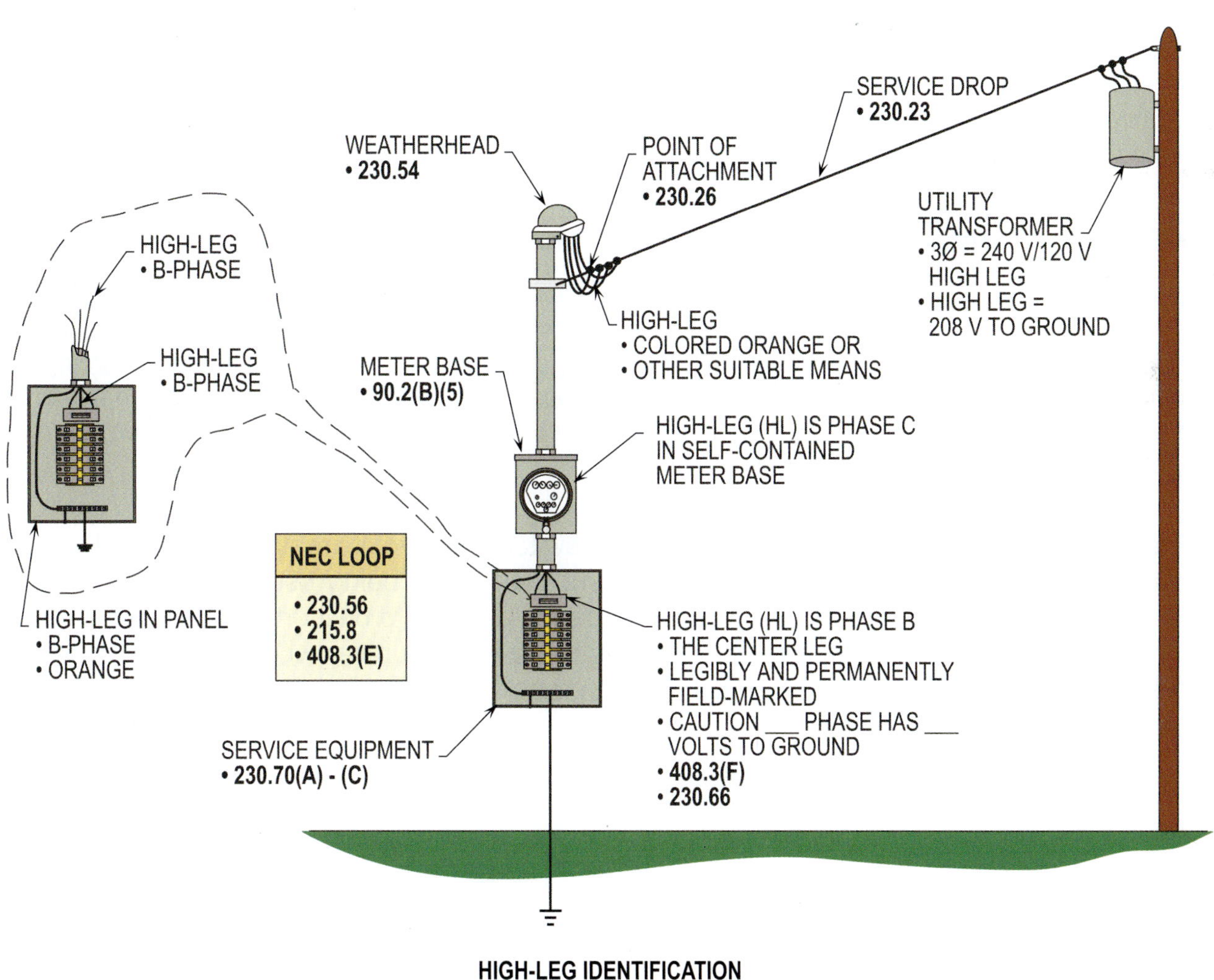

HIGH-LEG IDENTIFICATION
408.3(F)

Purpose of Change: A new subsection has been added to require the high-leg to be legibly and permanently field-marked in the switchboard or panelboard.

NEC Ch. 4 - Article 408
Part I - 408.4

Type of Change	Panel Action	UL	UL 508	API 500 - 1997	API 505 - 1997	OSHA - 1994
Revision	Accept	891 - 67	-	-	-	-
ROP	**ROC**	**NFPA 70E - 2004**	**NFPA 70B - 2006**	**NFPA 79 - 2007**	**NFPA**	**NEMA - 2006**
pg. 419 # 9-101	pg. 262 # 9-56	-	-	-	-	PB 1
log: 362 CMP: 9	log: 61	Submitter: Michael J. Johnston		2005 NEC: 408.4		IEC: -

2005 NEC - 408.4 Circuit Directory or Circuit Identification.

Every circuit and circuit modification shall be legibly identified as to its clear, evident, and specific purpose or use. The identification shall include sufficient detail to allow each circuit to be distinguished from all others. The identification shall be included in a circuit directory that is located on the face or inside of the panel door in the case of a panelboard, and located at each switch on a switchboard.

2008 NEC - 408.4 Circuit Directory or Circuit Identification.

Every circuit and circuit modification shall be legibly identified as to its clear, evident, and specific purpose or use. The identification shall include sufficient detail to allow each circuit to be distinguished from all others. Spare positions that contain unused overcurrent devices or switches shall be described accordingly. The identification shall be included in a circuit directory that is located on the face or inside of the panel door in the case of a panelboard, and located at each switch on a switchboard. No circuit shall be described in a manner that depends on transient conditions of occupancy.

Author's Comment: This change requires any spare switches or circuit breakers be labeled as "spare" to help identify all circuit breakers and switches, such as those being used and any spares. The comment has added a final sentence that does not permit a circuit to be idenified or a circuit directory to be labeled using transient identification. For example, a room number where the room has a label could be used as identification for the circuit, but a person's name for an office may be considered to be transient identification.

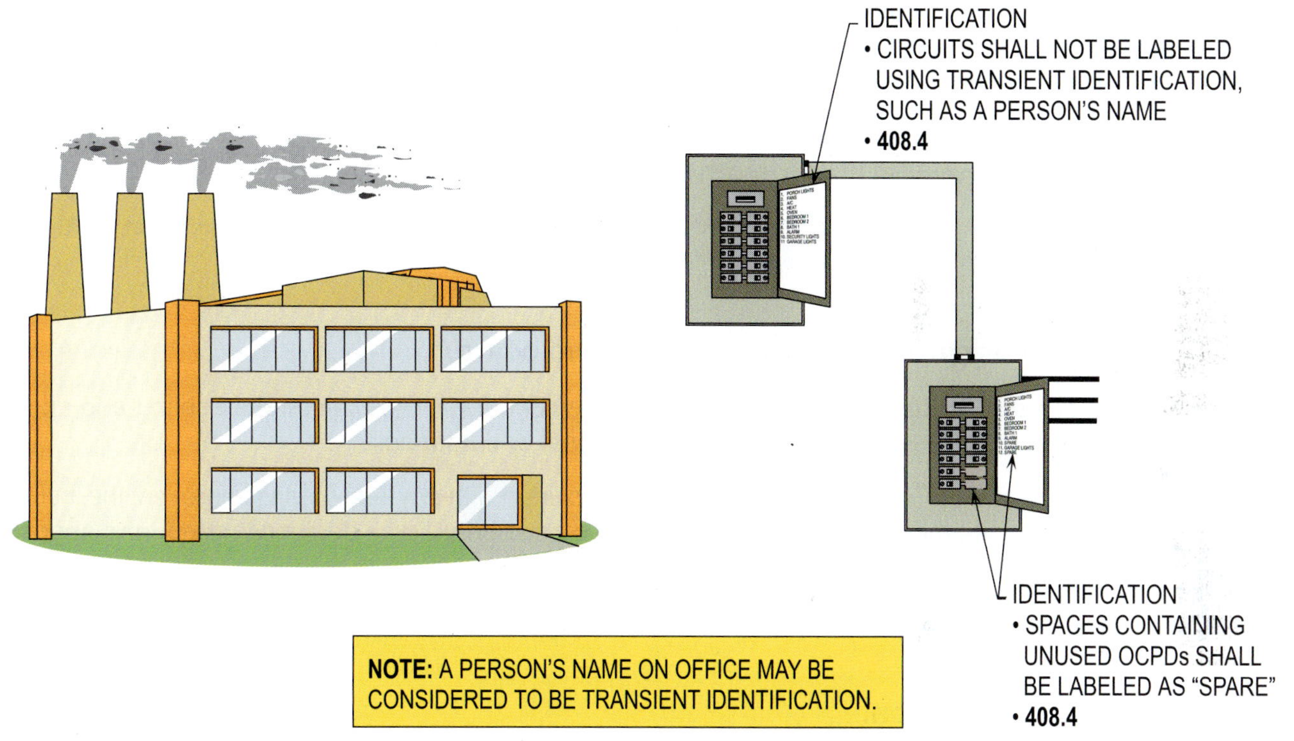
IDENTIFICATION
• CIRCUITS SHALL NOT BE LABELED USING TRANSIENT IDENTIFICATION, SUCH AS A PERSON'S NAME
• 408.4
IDENTIFICATION
• SPACES CONTAINING UNUSED OCPDs SHALL BE LABELED AS "SPARE"
• 408.4
NOTE: A PERSON'S NAME ON OFFICE MAY BE CONSIDERED TO BE TRANSIENT IDENTIFICATION.

**CIRCUIT DIRECTORY OR CIRCUIT IDENTIFICATION
408.4**

Purpose of Change: This revision clarifies that spare overcurrent protective devices shall be labeled accordingly and no circuit shall be described in a manner that depends on transient conditions of occupancy.

NEC Ch. 4 - Article 408
Part III - 408.36, Exception No. 1 through 3

Type of Change	Panel Action	UL	UL 508	API 500 - 1997	API 505 - 1997	OSHA - 1994
Revision	Accept in Principle	67	-	-	-	-
ROP	**ROC**	**NFPA 70E - 2004**	**NFPA 70B - 2006**	**NFPA 79 - 2007**	**NFPA**	**NEMA - 2006**
pg. 422 # 9-117	pg. 265 # 9-70	-	-	-	-	PB 1
log: 2643 CMP: 9	log: 2268	Submitter: Kevin J. Lippert		2005 NEC: 408.34; 408.35; 408.36		IEC: -

2005 NEC - ~~408.34 Classification of Panelboards.~~

~~Panelboards shall be classified for the purposes of this article as either lighting and appliance branch-circuit panelboards or power panelboards, based on their content. A lighting and appliance branch circuit is a branch circuit that has a connection to the neutral of the panelboard and that has overcurrent protection of 30 amperes or less in one or more conductors.~~

~~**(A) Lighting and Appliance Branch-Circuit Panelboard.** A lighting and appliance branch-circuit panelboard is one having more than 10 percent of its overcurrent devices protecting lighting and appliance branch circuits.~~

~~**(B) Power Panelboard.** A power panelboard is one having 10 percent or fewer of its overcurrent devices protecting lighting and appliance branch circuits.~~

408.35 Number of Overcurrent Devices on One Panelboard.

~~Not more than 42 overcurrent devices (other than those provided for in the mains) of a lighting and appliance branch-circuit panelboard shall be installed in any one cabinet or cutout box.~~

~~A lighting and appliance branch-circuit panelboard shall be provided with physical means to prevent the installation of more overcurrent devices than that number for which the panelboard was designed, rated, and approved.~~

~~For the purposes of this article, a 2-pole circuit breaker shall be considered two overcurrent devices; a 3-pole circuit breaker shall be considered three overcurrent devices.~~

408.36 Overcurrent Protection.

~~**(A) Lighting and Appliance Branch-Circuit Panelboard Individually Protected.** Each lighting and appliance branch-circuit panelboard shall be individually protected on the supply side by not more than two main circuit breakers or two sets of fuses having a combined rating not greater than that of the panelboard.~~

Exception No. 1: Individual protection ~~for a lighting and appliance panelboard shall not be required if the panelboard feeder has overcurrent protection not greater than the rating of the panelboard~~.

Exception No. 2: For existing installations, individual protection ~~for lighting and appliance branch-circuit panelboards~~ shall not be required ~~where such~~ panelboard~~s are~~ used as service equipment ~~in supplying~~ an individual residential occupancy.

~~**(B) Power Panelboard Protection.**~~ In addition to the requirements of 408.30, a ~~power~~ panelboard ~~with supply conductors that include a neutral, and having more than 10 percent of its overcurrent devices protecting branch circuits rated 30 amperes or less,~~ shall be protected by an overcurrent protective device having a rating not greater than that of the panelboard. This overcurrent protective device shall be located within or at any point on the supply side of the panelboard.

~~(C)~~ Snap Switches Rated at 30 Amperes or Less.

~~(D)~~ Supplied Through a Transformer.

~~(E)~~ Delta Breakers.

~~(F)~~ Back-Fed Devices.

2008 NEC - 408.36 Overcurrent Protection.

In addition to the requirement of 408.30, a panelboard shall be protected by an overcurrent protective device having a rating not greater than that of the panelboard. This overcurrent protective device shall be located within or at any point on the supply side of the panelboard.

Exception No. 1: Individual protection shall not be required for a panelboard used as service equipment with multiple disconnecting means in accordance with 230.71. In panelboards protected by three or more main circuit breakers or sets of fuses, the circuit breakers or sets of fuses shall not supply a second bus structure within the same panelboard assembly.

Exception No. 2: Individual protection shall not be required for a panelboard protected on its supply side by two main circuit breakers or two sets of fuses having a combined rating not greater than that of the panelboard. A panelboard constructed or wired under this exception shall not contain more than 42 overcurrent devices. For the purposes of determining the maximum of 42 overcurrent devices, a 2-pole or a 3- pole circuit breaker shall be considered as two or three overcurrent devices, respectively.

Exception No. 3: For existing panelboards, individual protection shall not be required for a panelboard used as service equipment for an individual residential occupancy.

(A) Snap Switches Rated at 30 Amperes or Less.

(B) Supplied Through a Transformer.

(C) Delta Breakers.

(D) Back-Fed Devices.

Author's Comment: By deleting Section **408.34**, the classification of panelboards as lighting and appliance panelboards or power panelboards has been deleted. Section **408.35** has also been deleted. In the past, power panels have not had the restriction of 42-overcurrent protective devices in a single panelboard. Now, lighting and appliance panelboards will no longer be limited to a maximum of 42 overcurrent protective devices.

Section **408.36, Exception No. 1** is based on the 2005 NEC text in Section **408.36(B), Exception**, which is intended to recognize a long-standing practice of permitting a small panel to be used as service equipment, with large line-to-line loads leaving at this point and a smaller feeder entering the building to supply what formerly was called a lighting and appliance branch circuit panelboard. The limitations in this exception prevent the extension of this limited practice to what could otherwise become a split-bus panelboard of unlimited size in the future. The six-disconnect limit echoes the customary service limitation in **230.71** and six disconnect limit in **225.33** for feeders.

Section **408.36, Exception No. 2** corresponds to the parent language in **408.36(A)**. Since prior practice effectively limited these panelboards to 42 circuits, the wording in the Panel action carries that limitation forward, but only for these split-bus panels.

Section **408.36, Exception No. 3** corresponds to the 2005 NEC edition **408.36(A), Exception No. 2**, and it continues without change. This exception still covers existing panelboards where individual protection shall not be required for a panelboard used as service equipment for an individual residential occupancy [older dwellings with existing split-bus panels are still acceptable].

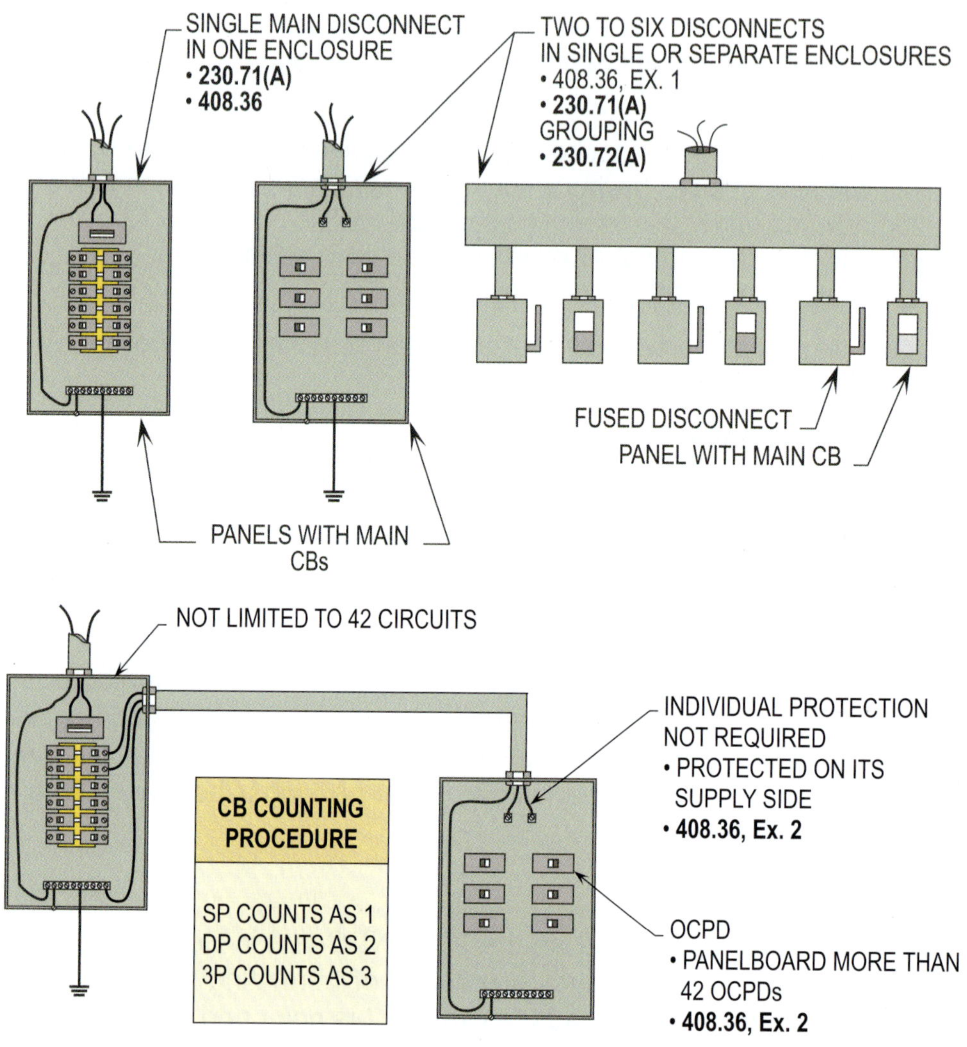

OVERCURRENT PROTECTION
408.36, Exception No. 1 through 3

Purpose of Change: This revision clarifies the permitted methods for providing protection of a panelboard and deletes the definition of power panelboards and lighting and appliance branch circuit panelboards.

NEC Ch. 4 - Article 408
Part IV - 408.54

Type of Change	Panel Action	UL	UL 508	API 500 - 1997	API 505 - 1997	OSHA - 1994
Relocated - Renumbered	Accept in Principle	67	-	-	-	-
ROP	**ROC**	**NFPA 70E - 2004**	**NFPA 70B - 2006**	**NFPA 79 - 2007**	**NFPA**	**NEMA - 2006**
pg. 425 # 9-127	pg. - # -	-	-	-	-	PB 1
log: 2948 CMP: 9	log: -	Submitter: Philip Simmons		2005 NEC: 408.35		IEC: -

2005 NEC - ~~408.35. Number of Overcurrent Devices on One Panelboard.~~

~~Not more than 42 overcurrent devices (other than those provided for in the mains) of a lighting and appliance branch-circuit panelboard shall be installed in any one cabinet or cutout box.~~

A ~~lighting and appliance branch-circuit~~ panelboard shall be provided with physical means to prevent the installation of more overcurrent devices than that number for which the panelboard was designed, rated, and ~~approved~~.

For the purposes of this ~~article~~, a 2-pole circuit breaker shall be considered two overcurrent devices; a 3-pole circuit breaker shall be considered three overcurrent devices.

2008 NEC - 408.54 Maximum Number of Overcurrent Devices.

A panelboard shall be provided with physical means to prevent the installation of more overcurrent devices than that number for which the panelboard was designed, rated, and listed. For the purposes of this section, a 2-pole circuit breaker or fusible switch shall be considered two overcurrent devices; a 3-pole circuit breaker or fusible switch shall be considered three overcurrent devices.

Author's Comment: This text was located in **408.35** and relocated to **Part IV** as new Section **408.54** because the text involves the construction of a panelboard and the physical restriction of the maximum number of overcurrent devices that the panel was designed, rated, and listed to contain. The maximum number of overcurrent devices is subject to the listing evaluation of the panel by the listing laboratory.

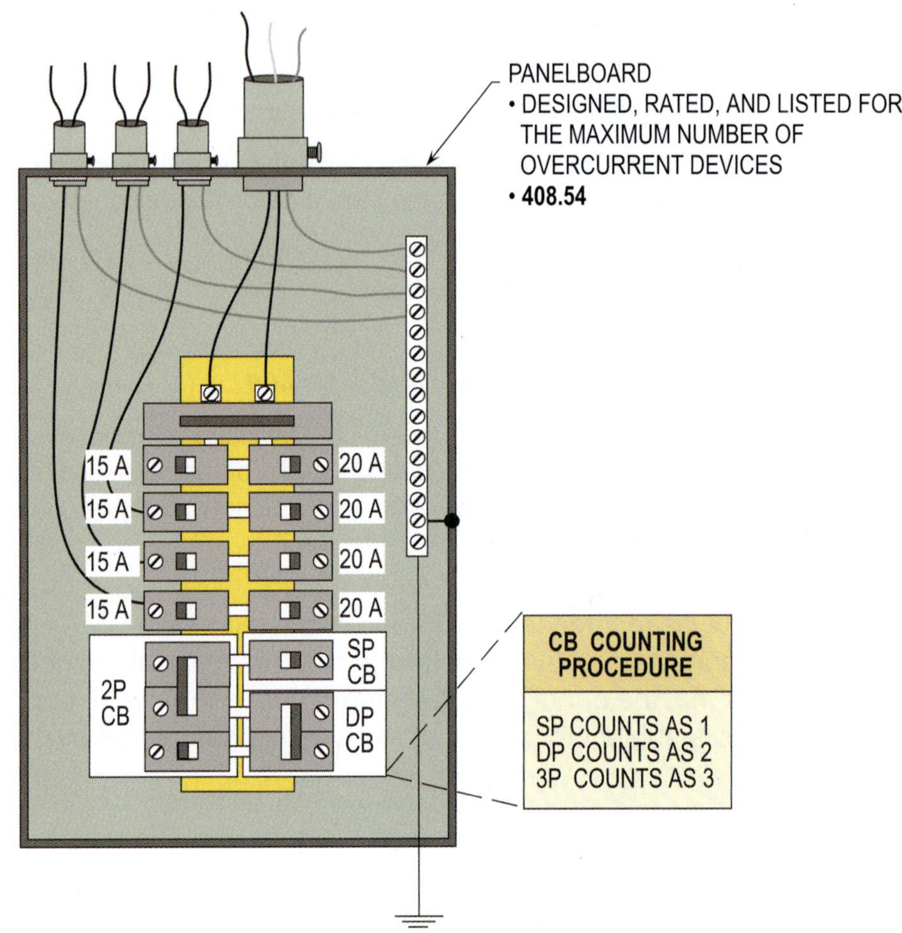

MAXIMUM NUMBER OF OVERCURRENT DEVICES
408.54

Purpose of Change: The maximum number of overcurrent devices that a panel was designed, rated, and listed to contain has been relocated to Part IV dealing with construction of panelboards.

NEC Ch. 4 - Article 408
Part IV - 408.55, Exception No. 1

Type of Change	Panel Action	UL	UL 508	API 500 - 1997	API 505 - 1997	OSHA - 1994
Revision	Accept	67	-	-	-	-
ROP	ROC	NFPA 70E - 2004	NFPA 70B - 2006	NFPA 79 - 2007	NFPA	NEMA - 2006
pg. 425 # 9-128a	pg. - # -	-	-	-	-	PB 1
log: CP 904 CMP: 9	log: -	Submitter: Code Making Panel 9		2005 NEC: 408.35		IEC: -

2005 NEC - 408.55 Wire-Bending Space in Panelboards.

The enclosure for a panelboard shall have the top and bottom wire-bending space sized in accordance with Table 312.6(B) for the largest conductor entering or leaving the enclosure. Side wire-bending space shall be in accordance with Table 312.6(A) for the largest conductor to be terminated in that space.

Exception No. 1: Either the top or bottom wire-bending space shall be permitted to be sized in accordance with Table 312.6(A) for a ~~lighting and appliance branch-circuit~~ panelboard rated 225 amperes or less.

2008 NEC - 408.55 Wire-Bending Space in Panelboards.

The enclosure for a panelboard shall have the top and bottom wire-bending space sized in accordance with Table 312.6(B) for the largest conductor entering or leaving the enclosure. Side wire-bending space shall be in accordance with Table 312.6(A) for the largest conductor to be terminated in that space.

Exception No. 1: Either the top or bottom wire-bending space shall be permitted to be sized in accordance with Table 312.6(A) for a panelboard rated 225 amperes or less and designed to contain not over 42 overcurrent devices. For the purposes of this exception, a 2-pole or a 3-pole circuit breaker shall be considered as two or three overcurrent devices, respectively.

Author's Comment: Code Making Panel 9 has removed the category of "lighting and appliance branch circuit panelboard" from **Article 408** by its action on Proposal 9-117. Because prior practice limited lighting and appliance panelboards to 42 circuits, the text added to the exception carries that limitation forward, along with information from former **408.35** defining how the circuit numbering is to be done. The change in this proposal correlates the provisions in this exception to the actions taken in Section **408.34**, **408.35**, and **408.36**.

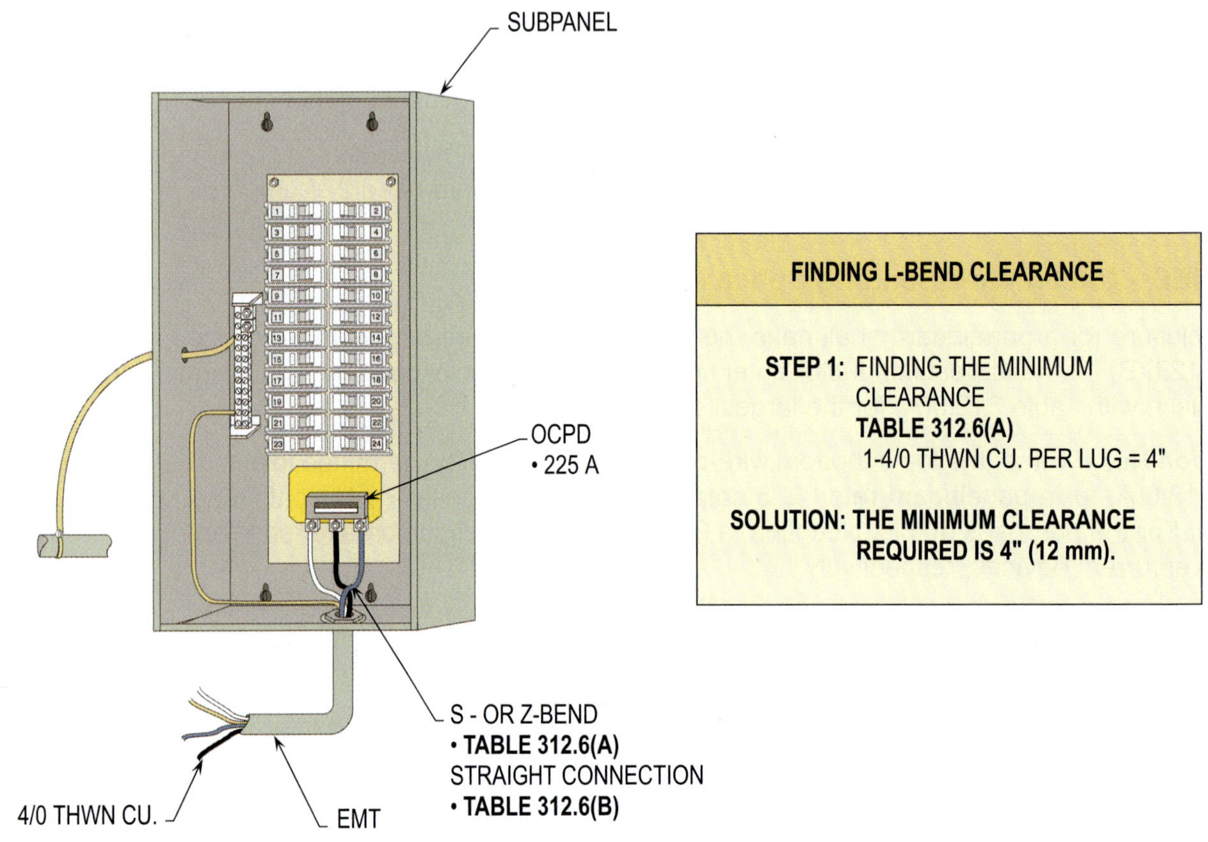

WIRE-BENDING SPACE IN PANELBOARDS
408.55, Exception No. 1

Purpose of Change: This revision clarifies that the exception shall not be applied for panelboards containing more than 42 overcurrent devices and rated more than 225 amperes.

NEC Ch. 4 - Article 408
Part IV - 408.58

Type of Change	Panel Action	UL	UL 508	API 500 - 1997	API 505 - 1997	OSHA - 1994
Relocated - Renumbered	Accept in Principle	67	-	-	-	-
ROP	**ROC**	**NFPA 70E - 2004**	**NFPA 70B - 2006**	**NFPA 79 - 2007**	**NFPA**	**NEMA - 2006**
pg. 424 # 9-123	pg. - # -	-	-	-	-	PB 1
log: 2946 CMP: 9	log: -	Submitter: Philip Simmons		2005 NEC: 408.30		IEC: -

2005 NEC - 408.30 General.

All panelboards shall have a rating not less than the minimum feeder capacity required for the load calculated in accordance with Article 220. ~~Panelboards shall be durably marked by the manufacturer with the voltage and the current rating and the number of phases for which they are designed and with the manufacturer's name or trademark in such a manner so as to be visible after installation, without disturbing the interior parts or wiring.~~

2008 NEC - 408.58 Panelboard Marking.

Panelboards shall be durably marked by the manufacturer with the voltage and the current rating and the number of phases for which they are designed and with the manufacturer's name or trademark in such a manner so as to be visible after installation, without disturbing the interior parts or wiring.

Author's Comment: The marking requirements formerly located in **408.30** were relocated to a new Section **408.58** in **Part IV** dealing with construction of panelboards because these markings are required to be installed by the manufacturer of the panelboards, not applied in the field. The existing **FPN** referencing **110.22** should probably be moved to **408.56** in the 2011 NEC since it deals with marking.

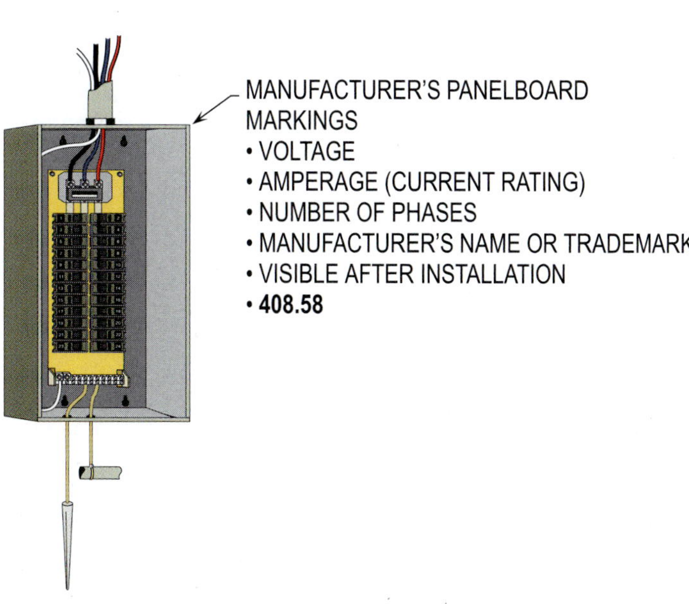

PANELBOARD MARKING
408.58

Purpose of Change: The marking requirements for panelboards have been relocated to Part IV dealing with construction of panelboards.

NEC Ch. 4 - Article 409
Part I - 409.2

Type of Change	Panel Action	UL	UL 508	API 500 - 1997	API 505 - 1997	OSHA - 1994
Revision	Accept	508A	-	-	-	-
ROP	**ROC**	**NFPA 70E - 2004**	**NFPA 70B - 2006**	**NFPA 79 - 2007**	**NFPA**	**NEMA - 2005**
pg. 426 # 11-3	pg. - # -	-	6.3.5	3.3.24	-	ICS 1
log: 1953 CMP: 11	log: -	Submitter: Vince Baclawski		2005 NEC: 409.2		IEC: -

2005 NEC - 409.2 Industrial Control Panel.

~~An assembly of a systematic and standard arrangement of two or more components such as motor controllers, overload relays, fused disconnect switches, and circuit breakers and related control devices such as pushbutton stations, selector switches, timers, switches, control relays, and the like with associated wiring, terminal blocks, pilot lights, and similar components. The industrial control panel does not include the controlled equipment.~~

2008 NEC - 409.2 Industrial Control Panel.

An assembly of two or more components consisting of one of the following:

(1) Power circuit components only, such as motor controllers, overload relays, fused disconnect switches, and circuit breakers

(2) Control circuit components only, such as pushbuttons, pilot lights, selector switches, timers, switches, control relays

(3) A combination of power and control circuit components

These components, with associated wiring and terminals, are mounted on or contained within an enclosure or mounted on a sub-panel. The industrial control panel does not include the controlled equipment.

Author's Comment: The approved method for determining the short circuit current rating of an industrial control panel is based only on power circuit components. However, an industrial control panel, consisting only of control circuit components, is not required to be marked with a short circuit current rating. The proposal adds clarification to the definition by recognizing that some panels may be constructed solely of control components. The last paragraph provides acceptable mounting locations and are specifically identified. The last paragraph is unchanged from the 2005 text.

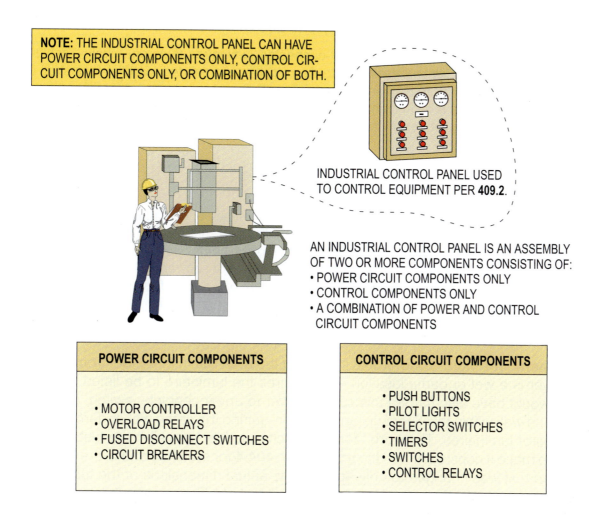

NOTE: THE INDUSTRIAL CONTROL PANEL CAN HAVE POWER CIRCUIT COMPONENTS ONLY, CONTROL CIRCUIT COMPONENTS ONLY, OR COMBINATION OF BOTH.

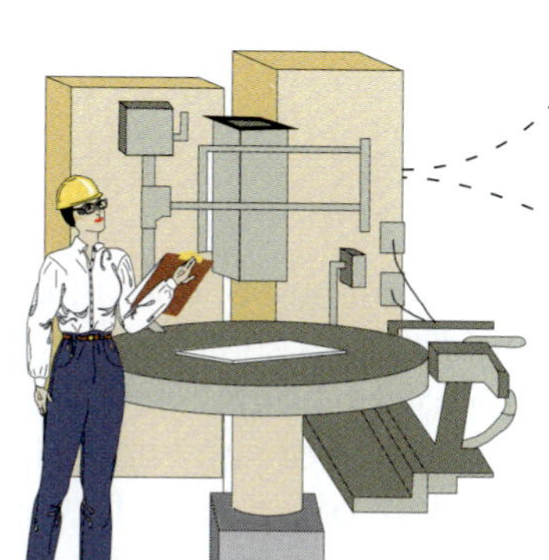

INDUSTRIAL CONTROL PANEL USED TO CONTROL EQUIPMENT PER **409.2**.

AN INDUSTRIAL CONTROL PANEL IS AN ASSEMBLY OF TWO OR MORE COMPONENTS CONSISTING OF:

- POWER CIRCUIT COMPONENTS ONLY
- CONTROL COMPONENTS ONLY
- A COMBINATION OF POWER AND CONTROL CIRCUIT COMPONENTS

POWER CIRCUIT COMPONENTS

- MOTOR CONTROLLER
- OVERLOAD RELAYS
- FUSED DISCONNECT SWITCHES
- CIRCUIT BREAKERS

CONTROL CIRCUIT COMPONENTS

- PUSH BUTTONS
- PILOT LIGHTS
- SELECTOR SWITCHES
- TIMERS
- SWITCHES
- CONTROL RELAYS

INDUSTRIAL CONTROL PANEL
409.2

Purpose of Change: This revision clarifies what constitutes an industrial control panel.

NEC Ch. 4 - Article 410
Part II - 410.10(D)

Type of Change	Panel Action	UL	UL 508	API 500 - 1997	API 505 - 1997	OSHA - 1994
Revision	Accept	1598	-	-	-	-
ROP	**ROC**	**NFPA 70E - 2004**	**NFPA 70B - 2006**	**NFPA 79 - 2007**	**NFPA**	**NEMA**
pg. 441 # 18-51	pg. 268 # 18-67	-	-	-	-	-
log: 1906 CMP: 18	log: 2220	Submitter: James W. Carpenter		2005 NEC: 410.4(D)		IEC: -

2005 NEC - ~~410.4~~ Luminaires (Fixtures) in Specific Locations.

(D) Bathtub and Shower Areas. No parts of cord-connected luminaires ~~(fixtures)~~, chain-, cable-, or cord-suspended-luminaires ~~(fixtures)~~, lighting track, pendants, or ceiling-suspended (paddle) fans shall be located within a zone measured 900 mm (3 ft) horizontally and 2.5 m (8 ft) vertically from the top of the bathtub rim or shower stall threshold. This zone is all encompassing and includes the ~~zone locations,~~ directly over the tub or shower stall. Luminaires ~~(lighting fixtures)~~ located ~~in this zone~~ shall be listed for damp or listed for wet locations where subject to shower spray.

2008 NEC - 410.10 Luminaires in Specific Locations.

(D) Bathtub and Shower Areas. No parts of cord-connected luminaires, chain-, cable-, or cord-suspended-luminaires, lighting track, pendants, or ceiling-suspended (paddle) fans shall be located within a zone measured 900 mm (3 ft) horizontally and 2.5 m (8 ft) vertically from the top of the bathtub rim or shower stall threshold. This zone is all encompassing and includes the space directly over the tub or shower stall. Luminaires located within the actual outside dimension of the bathtub or shower to a height of 2.5 m (8 ft) vertically from the top of the bathtub rim or shower threshold shall be marked for damp locations or marked for wet locations where subject to shower spray.

Author's Comment: The word "zone" would imply that the area within three feet from the edge of the bathtub or shower is a wet or damp location and requires the luminaire to be listed for wet or damp locations. This would have also carried this concept over to any receptacle, switch, or luminaire that was located on the wall within this 3 foot distance, thus requiring the same wet or damp location rating for these devices or luminaires. The word "zone" was changed to "space" in the last sentence of the first paragraph to make it consistent with the text used in **404.4** for switches and in the title for **406.8** for receptacles. The text was changed to only apply to the actual dimensions of the tub or shower to a height of 8 feet from the top of the bathtub rim or shower threshold. The comment deleted parenthetical "fixtures" and "lighting fixtures" and deleted listing of luminaires for damp and wet locations. Luminaires must be marked "damp" or "wet."

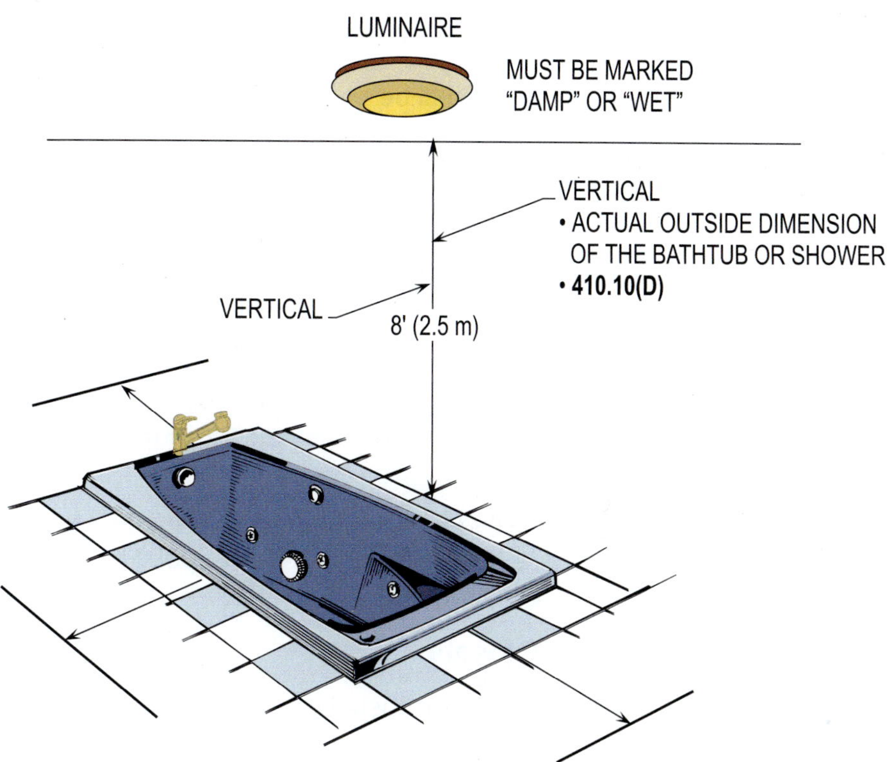

NOTE 1: NO CORD, CHAIN, OR CABLE TYPE LUMINAIRES SHALL BE INSTALLED PER **410.10(D)**.

NOTE 2: LUMINAIRES LOCATED IN THIS ACTUAL DIMENSION OF TUB OR SHOWER AND WITHIN 8 FEET SHALL BE MARKED FOR DAMP OR MARKED FOR WET LOCATIONS WHERE SUBJECT TO SHOWER SPRAY.

BATH AND SHOWER AREAS
410.10(D)

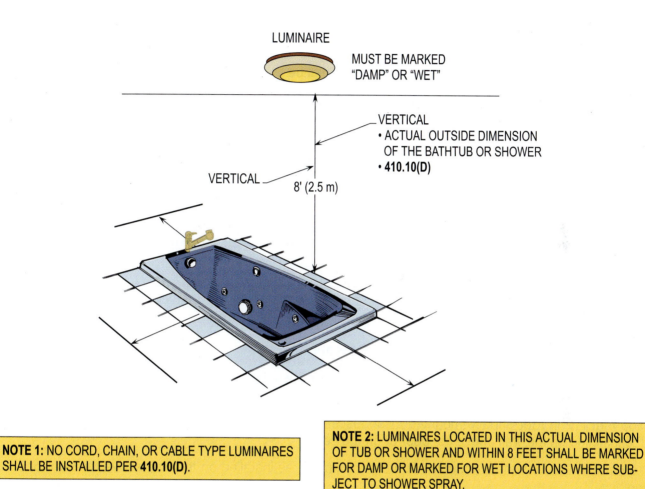

Purpose of Change: This revision clarifies that the luminaires located within the actual outside dimension of the bathtub or shower to a height of 8 ft. (2.5 m) vertically from the top of the bathtub rim or shower threshold to be marked "damp" or if subject to shower spray, marked "wet" location.

NEC Ch. 4 - Article 410
Part II - 410.16(C)(1) through (C)(5)

Type of Change	Panel Action	UL	UL 508	API 500 - 1997	API 505 - 1997	OSHA - 1994
Revision	Accept in Principle	1598	-	-	-	-
ROP	**ROC**	**NFPA 70E - 2004**	**NFPA 70B - 2006**	**NFPA 79 - 2007**	**NFPA**	**NEMA - 2001**
pg. 443 # 18-60	pg. 268 # 18-66	-	-	-	-	LE 4
log: 1352 CMP: 18	log: 1407 Submitter: Mike Holt			2005 NEC: 410.8(D)		IEC: -

2005 NEC - ~~410.8~~ Luminaires ~~(Fixtures)~~ in Clothes Closets.

~~(D)~~ Location. ~~Luminaires (fixtures) in clothes closets shall be permitted to be installed as follows:~~

(1) ~~Surface-mounted incandescent luminaires (fixtures) installed on the wall above the door or on the ceiling, provided there is a minimum clearance of 300 mm (12 in.) between the luminaire (fixture) and the nearest point of a storage space~~

(2) ~~Surface-mounted fluorescent luminaires (fixtures) installed on the wall above the door or on the ceiling, provided there is a minimum clearance of 150 mm (6 in.) between the luminaire (fixture) and the nearest point of a storage space~~

(3) ~~Recessed incandescent luminaires (fixtures) with a completely enclosed lamp installed in the wall or the ceiling, provided there is a minimum clearance of 150 mm (6 in.) between the luminaire (fixture) and the nearest point of a storage space~~

(4) ~~Recessed fluorescent luminaires (fixtures) installed in the wall or the ceiling, provided there is a minimum clearance of 150 mm (6 in.) between the luminaire (fixture) and the nearest point of a storage space~~

2008 NEC - 410.16 Luminaires in Clothes Closets.

(C) Location. The minimum clearance between luminaires installed in clothes closets and the nearest point of a storage space shall be as follows:

(1) 300 mm (12 in.) for surface-mounted incandescent or LED luminaires with a completely enclosed light source installed on the wall above the door or on the ceiling

(2) 150 mm (6 in.) for surface-mounted fluorescent luminaires installed on the wall above the door or on the ceiling

(3) 150 mm (6 in.) for recessed incandescent or LED luminaires with a completely enclosed light source installed in the wall or the ceiling

(4) 150 mm (6 in.) for recessed fluorescent luminaires installed in the wall or the ceiling

(5) Surface-mounted fluorescent or LED luminaires shall be permitted to be installed within the storage space where identified for this use

Author's Comment: The text was revised to make it easier to use and an additional (4) was added to deal with recessed fluorescent luminaires in a clothes closet. LED luminaires have also been added to the clearance requirements for luminaires in clothes closets. A new **(5)** has been added permitting surface-mounted fluorescent or LED luminaires to be installed within the storage space where identified for this use.

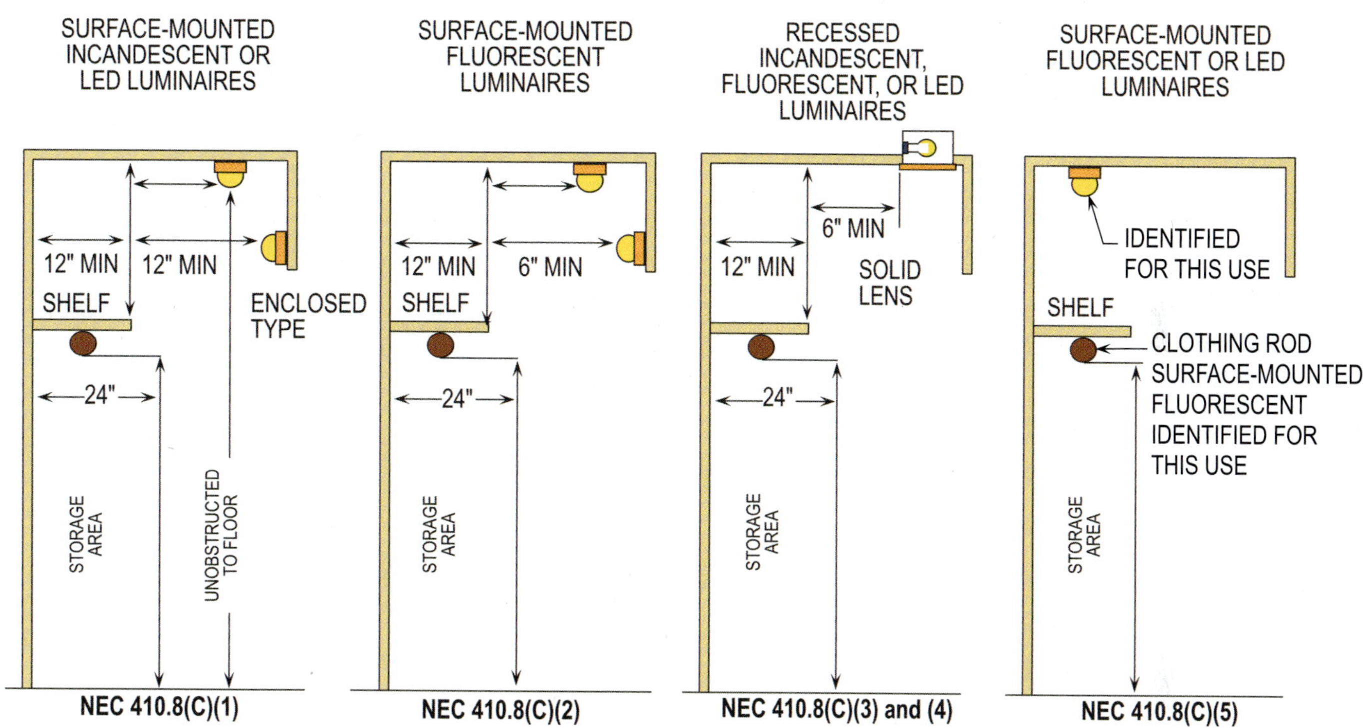

LUMINAIRES IN CLOTHES CLOSETS
410.16(C)(1) through (C)(5)

Purpose of Change: This revision clarifies the installation requirements for surface-mounted incandescent, fluorescent, or LED luminaires, and surface-mounted fluorescent or LED luminaires identified for the use where installed within the storage space.

NEC Ch. 4 - Article 410
Part VI - 410.62(C)(1)(2)c

Type of Change	Panel Action	UL	UL 508	API 500 - 1997	API 505 - 1997	OSHA - 1994
Revision	Accept	1598	-	-	-	-
ROP	**ROC**	**NFPA 70E - 2004**	**NFPA 70B - 2006**	**NFPA 79 - 2007**	**NFPA**	**NEMA**
pg. 446 # 18-78	pg. - # -	-	-	15.2.4	-	-
log: 3176 CMP: 18	log: - Submitter: Michael S. O'Boyle			2005 NEC: 410.30(C)(1)(2)c		IEC: -

2005 NEC - ~~410.30~~ Cord-Connected Lampholders and Luminaires.

(C) Electric-Discharge Luminaires.

(1) Cord-Connected Installation. A ~~listed~~ luminaire or a listed assembly shall be permitted to be cord connected if the following conditions apply:

(2) The flexible cord meets all the following:

a. Is visible for its entire length outside the luminaire

b. Is not subject to strain or physical damage

c. Is terminated in a grounding-type attachment plug cap or busway plug, or is a part of a listed assembly incorporating a manufactured wiring system connector in accordance with 604.6(C), or has a luminaire ~~(fixture)~~ assembly with a strain relief and canopy.

2008 NEC - 410.62 Cord-Connected Lampholders and Luminaires.

(C) Electric-Discharge Luminaires.

(1) Cord-Connected Installation. A luminaire or a listed assembly shall be permitted to be cord connected if the following conditions apply:

(1) The luminaire is located directly below the outlet or busway.

(2) The flexible cord meets all the following:

a. Is visible for its entire length outside the luminaire

b. Is not subject to strain or physical damage

c. Is terminated in a grounding-type attachment plug cap or busway plug, or is a part of a listed assembly incorporating a manufactured wiring system connector in accordance with 604.6(C), or has a luminaire assembly with a strain relief and canopy having a maximum 152 mm (6 in.) long section of raceway for attachment to an outlet box above a suspended ceiling.

Author's Comment: Electric-discharge luminaires are sometimes attached to grid members of suspended ceilings. To allow the canopy assembly to be positioned in line with the suspension hardware, the canopy needs to be mounted directly below a grid member. This is a problem because the ceiling grid member blocks placement of an outlet box flush with the ceiling surface. Running flexible cord unprotected through a hole in a suspended ceiling is a violation of **400.8**. A short length of raceway attached between a luminaire canopy and an outlet box would permit placement of the box on top of the ceiling grid. The raceway would protect the cord above the ceiling line. Section **400.8** allows flexible cord to be run in a raceway when specifically permitted elsewhere in the Code as this change would now permit this application.

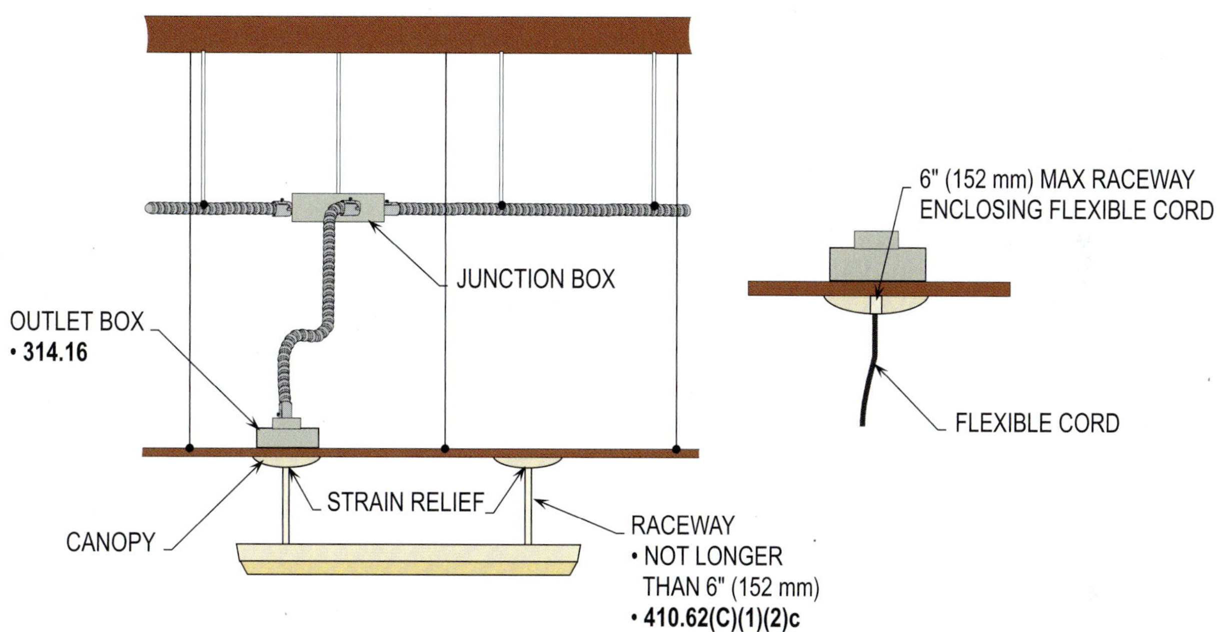

CORD-CONNECTED INSTALLATION
410.62(C)(1)(2)c

Purpose of Change: This revision clarifies that a luminaire assembly with strain relief and canopy can have a maximum 6 inch (152 mm) raceway enclosing a flexible cord installed to an outlet box above a suspended ceiling.

NEC Ch. 4 - Article 410
Part XI - 410.130(A)

Type of Change	Panel Action	UL	UL 508	API 500 - 1997	API 505 - 1997	OSHA - 1994
Revision	Accept	1598	-	-	-	-
ROP	**ROC**	**NFPA 70E - 2004**	**NFPA 70B - 2006**	**NFPA 79 - 2007**	**NFPA**	**NEMA**
pg. 448 / # 18-87	pg. - / # -	-	18.4	15.2.4	-	-
log: 2132 / CMP: 18	log: -	Submitter: Russell LeBlanc		2005 NEC: 410.73(A)		IEC: -

2005 NEC - ~~410.73~~ General.

(A) Open-Circuit Voltage of 1000 Volts or Less. Equipment for use with electric-discharge lighting systems and designed for an open-circuit voltage of 1000 volts or less shall be of a type ~~intended~~ for such service.

2008 NEC - 410.130 General.

(A) Open-Circuit Voltage of 1000 Volts or Less. Equipment for use with electric-discharge lighting systems and designed for an open-circuit voltage of 1000 volts or less shall be of a type identified for such service.

Author's Comment: The word "intended" was changed to "identified" to ensure that the equipment is identified for use with electric discharge lighting systems. "Identified" is defined as recognizable as suitable for the specific purpose, function, use, environment, application, and so forth.

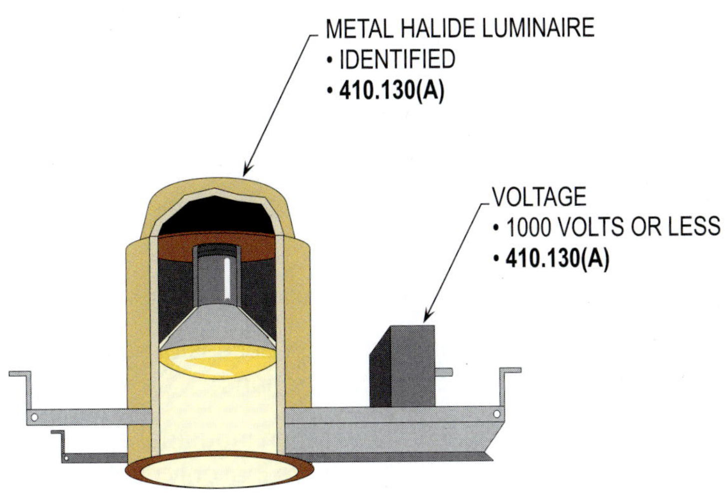

**OPEN-CIRCUIT VOLTAGE OF 1000 VOLTS OR LESS
410.130(A)**

Purpose of Change: This revision clarifies that the equipment is required to be identified not just intended for use with electric discharge lighting systems.

NEC Ch. 4 - Article 410
Part XI - 410.130(G)

66

Type of Change	Panel Action	UL	UL 508	API 500 - 1997	API 505 - 1997	OSHA - 1994
Revision	Accept	1598	-	-	-	-
ROP	ROC	NFPA 70E - 2004	NFPA 70B - 2006	NFPA 79 - 2007	NFPA	NEMA
pg. 449 # 18-90b	pg. 270 # 18-79	-	-	-	-	-
log: CP 1803 CMP: 18	log: 2091	Submitter: Code Making Panel 18		2005 NEC: 410.73(G)		IEC: C78.81 - 2005

2005 NEC - ~~410.73~~ General.

(G) Disconnecting Means. In indoor locations, other than dwellings and ~~assorted~~ accessory structures, fluorescent luminaires ~~(fixtures)~~ that utilize double-ended lamps and contain ballast(s) that can be serviced in place ~~or ballasted luminaires that are supplied from multiwire branch circuits and contain ballast(s) that can be serviced in place~~ shall have a disconnecting means either internal or external to each luminaire ~~(fixture), to disconnect simultaneously from the source of supply all conductors of the ballast, including the grounded conductor, if any.~~ The line side terminals of the disconnecting means shall be guarded. The disconnecting means shall be located so as to be accessible to qualified persons before servicing or maintaining the ballast. ~~This requirement shall become effective January 1, 2008.~~

Exception No. 1: A disconnecting means shall not be required for luminaires ~~(fixtures)~~ installed in hazardous (classified) location(s).

Exception No. 2: A disconnecting means shall not be required for emergency illumination required in 700.16.

Exception No. 3: For cord-and-plug-connected luminaires, an accessible separable or an accessible plug and receptacle shall be permitted to serve as the disconnecting means.

Exception No. 4: A disconnecting means shall not be required in industrial establishments with restricted public access where conditions of maintenance and supervision ensure that only qualified persons service the installation by written procedures.

Exception No. 5: Where more than one luminaire is installed and supplied by other than a mutlwire branch circuit, a disconnecting means shall not be required for every luminaire when the design of the installation includes ~~locally accessible disconnects~~, such that the illuminated space cannot be left in total darkness.

2008 NEC - 410.130 General.

(G) Disconnecting Means.

(1) General. In indoor locations, other than dwellings and associated accessory structures, fluorescent luminaires that utilize double-ended lamps and contain ballast(s) that can be serviced in place shall have a disconnecting means either internal or external to each luminaire. The line side terminals of the disconnecting means shall be guarded.

Exception No. 1: A disconnecting means shall not be required for luminaires installed in hazardous (classified) location(s).

Exception No. 2: A disconnecting means shall not be required for emergency illumination required in 700.16.

Exception No. 3: For cord-and-plug-connected luminaires, an accessible separable connector or an accessible plug and receptacle shall be permitted to serve as the disconnecting means.

Exception No. 4: A disconnecting means shall not be required in industrial establishments with restricted public access where conditions of maintenance and supervision ensure that only qualified persons service the installation by written procedures.

Exception No. 5: Where more than one luminaire is installed and supplied by other than a mutlwire branch circuit, a disconnecting means shall not be required for every luminaire when the design of the installation includes disconnecting means, such that the illuminated space cannot be left in total darkness.

(2) Multiwire Branch Circuits. When connected to multiwire branch circuits, the disconnecting means shall simultaneously break all the supply conductors to the ballast, including the grounded conductor.

(3) Location. The disconnecting means shall be located so as to be accessible to qualified persons before servicing or maintaining the ballast. Where the disconnecting means is external to the luminaire, it shall be a single device, and shall be attached to the luminaire or the luminaire shall be located within sight of the disconnecting means.

Author's Comment: The changes include removal of the effective date of January 1, 2008 and removing the disconnection of the grounded conductor with exception of multiwire-branch circuits. Where the location of the disconnect is an external disconnect, it must be within sight of the luminaire and must be a single device. The changes provide clarification that the rule applies only to double-ended fluorescent luminaires. Comment 18-79 has revised the last sentence of the recommendation regarding the location of the luminaire disconnecting means to clarify that either an external disconnecting means is required to be attached to the luminaire (could be attached to the luminaire and located above an accessible ceiling area) or that the disconnect is required to be located within sight of the luminaire. In addition, this section has been reorganized in order to place multiple requirements into separate subdivisions. This action also aids the user in understanding the application of the exceptions. Comment 18-80 has replaced "assorted" accessory structures with "associated" accessory structures.

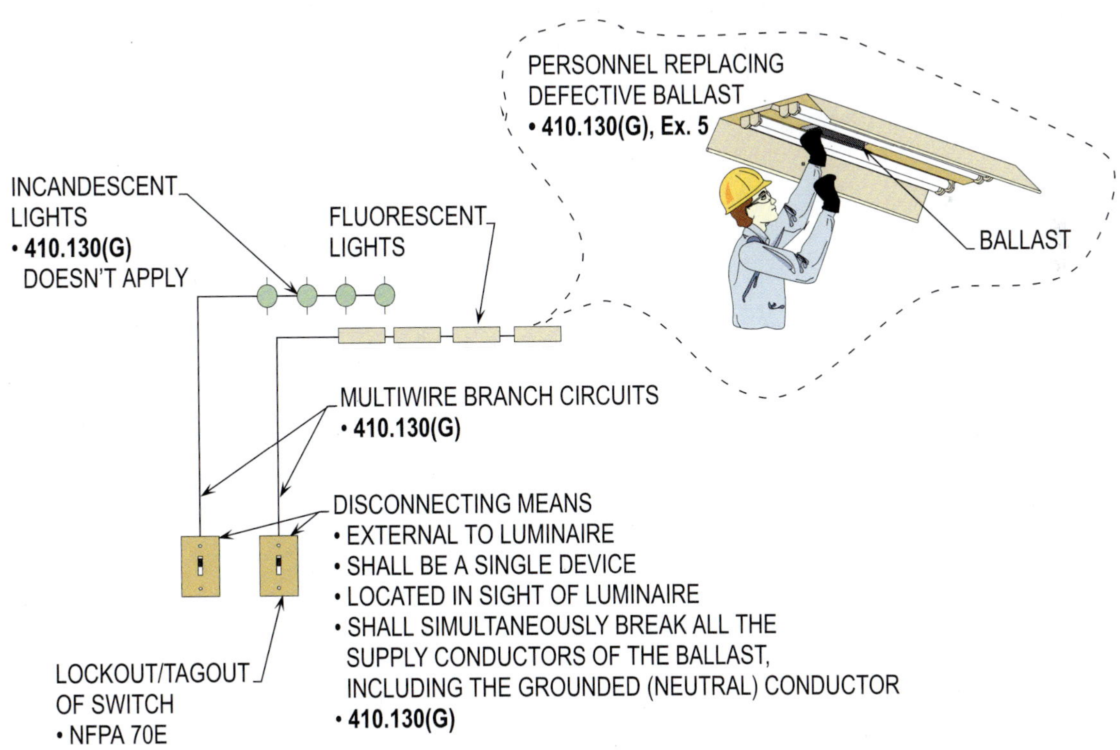

DISCONNECTING MEANS
410.130(G)

Purpose of Change: This revision clarifies that the disconnecting means for multiwire branch circuits shall simultaneously break all the supply conductors of the ballast, including the grounded (neutral) conductor.

NEC Ch. 4 - Article 410
Part XIV - 410.141(B)

Type of Change		Panel Action		UL	UL 508	API 500 - 1997	API 505 - 1997	OSHA - 1994
Revision		Accept		1598	-	-	-	-
ROP		**ROC**		**NFPA 70E - 2004**	**NFPA 70B - 2006**	**NFPA 79 - 2007**	**NFPA**	**NEMA**
pg. 451	# 18-100	pg. -	# -	-	-	-	-	-
log: 486	CMP: 18	log: -	Submitter: Michael J. Johnston			2005 NEC: 410.81(B)		IEC: -

2005 NEC - ~~410.81~~ Control.

(B) Within Sight or Locked Type. The switch or circuit breaker shall be located within sight from the luminaires ~~(fixtures)~~ or lamps, or it shall be permitted elsewhere if it is provided with a means for locking in the open position.

2008 NEC - 410.141 Control.

(B) Within Sight or Locked Type. The switch or circuit breaker shall be located within sight from the luminaires or lamps, or it shall be permitted elsewhere if it is provided with a means for locking in the open position. The provisions for locking or adding a lock to the disconnecting means must remain in place at the switch or circuit breaker whether the lock is installed or not. Portable means for adding a lock to the switch or circuit breaker shall not be permitted.

Author's Comment: The added text plus various other proposals on this same concept throughout the remainder of the NEC provides consistency for the lockout of disconnecting means throughout the NEC. The last sentence prohibiting a portable lockout device, will ensure the lockout device will remain at the circuit breaker or switch.

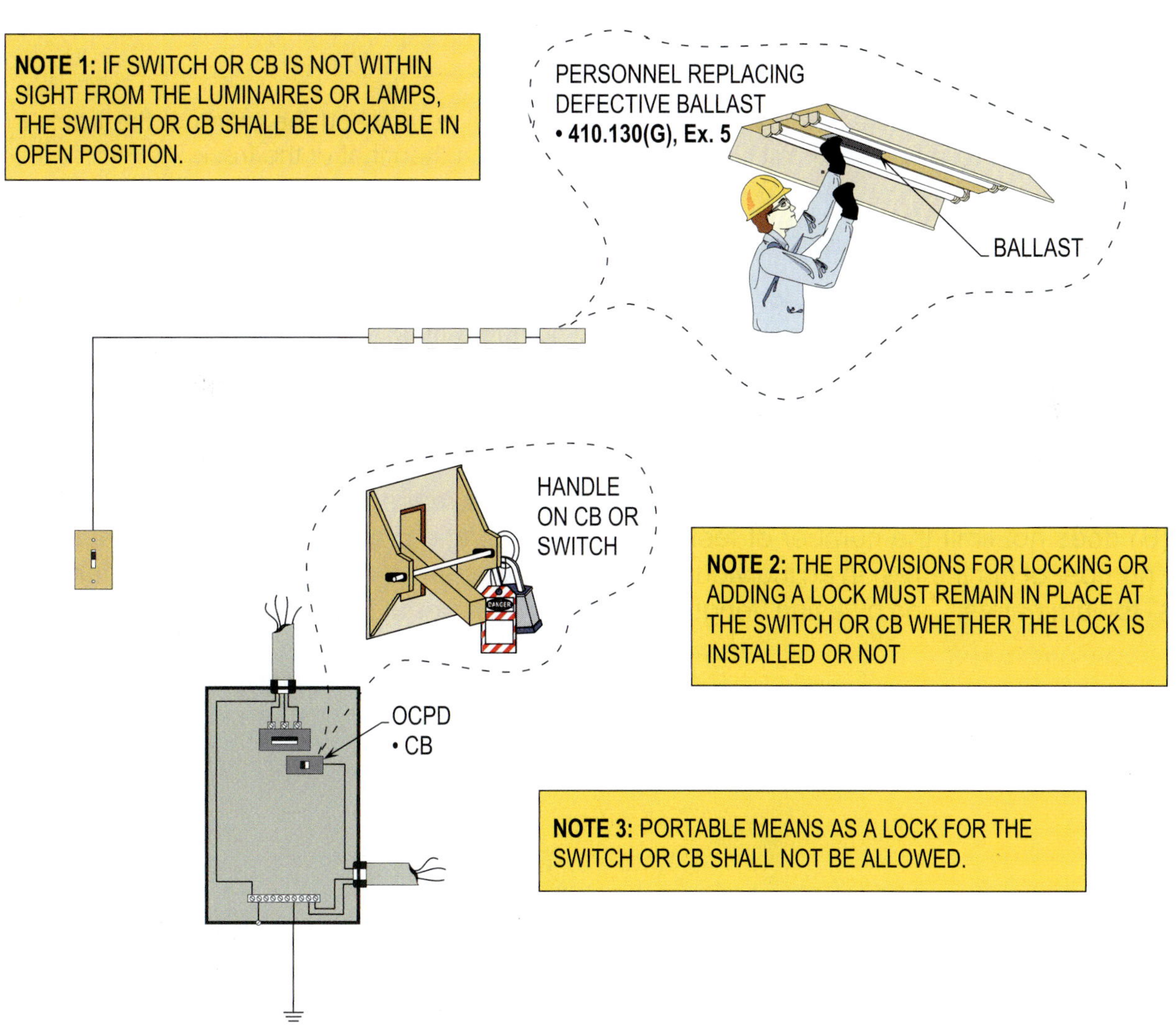

NOTE 1: IF SWITCH OR CB IS NOT WITHIN SIGHT FROM THE LUMINAIRES OR LAMPS, THE SWITCH OR CB SHALL BE LOCKABLE IN OPEN POSITION.
PERSONNEL REPLACING DEFECTIVE BALLAST
• 410.130(G), Ex. 5
BALLAST
HANDLE ON CB OR SWITCH
NOTE 2: THE PROVISIONS FOR LOCKING OR ADDING A LOCK MUST REMAIN IN PLACE AT THE SWITCH OR CB WHETHER THE LOCK IS INSTALLED OR NOT
OCPD
• CB
NOTE 3: PORTABLE MEANS AS A LOCK FOR THE SWITCH OR CB SHALL NOT BE ALLOWED.

WITHIN SIGHT OR LOCKED TYPE
410.141(B)

Purpose of Change: This revision clarifies that provisions for locking or adding a lock to the disconnecting means shall remain in place whether the lock is installed or not.

NEC Ch. 4 - Article 410
Part XV - 410.151(B), FPN

Type of Change	Panel Action	UL	UL 508	API 500 - 1997	API 505 - 1997	OSHA - 1994
New FPN	Accept	1598	-	-	-	-
ROP	**ROC**	**NFPA 70E - 2004**	**NFPA 70B - 2006**	**NFPA 79 - 2007**	**NFPA**	**NEMA**
pg. 444 / # 18-65	pg. 272 / # 18-89	-	-	-	-	-
log: 1670 / CMP: 18	log: 1944 / Submitter: Alcah Thompson			2005 NEC: 410.101(B)		IEC: -

2005 NEC - ~~410.101~~ Installation.

(B) Connected Load. The connected load on lighting track shall not exceed the rating of the track. Lighting track shall be supplied by a branch circuit having a rating not more than that of the track.

2008 NEC - 410.151 Installation.

(B) Connected Load. The connected load on lighting track shall not exceed the rating of the track. Lighting track shall be supplied by a branch circuit having a rating not more than that of the track.

FPN: The load calculation in 220.43(B) does not limit the length of track on a single branch circuit, and it does not limit the number of luminaires on a single track.

Author's Comment: The Fine Print Note has been added to make it clear that the load calculation in **220.43(B)** does not limit the number of feet of track on a single branch circuit and is not intended to limit the number of luminaires on an individual track. The comment omits statements of intent, changes "fixtures" to "luminaires," and by using "length" instead of "number of feet" avoids metrication problems.

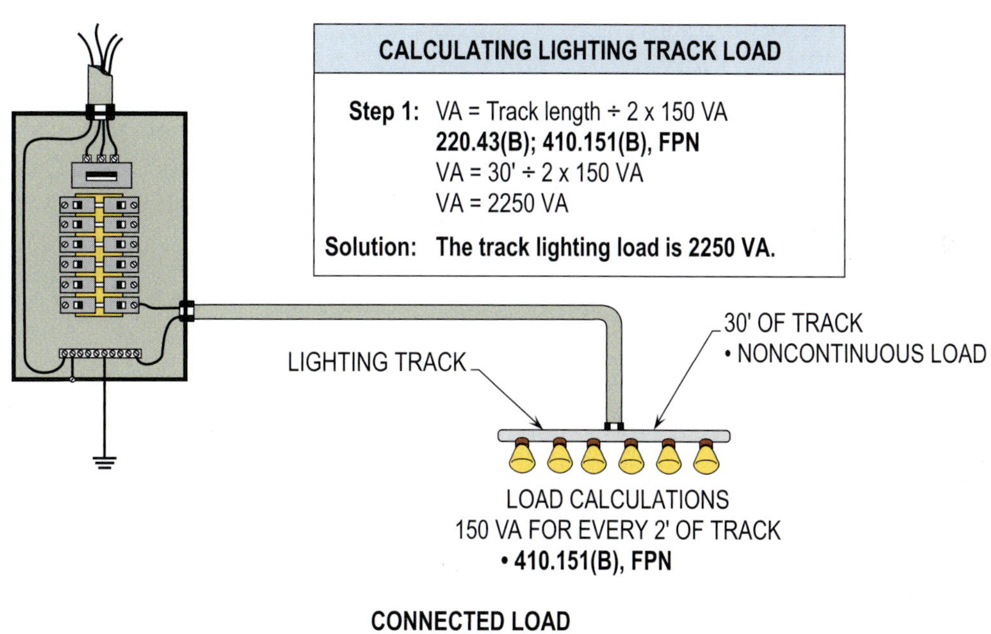

CALCULATING LIGHTING TRACK LOAD
Step 1: VA = Track length ÷ 2 x 150 VA
220.43(B); 410.151(B), FPN
VA = 30' ÷ 2 x 150 VA
VA = 2250 VA
Solution: The track lighting load is 2250 VA.
LIGHTING TRACK
30' OF TRACK
• NONCONTINUOUS LOAD
LOAD CALCULATIONS
150 VA FOR EVERY 2' OF TRACK
• 410.151(B), FPN

CONNECTED LOAD
410.151(B), FPN

Purpose of Change: A new FPN has been added to make it clear that the load calculation in **220.43(B)** does not limit the number of feet of track on a single branch circuit and is not intended to limit the number of luminaires on an individual track.

NEC Ch. 4 - Article 411
Part - 411.3(A) and (B)

Type of Change	Panel Action	UL	UL 508	API 500 - 1997	API 505 - 1997	OSHA - 1994
Revision	Accept	1598 - 2108	-	-	-	-
ROP	**ROC**	**NFPA 70E - 2004**	**NFPA 70B - 2006**	**NFPA 79 - 2007**	**NFPA**	**NEMA**
pg. 452 # 18-106	pg. 273 # 18-92	-	-	-	-	-
log: 2712 CMP: 18	log: 219 Submitter: Steven D. Holmes			2005 NEC: 411.3		IEC: -

2005 NEC - 411.3 Listing Required.

Lighting systems operating at 30 volts or less shall ~~be listed~~.

2008 NEC - 411.3 Listing Required.

Lighting systems operating at 30 volts or less shall comply with 411.3(A) or 411.3(B):

(A) Listed System. Lighting systems operating at 30 volts or less shall be listed as a complete system. The luminaires, power supply and luminaire fittings (including the exposed bare conductors) of an exposed bare conductor lighting system shall be listed for the use as part of the same identified lighting system.

(B) Assembly of Listed Parts. A lighting system assembled from the following listed parts shall be permitted.

(1) Low-voltage luminaires

(2) Low-voltage luminaire power supply

(3) Class 2 power supply

(4) Low-voltage luminaire fittings

(5) Cord (secondary circuit) that the luminaires and power supply are listed for use with

(6) Cable, conductors in conduit, or other fixed wiring method for the secondary circuit.

The luminaires, power supply, and luminaire fittings (including the exposed bare conductors) of an exposed bare conductor lighting system shall be listed for use as part of the same identified lighting system.

Author's Comment: This change has expanded the various types of lighting systems operating at 30 volts or less to either a listed low voltage lighting system or a lighting system assembled from listed components. The comment added headings as directed by the NEC Technical Correlating Committee and also revised and reorganized the text to clarify the application.

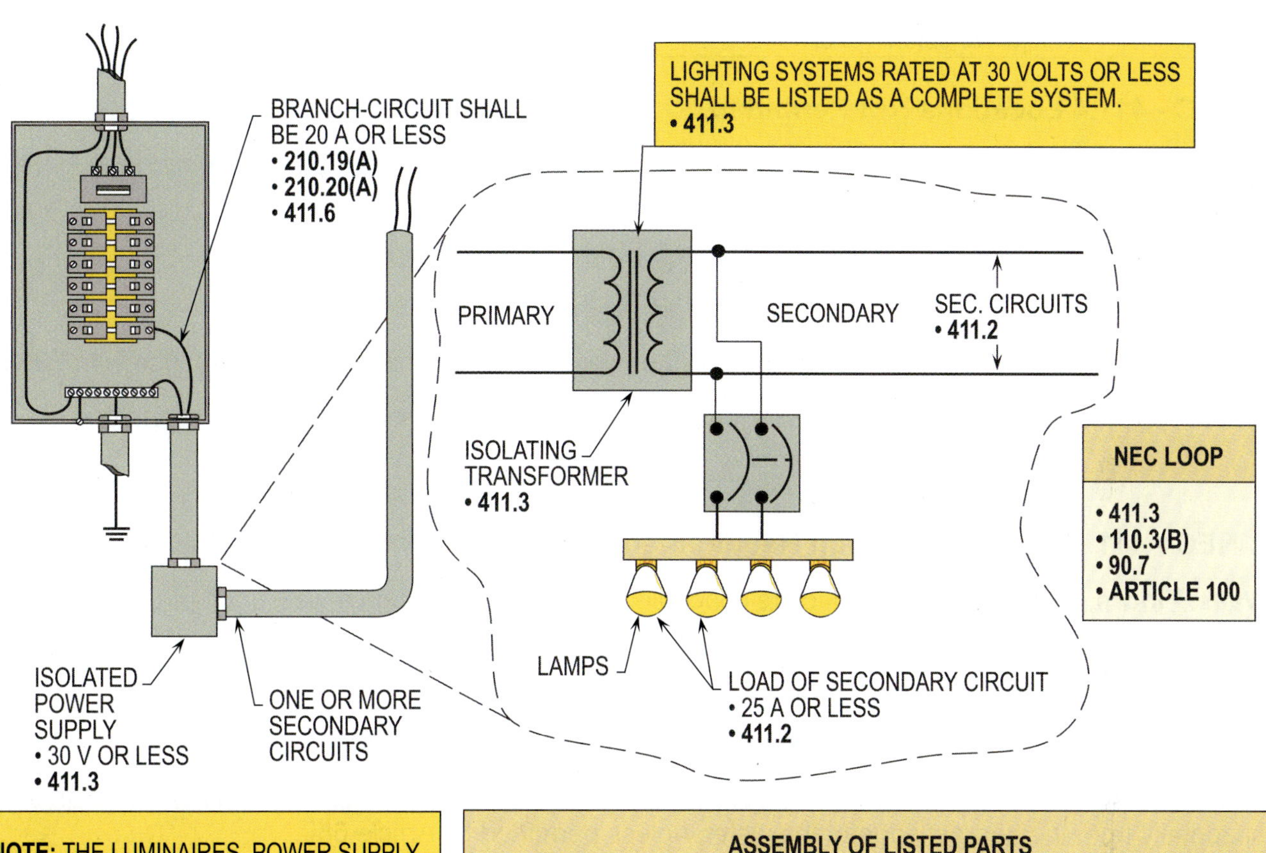

NOTE: THE LUMINAIRES, POWER SUPPLY, AND LUMINAIRE FITTINGS (INCLUDING THE EXPOSED BARE CONDUCTORS) OF AN EXPOSED BARE CONDUCTOR LIGHTING SYSTEM SHALL BE LISTED FOR THE USE AS PART OF THE SAME IDENTIFIED LIGHTING SYSTEM.

ASSEMBLY OF LISTED PARTS

- LOW VOLTAGE LUMINAIRES
- LOW VOLTAGE LUMINAIRE POWER SUPPLY
- CLASS 2 POWER SUPPLY
- LOW VOLTAGE LUMINAIRE FITTINGS
- CORD (SECONDARY CIRCUIT) THAT THE LUMINAIRES AND POWER SUPPLY ARE LISTED FOR USE WITH
- CABLE, CONDUCTORS IN CONDUIT, OR OTHER FIXED WIRING METHOD FOR THE SECONDARY CIRCUIT

LISTING REQUIRED
411.3(A) and (B)

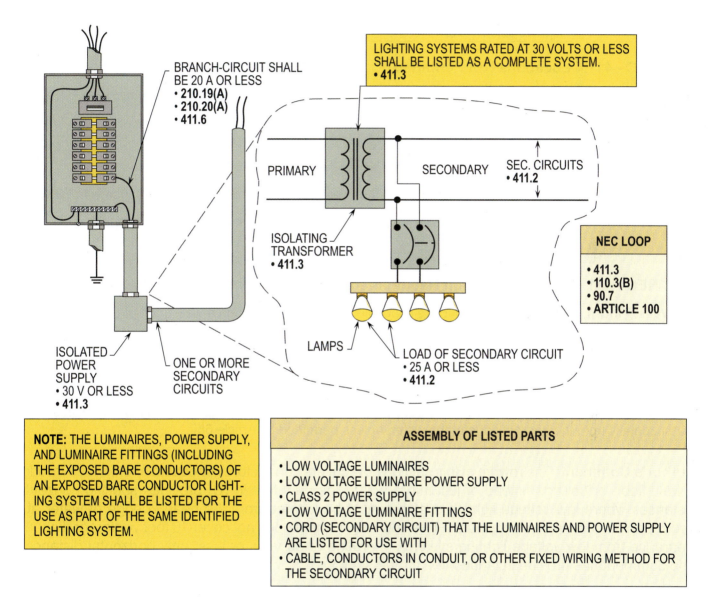

Purpose of Change: This revision clarifies that the various types of lighting systems operating at 30 volts or less are either a listed low voltage lighting system or a lighting system assembled from listed parts.

NEC Ch. 4 - Article 411
Part - 411.4(A) and (B)

Type of Change	Panel Action	UL	UL 508	API 500 - 1997	API 505 - 1997	OSHA - 1994
Revision	Accept	1598 - 2108	-	-	-	-
ROP	**ROC**	**NFPA 70E - 2004**	**NFPA 70B - 2006**	**NFPA 79 - 2007**	**NFPA**	**NEMA**
pg. 453 # 18-107	pg. 274 # 18-94	-	-	-	-	-
log: 2713 CMP: 18	log: 304	Submitter: Steven D. Holmes		2005 NEC: 411.4		IEC: -

2005 NEC - 411.4 ~~Locations Not Permitted.~~

~~Lighting systems operating at 30 volts or less shall not be installed in the locations described in 411.4(A) and 411.4(B).~~

(A) ~~Where~~ concealed or extended through a ~~building~~ wall ~~unless permitted in~~ (1) or (2):

(1) Installed using any of the wiring methods specified in Chapter 3

(2) Installed using wiring supplied by a listed Class 2 power source and installed in accordance with ~~725.52~~

(B) ~~Where~~ installed ~~within~~ 3 m (10 ft) ~~of pools, spas, fountains, or similar locations~~, unless permitted by Article 680.

2008 NEC - 411.4 Specific Location Requirements.

(A) Walls, Floors, and Ceilings. Conductors concealed or extended through a wall, floor, or ceiling shall be in accordance with (1) or (2):

(1) Installed using any of the wiring methods specified in Chapter 3

(2) Installed using wiring supplied by a listed Class 2 power source and installed in accordance with 725.130

(B) Pools, Spas, Fountains, and Similar Locations. Lighting systems shall be installed not less than 3 m (10 ft) horizontally from the nearest edge of the water, unless permitted by Article 680.

Author's Comment: The rewrite of this section now covers conductors concealed or extended through a wall, floor, or ceiling, instead of just into a wall. The new text in **(B)** provides specific horizontal distances where low voltage lighting systems must be located from the nearest edge of water at a pool, spa, fountain, or similar location. This new requirement does not cover 10 foot vertical or diagonal clearance from low voltage lighting systems operating at 30 volts or less, only horizontal distances. The comment has provided clarified text for the section.

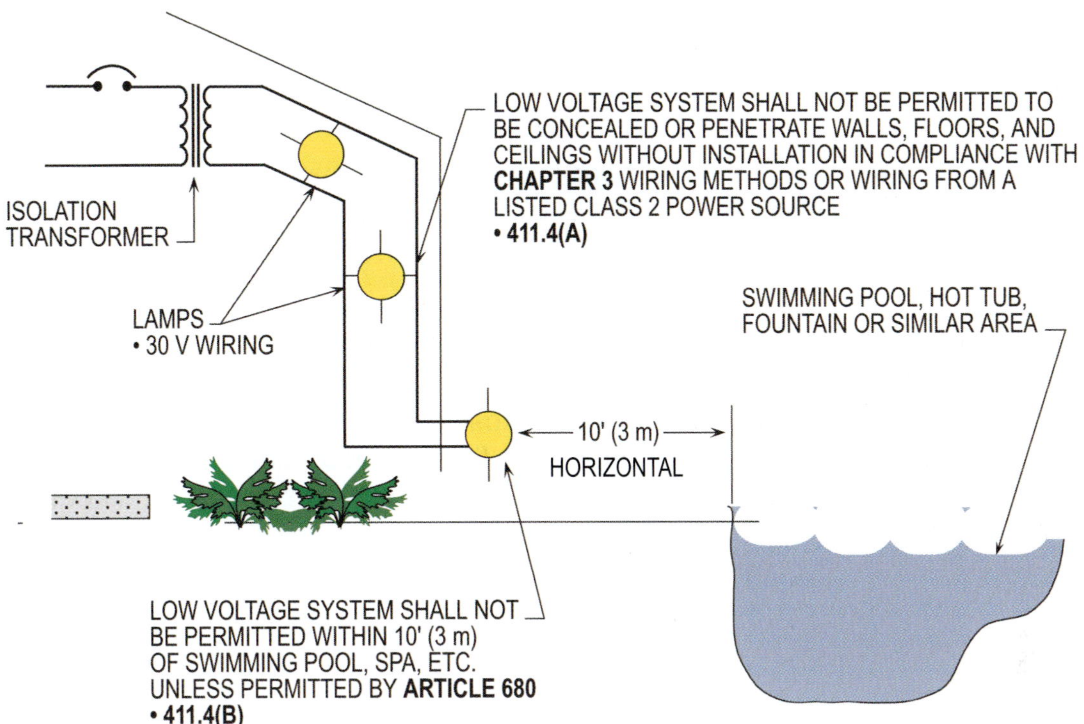
LOW VOLTAGE SYSTEM SHALL NOT BE PERMITTED TO BE CONCEALED OR PENETRATE WALLS, FLOORS, AND CEILINGS WITHOUT INSTALLATION IN COMPLIANCE WITH CHAPTER 3 WIRING METHODS OR WIRING FROM A LISTED CLASS 2 POWER SOURCE
• 411.4(A)
ISOLATION TRANSFORMER
LAMPS
• 30 V WIRING
SWIMMING POOL, HOT TUB, FOUNTAIN OR SIMILAR AREA
10' (3 m)
HORIZONTAL
LOW VOLTAGE SYSTEM SHALL NOT BE PERMITTED WITHIN 10' (3 m) OF SWIMMING POOL, SPA, ETC. UNLESS PERMITTED BY ARTICLE 680
• 411.4(B)

SPECIFIC LOCATION REQUIREMENTS
411.4(A) and (B)

Purpose of Change: This revision clarifies that low voltage lighting systems shall not be permitted to be concealed or penetrate walls, floors, and ceiling without installation in compliance with chapter 3 wiring methods or wiring from a listed Class 2 power source. Low voltage lighting systems shall not be permitted to be installed less than 10 ft. (3 m) horizontally from the nearest edge of the water, unless permitted by **Article 680**.

NEC Ch. 4 - Article 411
Part - 411.5(D)

Type of Change		Panel Action		UL	UL 508	API 500 - 1997	API 505 - 1997	OSHA - 1994
New Subsection		Accept in Principle in Part		1598 - 2108	-	-	-	-
ROP		ROC		NFPA 70E - 2004	NFPA 70B - 2006	NFPA 79 - 2007	NFPA	NEMA
pg. 454	# 18-109	pg. -	# -	-	-	-	-	-
log: 2715	CMP: 18	log: -	Submitter: Steven D. Holmes			2005 NEC: -		IEC: -

2008 NEC - 411.5 Secondary Circuits.

(D) Insulated Conductors. Exposed insulated secondary circuit conductors shall be of the type, and installed as, described in (1), (2), or (3):

(1) Class 2 cable supplied by a Class 2 power source and installed in accordance with Parts I and III of Article 725

(2) Conductors, cord, or cable of the listed system and installed not less than 2.1 m (7 ft) above the finished floor unless the system is specifically listed for a lower installation height

(3) Wiring methods described in Chapter 3

Author's Comment: New **411.5(D)** provides three different installation requirements for exposed secondary insulated conductors for low-voltage lighting systems operating at 30 volts or less. Chapter 3 wiring methods can be used. Conductors, cords, or cables that are part of the listed system can be installed at a height of not less than 7 feet above the finished floor, unless listed for a lower height. Finally, a Class 2 cable supplied by a Class 2 power system can be installed in accordance with **Parts I** and **III** of **Article 725**.

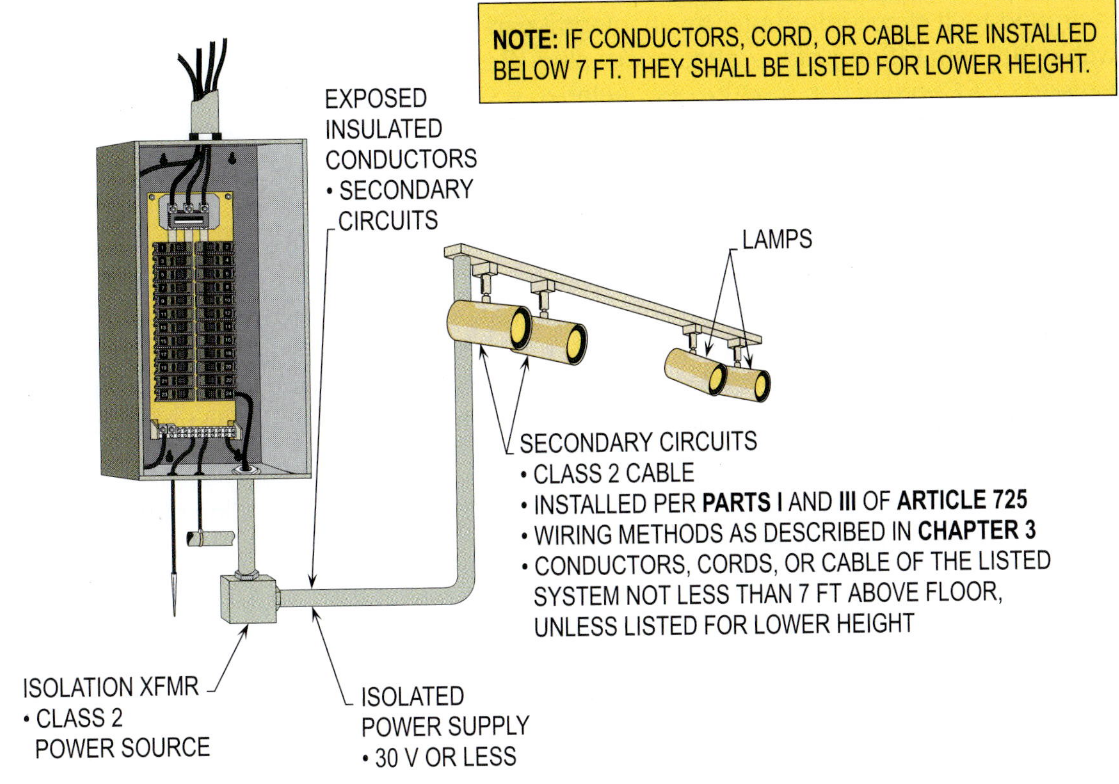

INSULATED CONDUCTORS
411.5(D)

Purpose of Change: A new subsection has been added to provide installation requirements for exposed secondary insulated conductors for low voltage lighting systems operating at 30 volts or less.

NEC Ch. 4 - Article 422
Part IV - 422.51 and FPN

Type of Change	Panel Action	UL	UL 508	API 500 - 1997	API 505 - 1997	OSHA - 1994
Revision	Accept in Part	541 - 751	-	-	-	-
ROP	**ROC**	**NFPA 70E - 2004**	**NFPA 70B - 2006**	**NFPA 79 - 2007**	**NFPA**	**NEMA**
pg. 462 # 17-27	pg. - # -	-	-	-	-	-
log: 1739 CMP: 17	log: -	Submitter: Daniel Cuddy		2005 NEC: 422.51		IEC: -

2005 NEC - 422.51 Cord-and-Plug-Connected Vending Machines.

Cord-and-plug-connected vending machines manufactured or remanufactured on or after January 1, 2005, shall include a ground-fault circuit-interrupter as an integral part of the attachment plug or located ~~in the power supply cord~~ within 300 mm (12 in.) of the attachment plug. ~~Cord-and-plug-connected vending machines not incorporating integral GFCI protection shall be connected to a GFCI protected outlet.~~

2008 NEC - 422.51 Cord-and-Plug-Connected Vending Machines.

Cord-and-plug-connected vending machines manufactured or remanufactured on or after January 1, 2005, shall include a ground-fault-circuit interrupter as an integral part of the attachment plug or be located within 300 mm (12 in) of the attachment plug. Older vending machines manufactured or remanufactured prior to January 1, 2005, shall be connected to a GFCI-protected outlet. For the purpose of this section, the term "vending machine" means any self-service device that dispenses products or merchandise without the necessity of replenishing the device between each vending operation and is designed to require insertion of a coin, paper currency, token, card, key, or receipt of payment by other means.

FPN: For further information, see ANSI/UL 541-2005, Standard for Refrigerated Vending Machines, or ANSI/UL 751-2005, Standard for Vending Machines.

Author's Comment: A definition of vending machine has been added to **422.51** to help the user determine what constitutes a vending machine. A requirement has been added for vending machines, manufactured or remanufactured before GFCI protection was required for vending machines, be connected to a GFCI-protected outlet. A Fine Print Note has been added identifying the UL Standards for listing of vending machines.

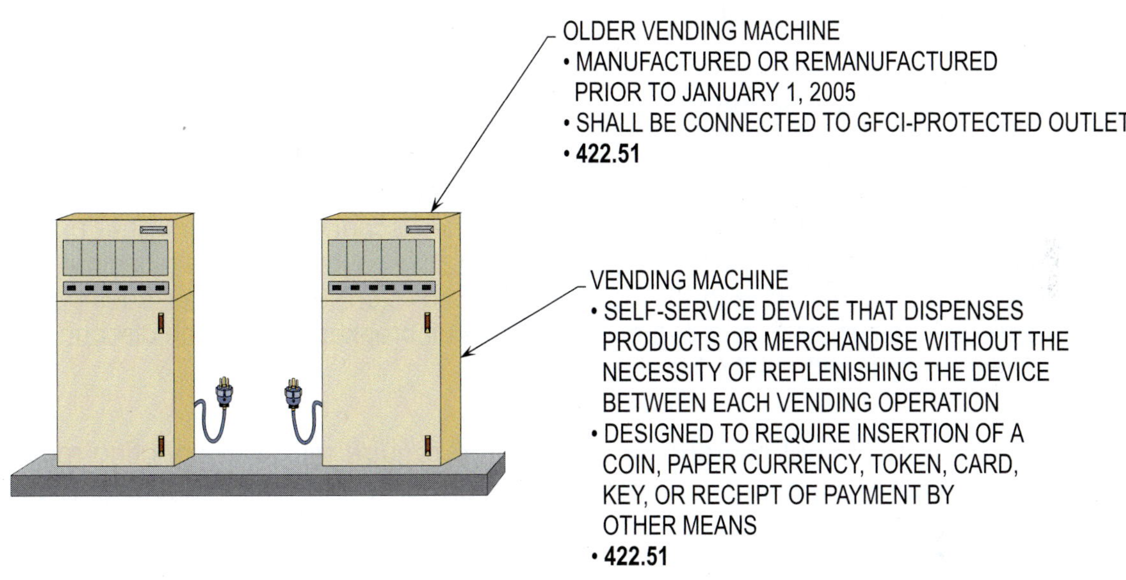
OLDER VENDING MACHINE
• MANUFACTURED OR REMANUFACTURED PRIOR TO JANUARY 1, 2005
• SHALL BE CONNECTED TO GFCI-PROTECTED OUTLET
• 422.51
VENDING MACHINE
• SELF-SERVICE DEVICE THAT DISPENSES PRODUCTS OR MERCHANDISE WITHOUT THE NECESSITY OF REPLENISHING THE DEVICE BETWEEN EACH VENDING OPERATION
• DESIGNED TO REQUIRE INSERTION OF A COIN, PAPER CURRENCY, TOKEN, CARD, KEY, OR RECEIPT OF PAYMENT BY OTHER MEANS
• 422.51

CORD-AND-PLUG CONNECTED VENDING MACHINES
422.51 and FPN

Purpose of Change: This revision clarifies that older vending machines manufactured or remanufactured before January 1, 2005, must be connected to a GFCI-protected outlet.

NEC Ch. 4 - Article 424
Part III - 424.19

Type of Change	Panel Action	UL	UL 508	API 500 - 1997	API 505 - 1997	OSHA - 1994
Revision	Accept	1995	-	-	-	1910.305(j)(3)(ii)
ROP	**ROC**	**NFPA 70E - 2004**	**NFPA 70B - 2006**	**NFPA 79 - 2007**	**NFPA**	**NEMA - 2003**
pg. 464 # 17-33	pg. - # -	420.10(D)(2)	12.5.2	5.5	-	250
log: 2036 CMP: 17	log: -	Submitter: James T. Dollard, Jr.		2005 NEC: 424.19		IEC: -

2005 NEC - 424.19 Disconnecting Means

Means shall be provided to disconnect the heater, motor controller(s), and supplementary overcurrent protective device(s) of all fixed electric space heating equipment from all ungrounded conductors. Where heating equipment is supplied by more than one source, the disconnecting means shall be grouped and marked.

2008 NEC - 424.19 Disconnecting Means

Means shall be provided to simultaneously disconnect the heater, motor controller(s), and supplementary overcurrent protective device(s) of all fixed electric space heating equipment from all ungrounded conductors. Where heating equipment is supplied by more than one source, the disconnecting means shall be grouped and marked. The disconnecting means specified in 424.19(A) and (B) shall have an ampere rating not less than 125 percent of the total load of the motors and the heaters. The provision for locking or adding a lock to the disconnecting means shall be installed on or at the switch or circuit breaker used as the disconnecting means and shall remain in place with or without the lock installed.

Author's Comment: An additional sentence was added to require the disconnecting means for the electric space-heating, with and without supplementary overcurrent protective devices, to be rated for not less than 125 percent of the total combined load of the motors and the heaters.

The requirement for locking or adding a lock to the disconnecting means of a heater has been added to provide safety for the installer or maintainer of the heating system. Installing this locking means or providing the locking mechanism will help ensure that the device is available where it becomes necessary to lock the disconnecting means to the "off" position.

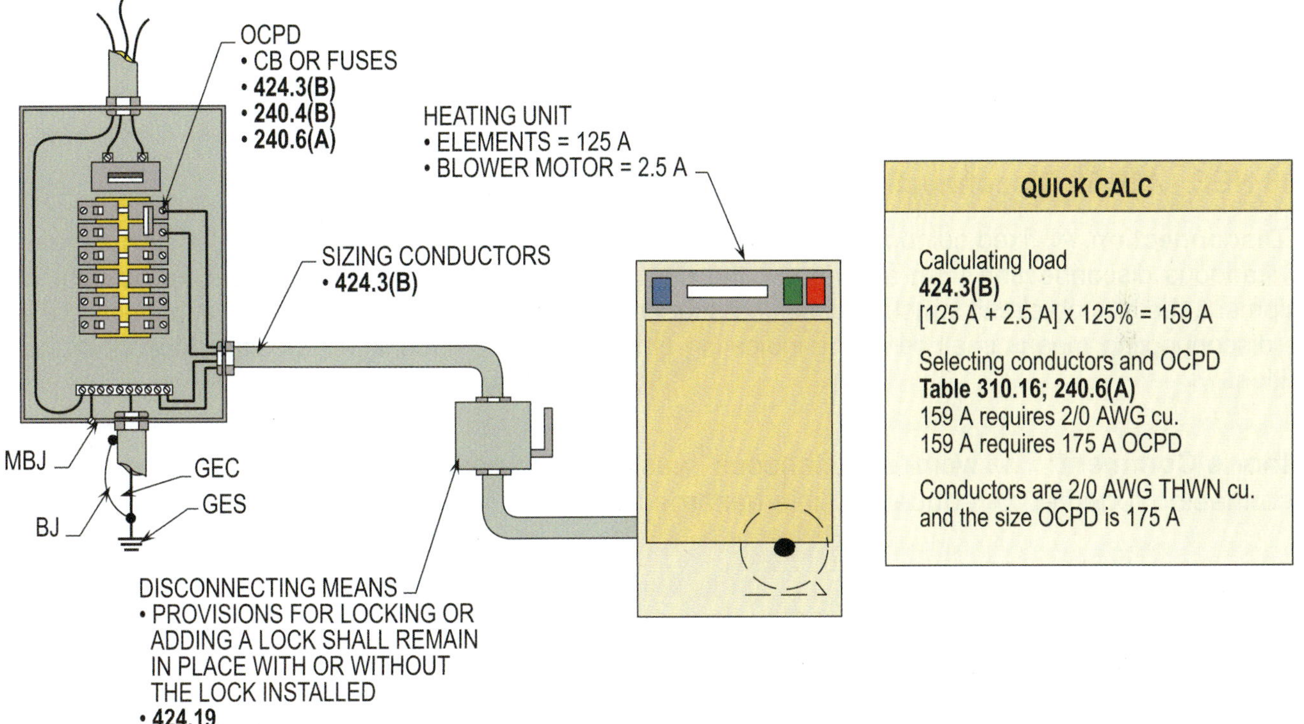

DISCONNECTING MEANS
424.19

Purpose of Change: This revision clarifies that the disconnecting means shall have an ampere rating not less than 125 percent of the total load of the motors and the heaters and have provisions for locking or adding a lock that shall remain in place with or without the lock installed.

NEC Ch. 4 - Article 426
Part VI - 426.50(A)

Type of Change		Panel Action		UL	UL 508	API 500 - 1997	API 505 - 1997	OSHA - 1994
Revision		Accept		1588	-	-	-	-
ROP		ROC		NFPA 70E - 2004	NFPA 70B - 2006	NFPA 79 - 2007	NFPA	NEMA - 2003
pg. 467	# 17-48	pg. -	# -	-	-	-	-	250
log: 3408	CMP: 17	log: -	Submitter: Frederic P. Hartwell			2005 NEC: 426.50(A)		IEC: -

2005 NEC - 426.50 Disconnecting Means.

(A) Disconnection. All fixed outdoor deicing and snow-melting equipment shall be provided with a means for disconnection from all ungrounded conductors. Where readily accessible to the user of the equipment, the branch-circuit switch or circuit breaker shall be permitted to serve as the disconnecting means. The disconnecting means shall be of the indicating type and be provided with a positive lockout in the "off" position.

2008 NEC - 426.50 Disconnecting Means.

(A) Disconnection. All fixed outdoor deicing and snow-melting equipment shall be provided with a means for simultaneous disconnection from all ungrounded conductors. Where readily accessible to the user of the equipment, the branch-circuit switch or circuit breaker shall be permitted to serve as the disconnecting means. The disconnecting means shall be of the indicating type and be provided with a positive lockout in the "off" position.

Author's Comment: The word "simultaneous" was added to ensure all ungrounded conductors were disconnected together to reduce possible hazards during servicing.

DISCONNECTING MEANS SHALL BE THE INDICATING TYPE AND CAPABLE OF BEING LOCKED IN THE OPEN POSITION

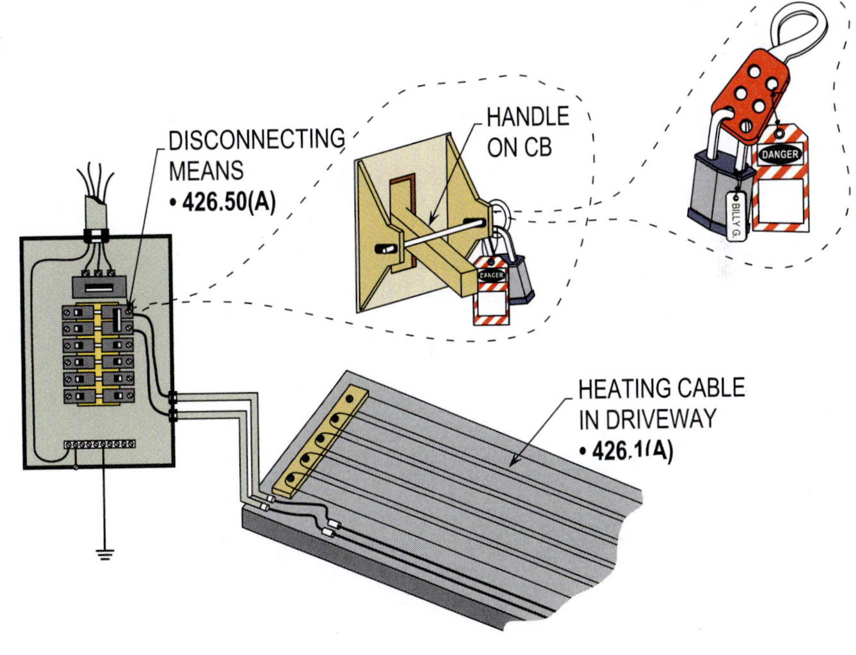

DISCONNECTION
426.50(A)

Purpose of Change: This revision clarifies that the disconnecting means shall simultaneously open all ungrounded conductors.

NEC Ch. 4 - Article 427
Part II - 427.13

Type of Change	Panel Action	UL	UL 508	API 500 - 1997	API 505 - 1997	OSHA - 1994
Revision	Accept	515	-	-	-	-
ROP	**ROC**	**NFPA 70E - 2004**	**NFPA 70B - 2006**	**NFPA 79 - 2007**	**NFPA**	**NEMA**
pg. 468 # 17-52a	pg. - # -	-	-	-	-	-
log: CP 1705 CMP: 17	log: -	Submitter: Code Making Panel 17		2005 NEC: 427.13		IEC: -

2005 NEC - 427.13 Identification.

The presence of electrically heated pipelines, vessels, or both, shall be evident by the posting of appropriate caution signs or markings at ~~frequent~~ intervals along the pipeline or vessel.

2008 NEC - 427.13 Identification.

The presence of electrically heated pipelines, vessels, or both, shall be evident by the posting of appropriate caution signs or markings at intervals not exceeding 6 m (20 ft) along the pipeline or vessel and on or adjacent to equipment in the piping system that requires periodic servicing.

Author's Comment: The word "frequent" was deleted because it is vague and unenforceable. A specific dimension of 20 feet or less for the location of caution signs or markings has been established along the pipeline or vessel and on or adjacent to the equipment where periodic servicing is required.

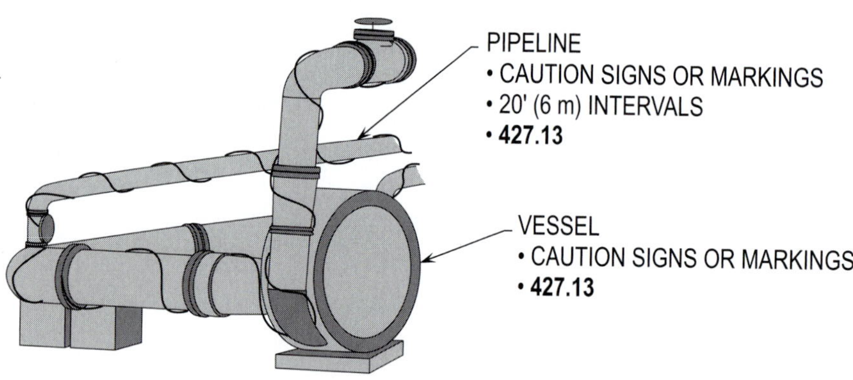

**IDENTIFICATION
427.13**

Purpose of Change: This revision clarifies that electrically heated pipeline, vessels, or both, shall have appropriate caution signs or markings at intervals not exceeding 20 ft (6 m).

NEC Ch. 4 - Article 430
Part III - 430.32(C)

Type of Change	Panel Action	UL	UL 508	API 500 - 1997	API 505 - 1997	OSHA - 1994
Revision	Accept	508	-	-	-	1910.305(j)(4)(iii)
ROP	**ROC**	**NFPA 70E - 2004**	**NFPA 70B - 2006**	**NFPA 79 - 2007**	**NFPA**	**NEMA - 2003**
pg. 475 # 11-37	pg. - # -	420.10(E)(2)(b)(f)	-	7.3.1.1	-	MG 1
log: 157 CMP: 11	log: -	Submitter: Steven Duritt		2005 NEC: 430.32(C)		IEC: -

2005 NEC - 430.32 Continuous-Duty Motors.

(C) Selection of Overload ~~Relay~~. Where the sensing element or setting of the overload ~~relay~~ selected in accordance with 430.32(A)(1) and 430.32(B)(1) is not sufficient to start the motor or to carry the load, higher size sensing elements or incremental settings shall be permitted to be used, provided the trip current of the overload ~~relay~~ does not exceed the following percentage of motor nameplate full-load current rating:

2008 NEC - 430.32 Continuous-Duty Motors.

(C) Selection of Overload Device. Where the sensing element or setting or sizing of the overload device selected in accordance with 430.32(A)(1) and 430.32(B)(1) is not sufficient to start the motor or to carry the load, higher size sensing elements or incremental settings or sizing shall be permitted to be used, provided the trip current of the overload device does not exceed the following percentage of motor nameplate full-load current rating:

Author's Comment: Section **430.32(C)** refers to all types of overload devices, not exclusively overload relays, so "relays" was replaced with "devices." The phrase "or sizing" was inserted because properly "sized" fuses can be used for motor overload protection in accordance with **430.36**.

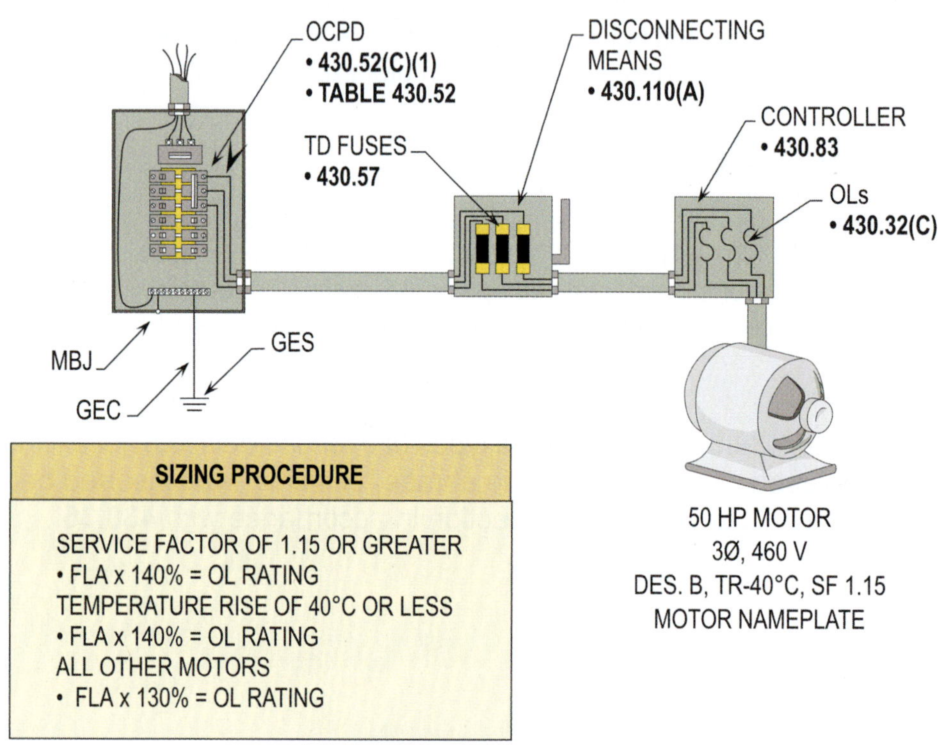

SELECTION OF OVERLOAD DEVICE
430.32(C)

Purpose of Change: This revision clarifies that this subsection refers to all types of overload devices.

NEC Ch. 4 - Article 430
Part IV - 430.52(C)(1), Exception No. 2

69

Type of Change	Panel Action	UL	UL 508	API 500 - 1997	API 505 - 1997	OSHA - 1994
Revision	Accept	-	-	-	-	1910.305(j)(4)(iii)
ROP	**ROC**	**NFPA 70E - 2004**	**NFPA 70B - 2006**	**NFPA 79 - 2007**	**NFPA**	**NEMA - 2003**
pg. 476 # 11-42	pg. - # -	420.10(E)(2)(b)(f)	-	7.3.1.1	-	MG 1
log: 2177 CMP: 11	log: -	Submitter: Dann Strube		2005 NEC: 430.52(C), Exception No. 2		IEC: -

2005 NEC - 430.52 Rating or Setting for Individual Motor Circuit.

(C) Rating or Setting.

(1) In Accordance with Table 430.52. A protective device that has a rating or setting not exceeding the value calculated according to the values given in Table 430.52 shall be used.

Exception No. 1: Where the values for branch-circuit short-circuit and ground-fault protective devices determined by Table 430.52 do not correspond to the standard sizes or ratings of fuses, nonadjustable circuit breakers, thermal protective devices, or possible settings of adjustable circuit breakers, a higher size, rating, or possible setting that does not exceed the next higher standard ampere rating shall be permitted.

Exception No. 2: Where the rating specified in Table 430.52 ~~as~~ modified by Exception No. 1 is not sufficient for the starting current of the motor:

(a) The rating of a nontime-delay fuse not exceeding 600 amperes or a time-delay Class CC fuse shall be permitted to be increased but shall in no case exceed 400 percent of the full-load current.

(b) The rating of a time-delay (dual-element) fuse shall be permitted to be increased but shall in no case exceed 225 percent of the full-load current.

(c) The rating of an inverse time circuit breaker shall be permitted to be increased but shall in no case exceed 400 percent for full-load currents of 100 amperes or less or 300 percent for full-load currents greater than 100 amperes.

(d) The rating of a fuse of 601–6000 ampere classification shall be permitted to be increased but shall in no case exceed 300 percent of the full-load current.

FPN: See Annex D, Example D8, and Figure 430.1.

2008 NEC - 430.52 Rating or Setting for Individual Motor Circuit.

(C) Rating or Setting.

(1) In Accordance with Table 430.52. A protective device that has a rating or setting not exceeding the value calculated according to the values given in Table 430.52 shall be used.

Exception No. 1: Where the values for branch-circuit short-circuit and ground-fault protective devices determined by Table 430.52 do not correspond to the standard sizes or ratings of fuses, nonadjustable circuit breakers, thermal protective devices, or possible settings of adjustable circuit breakers, a higher size, rating, or possible setting that does not exceed the next higher standard ampere rating shall be permitted.

Exception No. 2: Where the rating specified in Table 430.52, or the rating modified by Exception No. 1, is not sufficient for the starting current of the motor:

(a) The rating of a nontime-delay fuse not exceeding 600 amperes or a time-delay Class CC fuse shall be permitted to be increased but shall in no case exceed 400 percent of the full-load current.

(b) The rating of a time-delay (dual-element) fuse shall be permitted to be increased but shall in no case exceed 225 percent of the full-load current.

(c) The rating of an inverse time circuit breaker shall be permitted to be increased but shall in no case exceed 400 percent for full-load currents of 100 amperes or less or 300 percent for full-load currents greater than 100 amperes.

(d) The rating of a fuse of 601–6000 ampere classification shall be permitted to be increased but shall in no case exceed 300 percent of the full-load current.

FPN: See Annex D, Example D8, and Figure 430.1.

Author's Comment: Exception No. 2 has been modified to clarify where the rating specified in **Table 430.52** or as modified by **Exception No. 1** is not sufficient for starting the motor, then a higher rating specified in **(a) through (d)** can be used.

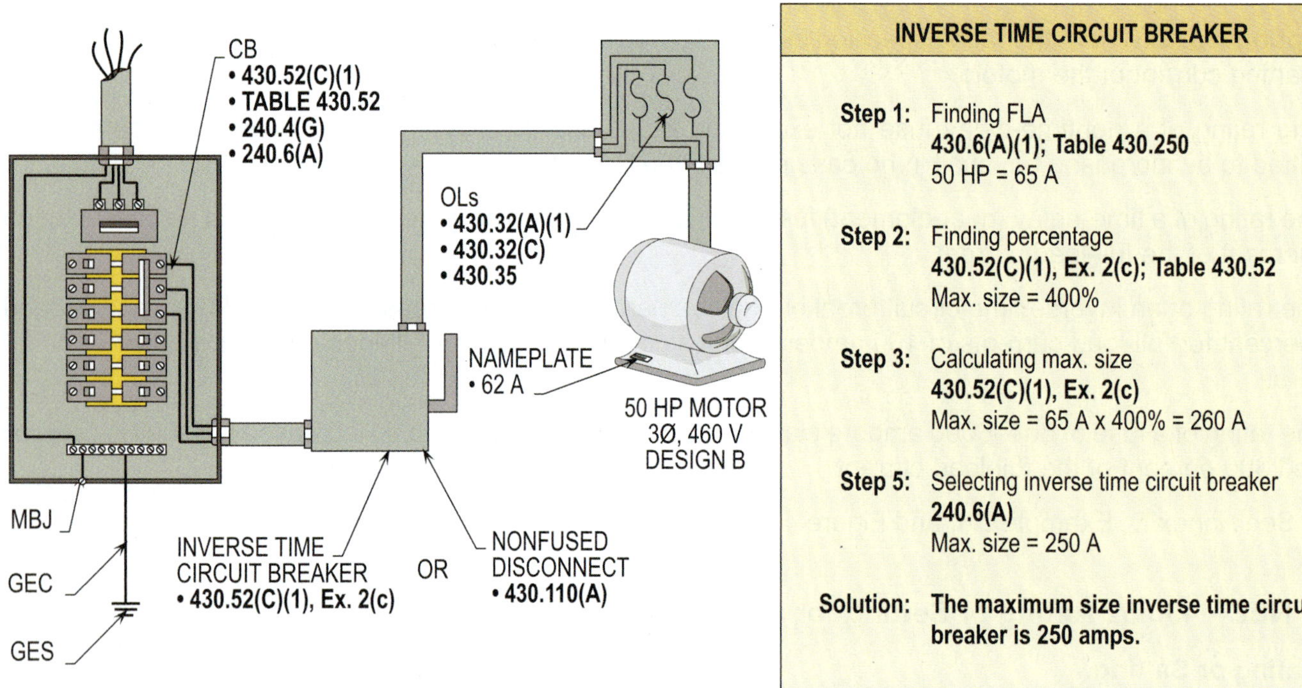

IN ACCORDANCE WITH TABLE 430.52
430.52(C)(1), Exception No. 2

Purpose of Change: This revision clarifies that where the rating specified in **Table 430.52** or the ratings modified by **Exception No. 1** is not sufficient for the starting current of the motor, the **Exception No. 2** shall be permitted to be applied.

NEC Ch. 4 - Article 430
Part VII - 430.81(A)

Type of Change	Panel Action	UL	UL 508	API 500 - 1997	API 505 - 1997	OSHA - 1994
Revision	Accept	1004	-	-	-	-
ROP	**ROC**	**NFPA 70E - 2004**	**NFPA 70B - 2006**	**NFPA 79 - 2007**	**NFPA**	**NEMA - 2003**
pg. 479 # 11-53a	pg. - # -	-	-	-	-	MG 1
log: CP 1100 CMP: 11	log: -	Submitter: Code Making Panel 11		2005 NEC: 430.81(A)		IEC: -

2005 NEC - 430.81 General.

Part VII is intended to require suitable controllers for all motors.

(A) Stationary Motor of 1/8 Horsepower or Less. For a stationary motor rated at 1/8 hp or less that is normally left running and is constructed so that it cannot be damaged by overload or failure to start, such as clock motors and the like, the branch-circuit ~~protective device~~ shall be permitted to serve as the controller.

2008 NEC - 430.81 General.

Part VII is intended to require suitable controllers for all motors.

(A) Stationary Motor of 1/8 Horsepower or Less. For a stationary motor rated at 1/8 hp or less that is normally left running and is constructed so that it cannot be damaged by overload or failure to start, such as clock motors and the like, the branch-circuit disconnecting means shall be permitted to serve as the controller.

Author's Comment: The branch-circuit protective device might not be able to act as the controller because the protective device may not be able to be opened by hand as a manual controller, whereas a disconnecting means can. Proposal 11-54 provides a new **Exception No. 1** in **430.87(A)** that permits small portable motors of 1/3 horsepower or less to be installed on the same branch circuit.

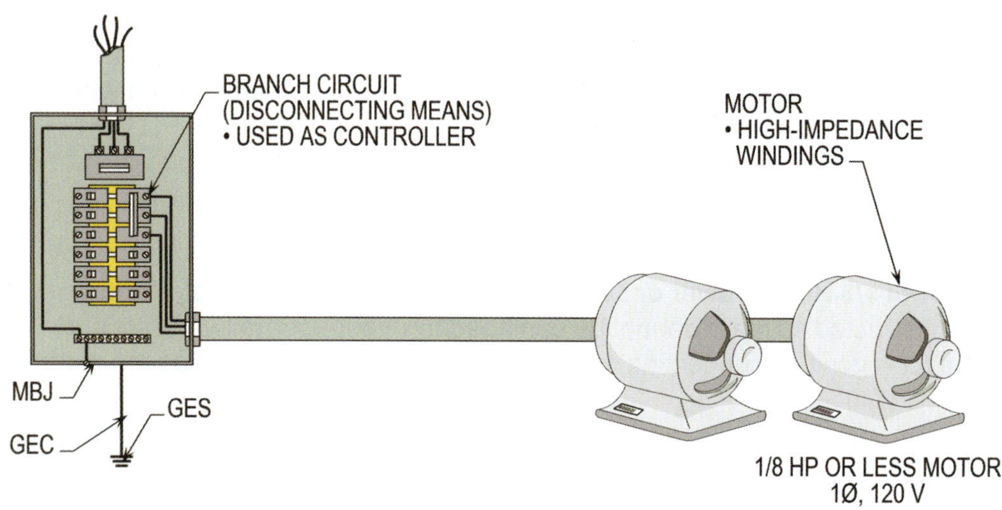

NOTE: A BRANCH CIRCUIT DISCONNECTING MEANS SERVING AS THE CONTROLLER AS ALLOWED IN NEW **430.87(A), EXCEPTION NO. 2** SHALL BE PERMITTED TO SERVE MORE THAN ONE MOTOR.

STATIONARY MOTOR OF 1/8 HORSEPOWER OR LESS
430.81(A)

Purpose of Change: This revision clarifies that the branch circuit disconnecting means shall be permitted to serve as motor controller.

NEC Ch. 4 - Article 430
Part IX - 430.102(B)(1) and (B)(2)

Type of Change	Panel Action	UL	UL 508	API 500 - 1997	API 505 - 1997	OSHA - 1994
Revision	Accept	-	-	-	-	1910.305(j)(4)(ii)
ROP	**ROC**	**NFPA 70E - 2004**	**NFPA 70B - 2006**	**NFPA 79 - 2007**	**NFPA**	**NEMA - 2003**
pg. 482 # 17-67	pg. - # 11-27	420.10(E)(2)(b)	7.4.2	5.5	-	MG 1
log: 2704 CMP: 17	log: 792	Submitter: Dorothy Kellogg		2005 NEC: 430.102(B)		IEC: -

2005 NEC - 430.102 Location.

(B) Motor. A disconnecting means shall be located within sight from the motor location and the driven machinery location.

Exception: The disconnecting means shall not be required ~~to be in sight from the motor and driven machinery location~~ under either condition (a) or (b), provided the disconnecting means required in accordance with 430.102(A) is individually capable of being locked in the open position. The provision for locking or adding a lock to the disconnecting means shall be installed on or at the switch or circuit breaker used as the disconnecting means and shall remain in place with or without the lock installed.

(a) Where such a location of the disconnecting means is impracticable or introduces additional or increased hazards to persons or property

(b) In industrial installations, with written safety procedures, where conditions of maintenance and supervision ensure that only qualified persons service the equipment

FPN No. 1: Some examples of increased or additional hazard include, but are not limited to, motors rated in excess of 100 hp, multimotor equipment, submersible motors, motors associated with adjustable speed drives, and motors located in hazardous (classified) locations.

FPN No. 2: For information on lockout/tagout procedures, see NFPA 70E-2004, Standard for Electrical Safety in the Workplace.

~~The disconnecting means required in accordance with 430.102(A) shall be permitted to serve as the disconnecting means for the motor if it is located in sight from the motor location and the driven machinery location.~~

2008 NEC - 430.102 Location.

(B) Motor. A disconnecting means shall be provided for a motor in accordance with (B)(1) or (B)(2).

(1) Separate Motor Disconnect. A disconnecting means for the motor shall be located in sight from the motor location and the driven machinery location.

(2) Controller Disconnect. The controller disconnecting means required in accordance with 430.102(A) shall be permitted to serve as the disconnecting means for the motor if it is in sight from the motor location and the driven machinery location.

Exception to (1) and (2): The disconnecting means for the motor shall not be required under either condition (1) or (2), provided the controller disconnecting means required in accordance with 430.102(A) is individually capable of being locked in the open position. The provision for locking or adding a lock to the controller disconnecting means shall be installed on or at the switch or circuit breaker used as the disconnecting means and shall remainin place with or without the lock installed.

(1) Where such a location of the disconnecting means for the motor is impracticable or introduces additional or increased hazards to persons or property

(2) In industrial installations, with written safety procedures, where conditions of maintenance and supervision ensure that only qualified persons service the equipment

FPN No. 1: Some examples of increased or additional hazards include, but are not limited to, motors rated in excess of 100 hp, multimotor equipment, submersible motors, motors associated with adjustable speed drives, and motors located in hazardous (classified) locations.

FPN No. 2: For information on lockout/tagout procedures, see NFPA 70E-2004, Standard for Electrical Safety in the Workplace.

Author's Comment: The last sentence of **430.102(B)** was causing NEC users to question the application of the exception and gave the impression of contradicting the exception. The last sentence is a continuation of the idea expressed in this section and is now placed as **(2)** to follow immediately after the first sentence. The remainder of the text provides an exception to the general rule and follows the complete general rule to eliminate confusion. The comment has added text and subsection titles to clarify which disconnecting means is being covered, the motor disconnect or the controller disconnect.

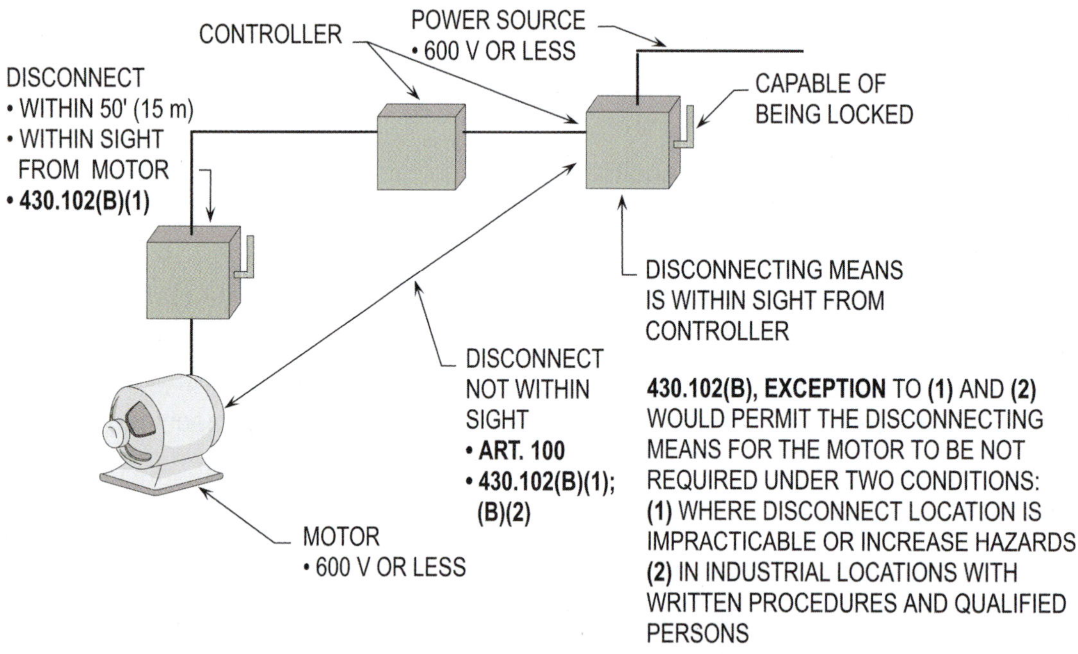

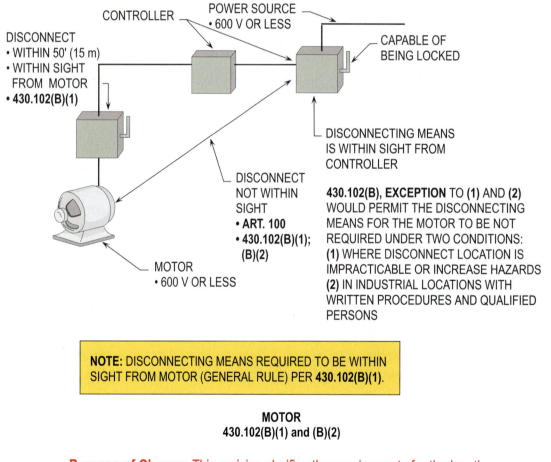

MOTOR
430.102(B)(1) and (B)(2)

Purpose of Change: This revision clarifies the requirements for the location of the disconnecting means for the motor by reformatting the text.

NEC Ch. 4 - Article 430
Part IX - 430.103

Type of Change		Panel Action		UL	UL 508	API 500 - 1997	API 505 - 1997	OSHA - 1994
Revision		Accept		-	-	-	-	1910.305(j)(4)(iii)
ROP		ROC		NFPA 70E - 2004	NFPA 70B - 2006	NFPA 79 - 2007	NFPA	NEMA - 2003
pg. 483	# 11-71	pg. -	# -	420.10(E)(2)(b)	7.4.2	5.5	-	MG 1
log: 1115	CMP: 11	log: -	Submitter: Daniel Leaf			2005 NEC: 430.103		IEC: -

2005 NEC - 430.103 Operation.

The disconnecting means shall open all ungrounded supply conductors and shall be designed so that no pole can be operated independently. The disconnecting means shall be permitted in the same enclosure with the controller.

FPN: See 430.113 for equipment receiving energy from more than one source.

2008 NEC - 430.103 Operation.

The disconnecting means shall open all ungrounded supply conductors and shall be designed so that no pole can be operated independently. The disconnecting means shall be permitted in the same enclosure with the controller. The disconnecting means shall be designed so that it cannot be closed automatically.

FPN: See 430.113 for equipment receiving energy from more than one source.

Author's Comment: This change will ensure that a disconnecting means cannot be re-closed automatically, such as would be the case with a time clock used as a disconnecting means. The time clock could turn the motor back on inadvertently as the clock progresses during its 24 hour operation.

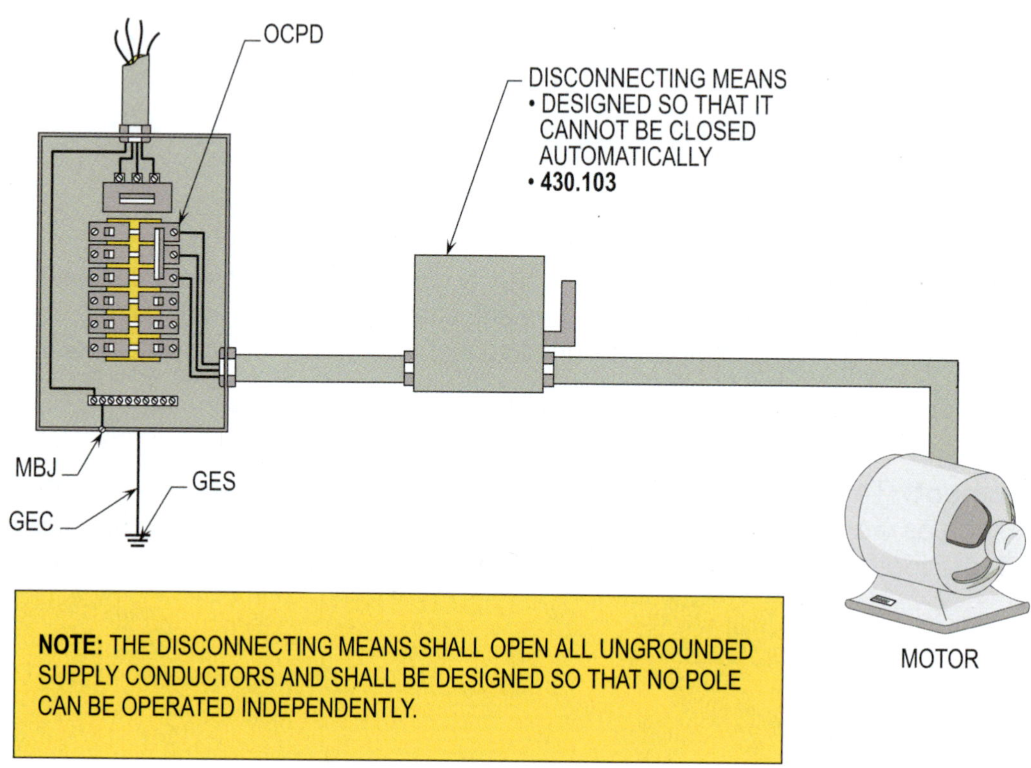
OCPD
DISCONNECTING MEANS
• DESIGNED SO THAT IT CANNOT BE CLOSED AUTOMATICALLY
• 430.103
MBJ
GEC
GES
MOTOR
NOTE: THE DISCONNECTING MEANS SHALL OPEN ALL UNGROUNDED SUPPLY CONDUCTORS AND SHALL BE DESIGNED SO THAT NO POLE CAN BE OPERATED INDEPENDENTLY.

OPERATION
430.103

Purpose of Change: This revision clarifies that the disconnecting means must not be closed automatically.

NEC Ch. 4 - Article 430
Part IX - 430.110(C)(1)

Type of Change		Panel Action		UL	UL 508	API 500 - 1997	API 505 - 1997	OSHA - 1994
Revision		Accept in Principle		-	-	-	-	-
ROP		ROC		NFPA 70E - 2004	NFPA 70B - 2006	NFPA 79 - 2007	NFPA	NEMA
pg. 484	# 11-76	pg. -	# -	-	-	-	-	-
log: 2131	CMP: 11	log: -	Submitter: Russell LeBlanc			2005 NEC: 430.110(C)(1)		IEC: -

2005 NEC - 430.110 Ampere Rating and Interrupting Capacity.

(C) For Combination Loads. Where two or more motors are used together or where one or more motors are used in combination with other loads, such as resistance heaters, and where the combined load may be simultaneous on a single disconnecting means, the ampere and horsepower ratings of the combined load shall be determined as follows.

(1) Horsepower Rating. The rating of the disconnecting means shall be determined from the sum of all currents, including resistance loads, at the full-load condition and also at the locked-rotor condition. The combined full-load current and the combined locked-rotor current so obtained shall be considered as a single motor for the purpose of this requirement as follows.

The full-load current equivalent to the horsepower rating of each motor shall be selected from Table 430.247, Table 430.248, Table 430.249, or Table 430.250. These full-load currents shall be added to the rating in amperes of other loads to obtain an equivalent full-load current for the combined load.

The locked-rotor current equivalent to the horsepower rating of each motor shall be selected from Table 430.251(A) or Table 430.251(B). The locked-rotor currents shall be added to the rating in amperes of other loads to obtain an equivalent locked-rotor current for the combined load. Where two or more motors or other loads cannot be started simultaneously, the largest sum of locked rotor currents of a motor or group of motors that can be started simultaneously and the full load currents of other concurrent loads shall be permitted to be used to determine the equivalent locked-rotor current for the simultaneous combined loads.

Exception: Where part of the concurrent load is resistance load, and where the disconnecting means is a switch rated in horsepower and amperes, the switch used shall be permitted to have a horsepower rating that is not less than the combined load of the motor(s), if the ampere rating of the switch is not less than the locked-rotor current of the motor(s) plus the resistance load.

2008 NEC - 430.110 Ampere Rating and Interrupting Capacity.

(C) For Combination Loads. Where two or more motors are used together or where one or more motors are used in combination with other loads, such as resistance heaters, and where the combined load may be simultaneous on a single disconnecting means, the ampere and horsepower ratings of the combined load shall be determined as follows.

(1) Horsepower Rating. The rating of the disconnecting means shall be determined from the sum of all currents, including resistance loads, at the full-load condition and also at the locked-rotor condition. The combined full-load current and the combined locked-rotor current so obtained shall be considered as a single motor for the purpose of this requirement as follows.

The full-load current equivalent to the horsepower rating of each motor shall be selected from Table 430.247, Table 430.248, Table 430.249, or Table 430.250. These full-load currents shall be added to the rating in amperes of other loads to obtain an equivalent full-load current for the combined load.

The locked-rotor current equivalent to the horsepower rating of each motor shall be selected from Table 430.251(A) or Table 430.251(B). The locked-rotor currents shall be added to the rating in amperes of other loads to obtain an equivalent locked-rotor current for the combined load. Where two or more motors or other loads cannot be started simultaneously, the largest sum of locked rotor currents of a motor or group of motors that can be started simultaneously and the full load currents of other concurrent loads shall be permitted to be used to determine the equivalent locked-rotor current for the simultaneous combined loads. In cases where different current ratings are obtained when applying these tables, the largest value obtained shall be used.

Exception: Where part of the concurrent load is resistance load, and where the disconnecting means is a switch rated in horsepower and amperes, the switch used shall be permitted to have a horsepower rating that is not less than the combined load of the motor(s), if the ampere rating of the switch is not less than the locked-rotor current of the motor(s) plus the resistance load.

Author's Comment: This sentence is used in **440.12(A)(2)** and is now being placed in **430.110(C)(1)** to explain that where different current ratings are obtained with multiple motors in a group installation, the largest current rating shall be used.

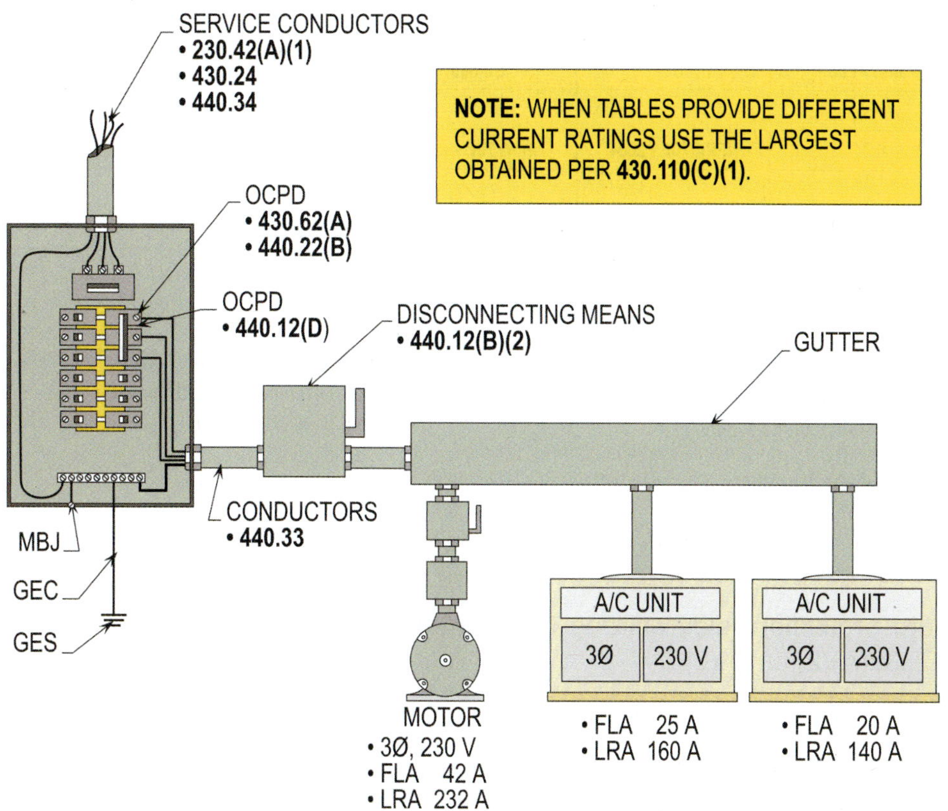

FINDING DISCONNECTING MEANS USING HORSEPOWER

Sizing disconnect using HP

Step 1: Calculating disconnect
440.12(A)(2); Table 430.250; 430.251(B)

FLA = HP	LRA = HP
42 A = 15	232 A = 15
25 A = 10	160 A = 10
20 A = 7 1/2	140 A = 10
87 A = 32 1/2	532 A = 35

Step 2: Selecting disconnect
440.12(A)(2)
Use the higher horsepower rating

Solution: **The disconnecting means is required to be rated at 35 HP.**

HORSEPOWER RATING
430.110(C)(1)

Purpose of Change: This revision clarifies that where different current ratings are obtained with multiple motors in a group installation, the largest current ratings shall be used.

NEC Ch. 4 - Article 430
Part X - 430.126(A)

Type of Change	Panel Action	UL	UL 508	API 500 - 1997	API 505 - 1997	OSHA - 1994
Revision	Accept	-	-	-	-	-
ROP	**ROC**	**NFPA 70E - 2004**	**NFPA 70B - 2006**	**NFPA 79 - 2007**	**NFPA**	**NEMA - 2006**
pg. 484 # 11-77	pg. - # -	-	21.22	7.3.1.2	-	ICS 7
log: 1951 CMP: 11	log: -	Submitter: Vince Baclawski		2005 NEC: 430.126(A)		IEC: -

2005 NEC - 430.126 Motor Overtemperature Protection.

(A) General. Adjustable Speed drive systems shall protect against motor overtemperature conditions. ~~Overtemperature~~ protection ~~is~~ in addition to the conductor protection required in 430.32. Protection shall be provided by one of the following means.

(1) Motor thermal protector in accordance with 430.32

(2) Adjustable speed drive ~~controller~~ with load and speed-sensitive overload protection and thermal memory retention upon shutdown or power loss

(3) Overtemperature protection relay utilizing thermal sensors embedded in the motor and meeting the requirements of 430.32(A)(2) or (B)(2)

(4) Thermal sensor embedded in the motor ~~that is~~ received and acted upon by an adjustable speed drive.

~~**(B) Motors with Cooling Systems.** Motors that utilize external forced air or liquid cooling systems shall be provided with protection that shall be continuously enabled or enabled automatically if the cooling system fails.~~

~~**FPN:** Protection against cooling system failure can take many forms. Some examples of protection against inoperative or failed cooling systems are direct sensing of the motor temperature as described in 430.32(A)(1), (A)(3), and (A)(4) or sensing of the presence or absence of the cooling media (flow or pressure sensing).~~

~~(C)~~ Multiple Motor Applications. For multiple motor application, individual motor overtemperature protection shall be provided.

FPN: The relationship between motor current and motor temperature changes when the motor is operated by an adjustable speed drive. When operated at reduced speed, overheating of motors may occur at current levels less than or equal to a motor's rated full load current. This is the result of reduced motor cooling when its shaft-mounted fan is operating less than rated nameplate RPM.

~~(D)~~ Automatic Restarting and Orderly Shutdown. The provisions of 430.43 and 430.44 shall apply to the motor over-temperature protection means.

2008 NEC - 430.126 Motor Overtemperature Protection.

(A) General. Adjustable Speed drive systems shall protect against motor overtemperature conditions where the motor is not rated to operate at the nameplate rated current over the speed range required by the application. This protection shall be provided in addition to the conductor protection required in 430.32. Protection shall be provided by one of the following means:

(1) Motor thermal protector in accordance with 430.32

(2) Adjustable speed drive system with load and speed-sensitive overload protection and thermal memory retention upon shutdown or power loss

Exception to (2): Thermal memory retention upon shutdown or power loss is not required for continuous duty loads.

(3) Overtemperature protection relay utilizing thermal sensors embedded in the motor and meeting the requirements of 430.32(A)(2) or (B)(2)

(4) Thermal sensor embedded in the motor whose communications are received and acted upon by an adjustable speed drive system.

FPN: The relationship between motor current and motor temperature changes when the motor is operated by an adjustable speed drive. In certain applications, overheating of motors can occur when operated at reduced speed, even at current levels less than a motor's rated full-load current. The overheating can be the result of reduced motor cooling when its shaft-mounted fan is operating less than rated nameplate RPM. As part of the analysis to determine whether overheating will occur, it is necessary to consider the continuous torque capability curves for the motor given the application requirements. This will assist in determining whether the motor overload protection will be able, on its own, to provide protection against overheating. These overheating protection requirements are only intended to apply to applications where an adjustable speed drive, as defined in 430.2, is used.

For motors that utilize external forced air or liquid cooling systems, over-temperature can occur if the cooling system is not operating. Although this issue is not unique to adjustable speed applications, externally cooled motors are most often encountered with such applications. In these instances, over-temperature protection using direct temperature sensing is recommended [i.e. 430.126(A)(1), (A)(3) or (A)(4)] or additional means should be provided to ensure that the cooling system is operating (flow or pressure sensing, interlocking of adjustable speed drive system and cooling system, etc.).

(B) Multiple Motor Applications. For multiple motor applications, individual motor overtemperature protection shall be provided, as provided in 430.126(A).

(C) Automatic Restarting and Orderly Shutdown. The provisions of 430.43 and 430.44 shall apply to the motor over-temperature protection means.

Author's Comment: The revision to **430.126(A)** reflects conditions where over-temperature protection is needed. The means to provide over-temperature protection are not limited to the controller or drive and can be part of the system. The new exception recognizes that the thermal memory retention are not required for continuous duty loads but mainly for protection of short-time, intermittent, periodic or varying duty loads.

The **FPN** has been relocated and revised as it more appropriately applies to all adjustable speed drives, not to only multiple motor applications; however, it does not apply in all applications. The additional text in the **FPN** is intended to provide guidance to help the authority having jurisdiction in considering the analysis of potential motor overheating when forced air or liquid cooling systems are used to obtain the desired motor torque capability over the speed range required by the application. When the forced cooling is present and functioning, over-temperature protection is provided by the overload protection mechanism.

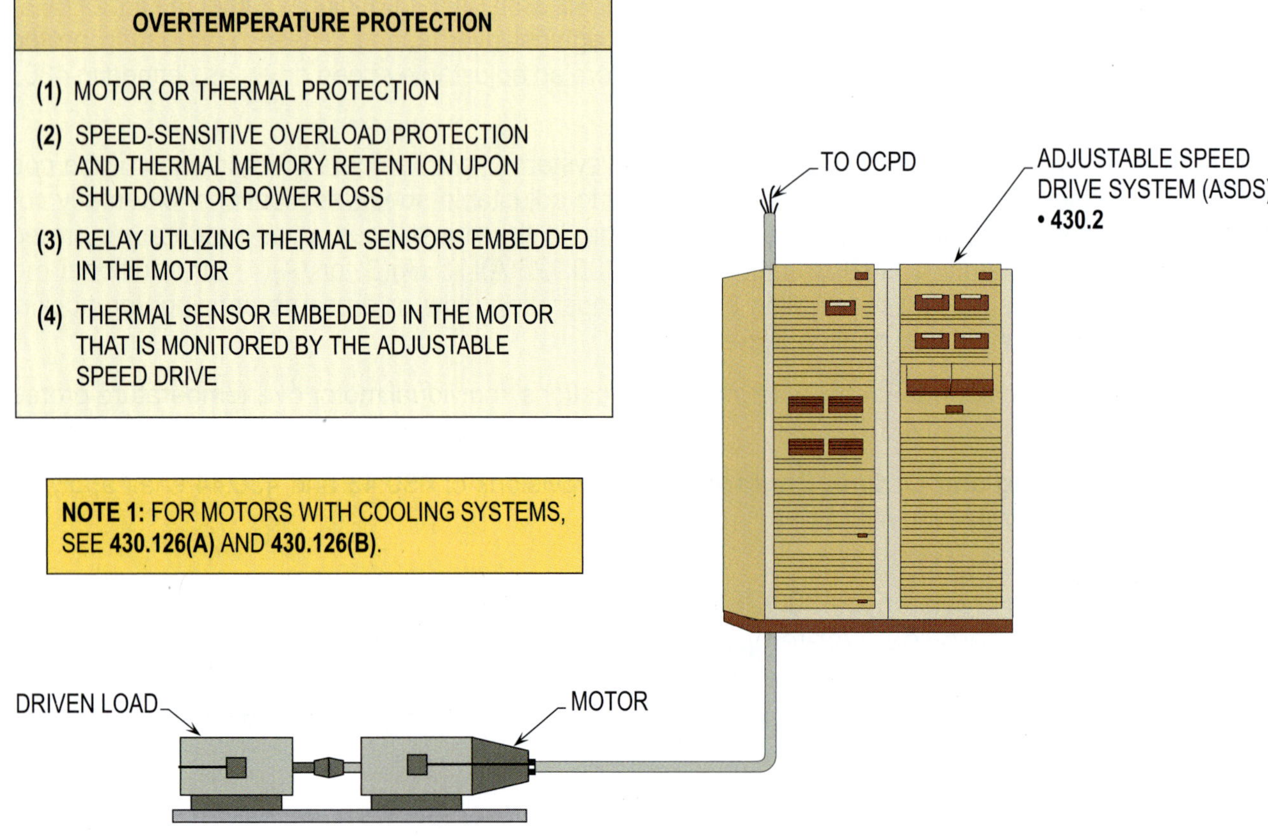

OVERTEMPERATURE PROTECTION
(1) MOTOR OR THERMAL PROTECTION
(2) SPEED-SENSITIVE OVERLOAD PROTECTION AND THERMAL MEMORY RETENTION UPON SHUTDOWN OR POWER LOSS
(3) RELAY UTILIZING THERMAL SENSORS EMBEDDED IN THE MOTOR
(4) THERMAL SENSOR EMBEDDED IN THE MOTOR THAT IS MONITORED BY THE ADJUSTABLE SPEED DRIVE
TO OCPD
ADJUSTABLE SPEED DRIVE SYSTEM (ASDS) • 430.2
NOTE 1: FOR MOTORS WITH COOLING SYSTEMS, SEE 430.126(A) AND 430.126(B).
DRIVEN LOAD
MOTOR
NOTE 2: TO ENSURE COOLING SYSTEM IS OPERATING, USE FLOW OR PRESSURE SENSING, INTERLOCKING OF ADJUSTABLE SPEED DRIVE SYSTEM, COOLING SYSTEM, ETC.

MOTOR OVERTEMPERATURE PROTECTION
430.126(A)

Purpose of Change: To include motor overtemperature protection for drive systems with excessive temperatures.

NEC Ch. 4 - Article 430
Part XI - 430.227

Type of Change	Panel Action	UL	UL 508	API 500 - 1997	API 505 - 1997	OSHA - 1994
Revision	Accept	-	-	-	-	-
ROP	**ROC**	**NFPA 70E - 2004**	**NFPA 70B - 2006**	**NFPA 79 - 2007**	**NFPA**	**NEMA**
pg. 486 # 11-83	pg. - # -	-	-	-	-	-
log: 2037 CMP: 11	log: -	Submitter: James T. Dollard, Jr.		2005 NEC: 430.227		IEC: -

2005 NEC - 430.227 Disconnecting Means.

The controller disconnecting means shall be capable of being locked in the open position.

2008 NEC - 430.227 Disconnecting Means.

The controller disconnecting means shall be capable of being locked in the open position. The provision for locking or adding a lock to the disconnecting means shall be installed on or at the switch or circuit breaker used as the disconnecting means and shall remain in place with or without the lock installed.

Author's Comment: The additional text provides consistency with other sections of the NEC for lockout/tagout procedures and also more clearly complies with the NFPA 70E and OSHA rules.

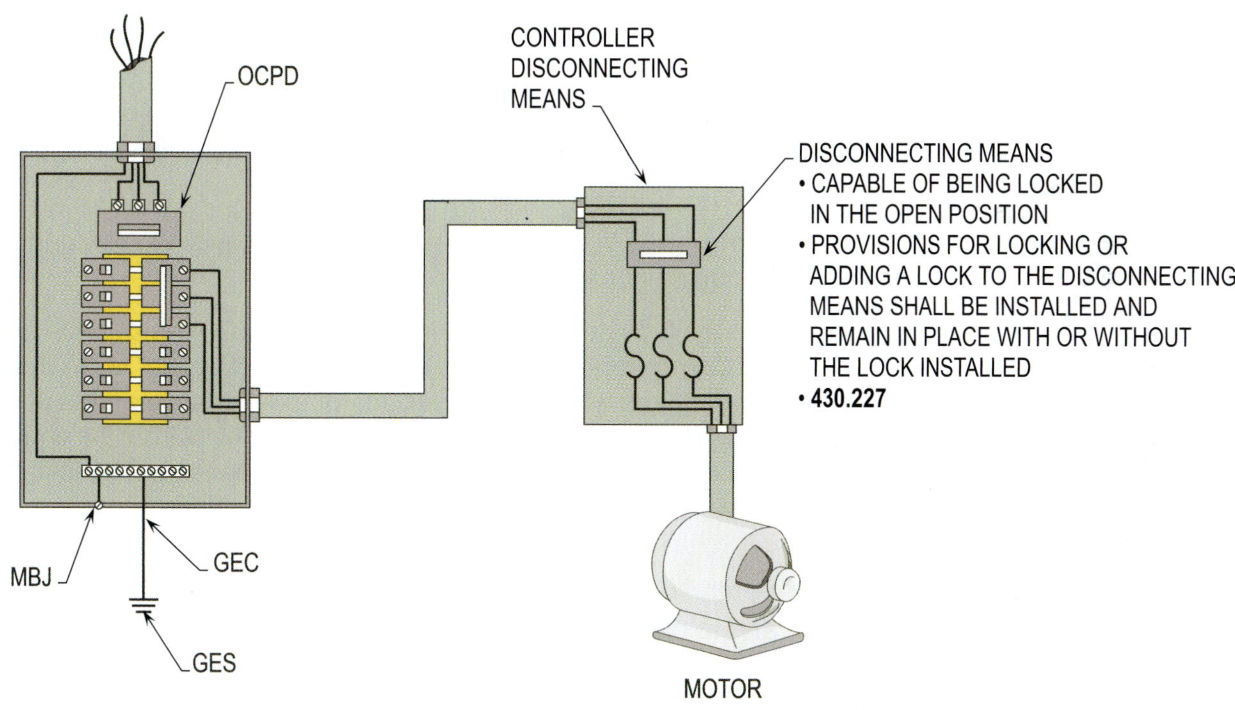

DISCONNECTING MEANS
430.227

Purpose of Change: This revision clarifies that provisions for locking or adding a lock to the disconnecting means shall be installed and remain in place with or without the lock installed.

NEC Ch. 4 - Article 440
Part II - 440.14, Exception No. 1

Type of Change	Panel Action	UL	UL 508	API 500 - 1997	API 505 - 1997	OSHA - 1994
Revision	Accept in Principle	-	-	-	-	1910.305(j)(3)(iii)
ROP	**ROC**	**NFPA 70E - 2004**	**NFPA 70B - 2006**	**NFPA 79 - 2007**	**NFPA**	**NEMA**
pg. 488 # 11-95	pg. - # -	420.10(D)(2)	7.4.2	5.5	-	-
log: 493 CMP: 11	log: -	Submitter: Michael J. Johnston		2005 NEC: 440.14, Exception No. 1		IEC: -

2005 NEC - 440.14 Location.

Disconnecting means shall be located within sight from and readily accessible from the air-conditioning or refrigerating equipment. The disconnecting means shall be permitted to be installed on or within the air-conditioning or refrigerating equipment.

The disconnecting means shall not be located on panels that are designed to allow access to the air-conditioning or refrigeration equipment.

Exception No. 1: Where the disconnecting means provided in accordance with 430.102(A) is capable of being locked in the open position, and the refrigerating or air-conditioning equipment is essential to an industrial process in a facility with written safety procedures, and where the conditions of maintenance and supervision ensure that only qualified persons service the equipment, a disconnecting means within sight from the equipment shall not be required. The provision for locking or adding a lock to the disconnecting means shall be ~~permanently~~ installed on or at the switch or circuit breaker ~~used as the disconnecting means~~.

2008 NEC - 440.14 Location.

Disconnecting means shall be located within sight from and readily accessible from the air-conditioning or refrigerating equipment. The disconnecting means shall be permitted to be installed on or within the air-conditioning or refrigerating equipment.

The disconnecting means shall not be located on panels that are designed to allow access to the air-conditioning or refrigeration equipment or to obscure the equipment nameplate(s).

Exception No. 1: Where the disconnecting means provided in accordance with 430.102(A) is capable of being locked in the open position, and the refrigerating or air-conditioning equipment is essential to an industrial process in a facility with written safety procedures, and where the conditions of maintenance and supervision ensure that only qualified persons service the equipment, a disconnecting means within sight from the equipment shall not be required. The provision for locking or adding a lock to the disconnecting means shall be installed on or at the switch or circuit breaker and shall remain in place with or without the lock installed.

Author's Comment: The word "permanently" was removed in **Exception No. 1** for consistency with the remainder of similar text in the NEC and to clarify that the provision for locking the disconnect in the open position does not have to be permanent, though it must remain in place with or without the lock installed. The text in the second paragraph of the main section was added to ensure that the disconnects are not mounted over the manufacturer's data nameplate, preventing safe access to pertinent information such as voltage, amperage, etc.

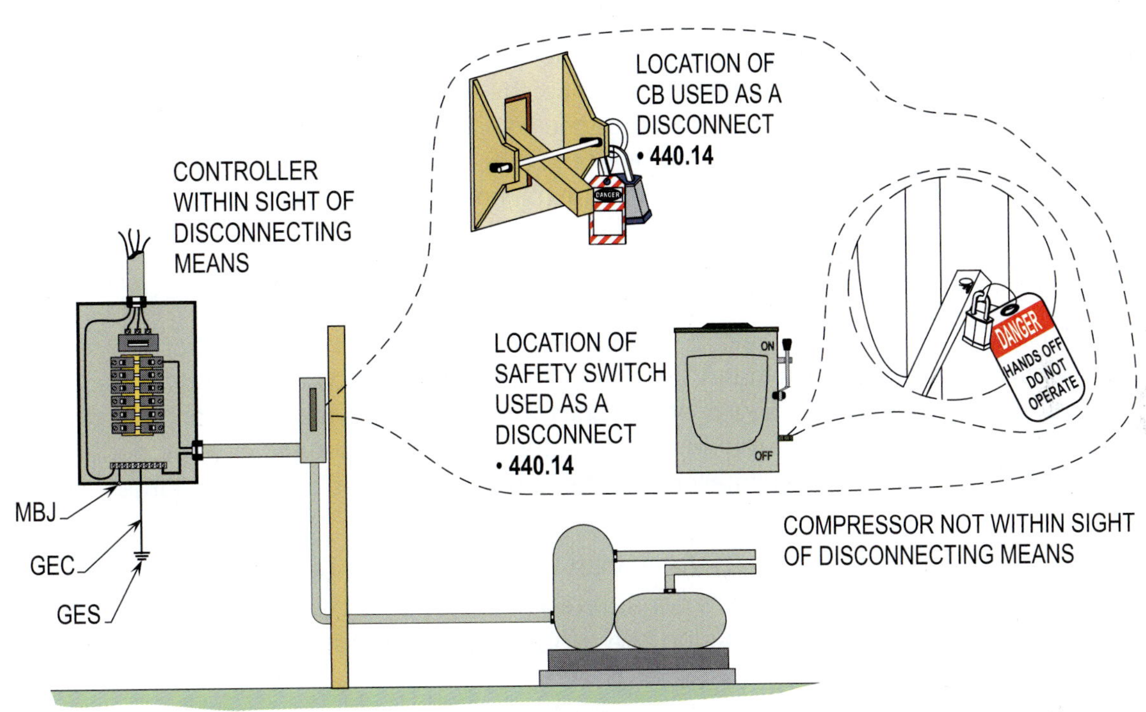

NOTE: PROVISIONS FOR LOCKING OR ADDING A LOCK TO THE DISCONNECTING MEANS SHALL BE INSTALLED ON THE SWITCH OR CB AND SHALL REMAIN IN PLACE WITH OR WITHOUT THE LOCKED INSTALLED PER **440.14, Ex. 1**.

LOCATION
440.14, Exception No. 1

Purpose of Change: This revision clarifies that provisions for locking or adding a lock to the disconnect means shall be installed and remain in place with or without the lock installed.

NEC Ch. 4 - Article 445
Part - 445.13

Type of Change	Panel Action	UL	UL 508	API 500 - 1997	API 505 - 1997	OSHA - 1994
Revision	Accept	-	-	-	-	-
ROP	**ROC**	**NFPA 70E - 2004**	**NFPA 70B - 2006**	**NFPA 79 - 2007**	**NFPA**	**NEMA - 2003**
pg. 491 # 13-7	pg. - # -	-	Annex J.4	-	-	MG 1
log: 2891 CMP: 13	log: -	Submitter: Mark R. Hilbert		2005 NEC: 445.13		IEC: -

2005 NEC - 445.13 Ampacity of Conductors.

The ampacity of the conductors from the generator terminals to the first distribution device(s) containing overcurrent protection shall not be less than 115 percent of the nameplate current rating of the generator. It shall be permitted to size the neutral conductors in accordance with 220.61. Conductors that must carry ground-fault currents shall not be smaller than that required by ~~250.24(C)~~. Neutral conductors of dc generators that must carry ground-fault currents shall not be smaller than the minimum required size of the largest conductor.

2008 NEC - 445.13 Ampacity of Conductors.

The ampacity of the conductors from the generator terminals to the first distribution device(s) containing overcurrent protection shall not be less than 115 percent of the nameplate current rating of the generator. It shall be permitted to size the neutral conductors in accordance with 220.61. Conductors that must carry ground-fault currents shall not be smaller than that required by 250.30(A). Neutral conductors of dc generators that must carry ground-fault currents shall not be smaller than the minimum required size of the largest conductor.

Author's Comment: Section **250.24(C)** deals with the sizing of the grounded conductor for services, however, generator output conductors are considered to be feeder conductors, not service conductors, so the more appropriate reference for sizing the grounded conductor is in **250.30(A)**.

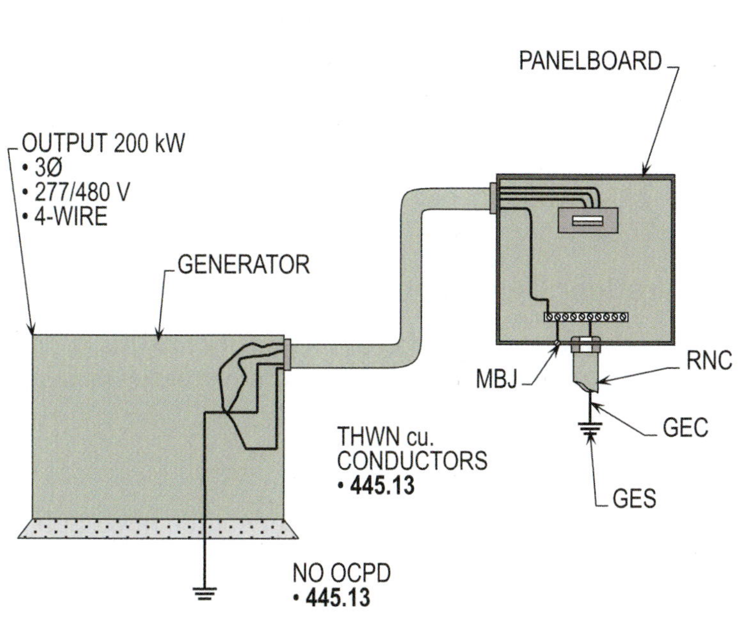

NOTE: CONDUCTORS THAT MUST CARRY GROUND-FAULT CURRENTS SHALL NOT BE SMALLER THAN THAT REQUIRED BY **250.30(A)**, AS WELL AS NEUTRAL CURRENT

Sizing Conductors To Panelboard

Step 1: Finding FLA of generator
FLA = 200 kW ÷ [480 V x 1.732]
FLA = 240.7 A

Step 2: Calculating FLA for conductors
445.13
240.7 A x 115% = 276.8 A

Step 3: Selecting conductors
Table 310.16
276.8 A requires 300 KCMIL cu.

Solution: **The THWN conductors are required to be 300 KCMIL copper since there is no OCPD at the generator.**

Sizing Grounded Conductor

Step 1: Finding grounded conductor
250.30(A)(8); Table 250.66
300 KCMIL requires 2 AWG cu.

Solution: **The size of the grounded conductor is required to be 2 AWG copper to carry fault current but could be larger based on calculation in accordance with 220.61 for neutral loads.**

AMPACITY OF CONDUCTORS
445.13

Purpose of Change: This revision clarifies that the grounded conductor shall be sized based on the requirements for feeders.

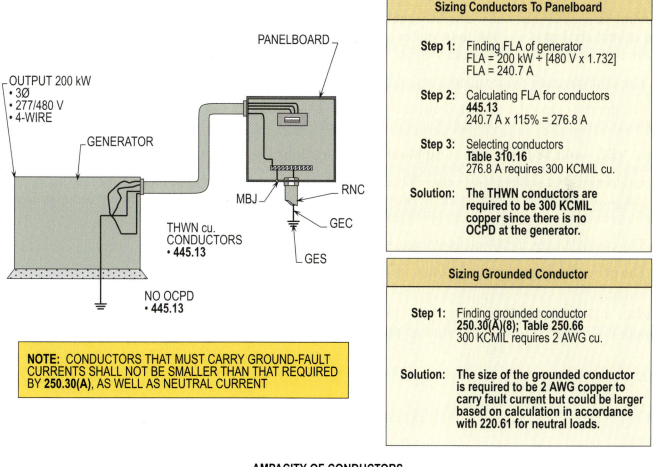

NEC Ch. 4 - Article 445
Part - 445.18

Type of Change		Panel Action		UL	UL 508	API 500 - 1997	API 505 - 1997	OSHA - 1994
Revision		Accept in Principle		-	-	-	-	-
ROP		**ROC**		**NFPA 70E - 2004**	**NFPA 70B - 2006**	**NFPA 79 - 2007**	**NFPA**	**NEMA - 2003**
pg. 491	# 13-9	pg. -	# -	-	Annex J.4	-	-	MG 1
log: 2121	CMP: 13	log: -	Submitter: Bud Swathwood			2005 NEC: 445.18		IEC: -

2005 NEC - 445.18 Disconnecting Means Required for Generators.

Generators shall be equipped with disconnect(s) by means of which the generator and all protective devices and control apparatus are able to be disconnected entirely from the circuits supplied by the generator except where both of the following conditions apply:

(1) The driving means for the generator can be readily shut down.

(2) The generator is not arranged to operate in parallel with another generator or other source of voltage.

2008 NEC - 445.18 Disconnecting Means Required for Generators.

Generators shall be equipped with disconnect(s), lockable in the open position by means of which the generator and all protective devices and control apparatus are able to be disconnected entirely from the circuits supplied by the generator except where both of the following conditions apply:

(1) The driving means for the generator can be readily shut down.

(2) The generator is not arranged to operate in parallel with another generator or other source of voltage.

Author's Comment: Requiring the generator disconnecting to be lockable in the open position will ensure the generator can be locked off for safe servicing of the feeder circuit and the equipment.

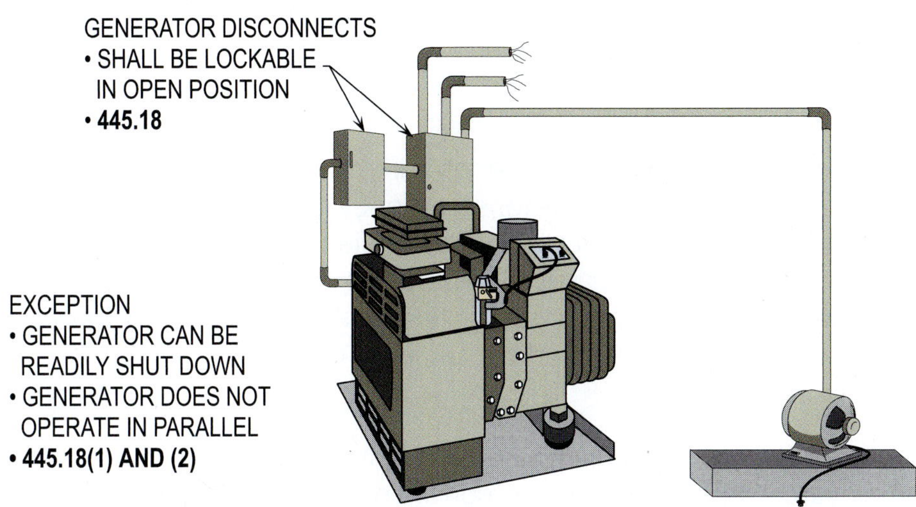
GENERATOR DISCONNECTS
• SHALL BE LOCKABLE IN OPEN POSITION
• **445.18**
EXCEPTION
• GENERATOR CAN BE READILY SHUT DOWN
• GENERATOR DOES NOT OPERATE IN PARALLEL
• **445.18(1) AND (2)**

DISCONNECTING MEANS REQUIRED FOR GENERATORS
445.18

Purpose of Change: This revision clarifies that the generator disconnect shall be capable of being lockable in the open position.

NEC Ch. 4 - Article 445
Part - 445.19

Type of Change	Panel Action	UL	UL 508	API 500 - 1997	API 505 - 1997	OSHA - 1994
New Section	Accept in Principle	1004	-	-	-	-
ROP	**ROC**	**NFPA 70E - 2004**	**NFPA 70B - 2006**	**NFPA 79 - 2007**	**NFPA**	**NEMA - 2003**
pg. 491 / # 13-8	pg. - / # -	-	Annex J.4	-	-	MG 1
log: 832 / CMP: 13	log: -	Submitter: Thomas H. Wood		2005 NEC: -		IEC: -

2008 NEC - 445.19 Generators Supplying Multiple Loads.

A single generator supplying more than one load, or multiple generators operating in parallel, shall be permitted to supply either of the following:

(1) A vertical switchboard with separate sections,

(2) Individual enclosures with overcurrent protection tapped from a single feeder for load separation and distribution.

Author's Comment: The supply tap box on generators is not generally designed or manufactured with multiple devices to serve separate circuits for emergency loads, fire pumps, legally required standby loads, and optional standby loads. This new section clarifies that the disconnect(s) may be provided using a single feeder from the generator set to separately mounted enclosed disconnects providing separation of the emergency and standby circuits or a distribution switchboard that separates emergency and standby load disconnects in different vertical sections.

Multiple generators operating in parallel are treated similarly downstream of the paralleling switchboard. Separately enclosed overcurrent devices or overcurrent devices mounted in separate vertical sections of a distribution switchboard provides physical separation of the different systems or branches of distribution and clarifies that the origin of the emergency, legally required standby, and optional standby systems is at the feeder overcurrent protection device, not at the generator terminals. A related proposal has been acted on for **700.9**.

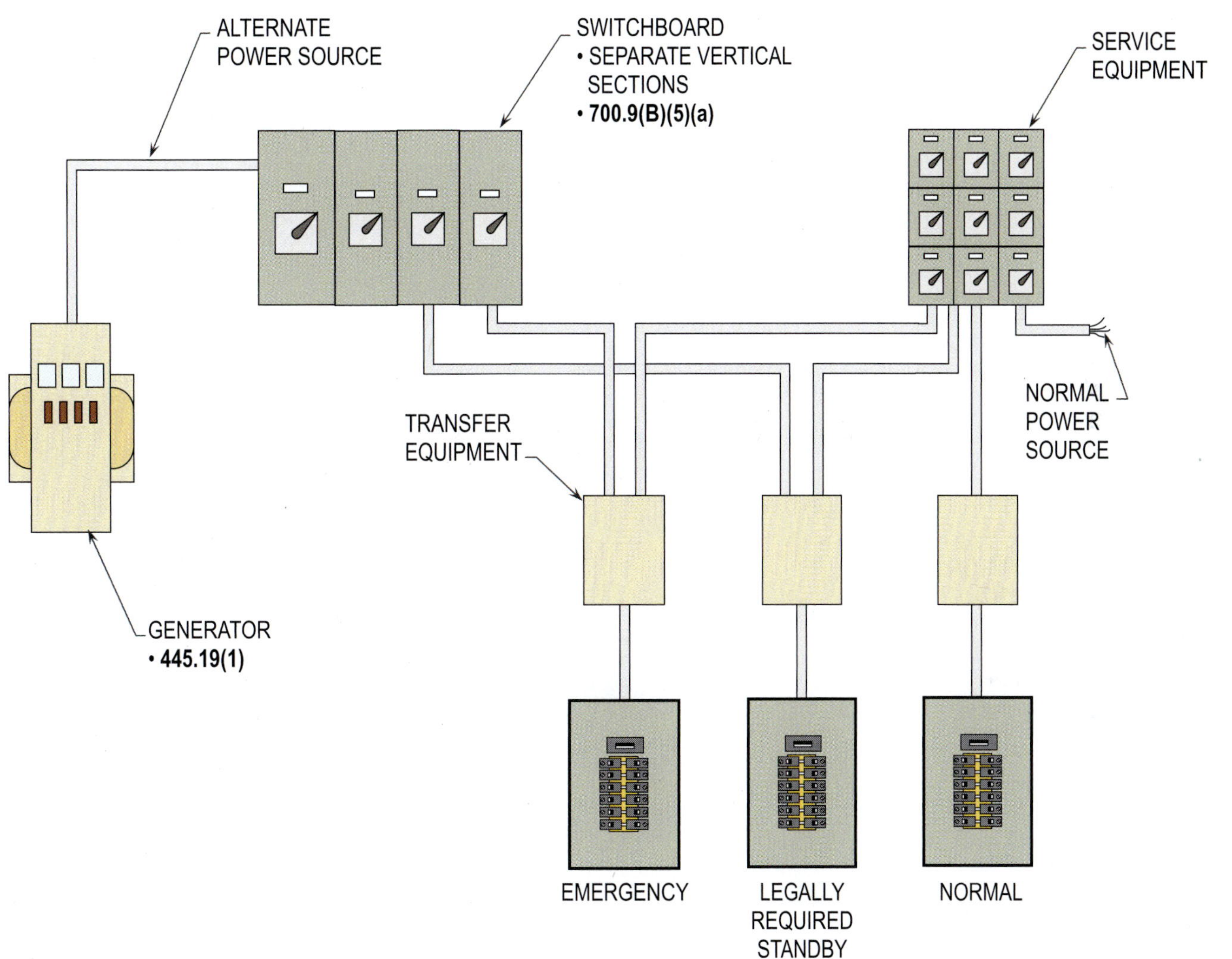

GENERATORS SUPPLYING MULTIPLE LOADS
445.19

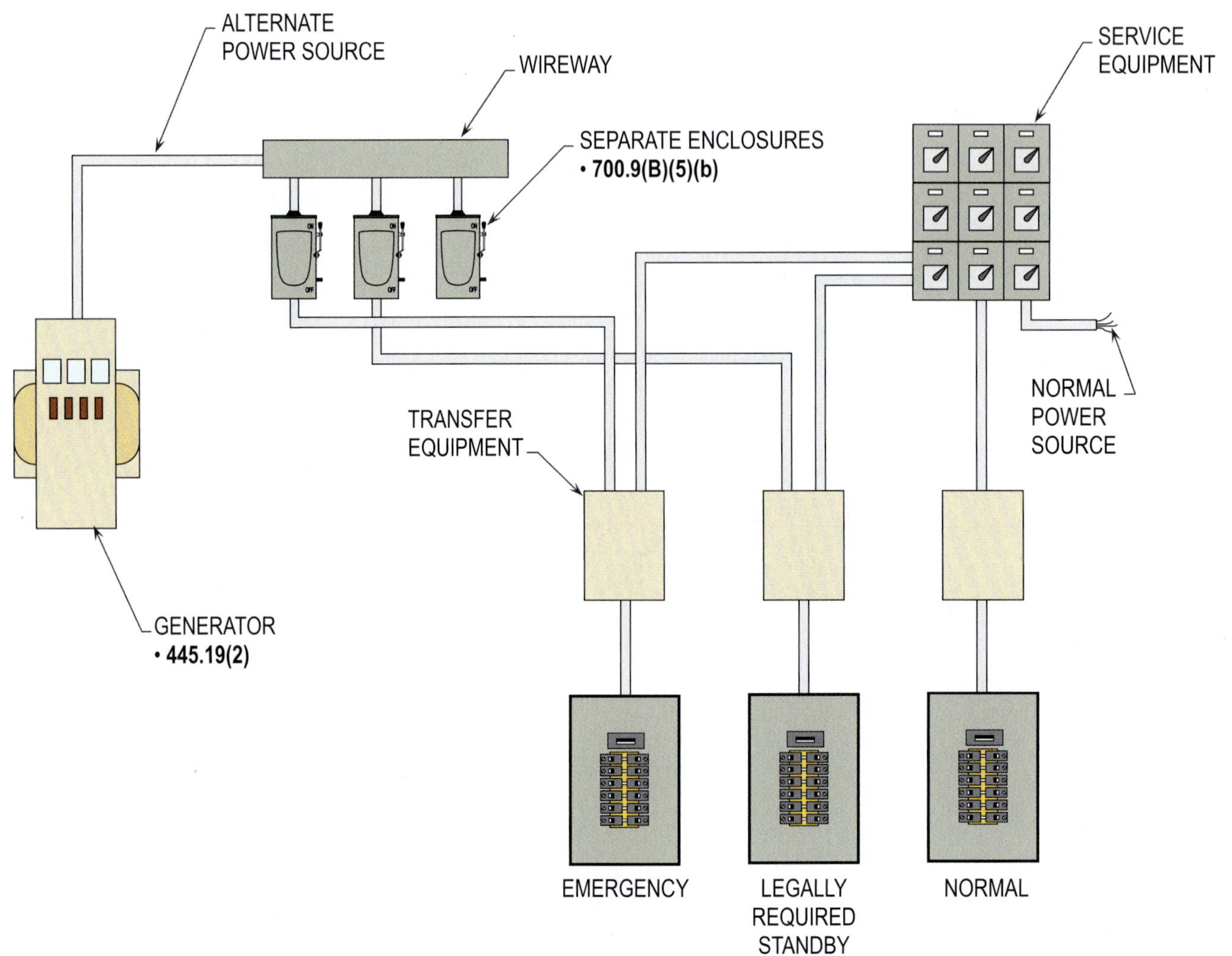

GENERATORS SUPPLYING MULTIPLE LOADS
445.19

Purpose of Change: A new section has been added to provide requirements for a single generator supplying more than one load or multiple generators operating in parallel.

NEC Ch. 4 - Article 450
Part I - 450.5(B)(2)(b), Exception

Type of Change	Panel Action	UL	UL 508	API 500 - 1997	API 505 - 1997	OSHA - 1994
Revision	Accept in Principle	506 - 5085	-	-	-	-
ROP	**ROC**	**NFPA 70E - 2004**	**NFPA 70B - 2006**	**NFPA 79 - 2007**	**NFPA**	**NEMA - 1992**
pg. 493 # 9-135	pg. - # -	-	-	7.2.7.1	-	ST 20
log: 492 CMP: 9	log: -	Submitter: David Murray		2005 NEC: 450.5(B)(2)		IEC: -

2005 NEC - 450.5 Grounding Autotransformers.

(B) Ground Reference for Fault Protection Devices. A grounding autotransformer used to make available a specified magnitude of ground-fault current for operation of a ground-responsive protective device on a 3-phase, 3-wire ungrounded system shall conform to 450.5(B)(1) and (B)(2).

(1) Rating. The autotransformer shall have a continuous neutral-current rating sufficient for the specified ground-fault current.

(2) Overcurrent Protection. An overcurrent protective device ~~of adequate short-circuit rating~~ that will open simultaneously all ungrounded conductors when it operates shall be applied in the grounding autotransformer branch circuit ~~and~~ shall be rated or set at a current not exceeding 125 percent of the autotransformer continuous per-phase current rating or 42 percent of the continuous-current rating of any series connected devices in the autotransformer neutral connection. Delayed tripping for temporary overcurrents to permit the proper operation of ground-responsive tripping devices on the main system shall be permitted but shall not exceed values that would be more than the short-time current rating of the grounding autotransformer or any series connected devices in the neutral connection thereto.

2008 NEC - 450.5 Grounding Autotransformers.

(B) Ground Reference for Fault Protection Devices. A grounding autotransformer used to make available a specified magnitude of ground-fault current for operation of a ground-responsive protective device on a 3-phase, 3-wire ungrounded system shall conform to 450.5(B)(1) and (B)(2).

(1) Rating. The autotransformer shall have a continuous neutral-current rating sufficient for the specified ground-fault current.

(2) Overcurrent Protection. Overcurrent protection shall comply with (a) and (b).

(a) Operation and Interrupting Rating. An overcurrent protective device having an interrupting rating in compliance with 110.9 and that will open simultaneously all ungrounded conductors when it operates shall be applied in the grounding autotransformer branch circuit.

(b) Ampere Rating. The overcurrent protection shall be rated or set at a current not exceeding 125 percent of the autotransformer continuous per-phase current rating or 42 percent of the continuous-current rating of any series connected devices in the autotransformer neutral connection. Delayed tripping for temporary overcurrents to permit the proper operation of ground-responsive tripping devices on the main system shall be permitted but shall not exceed values that would be more than the short-time current rating of the grounding autotransformer or any series connected devices in the neutral connection thereto.

Exception: For high-impedance grounded systems covered in 250.36, where the maximum ground-fault current is designed to be not more than 10 amperes, and where the grounding autotransformer and the grounding impedance are rated for continuous duty, an overcurrent device rated not more than 20 amperes that will simultaneously open all ungrounded conductors shall be permitted to be installed on the line side of the grounding autotransformer.

Author's Comment: Subsections have been established and titles have been added to **(B)** for user-friendliness. An exception has been added for high-impedance grounded systems where the maximum ground fault current is designed to be not more than 10 amperes. This will permit a grounding transformer to have an overcurrent device rated not more than 20 amperes to be installed on the line side of the grounding autotransformer as long as the overcurrent protective device simultaneously opens all ungrounded conductors.

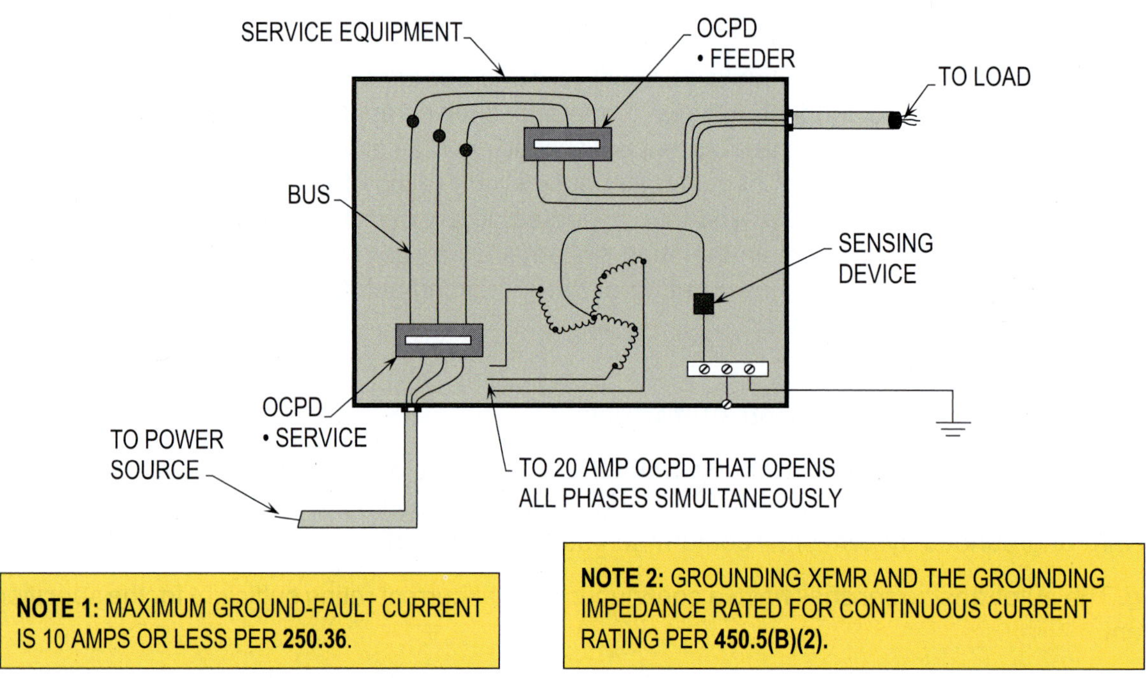

GROUND REFERENCE FOR FAULT PROTECTION DEVICES
450.5(B)(2), Exception

Purpose of Change: A new exception has been added to provide requirements for high impedance grounded systems where the maximum ground fault current is designed to be not more than 10 amperes.

NEC Ch. 4 - Article 490
Part III - 490.44(C)

Type of Change		Panel Action		UL	UL 508	API 500 - 1997	API 505 - 1997	OSHA - 1994
Revision		Accept in Principle		-	-	-	-	-
ROP		**ROC**		**NFPA 70E - 2004**	**NFPA 70B - 2006**	**NFPA 79 - 2007**	**IEEE - 2001**	**NEMA - 2000**
pg. 500	# 9-155	pg. -	# -	-	-	-	C37.20.3	SG 4
log: 2038	CMP: 9	log: -	Submitter: James T. Dollard, Jr.			2005 NEC: 490.44(C)		IEC: -

2005 NEC - 490.44 Fused Interrupter Switches.

(C) Switching Mechanism. The switching mechanism shall be arranged to be operated from a location outside the enclosure where the operator is not exposed to energized parts and shall be arranged to open all ungrounded conductors of the circuit simultaneously with one operation. Switches shall be capable of being locked in the open position.

2008 NEC - 490.44 Fused Interrupter Switches.

(C) Switching Mechanism. The switching mechanism shall be arranged to be operated from a location outside the enclosure where the operator is not exposed to energized parts and shall be arranged to open all ungrounded conductors of the circuit simultaneously with one operation. Switches shall be capable of being locked in the open position. The provisions for locking shall remain in place with or without the lock installed.

Author's Comment: The new last sentence will ensure that the lockout device will remain at the disconnecting means with or without the lock installed.

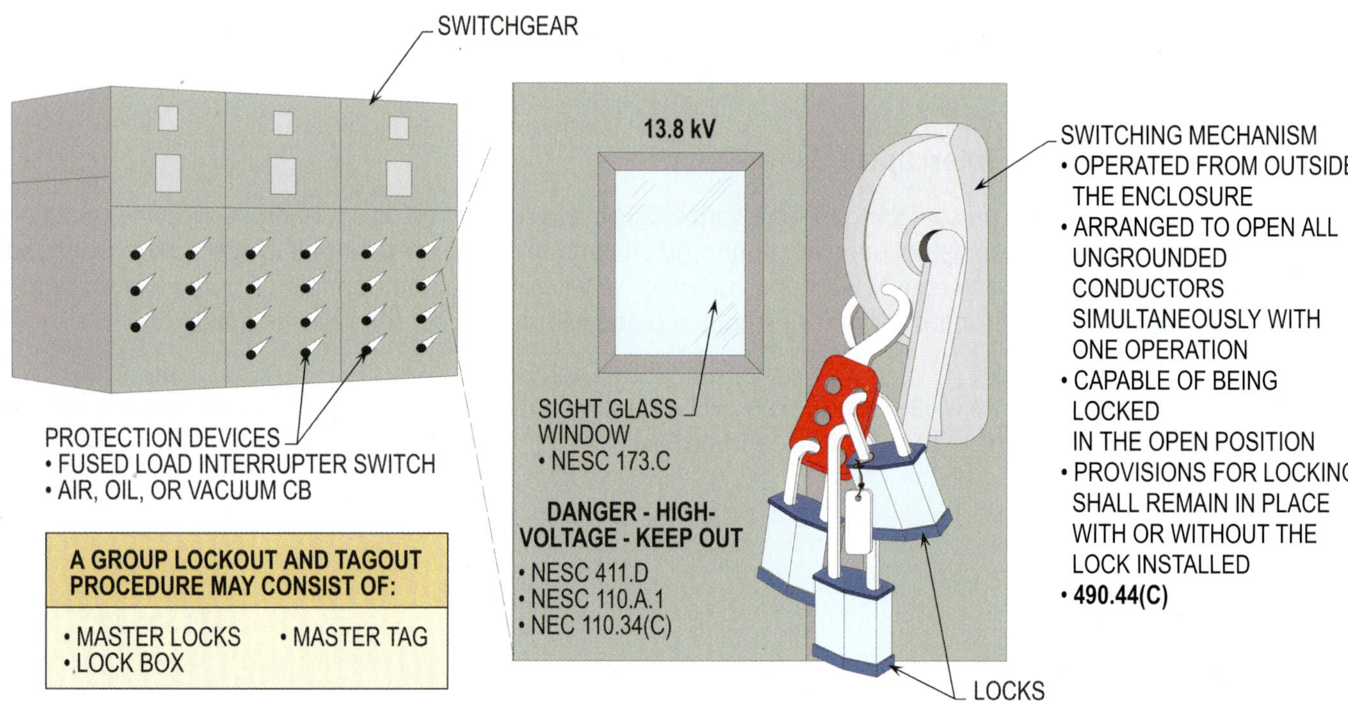

SWITCHING MECHANISM
490.44(C)

Purpose of Change: This revision clarifies that provisions for locking shall remain in place with or without the lock installed.

NEC Ch. 4 - Article 490
Part III - 490.46

Type of Change	Panel Action	UL	UL 508	API 500 - 1997	API 505 - 1997	OSHA - 1994
New Section	Accept in Principle	-	-	-	-	-
ROP	**ROC**	**NFPA 70E - 2004**	**NFPA 70B - 2006**	**NFPA 79 - 2007**	**IEEE - 2001**	**NEMA - 2000**
pg. 500 # 9-156	pg. - # -	-	-	-	C37.20.2	SG 4
log: 2039 CMP: 9	log: - Submitter: James T. Dollard, Jr.			2005 NEC: -		IEC: -

2005 NEC - ~~490.46~~ Metal Enclosed and Metal Clad Service Equipment.

Metal enclosed and metal clad switchgear installed as high-voltage service equipment shall include a ground bus for the connection of service cable shields and to facilitate the attachment of safety grounds for personnel protection. This bus shall be extended into the compartment where the service conductors are terminated.

2008 NEC - 490.46 Circuit Breaker Locking.

Circuit breakers shall be capable of being locked in the open position or, if they are installed in a drawout mechanism, that mechanism shall be capable of being locked in such a position that the mechanism cannot be moved into the connected position. In either case, the provision for locking shall remain in place with or without the lock.

490.47 Metal-Enclosed and Metal-Clad Service Equipment.

Metal-enclosed and metal-clad switchgear installed as high-voltage service equipment shall include a ground bus for the connection of service cable shields and to facilitate the attachment of safety grounds for personnel protection. This bus shall be extended into the compartment where the service conductors are terminated.

Author's Comment: This new section should help ensure that high voltage circuit breakers are capable of being locked in the open position or, if there is a drawout mechanism, that the mechanism is capable of being locked so that it cannot be moved back into the connected position. The lockout device must remain in place with or without the lock.

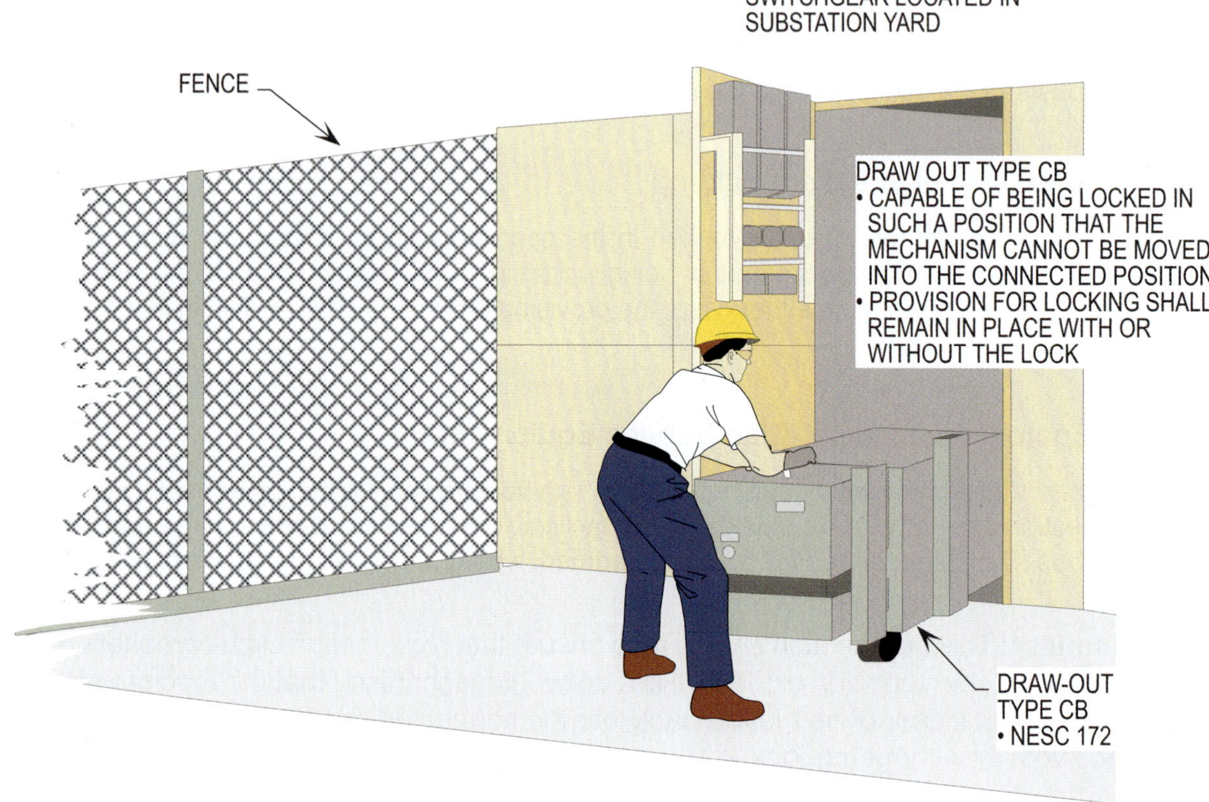
SWITCHGEAR LOCATED IN
SUBSTATION YARD
FENCE
DRAW OUT TYPE CB
• CAPABLE OF BEING LOCKED IN SUCH A POSITION THAT THE MECHANISM CANNOT BE MOVED INTO THE CONNECTED POSITION
• PROVISION FOR LOCKING SHALL REMAIN IN PLACE WITH OR WITHOUT THE LOCK
DRAW-OUT
TYPE CB
• NESC 172

CIRCUIT BREAKER LOCKING
490.46

Purpose of Change: A new section has been added to require high voltage circuit breakers to be locked in the open position or, if there is a draw-out mechanism, the mechanism shall be capable of being locked so that it cannot be moved into the connected position.

5

Special Occupancies

Chapter 5 of the NEC has always been known as the special chapter, covering specific subjects that have particular functions or purposes. **Chapter 5** deals with places or locations where people work, and such occupancies can have built-in electrical hazards.

The main function of this chapter is to protect personnel and equipment from electrical hazards that could occur in a particular work area, based upon the type of occupancy. For example, designers, installers and inspectors use the requirements of **Chapter 5** to design, install and inspect the electrical wiring methods and equipment located in hazardous areas.

Chapter 5 is special because it is based upon occupancies that prohibit a large portion of the design and installation techniques from falling under the requirements of **Chapters 1 through 4** in the NEC.

NEC Ch. 5 - Article 500
Part - 500.7(K)(1) through (K)(3)

Type of Change	Panel Action	UL	UL 508	API 500 - 1997	API 505 - 1997	OSHA - 1994
Revision	Accept in Principle	2075	-	6.5	6.8	1910.307(b)(2), Note
ROP	**ROC**	**NFPA 70E - 2004**	**NFPA 70B - 2006**	**NFPA 79 - 2007**	**NFPA**	**NEMA**
pg. 509 # 14-17	pg. - # -	-	23.1	-	-	-
log: 2495 CMP: 14	log: -	Submitter: Edward M. Briesch		2005 NEC: 500.7(K)		IEC: -

2005 NEC - 500.7(K) Combustible Gas Detection System.

A combustible gas detection system shall be permitted as a means of protection in industrial establishments with restricted public access and where the conditions of maintenance and supervision ensure that only qualified persons service the installation. ~~Gas detection equipment shall be listed for both the location in which it is installed and for detection of the specific gas or vapor to be encountered.~~ Where such a system is installed, equipment specified in 500.7(K)(1), (K)(2), or (K)(3) shall be permitted.

The type of combustible detection equipment, its listing, installation location(s), alarm and shutdown criteria, and calibration frequency shall be documented when combustible gas detectors are used as a protection technique.

FPN No. 1: For further information, see ANSI/ISA-12.13.01, *Performance Requirements, Combustible Gas Detectors.*

FPN No. 2: For further information, see ANSI/API RP 500, *Recommended Practice for Classification of Locations for Electrical Installations at Petroleum Facilities Classified for Class I, Division 1 or Division 2.*

FPN No. 3: For further information, see ANSI/ISA RP 12.13.02, *Installation, Operation, and Maintenance of Combustible Gas Detection Instruments.*

(1) Inadequate Ventilation. In a Class I, Division 1 location that is so classified due to inadequate ventilation, electrical equipment suitable for Class I, Division 2 locations shall be permitted.

(2) Interior of a Building. In a building located in, or with an opening into, a Class I, Division 2 location where the interior does not contain a source of flammable gas or vapor, electrical equipment for unclassified locations shall be permitted.

(3) Interior of a Control Panel. In the interior of a control panel containing instrumentation utilizing or measuring flammable liquids, gases, or vapors, electrical equipment suitable for Class I, Division 2 locations shall be permitted.

2008 NEC - 500.7(K) Combustible Gas Detection System.

A combustible gas detection system shall be permitted as a means of protection in industrial establishments with restricted public access and where the conditions of maintenance and supervision ensure that only qualified persons service the installation. Where such a system is installed, equipment specified in 500.7(K)(1), (K)(2), or (K)(3) shall be permitted.

The type of combustible detection equipment, its listing, installation location(s), alarm and shutdown criteria, and calibration frequency shall be documented when combustible gas detectors are used as a protection technique.

FPN No. 1: For further information, see ANSI/ISA-12.13.01-2003 (IEC 61779-1 through -5 Mod), *Performance Requirements, Combustible Gas Detectors, and ANSI/UL 2075, Gas and Vapor Detectors and Sensors.*

FPN No. 2: For further information, see ANSI/API RP 500, *Recommended Practice for Classification of Locations for Electrical Installations at Petroleum Facilities Classified for Class I, Division 1 or Division 2.*

FPN No. 3: For further information, see ANSI/ISA RP 12.13.02-2003 (IEC 61779-6 Mod), *Installation, Operation, and Maintenance of Combustible Gas Detection Instruments.*

(1) Inadequate Ventilation. In a Class I, Division 1 location that is so classified due to inadequate ventilation, electrical equipment suitable for Class I, Division 2 locations shall be permitted. Combustible gas detection equipment shall be listed for Class I, Division 1, for the appropriate material group, and for the detection of the specific gas or vapor to be encountered.

(2) Interior of a Building. In a building located in, or with an opening into, a Class I, Division 2 location where the interior does not contain a source of flammable gas or vapor, electrical equipment for unclassified locations shall be permitted. Combustible gas detection equipment shall be listed for Class I, Division 1 or Class I, Division 2, for the appropriate material group, and for the detection of the specific gas or vapor to be encountered.

(3) Interior of a Control Panel. In the interior of a control panel containing instrumentation utilizing or measuring flammable liquids, gases, or vapors, electrical equipment suitable for Class I, Division 2 locations shall be permitted. Combustible gas detection equipment shall be listed for Class I, Division 1, the appropriate material group, and for the detection of the specific gas or vapor to be encountered.

Author's Comment: The requirement for the combustible gas detection equipment to be listed for the location in which it is installed and for the detection of the specific gas or vapor encountered was deleted from the first paragraph to be placed in the subsection covering the specific location, such as the interior of the building.

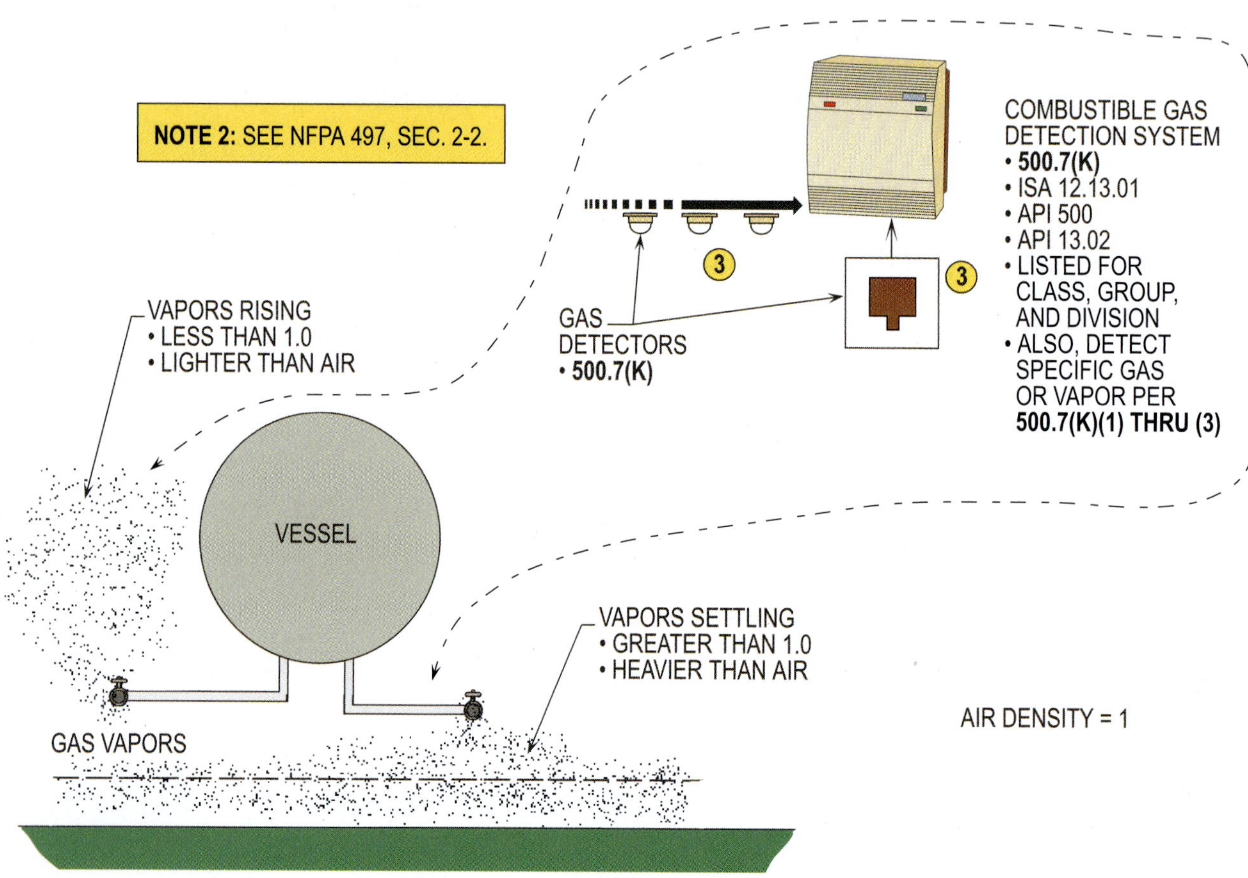

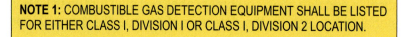

COMBUSTIBLE GAS DETECTION SYSTEM
500.7(K)(1) through (K)(3)

Purpose of Change: To provide requirements that more accurately address the rules for combustible gas detection systems.

NEC Ch. 5 - Article 500
Part - 500.8(A), FPN; 505.9(A), FPN; 506.9(A), FPN

Type of Change	Panel Action	UL	UL 508	API 500 - 1997	API 505 - 1997	OSHA - 1994
Revision	Accept	-	-	1.1.2	1.1.2	1910.307(b)(2)
ROP	**ROC**	**NFPA 70E - 2004**	**NFPA 70B - 2006**	**NFPA 79 - 2007**	**NFPA**	**NEMA**
pg. 509 # 14-18a	pg. - # -	440.4(E)	23.2.1	-	-	-
log: CP 1406 CMP: 14	log: -	Submitter: Code Making Panel 14		2005 NEC: 500.8; 505.9(A); 506.9(A)		IEC: -

2005 NEC - 500.8 Equipment.

Articles 500 through 504 require equipment construction and installation that ensure safe performance under conditions of proper use and maintenance.

FPN No. 1: It is important that inspection authorities and users exercise more than ordinary care with regard to installation and maintenance.

FPN No. 2: Since there is no consistent relationship between explosion properties and ignition temperature, the two are independent requirements.

FPN No. 3: Low ambient conditions require special consideration. Explosionproof or dust-ignitionproof equipment may not be suitable for use at temperatures lower than -25∞C (-13∞F) unless they are identified for low-temperature service. However, at low ambient temperatures, flammable concentrations of vapors may not exist in a location classified as Class I, Division 1 at normal ambient temperature.

~~(A)~~ Approval for Class and Properties.

(1) Equipment shall be identified not only for the class of location but also for the explosive, combustible, or ignitible properties of the specific gas, vapor, dust, fiber, or flyings that will be present. In addition, Class I equipment shall not have any exposed surface that operates at a temperature in excess of the ignition temperature of the specific gas or vapor. Class II equipment shall not have an external temperature higher than that specified in 500.8(C)(2). Class III equipment shall not exceed the maximum surface temperatures specified in 503.5.

FPN: Luminaires (lighting fixtures) and other heat-producing apparatus, switches, circuit breakers, and plugs and receptacles are potential sources of ignition and are investigated for suitability in classified locations. Such types of equipment, as well as cable terminations for entry into explosionproof enclosures, are available as listed for Class I, Division 2 locations. Fixed wiring, however, may utilize wiring methods that are not evaluated with respect to classified locations. Wiring products such as cable, raceways, boxes, and fittings, therefore, are not marked as being suitable for Class I, Division 2 locations. Also see 500.8(B)(6)(a).

Suitability of identified equipment shall be determined by any of the following:

(1) Equipment listing or labeling

(2) Evidence of equipment evaluation from a qualified testing laboratory or inspection agency concerned with product evaluation

(3) Evidence acceptable to the authority having jurisdiction such as a manufacturer's self-evaluation or an owner's engineering judgment

~~(B)~~ Marking.

~~(C)~~ Temperature.

~~(D)~~ Threading.

~~(E)~~ Fiber Optic Cable Assembly.

505.9 Equipment

(A) Suitability. Suitability of identified equipment shall be determined by one of the following:

(1) Equipment listing or labeling

(2) Evidence of equipment evaluation from a qualified testing laboratory or inspection agency concerned with product evaluation

(3) Evidence acceptable to the authority having jurisdiction such as a manufacturer's self-evaluation or an owner's engineering judgment

506.9 Equipment Requirements.

(A) Suitability. Suitability of identified equipment shall be determined by one of the following:

(1) Equipment listing or labeling

(2) Evidence of equipment evaluation from a qualified testing laboratory or inspection agency concerned with product evaluation

(3) Evidence acceptable to the authority having jurisdiction such as a manufacturer's self-evaluation or an owner's engineering judgment

2008 NEC - 500.8 Equipment.

Articles 500 through 504 require equipment construction and installation that ensure safe performance under conditions of proper use and maintenance.

FPN No. 1: It is important that inspection authorities and users exercise more than ordinary care with regard to installation and maintenance.

FPN No. 2: Since there is no consistent relationship between explosion properties and ignition temperature, the two are independent requirements.

FPN No. 3: Low ambient conditions require special consideration. Explosionproof or dust-ignitionproof equipment may not be suitable for use at temperatures lower than -25∞C (-13∞F) unless they are identified for low-temperature service. However, at low ambient temperatures, flammable concentrations of vapors may not exist in a location classified as Class I, Division 1 at normal ambient temperature.

(A) Suitability. Suitability of identified equipment shall be determined by one of the following:

(1) Equipment listing or labeling

(2) Evidence of equipment evaluation from a qualified testing laboratory or inspection agency concerned with product evaluation

(3) Evidence acceptable to the authority having jurisdiction such as a manufacturer's self-evaluation or an owner's engineering judgment.

FPN: Additional documentation for equipment may include certificates demonstrating compliance with applicable equipment standards, indicating special conditions of use, and other pertinent information.

(B) Approval for Class and Properties.

(1) Equipment shall be identified not only for the class of location but also for the explosive, combustible, or ignitible properties of the specific gas, vapor, dust, fiber, or flyings that will be present. In addition, Class I equipment shall not have any exposed surface that operates at a temperature in excess of the ignition temperature of the specific gas or vapor. Class II equipment shall not have an external temperature higher than that specified in 500.8(C)(2). Class III equipment shall not exceed the maximum surface temperatures specified in 503.5.

FPN: Luminaires (lighting fixtures) and other heat-producing apparatus, switches, circuit breakers, and plugs and receptacles are potential sources of ignition and are investigated for suitability in classified locations. Such types of equipment, as well as cable terminations for entry into explosionproof enclosures, are available as listed for Class I, Division 2 locations. Fixed wiring, however, may utilize wiring methods that are not evaluated with respect to classified locations. Wiring products such as cable, raceways, boxes, and fittings, therefore, are not marked as being suitable for Class I, Division 2 locations. Also see 500.8(B)(6)(a).

(C) Marking.

(D) Temperature.

(E) Threading.

(F) Fiber Optic Cable Assembly.

505.9 Equipment

(A) Suitability. Suitability of identified equipment shall be determined by one of the following:

(1) Equipment listing or labeling

(2) Evidence of equipment evaluation from a qualified testing laboratory or inspection agency concerned with product evaluation

(3) Evidence acceptable to the authority having jurisdiction such as a manufacturer's self-evaluation or an owner's engineering judgment

FPN: Additional documentation for equipment may include certificates demonstrating compliance with applicable equipment standards, indicating special conditions of use, and other pertinent information.

506.9 Equipment Requirements.

(A) Suitability. Suitability of identified equipment shall be determined by one of the following:

(1) Equipment listing or labeling

(2) Evidence of equipment evaluation from a qualified testing laboratory or inspection agency concerned with product evaluation

(3) Evidence acceptable to the authority having jurisdiction such as a manufacturer's self-evaluation or an owner's engineering judgment

FPN: Additional documentation for equipment may include certificates demonstrating compliance with applicable equipment standards, indicating special conditions of use, and other pertinent information.

Author's Comment: The text on suitability has been relocated from the final paragraph in former **500.8(A)** to a new subsection **500.8(A)** to help emphasize the importance of the suitability of equipment of identified equipment. A new fine print note has been added to **505.9(A)** and **506.9(A)** to acknowledge the current practice of providing certificates as part of the required documentation.

OSHA RULES PER 1910.302 THROUGH 1910.308. ALSO, SEE OSHA 1910.303(a), 1910.303(b)(1)(i) THROUGH (b)(vii) AND (b)(2).

NEC RULES PER **110.3(A)(1) THROUGH 110.3(A)(8), 110.3(B),** AND **110.2**

DESIGN BY:
• NEC • UL 508 • ANSI C SERIES • NFPA 79

NEC LOOP
• **90.7** • **110.3(A); (B)** • **110.2** • **500.8(A), FPN**

OSHA LOOP
• OSHA 1910.7 • OSHA 1910.303(a) • OSHA 1910.399

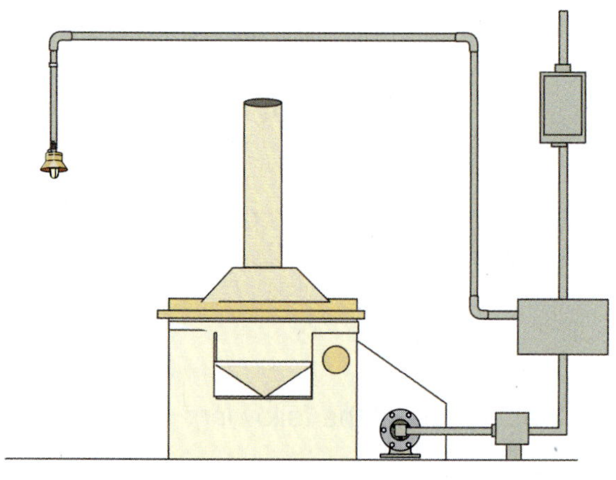

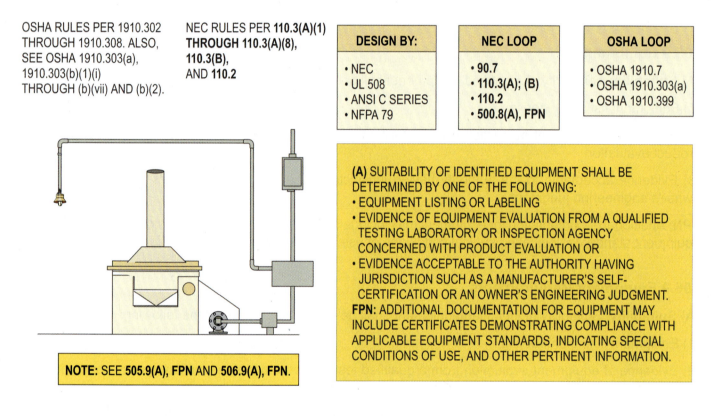

(A) SUITABILITY OF IDENTIFIED EQUIPMENT SHALL BE DETERMINED BY ONE OF THE FOLLOWING:
- EQUIPMENT LISTING OR LABELING
- EVIDENCE OF EQUIPMENT EVALUATION FROM A QUALIFIED TESTING LABORATORY OR INSPECTION AGENCY CONCERNED WITH PRODUCT EVALUATION OR
- EVIDENCE ACCEPTABLE TO THE AUTHORITY HAVING JURISDICTION SUCH AS A MANUFACTURER'S SELF-CERTIFICATION OR AN OWNER'S ENGINEERING JUDGMENT.

FPN: ADDITIONAL DOCUMENTATION FOR EQUIPMENT MAY INCLUDE CERTIFICATES DEMONSTRATING COMPLIANCE WITH APPLICABLE EQUIPMENT STANDARDS, INDICATING SPECIAL CONDITIONS OF USE, AND OTHER PERTINENT INFORMATION.

NOTE: SEE **505.9(A), FPN** AND **506.9(A), FPN**.

EQUIPMENT
500.8(A), FPN
505.9(A), FPN
506.9(A), FPN

Purpose of Change: To provide provisions for approving and accepting electrical equipment in hazardous (classified) locations.

NEC Ch. 5 - Article 501
Part II - 501.10(B)(1)(7)

Type of Change		Panel Action		UL	UL 508	API 500 - 1997	API 505 - 1997	OSHA - 1994
New Subdivision		Accept in Principle		-	-	-	-	1910.307(b)(2), Note
ROP		ROC		NFPA 70E - 2004	NFPA 70B - 2006	NFPA 79 - 2007	NFPA	NEMA
pg. 515	# 14-33a	pg. 305	# 14-8	440.4(A), FPN	23.2.10	-	-	-
log: CP 1402	CMP: 14	log: 548	Submitter: Donald W. Ankele			2005 NEC: -		IEC: -

2005 NEC - 501.10 Wiring Methods.

Wiring methods shall comply with 501.10(A) or 501.10(B).

(B) Class I, Division 2.

(1) General. In Class I, Division 2 locations, the following wiring methods shall be permitted:

(1) All wiring methods permitted in Article 501.10(A).

(2) Threaded rigid metal conduit, threaded steel intermediate metal conduit.

(3) Enclosed gasketed busways, enclosed gasketed wireways.

(4) Type PLTC cable in accordance with the provisions of Article 725, or in cable tray systems. PLTC shall be installed in a manner to avoid tensile stress at the termination fittings.

(5) Type ITC cable as permitted in 727.4.

(6) Type MI, MC, MV, or TC cable with termination fittings, or in cable tray systems and installed in a manner to avoid tensile stress at the termination fittings. Single conductor Type MV cables shall be shielded or metallic armored.

2008 NEC - 501.10 Wiring Methods.

Wiring methods shall comply with 501.10(A) or 501.10(B).

(B) Class I, Division 2.

(1) General. In Class I, Division 2 locations, the following wiring methods shall be permitted:

(1) All wiring methods permitted in Article 501.10(A).

(2) Threaded rigid metal conduit, threaded steel intermediate metal conduit.

(3) Enclosed gasketed busways, enclosed gasketed wireways.

(4) Type PLTC cable in accordance with the provisions of Article 725, or in cable tray systems. PLTC shall be installed in a manner to avoid tensile stress at the termination fittings.

(5) Type ITC cable as permitted in 727.4.

(6) Type MI, MC, MV, or TC cable with termination fittings, or in cable tray systems and installed in a manner to avoid tensile stress at the termination fittings. Single conductor Type MV cables shall be shielded or metallic armored.

<u>(7) In industrial establishments with restricted public access where the conditions of maintenance and supervision ensure that only qualified persons service the installation and where metallic conduit does not provide sufficient corrosion resistance, reinforced thermosetting resin conduit (RTRC), factory elbows, and associated fittings, all marked with suffix - XW, and Schedule 80 PVC Conduit, factory elbows, and associated fittings, shall be permitted.</u>

Where seals are required for boundary conditions as defined in 501.15(A)(4), the Division 1 wiring method shall extend into the Division 2 area to the seal which shall be located on the Division 2 side of the Division 1 - Division 2 boundary.

Author's Comment: This new addition recognizes the corrosion that can occur to metal conduit in some locations and provides permission to use RTRC and Schedule 80 PVC in an industrial establishment with restricted access where qualified personnel will service the installation, in these severe corrosive conditions. The comment added an "XW" suffix to the RTRC conduit, elbows, and fittings requirements to make it equivalent to the Schedule 80 PVC requirement.

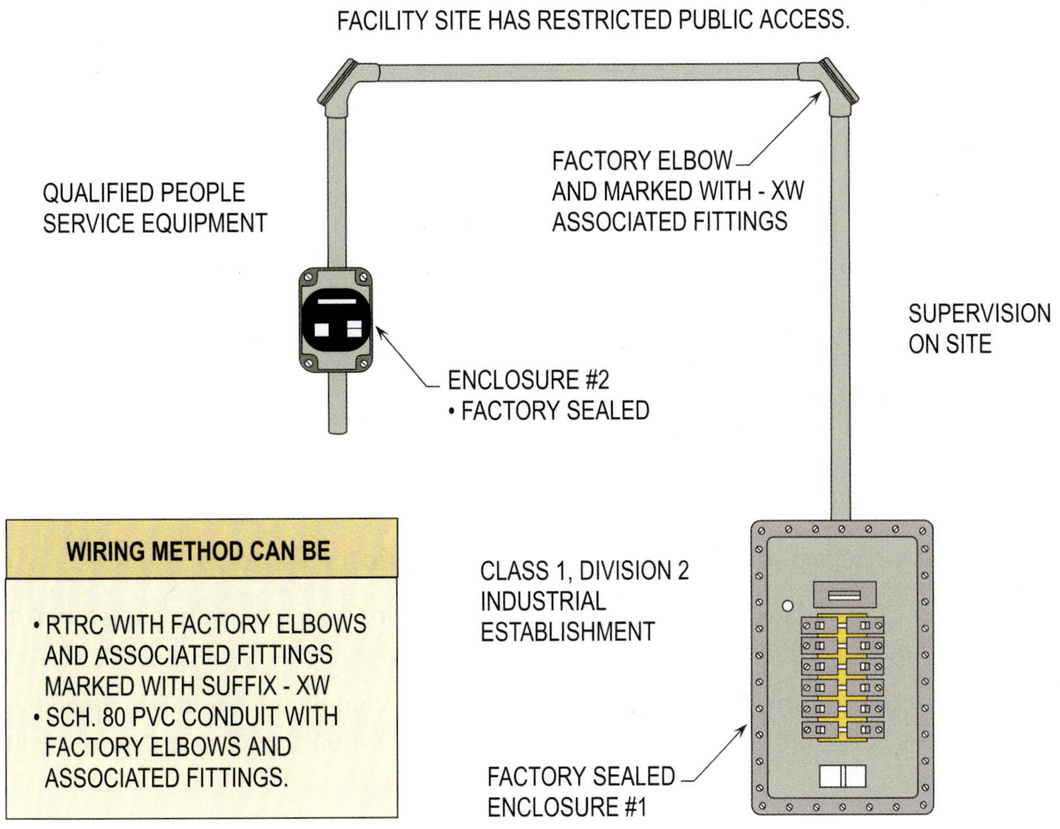

CLASS I, DIVISION 2
501.10(B)(1)(7)

Purpose of Change: To permit reinforced thermosetting resin conduit (RTRC) and Schedule 80 PVC in corrosion areas where RMC does not provide adequate corrosion resistance.

NEC Ch. 5 - Article 501
Part II - 501.30(B)

Type of Change	Panel Action	UL	UL 508	API 500 - 1997	API 505 - 1997	OSHA - 1994
Revision	Accept in Principle	-	-	-	-	1910.307(b)(2), Note
ROP	**ROC**	**NFPA 70E - 2004**	**NFPA 70B - 2006**	**NFPA 79 - 2007**	**NFPA**	**NEMA**
pg. 517 # 14-44	pg. - # -	440.4(G)	23.2.10	-	-	-
log: 1504 CMP: 14	log: -	Submitter: Daniel Leaf		2005 NEC: 501.30(B)		IEC: -

2005 NEC - 501.30 Grounding and Bonding, Class I, Divisions 1 and 2.

Wiring and equipment in Class I, Division 1 and 2 locations shall be grounded as specified in Article 250 and with the requirements in 501.30(A) and 501.30(B).

(B) Types of Equipment Grounding Conductors. ~~Where~~ flexible metal conduit ~~or~~ liquidtight flexible metal conduit ~~is used as permitted in 501.10(B) and is to be relied on to complete a sole equipment grounding path, it shall be installed with internal or external bonding jumpers in parallel with each conduit and complying with 250.102~~.

Exception: In Class I, Division 2 locations, the bonding jumper shall be permitted to be deleted where all of the following conditions are met:

(1) Listed liquidtight flexible metal conduit 1.8 m (6 ft) or less in length, with fittings listed for grounding, is used.

(2) Overcurrent protection in the circuit is limited to 10 amperes or less.

(3) The load is not a power utilization load.

2008 NEC - 501.30 Grounding and Bonding, Class I, Divisions 1 and 2.

Wiring and equipment in Class I, Division 1 and 2 locations shall be grounded as specified in Article 250 and with the requirements in 501.30(A) and 501.30(B).

(B) Types of Equipment Grounding Conductors. Flexible metal conduit and liquidtight flexible metal conduit shall not be used as the sole ground-fault current path. Where equipment bonding jumpers are installed, they shall comply with 250.102.

Exception: In Class I, Division 2 locations, the bonding jumper shall be permitted to be deleted where all of the following conditions are met:

(1) Listed liquidtight flexible metal conduit 1.8 m (6 ft) or less in length, with fittings listed for grounding, is used.

(2) Overcurrent protection in the circuit is limited to 10 amperes or less.

(3) The load is not a power utilization load.

Author's Comment: Flexible metal conduit and liquidtight flexible metal conduit cannot be used in Class I, Division 2 location as the sole ground-fault current path and the panel has revised the current text for clarity.

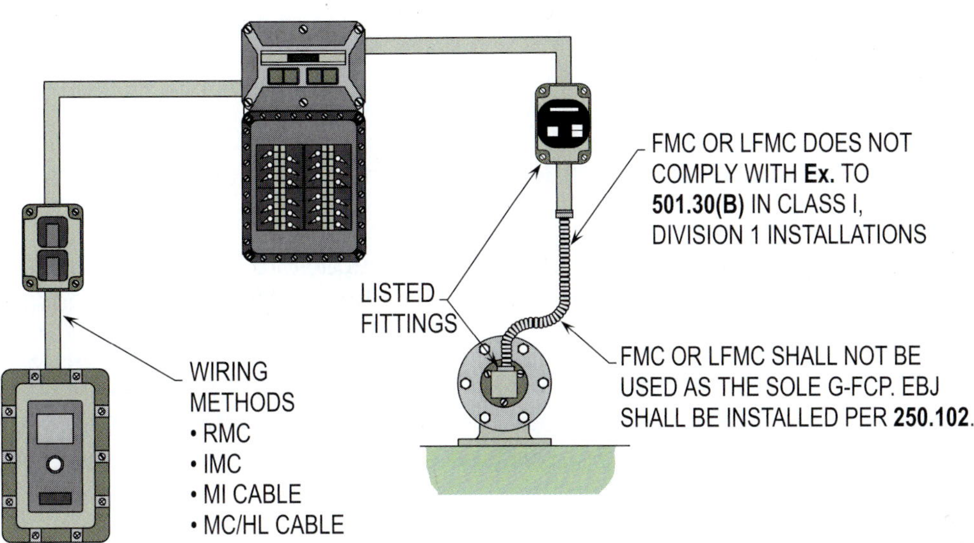
CLASS I, DIVISION 1 AND 2
FMC OR LFMC DOES NOT COMPLY WITH **Ex.** TO **501.30(B)** IN CLASS I, DIVISION 1 INSTALLATIONS
LISTED FITTINGS
FMC OR LFMC SHALL NOT BE USED AS THE SOLE G-FCP. EBJ SHALL BE INSTALLED PER **250.102**.
WIRING METHODS
• RMC
• IMC
• MI CABLE
• MC/HL CABLE

TYPES OF EQUIPMENT GROUNDING CONDUCTORS
501.30(B)

Purpose of Change: To clarify the use of flexible metal conduit (FMC) and liquidtight flexible metal conduit (LFMC) as an equipment grounding conductor in classified locations.

NEC Ch. 5 - Article 501
Part III - 501.100(A)(2)

Type of Change	Panel Action	UL	UL 508	API 500 - 1997	API 505 - 1997	OSHA - 1994
Revision	Accept	-	-	-	-	1910.307(b)(2), Note
ROP	**ROC**	**NFPA 70E - 2004**	**NFPA 70B - 2006**	**NFPA 79 - 2007**	**NFPA**	**NEMA**
pg. 518 # 14-46	pg. - # -	440.4(A), FPN	23.2.10	-	-	-
log: 550 CMP: 14	log: -	Submitter: Michael J. Johnston		2005 NEC: 501.100(A)(2)		IEC: -

2005 NEC - 501.100 Transformers and Capacitors.

(A) Class I, Division 1. In Class I, Division 1 locations, transformers and capacitors shall comply with 501.100(A)(1) and (A)(2).

(2) Not Containing Liquid That Will Burn. Transformers and capacitors that do not contain a liquid that will burn shall be installed in vaults complying with 501.100(A)(1) or be ~~approved~~ for Class I locations.

2008 NEC - 501.100 Transformers and Capacitors.

(A) Class I, Division 1. In Class I, Division 1 locations, transformers and capacitors shall comply with 501.100(A)(1) and (A)(2).

(2) Not Containing Liquid That Will Burn. Transformers and capacitors that do not contain a liquid that will burn shall be installed in vaults complying with 501.100(A)(1) or be identified for Class I locations.

Author's Comment: "Approved" is acceptable to the authority having jurisdiction and "identified" is marking the transformers and capacitors to be "recognizable as suitable for the specific purpose, function, use, environment, application, and so forth, where described in a particular Code requirement."

NOTE 1: XFMRS AND CAPACITORS SHALL BE "IDENTIFIED" WHICH MEANS ARE "RECOGNIZABLE" AS SUITABLE FOR THE SPECIFIC PURPOSE, ETC. PER **501.100(A)(2)**.

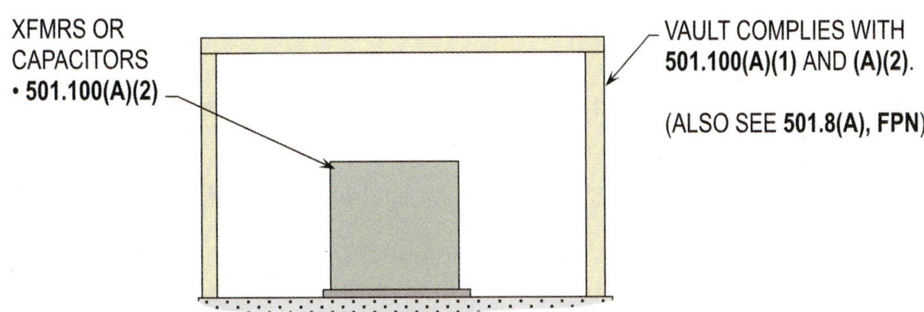

NEC LOOP	
LOCATION	**450.41**
WALLS, ROOFS AND FLOORS	**450.42**
DOORWAYS	**450.43**
VENT OPENINGS	**450.45**

NEC LOOP	
DRAINAGE	**450.46**
FOREIGN SYSTEMS	**450.47**
STORAGE IN VAULTS	**450.48**

NOTE 2: BEFORE INSTALLING XFMR'S OR CAPACITORS IN VAULTS LOCATED IN CLASS I, DIVISION 1 LOCATIONS, SEE ADDITIONAL RULES IN **501.100(A)(1)** AND **(A)(2)**.

TRANSFORMERS AND CAPACITORS
501.100(A)(2)

Purpose of Change: This revision clarifies that transformers and capacitors shall be marked as identified.

NEC Ch. 5 - Article 502
Part II - 502.30(B)

Type of Change	Panel Action	UL	UL 508	API 500 - 1997	API 505 - 1997	OSHA - 1994
Revision	Accept	-	-	-	-	1910.307(b)(2), Note
ROP	**ROC**	**NFPA 70E - 2004**	**NFPA 70B - 2006**	**NFPA 79 - 2007**	**NFPA**	**NEMA**
pg. 521 # 14-52a	pg. - # -	440.4(G)	23.2.10	-	-	-
log: CP 1404 CMP: 14	log: -	Submitter: Code Making Panel 14		2005 NEC: 502.30(B)		IEC: -

2005 NEC - 502.30 Grounding and Bonding, Class II, Divisions 1 and 2.

Wiring and equipment in Class II, Division 1 and 2 locations shall be grounded as specified in Article 250 and with the requirements in 502.30(A) and 502.30(B).

(B) Types of Equipment Grounding Conductors. ~~Where flexible conduit is used as permitted in 502.10, it shall be installed with internal or external bonding jumpers in parallel with each conduit and complying with 250.102.~~

Exception: In Class II, Division 2 locations, the bonding jumper shall be permitted to be deleted where all of the following conditions are met:

(1) Listed liquidtight flexible metal conduit 1.8 m (6 ft) or less in length, with fittings listed for grounding, is used.

(2) Overcurrent protection in the circuit is limited to 10 amperes or less.

(3) The load is not a power utilization load.

2008 NEC - 502.30 Grounding and Bonding, Class II, Divisions 1 and 2.

Wiring and equipment in Class II, Division 1 and 2 locations shall be grounded as specified in Article 250 and with the requirements in 502.30(A) and 502.30(B).

(B) Types of Equipment Grounding Conductors. Liquidtight flexible metal conduit shall not be used as the sole ground-fault current path. Where equipment bonding jumpers are installed, they shall comply with 250.102.

Exception: In Class II, Division 2 locations, the bonding jumper shall be permitted to be deleted where all of the following conditions are met:

(1) Listed liquidtight flexible metal conduit 1.8 m (6 ft) or less in length, with fittings listed for grounding, is used.

(2) Overcurrent protection in the circuit is limited to 10 amperes or less.

(3) The load is not a power utilization load.

Author's Comment: Liquidtight flexible metal conduit cannot be used in Class II, Division 2 location as the sole ground-fault current path and the panel has revised the current text for clarity.

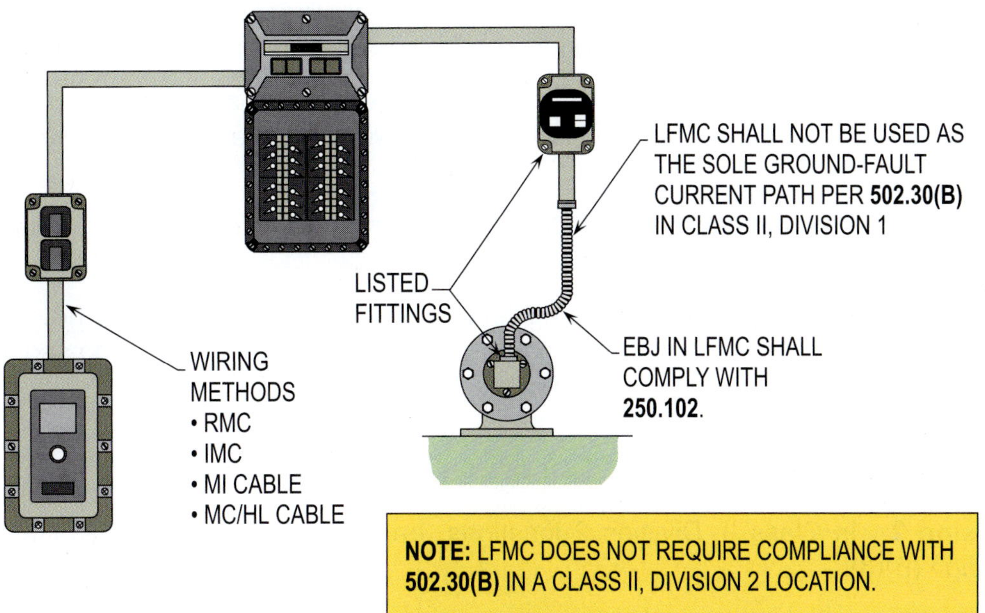
LFMC SHALL NOT BE USED AS THE SOLE GROUND-FAULT CURRENT PATH PER 502.30(B) IN CLASS II, DIVISION 1
LISTED FITTINGS
EBJ IN LFMC SHALL COMPLY WITH 250.102.
WIRING METHODS
• RMC
• IMC
• MI CABLE
• MC/HL CABLE
NOTE: LFMC DOES NOT REQUIRE COMPLIANCE WITH 502.30(B) IN A CLASS II, DIVISION 2 LOCATION.

TYPES OF EQUIPMENT GROUNDING CONDUCTORS
502.30(B)

Purpose of Change: To clarify the use of liquidtight flexible metal conduit (LFMC) as an equipment grounding conductor in Class II, Division 1 and 2 locations.

NEC Ch. 5 - Article 502
Part III - 502.120(B)(2)

Type of Change	Panel Action	UL	UL 508	API 500 - 1997	API 505 - 1997	OSHA - 1994
Revision	Accept	-	-	-	-	1910.307(b)(2), Note
ROP	**ROC**	**NFPA 70E - 2004**	**NFPA 70B - 2006**	**NFPA 79 - 2007**	**NFPA**	**NEMA**
pg. 522 # 14-64	pg. 310 # 14-28	440.4(A), FPN	23.2.10	-	-	-
log: 2500 CMP: 14	log: 1872	Submitter: Edward M. Briesch		2005 NEC: 502.120(B)(2)		IEC: -

2005 NEC - 502.120 Control Transformers and Resistors.

(B) Class II, Division 2. In Class II, Division 2 locations, transformers and resistors shall comply with 502.120(B)(1) through (B)(3).

(1) Switching Mechanisms. Switching mechanisms (including overcurrent devices) associated with control transformers, solenoids, impedance coils, and resistors shall be provided with dusttight enclosures.

(2) Coils and Windings. Where not located in the same enclosure with switching mechanisms, control transformers, solenoids, and impedance coils shall be provided with tight metal housings without ventilating openings.

(3) Resistors. Resistors and resistance devices shall have dust-ignitionproof enclosures identified for Class II locations.

Exception: Where the maximum normal operating temperature of the resistor will not exceed 120°C (248°F),nonadjustable resistors or resistors that are part of an automatically timed starting sequence shall be permitted to have enclosures complying with 502.120(B)(2).

2008 NEC - 502.120 Control Transformers and Resistors.

(B) Class II, Division 2. In Class II, Division 2 locations, transformers and resistors shall comply with 502.120(B)(1) through (B)(3).

(1) Switching Mechanisms. Switching mechanisms (including overcurrent devices) associated with control transformers, solenoids, impedance coils, and resistors shall be provided with dusttight enclosures.

(2) Coils and Windings. Where not located in the same enclosure with switching mechanisms, control transformers, solenoids, and impedance coils shall be provided with tight metal housings without ventilating openings or shall be installed in dusttight enclosures. Effective January 1, 2011, only dusttight enclosures shall be permitted.

(3) Resistors. Resistors and resistance devices shall have dust-ignitionproof enclosures identified for Class II locations.

Exception: Where the maximum normal operating temperature of the resistor will not exceed 120°C (248°F),nonadjustable resistors or resistors that are part of an automatically timed starting sequence shall be permitted to have enclosures complying with 502.120(B)(2).

Author's Comment: Since **502.10(B)(4)** requires all boxes and fittings in a Class II, Division 2 location to be dust-tight, text was added to require the same dust-tight enclosure for coils and windings, but this will only become effective on January 1, 2011 based on text added at the Comment stage. This deadline permits the manufacturers to transition their equipment to dusttight.

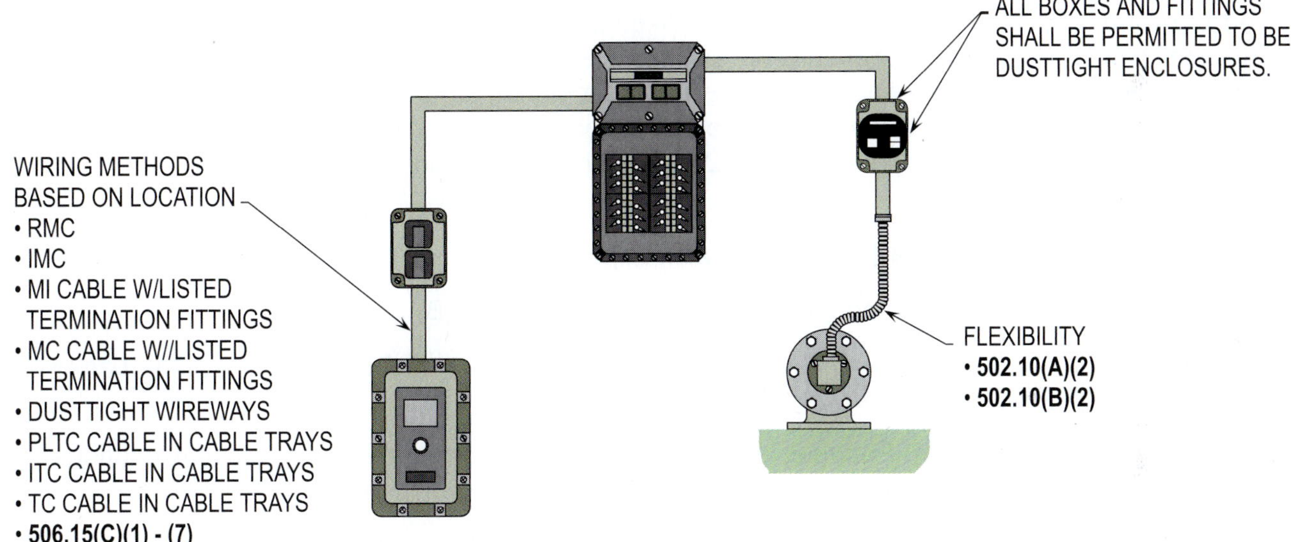

COILS AND WINDINGS
502.120(B)(2)

Purpose of Change: To permit dusttight enclosures to be used under certain conditions of use.

NEC Ch. 5 - Article 502
Part III - 502.130(B)(2)

Type of Change		Panel Action		UL	UL 508	API 500 - 1997	API 505 - 1997	OSHA - 1994
Revision		Accept		-	-	-	-	1910.307(b)(2), Note
ROP		**ROC**		**NFPA 70E - 2004**	**NFPA 70B - 2006**	**NFPA 79 - 2007**	**NFPA**	**NEMA**
pg. 523	# 14-69	pg. -	# -	440.4(A), FPN	23.2.10	-	-	-
log: 2498	CMP: 14	log: -	Submitter: Edward M. Briesch			2005 NEC: 502.130(B)(2)		IEC: -

2005 NEC - 502.130 Luminaires (Lighting Fixtures).

Luminaires (lighting fixtures) shall comply with 502.130(A) and 502.130(B).

(B) Class II, Division 2. In Class II, Division 2 locations, luminaires (lighting fixtures) shall comply with 502.130(B)(1) through (B)(5).

(1) Portable Lighting Equipment. Portable lighting equipment shall be identified for Class II locations. They shall be clearly marked to indicate the maximum wattage of lamps for which they are designed.

(2) Fixed Lighting. Luminaires (lighting fixtures) for fixed lighting, where not of a type identified for Class II locations, shall provide enclosures ~~for lamps and lampholders that shall be designed to minimize the deposit of dust on lamps and to prevent the escape of sparks, burning material, or hot metal.~~ Each fixture shall be clearly marked to indicate the maximum wattage of the lamp that shall be permitted without exceeding an exposed surface temperature in accordance with 500.8~~(C)~~(2) under normal conditions of use.

2008 NEC - 502.130 Luminaires (Lighting Fixtures).

Luminaires (lighting fixtures) shall comply with 502.130(A) and 502.130(B).

(B) Class II, Division 2. In Class II, Division 2 locations, luminaires (lighting fixtures) shall comply with 502.130(B)(1) through (B)(5).

(1) Portable Lighting Equipment. Portable lighting equipment shall be identified for Class II locations. They shall be clearly marked to indicate the maximum wattage of lamps for which they are designed.

(2) Fixed Lighting. Luminaires (lighting fixtures) for fixed lighting, where not of a type identified for Class II locations, shall be provided with dusttight enclosures. Each fixture shall be clearly marked to indicate the maximum wattage of the lamp that shall be permitted without exceeding an exposed surface temperature in accordance with 500.8(D)(2) under normal conditions of use.

Author's Comment: Section **502.10(B)(4)** requires all boxes and fittings in a Class II, Division 2 location to be dusttight so text was added to require luminaires to be dusttight rather than to merely minimize dust entrance.

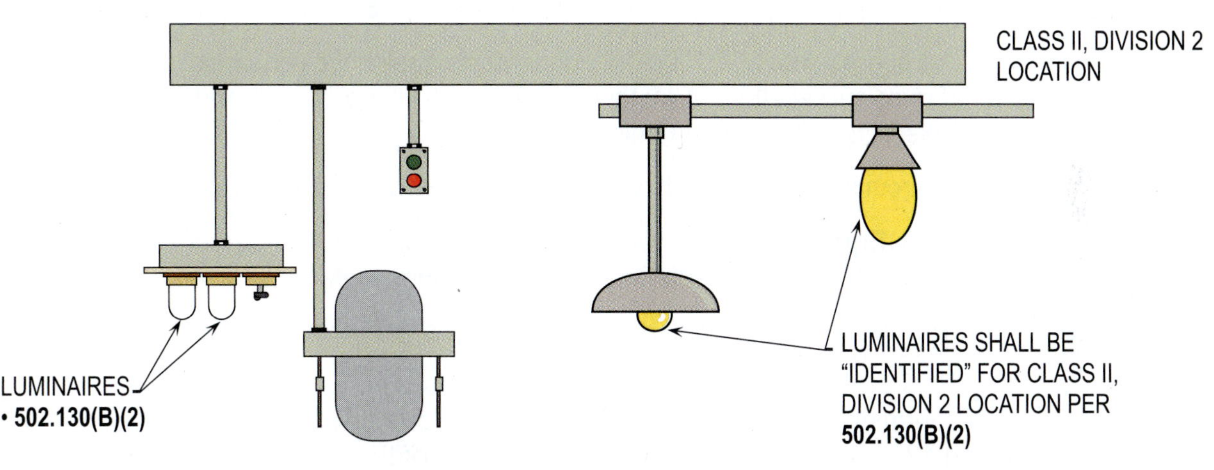
FOR REQUIREMENTS FOR BOXES AND FITTINGS, SEE **502.10(B)(4)**.
CLASS II, DIVISION 2 LOCATION
LUMINAIRES
• **502.130(B)(2)**
LUMINAIRES SHALL BE "IDENTIFIED" FOR CLASS II, DIVISION 2 LOCATION PER **502.130(B)(2)**
NOTE: ALL FIXED LUMINAIRES ARE REQUIRED TO BE DUSTTIGHT PER **502.130(B)(2)**.

FIXED LIGHTING
502.130(B)(2)

Purpose of Change: This revision clarifies the requirements for the use of dusttight enclosures in Class II, Division 2 location including fixed luminaires.

NEC Ch. 5 - Article 502
Part III - 502.150(B)(1)

Type of Change	Panel Action	UL	UL 508	API 500 - 1997	API 505 - 1997	OSHA - 1994
Revision	Accept	-	-	-	-	1910.307(b)(2), Note
ROP	**ROC**	**NFPA 70E - 2004**	**NFPA 70B - 2006**	**NFPA 79 - 2007**	**NFPA**	**NEMA**
pg. 524 # 14-73	pg. 311 # 14-33a	440.4(A), FPN	23.2.10	-	-	-
log: 2503 CMP: 14	log: CC 1400	Submitter: Edward M. Briesch		2005 NEC: 502.150(B)		IEC: -

2005 NEC - 502.150 Signaling, Alarm, Remote-Control, and Communications Systems; and Meters, Instruments, and Relays.

FPN: See Article 800 for rules governing the installation of communications circuits.

(B) Class II, Division 2. In Class II, Division 2 locations, signaling, alarm, remote-control, and communications systems; and meters, instruments, and relays shall comply with 502.150(B)(1) through (B)(5).

(1) Contacts. Enclosures shall comply with 502.150(A)(2), or contacts shall have tight metal enclosures designed to minimize the entrance of dust and shall have telescoping or tight- fitting covers and no openings through which, after installation, sparks or burning material might escape.

Exception: In nonincendive circuits, enclosures shall be permitted to be of the general-purpose type.

(2) Transformers and Similar Equipment. The windings and terminal connections of transformers, choke coils, and similar equipment shall ~~be provided with tight metal enclosures without ventilating openings~~.

(3) Resistors and Similar Equipment. Resistors, resistance devices, thermionic tubes, rectifiers, and similar equipment shall comply with ~~502.130(A)(3)~~.

Exception: Enclosures for thermionic tubes, nonadjustable resistors, or rectifiers for which maximum operating temperature will not exceed 120∞C (248∞F) shall be permitted to be of the general-purpose type.

(4) Rotating Machinery. Motors, generators, and other rotating electric machinery shall comply with 502.125(B).

~~**(5) Wiring Methods.** The wiring method shall comply with 502.10(B).~~

2008 NEC - 502.150 Signaling, Alarm, Remote-Control, and Communications Systems; and Meters, Instruments, and Relays.

FPN: See Article 800 for rules governing the installation of communications circuits.

(B) Class II, Division 2. In Class II, Division 2 locations, signaling, alarm, remote-control, and communications systems; and meters, instruments, and relays shall comply with 502.150(B)(1) through (B)(5).

(1) Contacts. Enclosures shall comply with 502.150(A)(2), or contacts shall have tight metal enclosures designed to minimize the entrance of dust and shall have telescoping or tight- fitting covers and no openings through which, after installation, sparks or burning material might escape or shall be installed in dusttight enclosures. Effective January 1, 2011, only dusttight enclosures shall be permitted.

Exception: In nonincendive circuits, enclosures shall be permitted to be of the general-purpose type.

(2) Transformers and Similar Equipment. The windings and terminal connections of transformers, choke coils, and similar equipment shall comply with 502.120(B)(2).

(3) Resistors and Similar Equipment. Resistors, resistance devices, thermionic tubes, rectifiers, and similar equipment shall comply with 502.120(B)(3).

(4) Rotating Machinery. Motors, generators, and other rotating electric machinery shall comply with 502.125(B).

Author's Comment: Section **502.10(B)(4)** requires all boxes and fittings in a Class II, Division 2 location to be dusttight so equipment **502.150** this section is also required to be dust-tight. The reference to **502.130(A)(3)** in **502.150(B)(3)** was changed to the proper reference of **502.120(B)(3)**. The comment correlates **502.150(B)** with previous actions on Comment 14-28 and removes some redundancy in the code.

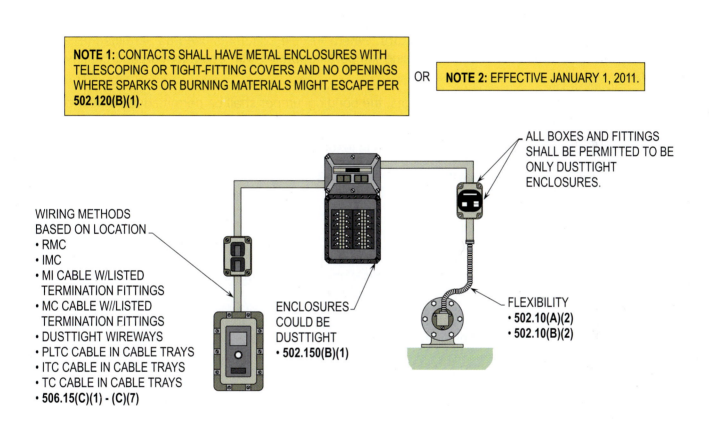

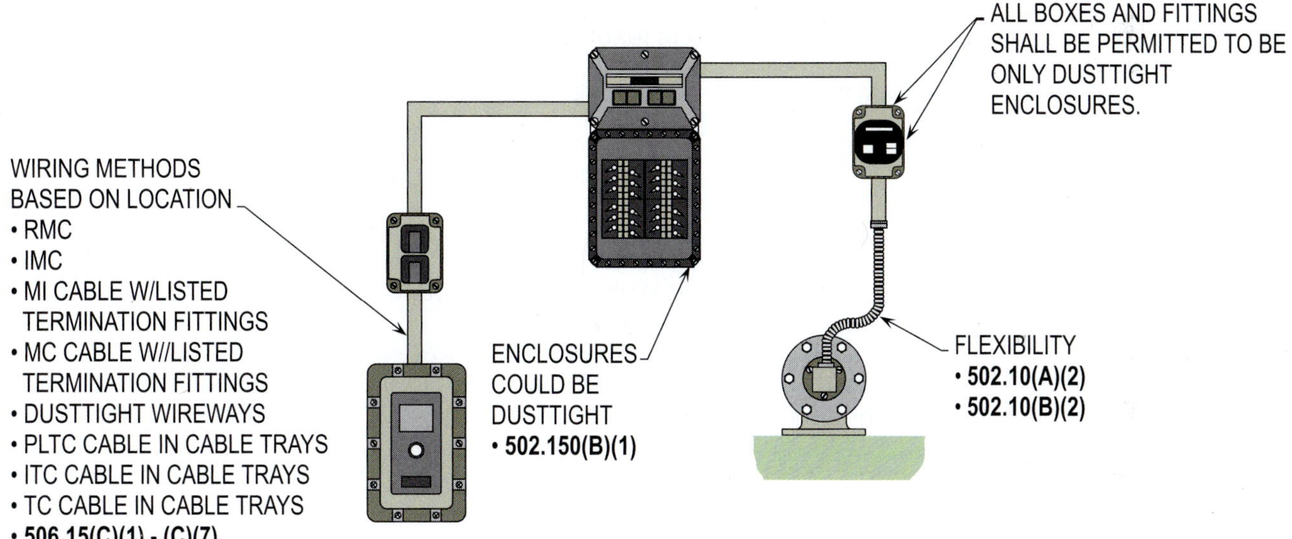

CLASS II, DIVISION 2
502.150(B)(1)

Purpose of Change: To permit dusttight enclosures to be used under certain conditions of use and will require only dusttight enclosures after January 1, 2011.

NEC Ch. 5 - Article 503
Part II - 503.30(B)

Type of Change	Panel Action	UL	UL 508	API 500 - 1997	API 505 - 1997	OSHA - 1994
Revision	Accept	-	-	-	-	1910.307(b)(2), Note
ROP	**ROC**	**NFPA 70E - 2004**	**NFPA 70B - 2006**	**NFPA 79 - 2007**	**NFPA**	**NEMA**
pg. 526 # 14-81a	pg. - # -	440.4(G)	23.2.10	-	-	-
log: CP 1405 CMP: 14	log: - Submitter: Code Making Panel 14			2005 NEC: 503.30(B)		IEC: -

2005 NEC - 503.30 Grounding and Bonding — Class III, Divisions 1 and 2.

Wiring and equipment in Class III, Division 1 and 2 locations shall be grounded as specified in Article 250 and with the following additional requirements in 503.30(A) and 503.30(B).

(B) Types of Equipment Grounding Conductors. ~~Where flexible conduit is used as permitted in 503.10, it shall be installed with internal or external bonding jumpers in parallel with each conduit and complying with 250.102.~~

Exception: In Class III, Division 1 and 2 locations, the bonding jumper shall be permitted to be deleted where all of the following conditions are met:

(1) Listed liquidtight flexible metal 1.8 m (6 ft) or less in length, with fittings listed for grounding, is used.

(2) Overcurrent protection in the circuit is limited to 10 amperes or less.

(3) The load is not a power utilization load.

2008 NEC - 503.30 Grounding and Bonding — Class III, Divisions 1 and 2.

Wiring and equipment in Class III, Division 1 and 2 locations shall be grounded as specified in Article 250 and with the following additional requirements in 503.30(A) and 503.30(B).

(B) Types of Equipment Grounding Conductors. Liquidtight flexible metal conduit shall not be used as the sole ground-fault current path. Where equipment bonding jumpers are installed, they shall comply with 250.102.

Exception: In Class III, Division 1 and 2 locations, the bonding jumper shall be permitted to be deleted where all of the following conditions are met:

(1) Listed liquidtight flexible metal 1.8 m (6 ft) or less in length, with fittings listed for grounding, is used.

(2) Overcurrent protection in the circuit is limited to 10 amperes or less.

(3) The load is not a power utilization load.

Author's Comment: Liquidtight flexible metal conduit cannot be used in Class III, Division 2 location as the sole ground-fault current path and the panel has revised the current text for clarity.

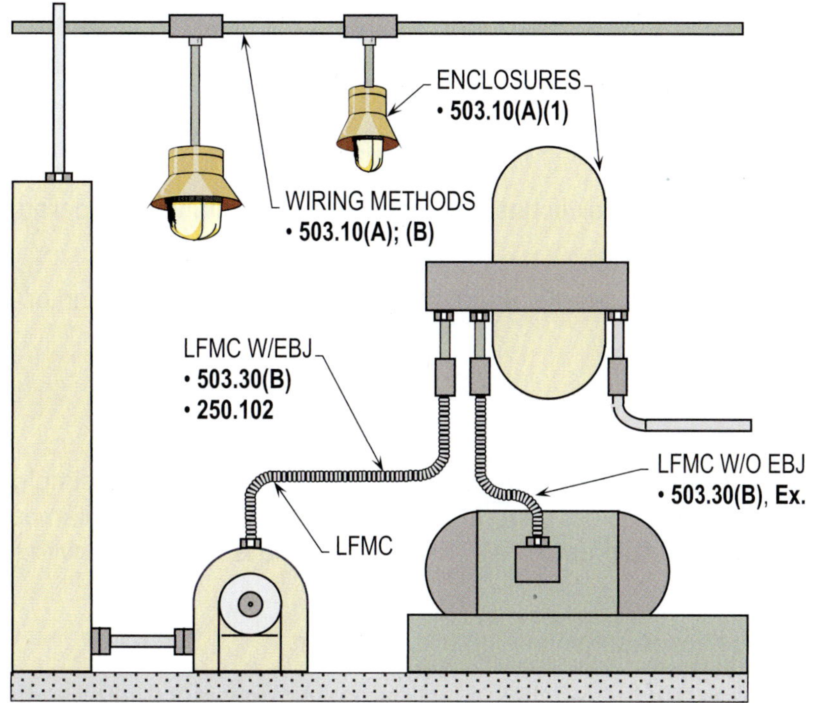

CLASS III, DIVISION 1 AND 2

TYPES OF EQUIPMENT GROUNDING CONDUCTORS
503.30(B)

Purpose of Change: To clarify the use of liquidtight flexible metal conduit (LFMC) in Class III, Division 1 and 2 locations.

NEC Ch. 5 - Article 504
Part - 504.2

73

Type of Change	Panel Action	UL	UL 508	API 500 - 1997	API 505 - 1997	OSHA - 1994
Revision	Accept	-	-	3.2.32	3.2.34	1910.307(b)(1)
ROP	**ROC**	**NFPA 70E - 2004**	**NFPA 70B - 2006**	**NFPA 79 - 2007**	**NFPA 497**	**NEMA**
pg. 527 # 14-86	pg. 313 # 14-39	440.2	23.2.10	-	4.1.5.1	-
log: 2457 CMP: 14	log: 127	Submitter: Eliana Beattie		2005 NEC: 504.2		IEC: -

2005 NEC - 504.2

Simple Apparatus. An electrical component or combination of components of simple construction with well-defined electrical parameters that does not generate more than 1.5 volts, 100 milliamps, and 25 milliwatts, or a passive component that does not dissipate more than 1.3 watts and is compatible with the intrinsic safety of the circuit in which it is used.

FPN: The following apparatus are examples of simple apparatus:

(a) Passive components, for example, switches, junction boxes, resistance temperature devices, and simple semiconductor devices such as LEDs

~~(b)~~ Sources of generated energy, for example, thermocouples and photocells, which do not generate more than 1.5 V, 100 mA, and 25 mW.

2008 NEC - 504.2

Simple Apparatus. An electrical component or combination of components of simple construction with well-defined electrical parameters that does not generate more than 1.5 volts, 100 milliamps, and 25 milliwatts, or a passive component that does not dissipate more than 1.3 watts and is compatible with the intrinsic safety of the circuit in which it is used.

FPN: The following apparatus are examples of simple apparatus:

(a) Passive components, for example, switches, junction boxes, resistance temperature devices, and simple semiconductor devices such as LEDs

(b) Sources of stored energy consisting of single components in simple circuits with well-defined parameters, for example, capacitors or inductors, whose values are considered when determining the overall safety of the system;

(c) Sources of generated energy, for example, thermocouples and photocells, which do not generate more than 1.5 V, 100 mA, and 25 mW.

Author's Comment: Single component stored energy sources with well defined parameters, such as small capacitors and inductors, can be simple apparatus. The Comment deleted the word "be" from the FPN as a clerical correction.

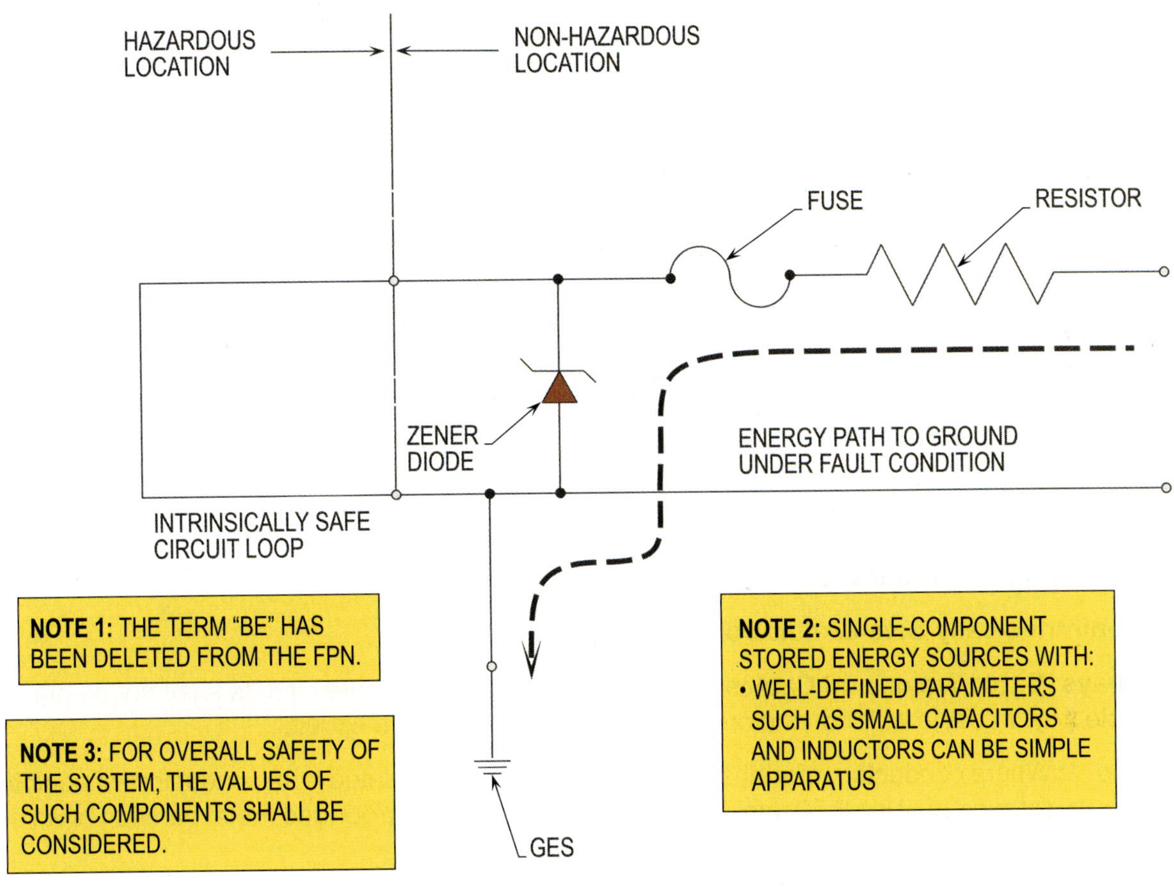
HAZARDOUS LOCATION
NON-HAZARDOUS LOCATION
FUSE
RESISTOR
ZENER DIODE
ENERGY PATH TO GROUND UNDER FAULT CONDITION
INTRINSICALLY SAFE CIRCUIT LOOP
NOTE 1: THE TERM "BE" HAS BEEN DELETED FROM THE FPN.
NOTE 3: FOR OVERALL SAFETY OF THE SYSTEM, THE VALUES OF SUCH COMPONENTS SHALL BE CONSIDERED.
GES
NOTE 2: SINGLE-COMPONENT STORED ENERGY SOURCES WITH:
• WELL-DEFINED PARAMETERS SUCH AS SMALL CAPACITORS AND INDUCTORS CAN BE SIMPLE APPARATUS

SIMPLE APPARATUS
504.2

Purpose of Change: To include single components in simple circuits well-defined parameters.

NEC Ch. 5 - Article 504
Part - 504.30(A)(1), Exceptions No. 3 and 4 and FPN

Type of Change	Panel Action	UL	UL 508	API 500 - 1997	API 505 - 1997	OSHA - 1994
New Ex.'s and FPN	Accept	-	-	3.2.32	3.2.34	1910.307(b)(1)
ROP	**ROC**	**NFPA 70E - 2004**	**NFPA 70B - 2006**	**NFPA 79 - 2007**	**NFPA 497**	**NEMA**
pg. 527 # 14-88a	pg. 313 # 14-41	440.2	23.2.10	-	4.1.5.1	-
log: CP 1407 CMP: 14	log: 635	Submitter: Code Making Panel 14		2005 NEC: 504.30(A)(1), Exception No. 3		IEC: -

2005 NEC - 504.30 Separation of Intrinsically Safe Conductors.

(A) From Nonintrinsically Safe Circuit Conductors.

(1) In Raceways, Cable Trays, and Cables. Conductors of intrinsically safe circuits shall not be placed in any raceway, cable tray, or cable with conductors of any nonintrinsically safe circuit.

Exception No. 1: Where conductors of intrinsically safe circuits are separated from conductors of nonintrinsically safe circuits by a distance of at least 50 mm (2 in.) and secured, or by a grounded metal partition or an approved insulating partition.

FPN: No. 20 gauge sheet metal partitions 0.91 mm (0.0359 in.) or thicker are generally considered acceptable.

Exception No. 2: Where either (1) all of the intrinsically safe circuit conductors or (2) all of the nonintrinsically safe circuit conductors are in grounded metal-sheathed or metal-clad cables where the sheathing or cladding is capable of carrying fault current to ground.

FPN: Cables meeting the requirements of Articles 330 and 332 are typical of those considered acceptable.

2008 NEC - 504.30 Separation of Intrinsically Safe Conductors.

(A) From Nonintrinsically Safe Circuit Conductors.

(1) In Raceways, Cable Trays, and Cables. Conductors of intrinsically safe circuits shall not be placed in any raceway, cable tray, or cable with conductors of any nonintrinsically safe circuit.

Exception No. 1: Where conductors of intrinsically safe circuits are separated from conductors of nonintrinsically safe circuits by a distance of at least 50 mm (2 in.) and secured, or by a grounded metal partition or an approved insulating partition.

FPN: No. 20 gauge sheet metal partitions 0.91 mm (0.0359 in.) or thicker are generally considered acceptable.

Exception No. 2: Where either (1) all of the intrinsically safe circuit conductors or (2) all of the nonintrinsically safe circuit conductors are in grounded metal-sheathed or metal-clad cables where the sheathing or cladding is capable of carrying fault current to ground.

FPN: Cables meeting the requirements of Articles 330 and 332 are typical of those considered acceptable.

Exception No. 3 : Intrinsically safe circuits in a Division 2 or Zone 2 location shall be permitted to be installed in a raceway, cable tray or cable along with nonincendive field wiring circuits when installed in accordance with 504.30(B).

Exception No. 4: Intrinsically safe circuits passing through a Division 2 or Zone 2 location to supply apparatus that is located in a Division 1, Zone 0, or Zone 1 location shall be permitted to be installed in a raceway, cable tray, or cable along with nonincendive field wiring circuits when installed in accordance with 504.30(B).

FPN: Nonincendive field wiring circuits are described in 501.10(B)(3), 502.10(B)(3), 503.10(B)(3), 505.15(C)(1)(g), and 506.15(C)(7).

Author's Comment: New **Exception No. 3** permits intrinsically safe circuit conductors and nonincendive circuit conductors in a Division 2 or Zone 2 location to be in the same raceway, cable tray or cable. Nonincendive circuits can be used in a Division 2 location so there isn't a reason to separate the two different system conductors. The Comment inserts a new **Exception No. 4** permitting intrinsically safe circuit and to be combined in the same raceway, cable tray or cable where passing through a Division 2 or Zone 2 to supply apparatus in a Class I, Division 1, Zone 0, or Zone 1 location.

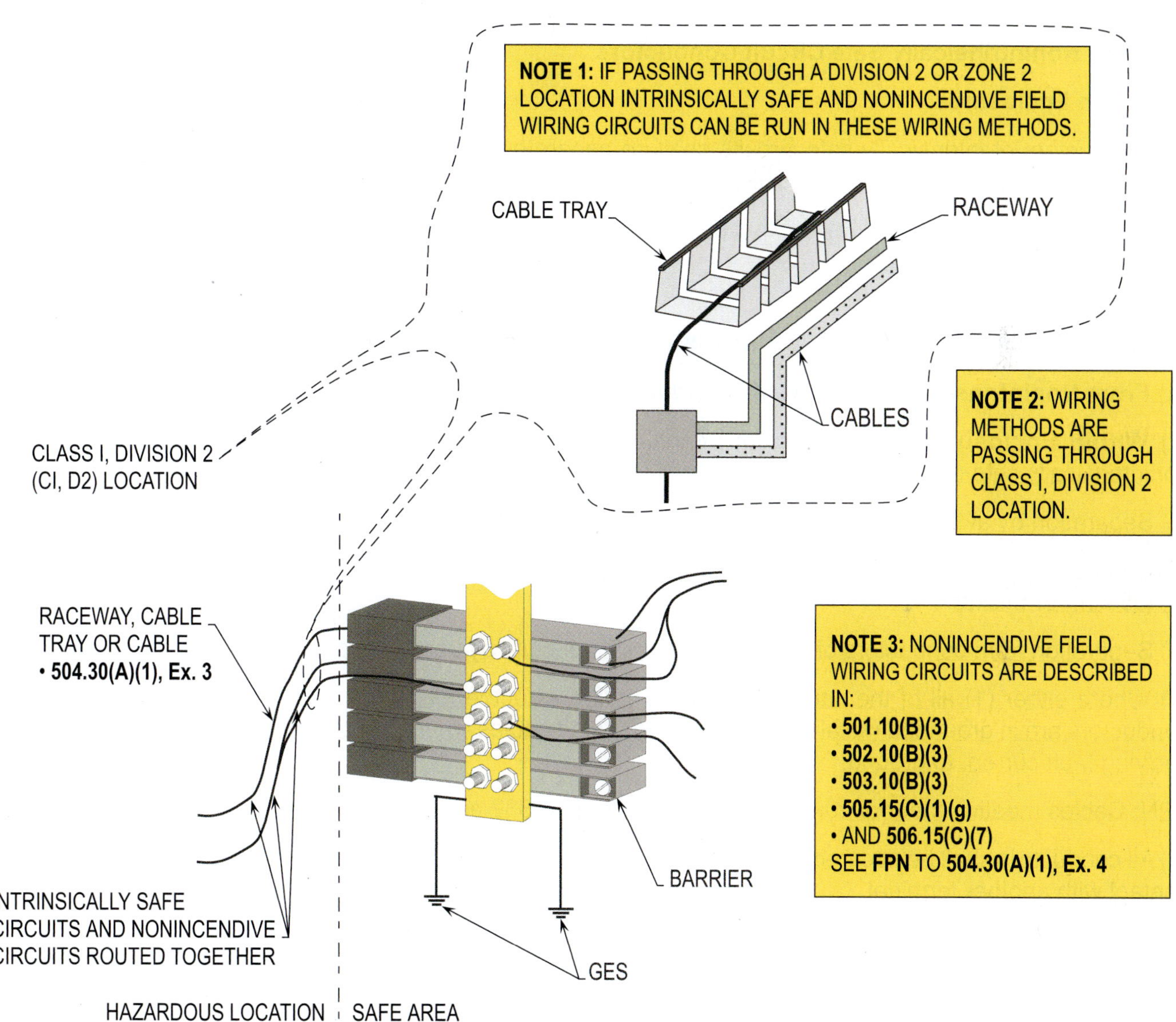

NOTE 1: IF PASSING THROUGH A DIVISION 2 OR ZONE 2 LOCATION INTRINSICALLY SAFE AND NONINCENDIVE FIELD WIRING CIRCUITS CAN BE RUN IN THESE WIRING METHODS.
CABLE TRAY
RACEWAY
CABLES
NOTE 2: WIRING METHODS ARE PASSING THROUGH CLASS I, DIVISION 2 LOCATION.
CLASS I, DIVISION 2 (CI, D2) LOCATION
RACEWAY, CABLE TRAY OR CABLE
• **504.30(A)(1), Ex. 3**
NOTE 3: NONINCENDIVE FIELD WIRING CIRCUITS ARE DESCRIBED IN:
• **501.10(B)(3)**
• **502.10(B)(3)**
• **503.10(B)(3)**
• **505.15(C)(1)(g)**
• AND **506.15(C)(7)**
SEE **FPN** TO **504.30(A)(1), Ex. 4**
BARRIER
INTRINSICALLY SAFE CIRCUITS AND NONINCENDIVE CIRCUITS ROUTED TOGETHER
GES
HAZARDOUS LOCATION
SAFE AREA

FROM NONINTRINSICALLY SAFE CIRCUIT CONDUCTORS
504.30(A)(1), Exceptions No. 3 and 4 and FPN

Purpose of Change: To provide requirements for running intrinsically safe circuits (ISCs) and nonincendive (NCCS) circuits together.

NEC Ch. 5 - Article 504
Part - 504.30(A)(2)

Type of Change	Panel Action	UL	UL 508	API 500 - 1997	API 505 - 1997	OSHA - 1994
Revision	Accept	-	-	3.2.32	3.2.34	1910.307(b)(1)
ROP	**ROC**	**NFPA 70E - 2004**	**NFPA 70B - 2006**	**NFPA 79 - 2007**	**NFPA 497**	**NEMA**
pg. 528 # 14-89	pg. 314 # 14-42	440.2	23.2.10	-	4.1.5.1	-
log: 2975 CMP: 14	log: 129	Submitter: Nicholas P. Ludlam		2005 NEC: 504.30(A)(2)		IEC: -

2005 NEC - 504.30 Separation of Intrinsically Safe Conductors.

(A) From Nonintrinsically Safe Circuit Conductors.

(2) Within Enclosures.

~~(1)~~ Conductors of intrinsically safe circuits shall be separated at least 50 mm (2 in.) from conductors of any nonintrinsically safe circuits ~~or as specified in 504.30(A)(2)~~.

~~(2)~~ All conductors shall be secured so that any conductor that might come loose from a terminal cannot come in contact with another terminal.

2008 NEC - 504.30 Separation of Intrinsically Safe Conductors.

(A) From Nonintrinsically Safe Circuit Conductors.

(2) Within Enclosures. Conductors of intrinsically safe circuits shall be separated from conductors of nonintrinsically safe circuits by one of the following means:

(1) Separation by at least 50 mm (2 in.) from conductors of any nonintrinsically safe circuits.

(2) Separation from conductors of nonintrinsically safe circuits by use of a grounded metal partition 0.91 mm (0.0359 in.) or thicker.

(3) Separation from conductors of nonintrinsically safe circuits by use of an approved insulating partition.

(4) Where either (1) all of the intrinsically safe circuit conductors or (2) all of the nonintrinsically safe circuit conductors are in grounded metal-sheathed or metal-clad cables where the sheathing or cladding is capable of carrying fault current to ground.

FPN: Cables meeting the requirements of Articles 330 and 332 are typical of those considered acceptable.

(5) All conductors shall be secured so that any conductor that might come loose from a terminal cannot come in contact with another terminal.

Author's Comment: The text was rearranged for clarity and the phrase "or as specified in **504.30(A)(2)**" was deleted to remove the circular reference. The Comment incorporated the Fine Print Note information into **504.30(A)(2)(2)** and deleted the Fine Print Note. It also made a minor editorial correction to the Fine Print Note in **504.30(A)(2)(5)**.

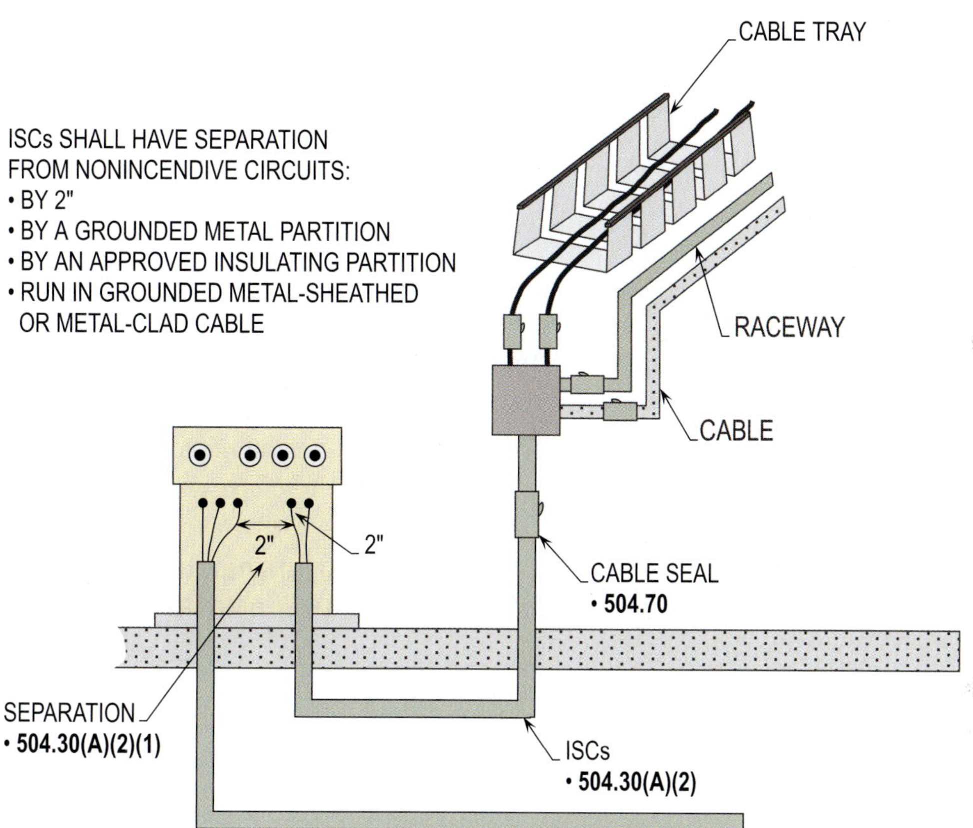
CABLE TRAY
ISCs SHALL HAVE SEPARATION
FROM NONINCENDIVE CIRCUITS:
• BY 2"
• BY A GROUNDED METAL PARTITION
• BY AN APPROVED INSULATING PARTITION
• RUN IN GROUNDED METAL-SHEATHED
OR METAL-CLAD CABLE
RACEWAY
CABLE
2"
2"
CABLE SEAL
• 504.70
SEPARATION
• 504.30(A)(2)(1)
ISCs
• 504.30(A)(2)

WITHIN ENCLOSURES
504.30(A)(2)

Purpose of Change: To clarify separation requirements between intrinsically safe conductors and nonincendive safe circuit conductors.

NEC Ch. 5 - Article 504
Part - 504.70

Type of Change	Panel Action	UL	UL 508	API 500 - 1997	API 505 - 1997	OSHA - 1994
Revision	Accept in Principle	-	-	3.2.32	3.2.34	1910.307(b)(1)
ROP	**ROC**	**NFPA 70E - 2004**	**NFPA 70B - 2006**	**NFPA 79 - 2007**	**NFPA 497**	**NEMA**
pg. 529 / # 14-97	pg. - / # -	440.2	23.2.10	-	4.1.5.1	-
log: 2504 / CMP: 14	log: -	Submitter: Edward M. Briesch		2005 NEC: 504.70		IEC: -

2005 NEC - 504.70 Sealing.

Conduits and cables that are required to be sealed by 501.15, 502.15, ~~and~~ 505.16, shall be sealed to minimize the passage of gases, vapors, or dusts. Such seals shall not be required to be explosionproof or flameproof.

Exception: Seals shall not be required for enclosures that contain only intrinsically safe apparatus, except as required by 501.15(F)(3).

2008 NEC - 504.70 Sealing.

Conduits and cables that are required to be sealed by 501.15, 502.15, 505.16, and 506.16 shall be sealed to minimize the passage of gases, vapors, or dusts. Such seals shall not be required to be explosionproof or flameproof but shall be identified for the purpose of minimizing passage of gases, vapors, or dusts under normal operating conditions and shall be accessible.

Exception: Seals shall not be required for enclosures that contain only intrinsically safe apparatus, except as required by 501.15(F)(3).

Author's Comment: The text in this section was changed to match the same requirements in **501.15(B)(2)** plus text was added to include Zone 20, 21, and 22 installations.

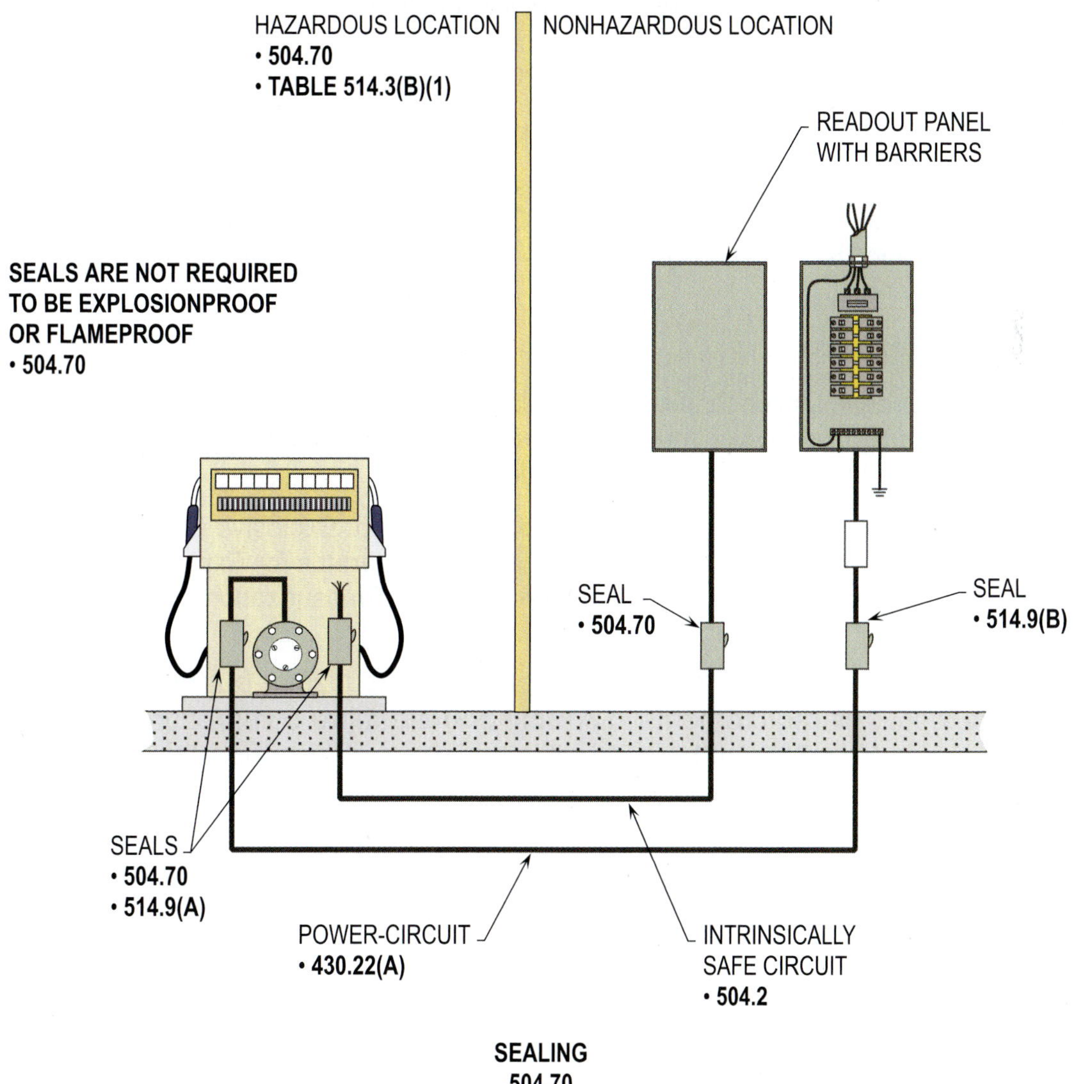

SEALING
504.70

Purpose of Change: This revision clarifies that seals do not have to be explosionproof or flameproof.

NEC Ch. 5 - Article 505
Part - 505.2

Type of Change	Panel Action	UL	UL 508	API 500 - 1997	API 505 - 1997	OSHA - 1994
Revision	Accept in Principle	-	-	-	-	1910.307(b)(1), Note
ROP	**ROC**	**NFPA 70E - 2004**	**NFPA 70B - 2006**	**NFPA 496**	**NFPA 497**	**NEMA**
pg. 529 # 14-100	pg. - # -	440.4(A), FPN	23.2.10	3.3.8	4.1.5.1	-
log: 3524 CMP: 14	log: -	Submitter: Sandra McCloskey		2005 NEC: 505.2		IEC: -

2005 NEC - 505.2 Definitions.

~~Purged and Pressurized~~. Type of protection for electrical equipment that uses the technique of guarding against the ingress of the external atmosphere, which may be explosive, into an enclosure by maintaining a protective gas therein at a pressure above that of the external atmosphere.

FPN No. 1: See NFPA 496-2003, Standard for Purged and Pressurized Enclosures for Electrical Equipment.

FPN ~~No. 2~~: See ~~IEC 60079-2-2000~~, Electrical Apparatus for Explosive Gas Atmospheres — Part 2: ~~Electrical Apparatus, Type of Protection "p"~~; and IEC 60079-13-1982, Electrical Apparatus for Explosive Gas Atmospheres — Part 13: Construction and Use of Rooms or Buildings Protected by Pressurization.

2008 NEC - 505.2 Definitions.

Pressurization "p". Type of protection for electrical equipment that uses the technique of guarding against the ingress of the external atmosphere, which may be explosive, into an enclosure by maintaining a protective gas therein at a pressure above that of the external atmosphere.

FPN: See ANSI/ISA 60079-2 (12.04.01)-2004, Electrical Apparatus for Explosive Gas Atmospheres — Part 2: Pressurized Enclosures "p"; and IEC 60079-13-1982, Electrical Apparatus for Explosive Gas Atmospheres — Part 13: Construction and Use of Rooms or Buildings Protected by Pressurization.

Author's Comment: Pressurization "p" always requires pressurizing, but does not require purging under all circumstances. The requirements in **Article 500** are more comprehensive in scope and are different from those in **Article 505** so the text in this definition was changed by deleting "Purged and Pressurized" and replacing it with "Pressurization "p". The first Fine Print Note was deleted and the second one revised since the ISA and IEC documents more appropriately cover pressurization.

NOTE: TYPE OF PROTECTION FOR ELECTRICAL EQUIPMENT THAT USES THE TECHNIQUE OF GUARDING AGAINST THE INGRESS OF THE EXTERNAL ATMOSPHERE, WHICH MAY BE EXPLOSIVE, INTO AN ENCLOSURE BY MAINTAINING A PROTECTIVE GAS THEREIN AT A PRESSURE ABOVE THAT OF THE EXTERNAL ATMOSPHERE.

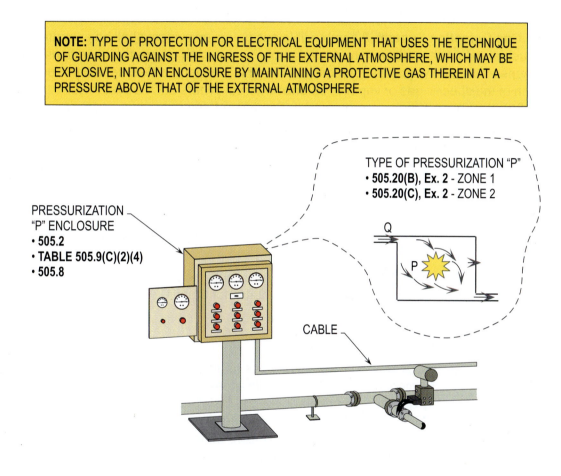

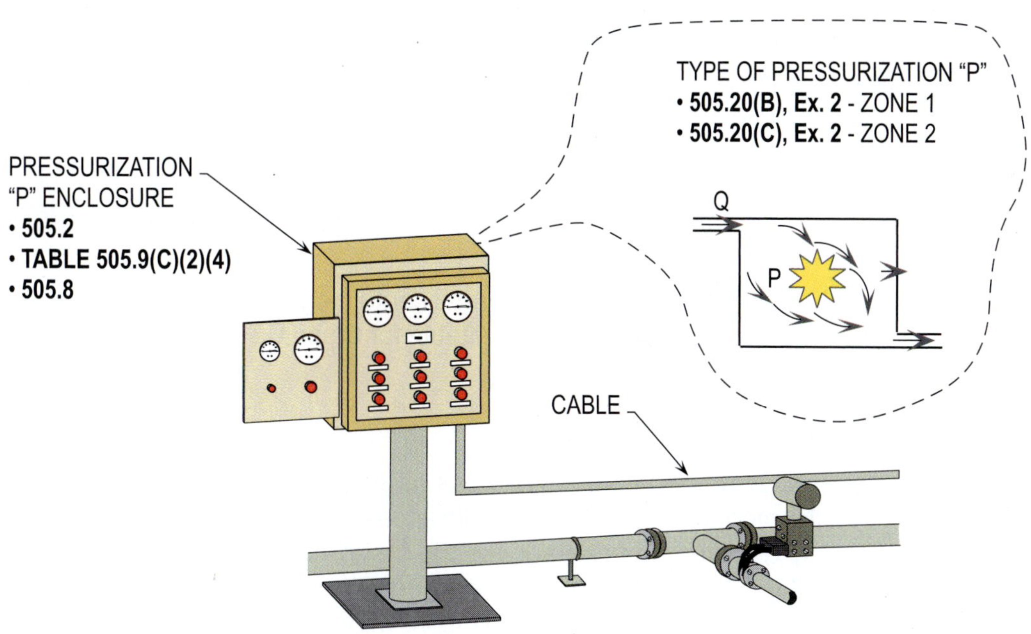

PRESSURIZATION "P"
505.2

Purpose of Change: Revised by deleting the term "purged and pressurized" and replacing it with the term "Pressurization "P".

NEC Ch. 5 - Article 505
Part - 505.8(K)(1) through (K)(3)

Type of Change	Panel Action	UL	UL 508	API 500 - 1997	API 505 - 1997	OSHA - 1994
Revision	Accept in Principle	-	-	6.5	6.8	1910.307(b)(2), Note
ROP	**ROC**	**NFPA 70E - 2004**	**NFPA 70B - 2006**	**NFPA 79 - 2007**	**NFPA**	**NEMA**
pg. 535 # 14-120	pg. - # -	-	23.1	-	-	-
log: 2494 CMP: 14	log: -	Submitter: Edward M. Briesch		2005 NEC: 505.8(I)		IEC: -

2005 NEC - 505.8 Protection Techniques.

~~(I)~~ Combustible Gas Detection System. A combustible gas detection system shall be permitted as a means of protection in industrial establishments with restricted public access and where the conditions of maintenance and supervision ensure that only qualified persons service the installation. ~~Gas detection equipment shall be listed for detection of the specific gas or vapor to be encountered.~~ Where such a system is installed, equipment specified in 505.8(I)(1), I(2), or I(3) shall be permitted. The type of detection equipment, its listing, installation location(s), alarm and shutdown criteria, and calibration frequency shall be documented when combustible gas detectors are used as a protection technique.

(1) Inadequate Ventilation. In a Class I, Zone 1 location that is so classified due to inadequate ventilation, electrical equipment suitable for Class I, Zone 2 locations shall be permitted.

(2) Interior of a Building. In a building located in, or with an opening into, a Class I, Zone 2 location where the interior does not contain a source of flammable gas or vapor, electrical equipment for unclassified locations shall be permitted.

(3) Interior of a Control Panel. In the interior of a control panel containing instrumentation utilizing or measuring flammable liquids, gases, or vapors, electrical equipment suitable for Class I, Zone 2 locations shall be permitted.

2008 NEC - 505.8 Protection Techniques.

(K) Combustible Gas Detection System. A combustible gas detection system shall be permitted as a means of protection in industrial establishments with restricted public access and where the conditions of maintenance and supervision ensure that only qualified persons service the installation. Where such a system is installed, equipment specified in 505.8(I)(1), I(2), or I(3) shall be permitted. The type of detection equipment, its listing, installation location(s), alarm and shutdown criteria, and calibration frequency shall be documented when combustible gas detectors are used as a protection technique.

(1) Inadequate Ventilation. In a Class I, Zone 1 location that is so classified due to inadequate ventilation, electrical equipment suitable for Class I, Zone 2 locations shall be permitted. Combustible gas detection equipment shall be listed for Class I, Zone 1, for the appropriate material group, and for the detection of the specific gas or vapor to be encountered.

(2) Interior of a Building. In a building located in, or with an opening into, a Class I, Zone 2 location where the interior does not contain a source of flammable gas or vapor, electrical equipment for unclassified locations shall be permitted. Combustible gas detection equipment shall be listed for Class I, Zone 1 or Class I, Zone 2, for the appropriate material group, and for the detection of the specific gas or vapor to be encountered.

(3) Interior of a Control Panel. In the interior of a control panel containing instrumentation utilizing or measuring flammable liquids, gases, or vapors, electrical equipment suitable for Class I, Zone 2 locations shall be permitted. Combustible gas detection equipment shall be listed for Class I, Zone 1, for the appropriate material group, and for the detection of the specific gas or vapor to be encountered.

Author's Comment: The 2005 NEC text was unclear with respect to the suitability of the gas detection equipment and the location in which it was installed. While the use of this technique permits equipment for Zone 2 or unclassified locations in a Zone 1 or 2 location respectively, with this change, the detection equipment itself is now required to be suitable for the actual classified location in which it is installed.

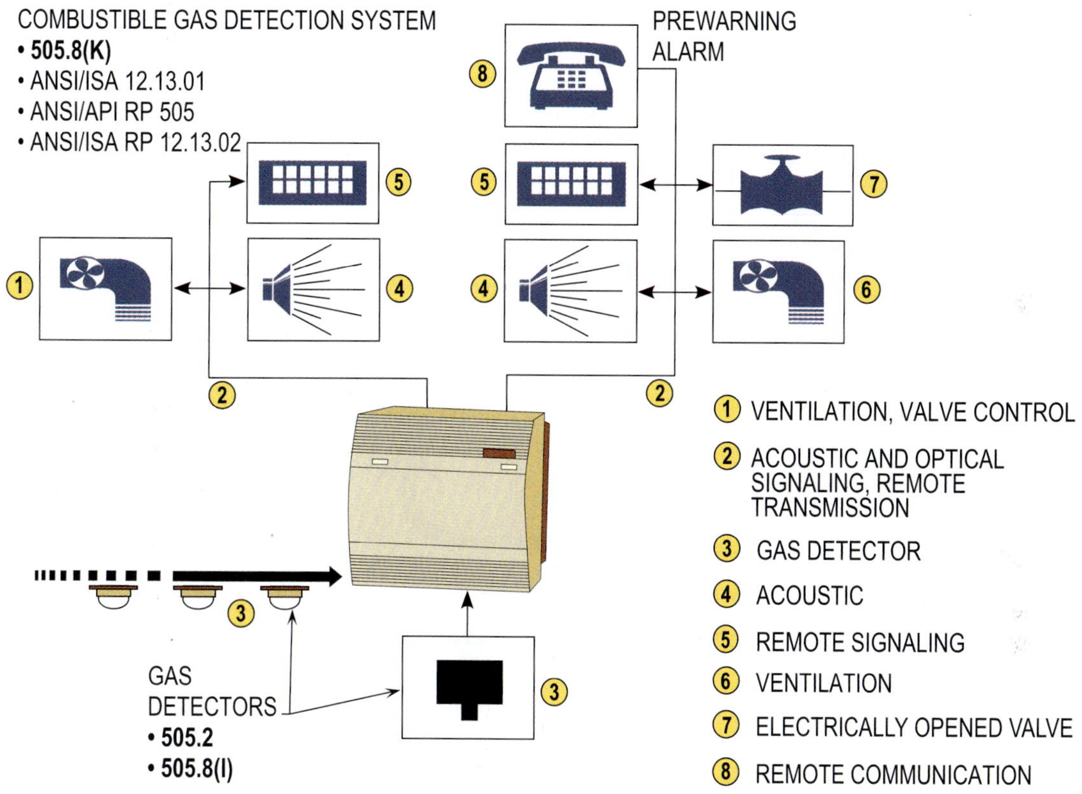

COMBUSTIBLE DETECTION SYSTEM
505.8(K)(1) through (K)(3)

Purpose of Change: To provide guidelines for the installation techniques when using combustible gas detection systems in order to install specific types of equipment.

NEC Ch. 5 - Article 505
Part - 505.9(F)

Type of Change		Panel Action		UL	UL 508	API 500 - 1997	API 505 - 1997	OSHA - 1994
New Subsection		Accept in Principle		-	-	1.1.2	1.1.2	1910.307(b)(2)
ROP		ROC		NFPA 70E - 2004	NFPA 70B - 2006	NFPA 79 - 2007	NFPA	NEMA
pg. 536	# 14-126	pg. -	# -	440.4(E)	23.2.1	-	-	-
log: 2565	CMP: 14	log: -	Submitter: Peter Schimmoeller			2005 NEC: -		IEC: -

2008 NEC - 505.9 Equipment.

(F) Fiber Optic Cable Assembly. Where a fiber optic cable assembly contains conductors that are capable of carrying current, the fiber optic cable assembly shall be installed in accordance with 505.15 and 505.16, as applicable.

Author's Comment: New **505.9(F)** provides installation criteria for fiber optic cable assemblies in Zone installations with similar wording for Division classifications.

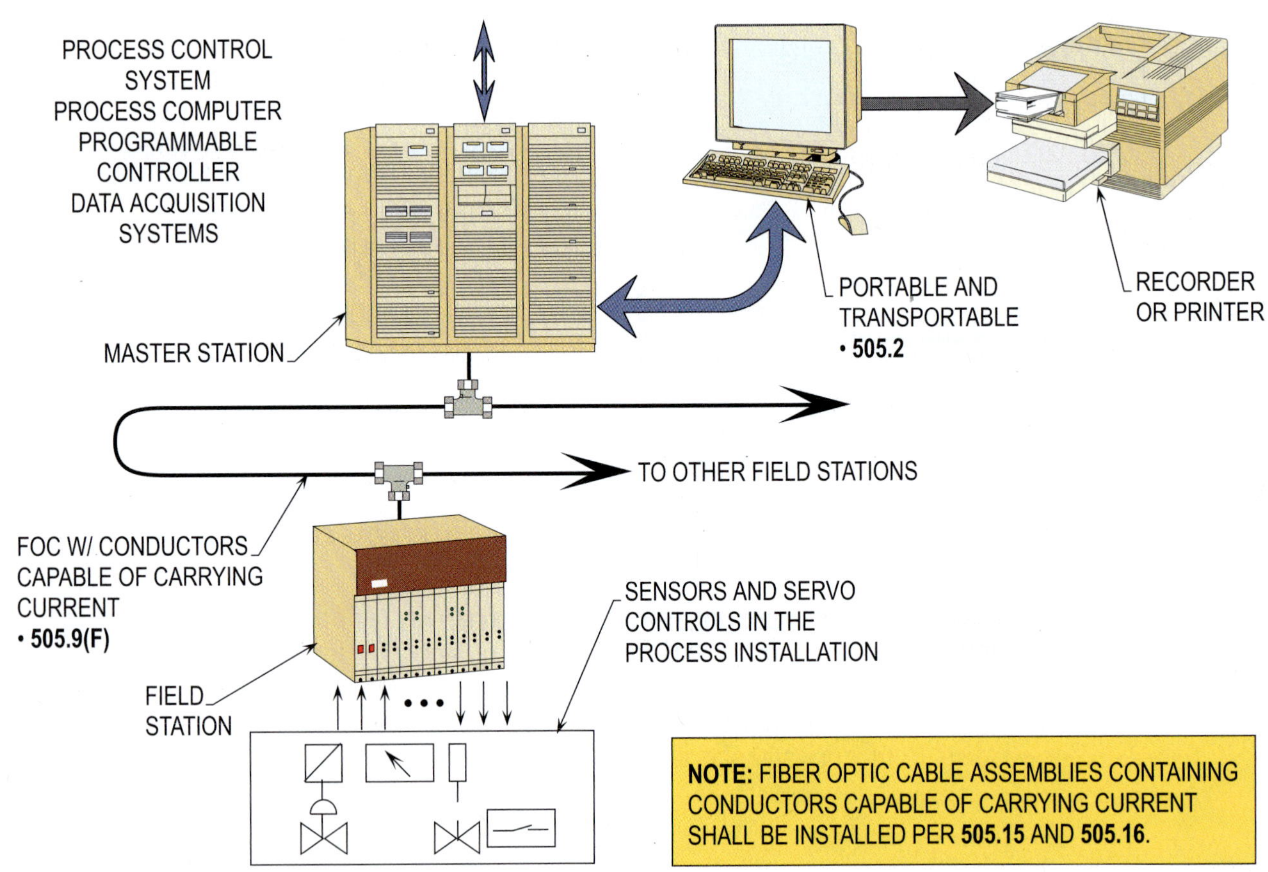

FIBER OPTIC CABLE ASSEMBLY
505.9(F)

Purpose of Change: To provide criteria for fiber optic cable (FOC) assemblies in zone installations.

NEC Ch. 5 - Article 505
Part - 505.17(6)

Type of Change	Panel Action	UL	UL 508	API 500 - 1997	API 505 - 1997	OSHA - 1994
New Subdivision	Accept	-	-	-	-	1910.307(b)(2), Note
ROP	**ROC**	**NFPA 70E - 2004**	**NFPA 70B - 2006**	**NFPA 79 - 2007**	**NFPA**	**NEMA**
pg. 538 # 14-133	pg. 317 # 14-56	440.4(A), FPN	23.2.10	-	-	-
log: 2480 CMP: 14	log: 135 Submitter: Donald W. Ankele			2005 NEC: -		IEC: -

2005 NEC - 505.17 Flexible Cords, Class I, Zones 1 and 2.

A flexible cord shall be permitted for connection between portable lighting equipment or other portable utilization equipment and the fixed portion of their supply circuit. Flexible cord shall also be permitted for that portion of the circuit where the fixed wiring methods of 505.15(B) cannot provide the necessary degree of movement for fixed and mobile electrical utilization equipment, in an industrial establishment where conditions of maintenance and engineering supervision ensure that only qualified persons install and service the installation, and the flexible cord is protected by location or by a suitable guard from damage. The length of the flexible cord shall be continuous. Where flexible cords are used, the cords shall comply with ~~all of~~ the following:

(1) Be of a type listed for extra-hard usage

(2) Contain, in addition to the conductors of the circuit, a grounding conductor complying with 400.23

(3) Be connected to terminals or to supply conductors in an approved manner

(4) Be supported by clamps or by other suitable means in such a manner that there will be no tension on the terminal connections

(5) Be provided with listed seals where the flexible cord enters boxes, fittings, or enclosures that are required to be explosionproof or flameproof

~~Exception: As provided in 505.16.~~

2008 NEC - 505.17 Flexible Cords, Class I, Zones 1 and 2.

A flexible cord shall be permitted for connection between portable lighting equipment or other portable utilization equipment and the fixed portion of their supply circuit. Flexible cord shall also be permitted for that portion of the circuit where the fixed wiring methods of 505.15(B) cannot provide the necessary degree of movement for fixed and mobile electrical utilization equipment, in an industrial establishment where conditions of maintenance and engineering supervision ensure that only qualified persons install and service the installation, and the flexible cord is protected by location or by a suitable guard from damage. The length of the flexible cord shall be continuous. Where flexible cords are used, the cords shall comply with ~~all of~~ the following:

(1) Be of a type listed for extra-hard usage

(2) Contain, in addition to the conductors of the circuit, a grounding conductor complying with 400.23

(3) Be connected to terminals or to supply conductors in an approved manner

(4) Be supported by clamps or by other suitable means in such a manner that there will be no tension on the terminal connections

(5) Be provided with listed seals where the flexible cord enters boxes, fittings, or enclosures that are required to be explosionproof or flameproof

<u>Exception to (5): As provided in 505.16.</u>

(6) Cord entering a increased safety "e" enclosure shall be terminated with a listed increased safety "e" cord connector.

FPN: See 400.7 for permitted uses of flexible cords.

Author's Comment: New **(6)** has been added to require a listed termination cord connector for an increased safety "e" enclosure. A Fine Print Note has been added to reference 400.7 for permitted uses of flexible cords. The NEC Technical Correlating Committee has deleted a proposed Fine Print Note referencing UL 2225 since this information more appropriately belongs in Annex A.

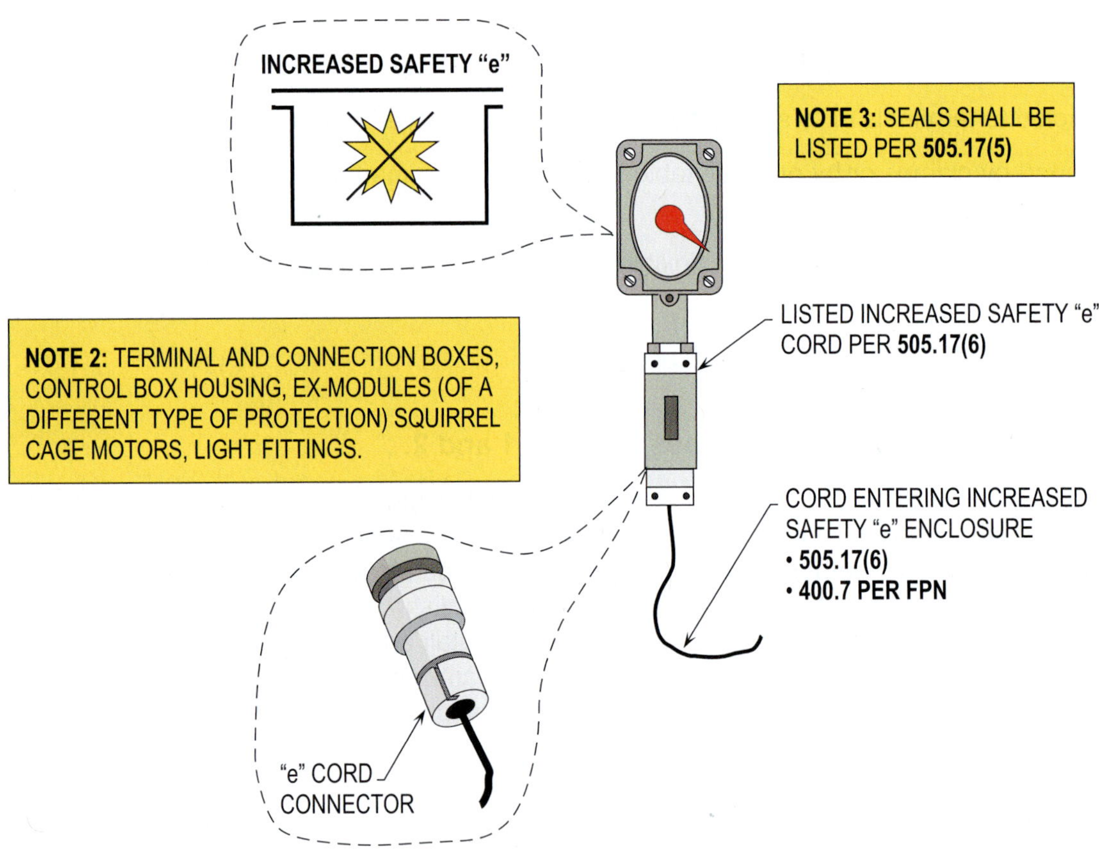
NOTE 1: TYPE OF PROTECTION APPLIED TO ELECTRICAL EQUIPMENT THAT DOES NOT PRODUCE ARCS OF SPARKS IN NORMAL SERVICE AND UNDER SPECIFIED ABNORMAL CONDITIONS, IN WHICH ADDITIONAL MEASURES ARE APPLIED SO AS TO GIVE INCREASED SECURITY AGAINST THE POSSIBILITY OF EXCESSIVE TEMPERATURES AND OF THE OCCURRENCE OF ARCS AND SPARKS.
INCREASED SAFETY "e"
NOTE 3: SEALS SHALL BE LISTED PER 505.17(5)
LISTED INCREASED SAFETY "e" CORD PER 505.17(6)
NOTE 2: TERMINAL AND CONNECTION BOXES, CONTROL BOX HOUSING, EX-MODULES (OF A DIFFERENT TYPE OF PROTECTION) SQUIRREL CAGE MOTORS, LIGHT FITTINGS.
CORD ENTERING INCREASED SAFETY "e" ENCLOSURE
• 505.17(6)
• 400.7 PER FPN
"e" CORD CONNECTOR

FLEXIBLE CORDS, CLASS I, ZONES 1 AND 2
505.17(6)

Purpose of Change: To require a listed termination cord connector for an increased safety "e" enclosure.

NEC Ch. 5 - Article 505
Part - 505.25(B)

Type of Change	Panel Action	UL	UL 508	API 500 - 1997	API 505 - 1997	OSHA - 1994
Revision	Accept in Principle	-	-	-	-	1910.307(b)(2), Note
ROP	**ROC**	**NFPA 70E - 2004**	**NFPA 70B - 2006**	**NFPA 79 - 2007**	**NFPA**	**NEMA**
pg. 539 # 14-138	pg. - # -	440.4(G)	23.2.10	-	-	-
log: 1505 CMP: 14	log: - Submitter: Daniel Leaf			2005 NEC: 505.25(B)		IEC: -

2005 NEC - 505.25 Grounding and Bonding.

Grounding and bonding shall comply with Article 250 and the requirements in 505.25(A) and 505.25(B).

(B) Types of Equipment Grounding Conductors. ~~Where~~ flexible metal conduit ~~or~~ liquidtight flexible metal conduit ~~is used as permitted in 505.15(C) and is to be relied on to complete a sole equipment grounding path, it shall be installed with internal or external bonding jumpers in parallel with each conduit and complying with 250.102~~.

Exception: In Class I, Zone 2 locations, the bonding jumper shall be permitted to be deleted where all of the following conditions are met:

(a) Listed liquidtight flexible metal conduit 1.8 m (6 ft) or less in length, with fittings listed for grounding, is used.

(b) Overcurrent protection in the circuit is limited to 10 amperes or less.

(c) The load is not a power utilization load.

2008 NEC - 505.25 Grounding and Bonding.

Grounding and bonding shall comply with Article 250 and the requirements in 505.25(A) and 505.25(B).

(B) Types of Equipment Grounding Conductors. Flexible metal conduit and liquidtight flexible metal conduit shall not be used as the sole ground-fault current path. Where equipment bonding jumpers are installed, they shall comply with 250.102.

Exception: In Class I, Zone 2 locations, the bonding jumper shall be permitted to be deleted where all of the following conditions are met:

(a) Listed liquidtight flexible metal conduit 1.8 m (6 ft) or less in length, with fittings listed for grounding, is used.

(b) Overcurrent protection in the circuit is limited to 10 amperes or less.

(c) The load is not a power utilization load.

Author's Comment: The text was changed to clarify that flexible metal conduit and liquidtight flexible metal conduit can not be used as the sole ground fault current path.

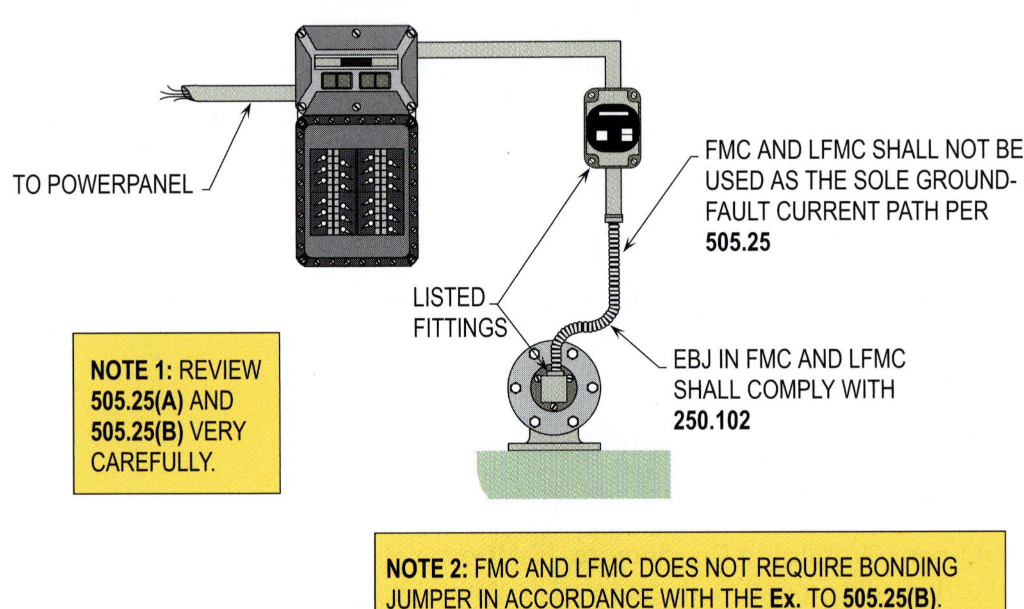
TO POWERPANEL
FMC AND LFMC SHALL NOT BE USED AS THE SOLE GROUND-FAULT CURRENT PATH PER **505.25**
LISTED FITTINGS
EBJ IN FMC AND LFMC SHALL COMPLY WITH **250.102**
NOTE 1: REVIEW **505.25(A)** AND **505.25(B)** VERY CAREFULLY.
NOTE 2: FMC AND LFMC DOES NOT REQUIRE BONDING JUMPER IN ACCORDANCE WITH THE **Ex.** TO **505.25(B)**.

TYPES OF EQUIPMENT GROUNDING CONDUCTORS
505.25(B)

Purpose of Change: To clarify the use of flexible metal conduit (FMC) and liquidtight flexible metal conduit (LFMC) in Class I, Zone 1 and 2 locations.

NEC Ch. 5 - Article 506
Part - 506.2

Type of Change	Panel Action	UL	UL 508	API 500 - 1997	API 505 - 1997	OSHA - 1994
New Definition	Accept in Part	-	-	-	-	1910.307(b)(2), Note
ROP	**ROC**	**NFPA 70E - 2004**	**NFPA 70B - 2006**	**NFPA 496**	**NFPA 497**	**NEMA**
pg. 539 # 14-139a	pg. 318 # 14-59	440.4(A), FPN	23.2.10	3.3.8	4.1.5.1	-
log: CP 1409 CMP: 14	log: 137	Submitter: Code Making Panel 14		2005 NEC: -		IEC: -

2008 NEC - 506.2 Definitions.

For purposes of this article, the following definitions apply.

Protection by Intrinsic Safety "iD". Type of protection where any spark or thermal effect is incapable of causing ignition of a mixture of combustible dust, fibers, or flyings in air under prescribed test conditions.

FPN: For additional information see ISA 61241-11 (12.10.06), Electrical Apparatus for use in Zone 20, Zone 21 and Zone 22 Hazardous (Classified) Locations- Protection by Intrinsic Safety "iD".

Protection by pressurization "pD". Type of protection that guards against the ingress of a mixture of combustible dust, fibers/flyings in air into an enclosure containing electrical equipment by providing and maintaining a protective gas atmosphere inside the enclosure at a pressure above that of the external atmosphere.

FPN: For additional information see, ISA 61241-2 (12.10.04), Electrical Apparatus for use in Zone 21 and Zone 22 Hazardous (Classified) Locations- Protection by Pressurization "pD".

Author's Comment: These definitions were added to Zones 20, 21, and 22 based on ISA standards ISA 61241-2 and 61241-11 for the recognized types of protection.

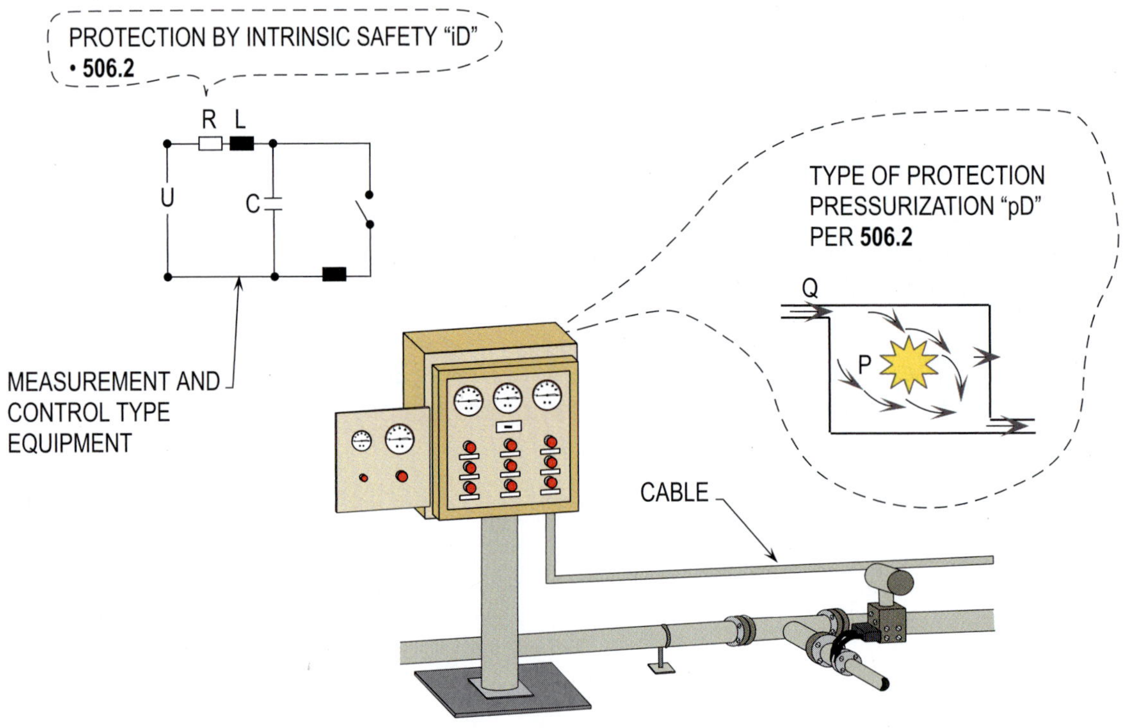

PROTECTION BY PRESSURIZATION "pD" AND INTRINSIC SAFETY "iD"
506.2

Purpose of Change: These definitions were added to zones 20, 21, and 22 for equipment used in combustible dust areas or areas with ignitible fibers or flyings.

NEC Ch. 5 - Article 506
Part - 506.9

Type of Change		Panel Action		UL	UL 508	API 500 - 1997	API 505 - 1997	OSHA - 1994
New Subsection		Accept		-	-	1.1.2	1.1.2	1910.307(b)(2)
ROP		ROC		NFPA 70E - 2004	NFPA 70B - 2006	NFPA 79 - 2007	NFPA	NEMA
pg. 542	# 14-153a	pg. 319	# 14-70	440.4(E)	23.2.1	-	-	-
log: CP 1411	CMP: 14	log: 915	Submitter: Code Making Panel 14			2005 NEC: -		IEC: -

2008 NEC - 506.9 Equipment Installation.

Table 506.9(C)(2)(2) Types of Protection Designation

Designation	Technique	Zone*
iaD	Protection by intrinsic safety	20
ibD	Protection by intrinsic safety	21
[iaD]	Associated apparatus	Unclassified**
[ibD]	Associated apparatus	Unclassified**
maD	Protection by encapsulation	20
mbD	Protection by encapsulation	21
pD	Protection by pressurization	21
tD	Protection by enclosures	21

* Does not address use where a combination of techniques is used.

** Associated apparatus is permitted to be installed in a hazardous (classified) location if suitably protected using another type of protection.

Author's Comment: This new table helps the user determine the type of protection permissible for each Zone with combustible dust or fibers/flyings.

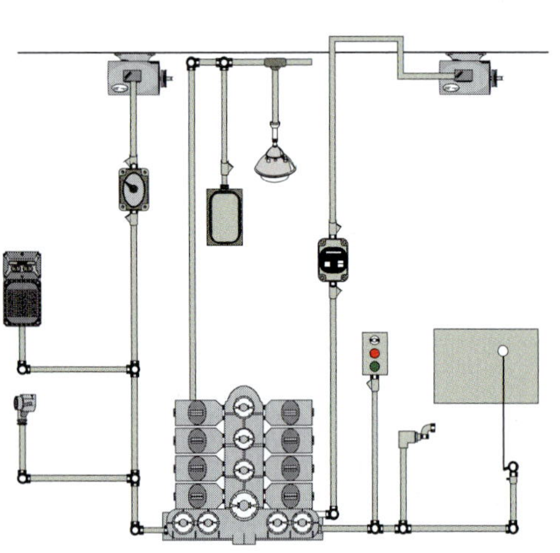

TABLE 506.9(C)(2)(2) TYPES OF PROTECTION DESIGNATION

DESIGNATION	TECHNIQUE	ZONE*
iaD	PROTECTION BY INTRINSIC SAFETY	20
ibD	PROTECTION BY INTRINSIC SAFETY	21
[iaD]	ASSOCIATED APPARATUS	UNCLASSIFIED**
[ibD]	ASSOCIATED APPARATUS	UNCLASSIFIED**
maD	PROTECTION BY ENCAPSULATION	20
mbD	PROTECTION BY ENCAPSULATION	21
pD	PROTECTION BY PRESSURIZATION	21
tD	PROTECTION BY ENCLOSURES	21

EQUIPMENT INSTALLATION
TABLE 506.9(C)(2)(2)

Purpose of Change: A new table was added to aid the user in selecting protection for each zone.

NEC Ch. 5 - Article 506
Part - 506.9(C)(1) and (C)(2)

Type of Change		Panel Action		UL	UL 508	API 500 - 1997	API 505 - 1997	OSHA - 1994
Revision		Accept		-	-	1.1.2	1.1.2	1910.307(b)(2)
ROP		**ROC**		**NFPA 70E - 2004**	**NFPA 70B - 2006**	**NFPA 79 - 2007**	**NFPA**	**NEMA**
pg. 541	# 14-150b	pg. -	# -	440.4(E)	23.2.1	-	-	-
log: CP 1410	CMP: 14	log: -	Submitter: Code Making Panel 14			2005 NEC: 506.9(C)(1) and (C)(2)		IEC: -

2005 NEC - 506.9 Equipment Requirements.

(C) Marking. Equipment identified for Class II, Division 1 or Class II, Division 2 shall, in addition to being marked in accordance with 500.8(B), be permitted to be marked with both of the following:

(1) Zone 20, 21, or 22 (as applicable)

(2) Temperature classification in accordance with 506.9(D)

2008 NEC - 506.9 Equipment Requirements.

(C) Marking.

(1) Division Equipment. Equipment identified for Class II, Division 1 or Class II, Division 2 shall, in addition to being marked in accordance with 500.8(B), be permitted to be marked with both of the following:

(1) Zone 20, 21, or 22 (as applicable)

(2) Temperature classification in accordance with 506.9(D)

(2) Zone Equipment. Equipment meeting one or more of the protection techniques described in 506.8 shall be marked with the following in the order shown:

(1) Symbol "AEx"

(2) Protection technique(s) in accordance with Table 506.209(C)(2)(2);

(3) Zone

(4) Temperature classification, marked as a temperature value, in degrees C, preceded by T;

(5) Ambient temperature marking in accordance with 506.9(D).

Author's Comment: A title has been added as marking for **(1)** for Division equipment and new **(2)** provides marking requirements for Zone equipment.

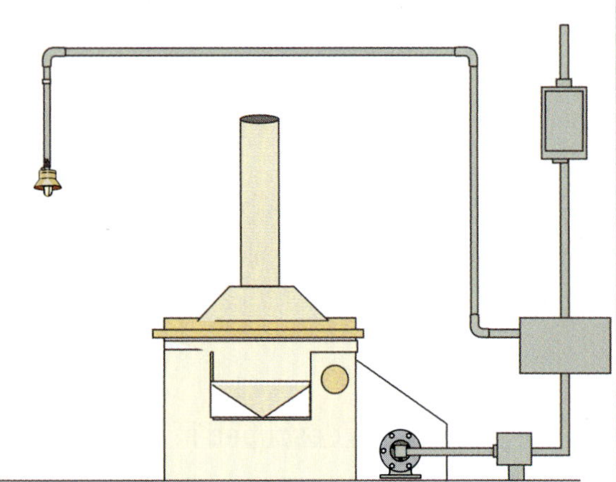

506.9(C) MARKINGS.
(1) DIVISION EQUIPMENT. EQUIPMENT IDENTIFIED FOR CLASS II, DIVISION 1 OR CLASS II, DIVISION 2 SHALL, IN ADDITION TO BEING MARKED IN ACCORDANCE WITH **500.8(B)**, BE PERMITTED TO BE MARKED WITH BOTH OF THE FOLLOWING:

- ZONE 20, 21, OR 22 (AS APPLICABLE)
- TEMPERATURE CLASSIFICATION IN ACCORDANCE WITH **506.9(D)**

(2) ZONE EQUIPMENT. EQUIPMENT MEETING ONE OR MORE OF THE PROTECTION TECHNIQUES DESCRIBED IN **506.8** SHALL BE MARKED WITH THE FOLLOWING IN THE ORDER SHOWN:

- SYMBOL "AEx"
- PROTECTION TECHNIQUE(S) IN ACCORDANCE WITH **TABLE 506.9(C)(2)(2)**:
- ZONE
- TEMPERATURE CLASSIFICATION, MARKED AS A TEMPERATURE VALUE, IN DEGREES C, PRECEDED BY T:
- AMBIENT TEMPERATURE MARKING IN ACCORDANCE WITH **506.9(D)**.

MARKING
506.9(C)(1) AND (C)(2)

Purpose of Change: New titles have been added to marked for (1) "Division Equipment" and (2) marking requirements for "Zone Equipment."

NEC Ch. 5 - Article 506
Part - 506.9(F)

Type of Change		Panel Action		UL	UL 508	API 500 - 1997	API 505 - 1997	OSHA - 1994
New Subsection		Accept in Principle		-	-	1.1.2	1.1.2	1910.307(b)(2)
ROP		ROC		NFPA 70E - 2004	NFPA 70B - 2006	NFPA 79 - 2007	NFPA	NEMA
pg. 541	# 14-151	pg. -	# -	440.4(E)	23.2.1	-	-	-
log: 2566	CMP: 14	log: -	Submitter: Peter Schimmoeller			2005 NEC: -		IEC: -

2008 NEC - 506.9 Equipment Requirements.

(F) Fiber Optic Cable Assembly. Where a fiber optic cable assembly contains conductors that are capable of carrying current, the fiber optic cable assembly shall be installed in accordance with 506.15 and 506.16, as applicable.

Author's Comment: New **506.9(F)** provides installation criteria for fiber optic cable assemblies in Zone installations with similar wording for Division classifications.

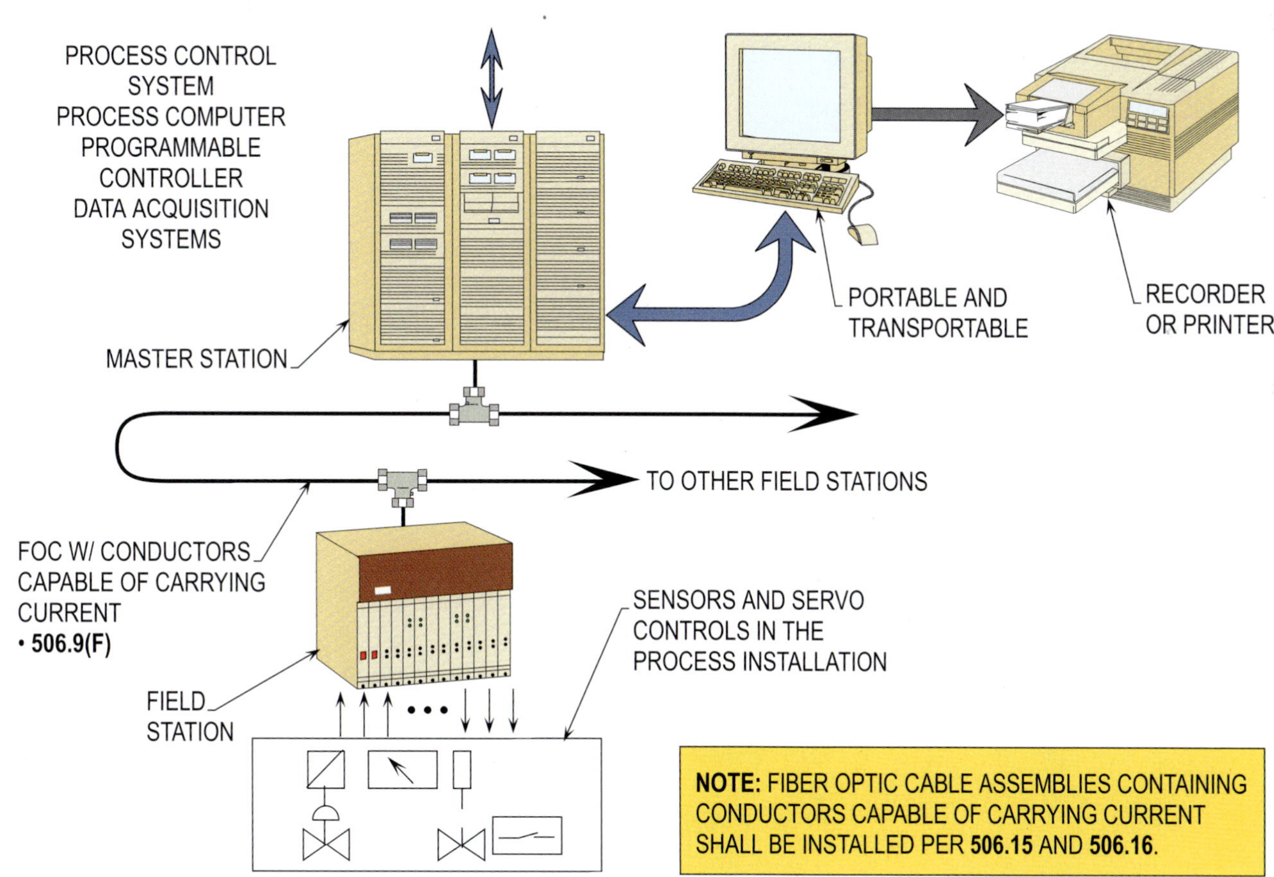

FIBER OPTIC CABLE ASSEMBLY
506.9(F)

Purpose of Change: To provide criteria for fiber optic cable (FOC) assemblies in zone installations.

NEC Ch. 5 - Article 506
Part - 506.25(B)

Type of Change	Panel Action	UL	UL 508	API 500 - 1997	API 505 - 1997	OSHA - 1994
Revision	Accept	-	-	-	-	1910.307(b)(2), Note
ROP	**ROC**	**NFPA 70E - 2004**	**NFPA 70B - 2006**	**NFPA 79 - 2007**	**NFPA**	**NEMA**
pg. 542 # 14-155a	pg. - # -	440.4(G)	23.2.10	-	-	-
log: CP 1403 CMP: 14	log: -	Submitter: Code Making Panel 14		2005 NEC: 506.25(B)		IEC: -

2005 NEC - 506.25 Grounding and Bonding.

Grounding and bonding shall comply with Article 250 and the requirements in 506.25(A) and 506.25(B).

(B) Types of Equipment Grounding Conductors. ~~Where flexible conduit is used as permitted in 506.15, it shall be installed with internal or external bonding jumpers in parallel with each conduit and complying with 250.102~~.

Exception: In Zone 22 locations, the bonding jumper shall be permitted to be deleted where all of the following conditions are met:

(1) Listed liquidtight flexible metal conduit 1.8 m (6 ft) or less in length, with fittings listed for grounding, is used.

(2) Overcurrent protection in the circuit is limited to 10 amperes or less.

(3) The load is not a power utilization load.

2008 NEC - 506.25 Grounding and Bonding.

Grounding and bonding shall comply with Article 250 and the requirements in 506.25(A) and 506.25(B).

(B) Types of Equipment Grounding Conductors. Liquidtight flexible metal conduit shall not be used as the sole ground-fault current path. Where equipment bonding jumpers are installed, they shall comply with 250.102.

Exception: In Zone 22 locations, the bonding jumper shall be permitted to be deleted where all of the following conditions are met:

(1) Listed liquidtight flexible metal conduit 1.8 m (6 ft) or less in length, with fittings listed for grounding, is used.

(2) Overcurrent protection in the circuit is limited to 10 amperes or less.

(3) The load is not a power utilization load.

Author's Comment: The text was changed to clarify that liquidtight flexible metal conduit can not be used as the sole ground fault current path.

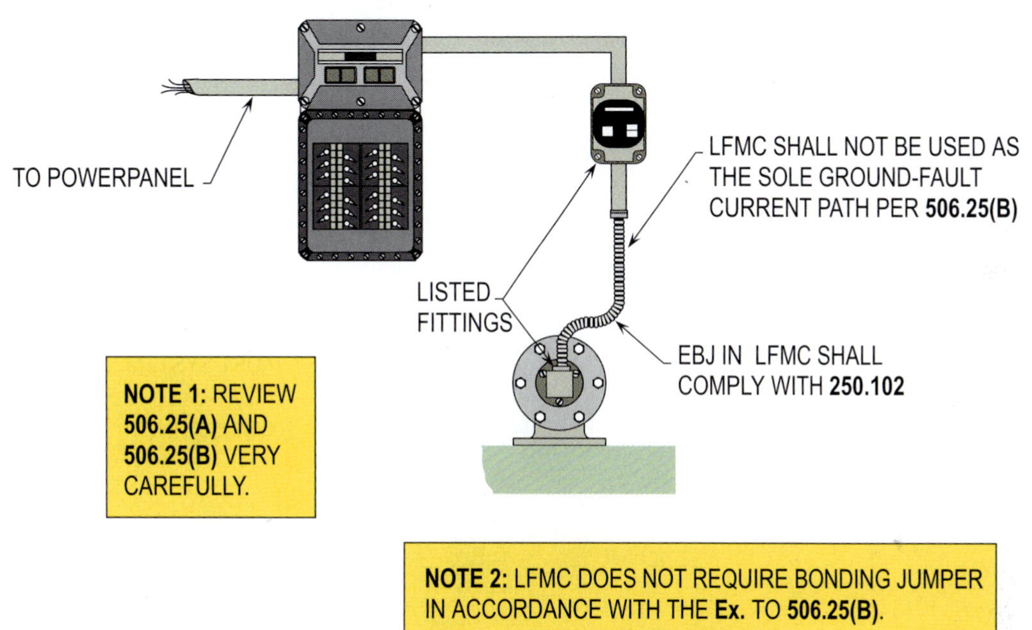
TO POWERPANEL
LFMC SHALL NOT BE USED AS THE SOLE GROUND-FAULT CURRENT PATH PER 506.25(B)
LISTED FITTINGS
EBJ IN LFMC SHALL COMPLY WITH 250.102
NOTE 1: REVIEW 506.25(A) AND 506.25(B) VERY CAREFULLY.
NOTE 2: LFMC DOES NOT REQUIRE BONDING JUMPER IN ACCORDANCE WITH THE Ex. TO 506.25(B).

TYPES OF EQUIPMENT GROUNDING CONDUCTORS
506.25(B)

Purpose of Change: To clarify the use of liquidtight flexible metal conduit (LFMC) in locations.

NEC Ch. 5 - Article 513
Part - 513.2

Type of Change		Panel Action		UL	UL 508	API 500 - 1997	API 505 - 1997	OSHA - 1994
New Definition		Accept		-	-	-	-	1910.307(b)(2), Note
ROP		ROC		NFPA 70E - 2004	NFPA 70B - 2006	NFPA 79 - 2007	NFPA	NEMA
pg. 545	# 14-165a	pg. -	# -	440.4(A), FPN	-	-	-	-
log: CP1413	CMP: 14	log: -	Submitter: Code Making Panel 14			2005 NEC: -		IEC: -

2008 NEC - 513.2 Definitions.

For the purpose of this article, the following definitions shall apply.

Aircraft Painting Hangar. An aircraft hangar constructed for the express purpose of spray/coating/dipping applications and provided with dedicated ventilation supply and exhaust.

Author's Comment: This definition has been added to **Article 513** as a specific type of aircraft hanger used to house aircraft during the painting process.

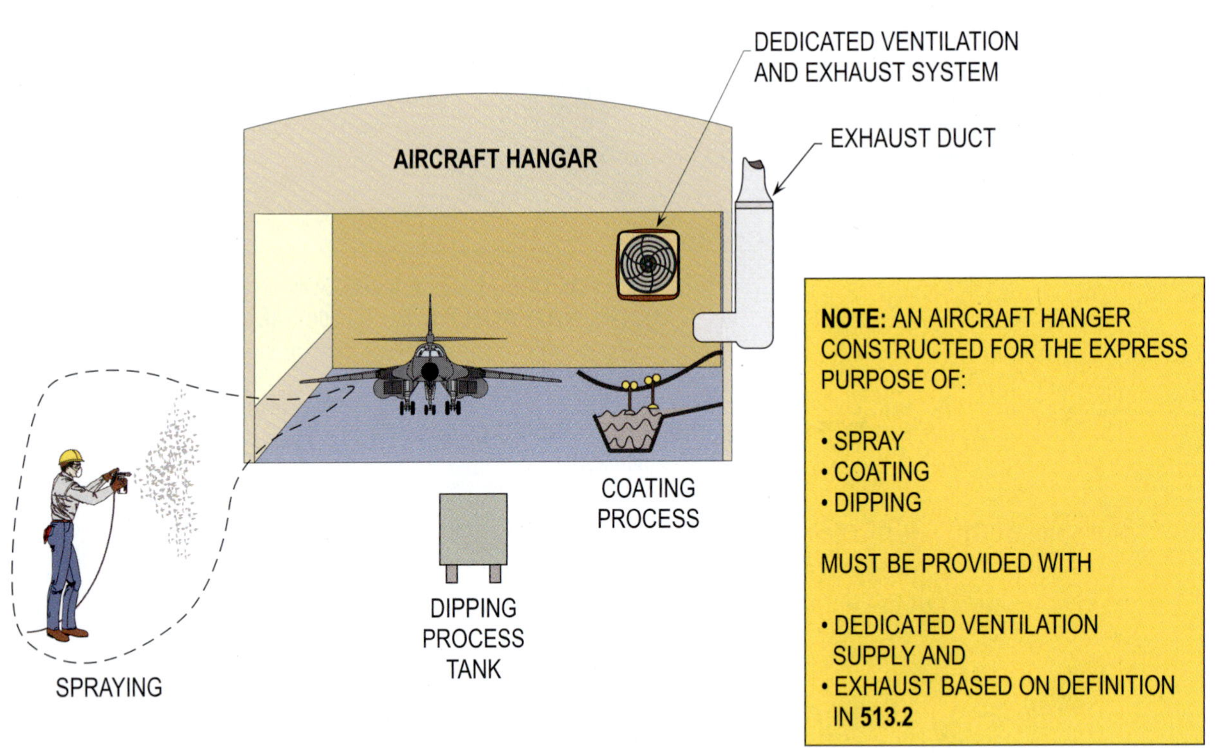

AIRCRAFT PAINTING HANGER
513.2

Purpose of Change: To include requirements for painting processes in aircraft hangars.

NEC Ch. 5 - Article 513
Part - 513.3(C)(2)

Type of Change		Panel Action		UL	UL 508	API 500 - 1997	API 505 - 1997	OSHA - 1994
Revision		Accept		-	-	-	-	1910.307(b)(2), Note
ROP		**ROC**		**NFPA 70E - 2004**	**NFPA 70B - 2006**	**NFPA 79 - 2007**	**NFPA**	**NEMA**
pg. 545	# 14-165b	pg. -	# -	440.4(A), FPN	-	-	-	-
log: CP 1412	CMP: 14	log: -	Submitter: Code Making Panel 14			2005 NEC: 513.3(C)		IEC: -

2005 NEC - 513.3 Classification of Locations.

(C) Vicinity of Aircraft. The area within 1.5 m (5 ft) horizontally from aircraft power plants or aircraft fuel tanks shall be classified as a Class I, Division 2 or Zone 2 locations that shall extend upward from the floor to a level 1.5 m (5 ft) above the upper surface of wings and of engine enclosures.

2008 NEC - 513.3 Classification of Locations.

(C) Vicinity of Aircraft.

(1) Aircraft Maintenance and Storage Hangars. The area within 1.5 m (5 ft) horizontally from aircraft power plants or aircraft fuel tanks shall be classified as a Class I, Division 2 or Zone 2 locations that shall extend upward from the floor to a level 1.5 m (5 ft) above the upper surface of wings and of engine enclosures.

(2) Aircraft Painting Hangars. The area within 3m (10ft) horizontally from aircraft surfaces from the floor to 3m (10ft) above the aircraft shall be classified as Class I, Division 1 or Class I, Zone 1. The area horizontally from aircraft surfaces between 3.0m (10ft) and 9.0m (30 ft) from the floor to 9.0m (30ft) above the aircraft surface shall be classified as Class I, Division 2 or Class I, Zone 2.

FPN: See NFPA 33-2007, Standard for Spray Application Using Flammable or Combustible Materials for information on ventilation and grounding for static protection in spray painting areas.

Author's Comment: A new subdivision **(2)** has been added to **513.3(C)**, covering classifications, dealing with aircraft painting hangers with the existing text in **513.3(C)** being labeled as **(1)** Aircraft Maintenance and Storage Hangers. NFPA 409-2005 has been revised to specifically separate the hazardous locations near aircraft for aircraft paint hangars from those of general maintenance. Aircraft paint hangars, while constructed like huge paint booths, do not have the same dimensional clearances found in traditional paint booths. The shape of the aircraft creates clearances far greater than that found in any other painting system. This creates a level of safety not found in traditional paint booths and supports hazardous location classification that is less than the entire hangar.

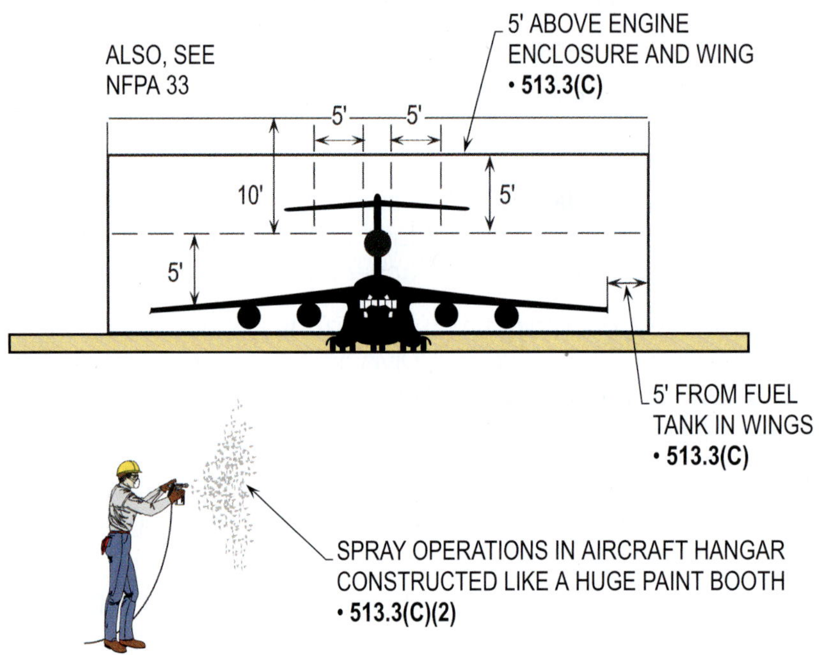

VICINITY OF AIRCRAFT
513.3(C)(2)

Purpose of Change: To include requirements for large aircraft hangars constructed like a paint booth.

NEC Ch. 5 - Article 517
Part I - 517.2

Type of Change		Panel Action		UL	UL 508	API 500 - 1997	API 505 - 1997	OSHA - 1994
New Definition		Accept		-	-	-	-	1910.308(b)
ROP		ROC		NFPA 70E - 2004	NFPA 70B - 2006	NFPA 79 - 2007	NFPA	NEMA
pg. 552	# 15-11	pg. -	# -	450.2	6.4.4.6(2)	-	-	-
log: 1781	CMP: 15	log: -	Submitter: Marvin J. Fischer			2005 NEC: -		IEC: -

2008 NEC - 517.2 Definitions

Emergency System. A system of circuits and equipment intended to supply alternate power to a limited number of prescribed functions vital to the protection of life and safety. [NFPA 99:3.3.41]

Author's Comment: The definition for "emergency system" has been added to **517.2** (it was deleted in the 2002 NEC) to clarify that it is a system of circuits and conductors to supply an alternate power source for prescribed functions vital to life and safety.

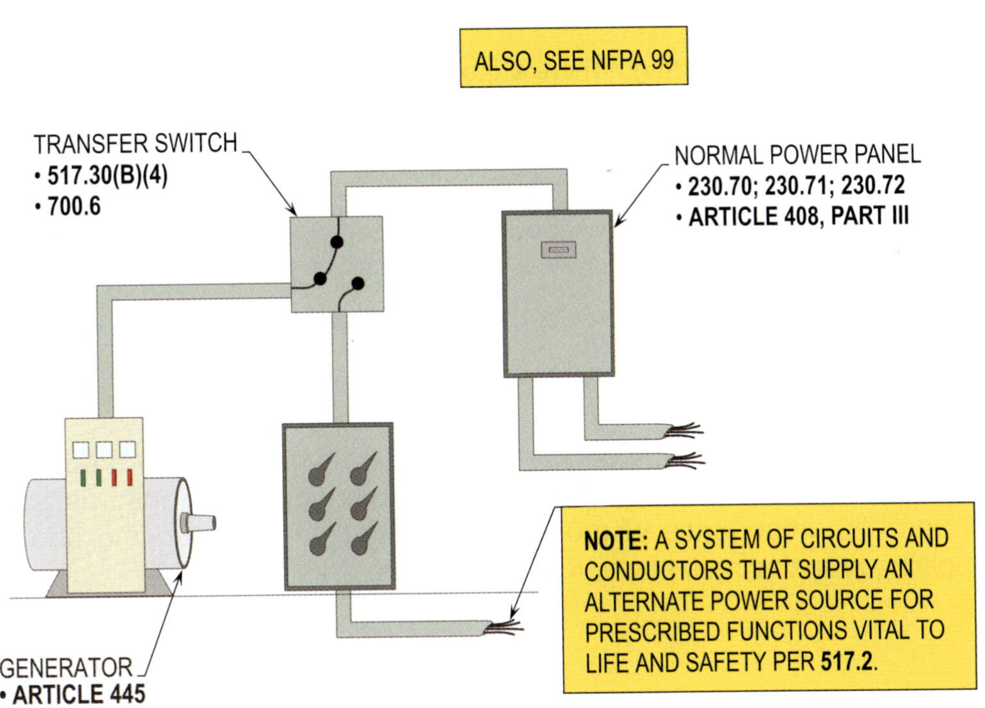

EMERGENCY SYSTEM
517.2

Purpose of Change: A new definition "Emergency System" was added to clarify its intended use.

NEC Ch. 5 - Article 517
Part II - 517.19(D)

Type of Change		Panel Action		UL	UL 508	API 500 - 1997	API 505 - 1997	OSHA - 1994
Revision		Accept		-	-	-	-	1910.304(f)(4)
ROP		ROC		NFPA 70E - 2004	NFPA 70B - 2006	NFPA 79 - 2007	NFPA	NEMA
pg. 557	# 15-38	pg. -	# -	410.10(A)	7.7	8.2.1.2	-	-
log: 1645	CMP: 15	log: -	Submitter: Michael J. Johnston			2005 NEC: 517.19(D)		IEC: -

2005 NEC - 517.19 Critical Care Areas.

(D) Panelboard Grounding. Where a grounded electrical distribution system is used and metal feeder raceway or Type MC or MI cable grounding of a panelboard or switchboard shall be ensured by one of the following means at each termination or junction point of the raceway or Type MC or MI cable:

(1) A grounding bushing and a continuous copper bonding jumper, sized in accordance with 250.122, with the bonding jumper connected to the junction enclosure or the ground bus of the panel

(2) Connection of feeder raceways or Type MC or MI cable to threaded hubs or bosses on terminating enclosures

(3) Other approved devices such as bonding-type locknuts or bushings.

2008 NEC - 517.19 Critical Care Areas.

(D) Panelboard Grounding and Bonding. Where a grounded electrical distribution system is used and metal feeder raceway or Type MC or MI cable that qualifies as an equipment grounding conductor in accordance with 250.118 is installed, grounding of a panelboard or switchboard shall be ensured by one of the following bonding means at each termination or junction point of the metal raceway or Type MC or MI cable:

(1) A grounding bushing and a continuous copper bonding jumper, sized in accordance with 250.122, with the bonding jumper connected to the junction enclosure or the ground bus of the panel

(2) Connection of feeder raceways or Type MC or MI cable to threaded hubs or bosses on terminating enclosures

(3) Other approved devices such as bonding-type locknuts or bushings.

Author's Comment: Adding the term "bonding" to panelboard grounding and adding text explaining that the raceway or cable qualifies as an equipment grounding conductor ensures that the user of this section understands the importance of providing both equipment grounding and bonding for the panelboard supplying branch circuit power to a critical care area.

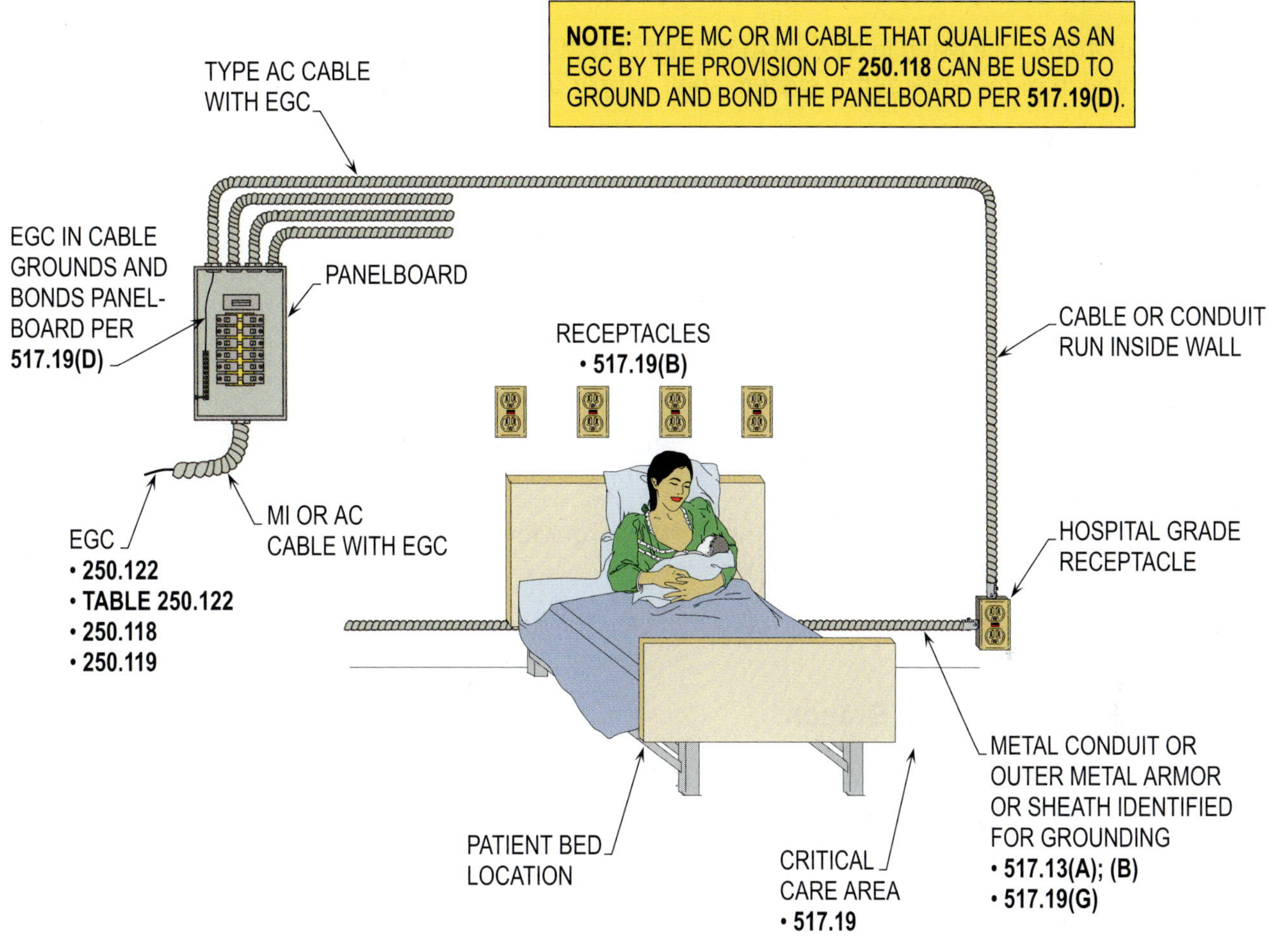

PANELBOARD GROUNDING AND BONDING
517.19(D)

Purpose of Change: The term bonding was added to clarify that both grounding and bonding is required at the panelboard.

NEC Ch. 5 - Article 517
Part III - 517.32(C), (E), and (F)

Type of Change		Panel Action		UL	UL 508	API 500 - 1997	API 505 - 1997	OSHA - 1994
New Subdivision		Accept in Principle		-	-	-	-	-
ROP		ROC		NFPA 70E - 2004	NFPA 70B - 2006	NFPA 79 - 2007	NFPA	NEMA
pg. 560	# 15-63	pg. -	# -	-	-	-	-	-
log: 2510	CMP: 15	log: -	Submitter: Hugh O. Nash, Jr.			2005 NEC: -		IEC: -

2005 NEC - 517.32 Life Safety Branch.

No function other than those listed in 517.32(A) through 517.32(G) shall be connected to the life safety branch. The life safety branch of the emergency system shall supply power for the following lighting, receptacles, and equipment.

(C) Alarm and Alerting Systems. Alarm and alerting systems including the following:

(1) Fire alarms

FPN: See NFPA 101-2003, Life Safety Code, Section 9.6 and 18.3.4.

(2) Alarms required for systems used for the piping of nonflammable medical gases

FPN: See NFPA 99-2002, Standard for Health Care Facilities, 4.4.2.2.2.2(3).

(D) Communications Systems. Hospital communications systems, where used for issuing instructions during emergency conditions.

(E) Generator Set Location. Task illumination battery charger for emergency battery-operated lighitng unit(s) and selected receptacles at the generator set location.

~~**(F)**~~ **Elevators.** Elevator cab lighting, control, communications, and signal systems.

2008 NEC - 517.32 Life Safety Branch.

No function other than those listed in 517.32(A) through 517.32(G) shall be connected to the life safety branch. The life safety branch of the emergency system shall supply power for the following lighting, receptacles, and equipment.

(C) Alarm and Alerting Systems. Alarm and alerting systems including the following:

(1) Fire alarms

FPN: See NFPA 101-2003, Life Safety Code, Section 9.6 and 18.3.4.

(2) Alarms required for systems used for the piping of nonflammable medical gases

FPN: See NFPA 99-2002, Standard for Health Care Facilities, 4.4.2.2.2.2(3).

(3) Mechanical, control, and other accessories required for effective life safety systems operation shall be permitted to be connected to the life safety branch.

(D) Communications Systems. Hospital communications systems, where used for issuing instructions during emergency conditions.

(E) Generator Set and Transfer Switch Locations. Task illumination battery charger for emergency battery-operated lighitng unit(s) and selected receptacles at the generator set and essential transfer switch locations. [**99:**4.4.2.2.2.2(5)]

(F) Generator Set Accessories. Generator set accessories as required for generator performance.

(G) **Elevators.** Elevator cab lighting, control, communications, and signal systems.

Author's Comment: HVAC controls, dampers, and certain motors are related to safety of life. The designer has the option to connect large motors to the equipment system where it is not practical to connect them to the life safety branch but dampers, controls, and accessories for HVAC systems are now permitted to be connected to the life safety branch.

Battery powered lighting units are necessary to provide lighting while working on transfer switches as well as generators so this was added to the life safety branch requirements.

A new subsection for generator accessories, such as day tank pumps and other critical equipment necessary for operation of the generator, permits these circuits to be supplied from the life safety branch. It will also help ensure the loads for the generator accessories are supplied from the same life safety panelboard.

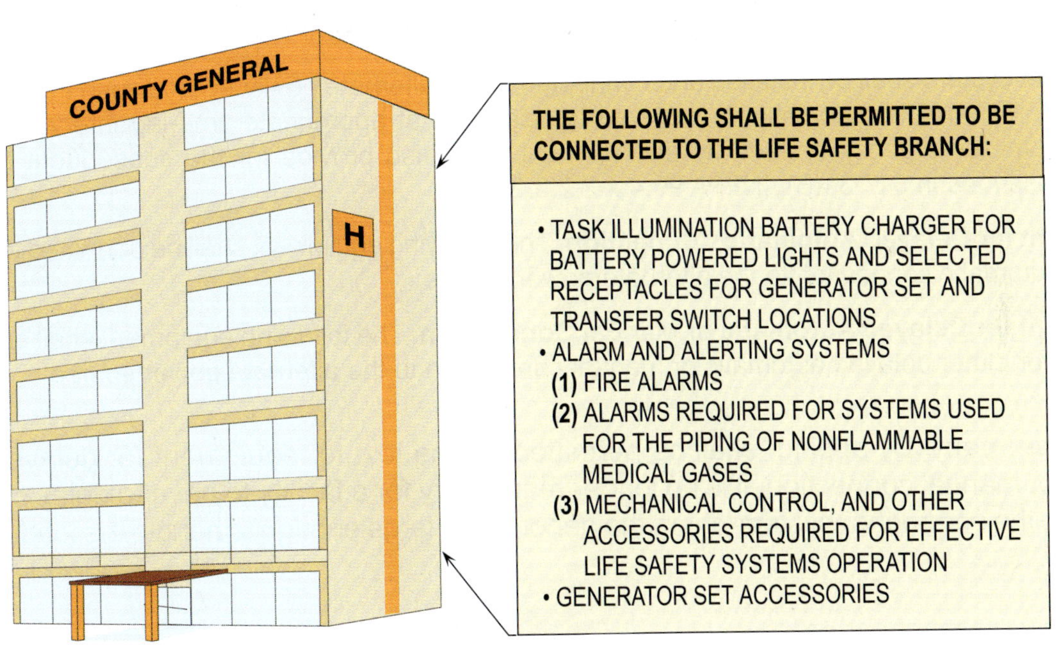
COUNTY GENERAL
H
THE FOLLOWING SHALL BE PERMITTED TO BE CONNECTED TO THE LIFE SAFETY BRANCH:
• TASK ILLUMINATION BATTERY CHARGER FOR BATTERY POWERED LIGHTS AND SELECTED RECEPTACLES FOR GENERATOR SET AND TRANSFER SWITCH LOCATIONS
• ALARM AND ALERTING SYSTEMS
(1) FIRE ALARMS
(2) ALARMS REQUIRED FOR SYSTEMS USED FOR THE PIPING OF NONFLAMMABLE MEDICAL GASES
(3) MECHANICAL CONTROL, AND OTHER ACCESSORIES REQUIRED FOR EFFECTIVE LIFE SAFETY SYSTEMS OPERATION
• GENERATOR SET ACCESSORIES

LIFE SAFETY BRANCH
517.32(C), (E) and (F)

Purpose of Change: There revisions clarify the items permitted to be connected to the life safety branch.

NEC Ch. 5 - Article 517
Part III - 517.34(A) and (B)

Type of Change	Panel Action	UL	UL 508	API 500 - 1997	API 505 - 1997	OSHA - 1994
Revision	Accept	-	-	-	-	-
ROP	**ROC**	**NFPA 70E - 2004**	**NFPA 70B - 2006**	**NFPA 79 - 2007**	**NFPA**	**NEMA**
pg. 561 # 15-67	pg. - # -	-	-	-	-	-
log: 2512 CMP: 15	log: -	Submitter: Hugh O. Nash, Jr.		2005 NEC: 517.34(A) and (B)		IEC: -

2005 NEC - 517.34 Equipment System Connection to Alternate Power Source.

The equipment system shall be installed and connected to the alternate power source such that the equipment described in 517.34(A) is automatically restored to operation at appropriate time-lag intervals following the energizing of the emergency system. Its arrangement shall also provide for the subsequent connection of equipment described in 517.34(B). [NFPA 99:4.4.2.2.3.2]

(A) Equipment for Delayed Automatic Connection. The following equipment shall be arranged for delayed automatic connection to the alternate power source:

(B) Equipment for Delayed Automatic or Manual Connection. The following equipment shall be arranged for either delayed automatic or manual connection to the alternate power source:

2008 NEC - 517.34 Equipment System Connection to Alternate Power Source.

The equipment system shall be installed and connected to the alternate power source such that the equipment described in 517.34(A) is automatically restored to operation at appropriate time-lag intervals following the energizing of the emergency system. Its arrangement shall also provide for the subsequent connection of equipment described in 517.34(B). [NFPA 99:4.4.2.2.3.2]

(A) Equipment for Delayed Automatic Connection. The following equipment shall be permitted to be arranged for delayed automatic connection to the alternate power source:

(B) Equipment for Delayed Automatic or Manual Connection. The following equipment shall be permitted to be arranged for either delayed automatic or manual connection to the alternate power source:

Author's Comment: By adding "shall be permitted" to the text for equipment for delayed automatic connection or manual connection, it is no longer mandatory for a facility to have to delay connection of equipment to the emergency system where the generator is large enough to handle the load immediately.

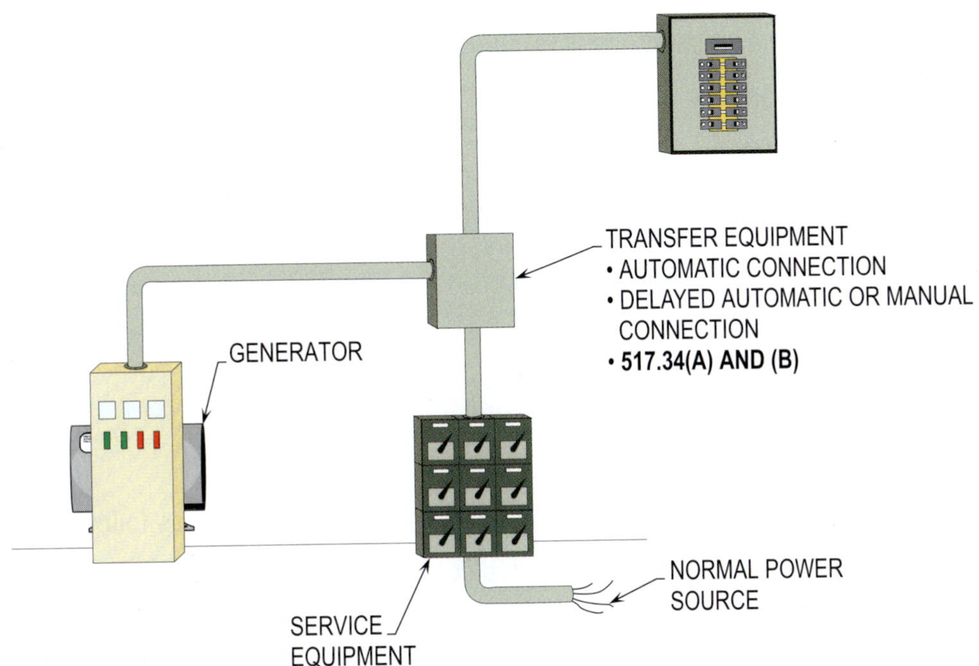

EQUIPMENT SYSTEM CONNECTION TO ALTERNATE POWER SOURCE 517.34(A) AND (B)

Purpose of Change: This revision clarifies that a facility shall be permitted to install equipment for delayed automatic connection, delayed automatic or manual connection, or non-delayed automatic connection.

NEC Ch. 5 - Article 517
Part VII - 517.160(A)(5)

Type of Change	Panel Action	UL	UL 508	API 500 - 1997	API 505 - 1997	OSHA - 1994
Revision	Accept in Principle	-	-	-	-	-
ROP	ROC	NFPA 70E - 2004	NFPA 70B - 2006	NFPA 79 - 2007	NFPA	NEMA
pg. 566 # 15-106	pg. - # -	-	-	-	-	-
log: 3656 CMP: 15	log: -	Submitter: Patricia Johnson		2005 NEC: 517.160(A)(5)		IEC: -

2005 NEC - 517.160 Isolated Power Systems.

(A) Installations.

(5) Conductor Identification. The isolated circuit conductors shall be identified as follows:

(1) Isolated Conductor No. 1 — Orange

(2) Isolated Conductor No. 2 — Brown

For 3-phase systems, the third conductor shall be identified as yellow. Where isolated circuit conductors supply 125-volt, single-phase, 15- and 20-ampere receptacles, the orange conductor(s) shall be connected to the terminal(s) on the receptacles that are identified in accordance with 200.10(B) for connection to the grounded circuit conductor.

2008 NEC - 517.160 Isolated Power Systems.

(A) Installations.

(5) Conductor Identification. The isolated circuit conductors shall be identified as follows:

(1) Isolated Conductor No. 1 — Orange with a distinctive colored stripe other than white, green or gray.

(2) Isolated Conductor No. 2 — Brown with a distinctive colored stripe other than white, green or gray.

For 3-phase systems, the third conductor shall be identified as yellow with a distinctive colored stripe other than white, green or gray. Where isolated circuit conductors supply 125-volt, single-phase, 15- and 20-ampere receptacles, the striped orange conductor(s) shall be connected to the terminal(s) on the receptacles that are identified in accordance with 200.10(B) for connection to the grounded circuit conductor.

Author's Comment: Since orange is used as a "high leg" color required by **110.15** plus orange and brown are also used for 480/277 volt wye colors, another use for the orange and brown conductors could cause confusion. The change to require striped insulation for the isolated circuit conductors provides proper identification based on the critical nature of these circuits in a manner that leaves no question as to their use.

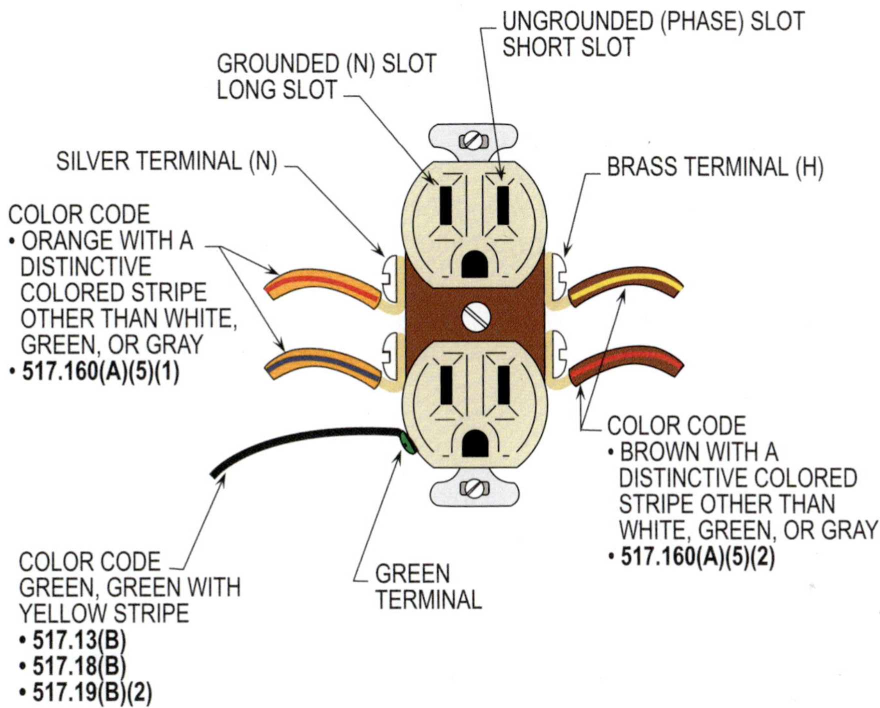

CONDUCTOR IDENTIFICATION
517.160(A)(5)

Purpose of Change: This revision clarifies that striped isolated conductors provides proper identification for isolated power systems.

NEC Ch. 5 - Article 518
Part - 518.3(B)

Type of Change		Panel Action		UL	UL 508	API 500 - 1997	API 505 - 1997	OSHA - 1994
Revision		Accept in Principle in Part		-	-	-	-	-
ROP		ROC		NFPA 70E - 2004	NFPA 70B - 2006	NFPA 79 - 2007	NFPA	NEMA
pg. 567	# 15-108	pg. -	# -	-	-	-	-	-
log: 1167	CMP: 15	log: -	Submitter: Daniel Leaf			2005 NEC: 518.3(A) and (B)		IEC: -

2005 NEC - 518.3 Other Articles.

(B) Temporary Wiring. In exhibition halls used for display booths, as in trade shows, the temporary wiring shall be installed in accordance with Article 590. Flexible cables and cords approved for hard or extra-hard usage shall be permitted to be laid on floors where protected from contact by the general public. The ground-fault circuit-interrupter requirements of 590.6 shall not apply.

Exception: Where conditions of supervision and maintenance ensure that only qualified persons will service the installation, flexible cords or cables identified in Table 400.4 for hard usage or extra-hard usage shall be permitted in cable trays used only for temporary wiring. All cords or cables shall be installed in a single layer. A permanent sign shall be attached to the cable tray at intervals not to exceed 7.5 m (25 ft). The sign shall read

CABLE TRAY FOR TEMPORARY WIRING ONLY

(C) Emergency Systems. Control of emergency systems shall comply with Article 700.

2008 NEC - 518.3 Other Articles.

(B) Temporary Wiring. In exhibition halls used for display booths, as in trade shows, the temporary wiring shall be permitted to be installed in accordance with Article 590. Flexible cables and cords approved for hard or extra-hard usage shall be permitted to be laid on floors where protected from contact by the general public. The ground-fault circuit-interrupter requirements of 590.6 shall not apply.

Exception: Where conditions of supervision and maintenance ensure that only qualified persons will service the installation, flexible cords or cables identified in Table 400.4 for hard usage or extra-hard usage shall be permitted in cable trays used only for temporary wiring. All cords or cables shall be installed in a single layer. A permanent sign shall be attached to the cable tray at intervals not to exceed 7.5 m (25 ft). The sign shall read

CABLE TRAY FOR TEMPORARY WIRING ONLY

(C) Emergency Systems. Control of emergency systems shall comply with Article 700.

Author's Comment: By the addition of the permissive text, "shall be permitted," to the first sentence in **518.3(B)**, exhibition halls may comply with all of the temporary wiring methods permitted in **Article 590**.

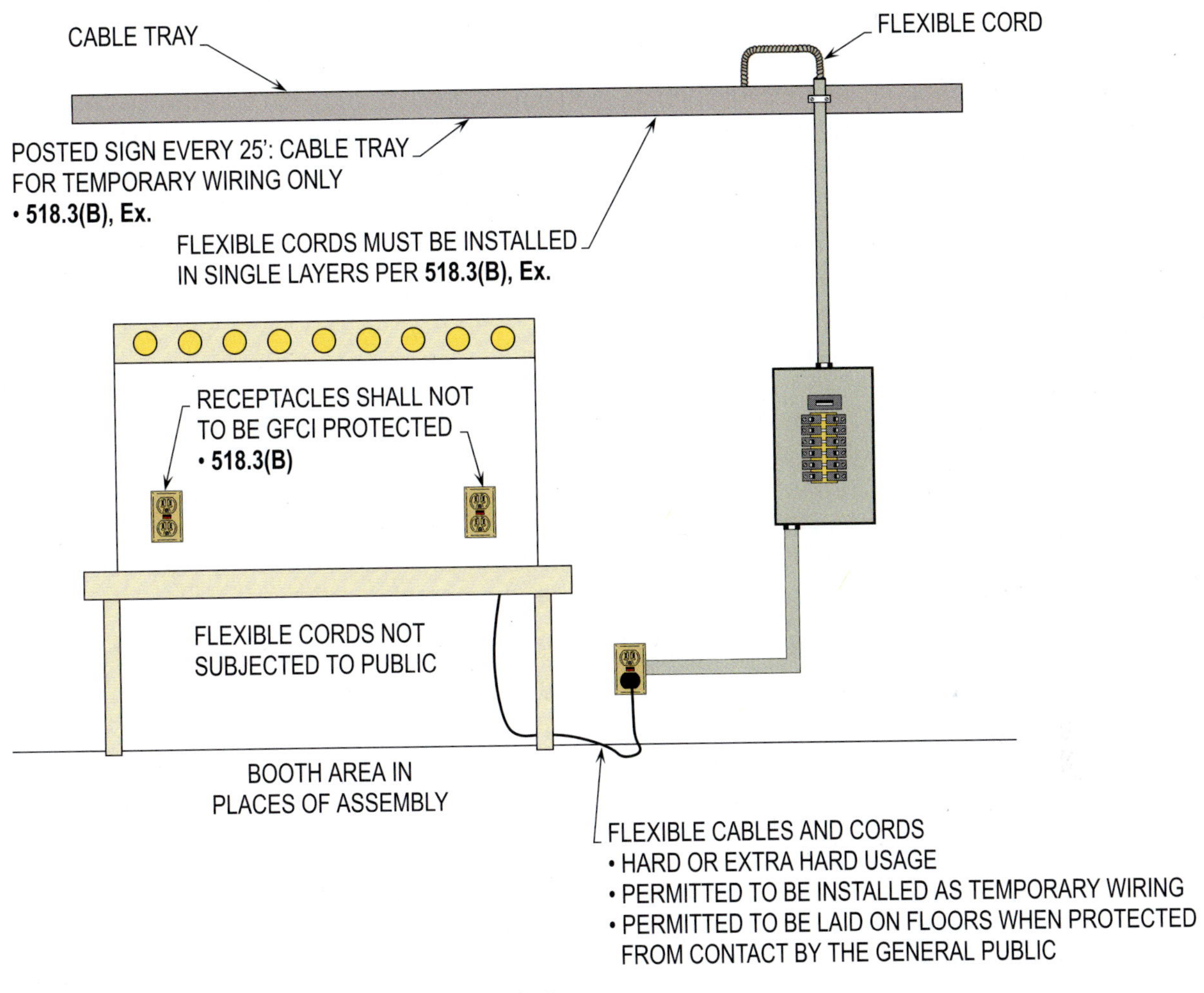

CABLE TRAY

FLEXIBLE CORD

POSTED SIGN EVERY 25': CABLE TRAY
FOR TEMPORARY WIRING ONLY
• **518.3(B), Ex.**

FLEXIBLE CORDS MUST BE INSTALLED
IN SINGLE LAYERS PER **518.3(B), Ex.**

RECEPTACLES SHALL NOT
TO BE GFCI PROTECTED

FLEXIBLE CORDS NOT
SUBJECTED TO PUBLIC

BOOTH AREA IN
PLACES OF ASSEMBLY

FLEXIBLE CABLES AND CORDS
• HARD OR EXTRA HARD USAGE
• PERMITTED TO BE INSTALLED AS TEMPORARY WIRING
• PERMITTED TO BE LAID ON FLOORS WHEN PROTECTED
 FROM CONTACT BY THE GENERAL PUBLIC

TEMPORARY WIRING
518.3(B)

Purpose of Change: This revision clarifies that temporary wiring installed in exhibit halls used for display booths shall be permitted as a temporary wiring method per **Article 590**.

NEC Ch. 5 - Article 518
Part - 518.5 and Exception

Type of Change	Panel Action	UL	UL 508	API 500 - 1997	API 505 - 1997	OSHA - 1994
Revision	Accept in Principle	-	-	-	-	-
ROP	**ROC**	**NFPA 70E - 2004**	**NFPA 70B - 2006**	**NFPA 79 - 2007**	**NFPA**	**NEMA**
pg. 569 # 15-116	pg. - # -	-	-	-	-	-
log: 3316 CMP: 15	log: -	Submitter: Steven R. Terry		2005 NEC: 518.5		IEC: -

2005 NEC - 518.5 Supply.

Portable switchboards and portable power distribution equipment shall be supplied only from listed power outlets of sufficient voltage and ampere rating. Such power outlets shall be protected by overcurrent devices. Such overcurrent devices and power outlets shall not be accessible to the general public. Provisions for connection of an equipment grounding conductor shall be provided. The neutral of feeders supplying solid-state, 3-phase, 4-wire ~~dimmer~~ systems shall be considered a current-carrying conductor.

2008 NEC - 518.5 Supply.

Portable switchboards and portable power distribution equipment shall be supplied only from listed power outlets of sufficient voltage and ampere rating. Such power outlets shall be protected by overcurrent devices. Such overcurrent devices and power outlets shall not be accessible to the general public. Provisions for connection of an equipment grounding conductor shall be provided. The neutral <u>conductor</u> of feeders supplying solid-state <u>phase-control</u>, 3-phase, 4-wire <u>dimming</u> systems shall be considered a current-carrying conductor <u>for purposes of derating</u>. <u>The neutral conductor of feeders supplying solid-state sine wave, 3-phase, 4-wire dimming systems shall not be considered a current-carrying conductor for the purpose of derating.</u>

<u>**Exception:** The neutral conductor of feeders supplying systems that use or may use both phase-control and sine-wave dimmers shall be considered as current-carrying for purposes of derating.</u>

<u>**FPN:** For definitions of solid state dimmer types, see 520.2.</u>

Author's Comment: A new class of listed solid state dimmer has been introduced to the professional performance lighting market: the Solid State Sine Wave dimmer. This type of dimmer varies the amplitude of the applied voltage wave form, without any of the nonlinear switching found in traditional phase control solid state dimmers. In the past, nonlinear phase control dimmers were the only type of readily available solid state dimmers. Since solid state sine wave dimmers are linear loads, they do not require the neutral of the feeder to the dimmers to be considered a current-carrying conductor. Wording has been added to clearly state the feeder requirements for both types of solid state dimmers that are now available, in order to avoid the confusion that might arise if only phase-control dimmers were mentioned in the requirement for neutral characteristics. The **FPN** directs the reader to **520.2** for the definition of both types of solid state dimmers since **Article 518** has no definitions section.

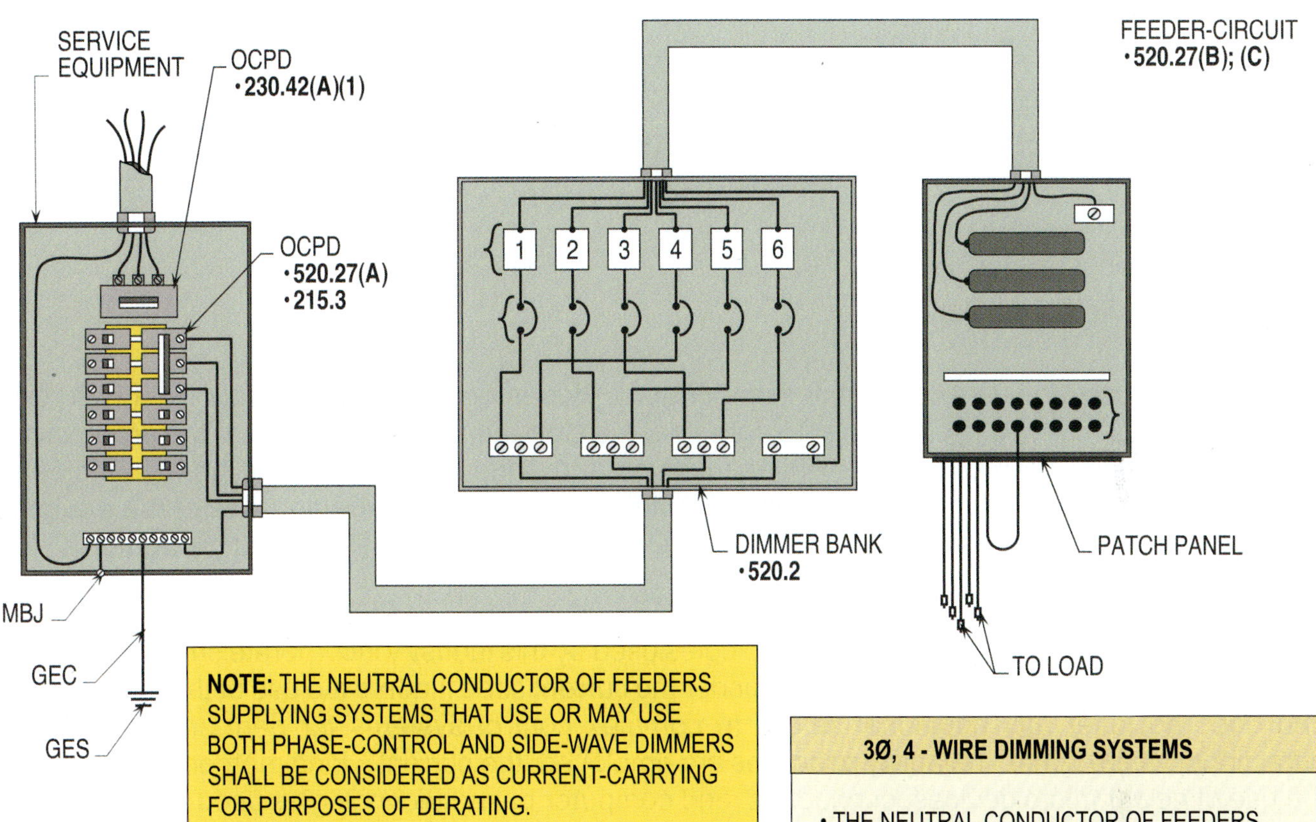

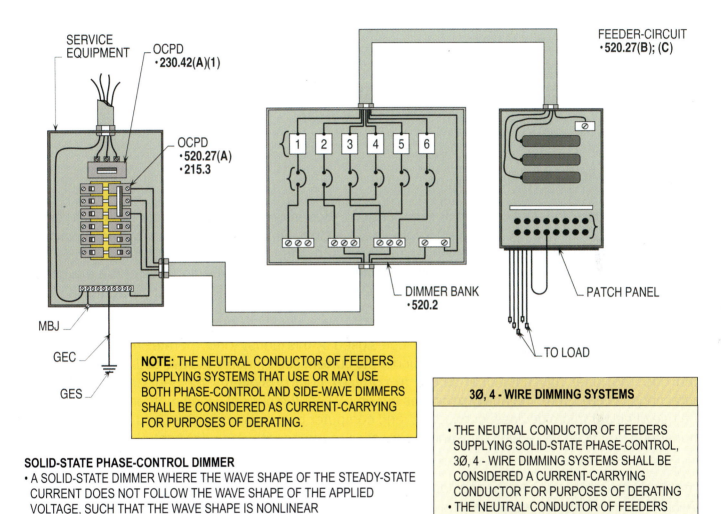

NOTE: THE NEUTRAL CONDUCTOR OF FEEDERS SUPPLYING SYSTEMS THAT USE OR MAY USE BOTH PHASE-CONTROL AND SIDE-WAVE DIMMERS SHALL BE CONSIDERED AS CURRENT-CARRYING FOR PURPOSES OF DERATING.

SOLID-STATE PHASE-CONTROL DIMMER

- A SOLID-STATE DIMMER WHERE THE WAVE SHAPE OF THE STEADY-STATE CURRENT DOES NOT FOLLOW THE WAVE SHAPE OF THE APPLIED VOLTAGE, SUCH THAT THE WAVE SHAPE IS NONLINEAR

SOLID-STATE SIDE-WAVE DIMMER

- A SOLID-STATE DIMMER WHERE THE WAVE SHAPE OF THE STEADY-STATE CURRENT FOLLOWS THE WAVE SHAPE OF THE APPLIED VOLTAGE, SUCH THAT THE WAVE SHAPE IS LINEAR

3Ø, 4 - WIRE DIMMING SYSTEMS

- THE NEUTRAL CONDUCTOR OF FEEDERS SUPPLYING SOLID-STATE PHASE-CONTROL, 3Ø, 4 - WIRE DIMMING SYSTEMS SHALL BE CONSIDERED A CURRENT-CARRYING CONDUCTOR FOR PURPOSES OF DERATING
- THE NEUTRAL CONDUCTOR OF FEEDERS SUPPLYING SOLID-STATE SIDE WAVE, 3Ø, 4 - WIRE DIMMING SYSTEMS SHALL NOT BE CONSIDERED A CURRENT-CARRYING CONDUCTOR.

SUPPLY
518.5 and Exception

Purpose of Change: This revision clarifies if the neutral conductor of feeders is considered a current-carrying conductor or not for different types of dimmer systems.

NEC Ch. 5 - Article 522
Parts I through III - Article 522

Type of Change		Panel Action		UL	UL 508	API 500 - 1997	API 505 - 1997	OSHA - 1994
New Article		Accept		-	-	-	-	-
ROP		ROC		NFPA 70E - 2004	NFPA 70B - 2006	NFPA 79 - 2007	NFPA	NEMA
pg. 573	# 15-121	pg. 335	# 15-64	-	-	-	-	-
log: 2111	CMP: 15	log: 150	Submitter: Joe Van Dam			2005 NEC: Article 522		IEC: -

2008 NEC - Article 522
Control Systems for Permanent Amusement Attractions.

I. General.

II. Control Circuits.

III. Control Circuit Wiring Methods.

Author's Comment: The Amusement Ride Industry presented several proposals for a new **Article 519** during the 2005 NEC cycle. Panel 15 agreed at the Comment Meeting to "hold" the proposals until a Task Force comprised of industry representatives and panel members could be assembled. The Task Force was assigned the responsibility of reviewing the unique characteristics of the equipment, new technology, and provisions for safety that are not already addressed in existing Code Articles.

Commercially available and standard technologies used by this industry incorporate smaller conductors. The limitation of larger conductors required under the current code hinders the use of these technologies and does not offer any additional safety. In complex control systems for permanent amusement attractions, where control reliability principles are employed, the current wiring methods limit or prevent the use of newer micro devices, connectors, and computer based technologies used for the increased monitoring, verification, redundancy and diagnostics of the apparatus under control.

The current restrictions require regular petitioning and work with the authorities having jurisdiction (AHJ) to allow alternate materials and methods appropriate to the application in order to attain the increased level of safety that can be offered to the public. This new article was written with the intent to be an inspection code, to provide direction for inspectors, and not intended to be a design manual (design is being addressed by the ASTM Std 2291). The NEC Technical Correlating Committee assigned this as new **Article 522** between **Article 520** and **Article 525** to allow for future expansion.

ARTICLE 522 - CONTROL SYSTEMS FOR PERMANENT AMUSEMENT ATTRACTIONS	
PART I - GENERAL	
SECTION	**SECTION HEADING**
522.1	SCOPE
522.2	DEFINITIONS
522.3	OTHER ARTICLES
522.5	VOLTAGE LIMITATIONS
522.7	MAINTENANCE
PART II - CONTROL CIRCUIT	
522.10	POWER SOURCES FOR CONTROL CIRCUITS
PART III - CONTROL CIRCUIT WIRING METHODS	
522.20	CONDUCTORS, BUSBARS, AND SLIP RINGS
522.21	CONDUCTOR SIZING
522.22	CONDUCTOR AMPACITY
522.23	OVERCURRENT PROTECTION FOR CONDUCTORS
522.24	CONDUCTORS OF DIFFERENT CIRCUITS IN THE SAME CABLE, CABLE TRAY, ENCLOSURE, OR RACEWAY
522.25	UNGROUNDED CONTROL CIRCUITS
522.28	CONTROL CIRCUITS IN WET LOCATIONS

CONTROL SYSTEMS FOR PERMANENT AMUSEMENT ATTRACTIONS ARTICLE 522

Purpose of Change: A new article has been added to address the requirements for control systems for permanent amusement attractions.

NEC Ch. 5 - Article 525
Part I - 525.5(B)(1) and (B)(2)

Type of Change	Panel Action	UL	UL 508	API 500 - 1997	API 505 - 1997	OSHA - 1994
Revision	Accept in Principle in Part	-	-	-	-	-
ROP	**ROC**	**NFPA 70E - 2004**	**NFPA 70B - 2006**	**NFPA 79 - 2007**	**NFPA**	**NEMA**
pg. 581 # 15-146	pg. - # -	430.11(B)	-	-	-	-
log: 2893 CMP: 15	log: -	Submitter: Mark R. Hilbert		2005 NEC: 525.5(A) and (B)		IEC: -

2005 NEC - 525.5 Overhead Conductor Clearances.

(B) Clearance to ~~Rides and Attractions~~. ~~Amusement rides and amusement attractions~~ shall be maintained not less than 4.5 m (15 ft) in any direction from overhead conductors operating at 600 volts or less, except for the conductors supplying the ~~amusement ride or attraction~~. ~~Amusement rides or attractions shall not be located under or within 4.5 m (15 ft) horizontally of conductors operating in excess of 600 volts.~~ Portable structures included in 525.3(D) shall ~~maintain a 6.9 m (22.5 ft) clearance in any direction from any part of portable structure.~~

~~Amusement rides or attractions~~.

2008 NEC - 525.5 Overhead Conductor Clearances.

(B) Clearance to Portable Structures.

(1) Under 600 Volts. Portable structures shall be maintained not less than 4.5 m (15 ft) in any direction from overhead conductors operating at 600 volts or less, except for the conductors supplying the portable structure. Portable structures included in 525.3(D) shall comply with Table 680.8.

(2) Over 600 Volts. Portable structures shall not be located under or within 4.5 m (15 ft) horizontally of conductors operating in excess of 600 volts.

Author's Comment: Changing the title and text using the term "portable structures," as it is defined in the new **525.2**, adds clarity to what type of equipment is included and expands the requirement to maintain clearances from overhead conductors for concessions, tents, power plants and similar equipment. This type of equipment was not considered under the previous language but the same hazards exist with overhead conductors. Revising this section adds clarity to the clearance requirements and resolves a conflict with the requirements of **Table 680.8, Parts A** and **C**. As previously written, the clearance requirement of 22.5 ft in **Table 680.8, Part A** had been reduced to 15 ft by Section **525.5(B)**. Conversely, the horizontal clearance requirement of **Table 680.8 Part C** has reduced the 15 ft requirement of **525.5(B)** to 10 ft.

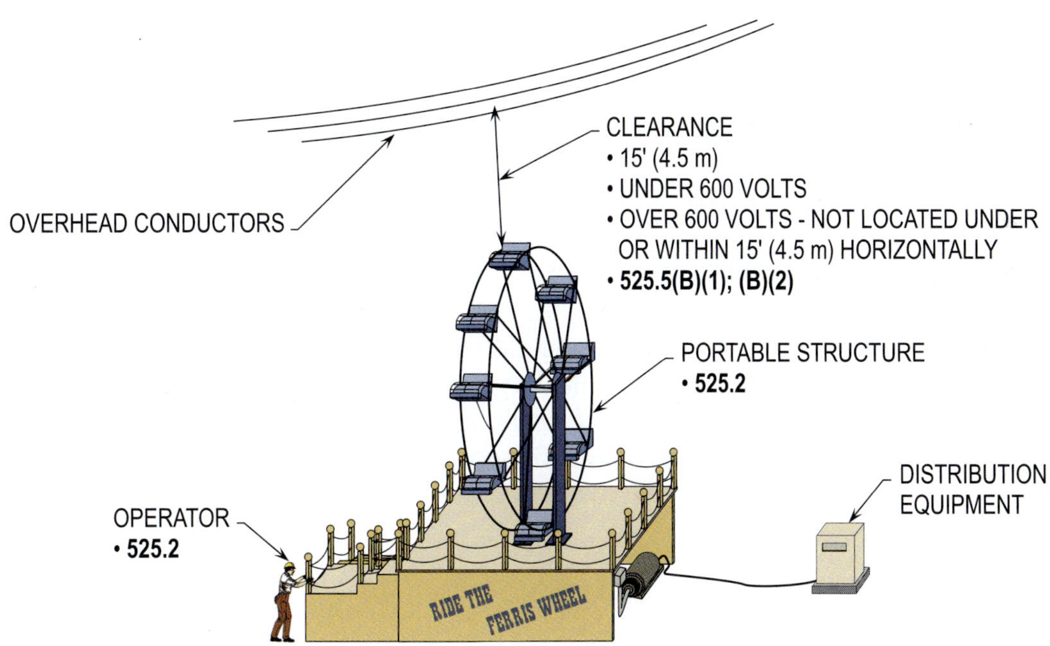

CLEARANCE TO PORTABLE STRUCTURES
525.5(B)(1) and (B)(2)

Purpose of Change: This revision clarifies the clearance requirements to portable structures.

NEC Ch. 5 - Article 547
Part - 547.5(F)

Type of Change		Panel Action		UL	UL 508	API 500 - 1997	API 505 - 1997	OSHA - 1994
Revision		Accept		-	-	-	-	-
ROP		ROC		NFPA 70E - 2004	NFPA 70B - 2006	NFPA 79 - 2007	NFPA	NEMA
pg. 590	# 19-18	pg. 341	# 19-13	-	-	-	-	-
log: 221	CMP: 19	log: 225	Submitter: Monte Ewing			2005 NEC: 547.5(F)		IEC: -

2005 NEC - 547.5 Wiring Methods.

(F) Separate Equipment Grounding Conductor. ~~Non-current-carrying metal parts of equipment, raceways, and other enclosures, where required to be grounded, shall be grounded by a copper equipment grounding conductor installed between the equipment and the building disconnecting means. If installed underground, the equipment grounding conductor shall be insulated or covered.~~

2008 NEC - 547.5 Wiring Methods.

(F) Separate Equipment Grounding Conductor. Where an equipment grounding conductor is installed within a location falling under the scope of Article 547 it shall be a copper conductor. Where an equipment grounding conductor is installed underground, it shall be insulated or covered copper.

Author's Comment: The text in **(F)** was revised to clarify that a separate copper equipment grounding conductor must be installed within an installation under Scope of **Article 547** and, if installed underground, must be a covered or insulated equipment grounding conductor to help protect the conductor. The Comment added a comma after the word "underground."

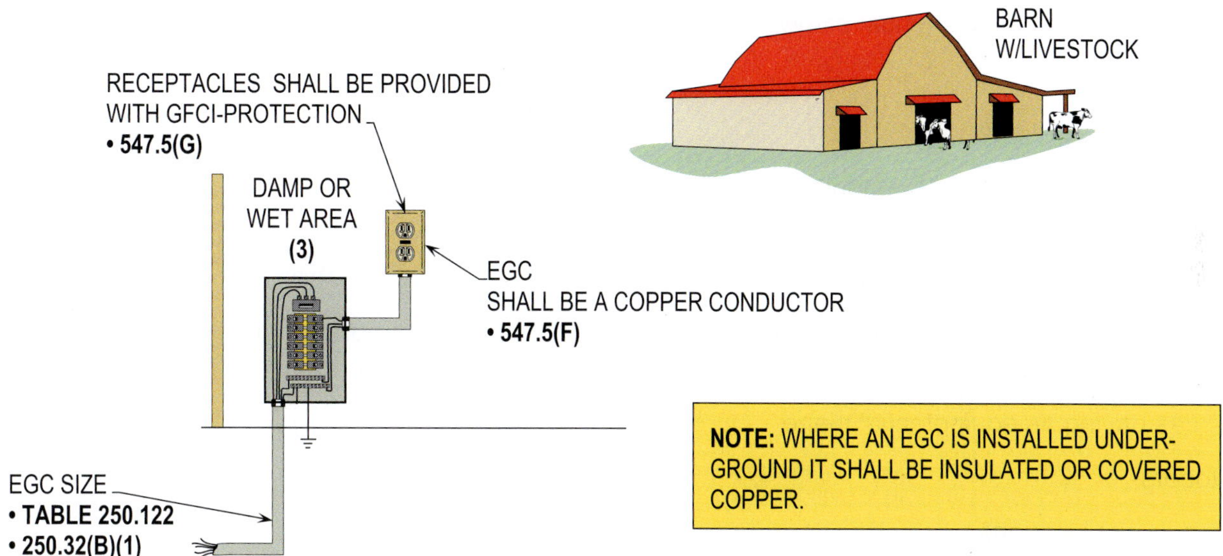

SEPARATE EQUIPMENT GROUNDING CONDUCTOR
547.5(F)

Purpose of Change: This revision clarifies that the equipment grounding conductor (EGC) shall be a copper conductor and must be insulated or covered where installed underground in locations under scope of **Article 547**.

NEC Ch. 5 - Article 547
Part - 547.5(G)

Type of Change		Panel Action		UL	UL 508	API 500 - 1997	API 505 - 1997	OSHA - 1994
Revision		Accept in Principle		-	-	-	-	-
ROP		ROC		NFPA 70E - 2004	NFPA 70B - 2006	NFPA 79 - 2007	NFPA	NEMA
pg. 590	# 19-21	pg. -	# -	-	-	-	-	-
log: 1130	CMP: 19	log: -	Submitter: Daniel Leaf			2005 NEC: 547.5(G)		IEC: -

2005 NEC - 547.5 Wiring Methods.

(G) Receptacles. All 125-volt, single-phase, 15- and 20-ampere general-purpose receptacles installed in the following locations shall have ground-fault circuit-interrupter protection for personnel:

(1) In areas having an equipotential plane

(2) Outdoors

(3) Damp or wet locations

(4) Dirt confinement areas for livestock

2008 NEC - 547.5 Wiring Methods.

(G) Receptacles. All 125-volt, single-phase, 15- and 20-ampere general-purpose receptacles installed in the following locations listed in (1) through (4) shall have ground-fault circuit-interrupter protection for personnel:

(1) In areas having an equipotential plane

(2) Outdoors

(3) Damp or wet locations

(4) Dirt confinement areas for livestock

GFCI protection shall not be required for an accessible receptacle supplying a dedicated load where a GFCI protected receptacle is located within 900 mm (3 ft) of the non-GFCI protected receptacle.

Author's Comment: The change in this section permits an accessible receptacle supplying a dedicated load to not have GFCI protection where a GFCI protected receptacle is located within 3 feet.

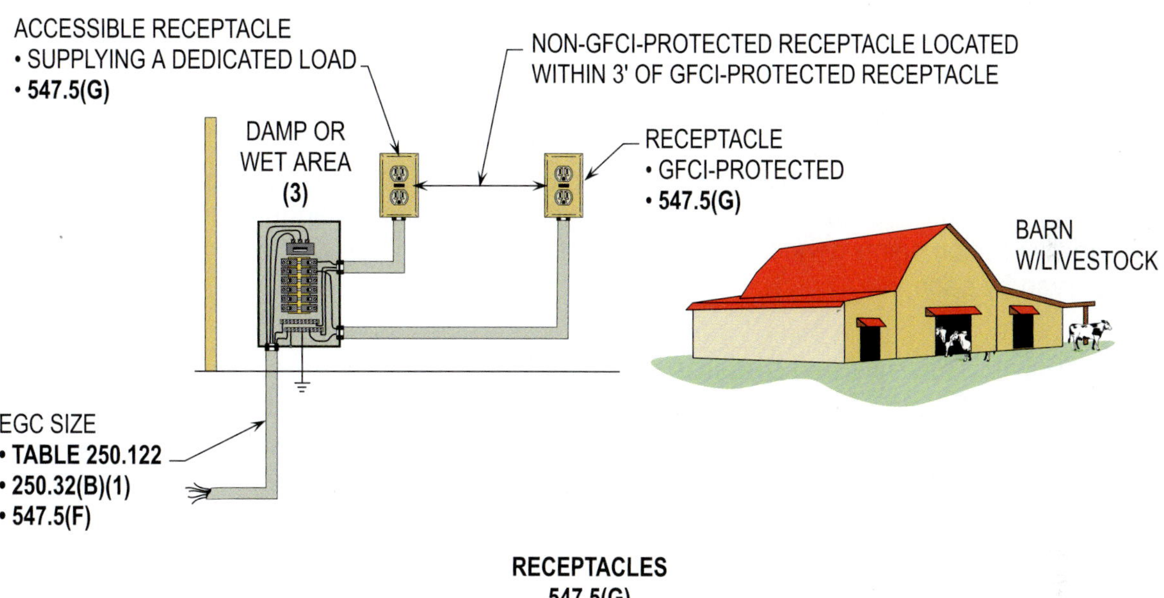

RECEPTACLES
547.5(G)

Purpose of Change: This revision clarifies that GFCI-protection shall not be required for an accessible receptacle supplying a dedicated load where a GFCI-protected receptacle is located within 3 ft. of the non-GFCI-protected receptacle.

NEC Ch. 5 - Article 547
Part - 547.9(E)

Type of Change	Panel Action	UL	UL 508	API 500 - 1997	API 505 - 1997	OSHA - 1994
Revision	Accept	-	-	-	-	-
ROP	**ROC**	**NFPA 70E - 2004**	**NFPA 70B - 2006**	**NFPA 79 - 2007**	**NFPA**	**NEMA**
pg. 592 # 19-29a	pg. - # -	-	-	-	-	-
log: CP 1903 CMP: 19	log: -	Submitter: Code Making Panel 19		2005 NEC: 547.9(E)		IEC: -

2005 NEC - 547.9 Electrical Supply to Building(s) or Structure(s) from a Distribution Point.

Overhead electrical supply shall comply with 547.9(A) and 547.9(B), or with 547.9(C). Underground electrical supply shall comply with 547.9(C) and 547.9(D).

2008 NEC - 547.9 Electrical Supply to Building(s) or Structure(s) from a Distribution Point.

Overhead electrical supply shall comply with 547.9(A) and 547.9(B), or with 547.9(C). Underground electrical supply shall comply with 547.9(C) and 547.9(D).

(E) Identification. Where a site is supplied by more than one service with any two services located a distance of 150 m (500 ft) or less apart, as measured in a straight line, a permanent plaque or directory shall be installed at each of these distribution points denoting the location of each of the other distribution points and the buildings or structures served by each.

Author's Comment: This new text requires a permanent plaque or directory to be installed at each distribution point denoting the location of each other distribution point and the buildings or structures served by each where the site is supplied by more than one service with any service located 500 feet or less apart from each other. This will ensure the proper equipment can be disconnected quickly in an emergency.

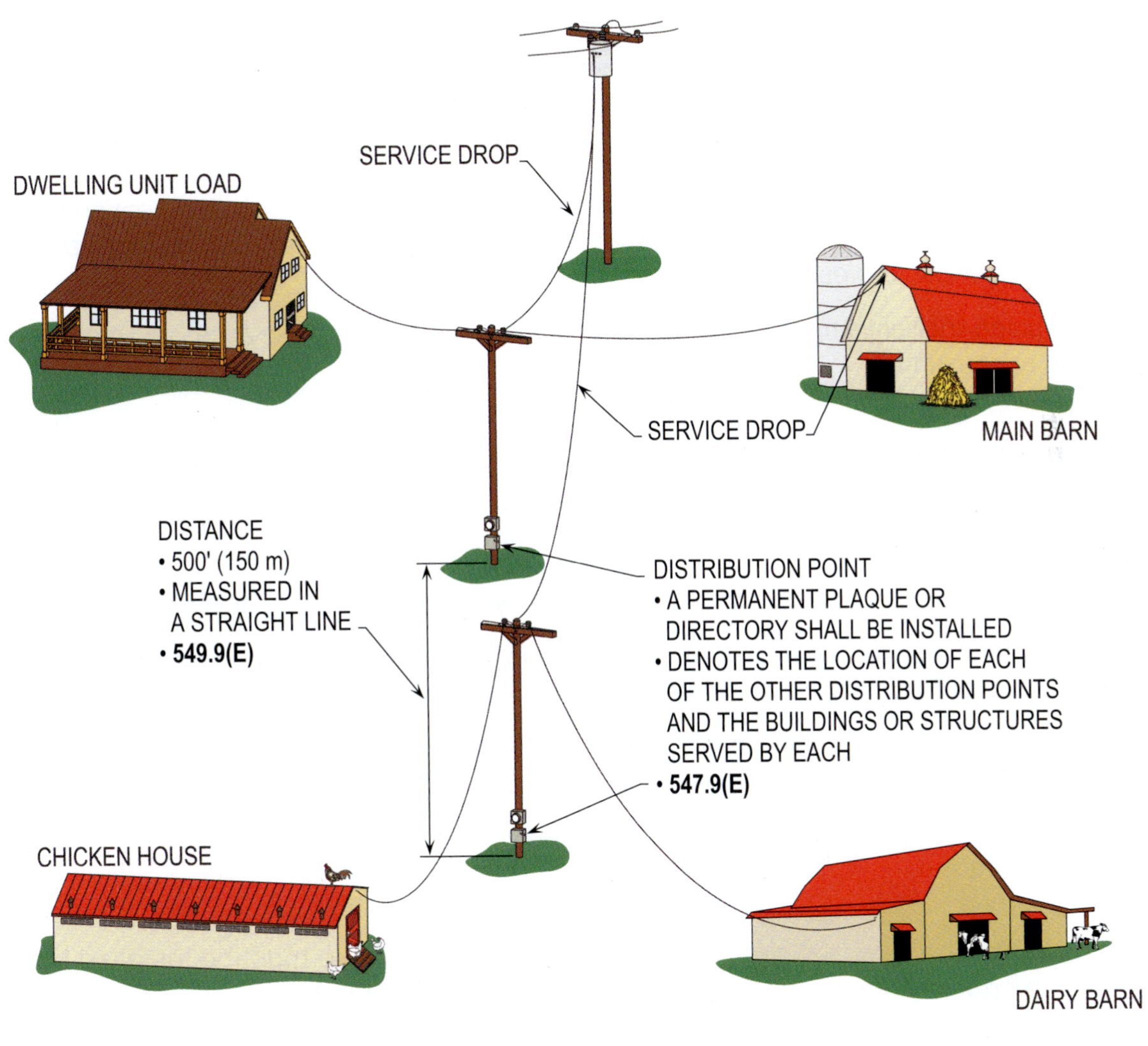

IDENTIFICATION
547.9(E)

Purpose of Change: A new subsection has been added to provide requirements for distribution points located 500 ft (150 m) or less apart.

NEC Ch. 5 - Article 547
Part - 547.10(A)(1) and (A)(2)

Type of Change	Panel Action	UL	UL 508	API 500 - 1997	API 505 - 1997	OSHA - 1994
Revision	Accept	-	-	-	-	-
ROP	**ROC**	**NFPA 70E - 2004**	**NFPA 70B - 2006**	**NFPA 79 - 2007**	**NFPA**	**NEMA**
pg. 594 # 19-36	pg. 341 # 19-18	-	-	-	-	-
log: 2595 CMP: 19	log: 2218	Submitter: Barry Bauman		2005 NEC: 547.10(A)		IEC: -

2005 NEC - 547.10 Equipotential Planes and Bonding of Equipotential Planes.

The installation and bonding of equipotential planes shall comply with 547.10(A) and 547.10(B). For the purposes of this section, the term livestock shall not include poultry.

(A) Where Required. Equipotential planes shall be installed in ~~all concrete floor~~ confinement areas ~~in livestock buildings, and in all outdoor confinement areas such as feedlots, containing~~ metallic equipment that may become energized and is accessible to livestock.

The equipotential plane shall encompass the area where the livestock stands while accessing metallic equipment that may become energized.

2008 NEC - 547.10 Equipotential Planes and Bonding of Equipotential Planes.

The installation and bonding of equipotential planes shall comply with 547.10(A) and (B). For the purposes of this section, the term livestock shall not include poultry.

(A) Where Required. Equipotential planes shall be installed where required in (A)(1) and (A)(2).

(1) Indoors. Equipotential planes shall be installed in confinement areas with concrete floors where metallic equipment is located that may become energized and is accessible to livestock.

(2) Outdoors. Equipotential planes shall be installed in concrete slabs where metallic equipment is located that may become energized and is accessible to livestock.

The equipotential plane shall encompass the area where the livestock stands while accessing metallic equipment that may become energized.

Author's Comment: The new text clarifies that an equipotential plane must be installed both indoors and outdoors with concrete floors or slabs where metallic equipment that could become energized is accessible to livestock.

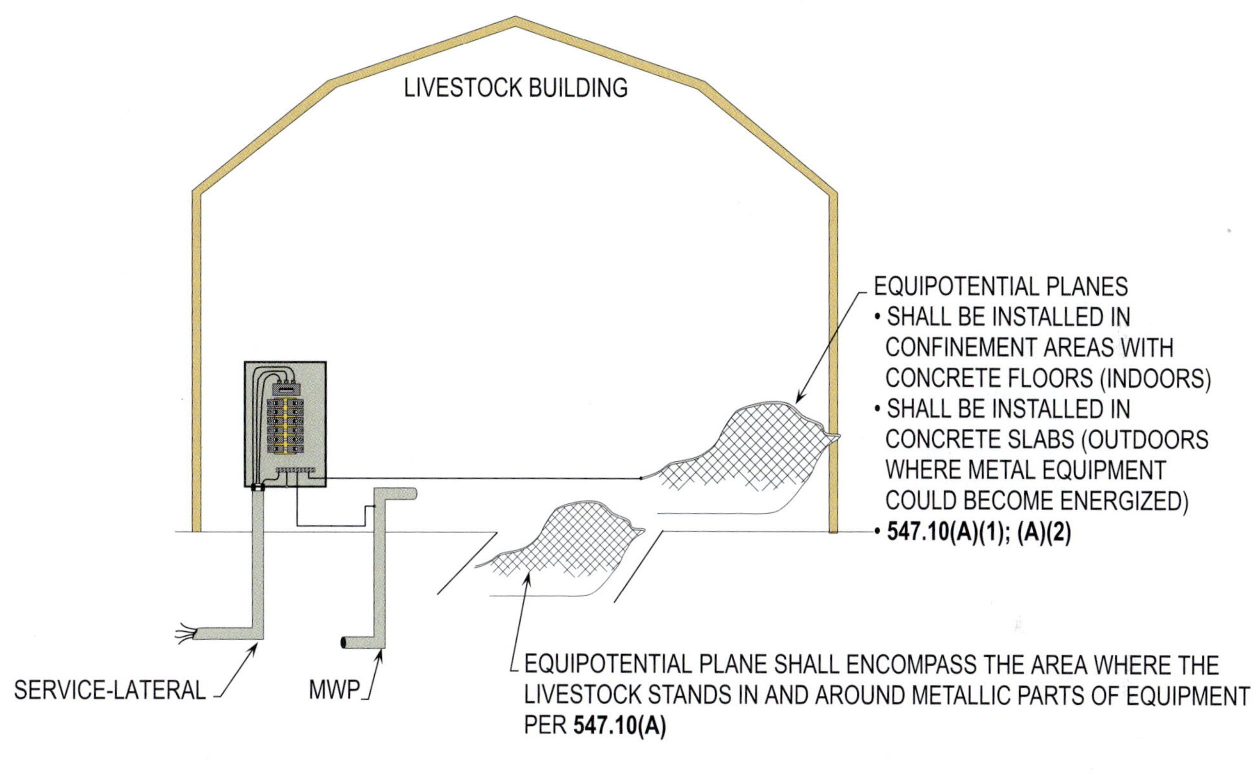
LIVESTOCK BUILDING
EQUIPOTENTIAL PLANES
• SHALL BE INSTALLED IN CONFINEMENT AREAS WITH CONCRETE FLOORS (INDOORS)
• SHALL BE INSTALLED IN CONCRETE SLABS (OUTDOORS WHERE METAL EQUIPMENT COULD BECOME ENERGIZED)
• 547.10(A)(1); (A)(2)
SERVICE-LATERAL
MWP
EQUIPOTENTIAL PLANE SHALL ENCOMPASS THE AREA WHERE THE LIVESTOCK STANDS IN AND AROUND METALLIC PARTS OF EQUIPMENT PER 547.10(A)

WHERE REQUIRED
547.10(A)(1) and (A)(2)

Purpose of Change: This revision clarifies that an equipotential plane must be installed both indoors and outdoors with concrete floors where metallic equipment that could become energized is accessible to livestock.

NEC Ch. 1 - Article 550
Part I - Article 100

Type of Change	Panel Action	UL	UL 508	API 500 - 1997	API 505 - 1997	OSHA - 1994
Relocated - Renumbered	Accept	-	-	-	-	-
ROP	**ROC**	**NFPA 70E - 2004**	**NFPA 70B - 2006**	**NFPA 79 - 2007**	**NFPA**	**NEMA**
pg. 13 # 19-1	pg. 14, 15 # 19-1, 19-2	-	-	-	-	-
log: CP 5 CMP: 1	log: 223	Submitter: Code Making Panel 19		2005 NEC: 550.2		IEC: -

2005 NEC - 550.2

Manufactured Home. A structure, transportable in one or more sections, that is ~~2.5 m~~ (8 body foot) or more in width or ~~12 m~~ (40 body foot) or more in length ~~in the traveling mode~~ or, when erected on site, is 30m² (320 ft²) or more; ~~which is~~ built on a chassis and designed to be used as a dwelling, with or without a permanent foundation, when connected ~~to the required utilities, including the plumbing, heating, air conditioning, and electrical systems contained~~ therein. Calculations used to determine the number of square meter (square feet) in a structure ~~will be~~ based on the structure's exterior dimensions, measured at the largest horizontal projections when erected on site. These dimensions include all expandable rooms, cabinets, and other projections containing interior space but do not include inside bay windows.

For the purpose of this *Code* and unless otherwise indicated, the term *mobile home* includes manufactured homes.

FPN No. 1: See the applicable building code for definition of the term *permanent foundation*.

FPN No. 2: See Part 3280, *Manufactured Home Construction and Safety Standards*, of the Federal Department of Housing and Urban Development, for additional information on the definition.

2008 NEC - Article 100

Manufactured Home. A structure, transportable in one or more sections, that<u>, in the traveling mode,</u> is <u>2.4 m</u> (8 foot body) or more in width or <u>12.2 m</u> (40 body foot) or more in length or, when erected on site, is 29.7m² (320 ft²) or more <u>and that</u> is built on a <u>permanent</u> chassis and designed to be used as a dwelling, with or without a permanent foundation, when connected therein. <u>The term manufactured home includes any structure that meets all the provisions of this paragraph except the size requirements and with respect to which the manufacturer voluntarily files a certification required by the regulatory agency, and except that such term does not include any self-propelled recreational vehicle.</u> Calculations used to determine the number of square meter (square feet) in a structure <u>are</u> based on the structure's exterior dimensions, measured at the largest horizontal projections when erected on site. These dimensions include all expandable rooms, cabinets, and other projections containing interior space, but do not include bay windows.

For the purpose of this *Code* and unless otherwise indicated, the term *mobile home* includes manufactured homes.

FPN No. 1: See the applicable building code for definition of the term *permanent foundation*.

FPN No. 2: See Part 3280, *Manufactured Home Construction and Safety Standards*, of the Federal Department of Housing and Urban Development, for additional information on the definition.

Author's Comment: This definition has been relocated from **550.2** to **Article 100** to assist the user by providing consistent meaning of defined terms throughout the National Fire Codes. The comments provided editorial corrections.

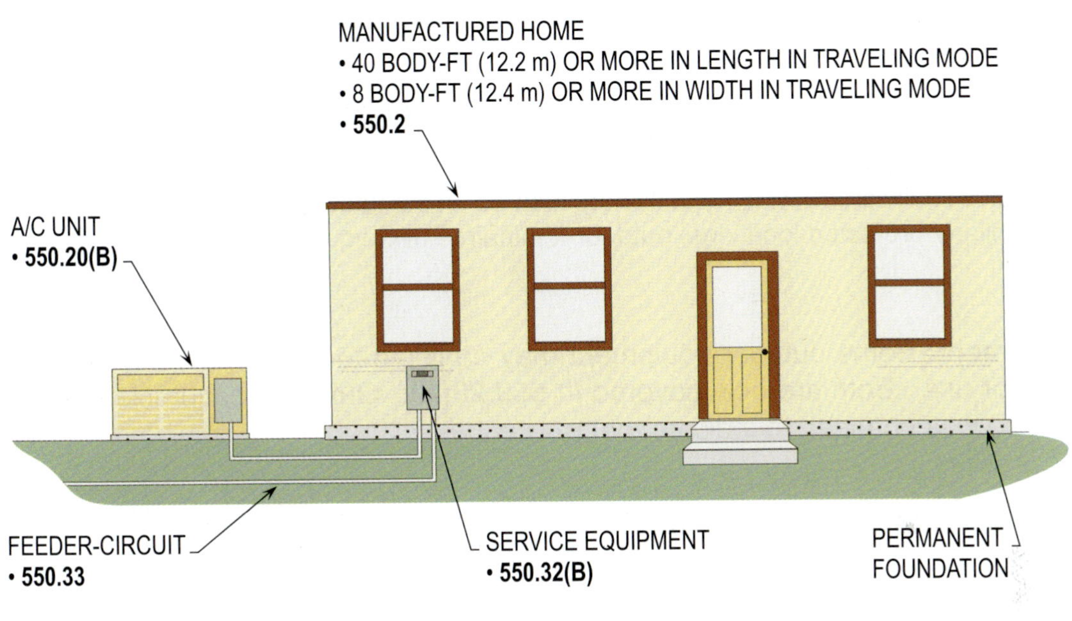

MANUFACTURED HOME
550.2

Purpose of Change: This definition has been revised to assist the user by providing consistent meaning of defined terms throughout the National Fire Codes.

NEC Ch. 5 - Article 550
Part - 550.20(A)

Type of Change	Panel Action	UL	UL 508	API 500 - 1997	API 505 - 1997	OSHA - 1994
Revision	Accept in Part	-	-	-	-	-
ROP	**ROC**	**NFPA 70E - 2004**	**NFPA 70B - 2006**	**NFPA 79 - 2007**	**NFPA**	**NEMA**
pg. 599 # 19-60	pg. - # -	-	-	-	-	-
log: 1180 CMP: 19	log: -	Submitter: Daniel Leaf		2005 NEC: 550.20(A)		IEC: -

2005 NEC - 550.20 Outdoor Outlets, Luminaires ~~(Fixtures)~~, Air-Cooling Equipment, and So Forth.

(A) Listed for Outdoor Use. Outdoor luminaires ~~(fixtures)~~ and equipment shall be listed for outdoor use. Outdoor receptacles ~~or convenience outlets~~ shall ~~be of a gasketed-cover type for use in wet locations~~. Where located on the underside of the home or located under roof extensions or similarly protected locations, outdoor luminaires ~~(fixtures)~~ and equipment shall be listed for use in damp locations.

2008 NEC - 550.20 Outdoor Outlets, Luminaires, Air-Cooling Equipment, and So Forth.

(A) Listed for Outdoor Use. Outdoor luminaires and equipment shall be listed for wet locations or outdoor use. Outdoor receptacles shall comply with 406.8. Where located on the underside of the home or located under roof extensions or similarly protected locations, outdoor luminaires and equipment shall be listed for use in damp locations.

Author's Comment: Some outdoor equipment may be listed for wet locations, whereas others are listed for outdoor use. Both are now covered in **550.20(A)**. Outdoor receptacles must comply with **406.8** so the text was changed to indicate that reference.

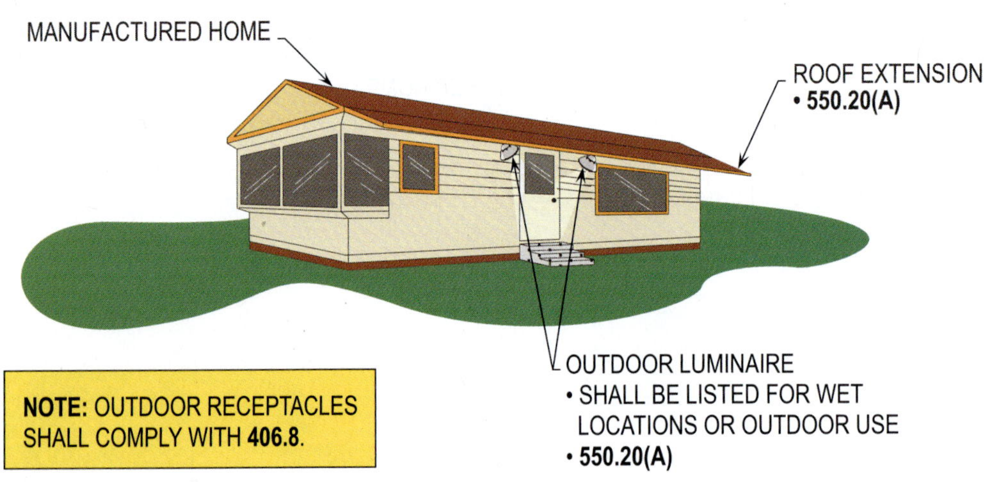

LISTED FOR OUTDOOR USE
550.20(A)

Purpose of Change: This revision clarifies that outdoor luminaires and equipment shall be listed for wet locations or outdoor use.

NEC Ch. 5 - Article 550
Part II - 550.25(B)

Type of Change	Panel Action	UL	UL 508	API 500 - 1997	API 505 - 1997	OSHA - 1994
Revision	Accept	-	-	-	-	-
ROP	ROC	NFPA 70E - 2004	NFPA 70B - 2006	NFPA 79 - 2007	NFPA	NEMA
pg. 600 # 19-64	pg. - # -	-	-	-	-	-
log: 1357 CMP: 19	log: -	Submitter: Mike Holt		2005 NEC: 550.25(B)		IEC: -

2005 NEC - 550.25 Arc-Fault Circuit-Interrupter Protection.

(B) Bedrooms of Mobile Homes and Manufactured Homes. All branch circuits that supply ~~125-volt, single-phase,~~ 15- and 20-ampere outlets installed in bedrooms of mobile homes and manufactured homes shall ~~be protected by arc-fault circuit interrupter(s)~~.

2008 NEC - 550.25 Arc-Fault Circuit-Interrupter Protection.

(B) Bedrooms of Mobile Homes and Manufactured Homes. All 120-volt branch circuits that supply 15- and 20-ampere outlets installed in bedrooms of mobile homes and manufactured homes shall comply with 210.12(B).

Author's Comment: The voltage was changed from the 125-volt rating of the receptacle to the actual 120-volt rating of the branch circuit. The requirement for AFCI protection was referenced to **210.12**.

NOTE: FOR SIMPLICITY ONLY BEDROOMS AND BATHROOMS ARE SHOWN.

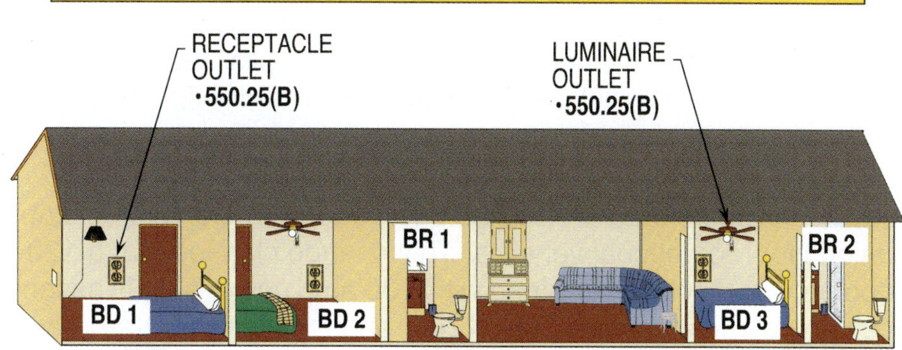

AFCI-PROTECTION
- SHALL BE PROVIDED FOR ALL 120-VOLT BRANCH CIRCUITS THAT SUPPLY 15- AND 20-AMPERE OUTLETS INSTALLED IN BEDROOMS OF MOBILE HOMES AND MANUFACTURED HOMES
- **550.25(B)**

**BEDROOMS OF MOBILE HOMES AND MANUFACTURED HOMES
550.25(B)**

Purpose of Change: This revision clarifies that AFCI-protection shall be provided for bedrooms in mobile homes and manufactured homes and must comply with **210.12(B)**.

NEC Ch. 5 - Article 551
Part IV - 551.42(C)

Type of Change		Panel Action		UL	UL 508	API 500 - 1997	API 505 - 1997	OSHA - 1994
Revision		Accept		-	-	-	-	-
ROP		ROC		NFPA 70E - 2004	NFPA 70B - 2006	NFPA 79 - 2007	NFPA	NEMA
pg. 603	# 19-78	pg. 343	# 19-29	-	-	-	-	-
log: 2305	CMP: 19	log: 2212	Submitter: Kent Perkins			2005 NEC: 551.42(C)		IEC: -

2005 NEC - 551.42 Branch Circuits Required.

Each recreational vehicle containing a 120-volt electrical system shall contain one of the following.

(C) Two to Five 15- or 20-Ampere Circuits. A maximum of five 15- or 20-ampere circuits to supply lights, receptacle outlets, and fixed appliances shall be permitted. Such recreational vehicles shall be equipped with a distribution panelboard rated at 120 volts maximum with a 30-ampere rated main power supply assembly. Not more than two 120-volt thermostatically controlled appliances (i.e., air conditioner and water heater) shall be installed in such systems unless appliance isolation switching, energy management systems, or similar methods are used.

2008 NEC - 551.42 Branch Circuits Required.

Each recreational vehicle containing a 120-volt electrical system shall contain one of the following.

(C) Two to Five 15- or 20-Ampere Circuits. A maximum of five 15- or 20-ampere circuits to supply lights, receptacle outlets, and fixed appliances shall be permitted. Such recreational vehicles shall be permitted to be equipped with distribution panelboards rated 120 volts maximum or 120/240 volts maximum and listed for 30-ampere applications supplied by the appropriate power supply assemblies. Such recreational vehicles shall be equipped with a distribution panelboard rated at 120 volts maximum with a 30-ampere rated main power supply assembly. Not more than two 120-volt thermostatically controlled appliances (i.e., air conditioner and water heater) shall be installed in such systems unless appliance isolation switching, energy management systems, or similar methods are used.

Author's Comment: A new second sentence has been added recognizing distribution panelboards with 30 amp applications for recreational vehicles. This change clarifies that a 120-volt only or a 120/240 volt rated panelboard would still be permitted with a 30 amp power supply assembly. Any reference to a 50 amp application was deleted since **551.42(C)** only applies to 30-amp applications. 50 amp requirements are covered under **551.42(D)**.

TWO TO FIVE 15- OR 20-AMPERE CIRCUITS
551.42(C)

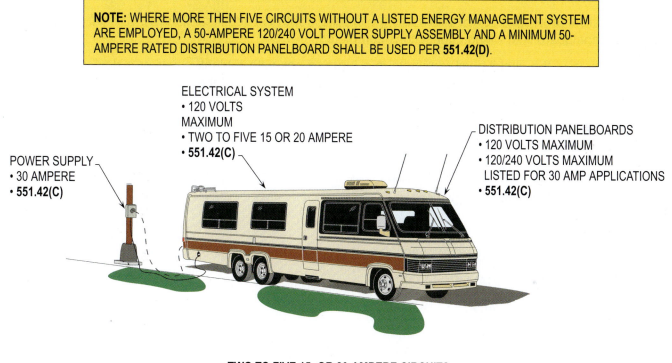

Purpose of Change: This revision clarifies that a 120 volt or 120/240 volt rated distribution panelboards shall be permitted with a 30 amp power supply assembly.

NEC Ch. 5 - Article 551
Part IV - 551.47(S)

Type of Change		Panel Action		UL	UL 508	API 500 - 1997	API 505 - 1997	OSHA - 1994
New Subsection		Accept		-	-	-	-	-
ROP		ROC		NFPA 70E - 2004	NFPA 70B - 2006	NFPA 79 - 2007	NFPA	NEMA
pg. 606	# 19-89	pg. 345	# 19-35	-	-	-	-	-
log: 2311	CMP: 19	log: 646	Submitter: Kent Perkins			2005 NEC: -		IEC: -

2008 NEC - 551.47 Wiring Methods.

(S) Prewiring for Other Circuits. Prewiring installed for the purpose of installing other appliances or devices shall comply with the applicable portions of this article and the following:

(1) An overcurrent protection device with a rating compatible with the circuit conductors shall be installed in the distribution panelboard with wiring connections completed.

(2) The load end of the circuit shall terminate in a junction box with a blank cover or a device listed for the purpose. Where a junction box with blank cover is used, the free ends of the conductors shall be adequately capped or taped.

(3) A label conforming to 551.46(D) shall be placed on or adjacent to the junction box or device listed for the purpose and shall read as follows:

THIS CONNECTION IS FOR ____ RATED ____

VOLT AC, 60HZ, ____ AMPERES MAXIMUM. DO

NOT EXCEED CIRCUIT RATING.

AN AMPERE RATING NOT TO EXCEED 80 PERCENT

OF THE CIRCUIT RATING SHALL BE LEGIBLY

MARKED IN THE BLANK SPACE.

Author's Comment: This new section provides installation requirements for pre-wiring of circuits for other appliances or devices within an RV.

LABEL SHALL READ:
- THIS CONNECTION IS FOR ____ RATED ____ VOLT AC, 60 HZ, ____ AMPERES MAXIMUM DO NOT EXCEED CIRCUITS RATING
- AN AMPERE RATING NOT TO EXCEED 80 PERCENT OF THE CIRCUIT RATING SHALL BE LEGIBLY MARKED IN THE BLANK SPACE
- **551.47(S)(3)**

PREWIRING FOR OTHER CIRCUITS

- OCPD WITH A RATING COMPATIBLE WITH THE CIRCUIT CONDUCTORS SHALL BE INSTALLED IN THE DISTRIBUTION PANELBOARD WITH WIRING CONNECTIONS COMPLETED.
- THE LOAD END OF THE CIRCUIT SHALL TERMINATE IN A JUNCTION BOX WITH A BLANK COVER OR A DEVICE LISTED FOR THE PURPOSE
- A LABEL SHALL BE PLACED ON OR ADJACENT TO THE JUNCTION BOX OR DEVICE LISTED FOR THE PURPOSE

PREWIRING FOR OTHER CIRCUITS
551.47(S)

Purpose of Change: A new subsection has been added to provide installation requirements for prewiring of circuits for other appliances or devices within an RV.

NEC Ch. 5 - Article 555
Part - 555.9

Type of Change	Panel Action	UL	UL 508	API 500 - 1997	API 505 - 1997	OSHA - 1994
Revision	Accept in Principle	-	-	-	-	-
ROP	**ROC**	**NFPA 70E - 2004**	**NFPA 70B - 2006**	**NFPA 79 - 2007**	**NFPA**	**NEMA**
pg. 612 # 19-117	pg. 346 # 19-46	-	-	-	-	-
log: 134 CMP: 19	log: 385 Submitter: David Fecitt			2005 NEC: 555.9		IEC: -

2005 NEC - 555.9 Electrical Connections.

~~All~~ electrical connections shall be located at least 305 mm (12 in.) above the deck of a floating pier. All electrical connections shall be located at least 305 mm (12 in.) above the deck of a fixed pier, but not below the electrical datum plane.

2008 NEC - 555.9 Electrical Connections.

Electrical connections shall be located at least 305 mm (12 in.) above the deck of a floating pier. Conductor splices, within approved junction boxes, utilizing sealed wire connector systems listed and identified for submersion shall be permitted where located above the waterline, but below the electrical datum field for floating piers. All electrical connections shall be located at least 305 mm (12 in.) above the deck of a fixed pier, but not below the electrical datum plane.

Author's Comment: The added sentence recognizes that electrical conductor splices can be installed in a Type 6P box that is designed to be directly immersed in water as long as the connectors are listed for direct immersion also. The Comment deleted the words "Type 6P" because a Type 6P box may not be needed in every case, however, a junction box suitable for the location is still required for splicing. In lieu of the 6P box, the Comment added the requirement for sealed wire connector systems to be lised and identified for submersion. Not all listed sealed wire connector systems provide the same degree of protection from moisture ingress.

NOTE 1: ELECTRICAL CONDUCTORS CAN BE SPLICED IN A TYPE 6P BOX OR A JUNCTION BOX THAT'S SUITABLE FOR THE LOCATION PER **555.9**.

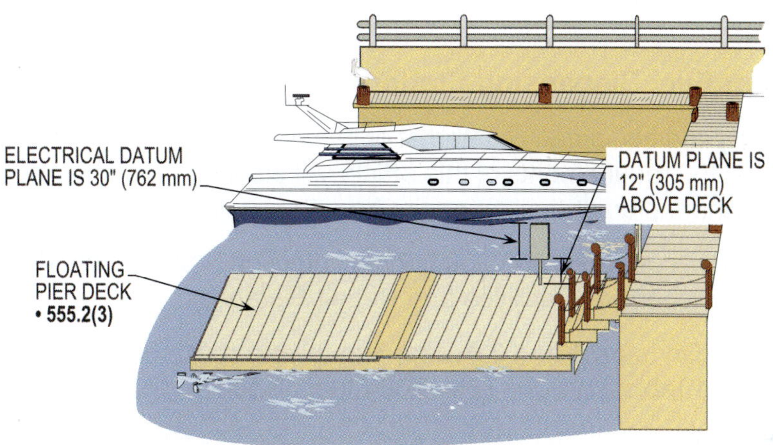

NOTE 2: DESIGNED TYPE 6P BOX CAN BE DIRECTLY IMMERSED IN WATER AS LONG AS CONNECTORS ARE LISTED FOR SUCH USE PER **555.9**.

ELECTRICAL CONNECTIONS
555.9

Purpose of Change: To provide requirements for selecting boxes and and conductor splices based on condition of use.

NEC Ch. 5 - Article 555
Part - 555.21(B)

82

Type of Change	Panel Action	UL	UL 508	API 500 - 1997	API 505 - 1997	OSHA - 1994
Revision	Accept	-	-	-	-	-
ROP	ROC	NFPA 70E - 2004	NFPA 70B - 2006	NFPA 79 - 2007	NFPA	NEMA
pg. 614 # 19-127	pg. 347 # 19-48	-	-	-	-	-
log: 2950 CMP: 19	log: 1402 Submitter: Philip Simmons			2005 NEC: 555.21		IEC: -

2005 NEC - 555.21 Motor Fuel Dispensing Stations — Hazardous (Classified) Locations.

Electrical wiring and equipment located at or serving motor fuel dispensing ~~stations~~ shall comply with Article 514 in addition to the requirements of this article. All electrical wiring for power and lighting shall be installed on the side of the wharf, pier, or dock opposite from the liquid piping system.

FPN: For additional information, see NFPA 303-2000, Fire Protection Standard for Marinas and Boatyards, and NFPA 30A-2003, Motor Fuel Dispensing Facilities and Repair Garages.

2008 NEC - 555.21 Motor Fuel Dispensing Stations — Hazardous (Classified) Locations.

(A) General. Electrical wiring and equipment located at or serving motor fuel dispensing locations shall comply with Article 514 in addition to the requirements of this article. All electrical wiring for power and lighting shall be installed on the side of the wharf, pier, or dock opposite from the liquid piping system.

FPN: For additional information, see NFPA 303-2000, Fire Protection Standard for Marinas and Boatyards, and NFPA 30A-2003, Motor Fuel Dispensing Facilities and Repair Garages.

(B) Classification of Class I, Division 1 and 2 Areas. The following criteria shall be used for the purposes of applying Table 514.3(B)(1) and Table 514.3(B)(2) to motor fuel dispensing equipment on floating or fixed piers, wharfs, or docks.

(1) Closed Construction. Where the construction of floating docks, piers. or wharfs is closed so that there is no space between the bottom of the dock, pier, or wharf and the water, such as concrete enclosed expanded foam or similar construction, and having integral service boxes with supply chases, the following shall apply:

(a) The space above the surface of the floating dock, pier, or wharf shall be a Class I, Division 2 location with distances as identified in Table 514.3(B)(1) Dispenser and Outdoor.

(b) The space below the surface of the floating dock, pier, or wharf having areas or enclosures such as tubs, voids, pits, vaults, boxes, depressions, fuel piping chases, or similar spaces where flammable liquid or vapor can accumulate shall be a Class I, Division 1 location.

Exception No. 1: Dock, pier, or wharf sections that do not support fuel dispensers and abut but are 6.0 m (20 feet) or more from dock sections that support fuel dispenser(s) shall be permitted to be Class I, Division 2 where documented air space is provided between dock sections to permit flammable liquids or vapors to dissipate and not travel to these dock sections. Such documentation shall comply with 500.4(A).

Exception No. 2: Dock, pier, or wharf sections that do not support fuel dispensers and do not directly abut sections that support fuel dispensers shall be permitted to be unclassified where documented air space is provided and where flammable liquids or vapors can not travel to these dock sections. Such documentation shall comply with 500.4(A).

FPN: See 500.4(A) for documentation requirements.

(2) Open Construction. Where the construction of piers, wharfs, or docks is open, such as decks built on stringers supported by pilings, floats, pontoons or similar construction, the following shall apply:

(a) The area 450 mm (18 in) above the surface of the dock, pier, or wharf and extending 6.0 m (20 ft) horizontally in all directions from the outside edge of the dispenser and down to the water level shall be Class I, Division 2.

(b) Enclosures such as tubs, voids, pits, vaults, boxes, depressions, piping chases, or similar spaces where flammable liquids or vapors can accumulate within 6.0 m (20 ft) of the dispenser shall be a Class I, Division 1 location.

Author's Comment: This revision provides area classification requirements of Class I, Division 1 and 2 locations for motor fuel dispensing areas and equipment for those facilities that may be on floating docks or piers with both closed construction and open construction. This change provides coverage of open and closed construction methods of docks, marinas, and piers.

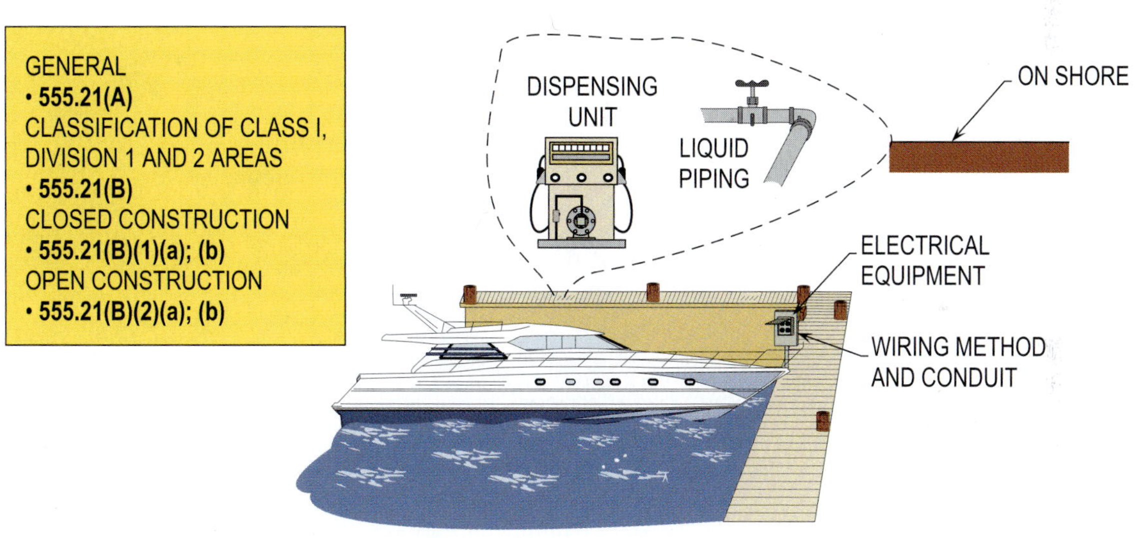

CLASSIFICATION OF CLASS I, DIVISION 1 AND 2 AREAS 555.21(B)

Purpose of Change: To include requirements to provide coverage of open and closed construction methods for docks, marinas, and piers.

6

Special Equipment

Chapter 6 has always been utilized by designers, installers and inspectors when dealing with special equipment that is located and used in specific occupancies.

Special equipment, due to the specialized rules and regulations that are needed for their safe design and installation, cannot be located in **Chapter 4**, which covers Equipment for General Use. Specialized equipment ranges from such diverse equipment as electric signs and outline lighting to elevators to electric vehicle charging.

If an engineer is designing an electrical system where elevators are to be installed in an office building, **Article 620** will contain the requirements necessary for such specialized equipment and its associated apparatus. Note, anytime special equipment is designed, installed and inspected, one of the 600 (series) Article numbers of **Chapter 6** will be selected, based upon the specialized equipment involved.

For example, an electrician designing an electrical circuit for a welder must consult **Article 630** and comply with the requirements pertaining to the type of welder being used.

For those users interested in the requirements necessary to design and install disconnects for deenergizing power circuits supplying cranes, they must utilize **Article 610** to find these rule.

NEC Ch. 6 - Article 600
Part I - 600.2

Type of Change	Panel Action	UL	UL 508	API 500 - 1997	API 505 - 1997	OSHA - 1994
Revision	Accept in Principle	48	-	-	-	1910.399(a)
ROP	**ROC**	**NFPA 70E - 2004**	**NFPA 70B - 2006**	**NFPA 79 - 2007**	**NFPA**	**NEMA**
pg. 633 / # 18-111	pg. - / # -	Article 100	-	-	-	-
log: 526 / CMP: 18	log: - / Submitter: Michael J. Johnston			2005 NEC: 600.2		IEC: -

2005 NEC - 600.2 Definitions.

Section Sign. A sign or outline lighting system, shipped as subassemblies, that requires field-installed wiring between the subassemblies to complete the overall sign.

2008 NEC - 600.2 Definitions.

Section Sign. A sign or outline lighting system, shipped as subassemblies, that requires field-installed wiring between the subassemblies to complete the overall sign. The subassemblies are either physically joined to form a single sign unit or are installed as separate remote parts of an overall sign.

Author's Comment: Section sign subassemblies may be physically joined to form a single sign unit, or may be installed as separate remote parts of an overall sign. The revision to this definition of section sign clarifies that the multiple parts of a section sign are referred to as subassemblies and the only field wiring involved are the interconnections and wiring between subassemblies or field-installed wiring between the subassemblies remote from one another for overall sign and connection of the subassemblies to the power source.

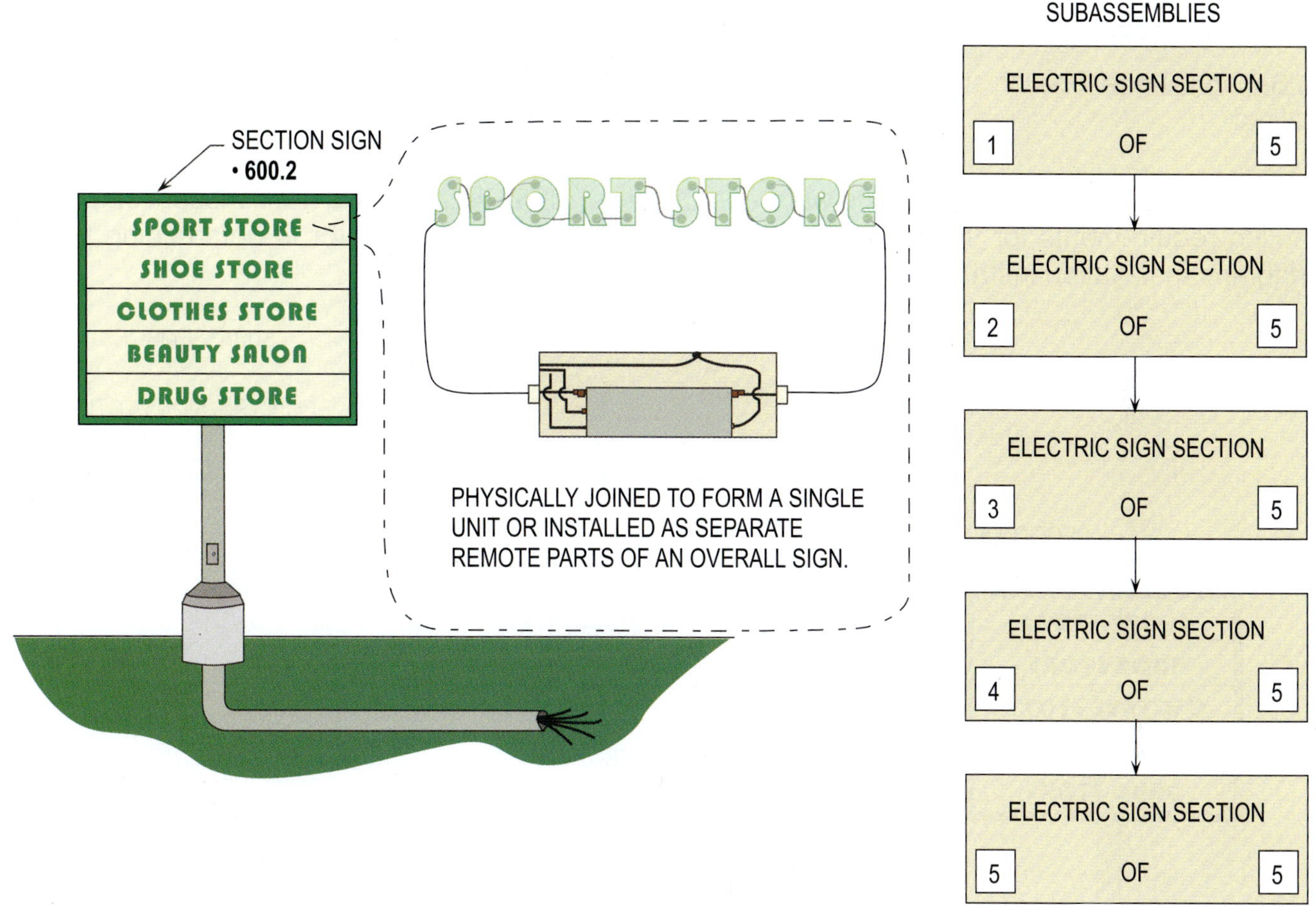

SECTION SIGN
600.2

Purpose of Change: This revision clarifies that section sign subassemblies are either physically joined to form a single sign unit or are installed as separate remote parts of an overall sign.

NEC Ch. 6 - Article 600
Part I - 600.4(C)

Type of Change	Panel Action	UL	UL 508	API 500 - 1997	API 505 - 1997	OSHA - 1994
New Subsection	Accept in Principle	48	-	-	-	-
ROP	**ROC**	**NFPA 70E - 2004**	**NFPA 70B - 2006**	**NFPA 79 - 2007**	**NFPA**	**NEMA**
pg. 633 # 18-113	pg. - # -	-	-	-	-	-
log: 2861 CMP: 18	log: -	Submitter: Randall K. Wright		2005 NEC: -		IEC: -

2008 NEC - 600.4 Markings.

(C) Section Signs. Section signs shall be marked to indicate that field-wiring and installation instructions are required.

Author's Comment: Sectional signs were added in the 2005 NEC and this new subsection provides marking requirements for these special signs to aid the electrical inspector in determining the filed wiring and installation instructions.

SUBASSEMBLIES

ELECTRIC SIGN SECTION 1 OF 5

ELECTRIC SIGN SECTION 2 OF 5

ELECTRIC SIGN SECTION 3 OF 5

ELECTRIC SIGN SECTION 4 OF 5

ELECTRIC SIGN SECTION 5 OF 5

SPORT STORE

SECTION SIGN
• 600.2

SPORT STORE
SHOE STORE
CLOTHES STORE
BEAUTY SALON
DRUG STORE

ELECTRIC SIGN SECTION 1 OF 5

SECTION SIGNS SHALL BE MARKED TO INDICATE THAT FIELD-WIRING AND INSTALLATION INSTRUCTIONS ARE REQUIRED

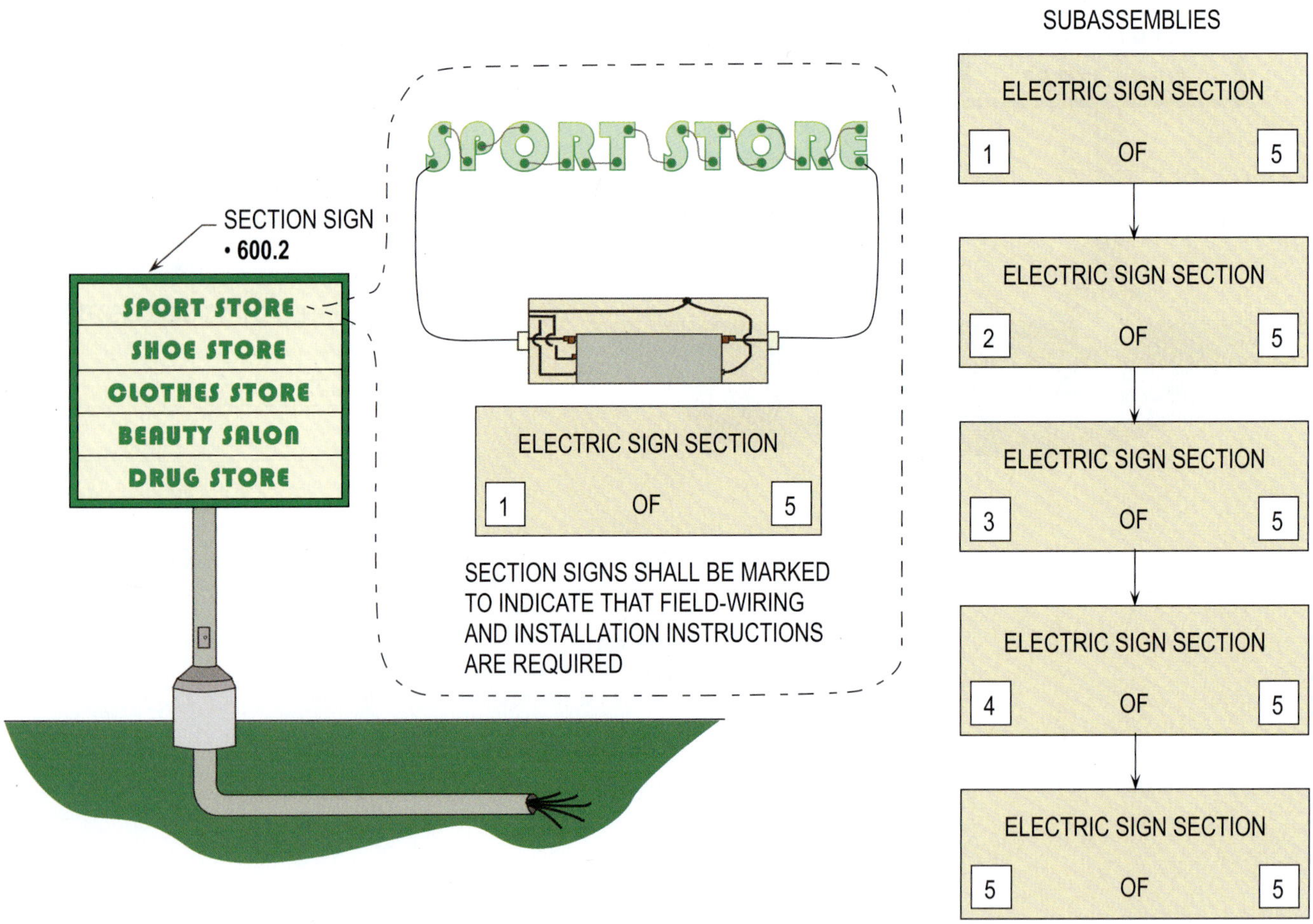

SECTION SIGNS
600.4(C)

Purpose of Change: A new section has been added to require section signs to be marked to indicate that field-wiring and installation instructions are required.

NEC Ch. 6 - Article 600
Part I - 600.5(C)(3)

Type of Change		Panel Action		UL	UL 508	API 500 - 1997	API 505 - 1997	OSHA - 1994
Revision		Accept		-	-	-	-	-
ROP		ROC		NFPA 70E - 2004	NFPA 70B - 2006	NFPA 79 - 2007	NFPA	NEMA
pg. 634	# 18-115	pg. -	# -	-	-	-	-	-
log: 520	CMP: 18	log: -	Submitter: Michael J. Johnston			2005 NEC: 600.6(C)(3)		IEC: -

2005 NEC - 600.5 Branch Circuits.

(C) Wiring Methods. Wiring methods used to supply signs shall comply with 600.5(C)(1), (C)(2), and (C)(3).

(3) Metal Poles. Metal poles used to support signs shall be permitted to enclose supply conductors, provided the poles and conductors are installed in accordance with 410.15(B).

2008 NEC - 600.5 Branch Circuits.

(C) Wiring Methods. Wiring methods used to supply signs shall comply with 600.5(C)(1), (C)(2), and (C)(3).

(3) Metal or Nonmetallic Poles. Metal or nonmetallic poles used to support signs shall be permitted to enclose supply conductors, provided the poles and conductors are installed in accordance with 410.30(B).

Author's Comment: Nonmetallic poles have been added to support signs and can be used to enclose supply conductors.

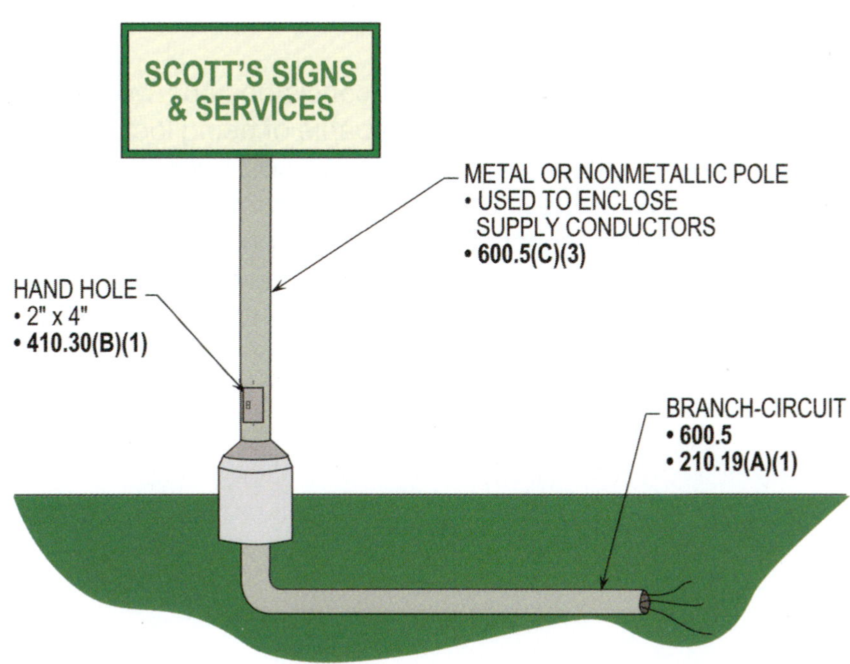

METAL OR NONMETALLIC POLES
600.5(C)(3)

Purpose of Change: This revision clarifies that nonmetallic poles shall be permitted to be used to enclose supply conductors.

NEC Ch. 6 - Article 600
Part I - 600.6(A)(1)

Type of Change	Panel Action	UL	UL 508	API 500 - 1997	API 505 - 1997	OSHA - 1994
Revision	Accept	-	-	-	-	1910.306(a)(1)
ROP	ROC	NFPA 70E - 2004	NFPA 70B - 2006	NFPA 79 - 2007	NFPA	NEMA
pg. 634 # 18-118	pg. - # -	430.1(B)(1)	-	-	-	-
log: 355 CMP: 18	log: - Submitter: Michael J. Johnston			2005 NEC: 600.6(A)(1)		IEC: -

2005 NEC - 600.6 Disconnects.

(A) Location.

(1) Within Sight of the Sign. The disconnecting means shall be within sight of the sign or outline lighting system that it controls. Where the disconnecting means is out of the line of sight from any section that may be energized, the disconnecting means shall be capable of being locked in the open position.

2008 NEC - 600.6 Disconnects.

(A) Location.

(1) Within Sight of the Sign. The disconnecting means shall be within sight of the sign or outline lighting system that it controls. Where the disconnecting means is out of the line of sight from any section that may be energized, the disconnecting means shall be capable of being locked in the open position. The provision for locking or adding a lock to the disconnecting means must remain in place at the switch or circuit breaker whether the lock is installed or not. Portable means for adding a lock to the switch or circuit breaker shall not be permitted.

Author's Comment: The change in wording provides consistency between other similar rules in the NEC that also call for the disconnecting means to be capable of being locked in the open position. The phrase "capable of being locked in the open position" is used over 25 times in the NEC and the purpose is the same in every instance. Electrical safety rules for the worker should be consistent and the wording and requirements are now consistent where this phrase is used. The last sentence was added since some of the portable units available for snapping on to circuit breakers do remain with the switch or circuit breaker after they are installed on the breakers when the lock is not installed, but they are portable.

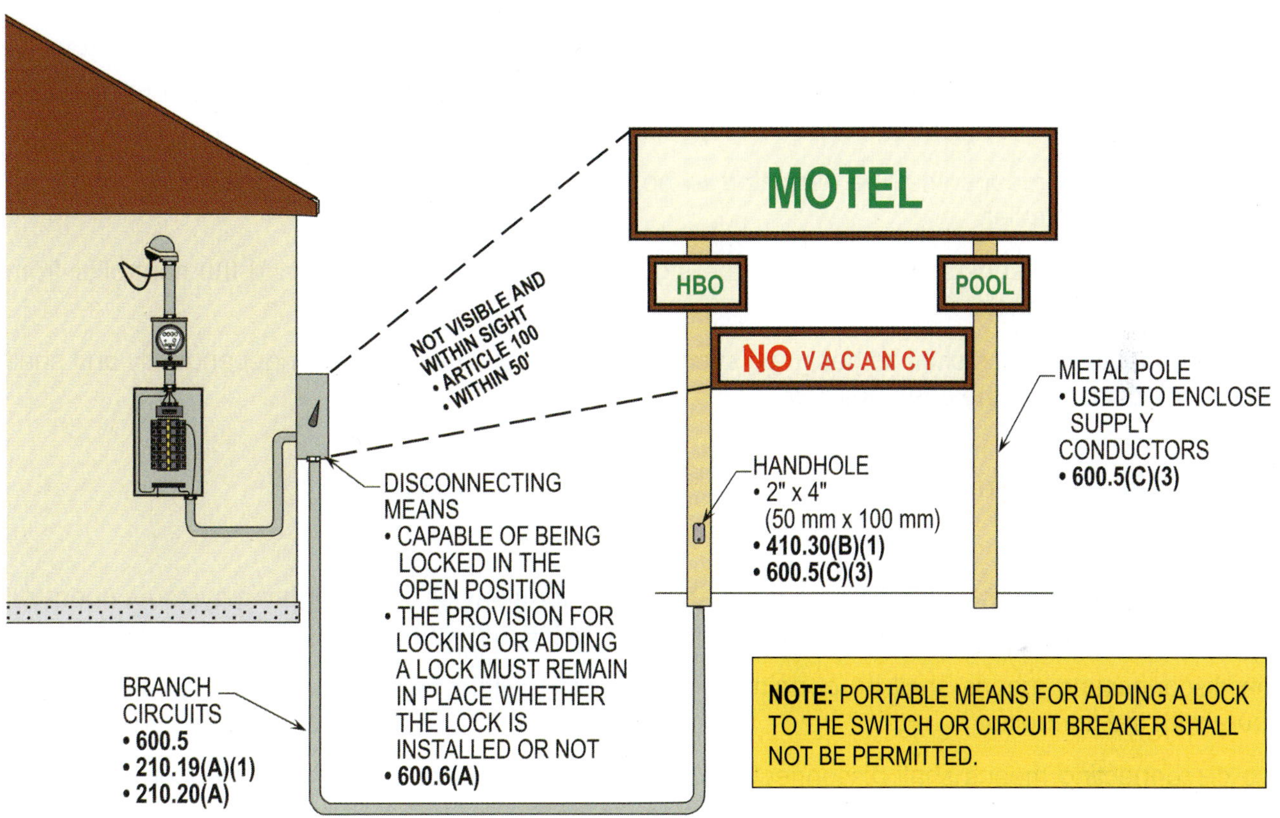

WITHIN SIGHT OF THE SIGN
600.6(A)(1)

Purpose of Change: This revision clarifies the provisions for locking or adding a lock to the disconnecting means.

NEC Ch. 6 - Article 600
Part I - 600.6(A)(2)(3)

Type of Change	Panel Action	UL	UL 508	API 500 - 1997	API 505 - 1997	OSHA - 1994
Revision	Accept	-	-	-	-	1910.306(a)(1)
ROP	**ROC**	**NFPA 70E - 2004**	**NFPA 70B - 2006**	**NFPA 79 - 2007**	**NFPA**	**NEMA**
pg. 634 # 18-120	pg. - # -	430.1(B)(2)	-	-	-	-
log: 360 CMP: 18	log: -	Submitter: Michael J. Johnston		2005 NEC: 600.6(A)(2)(3)		IEC: -

2005 NEC - 600.6 Disconnects.

(A) Location.

(2) Within Sight of the Controller. The following shall apply for signs or outline lighting systems operated by electronic or electromechanical controllers located external to the sign or outline lighting system:

(1) The disconnecting means shall be permitted to be located within sight of the controller or in the same enclosure with the controller.

(2) The disconnecting means shall disconnect the sign or outline lighting system and the controller from all ungrounded supply conductors.

(3) The disconnecting means shall be designed such that no pole can be operated independently and shall be capable of being locked in the open position.

2008 NEC - 600.6 Disconnects.

(A) Location.

(2) Within Sight of the Controller. The following shall apply for signs or outline lighting systems operated by electronic or electromechanical controllers located external to the sign or outline lighting system:

(1) The disconnecting means shall be permitted to be located within sight of the controller or in the same enclosure with the controller.

(2) The disconnecting means shall disconnect the sign or outline lighting system and the controller from all ungrounded supply conductors.

(3) The disconnecting means shall be designed such that no pole can be operated independently and shall be capable of being locked in the open position. The provisions for locking or adding a lock to the disconnecting means must remain in place at the switch or circuit breaker whether the lock is installed or not. Portable means for adding a lock to the switch or circuit breaker shall not be permitted.

Author's Comment: The change in wording provides consistency between other similar rules in the NEC that also call for the disconnecting means to be capable of being locked in the open position. The phrase "capable of being locked in the open position" is used over 25 times in the NEC and the purpose is the same in every instance. Electrical safety rules for the worker should be consistent and the wording and requirements should be consistent where this phrase is used. The last sentence was added since some of the portable units available for snapping on to circuit breakers do remain with the switch or circuit breaker after they are installed on the breakers when the lock is not installed, but they are portable.

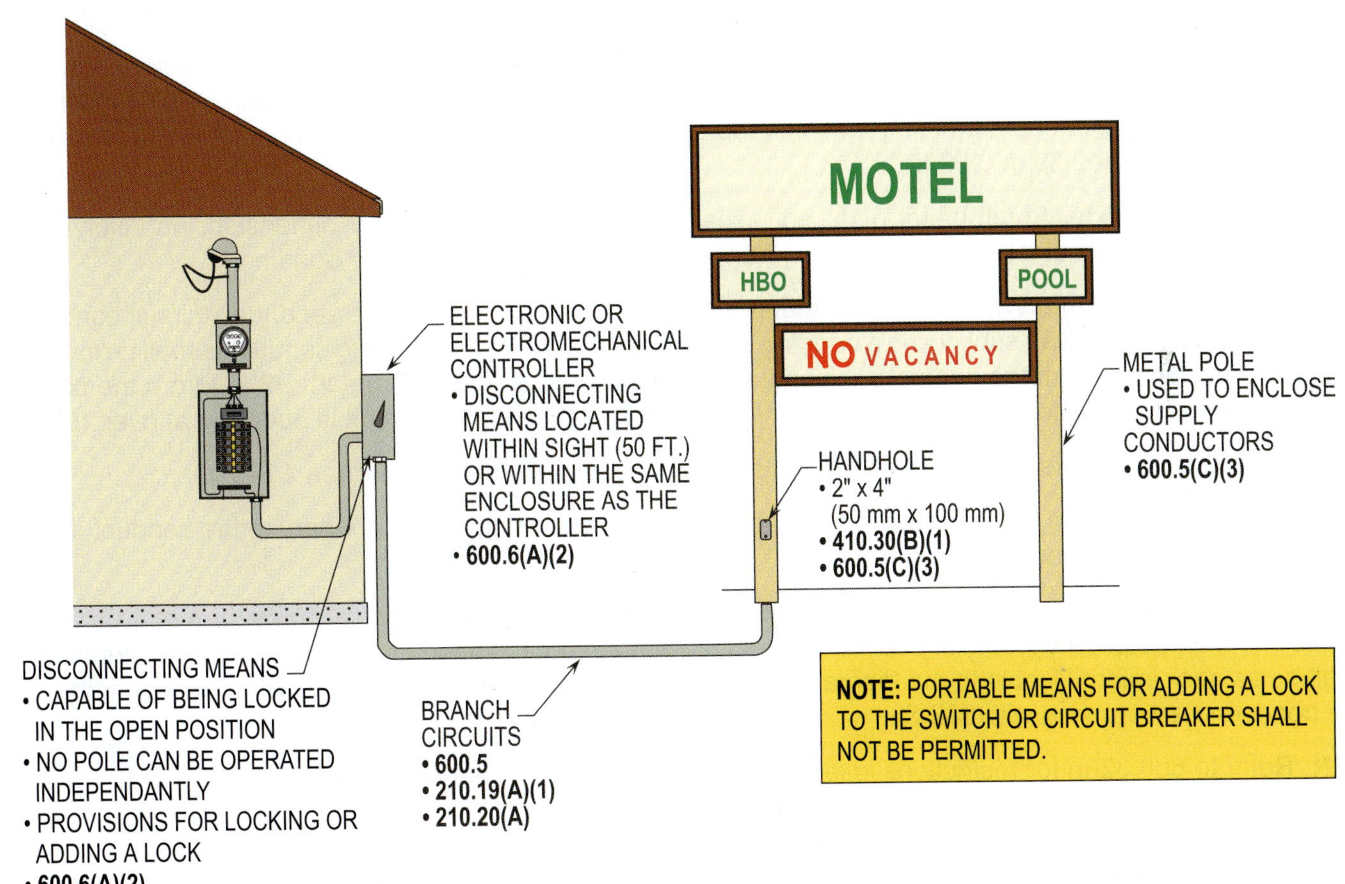

WITHIN SIGHT OF THE CONTROLLER
600.6(A)(2)(3)

Purpose of Change: This revision clarifies the provisions for locking or adding a lock to the disconnecting means.

NEC Ch. 6 - Article 600
Part I - 600.7(A) and (B)

Type of Change	Panel Action	UL	UL 508	API 500 - 1997	API 505 - 1997	OSHA - 1994
Revision	Accept	467	-	-	-	-
ROP	ROC	NFPA 70E - 2004	NFPA 70B - 2006	NFPA 79 - 2007	NFPA	NEMA
pg. 635 # 18-123	pg. 355 # 18-95a	-	-	-	-	-
log: 338 CMP: 18	log: CC 1500	Submitter: Michael J. Johnston		2005 NEC: 600.7		IEC: -

2005 NEC - 600.7 Grounding.

~~(A)~~ **Flexible Metal Conduit Length.** Listed flexible metal conduit or listed liquidtight flexible metal conduit that encloses the secondary circuit conductor from a transformer or power supply for use with ~~electric discharge~~ tubing shall be permitted as a bonding means if the total accumulative length of the conduit in the secondary circuit does not exceed 30 m (100 ft).

~~(B)~~ **Small Metal Parts.** Small metal parts not exceeding 50 mm (2 in.) in any dimension, not likely to be energized, and spaced at least 19 mm (3/4 in.) from neon tubing shall not require bonding.

~~(C)~~ **Nonmetallic Conduit.** Where listed nonmetallic conduit is used to enclose the secondary circuit conductor from a transformer or power supply and a bonding conductor is required, the bonding conductor shall be installed separate and remote from the nonmetallic conduit and be spaced at least 38 mm (1 1/2 in.) from the conduit when the circuit is operated at 100 Hz or less or 45 mm (1 3/4 in.) when the circuit is operated at over 100 Hz.

~~(D)~~ **Bonding Conductors.** Bonding conductors shall be copper and not smaller than 14 AWG.

~~(E)~~ **Metal Building Parts.** Metal parts of a building shall not be permitted as a secondary return conductor or an equipment grounding conductor.

~~(F)~~ **Signs in Fountains.** Signs or outline lighting installed inside a fountain shall have all metal parts ~~and~~ bonded to the equipment grounding conductor for the fountain recirculating system. The bonding connection shall be as near as practicable to the fountain and shall be permitted to be made to metal piping systems that are bonded in accordance with 680.53.

FPN: Refer to 600.32(J) for restrictions in length of high-voltage secondary conductors.

2008 NEC - 600.7 Grounding and Bonding.

(A) Grounding.

(1) Equipment Grounding. Signs and metal equipment of outline lighting systems shall be grounded by connection to the equipment grounding conductor of the supply branch circuit(s) or feeder using any of the types of equipment grounding conductors specified in 250.118.

Exception: portable cord-connected signs shall not be required to be connected to the equipment grounding conductor where protected by a system of double insulation or its equivalent. Double insulated equipment shall be distinctively marked.

(2) Size of Equipment Grounding Conductor. The equipment grounding conductor size shall be in accordance with 250.122 based on the rating of the overcurrent device protecting the branch circuit or feeder conductors supplying the sign or equipment.

(3) Connections. Equipment grounding conductor connections shall be made in accordance with 250.130 and in a method specified in 250.8.

(4) Auxiliary Grounding Electrode. Auxiliary grounding electrode(s) shall be permitted for electric signs and equipment outline lighting systems covered by this article and shall meet the requirements of 250.54.

(5) Metal Building Parts. Metal parts of a building shall not be permitted as a secondary return conductor or an equipment grounding conductor.

(B) Bonding.

(1) Bonding of Metal Parts. Metal parts and equipment of signs and outline lighting systems shall be bonded together and to the associated transformer or power supply equipment grounding conductor of the branch circuit or feeder supplying the sign or outline lighting system and shall meet the requirements of 250.90.

(2) Bonding Connections. Bonding connections shall be made in accordance with 250.8.

(3) Metal Building Parts. Metal parts of a building shall not be permitted to be used as a means for bonding metal parts and equipment of signs or outline lighting systems together or to the transformer or power supply equipment grounding conductor of the supply circuit.

(4) Flexible Metal Conduit Length. Listed flexible metal conduit or listed liquidtight flexible metal conduit that encloses the secondary circuit conductor from a transformer or power supply for use with neon tubing shall be permitted as a bonding means if the total accumulative length of the conduit in the secondary circuit does not exceed 30 m (100 ft).

(5) Small Metal Parts. Small metal parts not exceeding 50 mm (2 in.) in any dimension, not likely to be energized, and spaced at least 19 mm (3/4 in.) from neon tubing shall not require bonding.

(6) Nonmetallic Conduit. Where listed nonmetallic conduit is used to enclose the secondary circuit conductor from a transformer or power supply and a bonding conductor is required, the bonding conductor shall be installed separate and remote from the nonmetallic conduit and be spaced at least 38 mm (1 1/2 in.) from the conduit when the circuit is operated at 100 Hz or less or 45 mm (1 3/4 in.) when the circuit is operated at over 100 Hz.

(7) Bonding Conductors. Bonding conductors shall comply with (a) and (b).

(a) Bonding conductors shall be copper and not smaller than 14 AWG.

(b) Bonding conductors installed externally of a sign or raceway shall be protected from physical damage.

(8) Signs in Fountains. Signs or outline lighting installed inside a fountain shall have all metal parts bonded to the equipment grounding conductor of the branch circuit for the fountain recirculating system. The bonding connection shall be as near as practicable to the fountain and shall be permitted to be made to metal piping systems that are bonded in accordance with 680.53.

FPN: Refer to 600.32(J) for restrictions in length of high-voltage secondary conductors.

Author's Comment: The revision divides this section into two parts "Grounding" and "Bonding," and provides clear direction for users and enforcement as to what bonding is intended to accomplish and where the bonding conductors or jumpers are required to be connected. The panel added an **exception** to **600.7(A)(1)** to permit cord-connected double-insulated signs and changed the term from "electric discharge" to "neon". The Comment added titles to each subsection and subdivided the last subsection for clarity.

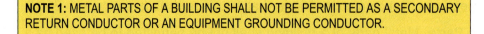

NOTE 1: METAL PARTS OF A BUILDING SHALL NOT BE PERMITTED AS A SECONDARY RETURN CONDUCTOR OR AN EQUIPMENT GROUNDING CONDUCTOR.

NOTE 2: METAL PARTS OF A BUILDING SHALL NOT BE PERMITTED TO BE USED AS A MEANS FOR BONDING METAL PARTS OR OUTLINE LIGHTING SYSTEMS TOGETHER OR TO THE TRANSFORMER OR POWER SUPPLY EQUIPMENT GROUNDING CONDUCTOR.

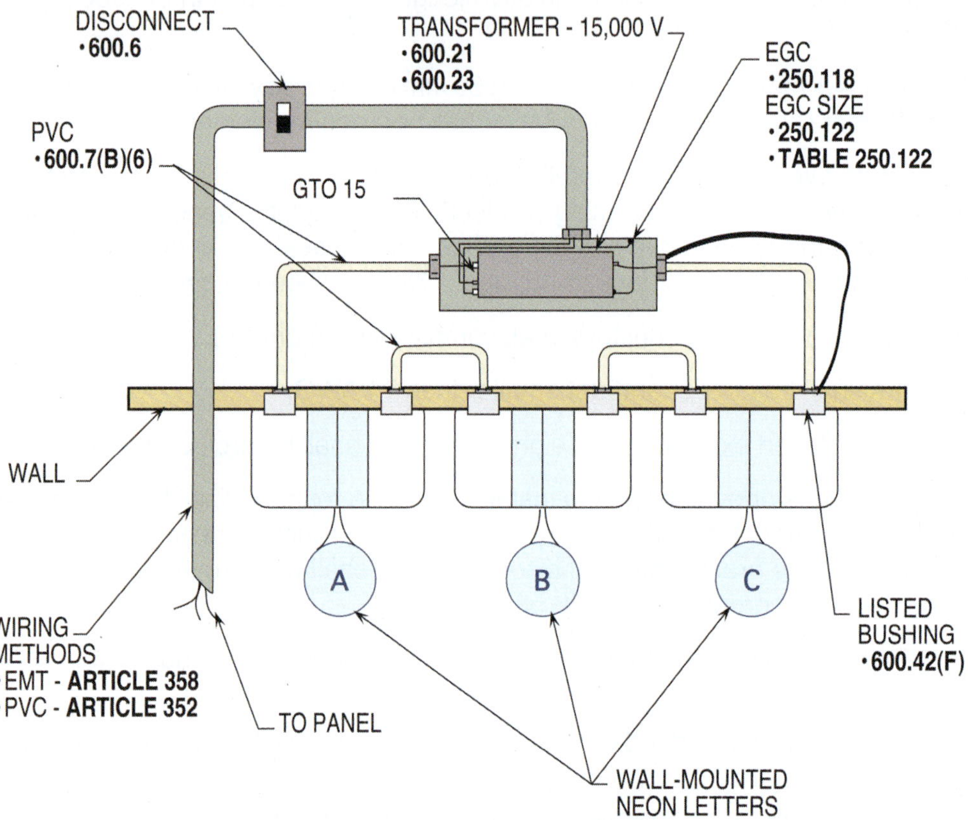

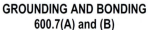

GROUNDING AND BONDING
600.7(A) and (B)

Purpose of Change: This revision divides this section into two parts, "grounding" and "bonding" and provides clear direction to what grounding and bonding is intended to accomplish.

NEC Ch. 6 - Article 600
Part I - 600.12(A) through (C)

Type of Change		Panel Action		UL	UL 508	API 500 - 1997	API 505 - 1997	OSHA - 1994
Revision		Accept		-	-	-	-	-
ROP		ROC		NFPA 70E - 2004	NFPA 70B - 2006	NFPA 79 - 2007	NFPA	NEMA
pg. 638	# 18-133	pg. 358	# 18-103	-	-	-	-	-
log: 856	CMP: 18	log: 292	Submitter: Michael J. Johnston			2005 NEC: 600.12		IEC: -

2005 NEC - 600.12 Field-Installed Secondary Wiring.

~~The~~ field ~~section signs shall comply with 600.31 if 1000 volts or less, or with 600.32 if over 1000 volts.~~

2008 NEC - 600.12 Field-Installed Secondary Wiring.

Field-installed secondary circuit wiring for electric signs and outline lighting systems shall be in accordance with 600.12(A), (B), or (C).

(A) 1000 Volts or Less. Secondary circuit wiring of 1000 volts or less shall comply with 600.31.

(B) Over 1000 Volts. Secondary circui wiringt of over 1000 volts shall comply with 600.32.

(C) Less Than 50 Volts. Secondary circuit wiring less than 50 volts shall be installed in accordance with either of the following:

(1) Any wiring method included in Chapter 3 suitable for the conditions.

(2) Where the power source complies with the requirements in 725.121, wiring methods shall be permitted to be installed in accordance with 725.130(A) or (B).

Author's Comment: This section was originally intended in the 2005 NEC to permit section signs to be installed with secondary wiring similar to the wiring for field installed skeleton since there often was neon lighting within each sectional sign. This concept has been expanded for the 2008 NEC to also include less than 50 volt power sources wired by Chapter 3 wiring methods or by Class 2 wiring systems where using a Class 2 power supply. The Comment provided the following information: Just limiting the power source to less than 50 volts does not cover the power sources that may exceed the current limitations outlined in **Tables 11(A)** and **11(B)** in **Chapter 9**.

For example, **725.41** permits Class 1 power limited circuits to be 30 volts and 1000 volt-amperes with a current rating of 33.33 amps. Where a power-limited Class 1 power source is other than a transformer, the maximum output or VA max can be as high as 2500 volt-amperes with a current peak of 83.33 amps. A Class 2 wiring method would not be acceptable for circuits with amperage and volt-ampere levels in these ranges.

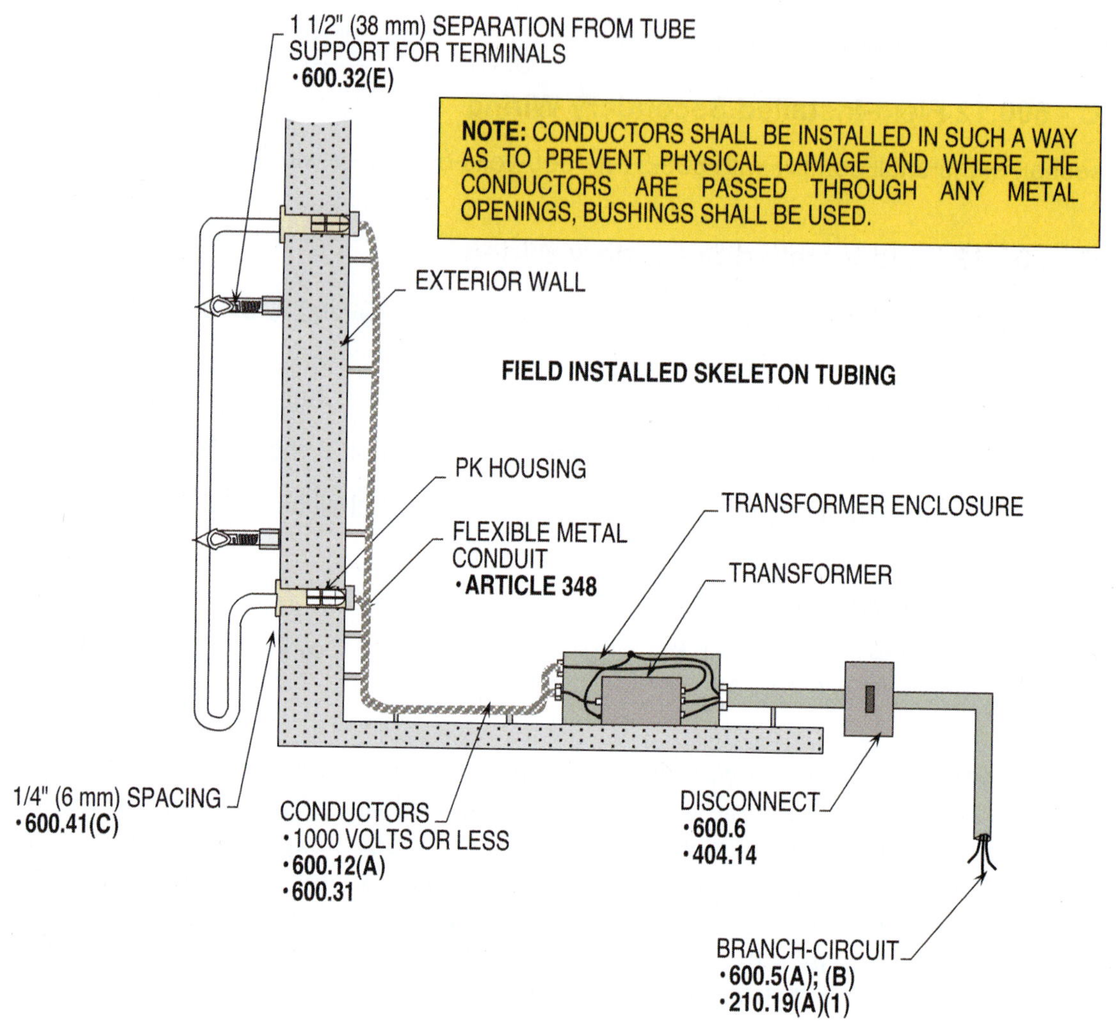

FIELD-INSTALLED SECONDARY WIRING
600.12(A) through (C)

Purpose of Change: This revision clarifies the installation requirements for field-installed secondary wiring at 1000 volts or less, over 1000 volts, and less than 50 volts.

NEC Ch. 6 - Article 600
Part I - 600.21(E)

Type of Change	Panel Action	UL	UL 508	API 500 - 1997	API 505 - 1997	OSHA - 1994
Revision	Accept	48 - 2161	-	-	-	-
ROP	**ROC**	**NFPA 70E - 2004**	**NFPA 70B - 2006**	**NFPA 79 - 2007**	**NFPA**	**NEMA**
pg. 638 / # 18-136	pg. - / # -	-	-	-	-	-
log: 660 / CMP: 18	log: -	Submitter: Leon Przybyla		2005 NEC: 600.21(E)		IEC: -

2005 NEC - 600.21 Ballasts, Transformers, and Electronic Power Supplies.

(E) Attic and Soffit Locations. Ballasts, transformers, and electronic power supplies shall be permitted to be located in attics and soffits, provided there is an access door at least 900 mm by ~~600~~ mm (~~3 ft by 2 ft~~) and a passageway of at least 900 mm (3 ft) high by 600 mm (2 ft) wide with a suitable permanent walkway at least 300 mm (12 in.) wide extending from the point of entry to each component.

2008 NEC - 600.21 Ballasts, Transformers, and Electronic Power Supplies.

(E) Attic and Soffit Locations. Ballasts, transformers, and electronic power supplies shall be permitted to be located in attics and soffits, provided there is an access door at least 900 mm by 562.5 mm (36 in. by 22 1/2 in.) and a passageway of at least 900 mm (3 ft) high by 600 mm (2 ft) wide with a suitable permanent walkway at least 300 mm (12 in.) wide extending from the point of entry to each component. At least one lighting outlet containing a switch or controlled by a wall switch shall be installed in such spaces. At least one point of control shall be at the usual point of entry to these spaces. The lighting outlet shall be provided at or near the equipment requiring servicing.

Author's Comment: Similar to requirements in **210.70(C)** for illumination in attic and soffit areas where equipment is located, the same type of requirement for illumination has been inserted in **600.21(E)** to ensure sufficient lighting to work on the sign lighting equipment.

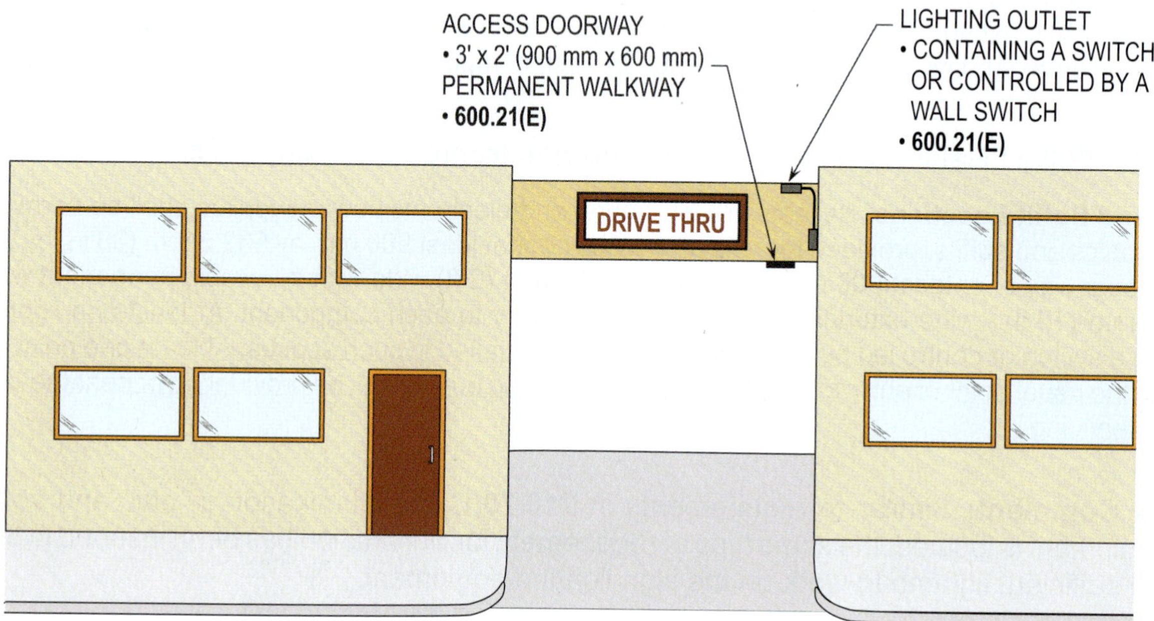

ATTIC AND SOFFIT LOCATIONS
600.21(E)

Purpose of Change: This revision clarifies that at least one lighting outlet containing a switch or controlled by a wall switch shall be installed in attic and soffit locations.

NEC Ch. 6 - Article 600
Part I - 600.24(A) through (C)

Type of Change		Panel Action		UL	UL 508	API 500 - 1997	API 505 - 1997	OSHA - 1994
Revision		Accept		1310	-	-	-	-
ROP		ROC		NFPA 70E - 2004	NFPA 70B - 2006	NFPA 79 - 2007	NFPA	NEMA
pg. 639	# 18-139	pg. 359	# 18-104	-	-	-	-	-
log: 855	CMP: 18	log: 1198	Submitter: Michael J. Johnston			2005 NEC: 600.24		IEC: -

2005 NEC - 600.24 Class 2 Power Sources.

~~In addition to the requirements of Article 600,~~ signs and outline lighting systems supplied by Class 2 transformers, power supplies, and power sources shall comply with ~~725.41~~.

2008 NEC - 600.24 Class 2 Power Sources.

Signs and outline lighting systems supplied by Class 2 transformers, power supplies, and power sources shall comply with <u>the applicable requirements of Article 600 and 600.24(A), (B), and (C):</u>

<u>**(A) Listing.** Class 2 Power supplies and power sources shall be listed for use with electric signs and outline lighting systems and shall comply with 725.121.</u>

<u>**(B) Grounding.** Metal parts of signs and outline lighting systems shall be grounded and bonded in accordance with 600.7.</u>

<u>**(C) Secondary Wiring.** Secondary wiring from Class 2 power sources shall comply with 600.12(C).</u>

Author's Comment: Not only must signs and outline lighting systems comply with the requirements for listing as a Class 2 power source but they must be grounded if required by **250.20(A)**. Metal parts of signs and outline lighting systems must also be grounded in accordance with **600.7** and **250.112(G)**. The secondary wiring must also comply with new **600.12(C)** by either wiring with **Chapter 3** wiring methods or **725.121** and **725.130(A)** or **(B)**.

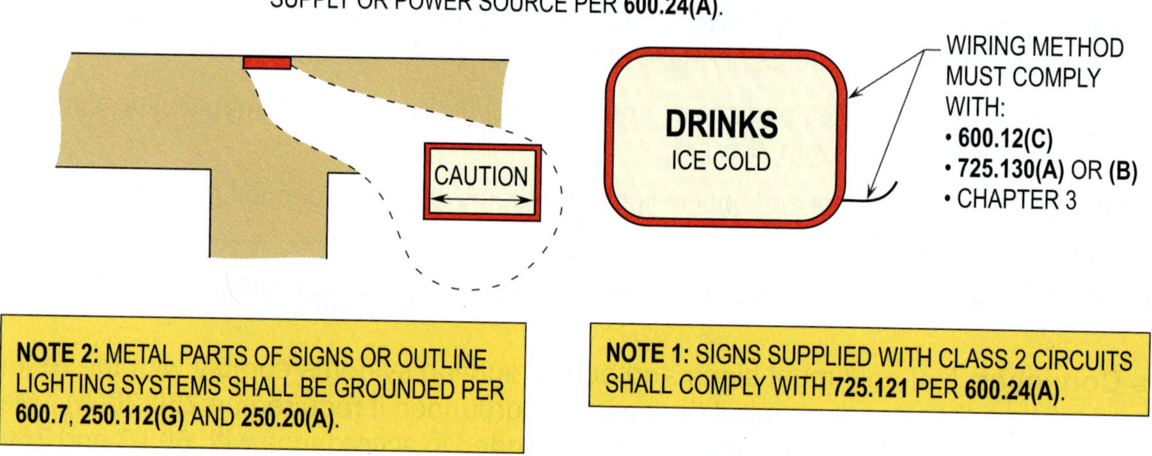
SIGNS SUPPLIED BY A LISTED CLASS 2 XFMR POWER SUPPLY OR POWER SOURCE PER **600.24(A)**.
CAUTION
DRINKS
ICE COLD
WIRING METHOD MUST COMPLY WITH:
• **600.12(C)**
• **725.130(A)** OR **(B)**
• CHAPTER 3
NOTE 2: METAL PARTS OF SIGNS OR OUTLINE LIGHTING SYSTEMS SHALL BE GROUNDED PER **600.7**, **250.112(G)** AND **250.20(A)**.
NOTE 1: SIGNS SUPPLIED WITH CLASS 2 CIRCUITS SHALL COMPLY WITH **725.121** PER **600.24(A)**.

CLASS 2 POWER SOURCES
600.24 (A) through (C)

Purpose of Change: This revision clarifies the requirements for electric signs and outline lighting systems supplied by Class 2 transfomers, power supplies and power sources.

NEC Ch. 6 - Article 600
Part II - 600.32(K)

Type of Change		Panel Action		UL	UL 508	API 500 - 1997	API 505 - 1997	OSHA - 1994
New Subsection		Accept in Principle		814 - 879	-	-	-	-
ROP		ROC		NFPA 70E - 2004	NFPA 70B - 2006	NFPA 79 - 2007	NFPA	NEMA
pg. 642	# 18-154	pg. -	# -	-	-	-	-	-
log: 407	CMP: 18	log: -	Submitter: Michael J. Johnston			2005 NEC: 600.32(K)		IEC: -

2008 NEC - 600.32 Neon Secondary Circuit Conductors, Over 1000 Volts, Nominal.

(K) Splices. Splices in high-voltage secondary circuit conductors shall be made in listed enclosures rated over 1000 volts. Splice enclosures shall be accessible after installation and listed for the location where they are installed.

Author's Comment: New **600.32(K)** requires splices in high voltage circuit conductors to be made in listed enclosures rated over 1000 volts. These enclosures must be accessible after installation and listed for the location where installed.

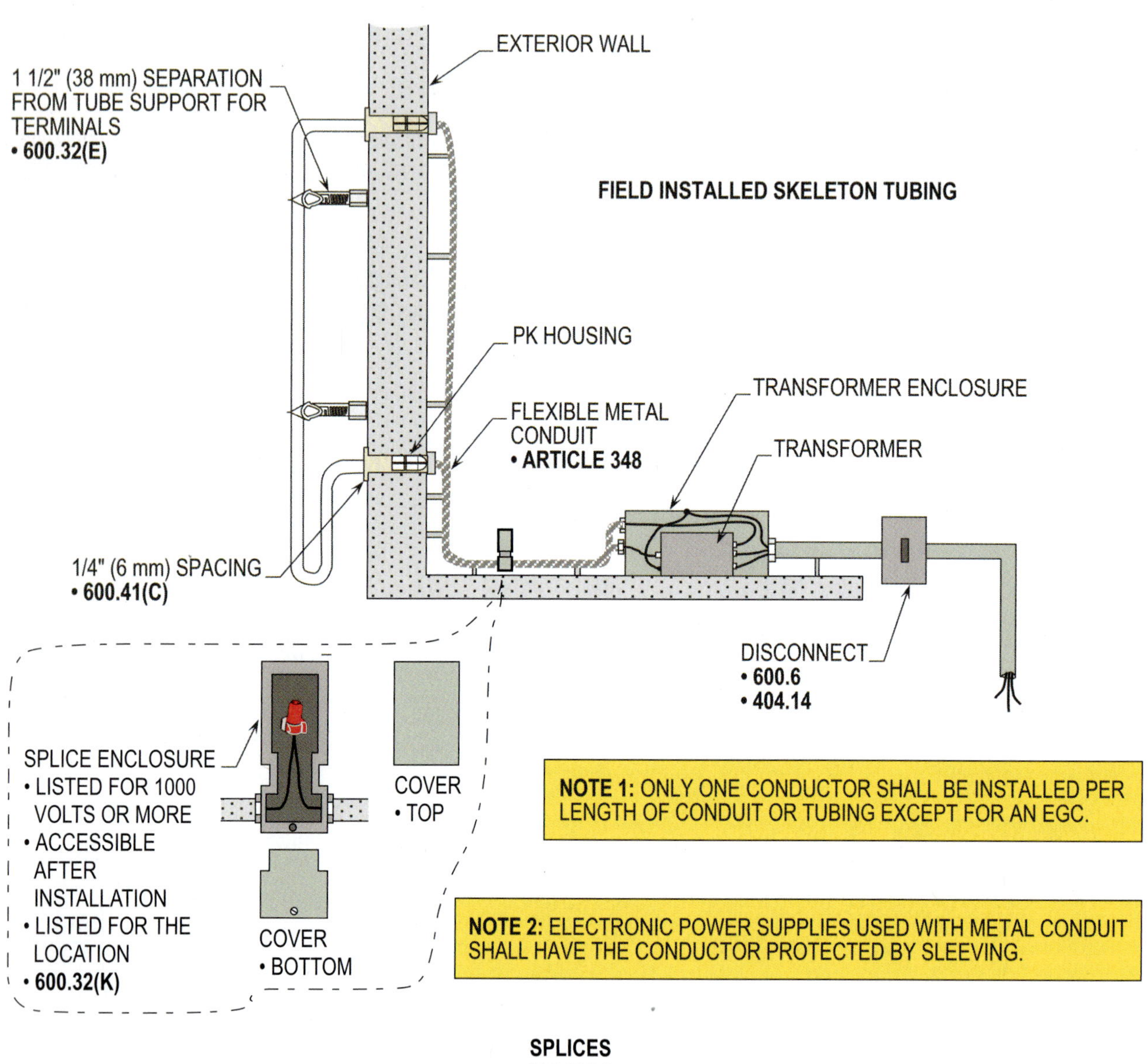

SPLICES
600.32(K)

Purpose of Change: A new subsection has been added to require splices in high-voltage secondary circuit conductors to be made in listed enclosures rated over 1000 volts.

NEC Ch. 6 - Article 600
Part II - 600.41(B)

Type of Change		Panel Action		UL	UL 508	API 500 - 1997	API 505 - 1997	OSHA - 1994
Revision		Accept		48	-	-	-	-
ROP		ROC		NFPA 70E - 2004	NFPA 70B - 2006	NFPA 79 - 2007	NFPA	NEMA
pg. 642	# 18-156	pg. -	# -	-	-	-	-	-
log: 530	CMP: 18	log: -	Submitter: Michael J. Johnston			2005 NEC: 600.41(B)		IEC: -

2005 NEC - 600.41 Neon Tubing.

(C) Spacing. A spacing of not less than 6 mm (1/4 in.) shall be maintained between the tubing and the nearest surface, other than its support.

2008 NEC - 600.41 Neon Tubing.

(B) Support. Tubing shall be supported by listed tube supports. The neon tubing shall be supported within 150 mm (6 in.) from the electrode connection.

Author's Comment: The new sentence added to neon tubing support now requires the support to be within 6 inches from the electrode connection.

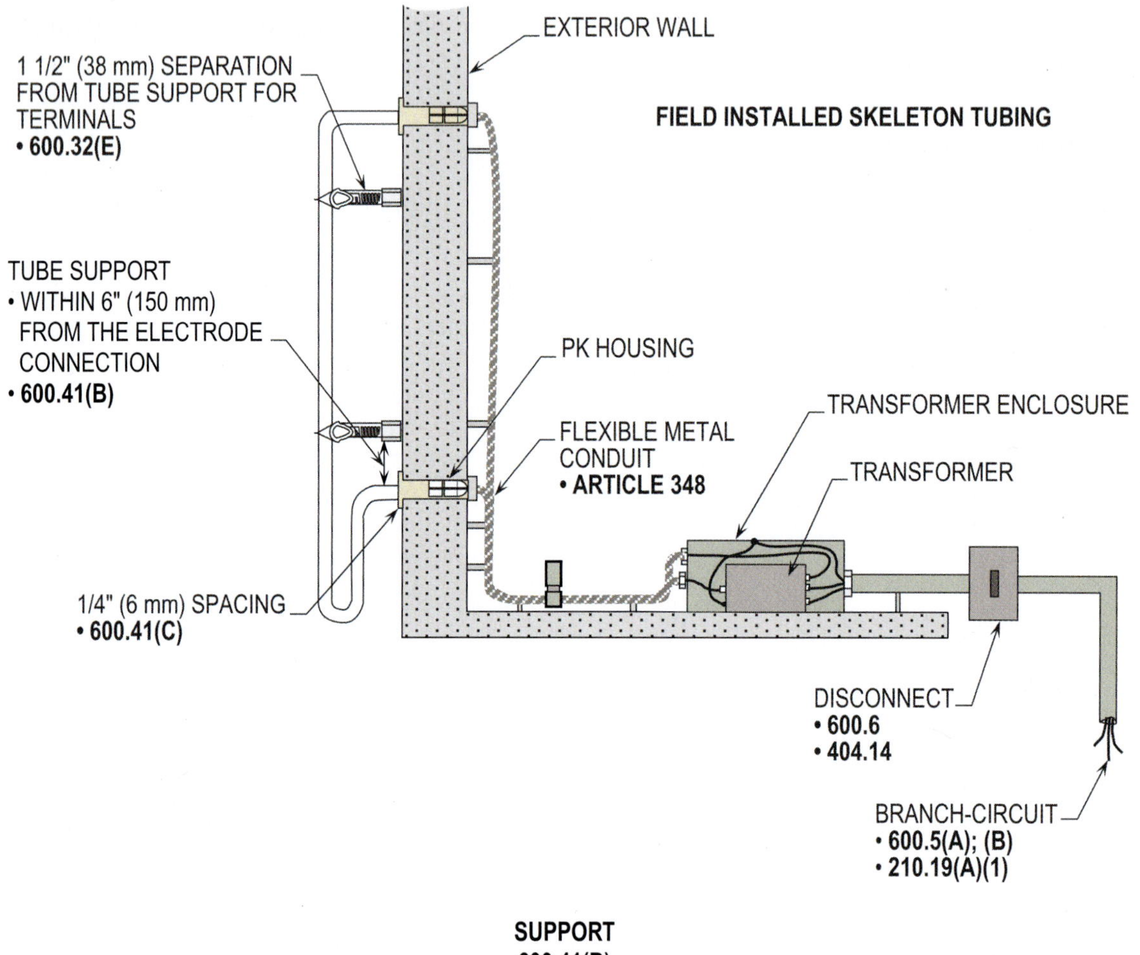

SUPPORT
600.41(B)

Purpose of Change: This revision clarifies that neon tubing shall be supported within 6 in. (150 mm) from the electrode connection.

NEC Ch. 6 - Article 600
Part II - 600.41(D)

Type of Change	Panel Action	UL	UL 508	API 500 - 1997	API 505 - 1997	OSHA - 1994
New Subsection	Accept	48	-	-	-	-
ROP	**ROC**	**NFPA 70E - 2004**	**NFPA 70B - 2006**	**NFPA 79 - 2007**	**NFPA**	**NEMA**
pg. 642 # 18-157	pg. - # -	-	-	-	-	-
log: 529 CMP: 18	log: -	Submitter: Michael J. Johnston		2005 NEC: 600.41(D)		IEC: -

2008 NEC - 600.41 Neon Tubing.

(D) Protection. Field-installed skeleton tubing shall not be subject to physical damage. Where the tubing is readily accessible to other than qualified persons, field-installed skeleton tubing shall be provided with suitable guards or protected by other approved means.

Author's Comment: New **(D)** requires field installed skeleton tubing to be installed where not subject to physical damage. Where readily accessible to other than qualified persons, suitable guards must be installed or it must be protected by some other means.

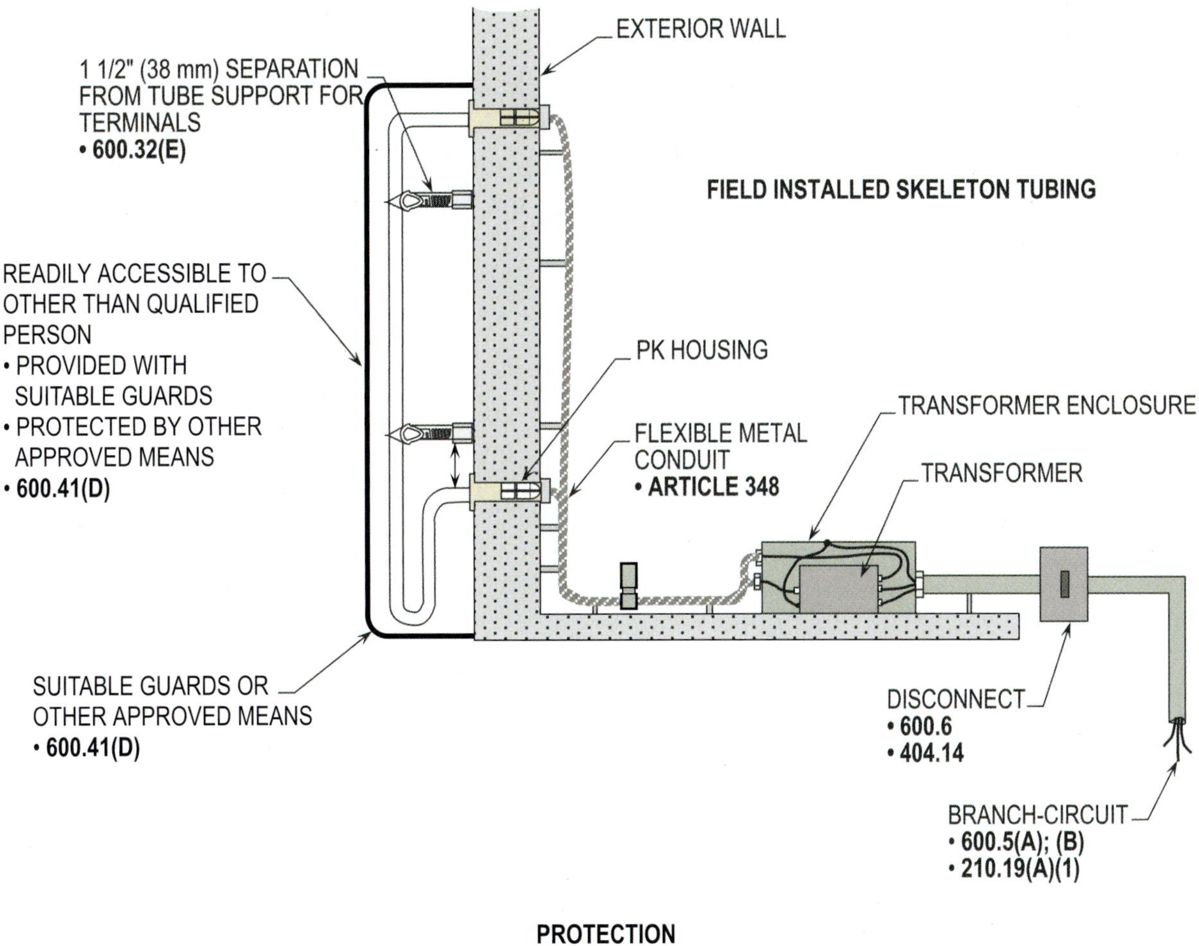

PROTECTION
600.41(D)

Purpose of Change: A new subsection has been added to require field-installed skeleton tubing to be protected if exposed to physical damage or readily accessible to other than qualified persons.

NEC Ch. 6 - Article 600
Part II - 600.42(G)(1) and (G)(2)

Type of Change	Panel Action	UL	UL 508	API 500 - 1997	API 505 - 1997	OSHA - 1994
Revision	Accept	48 - Suject 978B	-	-	-	-
ROP	**ROC**	**NFPA 70E - 2004**	**NFPA 70B - 2006**	**NFPA 79 - 2007**	**NFPA**	**NEMA**
pg. 643 / # 18-160	pg. - / # -	-	-	-	-	-
log: 361 / CMP: 18	log: -	Submitter: Michael J. Johnston		2005 NEC: 600.42(G)		IEC: -

2005 NEC - 600.42 Electrode Connections.

(G) Electrode Enclosures. Electrode enclosures shall be listed.

2008 NEC - 600.42 Electrode Connections.

(G) Electrode Enclosures. Electrode enclosures shall be listed.

(1) Dry Locations. Electrode enclosures that are listed for use in dry, damp, or wet locations shall be permitted to be installed and used in such locations.

(2) Damp and Wet Locations. Electrode enclosures installed in damp and wet locations shall be specifically listed and identified for use in such locations.

FPN: See Section 110.3(B) covering installation and use of electrical equipment.

Author's Comment: Listing requirements for dry, damp, and wet location electrode enclosures has been added to help the installer/technician understand the listing requirements for various locations where the electrode enclosures will be installed.

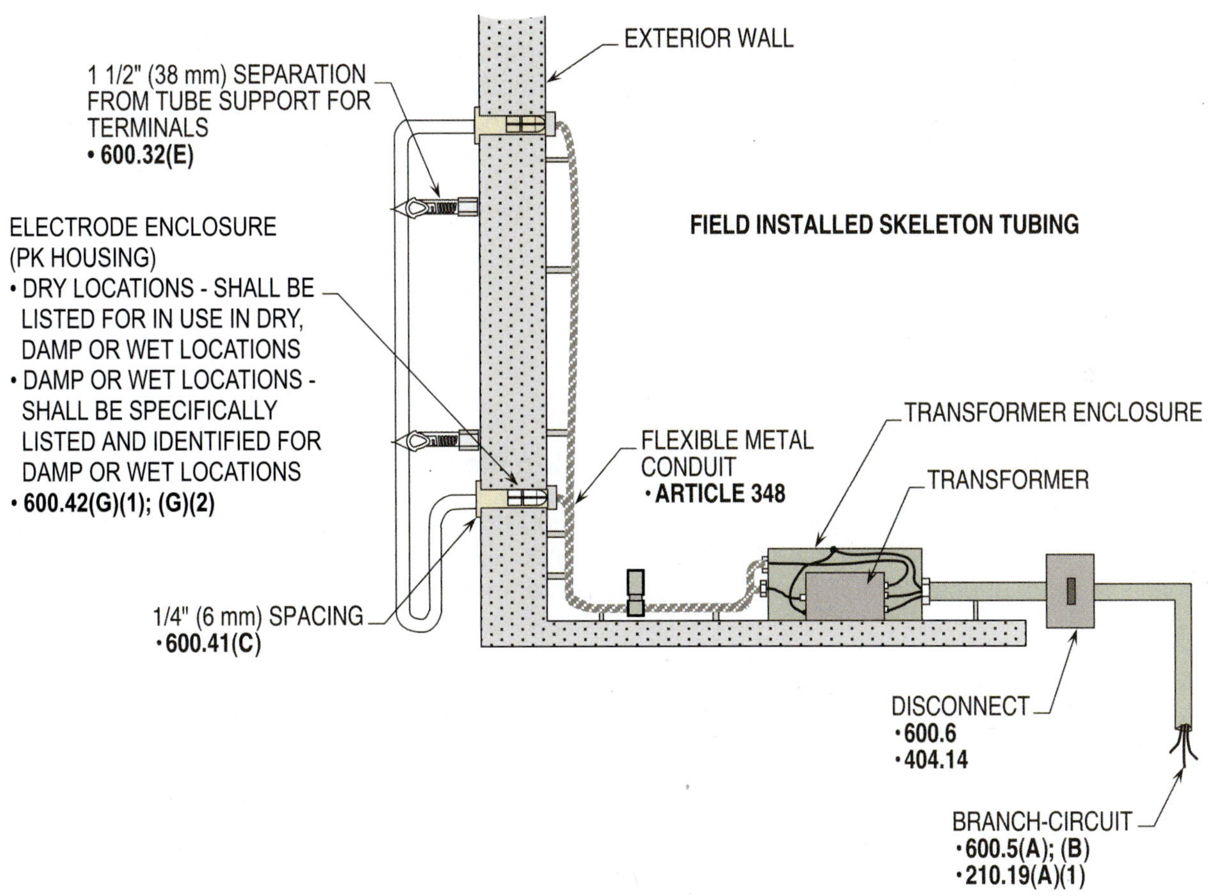

ELECTRODE ENCLOSURES
600.42(G)(1) and (G)(2)

Purpose of Change: This revision clarifies that electrode enclosures shall be listed for use in dry locations and specifically listed and identified for damp or wet locations.

NEC Ch. 6 - Article 610
Part IV - 610.31(2)

Type of Change		Panel Action		UL	UL 508	API 500 - 1997	API 505 - 1997	OSHA - 1994
Revision		Accept in Principle		-	-	-	-	1910.306(b)(1)(i)
ROP		ROC		NFPA 70E - 2004	NFPA 70B - 2006	NFPA 79 - 2007	NFPA	NEMA - 2005
pg. 647	# 12-9	pg. 364	# 12-4	430.2(A)(1)	-	-	-	ICS 8
log: 359	CMP: 12	log: 1852	Submitter: Michael J. Johnston			2005 NEC: 610.31		IEC: -

2005 NEC - 610.31 Runway Conductor Disconnecting Means.

A disconnecting means that has a continuous ampere rating not less than that calculated in 610.14(E) and 610.14(F) shall be provided between the runway contact conductors and the power supply. Such disconnecting means shall consist of a motor-circuit switch, circuit breaker, or molded case switch. This disconnecting means shall be as follows:

(1) Readily accessible and operable from the ground or floor level

(2) Capable of being locked in the open position.

(3) Open all ungrounded conductors simultaneously

(4) Placed within view of the runway contact conductors

2008 NEC - 610.31 Runway Conductor Disconnecting Means.

A disconnecting means that has a continuous ampere rating not less than that calculated in 610.14(E) and 610.14(F) shall be provided between the runway contact conductors and the power supply. Such disconnecting means shall consist of a motor-circuit switch, circuit breaker, or molded case switch. This disconnecting means shall be as follows:

(1) Readily accessible and operable from the ground or floor level

(2) Capable of being locked in the open position. <u>The provision for locking or adding a lock to the disconnecting means shall be installed on or at the switch or circuit breaker used as the disconnecting means and shall remain in place with or without the lock installed. Portable means for adding a lock to the switch or circuit breaker shall not be permitted as the means required to be installed at and remain with the equipment.</u>

(3) Open all ungrounded conductors simultaneously

(4) Placed within view of the runway contact conductors

Author's Comment: A series of changes have been accepted to provide standardized wording. The requirement for locking or adding a lock to the disconnecting means for a crane or hoist has been added to provide safety for the installer or maintainer of the system. Installing this locking means or providing the locking mechanism will help ensure that the device is available when it becomes necessary to lock the disconnecting means to the "off" position. Based on the Proposal and the Comment, an attachment device, such as scissors or multiple lock hasp, that provides a means to attach multiple locks to the locking device on the disconnect is not prohibited. Instead of inserting one lock in the locking device, the multiple lock hasp is inserted and multiple locks can be used. The intention of prohibiting portable means for adding a lock, as noted in the Comment, is to ensure that a provision for locking or adding a lock to the disconnecting means is permanent.

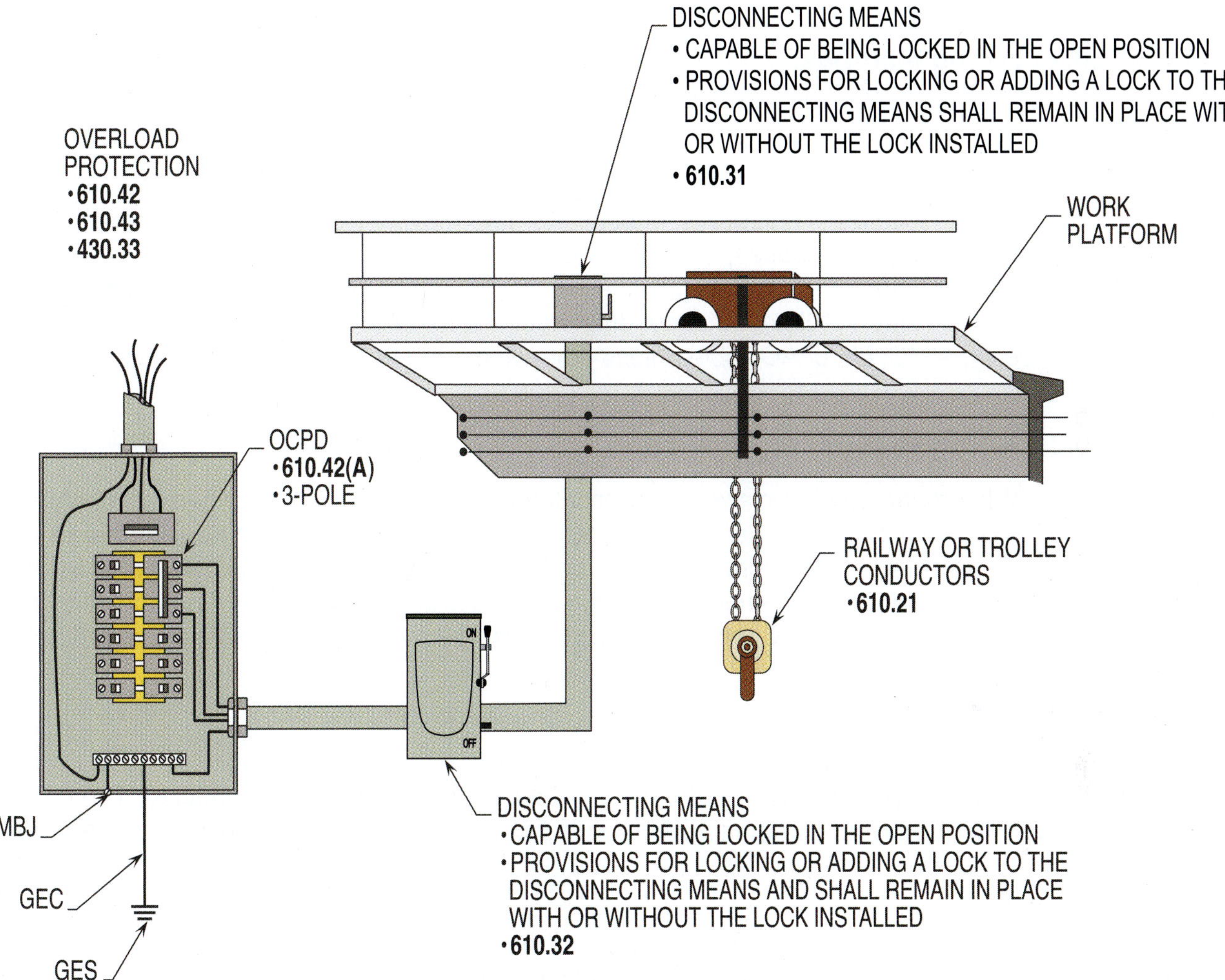

RUNWAY CONDUCTOR DISCONNECTING MEANS
610.31(2)

Purpose of Change: This revision clarifies the requirements for runway conductor disconnecting means and the crane and monorail hoists disconnecting means.

NEC Ch. 6 - Article 620
Part III - 620.21(A)(1)(e)

Type of Change	Panel Action	UL	UL 508	API 500 - 1997	API 505 - 1997	OSHA - 1994
New Subdivision	Accept in Principle	-	-	-	-	-
ROP	ROC	NFPA 70E - 2004	NFPA 70B - 2006	NFPA 79 - 2007	NFPA	NEMA
pg. 650 # 12-25	pg. 365 # 12-9	-	-	-	-	-
log: 3096 CMP: 12	log: 75	Submitter: Joseph A. Hertel		2005 NEC: -		IEC: -

2005 NEC - 620.21 Wiring Methods.

(A) Elevators.

(1) Hoistways.

Conductors and optical fibers located in hoistways, in escalator and moving walk wellways, in ~~wheelchair~~ lifts, stairway chairlift runways, machinery spaces, control spaces, in or on cars, in machine rooms and control rooms, not including the traveling cables connecting the car or counterweight and hoistway wiring, shall be installed in rigid metal conduit, intermediate metal conduit, electrical metallic tubing, rigid nonmetallic conduit, or wireways, or shall be Type MC, MI, or AC cable unless otherwise permitted in 620.21(A) through 620.21(C).

2008 NEC - 620.21 Wiring Methods.

Conductors and optical fibers located in hoistways, in escalator and moving walk wellways, in platform lifts, stairway chairlift runways, machinery spaces, control spaces, in or on cars, in machine rooms and control rooms, not including the traveling cables connecting the car or counterweight and hoistway wiring, shall be installed in rigid metal conduit, intermediate metal conduit, electrical metallic tubing, rigid nonmetallic conduit, or wireways, or shall be Type MC, MI, or AC cable unless otherwise permitted in 620.21(A) through 620.21(C).

(A) Elevators.

(1) Hoistways.

(e) A sump pump or oil recovery pump located in the pit shall be permitted to be cord connected. The cord shall be a hard usage oil-resistant type, of a length not to exceed 1.8 m (6 ft), and shall be located to be protected from physical damage.

Author's Comment: A new subdivision has been added that permits cord-connected sump pump and oil recovery pump equipment located in the pit but sets the limit on the length of the cord. The Comment changed the exception into positive text as a new **620.21(A)(1)(e)**.

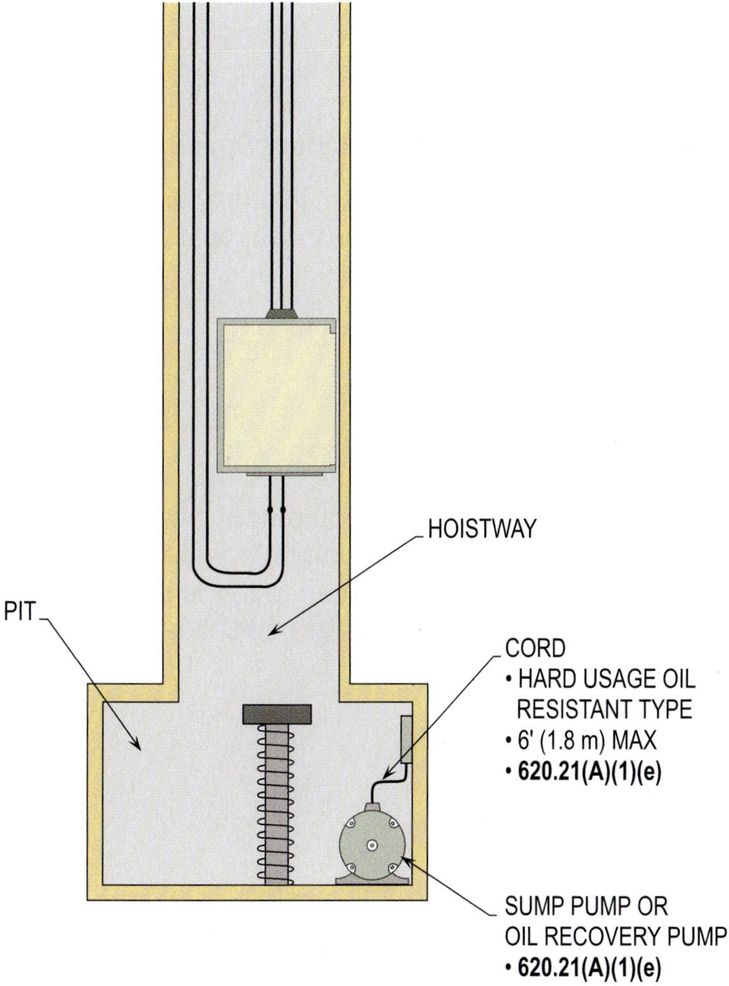

WIRING METHODS
620.21(A)(1)(e)

Purpose of Change: A new subdivision has been added to permit a sump pump or oil recovery pump to be cord- and plug-connected.

NEC Ch. 6 - Article 620
Part III - 620.21(A)(3)(e)

Type of Change		Panel Action		UL	UL 508	API 500 - 1997	API 505 - 1997	OSHA - 1994
New Subdivision		Accept		62	-	-	-	-
ROP		**ROC**		**NFPA 70E - 2004**	**NFPA 70B - 2006**	**NFPA 79 - 2007**	**NFPA**	**NEMA**
pg. 652	# 12-31	pg. 365	# 12-11	-	-	-	-	-
log: 1214	CMP: 12	log: 76	Submitter: Andy Juhasz			2005 NEC: -		IEC: -

2008 NEC - 620.21 Wiring Methods.

(A) Elevators.

(3) Within Machine Rooms, Control Rooms, and Machinery Spaces and Control Spaces.

(e) Flexible cords and cables in lengths not to exceed 1.8 m (6 ft) that are of a flame-retardant type and located to be protected from physical damage shall be permitted in these rooms and spaces without being installed in a raceway. They shall be part of the following:

(1) Listed equipment,

(2) A driving machine, or

(3) A driving machine brake.

Author's Comment: The same wiring methods for equipment located in the hoistway, on the car or on the counterweight is afforded to the same types equipment located in machine rooms, control rooms, machinery spaces and control spaces. The final sentence in the new text has been incorporated into the main text rather than being left at the end of the list of applications.

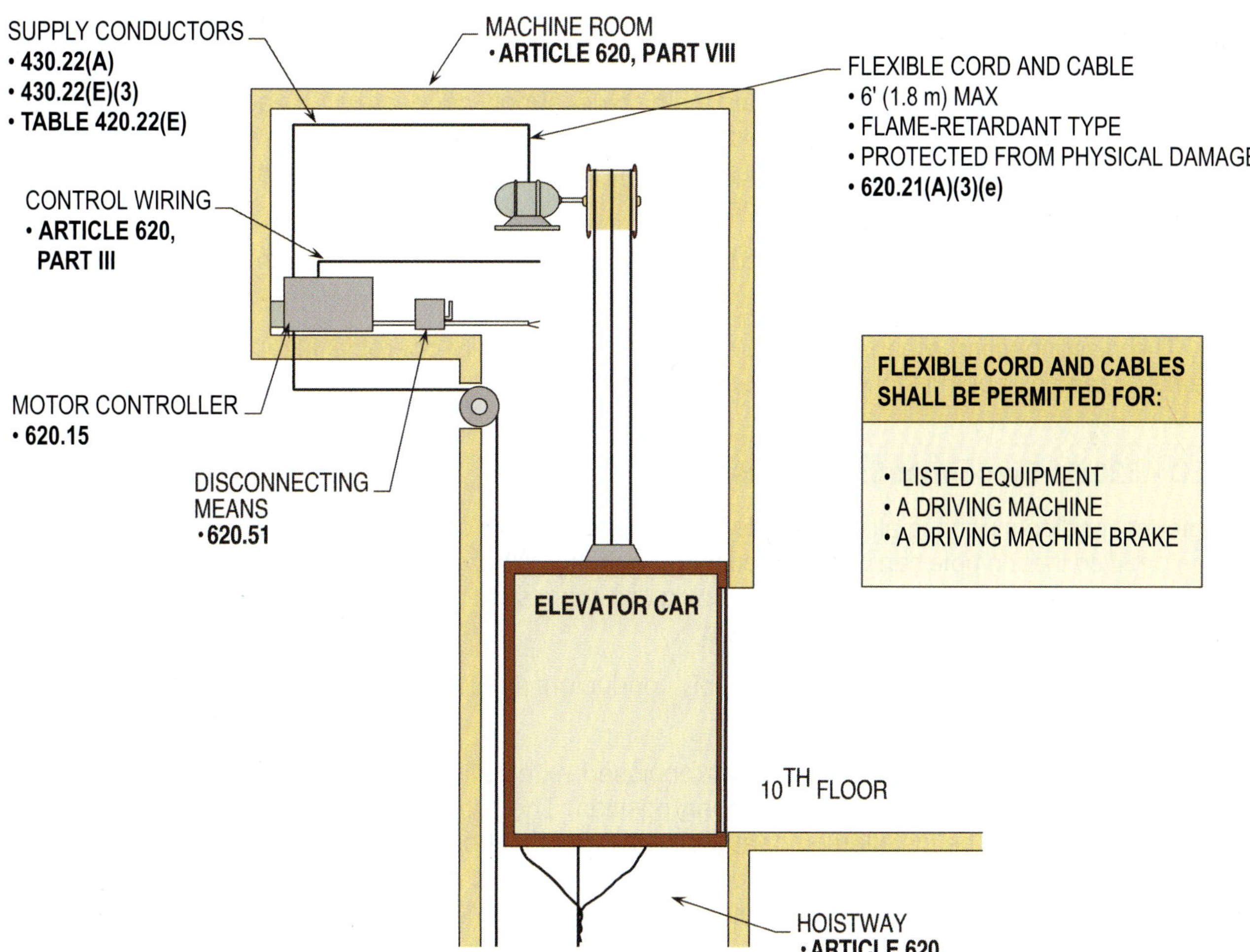

WITHIN MACHINE ROOMS, CONTROL ROOMS, AND MACHINERY SPACES AND CONTROL SPACES 620.21(A)(3)(e)

Purpose of Change: A new subdivision has been added to permit flexible cords and cables to be installed within machine rooms, control rooms and machinery spaces and control spaces.

NEC Ch. 6 - Article 620
Part VI - 620.51(A)

Type of Change	Panel Action	UL	UL 508	API 500 - 1997	API 505 - 1997	OSHA - 1994
Revision	Accept in Principle	-	-	-	-	1910.306(c)(1)
ROP	**ROC**	**NFPA 70E - 2004**	**NFPA 70B - 2006**	**NFPA 79 - 2007**	**NFPA**	**NEMA**
pg. 655 # 12-51	pg. 366 # 12-17	430.3(A)(1)	-	-	-	-
log: 352 CMP: 12	log: 553	Submitter: Michael J. Johnston		2005 NEC: 620.51(A)		IEC: -

2005 NEC - 620.51 Disconnecting Means.

A single means for disconnecting all ungrounded main power supply conductors for each unit shall be provided and be designed so that no pole can be operated independently. Where multiple driving machines are connected to a single elevator, escalator, moving walk, or pumping unit, there shall be one disconnecting means to disconnect the motor(s) and control valve operating magnets.

The disconnecting means for the main power supply conductors shall not disconnect the branch circuit required in 620.22, 620.23, and 620.24.

(A) Type. The disconnecting means shall be an enclosed externally operable fused motor circuit switch or circuit breaker capable of being locked in the open position. The disconnecting means shall be a listed device.

2008 NEC - 620.51 Disconnecting Means.

A single means for disconnecting all ungrounded main power supply conductors for each unit shall be provided and be designed so that no pole can be operated independently. Where multiple driving machines are connected to a single elevator, escalator, moving walk, or pumping unit, there shall be one disconnecting means to disconnect the motor(s) and control valve operating magnets.

The disconnecting means for the main power supply conductors shall not disconnect the branch circuit required in 620.22, 620.23, and 620.24.

(A) Type. The disconnecting means shall be an enclosed externally operable fused motor circuit switch or circuit breaker capable of being locked in the open position. <u>The provision for locking or adding a lock to the disconnecting means shall be installed on or at the switch or circuit breaker used as the disconnecting means and shall remain in place with or without the lock installed. Portable means for adding a lock to the switch or circuit breaker shall not be permitted as the means installed at and remain with the equipment. The disconnecting means shall be a listed device.</u>

Author's Comment: A series of changes have been accepted to provide standardized wording. The requirement for locking or adding a lock to the disconnecting means for the elevator motor has been added to provide safety for the installer or maintainer of the system. Installing this locking means or providing the locking mechanism will help ensure that the device is available when it becomes necessary to lock the disconnecting means to the "off" position. The Comment added text that prohibits a portable means as being the locking means left on or at the equipment, thus requiring a permanent locking means to be installed.

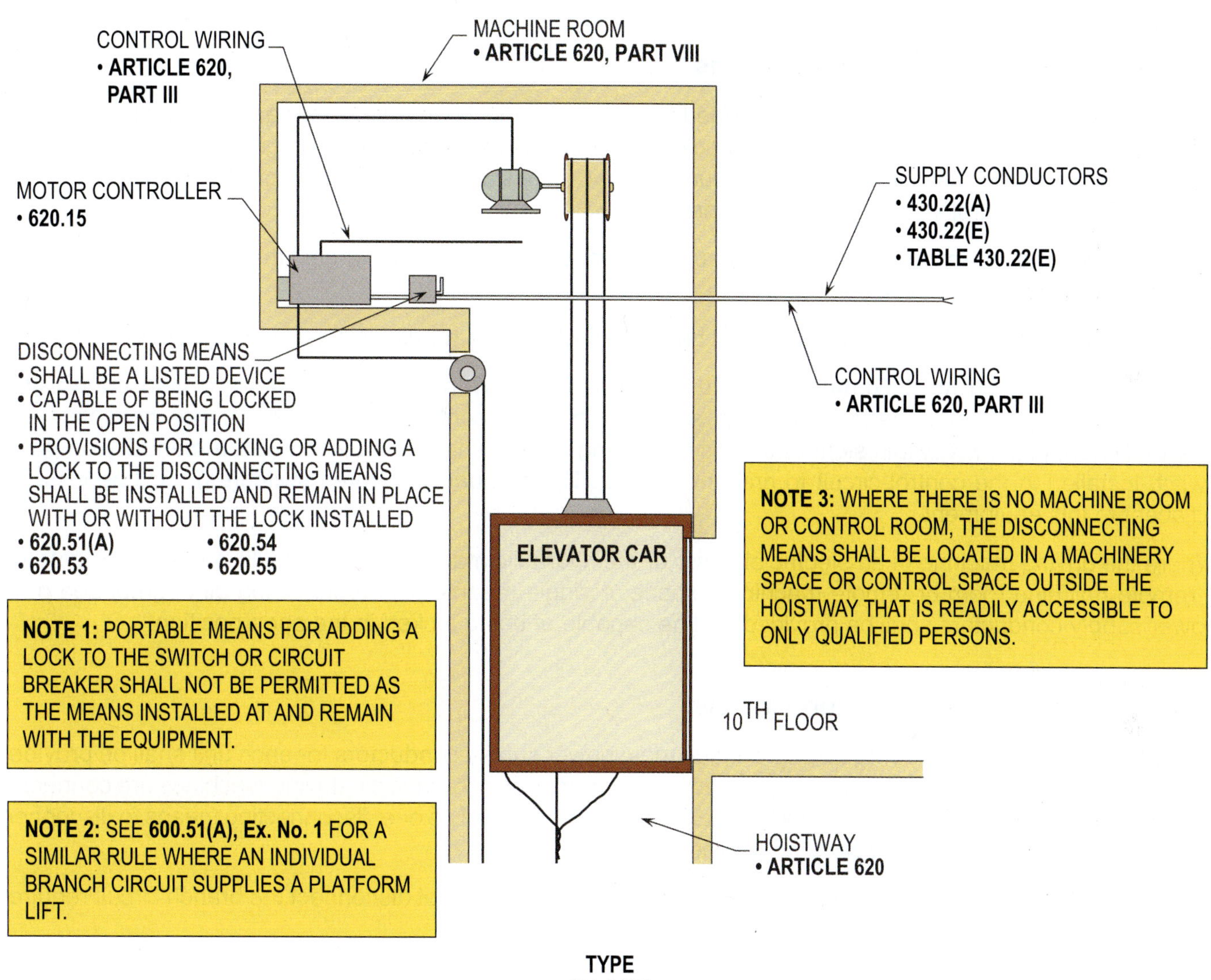
CONTROL WIRING
• ARTICLE 620, PART III
MACHINE ROOM
• ARTICLE 620, PART VIII
MOTOR CONTROLLER
• 620.15
SUPPLY CONDUCTORS
• 430.22(A)
• 430.22(E)
• TABLE 430.22(E)
DISCONNECTING MEANS
• SHALL BE A LISTED DEVICE
• CAPABLE OF BEING LOCKED IN THE OPEN POSITION
• PROVISIONS FOR LOCKING OR ADDING A LOCK TO THE DISCONNECTING MEANS SHALL BE INSTALLED AND REMAIN IN PLACE WITH OR WITHOUT THE LOCK INSTALLED
• 620.51(A)
• 620.53
• 620.54
• 620.55
CONTROL WIRING
• ARTICLE 620, PART III
NOTE 3: WHERE THERE IS NO MACHINE ROOM OR CONTROL ROOM, THE DISCONNECTING MEANS SHALL BE LOCATED IN A MACHINERY SPACE OR CONTROL SPACE OUTSIDE THE HOISTWAY THAT IS READILY ACCESSIBLE TO ONLY QUALIFIED PERSONS.
ELEVATOR CAR
NOTE 1: PORTABLE MEANS FOR ADDING A LOCK TO THE SWITCH OR CIRCUIT BREAKER SHALL NOT BE PERMITTED AS THE MEANS INSTALLED AT AND REMAIN WITH THE EQUIPMENT.
10TH FLOOR
NOTE 2: SEE 600.51(A), Ex. No. 1 FOR A SIMILAR RULE WHERE AN INDIVIDUAL BRANCH CIRCUIT SUPPLIES A PLATFORM LIFT.
HOISTWAY
• ARTICLE 620

TYPE
620.51(A)

Purpose of Change: This revision clarifies the provisions for locking or adding a lock to the disconnecting means.

NEC Ch. 6 - Article 620
Part VI - 620.51(C)(1)

Type of Change	Panel Action	UL	UL 508	API 500 - 1997	API 505 - 1997	OSHA - 1994
Revision	Accept in Principle	-	-	-	-	1910.306(c)(1)
ROP	**ROC**	**NFPA 70E - 2004**	**NFPA 70B - 2006**	**NFPA 79 - 2007**	**NFPA**	**NEMA**
pg. 656 # 12-58	pg. 367 # 12-22	430.3(A)(3)	-	-	-	-
log: 1226 CMP: 12	log: 1326 Submitter: Andy Juhasz			2005 NEC: 620.51(C)(1)		IEC: -

2005 NEC - 620.51 Disconnecting Means.

A single means for disconnecting all ungrounded main power supply conductors for each unit shall be provided and be designed so that no pole can be operated independently. Where multiple driving machines are connected to a single elevator, escalator, moving walk, or pumping unit, there shall be one disconnecting means to disconnect the motor(s) and control valve operating magnets.

The disconnecting means for the main power supply conductors shall not disconnect the branch circuit required in 620.22, 620.23, and 620.24.

(C) Location. The disconnecting means shall be located where it is readily accessible to qualified persons.

(1) On Elevators Without Generator Field Control. On elevators without generator field control, the disconnecting means shall be located within sight of the motor controller. Driving machines or motion and operation controllers not within sight of the disconnecting means shall be provided with a manually operated switch installed in the control circuit to prevent starting. The manually operated switch(es) shall be installed adjacent to this equipment.

Where the driving machine of an electric elevator or the hydraulic machine of a hydraulic elevator is located in a remote machine room or remote machinery space, a single means for disconnecting all ungrounded main power supply conductors shall be provided and be capable of being locked in the open position.

2008 NEC - 620.51 Disconnecting Means.

A single means for disconnecting all ungrounded main power supply conductors for each unit shall be provided and be designed so that no pole can be operated independently. Where multiple driving machines are connected to a single elevator, escalator, moving walk, or pumping unit, there shall be one disconnecting means to disconnect the motor(s) and control valve operating magnets.

The disconnecting means for the main power supply conductors shall not disconnect the branch circuit required in 620.22, 620.23, and 620.24.

(C) Location. The disconnecting means shall be located where it is readily accessible to qualified persons.

(1) On Elevators Without Generator Field Control. On elevators without generator field control, the disconnecting means shall be located within sight of the motor controller. Where the driving machine of an electric elevator or the hydraulic machine of a hydraulic elevator is located in a remote machine room or remote machinery space, a single means for disconnecting all ungrounded main power supply conductors shall be provided and be capable of being locked in the open position. Where the motor controller is located in the elevator hoistway, the disconnecting means required by 620.51(A) shall be located in a machinery space, machine room, control space or control room outside the hoistway; and an additional, non-fused enclosed externally operable motor circuit switch capable of being locked in the open position to disconnect all ungrounded main power supply conductors shall be located within sight of the motor controller. The additional switch shall be a listed device and shall comply with 620.91(C).

The provision for locking or adding a lock to the disconnecting means, required by this section, shall be installed on or at the switch or circuit breaker used as the disconnecting means and shall remain in place with or without the lock installed. Portable means for adding a lock to the switch or circuit breaker shall not be permitted.

Driving machines or motion and operation controllers not within sight of the disconnecting means shall be provided with a manually operated switch installed in the control circuit to prevent starting. The manually operated switch(es) shall be installed adjacent to this equipment.

Author's Comment: Where the motor controller is located in the hoistway, a disconnecting means must now be accessible from outside the hoistway. That means an additional non-fused disconnecting means is now necessary within sight of the motor controller for use by and protection of elevator personnel working on equipment. The added text is also consistent with **430.102(B), Exception** and the action on Proposal 12-9 to provide consistency with all the lockout requirements within the Code. The Comment added a requirement that, where the motor controller for the elevator is located in the elevator hoistway, the additional disconnecting means must be listed and must disconnect power from any emergency or standby power systems in accordance with **620.91(C)**.

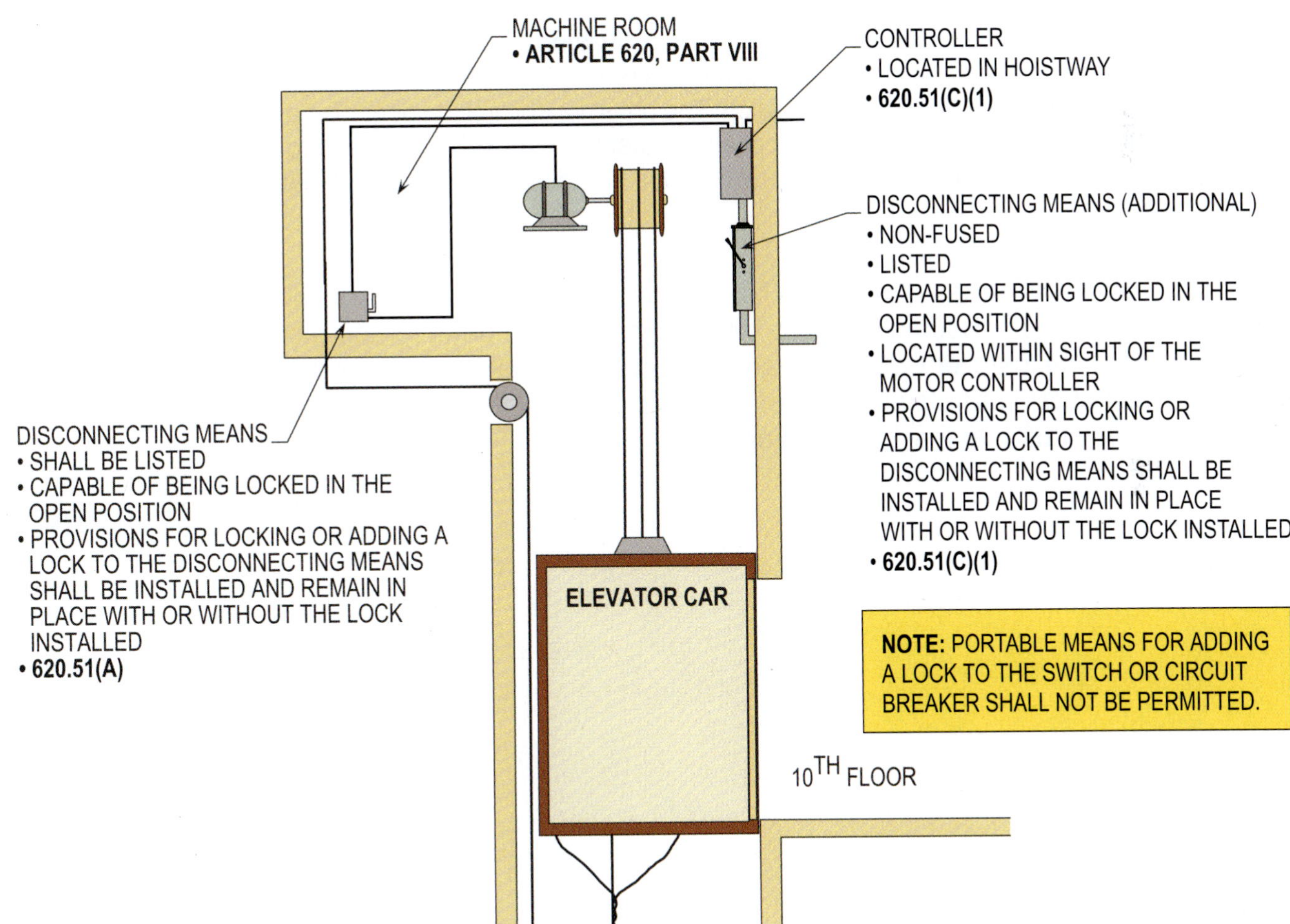

ON ELEVATORS WITHOUT GENERATOR FIELD CONTROL
620.51(C)(1)

Purpose of Change: This revision clarifies that where the motor controller is located in the elevator hoistway, the disconnecting means required by **620.51(A)** shall be located in a machinery space, machine room, control space or control room outside the hoistway; and as additional disconnecting means (non-fused) shall be located within sight of the motor controller.

NEC Ch. 6 - Article 625
Part III - 625.23

Type of Change		Panel Action		UL	UL 508	API 500 - 1997	API 505 - 1997	OSHA - 1994
Revision		Accept in Principle		-	-	-	-	-
ROP		ROC		NFPA 70E - 2004	NFPA 70B - 2006	NFPA 79 - 2007	NFPA	NEMA
pg. 660	# 12-77	pg. -	# -	-	-	-	-	-
log: 350	CMP: 12	log: -	Submitter: Michael J. Johnston			2005 NEC: 625.23		IEC: -

2005 NEC - 625.23 Disconnecting Means.

For electric vehicle supply equipment rated more than 60 amperes or more than 150 volts to ground, the disconnecting means shall be provided and installed in a readily accessible location. The disconnecting means shall be capable of being locked in the open position.

2008 NEC - 625.23 Disconnecting Means.

For electric vehicle supply equipment rated more than 60 amperes or more than 150 volts to ground, the disconnecting means shall be provided and installed in a readily accessible location. The disconnecting means shall be capable of being locked in the open position. <u>The provision for locking or adding a lock to the disconnecting means shall be installed on or at the switch or circuit breaker used as the disconnecting means and shall remain in place with or without the lock installed. Portable means for adding a lock to the switch or circuit breaker shall not be permitted.</u>

Author's Comment: A series of changes have been accepted to provide standardized wording. The requirement for locking or adding a lock to the disconnecting means for the motor has been added to provide safety for the installer or maintainer of the system. Installing this locking means or providing the locking mechanism will help ensure that the device is available when it becomes necessary to lock the disconnecting means to the "off" position.

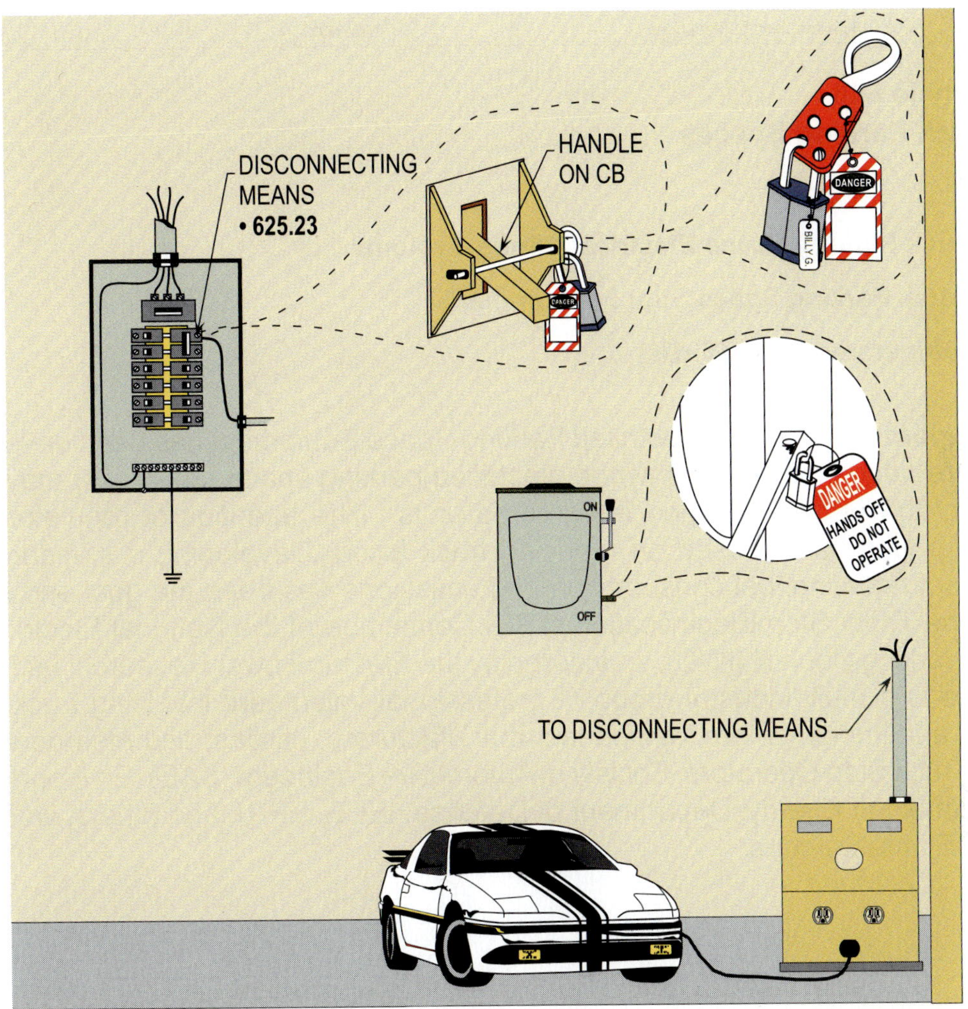

NOTE: THE MEANS TO PLACE LOCK SHALL REMAIN IN PLACE ON THE DISCONNECTING MEANS AND PORTABLE MEANS SHALL NOT BE USED PER **625.23**.

DISCONNECTING MEANS
625.23

Purpose of Change: This revision clarifies that provisions for locking or adding a lock to the disconnecting means shall remain in place with or without the lock installed.

NEC Ch. 6 - Article 626
Parts I thru IV - 626.1 through 626.32

88

Type of Change		Panel Action		UL	UL 508	API 500 - 1997	API 505 - 1997	OSHA - 1994
New Article		Accept in Principle		-	-	-	-	-
ROP		ROC		NFPA 70E - 2004	NFPA 70B - 2006	NFPA 79 - 2007	NFPA	NEMA
pg. 661	# 12-81	pg. 373	# 12-44	-	-	-	-	-
log: 1650	CMP: 12	log: 2025	Submitter: Juan C. Menendez			2005 NEC: -		IEC: -

2008 NEC - Article 626
Electrified Truck Parking Spaces

I. General

II. Electrified Truck Parking Space Electrical Wiring Systems

III. Electrified Truck Parking Space Supply Equipment

IV. Transport Refrigerated Units (TRUs)

Author's Comment: Article 626 Electrified Parking Space Equipment has been developed to identify the infrastructure needs for systems where electrified parking space equipment may be installed for both heavy duty trucks and transport refrigerated units. The attention of regulatory agencies and environmental groups has focused on reducing truck idling. Developing a standardized, safe and efficient means of reducing fuel consumption and emissions has been the goal and **Article 626** was developed by the Truck Stop Electrification (TSE) Committee of the National Electric Transportation Infrastructure Working Council (IWC), sponsored by the Electric Power Research Institute (EPRI). The TSE Committee is a multi-industry group of professional volunteers involving truck manufacturers, TSE designers and implementers, component manufacturers, utilities, and members of the National Association of Truck Stop Operators, Society of Automotive Engineers (SAE), Environmental Protection Agency, Department of Energy, Department of Defense, IEEE, EPRI, and others, working together to develop the TSE infrastructure.

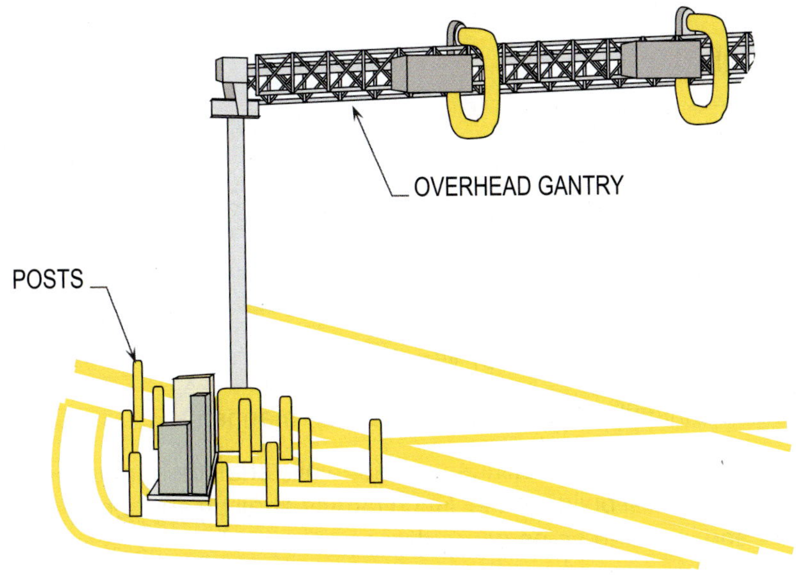

ARTICLE 626 - ELECTRIFIED TRUCK PARKING SPACES	
PART I - GENERAL	
SECTION	**HEADING**
626.1	SCOPE
626.2	DEFINITIONS
626.3	OTHER ARTICLES
626.4	GENERAL REQUIREMENTS
PART II - ELECTRIFIED TRUCK PARKING SPACE ELECTRICAL WIRING SYSTEMS	
626.10	BRANCH CIRCUIT
626.11	FEEDER AND SERVICE LOAD CALCULATIONS
PART III - ELECTRIFIED TRUCK PARKING SPACE SUPPLY EQUIPMENT	
626.22	WIRING METHODS AND MATERIALS
626.23	OVERHEAD GANTRY OR CABLE MANAGEMENT SYSTEM
626.24	ELECTRIFIED TRUCK PARKING SPACE SUPPLY EQUIPMENT CONNECTION MEANS
626.25	SEPARABLE POWER-SUPPLY CABLE ASSEMBLY
626.26	LOSS OF PRIMARY POWER
626.27	INTERACTIVE SYSTEMS
PART IV - TRANSPORT REFRIGERATED UNITS (TRUs)	
626.30	TRANSPORT REFRIGERATED UNITS
626.31	DISCONNECTING MEANS AND RECEPTACLES
626.32	SEPARABLE POWER SUPPLY CABLE ASSEMBLY

ELECTRIFIED TRUCK PARKING SPACES
626.1 through 626.32

Purpose of Change: A new article has been added to address requirements for electrified truck parking spaces.

NEC Ch. 6 - Article 640
Part I - 640.6(A) through (D)

Type of Change	Panel Action	UL	UL 508	API 500 - 1997	API 505 - 1997	OSHA - 1994
Revision	Accept in Principle	-	-	-	-	-
ROP	ROC	NFPA 70E - 2004	NFPA 70B - 2006	NFPA 79 - 2007	NFPA	NEMA
pg. 676 # 12-94	pg. 386 # 12-67	-	-	-	-	-
log: 3049 CMP: 12	log: 545	Submitter: Harold C. Ohde		2005 NEC: 640.6		IEC: -

2005 NEC - 640.6 Mechanical Execution of Work.

Equipment ~~and~~ cables shall be installed in a neat workmanlike manner.

Cables installed exposed on the surface of ceilings and sidewalls shall be supported in such a manner that the cables will not be damaged by normal building use. Such cables shall be ~~supported~~ by straps, staples, hangers, or similar fittings designed and installed so as not to damage the cable. The installation shall also comply with 300.4~~(D)~~ and ~~300.11~~.

2008 NEC - 640.6 Mechanical Execution of Work.

(A) Neat and Workmanlike Manner. Audio signal processing, amplification, and reproduction equipment, cables, and circuits shall be installed in a neat and workmanlike manner.

(B) Installation of Audio Distribution Cables. Cables installed exposed on the surface of ceilings and sidewalls shall be supported in such a manner that the audio distribution cables will not be damaged by normal building use. Such cables shall be secured by straps, staples, cable ties, hangers, or similar fittings designed and installed so as not to damage the cable. The installation shall also comply with 300.4 and (A).

(C) Abandoned Audio Distribution Cables. The accessible portion of abandoned audio distribution cables shall be removed.

(D) Installed Audio Distribution Cable Identified for Future Use.

(1) Cables identified for future use shall be marked with a tag of sufficient durability to withstand the environment involved.

(2) Cable tags shall have the following information:

(a) Date cable was identified for future use

(b) Date of intended use

(c) Information relating to the intended future use of cable

Author's Comment: Section **640.6** has been reformatted with four new individual subsections. New **(C)** is the text from **640.3(A)** requiring abandoned cables to be removed. New **(D)** provides marking requirements for audio cables identified for future use and requires the tag to be impervious to temperature, dampness, in other words "able to withstand the environment involved." The marking tags must have date of marking, date of future use, and reason for future use. The Comment amended the proposal by requiring the tag to be of sufficient durability to withstand the environment in which it is installed but deleted the requirements for the tag to be impervious to the effects of temperature and dampness or the effects of gnawing by rodents.

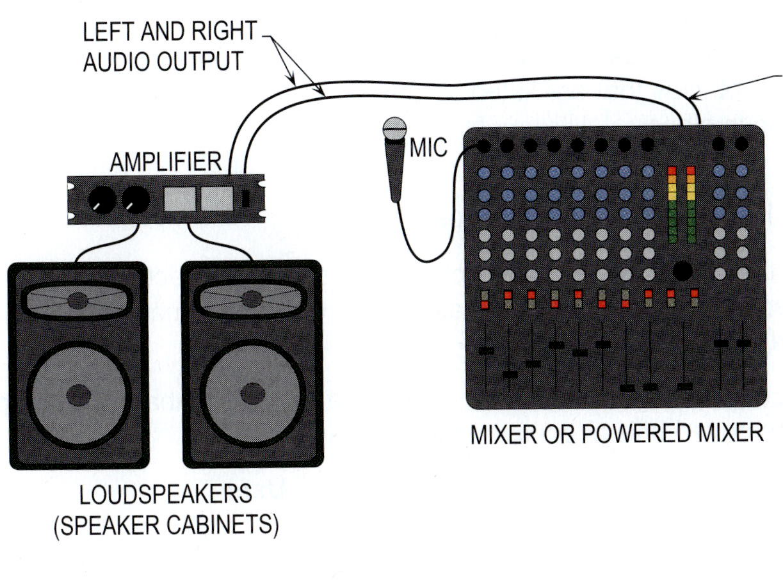

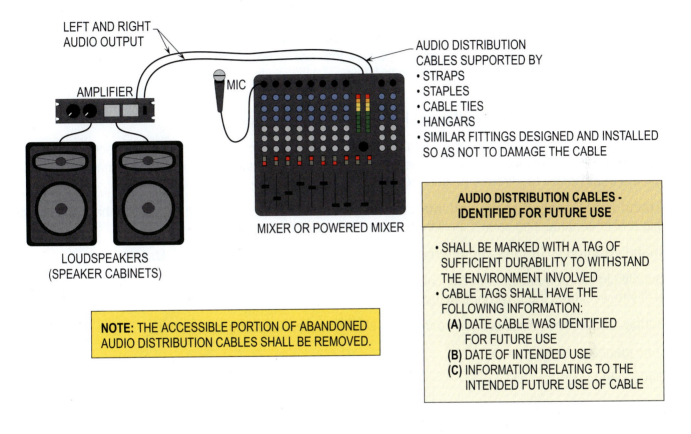

AUDIO DISTRIBUTION CABLES SUPPORTED BY

- STRAPS
- STAPLES
- CABLE TIES
- HANGARS
- SIMILAR FITTINGS DESIGNED AND INSTALLED SO AS NOT TO DAMAGE THE CABLE

AUDIO DISTRIBUTION CABLES - IDENTIFIED FOR FUTURE USE

- SHALL BE MARKED WITH A TAG OF SUFFICIENT DURABILITY TO WITHSTAND THE ENVIRONMENT INVOLVED
- CABLE TAGS SHALL HAVE THE FOLLOWING INFORMATION:
  - **(A)** DATE CABLE WAS IDENTIFIED FOR FUTURE USE
  - **(B)** DATE OF INTENDED USE
  - **(C)** INFORMATION RELATING TO THE INTENDED FUTURE USE OF CABLE

NOTE: THE ACCESSIBLE PORTION OF ABANDONED AUDIO DISTRIBUTION CABLES SHALL BE REMOVED.

MECHANICAL EXECUTION OF WORK
640.6(A) through (D)

Purpose of Change: This revision has reformatted section **640.6** into four new subsections.

NEC Ch. 6 - Article 645
Part - 645.5(F) and (G)

Type of Change		Panel Action		UL	UL 508	API 500 - 1997	API 505 - 1997	OSHA - 1994
New Subsection		Accept in Principle		60950	-	-	-	-
ROP		ROC		NFPA 70E - 2004	NFPA 70B - 2006	NFPA 79 - 2007	NFPA	NEMA
pg. 682	# 12-116	pg. 389	# 12-80	-	-	-	-	-
log: 2649	CMP: 12	log: 546	Submitter: Robert W. Jensen			2005 NEC: -		IEC: -

2005 NEC - 645.5 Supply Circuits and Interconnecting Cables.

(6) Abandoned cables shall be removed unless contained in metal raceways.

(E) Securing in Place. Power cables; communications cables; connecting cables; interconnecting cables; and associated boxes, connectors, plugs, and receptacles that are listed as part of, or for, information technology equipment shall not be required to be secured in place.

2008 NEC - 645.5 Supply Circuits and Interconnecting Cables.

(E) Securing in Place. Power cables; communications cables; connecting cables; interconnecting cables; and associated boxes, connectors, plugs, and receptacles that are listed as part of, or for, information technology equipment shall not be required to be secured in place.

(F) Abandoned Supply Circuits and Interconnecting Cables. The accessible portion of abandoned supply circuits and interconnecting cables shall be removed unless contained in a metal raceway.

(G) Installed Supply Circuits and Interconnecting Cables Identified for Future Use.

(1) Supply circuits and interconnecting cables identified for future use shall be marked with a tag of sufficient durability to withstand the environment involved.

(2) Supply circuits and interconnecting cable tags shall have the following information:

(a) Date identified for future use

(b) Date of intended use

(c) Information relating to the intended future use

Author's Comment: The abandoned cable text has been moved from **645.5(D)(6)** to new subsection **645.5(F)** and new **645.5(G)** has been added for the requirements of marking cables for future use. The Comment amended the proposal by requiring the tag to be of sufficient durability to withstand the environment in which it is installed but deleted the requirements for the tag to be impervious to the effects of temperature and dampness or the effects of gnawing by rodents.

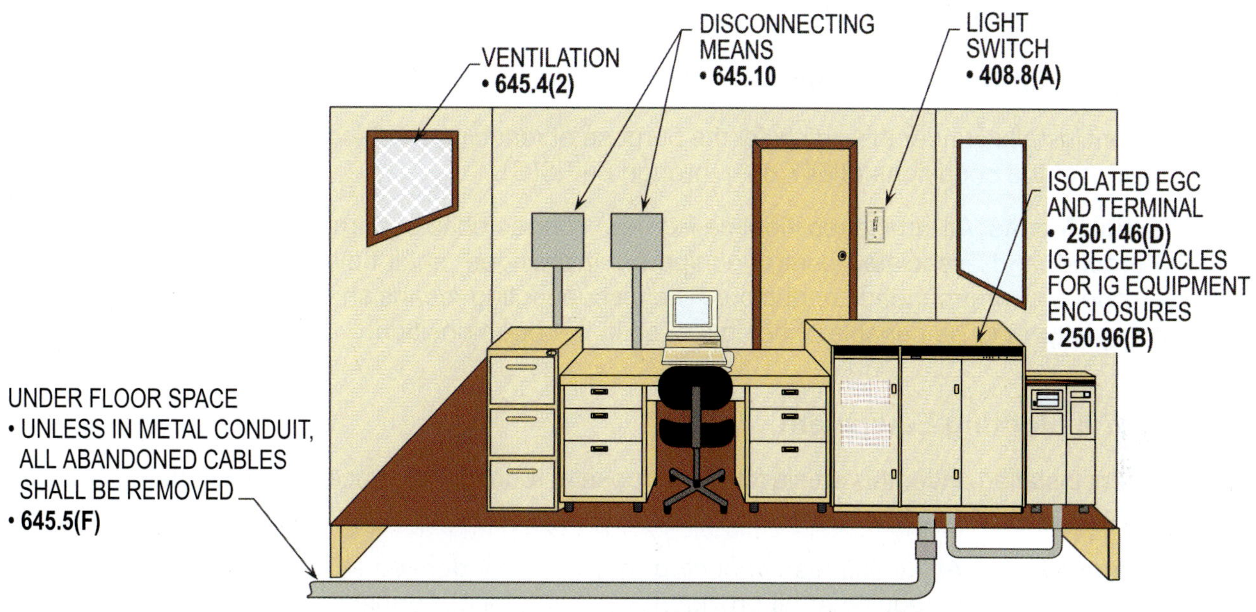

INSTALLED SUPPLY CIRCUITS AND INTERCONNECTING CABLES - IDENTIFIED FOR FUTURE USE

- SHALL BE MARKED WITH A TAG OF SUFFICIENT DURABILITY TO WITHSTAND THE ENVIRONMENT
- CABLE TAGS SHALL HAVE FOLLOWING INFORMATION

(A) DATE IDENTIFIED FOR FUTURE USE
(B) DATE OF INTENDED USE
(C) INFORMATION RELATING TO THE INTENDED USE

SUPPLY CIRCUITS AND INTERCONNECTING CABLES
645.5(F) and (G)

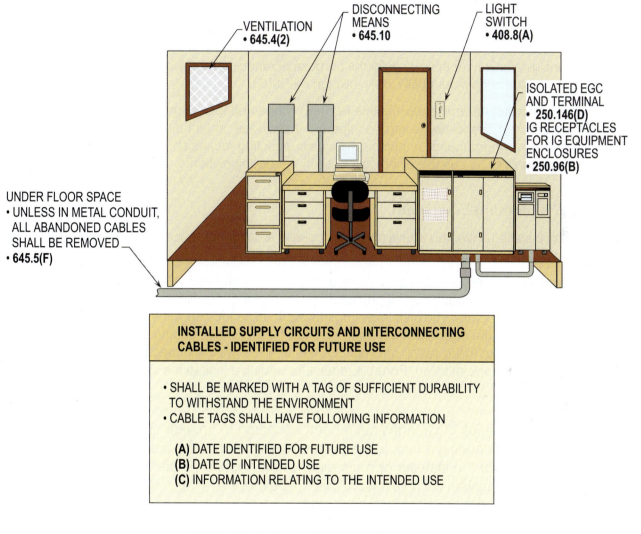

Purpose of Change: Section **645.5(D)(6)** has been moved to new subsection **645.5(F)**. A new subsection has been added to clarify the marking requirements of installed supply circuits and interconnecting cables for future use.

NEC Ch. 6 - Article 647
Part - 647.8(A)

Type of Change	Panel Action	UL	UL 508	API 500 - 1997	API 505 - 1997	OSHA - 1994
Revision	Accept in Principle	-	-	-	-	-
ROP	ROC	NFPA 70E - 2004	NFPA 70B - 2006	NFPA 79 - 2007	NFPA	NEMA
pg. 685 # 12-129	pg. - # -	-	-	-	-	-
log: 349 CMP: 12	log: -	Submitter: Michael J. Johnston		2005 NEC: 647.8(A)		IEC: -

2005 NEC - 647.8 Lighting Equipment.

Lighting equipment installed under this article for the purpose of reducing electrical noise originating from lighting equipment shall meet the conditions of 647.8(A) through ~~647.8~~(C).

(A) Disconnecting Means. All luminaires ~~(lighting fixtures)~~ connected to separately derived systems operating at 60 volts to ground, and associated control equipment if provided, shall have a disconnecting means that simultaneously opens all ungrounded conductors. The disconnecting means shall be located within sight of the luminaire ~~(lighting fixture)~~ or be capable of being locked in the open position.

2008 NEC - 647.8 Lighting Equipment.

Lighting equipment installed under this article for the purpose of reducing electrical noise originating from lighting equipment shall meet the conditions of 647.8(A) through (C).

(A) Disconnecting Means. All luminaires connected to separately derived systems operating at 60 volts to ground, and associated control equipment if provided, shall have a disconnecting means that simultaneously opens all ungrounded conductors. The disconnecting means shall be located within sight of the luminaire or be capable of being locked in the open position. The provision for locking or adding a lock to the disconnecting means shall be installed on or at the switch or circuit breaker used as the disconnecting means and shall remain in place with or without the lock installed. Portable means for adding a lock to the switch or circuit breaker shall not be permitted.

Author's Comment: A series of changes have been accepted to provide standardized wording. The requirement for locking or adding a lock to the disconnecting means for the elevator motor has been added to provide safety for the installer or maintainer of the system. Installing this locking means or providing the locking mechanism will help ensure that the device is available when it becomes necessary to lock the disconnecting means to the "off" position.

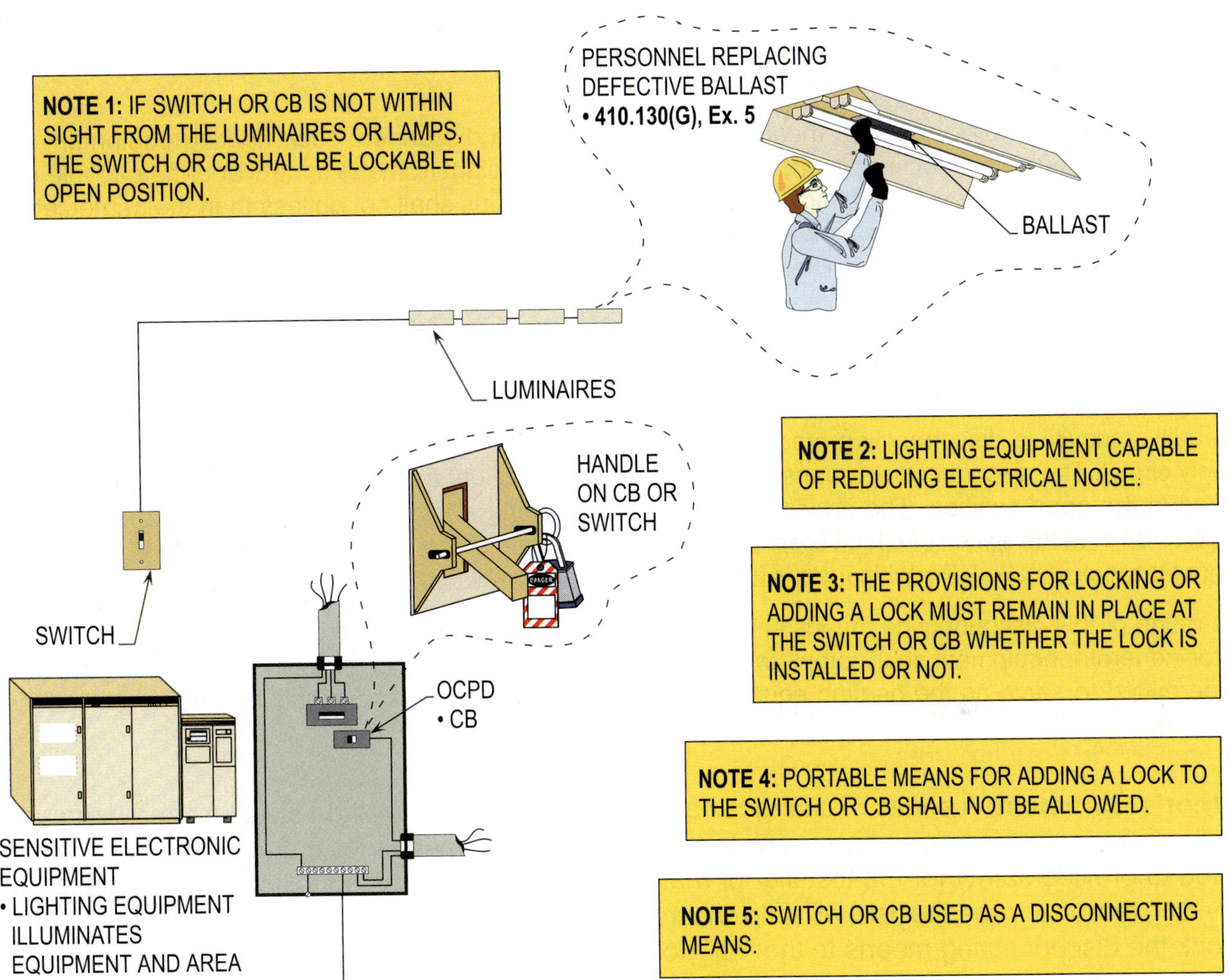

NOTE 1: IF SWITCH OR CB IS NOT WITHIN SIGHT FROM THE LUMINAIRES OR LAMPS, THE SWITCH OR CB SHALL BE LOCKABLE IN OPEN POSITION.
PERSONNEL REPLACING DEFECTIVE BALLAST
• 410.130(G), Ex. 5
BALLAST
LUMINAIRES
HANDLE ON CB OR SWITCH
NOTE 2: LIGHTING EQUIPMENT CAPABLE OF REDUCING ELECTRICAL NOISE.
NOTE 3: THE PROVISIONS FOR LOCKING OR ADDING A LOCK MUST REMAIN IN PLACE AT THE SWITCH OR CB WHETHER THE LOCK IS INSTALLED OR NOT.
SWITCH
OCPD
• CB
NOTE 4: PORTABLE MEANS FOR ADDING A LOCK TO THE SWITCH OR CB SHALL NOT BE ALLOWED.
SENSITIVE ELECTRONIC EQUIPMENT
• LIGHTING EQUIPMENT ILLUMINATES EQUIPMENT AND AREA
NOTE 5: SWITCH OR CB USED AS A DISCONNECTING MEANS.

DISCONNECTING MEANS
647.8(A)

Purpose of Change: This revision clarifies that provisions for locking or adding a lock to the disconnecting means shall remain in place with or without the lock installed.

NEC Ch. 6 - Article 665
Part I - 665.12

Type of Change		Panel Action		UL	UL 508	API 500 - 1997	API 505 - 1997	OSHA - 1994
Revision		Accept in Principle		-	-	-	-	1910.306(g)(2)(vi)
ROP		ROC		NFPA 70E - 2004	NFPA 70B - 2006	NFPA 79 - 2007	NFPA	NEMA
pg. 686	# 12-136	pg. -	# -	430.7(B)(6)	-	-	-	-
log: 348	CMP: 12	log: -	Submitter: Michael J. Johnston			2005 NEC: 665.12		IEC: -

2005 NEC - 665.12 Disconnecting Means.

A readily accessible disconnecting means shall be provided to disconnect each heating equipment from its supply circuit. The disconnecting means shall be located within sight from the controller or be capable of being locked in the open position. The rating of this disconnecting means shall not be less than the nameplate rating of the heating equipment. Motor-generator equipment shall comply with Article 430, Part IX. The supply circuit disconnecting means shall be permitted to serve as the heating equipment disconnecting means where only one heating equipment is supplied.

2008 NEC - 665.12 Disconnecting Means.

A readily accessible disconnecting means shall be provided to disconnect each heating equipment from its supply circuit. The disconnecting means shall be located within sight from the controller or be capable of being locked in the open position. The provision for locking or adding a lock to the disconnecting means shall be installed on or at the switch or circuit breaker used as the disconnecting means and shall remain in place with or without the lock installed. Portable means for adding a lock to the switch or circuit breaker shall not be permitted.

The rating of this disconnecting means shall not be less than the nameplate rating of the heating equipment. Motor-generator equipment shall comply with Article 430, Part IX. The supply circuit disconnecting means shall be permitted to serve as the heating equipment disconnecting means where only one heating equipment is supplied.

Author's Comment: A series of changes have been accepted to provide standardized wording. The requirement for locking or adding a lock to the disconnecting means for the elevator motor has been added to provide safety for the installer or maintainer of the system. Installing this locking means or providing the locking mechanism will help ensure that the device is available when it becomes necessary to lock the disconnecting means to the "off" position.

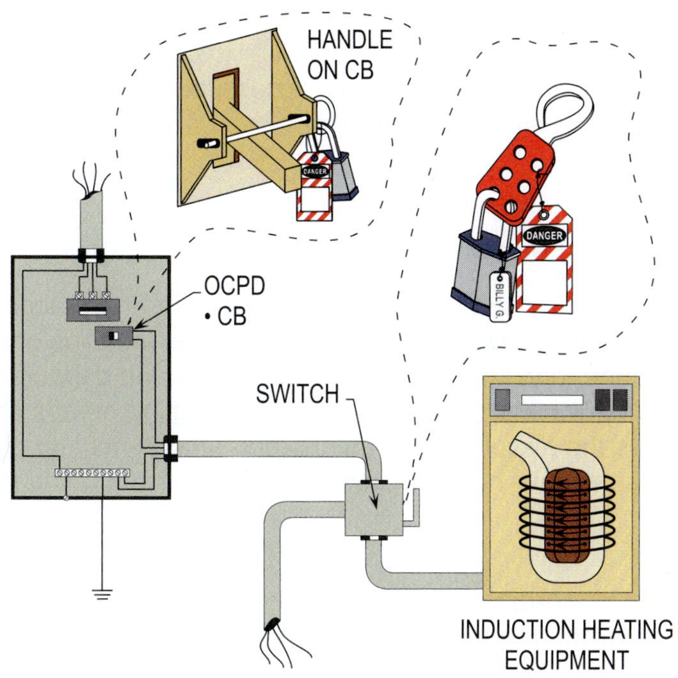

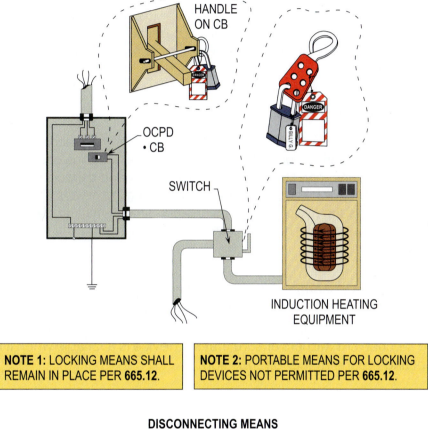

NOTE 1: LOCKING MEANS SHALL REMAIN IN PLACE PER **665.12**.	**NOTE 2:** PORTABLE MEANS FOR LOCKING DEVICES NOT PERMITTED PER **665.12**.

DISCONNECTING MEANS
665.12

Purpose of Change: This revision clarifies that provisions for locking or adding a lock to the disconnecting means shall remain in place or without the lock installed.

NEC Ch. 6 - Article 675
Part I - 675.8(B)

Type of Change		Panel Action		UL	UL 508	API 500 - 1997	API 505 - 1997	OSHA - 1994
Revision		Accept		-	-	-	-	1910.307(i)(2)
ROP		**ROC**		**NFPA 70E - 2004**	**NFPA 70B - 2006**	**NFPA 79 - 2007**	**NFPA**	**NEMA**
pg. 689	# 19-137	pg. -	# -	430.9(B)	-	-	-	-
log: 2046	CMP: 19	log: -	Submitter: James T. Dollard, Jr.			2005 NEC: 675.8(B)		IEC: -

2005 NEC - 675.8 Disconnecting Means.

(B) Main Disconnecting Means. The main disconnecting means for the machine shall provide overcurrent protection, shall be at the point of connection of electrical power to the machine or shall be visible and not more than 15 m (50 ft) from the machine, and shall be readily accessible and capable of being locked in the open position. This disconnecting means shall have a horsepower and current rating not less than required for the main controller.

2008 NEC - 675.8 Disconnecting Means.

(B) Main Disconnecting Means. The main disconnecting means for the machine shall provide overcurrent protection, shall be at the point of connection of electrical power to the machine or shall be visible and not more than 15 m (50 ft) from the machine, and shall be readily accessible and capable of being locked in the open position. The provision for locking or adding a lock to the disconnecting means shall be installed on or at the switch or circuit breaker used as the disconnecting means and shall remain in place with or without the lock installed. This disconnecting means shall have a horsepower and current rating not less than required for the main controller.

Author's Comment: A series of changes have been accepted to provide standardized wording. The requirement for locking or adding a lock to the disconnecting means for the elevator motor has been added to provide safety for the installer or maintainer of the system. Installing this locking means or providing the locking mechanism will help ensure that the device is available when it becomes necessary to lock the disconnecting means to the "off" position.

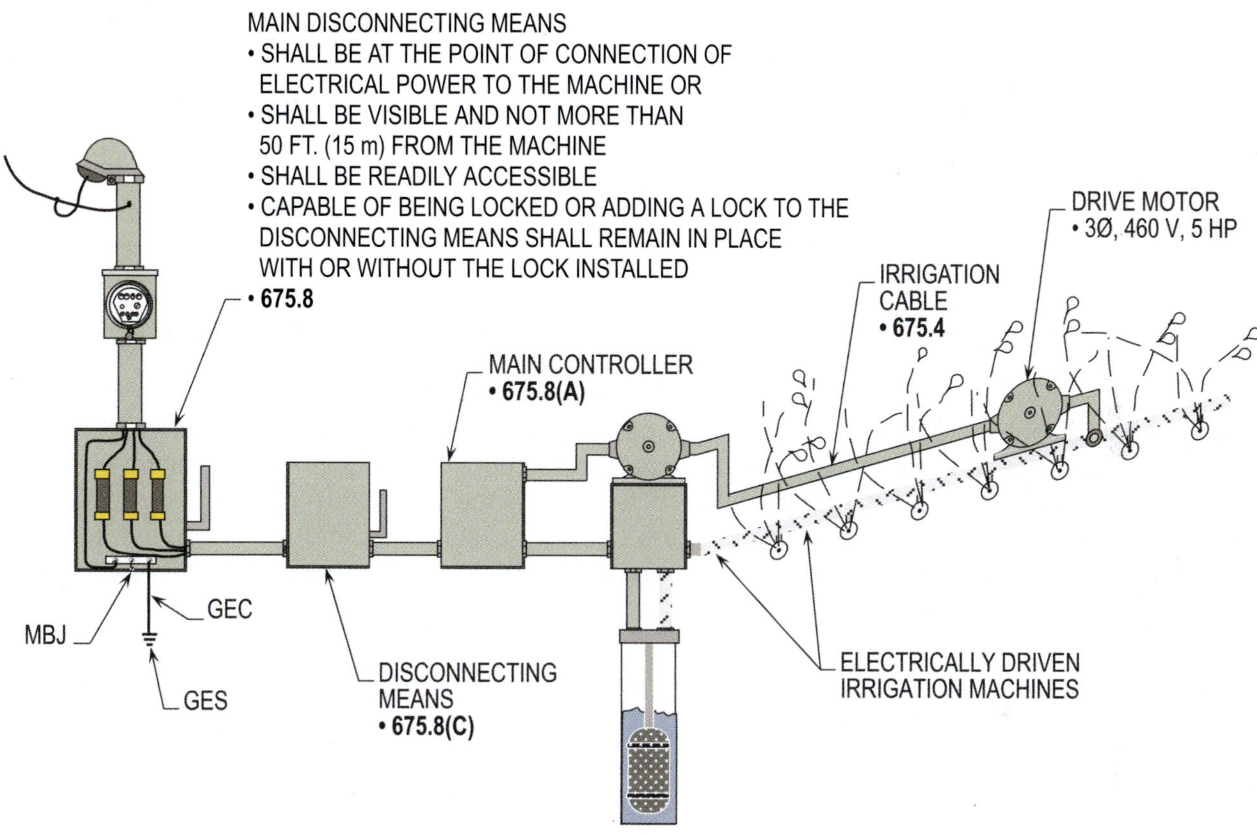
MAIN DISCONNECTING MEANS
• SHALL BE AT THE POINT OF CONNECTION OF ELECTRICAL POWER TO THE MACHINE OR
• SHALL BE VISIBLE AND NOT MORE THAN 50 FT. (15 m) FROM THE MACHINE
• SHALL BE READILY ACCESSIBLE
• CAPABLE OF BEING LOCKED OR ADDING A LOCK TO THE DISCONNECTING MEANS SHALL REMAIN IN PLACE WITH OR WITHOUT THE LOCK INSTALLED
• 675.8
DRIVE MOTOR
• 3Ø, 460 V, 5 HP
IRRIGATION CABLE
• 675.4
MAIN CONTROLLER
• 675.8(A)
GEC
MBJ
GES
DISCONNECTING MEANS
• 675.8(C)
ELECTRICALLY DRIVEN IRRIGATION MACHINES

NOTE: PROVISIONS FOR LOCKING OR ADDING A LOCK FOR ACCESS TO INTERNAL EQUIPMENT SHALL BE INSTALLED ON OR AT THE SWITCH OR CIRCUIT BREAKER AND SHALL REMAIN IN PLACE WITH OR WITHOUT THE LOCK INSTALLED.

MAIN DISCONNECTING MEANS
675.8(B)

Purpose of Change: This revision clarifies the provisions for locking or adding a lock to the main disconnecting means.

NEC Ch. 6 - Article 680
Part I - 680.10

Type of Change	Panel Action	UL	UL 508	API 500 - 1997	API 505 - 1997	OSHA - 1994
Revision	Accept	-	-	-	-	-
ROP	**ROC**	**NFPA 70E - 2004**	**NFPA 70B - 2006**	**NFPA 79 - 2007**	**NFPA**	**NEMA**
pg. 695 # 17-76	pg. - # -	-	-	-	-	-
log: 1634 CMP: 17	log: - Submitter: L. Keith Lofland			2005 NEC: 680.10		IEC: -

2005 NEC - 680.10 Underground Wiring Location.

Underground wiring shall not be permitted under the pool or within the area extending 1.5 m (5 ft) horizontally from the inside wall of the pool unless this wiring is necessary to supply pool equipment permitted by this article. Where space limitations prevent wiring from being routed a distance 1.5 m (5 ft) or more from the pool, such wiring shall be permitted where installed in rigid metal conduit, intermediate metal conduit, or a nonmetallic raceway system. All metal conduit shall be corrosion resistant and suitable for the location. The minimum ~~burial~~ depth shall be as given in Table 680.10.

2008 NEC - 680.10 Underground Wiring Location.

Underground wiring shall not be permitted under the pool or within the area extending 1.5 m (5 ft) horizontally from the inside wall of the pool unless this wiring is necessary to supply pool equipment permitted by this article. Where space limitations prevent wiring from being routed a distance 1.5 m (5 ft) or more from the pool, such wiring shall be permitted where installed in <u>complete raceway systems of</u> rigid metal conduit, intermediate metal conduit, or a nonmetallic raceway system. All metal conduit shall be corrosion resistant and suitable for the location. The minimum <u>cover</u> depth shall be as given in Table 680.10.

Author's Comment: The main requirements of **680.10** requires underground wiring systems to be located a minimum of 1.5 m (5 ft) horizontally from the inside wall of a pool. Where space limitations prevent this, an allowance inside the restricted 1.5 m (5 ft) distance is permitted when employing certain complete raceway systems within the restricted 1.5 m (5 ft) area of the pool.

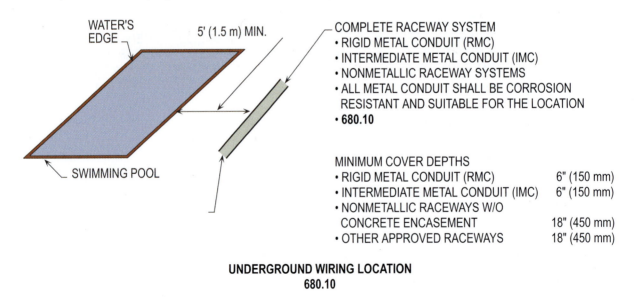

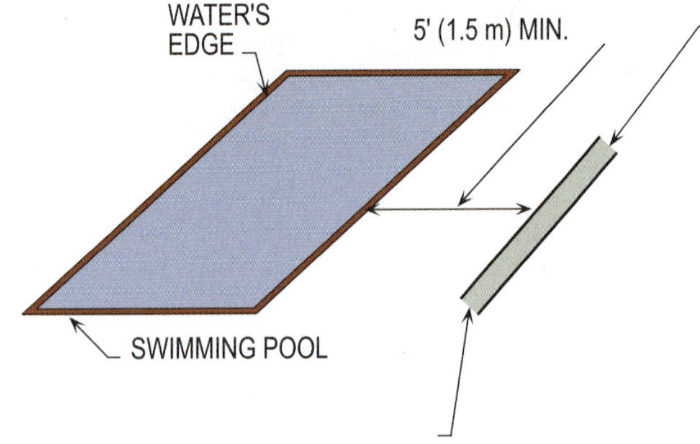

COMPLETE RACEWAY SYSTEM
- RIGID METAL CONDUIT (RMC)
- INTERMEDIATE METAL CONDUIT (IMC)
- NONMETALLIC RACEWAY SYSTEMS
- ALL METAL CONDUIT SHALL BE CORROSION RESISTANT AND SUITABLE FOR THE LOCATION
- **680.10**

MINIMUM COVER DEPTHS
- RIGID METAL CONDUIT (RMC) 6" (150 mm)
- INTERMEDIATE METAL CONDUIT (IMC) 6" (150 mm)
- NONMETALLIC RACEWAYS W/O CONCRETE ENCASEMENT 18" (450 mm)
- OTHER APPROVED RACEWAYS 18" (450 mm)

UNDERGROUND WIRING LOCATION
680.10

Purpose of Change: This revision clarifies that underground wiring systems installed inside the restricted 5 ft. (1.5 m) distance shall be permitted when employing a complete raceway stystem of specified raceways.

NEC Ch. 6 - Article 680
Part I - 680.12

Type of Change	Panel Action	UL	UL 508	API 500 - 1997	API 505 - 1997	OSHA - 1994
Revision	Accept in Principle	-	-	-	-	-
ROP	**ROC**	**NFPA 70E - 2004**	**NFPA 70B - 2006**	**NFPA 79 - 2007**	**NFPA**	**NEMA**
pg. 696 # 17-80	pg. 395 # 17-71	-	-	-	-	-
log: 1635 CMP: 17	log: 1950	Submitter: L. Keith Lofland		2005 NEC: 680.12		IEC: -

2005 NEC - 680.12 Maintenance Disconnecting Means.

One or more means to disconnect all ungrounded conductors shall be provided for all utilization equipment other than lighting. Each means shall be readily accessible and within sight from its equipment.

2008 NEC - 680.12 Maintenance Disconnecting Means.

One or more means to simultaneously disconnect all ungrounded conductors shall be provided for all utilization equipment other than lighting. Each means shall be readily accessible and within sight from its equipment, and shall be located at least 1.5 m (5 ft) horizontally from the inside walls of a pool, spa, or hot tub unless separated from the open water by a permanently installed barrier that provides a 1.5 m (5 ft) to reach path or greater. This horizontal distance is to be measured from the water's edge along the shortest path required to reach the disconnect.

Author's Comment: The maintenance disconnecting means was required to be located at least 1.5 m (5 ft) horizontally from the inside wall of a pool in the 1999 edition of the National Electrical Code. During the reorganization of **Article 680** for the 2002 NEC, this requirement was removed from **680.12**. The action of this proposal reinserts the text requiring the disconnecting means to be located at least 5 feet horizontally from the inside walls of the pool, spa, or hot tub, unless the disconnect is separated from open water by a permanently installed barrier that would require a reach of 5 feet or more.

For example, installing a brick or block wall at the edge of the water that is high enough to require a reach of five feet or more to reach the disconnect would satisfy the requirement.

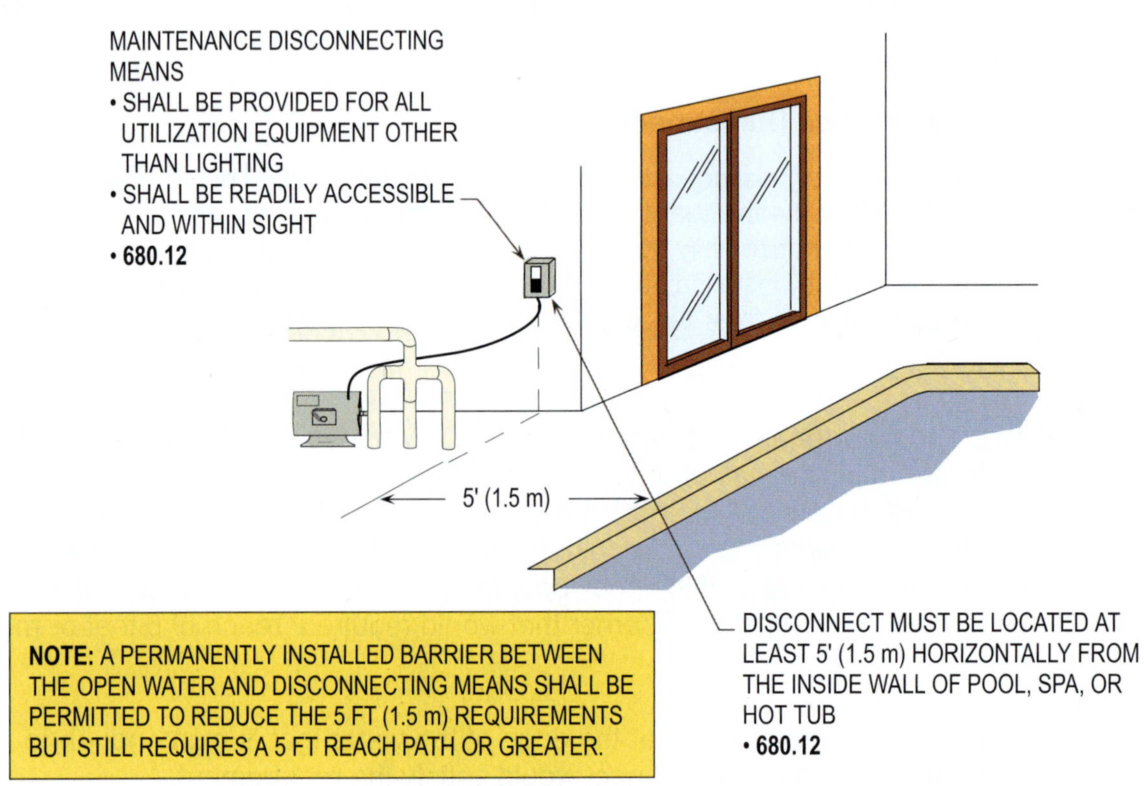

MAINTENANCE DISCONNECTING MEANS
• SHALL BE PROVIDED FOR ALL UTILIZATION EQUIPMENT OTHER THAN LIGHTING
• SHALL BE READILY ACCESSIBLE AND WITHIN SIGHT
• 680.12
5' (1.5 m)
DISCONNECT MUST BE LOCATED AT LEAST 5' (1.5 m) HORIZONTALLY FROM THE INSIDE WALL OF POOL, SPA, OR HOT TUB
• 680.12
NOTE: A PERMANENTLY INSTALLED BARRIER BETWEEN THE OPEN WATER AND DISCONNECTING MEANS SHALL BE PERMITTED TO REDUCE THE 5 FT (1.5 m) REQUIREMENTS BUT STILL REQUIRES A 5 FT REACH PATH OR GREATER.

MAINTENANCE DISCONNECTING MEANS
680.12

Purpose of Change: This revision clarifies the distance for the disconnecting means from open water.

NEC Ch. 6 - Article 680
Part II - 680.22(A)(1) through (A)(3)

Type of Change	Panel Action	UL	UL 508	API 500 - 1997	API 505 - 1997	OSHA - 1994
Revision	Accept	-	-	-	-	1910.307(j)(2)(i)
ROP	**ROC**	**NFPA 70E - 2004**	**NFPA 70B - 2006**	**NFPA 79 - 2007**	**NFPA**	**NEMA**
pg. 697 # 17-85a	pg. - # -	430.10(B)	-	-	-	-
log: CP 1707 CMP: 17	log: -	Submitter: Code Making Panel 17		2005 NEC: 680.22		IEC: -

2005 NEC - 680.22 Area Lighting, Receptacles, and Equipment.

(A) Receptacles.

(1) Circulation and Sanitation System, Location. Receptacles that provide power for water-pump motors or for other loads directly related to the circulation and sanitation system shall be located at least 3.0 m (10 ft) from the inside walls of the pool, or not less than ~~1.5 m (5 ft)~~ from the inside walls of the pool if they meet all of the following conditions:

(1) Consist of single receptacles

(2) Employ a locking configuration

(3) Are of the grounding type

(4) Have GFCI protection

(2) Other Receptacles, Location. Other receptacles shall be not less than ~~3.0 m (10 ft)~~ from the inside walls of a pool.

(3) Dwelling Unit(s). Where a permanently installed pool is installed at a dwelling unit(s), no fewer than one 125-volt 15- or 20-ampere receptacle on a general-purpose branch circuit shall be located not less than ~~3.0 m (10 ft)~~ from, and not more than 6.0 m (20 ft) from, the inside wall of the pool. This receptacle shall be located not more than 2.0 m (6 ft 6 in.) above the floor, platform, or grade level serving the pool.

~~**(4) Restricted Space.** Where a pool is within 3.0 m (10 ft) of a dwelling and the dimensions of the lot preclude meeting the required clearances, not more than one receptacle outlet shall be permitted if not less than 1.5 m (5 ft) measured horizontally from the inside wall of the pool.~~

2008 NEC - 680.22 Area Lighting, Receptacles, and Equipment.

(A) Receptacles.

(1) Circulation and Sanitation System, Location. Receptacles that provide power for water-pump motors or for other loads directly related to the circulation and sanitation system shall be located at least 3.0 m (10 ft) from the inside walls of the pool, or not less than 1.83 m (6 ft) from the inside walls of the pool if they meet all of the following conditions:

(1) Consist of single receptacles

(2) Employ a locking configuration

(3) Are of the grounding type

(4) Have GFCI protection

(2) Other Receptacles, Location. Other receptacles shall be not less than 1.83 m.(6 ft.) from the inside walls of a pool.

(3) Dwelling Unit(s). Where a permanently installed pool is installed at a dwelling unit(s), no fewer than one 125-volt 15- or 20-ampere receptacle on a general-purpose branch circuit shall be located not less than 1.83 m (6 ft) from, and not more than 6.0 m (20 ft) from, the inside wall of the pool. This receptacle shall be located not more than 2.0 m (6 ft 6 in.) above the floor, platform, or grade level serving the pool.

Author's Comment: The measurements for receptacle locations relative to distance to water was changed from 5 ft to 6 ft and 10 ft to 6 ft to ensure consistency throughout **Article 680**. The 10 ft measurement has been in the Code for many years, previous to the introduction of GFCI devices. The panel determined that a measurement of 6 ft is sufficient and the 6 ft measurement correlates with standard power supply cord lengths.

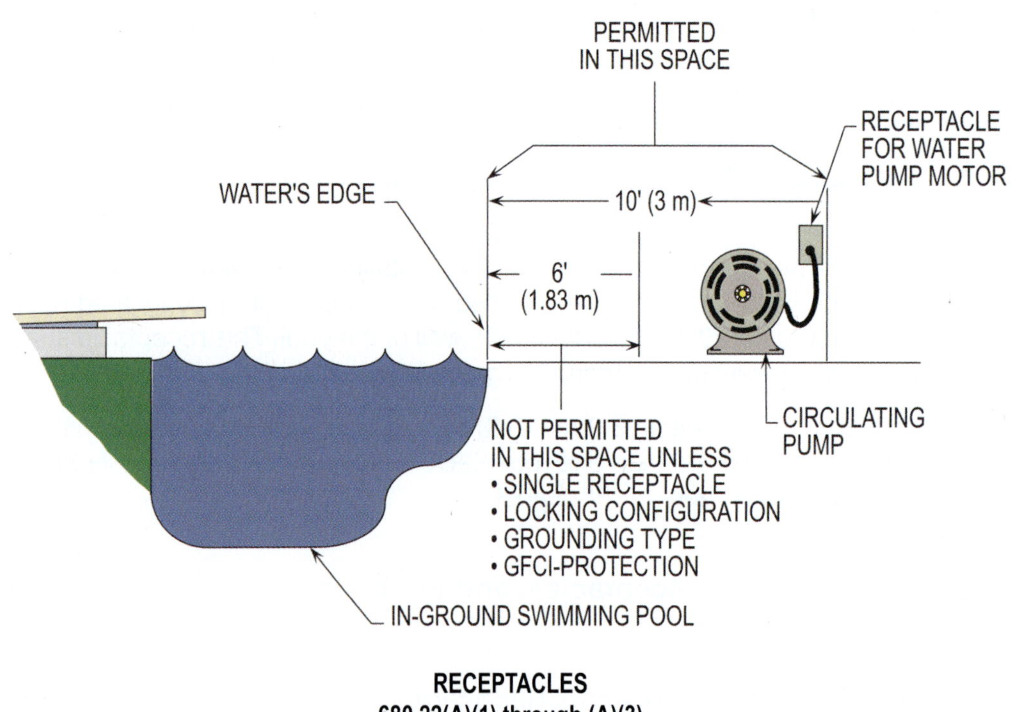

RECEPTACLES
680.22(A)(1) through (A)(3)

Purpose of Change: This revision clarifies that receptacles for circulation and sanitation systems, other receptacles and for dwelling units shall be located not less than 6 ft. (1.83) from the inside walls of the pool.

NEC Ch. 6 - Article 680
Part II - 680.22(E) and FPN

Type of Change		Panel Action		UL	UL 508	API 500 - 1997	API 505 - 1997	OSHA - 1994
New Subsection		Accept in Principle		-	-	-	-	-
ROP		ROC		NFPA 70E - 2004	NFPA 70B - 2006	NFPA 79 - 2007	NFPA	NEMA
pg. 699	# 17-96	pg. -	# -	-	-	-	-	-
log: 389	CMP: 17	log: -	Submitter: Bryan P. Holland			2005 NEC: -		IEC: -

2008 NEC - 680.22 Area Lighting, Receptacles, and Equipment.

(E) Other Outlets. Other outlets shall be not less than 3.0 m (10 ft) from the inside walls of the pool. Measurements shall be determined in accordance with 680.22(A)(5).

FPN: Other outlets may include, but are not limited to, remote-control, signaling, fire alarm, and communications circuits.

Author's Comment: A safety hazard could arise from the use of other equipment, such as remote-control, signaling, fire alarm, and communications circuits, used in the vicinity of a pool. The typical pool user may not be aware of the potential hazards associated with communication and other "low-voltage" equipment and want to use this equipment near or at the pool if an outlet is also near or at the pool area. This restriction provides reasonable protection by requiring distance from a pool at which these signaling, fire alarm, communications, and other "low voltage" outlets may be installed.

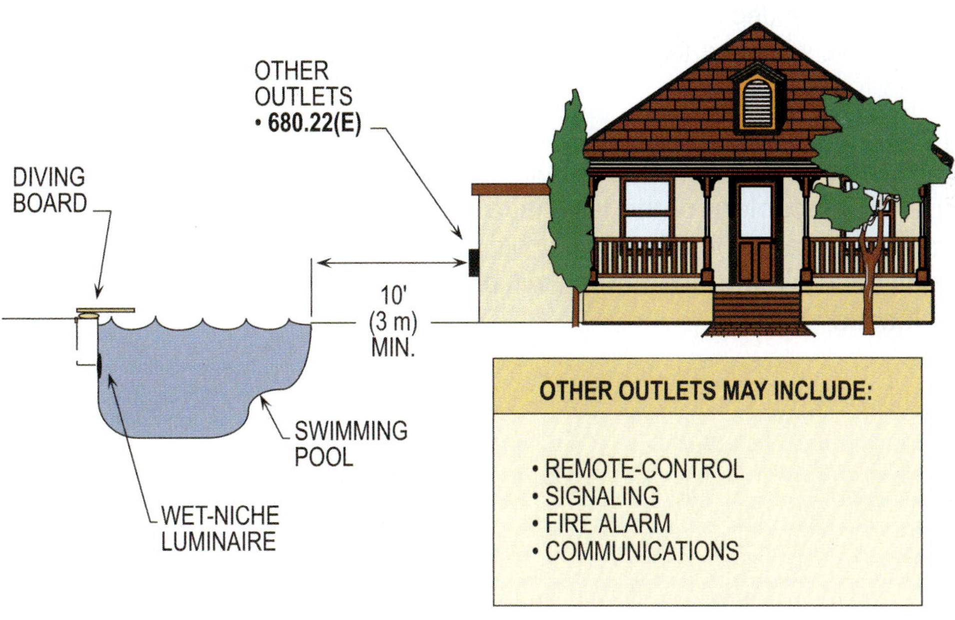

OTHER OUTLETS
680.22(E) and FPN

Purpose of Change: A new subsection has been added to require other outlets to be installed not less than 10 ft. (3 m) from the inside walls of the pool.

NEC Ch. 6 - Article 680
Part II - 680.23(B)(6)

Type of Change		Panel Action		UL	UL 508	API 500 - 1997	API 505 - 1997	OSHA - 1994
Revision		Accept in Principle		-	-	-	-	1910.307(j)(4)(i)
ROP		ROC		NFPA 70E - 2004	NFPA 70B - 2006	NFPA 79 - 2007	NFPA	NEMA
pg. 702	# 17-103	pg. -	# -	430.10(E)(2)	-	-	-	-
log: 2786	CMP: 17	log: -	Submitter: Steven D. Holmes		2005 NEC: 680.23(B)(6)			IEC: -

2005 NEC - 680.23 Underwater Luminaires ~~(Lighting Fixtures)~~.

This section covers all luminaires ~~(lighting fixtures)~~ installed below the normal water level of the pool.

(B) Wet-Niche Luminaires ~~(Fixtures)~~.

(6) Servicing. All wet-niche luminaires shall be removable from the water for ~~relamping or normal maintenance. Luminaires shall be installed in such a manner that personnel can reach the luminaire for relamping, maintenance, or inspection while on the deck or equivalently dry location~~

2008 NEC - 680.23 Underwater Luminaires.

This section covers all luminaires installed below the normal water level of the pool.

(B) Wet-Niche Luminaires.

(6) Servicing. All wet-niche luminaires shall be removable from the water for inspection, relamping, or other maintenance. The forming shell location and length of cord in the forming shell shall permit personnel to place the removed luminaire on the deck or other dry location for such maintenance. The luminaire maintenance location shall be accessible without entering or going in the pool water.

Author's Comment: The requirements for inspection, relamping, or other maintenance of underwater wet niche luminaires has been revised by requiring the forming shell location and length of cord in the forming shell to permit placement of the luminaire on the deck or other dry location for maintenance and the luminaire maintenance location must be accessible without entering or going into the pool water. This ensures the maintenance person will be relatively dry and will be working on the luminaire at a relatively dry location.

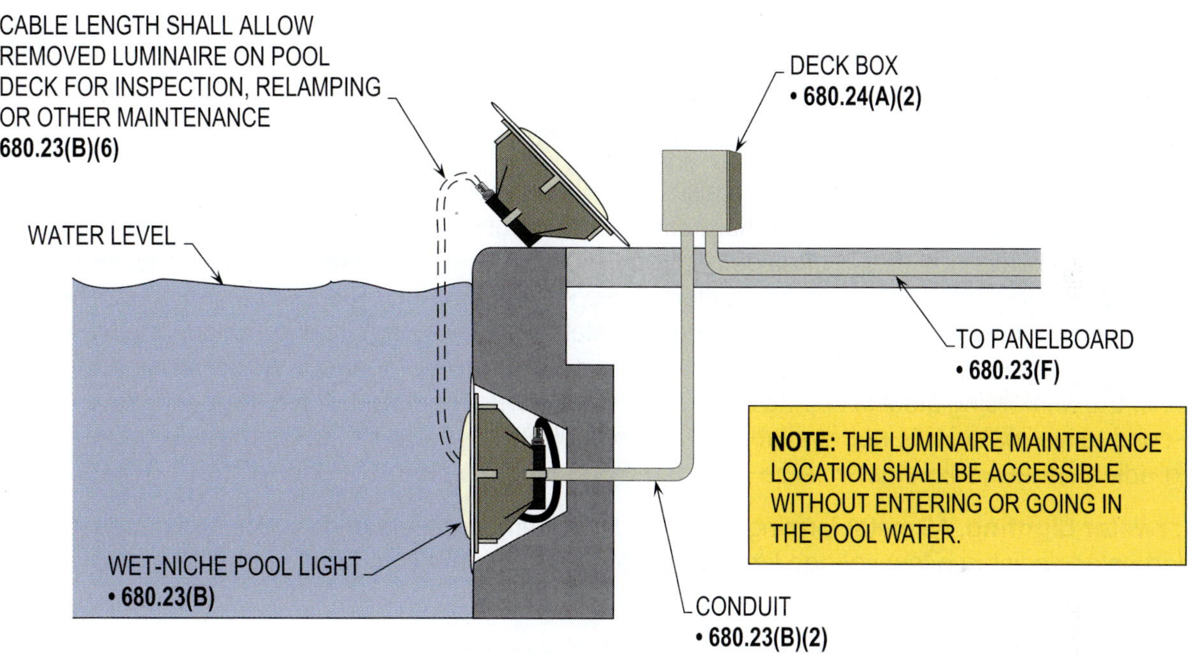

SERVICING
680.23(B)(6)

Purpose of Change: This revision clarifies the requirements for inspections, relamping or other maintenance for underwater wet-niche luminaires.

NEC Ch. 6 - Article 680
Part II - 680.26(A) through (C)

Type of Change	Panel Action	UL	UL 508	API 500 - 1997	API 505 - 1997	OSHA - 1994
Revision	Accept in Principle	-	-	-	-	-
ROP	**ROC**	**NFPA 70E - 2004**	**NFPA 70B - 2006**	**NFPA 79 - 2007**	**NFPA**	**NEMA**
pg. 705 # 17-114a	pg. 400 # 17-92	-	-	-	-	-
log: CP 1708 CMP: 17	log: 950	Submitter: Code Making Panel 17		2005 NEC: 680.25(A) and (B)		IEC: -

2005 NEC - 680.26 Equipotential Bonding.

(A) Performance. The equipotential bonding required by this section shall be installed to ~~eliminate~~ voltage gradients in the pool area ~~as prescribed~~.

FPN: The 8 AWG or larger solid copper bonding conductor shall not be required to be extended or attached to any remote panelboard, service equipment, or any electrode.

(B) Bonded Parts. The parts specified in 680.26(B)(1) through ~~(B)(5)~~ shall be bonded.

~~(1)~~ **Metallic ~~Structural~~ Components.** All metallic parts of the pool structure, including reinforcing metal. Where reinforcing steel is encapsulated with a nonconductive compound ~~it~~ shall not be required to be bonded. ~~of the pool shell, coping stones, and deck, shall be bonded. The usual steel tie wires shall be considered suitable for bonding the reinforcing steel together, and welding or special clamping shall not be required. These tie wires shall be made tight. If reinforcing steel is effectively insulated by an encapsulating nonconductive compound at the time of manufacture and installation, it shall not be required to be bonded. Where reinforcing steel of the pool shell or the reinforcing steel of coping stones and deck is encapsulated with a nonconductive compound or another conductive material is not available, provisions shall be made for an alternative means to eliminate voltage gradients that would otherwise be provided by unencapsulated, bonded reinforcing steel.~~

~~(2)~~ **Underwater Lighting.** All metal forming shells and mounting brackets of no-niche luminaires ~~(fixtures)~~ shall be bonded ~~unless a~~ listed ~~a~~ low-voltage lighting systems with nonmetallic forming shells not requiring bonding ~~is used~~.

~~(3)~~ **Metal Fittings.** All metal fittings within or attached to the pool structure shall be bonded. Isolated parts that are not over 100 mm (4 in.) in any dimension and do not penetrate into the pool structure more than 25 mm (1 in.) shall not require bonding.

~~(4)~~ **Electrical Equipment.** Metal parts of electrical equipment associated with the pool water circulating system, including pump motors and metal parts of equipment associated with pool covers, including electric motors, shall be bonded.

Accessible metal parts of listed equipment incorporating an approved system of double insulation ~~and providing a means for grounding internal nonaccessible, non–current-carrying metal parts shall not be bonded by a direct connection to the equipotential bonding grid~~.

Where a double-insulated water pump motor is installed under the provisions of this rule, a solid 8 AWG copper conductor that is of sufficient length to make a bonding connection to a replacement motor shall be extended from the bonding grid to an accessible point in the ~~motor~~ vicinity. Where there is no connection between the swimming pool bonding grid and the equipment grounding system for the premises, this bonding conductor shall be connected to the equipment grounding conductor of the motor circuit.

~~(5)~~ **Metal Wiring Methods and Equipment.** Metal sheathed cables and raceways, metal piping, and all fixed metal parts ~~that are within the following distances of the pool, except~~ those separated from the pool by a permanent barrier~~,~~ shall be bonded:

~~(1) Within~~ 1.5 m (5 ft) horizontally of the inside walls of the pool.

~~(2) Within~~ 3.7 m (12 ft) measured vertically above the maximum water level of the pool, or any observation stands, towers, or platforms, or any diving structures.

(C) Equipotential Bonding Grid. The parts specified in 680.26(B) shall be connected to an equipotential bonding grid with a solid copper conductor, insulated, covered, or bare, not smaller than 8 AWG or rigid metal conduit of brass or other identified corrosion-resistant metal conduit. Connection shall be made by exothermic welding or by listed pressure connectors or clamps that are labeled as being suitable for the purpose and are of stainless steel, brass, copper, or copper alloy. The equipotential ~~common~~ bonding grid shall extend under paved walking surfaces for 1 m (3 ft) horizontally beyond the inside walls of the pool and shall be permitted to be any of the following:

(1) Structural Reinforcing Steel. The structural reinforcing steel of a concrete pool where the reinforcing rods are bonded together by the usual steel tie wires or the equivalent

(2) Bolted or Welded Metal Pools. The wall of a bolted or welded metal pool

(3) Alternate Means. This system shall be permitted to be constructed as specified in (a) through (c):

a. Materials and Connections. The grid shall be constructed of minimum 8 AWG bare solid copper conductors. Conductors shall be bonded to each other at all points of crossing. Connections shall be made as required by 680.26(D).

b. Grid Structure. The equipotential bonding grid shall cover the contour of the pool and the pool deck extending 1 m (3 ft) horizontally from the inside walls of the pool. The equipotential bonding grid shall be arranged in a 300 mm (12 in.) by 300 mm (12 in.) network of conductors in a uniformly spaced perpendicular grid pattern with tolerance of 100 mm (4 in.).

c. Securing. The below-grade grid shall be secured within or under the pool and deck media.

(D) Connections. Where structural reinforcing steel or the walls of bolted or welded metal pool structures are used as an equipotential bonding grid for nonelectrical parts, the connections shall be made in accordance with 250.8.

(E) Pool Water Heaters. For pool water heaters rated at more than 50 amperes and having specific instructions regarding bonding and grounding, only those parts designated to be bonded shall be bonded and only those parts designated to be grounded shall be grounded.

2008 NEC - 680.26 Equipotential Bonding.

(A) Performance. The equipotential bonding required by this section shall be installed to reduce voltage gradients in the pool area.

(B) Bonded Parts. The parts specified in 680.26(B)(1) through (B)(7) shall be bonded together using solid copper conductors, insulated, covered, or bare, not smaller than 8 AWG or with rigid metal conduit of brass or other identified corrosion-resistant metal. Connections to bonded parts shall be made in accordance with 250.8. An 8 AWG or larger solid copper bonding conductor provided to reduce voltage gradients in the pool area shall not be required to be extended or attached to remote panelboards, service equipment, or electrodes.

(1) Conductive Pool Shells. Bonding to conductive pool shells shall be provided as specified in 680.26(B)(1)(a) or (B)(1)(b). Poured concrete, pneumatically applied or sprayed concrete, and concrete block with painted or plastered coatings shall all be considered conductive materials due to water permeability and porosity. Vinyl liners and fiberglass composite shells shall be considered to be nonconductive materials.

(a) Structural Reinforcing Steel. Unencapsulated structural reinforcing steel shall be bonded together by steel tie wires or the equivalent. Where structural reinforcing steel is encapsulated in a nonconductive compound, copper conductor grid shall be installed in accordance with 680.26(B)(1)(b).

(b) Copper Conductor Grid. A copper conductor grid shall be provided and shall comply with (b)(1) through (b)(4):

(1) Be constructed of minimum 8 AWG bare solid copper conductors bonded to each other at all points of crossing.

(2) Conform to the contour of the pool and the pool deck.

(3) Be arranged in a 300 mm (12 in.) by 300 mm (12 in.) network of conductors in a uniformly spaced perpendicular grid pattern with a tolerance of 100 mm (4 in.).

(4) Be secured within or under the pool no more than 150 mm (6 in.) from the outer contour of the pool shell.

(2) Perimeter Surfaces. The perimeter surface shall extend for 1 m (3 ft) horizontally beyond the inside walls of the pool and shall include unpaved surfaces as well as poured concrete and other types of paving. Bonding to perimeter surfaces shall be provided as specified in 680.26(B)(2)(a) or (2)(b), and shall be attached to the pool reinforcing steel or copper conductor grid at a minimum of four (4) points uniformly spaced around the perimeter of the pool. For nonconductive pool shells, bonding at four points shall not be required.

(a) Structural Reinforcing Steel. Structural reinforcing steel shall be bonded in accordance with 680.26(B)(1)(a).

(b) Alternate Means. Where structural reinforcing steel is not available or is encapsulated in a nonconductive compound, copper conductor(s) shall be utilized where the following requirements are met:

(1) At least one minimum 8 AWG bare solid copper conductor shall be provided.

(2) The conductors shall follow the contour of the perimeter surface.

(3) Only listed approved splices shall be permitted.

(4) The required conductor shall be 450 to 600 mm (18 to 24 in.) from the inside walls of the pool.

(5) The required conductor shall be secured within or under the perimeter surface 100 mm to 150 mm (4 in. to 6 in.) below the subgrade.

(3) Metallic Components. All metallic parts of the pool structure, including reinforcing metal not addressed in 680.26(B)(1)(a) shall be bonded. Where reinforcing steel is encapsulated with a nonconductive compound, the reinforcing steel shall not be required to be bonded.

(4) Underwater Lighting. All metal forming shells and mounting brackets of no-niche luminaires shall be bonded.

Exception: Listed low-voltage lighting systems with nonmetallic forming shells shall not require bonding.

(5) Metal Fittings. All metal fittings within or attached to the pool structure shall be bonded. Isolated parts that are not over 100 mm (4 in.) in any dimension and do not penetrate into the pool structure more than 25 mm (1 in.) shall not require bonding.

(6) Electrical Equipment. Metal parts of electrical equipment associated with the pool water circulating system, including pump motors and metal parts of equipment associated with pool covers, including electric motors, shall be bonded.

Exception: Metal parts of listed equipment incorporating an approved system of double insulation, shall not be bonded.

(a) Double-Insulated Water Pump Motors. Where a double-insulated water pump motor is installed under the provisions of this rule, a solid 8 AWG copper conductor of sufficient length to make a bonding connection to a replacement motor shall be extended from the bonding grid to an accessible point in the vicinity of the pool pump motor. Where there is no connection between the swimming pool bonding grid and the equipment grounding system for the premises, this bonding conductor shall be connected to the equipment grounding conductor of the motor circuit.

(b) Pool Water Heaters. For pool water heaters rated at more than 50 amperes and having specific instructions regarding bonding and grounding, only those parts designated to be bonded shall be bonded and only those parts designated to be grounded shall be grounded.

(7) Metal Wiring Methods and Equipment. Metal-sheathed cables and raceways, metal piping, and all fixed metal parts shall be bonded:

Exception No. 1: Those separated from the pool by a permanent barrier shall not be required to be bonded.

Exception No. 2: Those greater than 1.5 m (5 ft) horizontally of the inside walls of the pool shall not be required to be bonded.

Exception No. 3: Those greater than 3.7 m (12 ft) measured vertically above the maximum water level of the pool, or as measured vertically above any observation stands, towers, or platforms, or any diving structures shall not be required to be bonded.

(C) Pool Water. an intentional bond of a minimum conductive surface area of 5806 mm^2 (9 in.^{2}) shall be installed in contact with the pool water. This bond shall be permitted to consist of parts that are required to be bonded in 680.26(B).

Author's Comment: The text dealing with equipotential bonding has been revised for clarity.

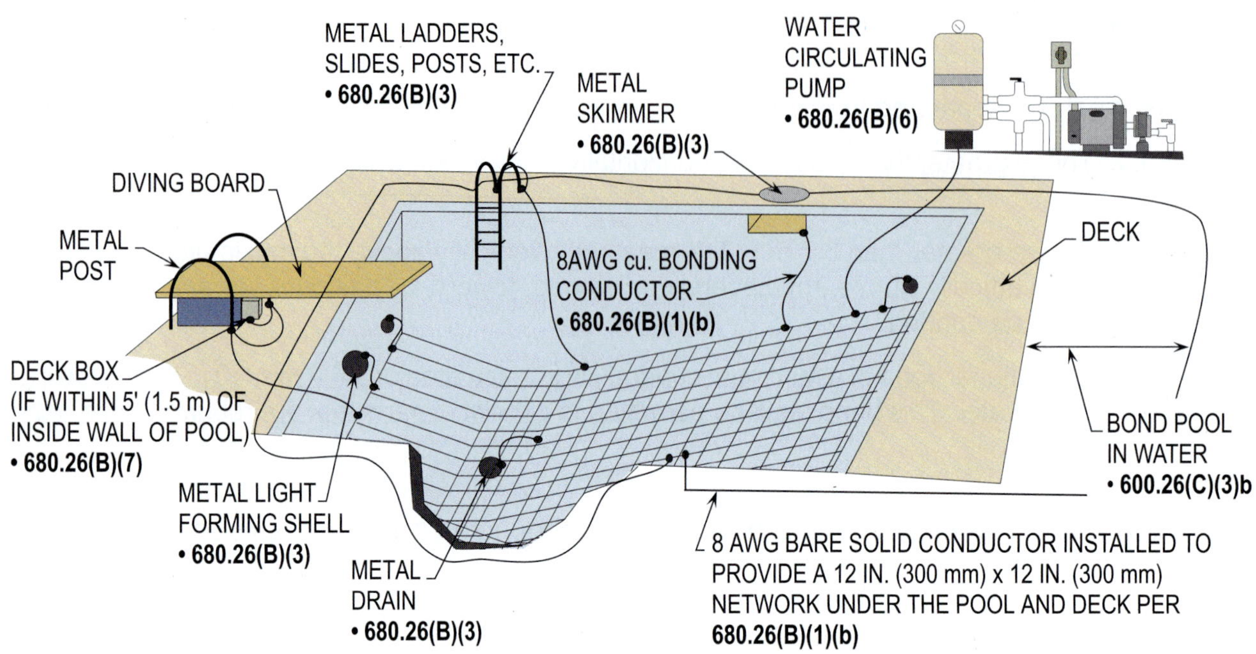

CONDUCTIVE POOL SHELLS	ALTERNATE MEANS
• POURED CONCRETE • PNEUMATICALLY APPLIED OR SPRAYED CONCRETE • CONCRETE BLOCK WITH PAINTED OR PLASTERED COATINGS	• VINYL LINERS • FIBERGLASS COMPOSITE

COPPER CONDUCTOR GRID	ALTERNATE MEANS
• BE CONSTRUCTED OF MINIMUM 8 AWG BARE SOLID COPPER CONDUCTORS BONDED TO EACH OTHER AT ALL POINTS OF CROSSING • CONFORM TO THE CONTOUR OF THE POOL AND THE POOL DECK • BE ARRANGED IN A 12 IN. (300 mm) BY 12 IN. (300 mm) NETWORK OF CONDUCTORS IN A UNIFORMLY SPACED PERPENDICULAR GRID PATTERN WITH TOLERANCE OF 4 IN. (100 mm)	• AT LEAST ONE MINIMUM 8 AWG BARE SOLID COPPER CONDUCTOR SHALL BE PROVIDED • THE CONDUCTORS SHALL FOLLOW THE CONTOUR OF THE PERIMETER SURFACE • THE REQUIRED CONDUCTOR SHALL BE 18 TO 24 IN. (450 mm TO 600 mm) FROM THE INSIDE WALL OF THE POOL • THE REQUIRED CONDUCTOR SHALL BE SECURED WITHIN OR UNDER THE PERIMETER SURFACE 4 TO 6 IN. (100 mm TO 150 mm) BELOW THE SUBGRADE

EQUIPOTENTIAL BONDING
680.26(A) through (C)

Purpose of Change: This revision clarifies the equipotential bonding requirements.

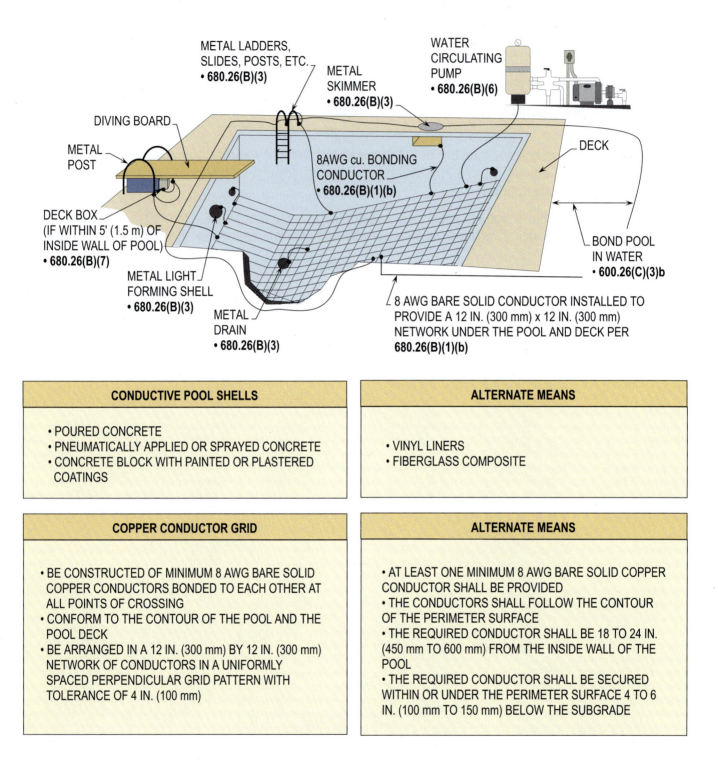

NEC Ch. 6 - Article 680
Part III - 680.31

92

Type of Change	Panel Action	UL	UL 508	API 500 - 1997	API 505 - 1997	OSHA - 1994
Revision	Accept in Principle in Part	1081	-	-	-	-
ROP	**ROC**	**NFPA 70E - 2004**	**NFPA 70B - 2006**	**NFPA 79 - 2007**	**NFPA**	**NEMA - 2003**
pg. 714 # 17-135	pg. - # -	-	-	-	-	MG 1
log: 3640 CMP: 17	log: -	Submitter: Aaron B. Chase		2005 NEC: 680.31		IEC: -

2005 NEC - 680.31 Pumps.

A cord-connected pool filter pump shall incorporate an approved system of double insulation or its equivalent and shall be provided with means for grounding only the internal and nonaccessible non–current-carrying metal parts of the appliance. The means for grounding shall be an equipment grounding conductor run with the power-supply conductors in the flexible cord that is properly terminated in a grounding-type attachment plug having a fixed grounding contact member.

2008 NEC - 680.31 Pumps.

A cord-connected pool filter pump shall incorporate an approved system of double insulation or its equivalent and shall be provided with means for grounding only the internal and nonaccessible non–current-carrying metal parts of the appliance.

The means for grounding shall be an equipment grounding conductor run with the power-supply conductors in the flexible cord that is properly terminated in a grounding-type attachment plug having a fixed grounding contact member.

Cord-connected pool filter pumps shall be provided with a ground-fault circuit interrupter that is an integral part of the attachment plug or located in the power supply cord within 300 mm (12 in.) of the attachment plug.

Author's Comment: Cord-connected pool filter pumps for storable pools must now be protected with a GFCI that is an integral part of the attachment plug or located in the power supply cord within 12 inches of the attachment plug.

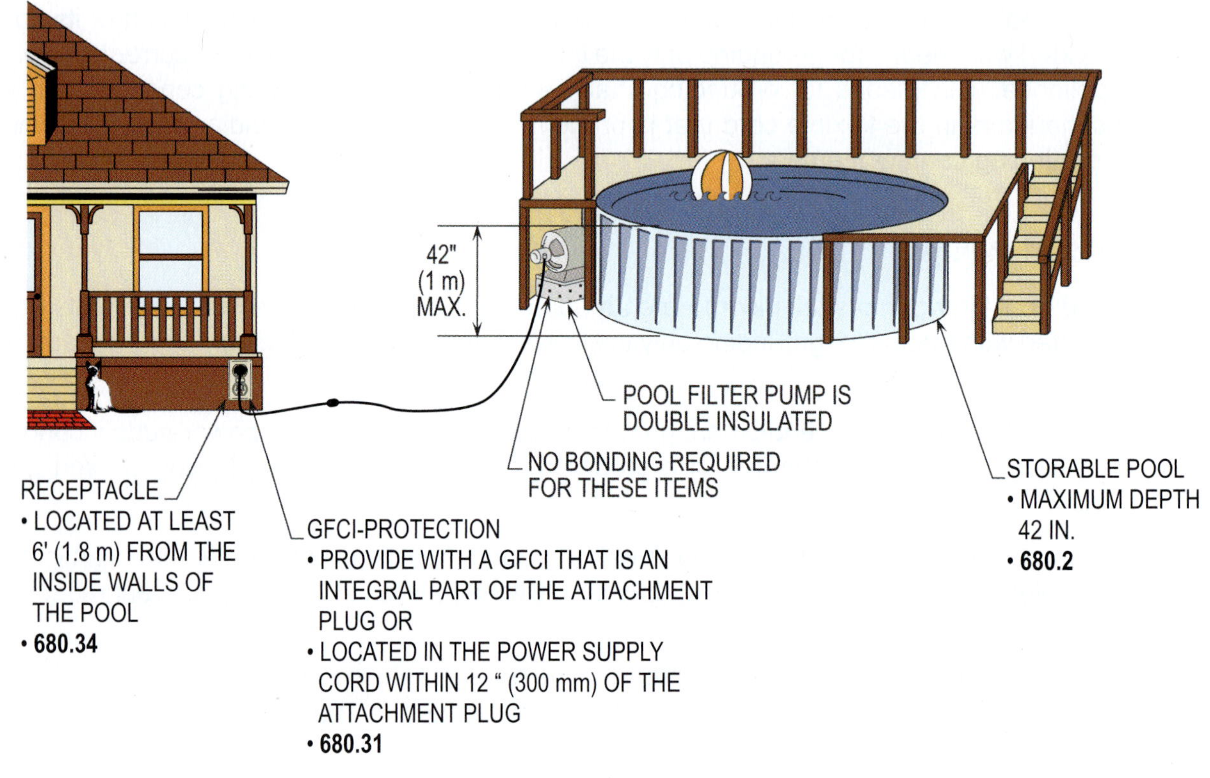

PUMPS
680.31

Purpose of Change: This revision clarifies that storable pool cord-connected filter pumps shall be protected with a GFCI that is an integral part of the attachment plug or located in the power supply cord within 12 inches of the attachment plug.

NEC Ch. 6 - Article 680
Part IV - 680.43(D), Exception No. 2

Type of Change	Panel Action	UL	UL 508	API 500 - 1997	API 505 - 1997	OSHA - 1994
Revision	Accept	1081	-	-	-	-
ROP	**ROC**	**NFPA 70E - 2004**	**NFPA 70B - 2006**	**NFPA 79 - 2007**	**NFPA**	**NEMA**
pg. 717 # 17-144	pg. - # -	-	-	-	-	-
log: 2702 CMP: 17	log: - Submitter: Gary L. Siggins			2005 NEC: -		IEC: -

2005 NEC - 680.43 Indoor Installations.

A spa or hot tub installed indoors shall comply with the provisions of Parts I and II of this article except as modified by this section and shall be connected by the wiring methods of Chapter 3.

Exception: Listed spa and hot tub packaged units rated 20 amperes or less shall be permitted to be cord-and-plug connected to facilitate the removal or disconnection of the unit for maintenance and repair.

(D) Bonding. The following parts shall be bonded together:

(1) All metal fittings within or attached to the spa or hot tub structure

(2) Metal parts of electrical equipment associated with the spa or hot tub water circulating system, including pump motors

(3) Metal ~~conduit~~ and metal piping that are within 1.5 m (5 ft) of the inside walls of the spa or hot tub and that are not separated from the spa or hot tub by a permanent barrier

(4) All metal surfaces that are within 1.5 m (5 ft) of the inside walls of the spa or hot tub and that are not separated from the spa or hot tub area by a permanent barrier

Exception: Small conductive surfaces not likely to become energized, such as air and water jets and drain fittings, where not connected to metallic piping, towel bars, mirror frames, and similar nonelectrical equipment, shall not be required to be bonded.

(5) Electrical devices and controls that are not associated with the spas or hot tubs and that are located not less than 1.5 m (5 ft) from such units; otherwise they shall be bonded to the spa or hot tub system.

2008 NEC - 680.43 Indoor Installations.

A spa or hot tub installed indoors shall comply with the provisions of Parts I and II of this article except as modified by this section and shall be connected by the wiring methods of Chapter 3.

Exception: Listed spa and hot tub packaged units rated 20 amperes or less shall be permitted to be cord-and-plug connected to facilitate the removal or disconnection of the unit for maintenance and repair.

(D) Bonding. The following parts shall be bonded together:

(1) All metal fittings within or attached to the spa or hot tub structure

(2) Metal parts of electrical equipment associated with the spa or hot tub water circulating system, including pump motors

(3) Metal raceway and metal piping that are within 1.5 m (5 ft) of the inside walls of the spa or hot tub and that are not separated from the spa or hot tub by a permanent barrier

(4) All metal surfaces that are within 1.5 m (5 ft) of the inside walls of the spa or hot tub and that are not separated from the spa or hot tub area by a permanent barrier

Exception No. 1: Small conductive surfaces not likely to become energized, such as air and water jets and drain fittings, where not connected to metallic piping, towel bars, mirror frames, and similar nonelectrical equipment, shall not be required to be bonded.

Exception No. 2: Metal parts of electrical equipment associated with the water circulating system, including pump motors that are part of a listed self-contained spa or hot tub.

(5) Electrical devices and controls that are not associated with the spas or hot tubs and that are located not less than 1.5 m (5 ft) from such units; otherwise, they shall be bonded to the spa or hot tub system.

Author's Comment: A new **Exception No. 2** was added to **680.43(D)**. Metal parts of electrical equipment associated with the water circulating system, including pump motors that are part of a listed self-contained spa or hot tub need not be bonded since grounding and bonding in listed self-contained spas can be evaluated and controlled as part of the listing. More options are available for listed self-contained spas or hot tubs than there are for field assembled spas.

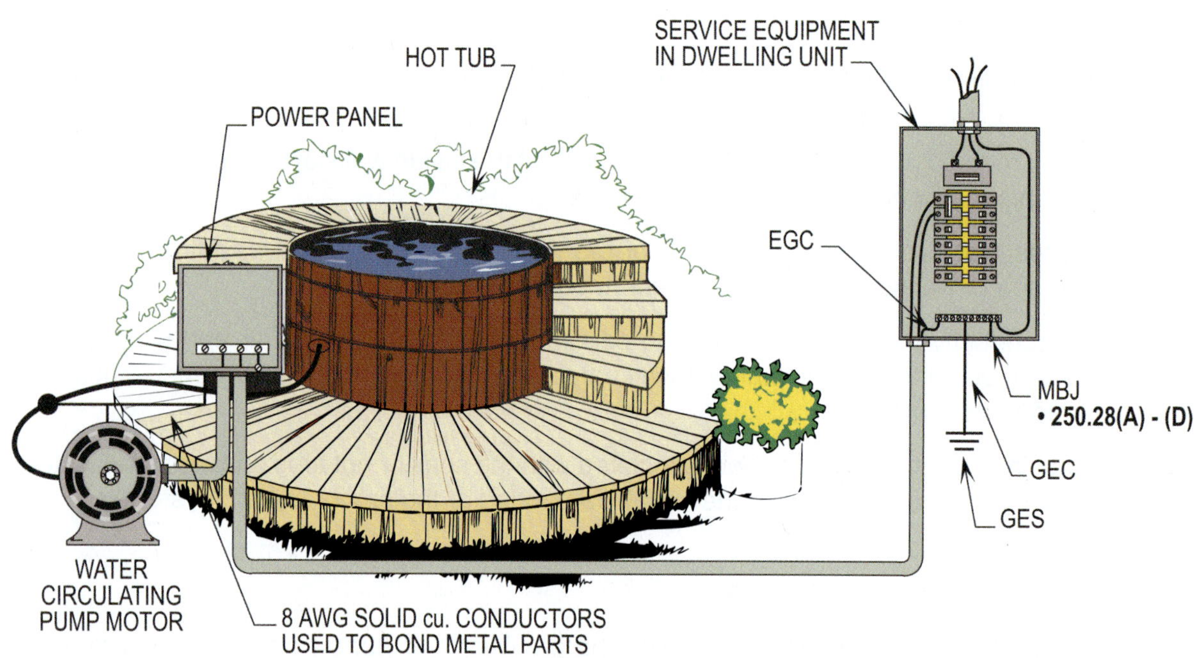

NOTE: METAL PARTS OF ELECTRICAL EQUIPMENT ASSOCIATED WITH THE WATER CIRCULATING SYSTEMS, INCLUDING PUMP MOTORS THAT ARE PART OF A LISTED SELF-CONTAINED SPA OR HOT TUB SHALL NOT BE REQUIRED TO BE BONDED.

BONDING
680.43(D), Exception No. 2

Purpose of Change: A new exception has been added to **680.43(D)** to clarify that metal parts of electrical equipment associated with the water circulating systems, including pump motors, that are part of a listed self-contained spa or hot tub shall not be required to be bonded.

NEC Ch. 6 - Article 680
Part VII - 680.71

Type of Change		Panel Action		UL	UL 508	API 500 - 1997	API 505 - 1997	OSHA - 1994
Revision		Accept		-	-	-	-	-
ROP		ROC		NFPA 70E - 2004	NFPA 70B - 2006	NFPA 79 - 2007	NFPA	NEMA
pg. 720	# 17-165	pg. 404	# 17-106	-	-	-	-	-
log: 2547	CMP: 17	log: 1660	Submitter: Robert P. McGann			2005 NEC: 680.71		IEC: -

2005 NEC - 680.71 Protection.

Hydromassage bathtubs and their associated electrical components shall be protected by a ground-fault circuit interrupter. All 125-volt, single-phase receptacles not exceeding 30 amperes and located within ~~1.5 m (5 ft)~~ measured horizontally of the inside walls of a hydromassage tub shall be protected by a ground-fault circuit interrupter(s).

2008 NEC - 680.71 Protection.

Hydromassage bathtubs and their associated electrical components shall be <u>on an individual branch circuit(s) and</u> protected by a <u>readily accessible</u> ground-fault circuit interrupter. All 125-volt, single-phase receptacles not exceeding 30 amperes and located within <u>1.83 m (6 ft)</u> measured horizontally of the inside walls of a hydromassage tub shall be protected by a ground-fault circuit interrupter(s).

Author's Comment: Installers are tapping into existing hallway, living room, or bedroom circuits and this change now requires the hydromassage bathtub and the associated components to be on an individual branch circuit(s) and protected by a readily accessible GFCI. The term "dedicated circuit" is not defined in the NEC but an "individual branch circuit" is defined.

For example, an installer may interpret "dedicated" as primarily serving the hydromassage bathtub while also providing limited lighting in the area.

The measurements for recepacle locations relative to distance to water was changed from 5 ft to 6 ft to ensure consistency throughout Article 680. The panel determined that a measurement of 6 ft is sufficient and the 6 ft correlates with standard power supply cord lengths.

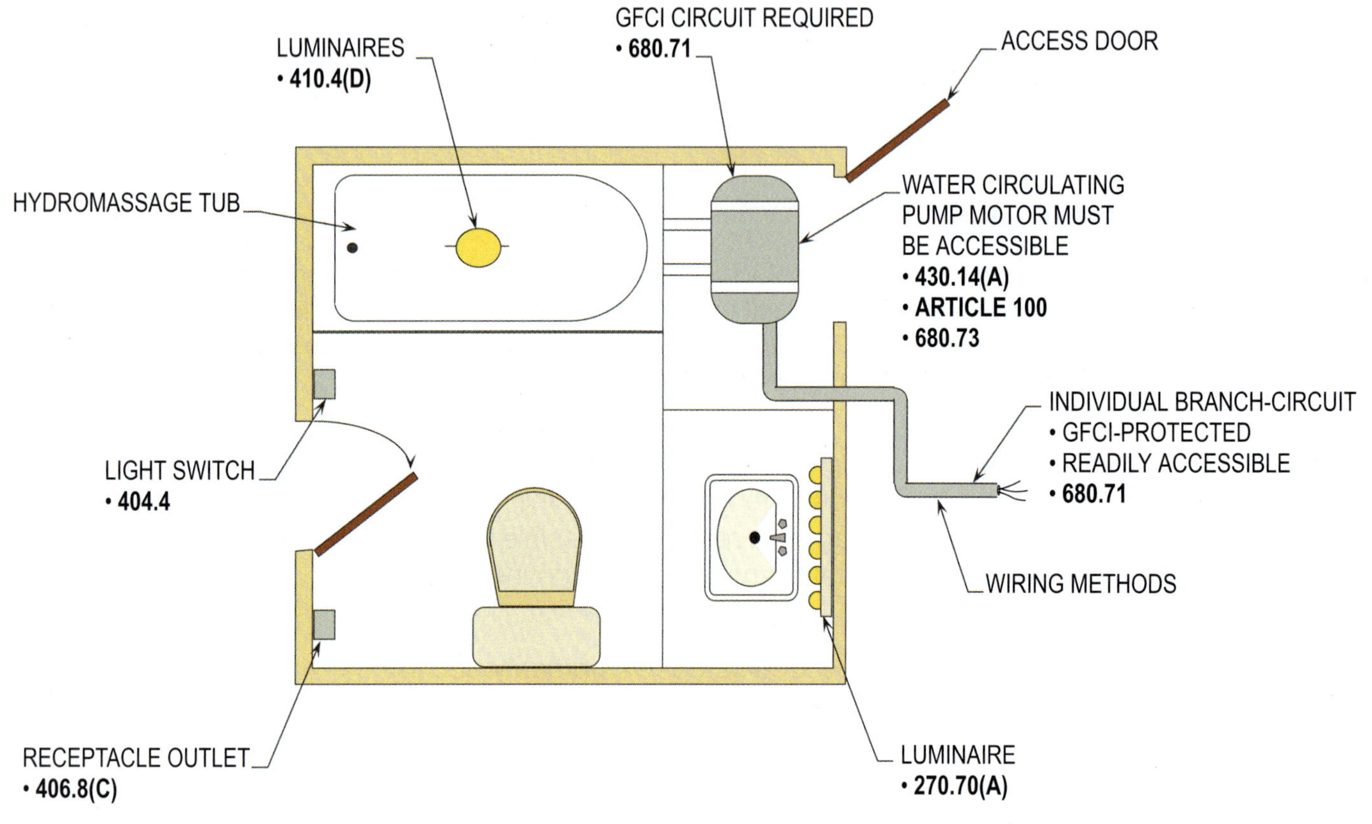

PROTECTION
680.71

Purpose of Change: This revision clarifies that hydromassage bathtubs and thier associated electrical components shall be on an individual branch-circuit(s) and protected by a readily accessible GFCI.

NEC Ch. 6 - Article 680
Part VII - 680.74

Type of Change	Panel Action	UL	UL 508	API 500 - 1997	API 505 - 1997	OSHA - 1994
Revision	Accept	-	-	-	-	-
ROP	**ROC**	**NFPA 70E - 2004**	**NFPA 70B - 2006**	**NFPA 79 - 2007**	**NFPA**	**NEMA**
pg. 720 # 17-166	pg. 404 # 17-110	-	-	-	-	-
log: 347 CMP: 17	log: 215	Submitter: Michael J. Johnston		2005 NEC: 680.74		IEC: -

2005 NEC - 680.74 Bonding.

All metal piping systems and all grounded metal parts in contact with the circulating water shall be bonded together using a copper bonding jumper, insulated, covered, or bare, not smaller that 8 AWG ~~solid~~.

2008 NEC - 680.74 Bonding.

All metal piping systems and all grounded metal parts in contact with the circulating water shall be bonded together using a soild copper bonding jumper, insulated, covered, or bare, not smaller that 8 AWG. The bonding jumper shall be connected to the terminal on the circulating pump motor that is intended for this purpose. The bonding jumper shall not be required to be connected to a double insulated circulating pump motor. The 8 AWG or larger solid copper bonding jumper shall be required for equipotential bonding in the area of the hydromassage bathtub and shall not be required to be extended or attached to any remote panelboard, service equipment, or any electrode.

Author's Comment: All the metal piping systems and all grounded metal parts in contact with the circulating water must be bonded together. The new text requires the bonding jumper to be connected to the bonding terminal on the circulating pump motor, unless the circulating motor is a double insulated circulating pump motor. The bonding jumpers are not required to be extended or attached to any remote panelboard, service equipment, or any electrode.

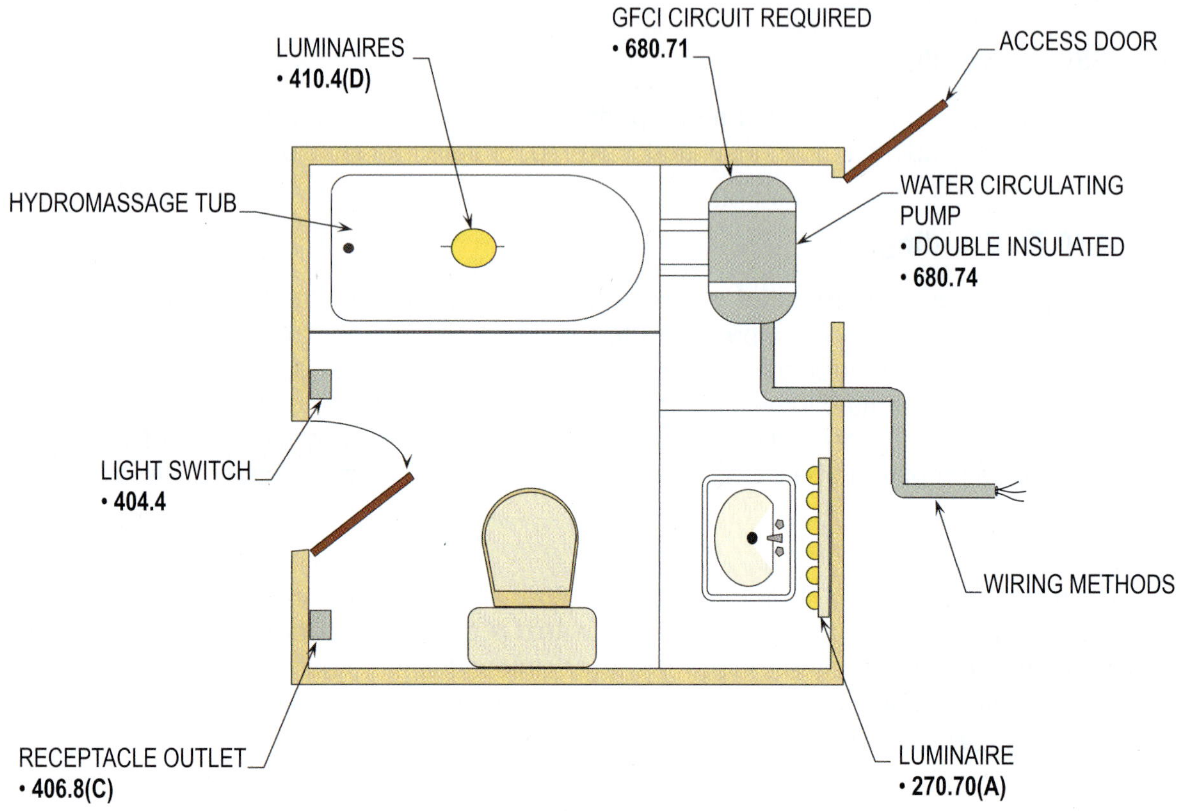

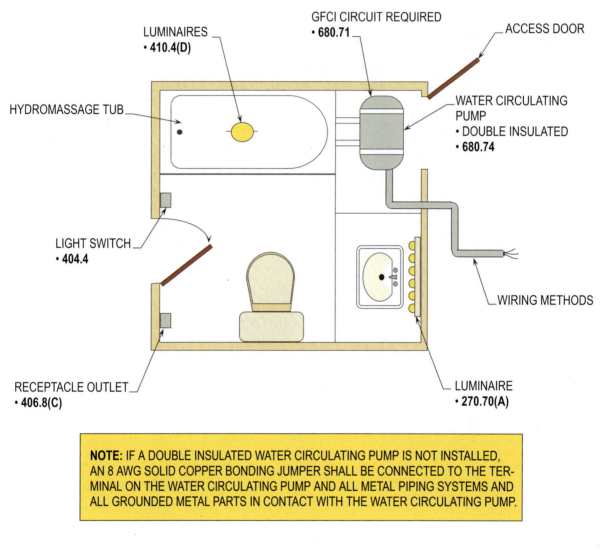

NOTE: IF A DOUBLE INSULATED WATER CIRCULATING PUMP IS NOT INSTALLED, AN 8 AWG SOLID COPPER BONDING JUMPER SHALL BE CONNECTED TO THE TERMINAL ON THE WATER CIRCULATING PUMP AND ALL METAL PIPING SYSTEMS AND ALL GROUNDED METAL PARTS IN CONTACT WITH THE WATER CIRCULATING PUMP.

BONDING
680.74

Purpose of Change: This revision clarifies the bonding requirements for the water circulating pump to a hydromassage bathtub.

NEC Ch. 6 - Article 690
Part I - 690.4(D)

Type of Change	Panel Action	UL	UL 508	API 500 - 1997	API 505 - 1997	OSHA - 1994
Revision	Accept in Principle	1741 - 1703	-	-	-	-
ROP	**ROC**	**NFPA 70E - 2004**	**NFPA 70B - 2006**	**NFPA 79 - 2007**	**NFPA**	**NEMA**
pg. 733 # 13-21	pg. - # -	-	-	-	-	-
log: 2083 CMP: 13	log: -	Submitter: John C. Wiles		2005 NEC: 690.4(D)		IEC: -

2005 NEC - 690.4 Installation.

(D) Equipment. Inverters, ~~or~~ motor generators, shall be identified ~~for use in solar photovoltaic systems~~.

2008 NEC - 690.4 Installation.

(D) Equipment. Inverters, motor generators, photovoltaic modules, photovoltaic panels, ac photovoltaic modules, source-circuit combiners, and charge controllers intended for use in photovoltaic power systems shall be identified and listed for the application.

Author's Comment: The complexity of PV system equipment and PV system designs require that the PV modules, the combiner boxes, the charge controllers, and the inverters be identified for use in PV systems and examined for safety (listed) by a third party to standards established by Underwriters Laboratory (UL). Electrical inspectors, sometimes faced with systems containing unlisted equipment in these categories, may request that this complex equipment be specifically listed or field evaluated to assist them in approving the installation. Also the identification and proper listing of these components rules out the inappropriate use of electrical components, such as inverters and charge controllers that may be listed to inappropriate standards, such as telecom and marine standards.

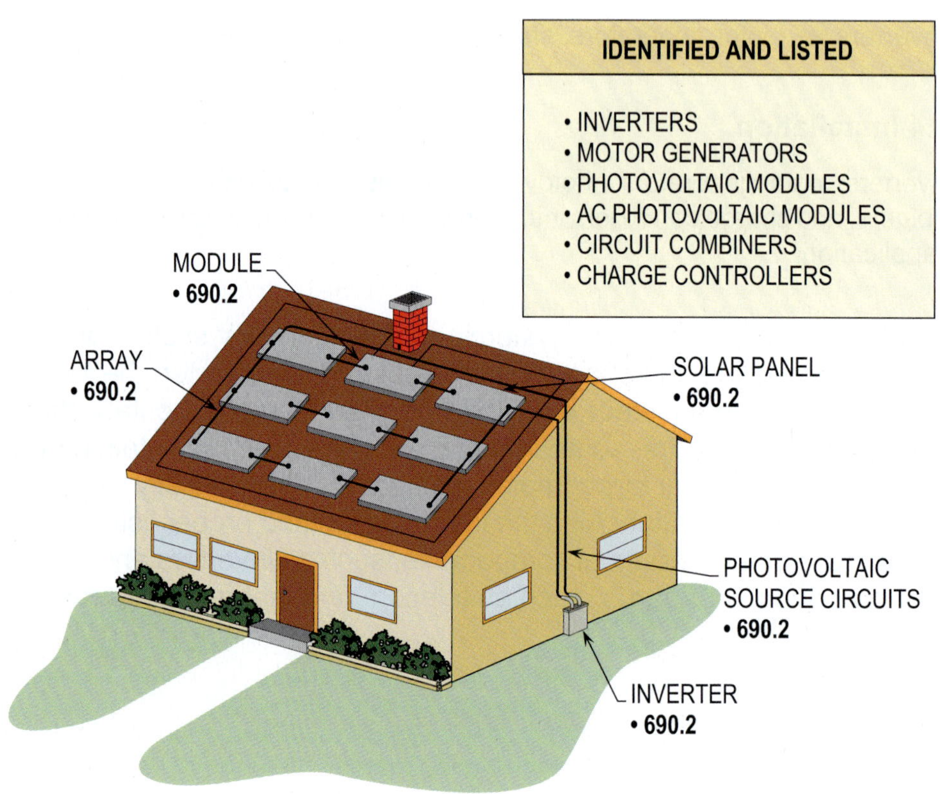

EQUIPMENT
690.4(D)

Purpose of Change: This revision clarifies the components that shall be listed and identified for photovoltaic systems.

NEC Ch. 6 - Article 690
Part I - 690.5(A) through (C)

Type of Change	Panel Action	UL	UL 508	API 500 - 1997	API 505 - 1997	OSHA - 1994
Revision	Accept	-	-	-	-	-
ROP	**ROC**	**NFPA 70E - 2004**	**NFPA 70B - 2006**	**NFPA 79 - 2007**	**NFPA**	**NEMA**
pg. 733 # 13-22	pg. 407 # 13-29	-	-	-	-	-
log: 2060 CMP: 13	log: 287	Submitter: John C. Wiles		2005 NEC: 690.5		IEC: -

2005 NEC - 690.5 Ground-Fault Protection.

~~Roof-mounted dc photovoltaic arrays located on dwellings shall be provided with dc ground-fault protection to reduce fire hazards.~~

(A) Ground-Fault Detection and Interruption. The ground-fault protection device or system shall be capable of detecting a ground fault, interrupting the flow of fault current, and providing an indication of the fault.

~~**(B) Disconnection of Conductors.** The ungrounded conductors of the faulted source circuit shall be automatically disconnected. If the grounded conductors of the faulted source circuit are disconnected to comply with the requirements of 690.5(A), all conductors of the faulted source circuit shall be opened automatically and simultaneously. Opening the grounded conductor of the array or opening the faulted sections of the array shall be permitted to interrupt the ground-fault current path.~~

(C) Labels and Markings. ~~Labels and markings~~ shall be applied near the ground-fault indicator at a visible location stating ~~that, if a ground fault is indicated, the normally grounded conductors may be energized and ungrounded~~.

2008 NEC - 690.5 Ground-Fault Protection.

Grounded dc photovoltaic arrays shall be provided with dc ground-fault protection meeting the requirements of 690.5 (A) through (C) to reduce fire hazards. Ungrounded dc photovoltaic arrays shall comply with 690.35.

Exception 1: Ground-mounted or pole-mounted photovoltaic arrays with not more than two paralleled source circuits and with all dc source and dc output circuits isolated from buildings shall be permitted without ground-fault protection.

Exception 2: PV arrays installed at other than dwelling units shall be permitted without ground-fault protection where the equipment grounding conductors are sized in accordance with 690.45.

(A) Ground-Fault Detection and Interruption. The ground-fault protection device or system shall be capable of detecting a ground fault current, interrupting the flow of fault current, and providing an indication of the fault. Automatically opening the grounded conductor of the faulted circuit to interrupt the ground-fault current path shall be permitted. If a grounded conductor is opened to interrupt the ground-fault current path, all conductors of the faulted circuit shall be automatically and simultaneously opened.

Manual operation of the main PV dc disconnect shall not activate the ground-fault protection device or result in grounded conductors becoming ungrounded.

(B) Isolating Faulted Circuits. The faulted circuits shall be isolated by one of the following methods:

(1) The ungrounded conductors of the faulted circuit shall be automatically disconnected.

(2) The inverter or charge controller fed by the faulted circuit shall automatically cease to supply power to output circuits.

(C) Labels and Markings. A warning label shall appear on the utility-interactive inverter or be applied by the installer near the ground-fault indicator at a visible location stating the following:

WARNING ELECTRIC SHOCK HAZARD

IF A GROUND FAULT IS INDICATED, NORMALLY GROUNDED CONDUCTORS MAY BE UNGROUNDED AND ENERGIZED

When the photovoltaic system also has batteries, the same warning shall also be applied by the installer in a visible location at the batteries.

Author's Comment: The current-limited characteristic of PV modules, sub arrays, and arrays and the ability to generate sustained ground-fault currents require that these ground-fault currents be sensed and interrupted at current levels depending on the system size to eliminate the need for significant over sizing of equipment grounding conductors. The text limiting ground fault protection to just roof mounted arrays on dwellings has been deleted to now apply to all arrays in any location and at any installation, regardless of the occupancy.

The added second paragraph was moved from 690.5(B) to 690.5(A) because it describes the various optional methods of interrupting the fault current as required by 690.5(A). The addition of the third paragraph establishes requirements that prevent interconnecting the main dc PV disconnect with the ground-fault protection device that could leave the PV array ungrounded when the PV disconnect was opened manually during normal service operations or in other situations. The intent is to keep a grounded PV system solidly grounded under all normal operating conditions including when the main PV or other disconnect is opened. An exception was added to permit listed equipment that is capable of interrupting the flow of fault current to be used. In addition, the requirement for a redundant disconnect was changed to a labeling requirement, as no known safety hazard exists.

A ground fault in a PV system and the response of the required ground-fault protection device modifies the normal grounding of the system. The need for an appropriate warning label on the inverter containing the ground-fault device or an installer-applied label on those systems with a separate ground-fault device is now clearly defined. Requiring specific wording and a second label on the battery banks (if installed and usually operating at or less than 48 volts, nominal) increases user/operator awareness and is now required since battery banks may also become ungrounded during ground faults.

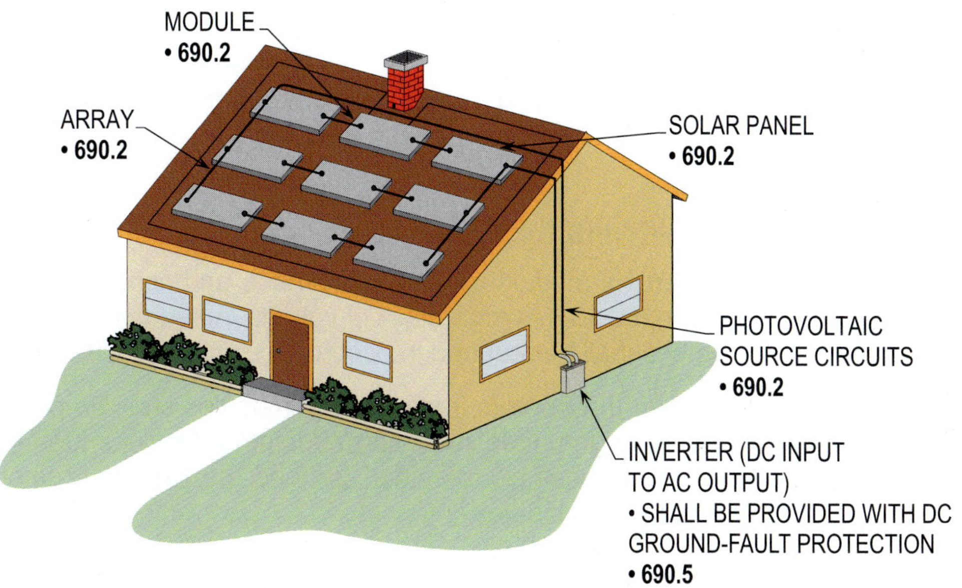

GROUND-FAULT DETECTION AND INTERRUPTION
• SHALL BE CAPABLE OF DETECTING A GROUND-FAULT CURRENT • INTERRUPTING THE FLOW OF CURRENT • PROVIDING AN INDICATION OF THE FAULT • AUTOMATICALLY OPENING THE GROUNDED CONDUCTOR OF THE FAULTED CIRCUIT SHALL BE PERMITTED • IF A GROUNDED CONDUCTOR IS OPENED TO INTERRUPT THE GROUND-FAULT CURRENT PATH, ALL CONDUCTORS SHALL BE AUTOMATICALLY AND SIMULTANEOUSLY OPENED.

LABELS AND MARKINGS
• A WARNING LABEL SHALL APPEAR ON THE UTILITY-INTERACTIVE INVERTER OR BE APPLIED BY THE INSTALLER NEAR THE GROUND-FAULT INDICATOR AT A VISIBLE LOCATION STATING THE FOLLOWING • WARNING ELECTRIC SHOCK HAZARD IF A GROUND-FAULT IS INDICATED, NORMALLY GROUNDED CONDUCTORS MAY BE UNGROUNDED AND ENERGIZED

GROUND-FAULT PROTECTION
690.5(A) through (C)

Purpose of Change: This revision clarifies the ground-fault protection requirements for solar photovoltaic systems.

NEC Ch. 6 - Article 690
Part V - 690.43

Type of Change	Panel Action	UL	UL 508	API 500 - 1997	API 505 - 1997	OSHA - 1994
Revision	Accept	-	-	-	-	-
ROP	**ROC**	**NFPA 70E - 2004**	**NFPA 70B - 2006**	**NFPA 79 - 2007**	**NFPA**	**NEMA**
pg. 741 # 13-47	pg. 412 # 13-57	-	-	-	-	-
log: 2074 CMP: 13	log: 88	Submitter: John C. Wiles		2005 NEC: 690.43		IEC: -

2005 NEC - 690.43 Equipment Grounding.

Exposed non–current-carrying metal parts of module frames, equipment, and conductor enclosures shall be grounded in accordance with 250.134 or 250.136(A) regardless of voltage.

2008 NEC - 690.43 Equipment Grounding.

Exposed non–current-carrying metal parts of module frames, equipment, and conductor enclosures shall be grounded in accordance with 250.134 or 250.136(A) regardless of voltage. An equipment grounding conductor between a PV array and other equipment shall be required in accordance with 250.110.

Devices listed and identified for grounding the metallic frames of PV modules shall be permitted to bond the exposed metallic frames of PV modules to grounded mounting structures. Devices identified and listed for bonding the metallic frames of PV modules shall be permitted to bond the exposed metallic frames of PV modules to the metallic frames of adjacent PV modules.

Equipment-grounding conductors for the PV array and structure (where installed) shall be contained within the same raceway or cable, or otherwise run with the PV array circuit conductors when those circuit conductors leave the vicinity of the PV array.

Author's Comment: The second paragraph was added because module grounding clips and other devices are being developed and listed that will effectively penetrate the oxide or anodizing on aluminum framed PV modules and ground them to grounded PV array mounting structures or effectively bond them to adjacent PV modules which, in turn, may be grounded. The existing grounding and bonding requirements in **250.134** or **250.136** do not specifically or generally allow the use of such devices. Nor do they prohibit the use of such devices.

The third paragraph was added because **250.134(B), Exception 2** allows dc equipment-grounding conductors to be routed separately from the circuit conductors. The Comments provide NEC Style Manual changes.

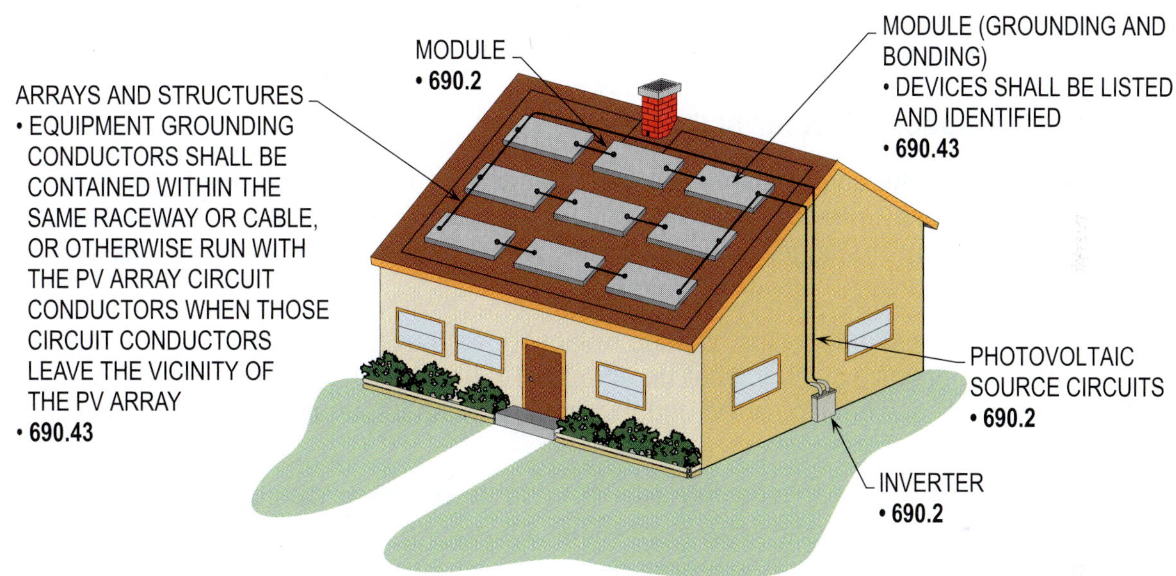
MODULE
• 690.2
ARRAYS AND STRUCTURES
• EQUIPMENT GROUNDING CONDUCTORS SHALL BE CONTAINED WITHIN THE SAME RACEWAY OR CABLE, OR OTHERWISE RUN WITH THE PV ARRAY CIRCUIT CONDUCTORS WHEN THOSE CIRCUIT CONDUCTORS LEAVE THE VICINITY OF THE PV ARRAY
• 690.43
MODULE (GROUNDING AND BONDING)
• DEVICES SHALL BE LISTED AND IDENTIFIED
• 690.43
PHOTOVOLTAIC SOURCE CIRCUITS
• 690.2
INVERTER
• 690.2

EQUIPMENT GROUNDING
690.43

Purpose of Change: This revision clarifies the grounding and bonding requirements for solar photovoltaic systems.

NEC Ch. 6 - Article 690
Part VII - 690.62

Type of Change		Panel Action		UL	UL 508	API 500 - 1997	API 505 - 1997	OSHA - 1994
Revision		Accept		-	-	-	-	-
ROP		ROC		NFPA 70E - 2004	NFPA 70B - 2006	NFPA 79 - 2007	NFPA	NEMA
pg. 745	# 13-58	pg. 416	# 13-74	-	-	-	-	-
log: 2096	CMP: 13	log: 95	Submitter: John C. Wiles			2005 NEC: 690.62		IEC: -

2005 NEC - 690.62 Ampacity of Neutral Conductor.

If a single-phase, 2-wire inverter output is connected to the neutral and one ungrounded conductor (only) of a 3-wire system or of a 3-phase, 4-wire wye-connected system, the maximum load connected between the neutral and any one ungrounded conductor plus the inverter output rating shall not exceed the ampacity of the neutral conductor.

2008 NEC - 690.62 Ampacity of Neutral Conductor.

If a single-phase, 2-wire inverter output is connected to the neutral conductor and one ungrounded conductor (only) of a 3-wire system or of a 3-phase, 4-wire wye-connected system, the maximum load connected between the neutral conductor and any one ungrounded conductor plus the inverter output rating shall not exceed the ampacity of the neutral conductor.

A conductor used solely for instrumentation, voltage detection, or phase detection, and connected to a single phase or 3-phase utility-interactive inverter, shall be permitted to be sized at less than the ampacity of the other current-carrying conductors and shall be sized equal to or larger than the equipment grounding conductor.

Author's Comment: A 14 AWG for instrumentation, voltage, or phase detection may not be sufficient to carry fault current from the branch circuit breaker. Changing the requirement that the minimum conductor size be no smaller than the equipment grounding conductor ensures that this instrumentation conductor will be able to carry fault current and trip the circuit overcurrent device. The comment clarified the text to conform to the NEC Style Manual. The last part of the sentence was changed to clarify that smaller conductors are permitted in certain cases.

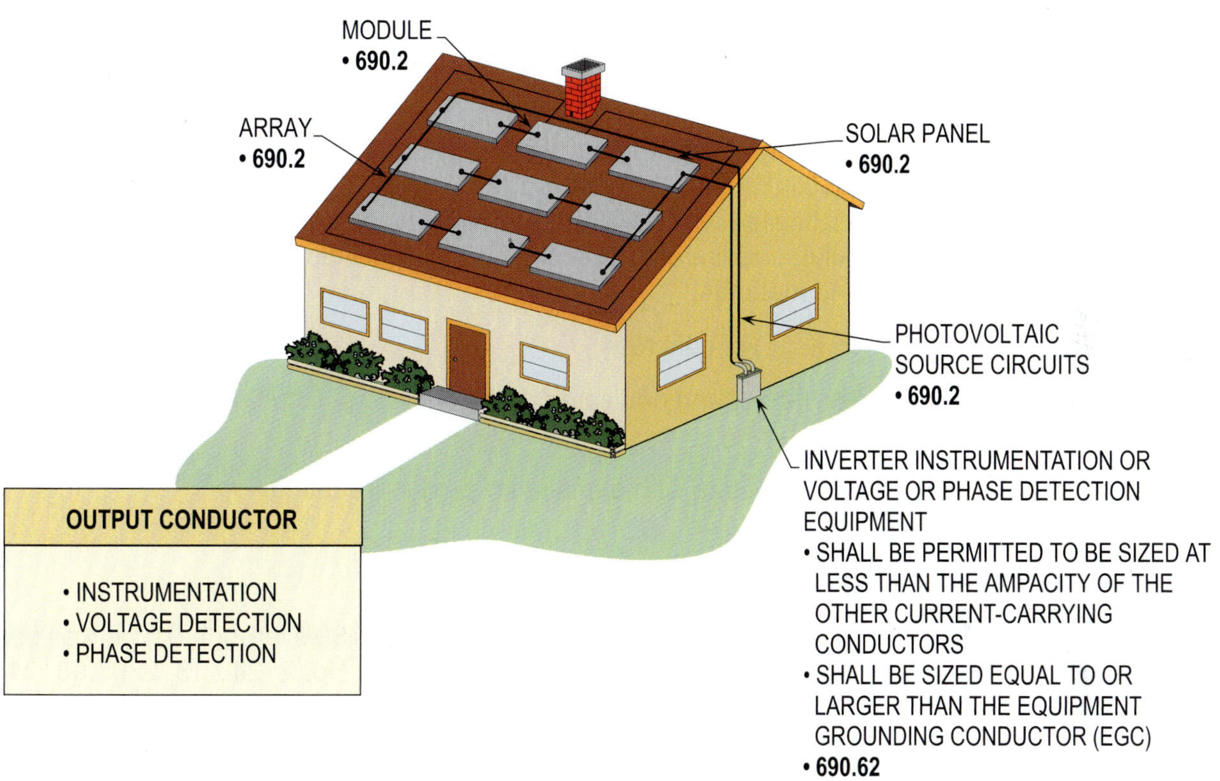

AMPACITY OF NEUTRAL CONDUCTOR
690.62

Purpose of Change: This revision clarifies the sizing requirements for a conductor used solely for instrumentation, voltage detection or phase detection and connected to a single phase or 3-phase utility-interactive inverter.

NEC Ch. 6 - Article 695
Part - 695.6(B)(2) and (B)(3)

93

Type of Change	Panel Action	UL	UL 508	API 500 - 1997	API 505 - 1997	OSHA - 1994
Revision	Accept	-	-	-	-	-
ROP	**ROC**	**NFPA 70E - 2004**	**NFPA 70B - 2006**	**NFPA 79 - 2007**	**NFPA**	**NEMA**
pg. 764 # 13-99	pg. - # -	-	-	-	-	-
log: 3018 CMP: 13	log: -	Submitter: James Conrad		2005 NEC: 695.6(B)(2) and (B)(3)		IEC: -

2005 NEC - 695.6 Power Wiring.

Power circuits and wiring methods shall comply with the requirements in 695.6(A) through (H), and as permitted in 230.90(A), Exception No. 4; 230.94, Exception No. 4; 230.95, Exception No. 2; 240.13; 230.208; 240.4(A); and 430.31.

(B) Circuit Conductors. Fire pump supply conductors on the load side of the final disconnecting means and overcurrent device(s) permitted by 695.4(B) shall be kept entirely independent of all other wiring. They shall supply only loads that are directly associated with the fire pump system, and they shall be protected to resist potential damage by fire, structural failure, or operational accident. They shall be permitted to be routed through a building(s) using one of the following methods:

(1) Be encased in a minimum 50 mm (2 in.) of concrete

(2) Be ~~within an enclosed construction 1-hour~~ and dedicated to the fire pump circuit(s). ~~and having a minimum of a 1-hour fire resistive rating~~.

(3) Be a listed electrical circuit protective system with a minimum ~~1-hour~~ fire rating

2008 NEC - 695.6 Power Wiring.

Power circuits and wiring methods shall comply with the requirements in 695.6(A) through (H), and as permitted in 230.90(A), Exception No. 4; 230.94, Exception No. 4; 230.95, Exception No. 2; 240.13; 230.208; 240.4(A); and 430.31.

(B) Circuit Conductors. Fire pump supply conductors on the load side of the final disconnecting means and overcurrent device(s) permitted by 695.4(B) shall be kept entirely independent of all other wiring. They shall supply only loads that are directly associated with the fire pump system, and they shall be protected to resist potential damage by fire, structural failure, or operational accident. They shall be permitted to be routed through a building(s) using one of the following methods:

(1) Be encased in a minimum 50 mm (2 in.) of concrete

(2) Be protected by a fire-rated assembly listed to achieve a minimum fire rating of 2-hours and dedicated to the fire pump circuit(s).

(3) Be a listed electrical circuit protective system with a minimum 2-hour fire rating

FPN: UL guide information for electrical protective systems (FHIT) contains information on proper installation requirements to maintain the fire rating.

Author's Comment: These proposals harmonize **695.6(B)(2)** with **700.9(D)(4)** by requiring the enclosure or assembly to be listed. Listed fire-rated assemblies have been tested and are described in the UL Fire Resistance Directory or other Listing Directories. In addition, NFPA 72 requires 2-hour survivability of the notification circuits and the interconnecting wiring of the fire command center. To achieve this 2-hour rating you can use 2-hour rated cables, a 2-hour fire-rated enclosure or, if permitted by the AHJ, the automatic sprinkler system for the building. If an electric driven fire pump is supplying the sprinkler system, then the fire pump circuits should be protected for 2-hours ensuring the operation of the sprinkler system for 2-hours to ensure survivability of the fire pump circuits.

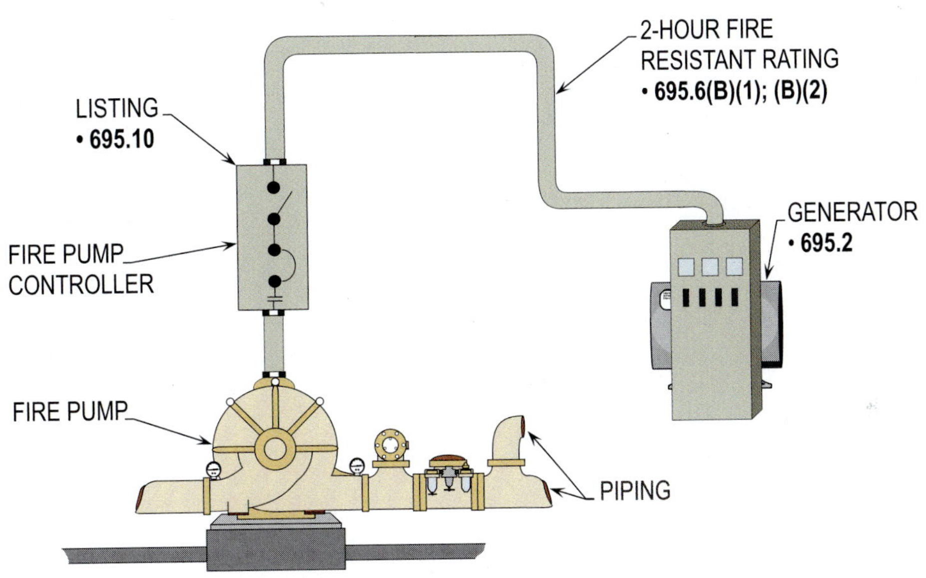

CIRCUIT CONDUCTORS
695.6(B)(2) and (B)(3)

Purpose of Change: This revision clarifies that fire pump supply conductors shall be protected by a 2-hour fire resistance rating if not encased in a minimum of 2 in. (50 mm) of concrete or by a listed electrical protective system with a 2-hour minimum fire rating.

7

Special Conditions

The requirements of **Chapter 7** have been used to design and install the electrical elements that are related to emergency and other special condition power sources. Emergency systems and their power sources are mandated by codes other than the NEC. The NEC establishes the provisions for the installation of emergency systems, once these systems are required. **Article 725** covers remote control, signaling, and power limited circuits. **Article 760** provides wiring requirements for fire alarm systems. Fiber optics are provided in **Article 770**.

Emergency systems must operate automatically and supply illumination to designated areas and power to designated equipment, and their wiring must not be run with the normal power system. Legally required standby systems must also operate automatically, but unlike emergency systems, their wiring can be run with the wiring of the normal power system. Optional standby systems are not required to operate automatically, and their wiring can also be run with the wiring of the normal power system.

Article and Sections of the 700 series govern wiring methods and equipment related to special conditions and use.

NEC Ch. 7 - Article 700
Part I - 700.6(C)

Type of Change		Panel Action		UL	UL 508	API 500 - 1997	API 505 - 1997	OSHA - 1994
Revision		Accept		1008	-	-	-	1910.308(b)(1)
ROP		**ROC**		**NFPA 70E - 2004**	**NFPA 70B - 2006**	**NFPA 79 - 2007**	**NFPA**	**NEMA**
pg. 768	# 13-117	pg. -	# -	450.2(A)	6.4.4.6(2)	-	-	-
log: 2369	CMP: 13	log: -	Submitter: Lawrence A. Bey			2005 NEC: 700.6(C)		IEC: -

2005 NEC - 700.6 Transfer Equipment.

(C) Automatic Transfer Switches. Automatic transfer switches shall be electrically operated and mechanically held.

2008 NEC - 700.6 Transfer Equipment.

(C) Automatic Transfer Switches. Automatic transfer switches shall be electrically operated and mechanically held. Automatic transfer switches, rated 600 VAC and below, shall be listed for emergency system use.

Author's Comment: A new sentence has been added to **700.6(C)** requiring automatic transfer switches to be listed for emergency system use.

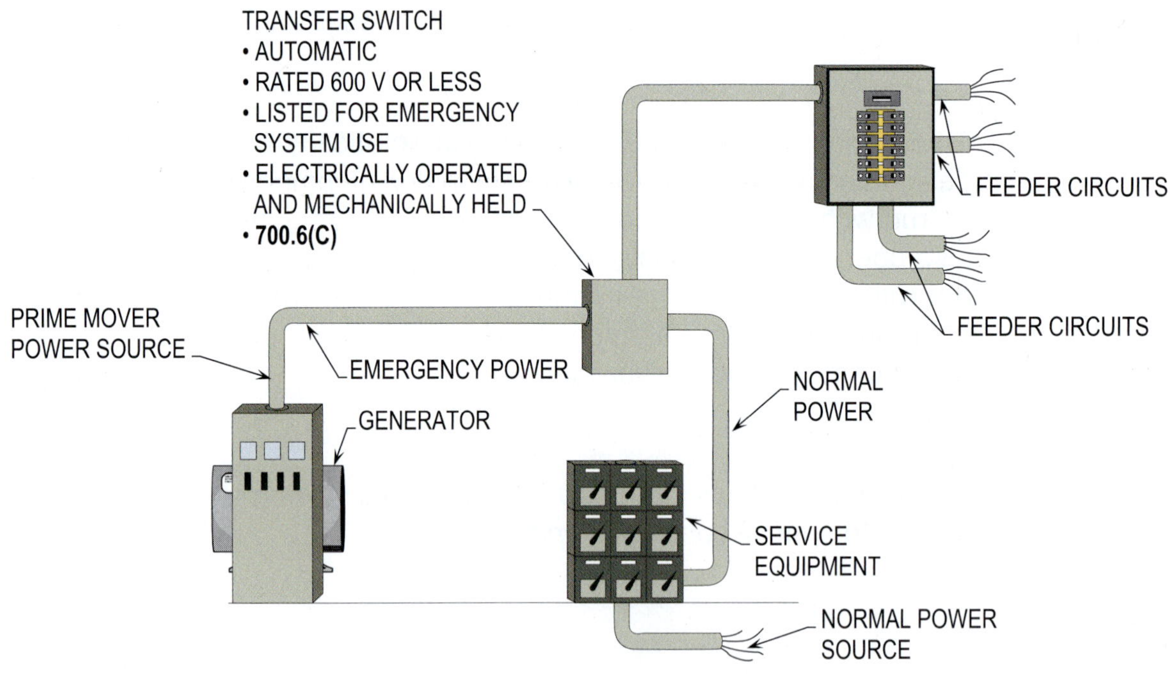

AUTOMATIC TRANSFER SWITCHES
700.6(C)

Purpose of Change: This revision clarifies that automatic transfer switches, rated 600 volts or less, are required to be listed for emergency system use.

NEC Ch. 7 - Article 700
Part II - 700.9(B)(5)

Type of Change	Panel Action	UL	UL 508	API 500 - 1997	API 505 - 1997	OSHA - 1994
New Subdivision	Accept	-	-	-	-	1910.308(b)(1)
ROP	ROC	NFPA 70E - 2004	NFPA 70B - 2006	NFPA 79 - 2007	NFPA	NEMA
pg. 768 # 13-118	pg. 444 # 13-156	450.2(A)	6.4.4.6(2)	-	-	-
log: 833 CMP: 13	log: 110	Submitter: Thomas H. Wood		2005 NEC: 700.9(B)		IEC: -

2005 NEC - 700.9 Wiring, Emergency System.

(B) Wiring. Wiring of two or more emergency circuits supplied from the same source shall be permitted in the same raceway, cable, box, or cabinet. Wiring from an emergency source or emergency source distribution overcurrent protection to emergency loads shall be kept entirely independent of all other wiring and equipment, unless otherwise permitted in (1) through ~~(4)~~:

(1) Wiring from the normal power source located in transfer equipment enclosures

(2) Wiring supplied from two sources in exit or emergency luminaires ~~(lighting fixtures)~~

(3) Wiring from two sources in a common junction box, attached to exit or emergency luminaires ~~(lighting fixtures)~~

(4) Wiring within a common junction box attached to unit equipment, containing only the branch circuit supplying the unit equipment and the emergency circuit supplied by the unit equipment

2008 NEC - 700.9 Wiring, Emergency System.

(B) Wiring. Wiring of two or more emergency circuits supplied from the same source shall be permitted in the same raceway, cable, box, or cabinet. Wiring from an emergency source or emergency source distribution overcurrent protection to emergency loads shall be kept entirely independent of all other wiring and equipment, unless otherwise permitted in (1) through (5):

(1) Wiring from the normal power source located in transfer equipment enclosures

(2) Wiring supplied from two sources in exit or emergency luminaires

(3) Wiring from two sources in a common junction box, attached to exit or emergency luminaires

(4) Wiring within a common junction box attached to unit equipment, containing only the branch circuit supplying the unit equipment and the emergency circuit supplied by the unit equipment

(5) Wiring from an emergency source to supply any combination of emergency, legally required, or optional loads in accordance with (a), (b) and (c).

(a) From separate vertical switchboard sections, with or without a common bus, or from individual disconnects mounted in separate enclosures.

(b) The common bus or separate sections of the switchboard or the individual enclosures shall be permitted to be supplied by single or multiple feeders without overcurrent protection at the source.

Exception to (5)(b). Overcurrent protection shall be permitted at the source or for the equipment, provided the overcurrent protection is selectively coordinated with the down-stream overcurrent protection.

(c) Legally required and optional standby circuits shall not originate from the same vertical switchboard section, panelboard enclosure or individual disconnect enclosure as emergency circuits.

Author's Comment: A new subsection **(5)** has been added to provide wiring requirements for emergency circuits. Emergency systems must originate from separate sections of a vertical switchboard or from individual disconnects mounted in separate enclosures. Common bus, separate sections of the switchboard, or individual disconnects in separate enclosures are permitted to be supplied by single or multiple feeders without overcurrent protection at the source. Finally, the legally required and optional standby circuits shall not originate from the same vertical switchboard section, panelboard enclosure or individual disconnect enclosure as emergency circuits. The comment changed the text in 700.9(B)(5)(a) to comply with NEC Style Manual.

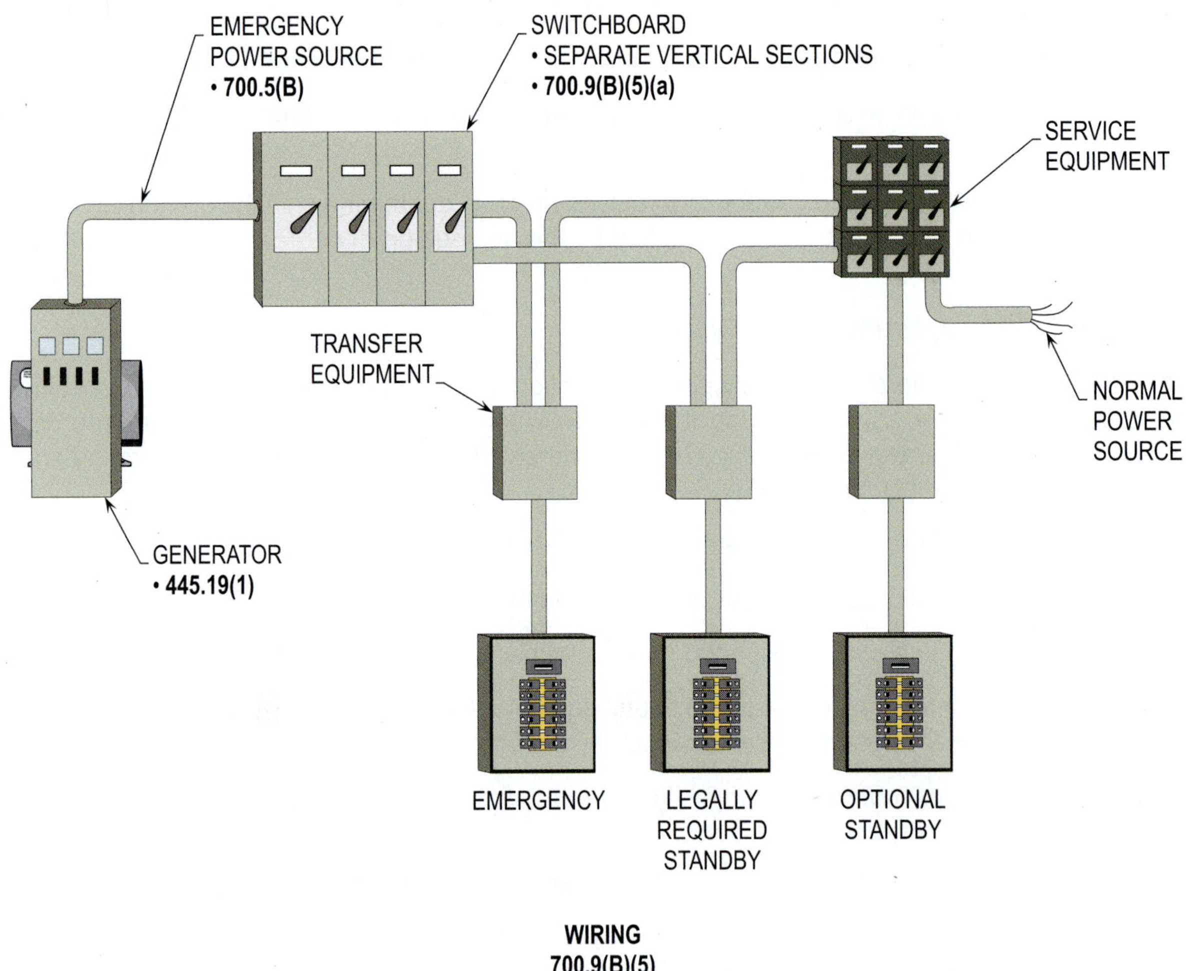

WIRING
700.9(B)(5)

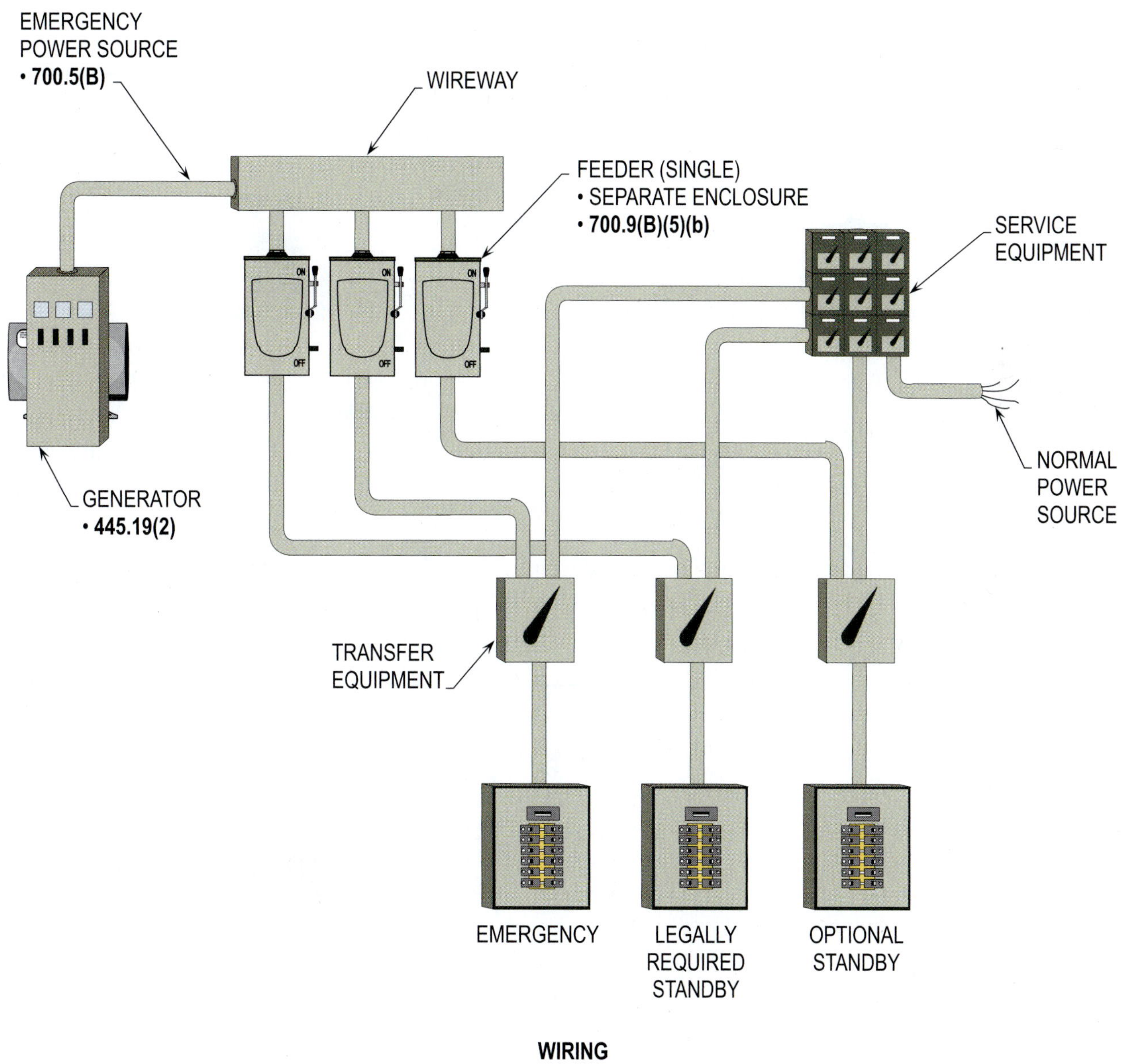
EMERGENCY
POWER SOURCE
• 700.5(B)
WIREWAY
FEEDER (SINGLE)
• SEPARATE ENCLOSURE
• 700.9(B)(5)(b)
ON
OFF
SERVICE
EQUIPMENT
NORMAL
POWER
SOURCE
GENERATOR
• 445.19(2)
TRANSFER
EQUIPMENT
EMERGENCY
LEGALLY
REQUIRED
STANDBY
OPTIONAL
STANDBY

WIRING
700.9(B)(5)

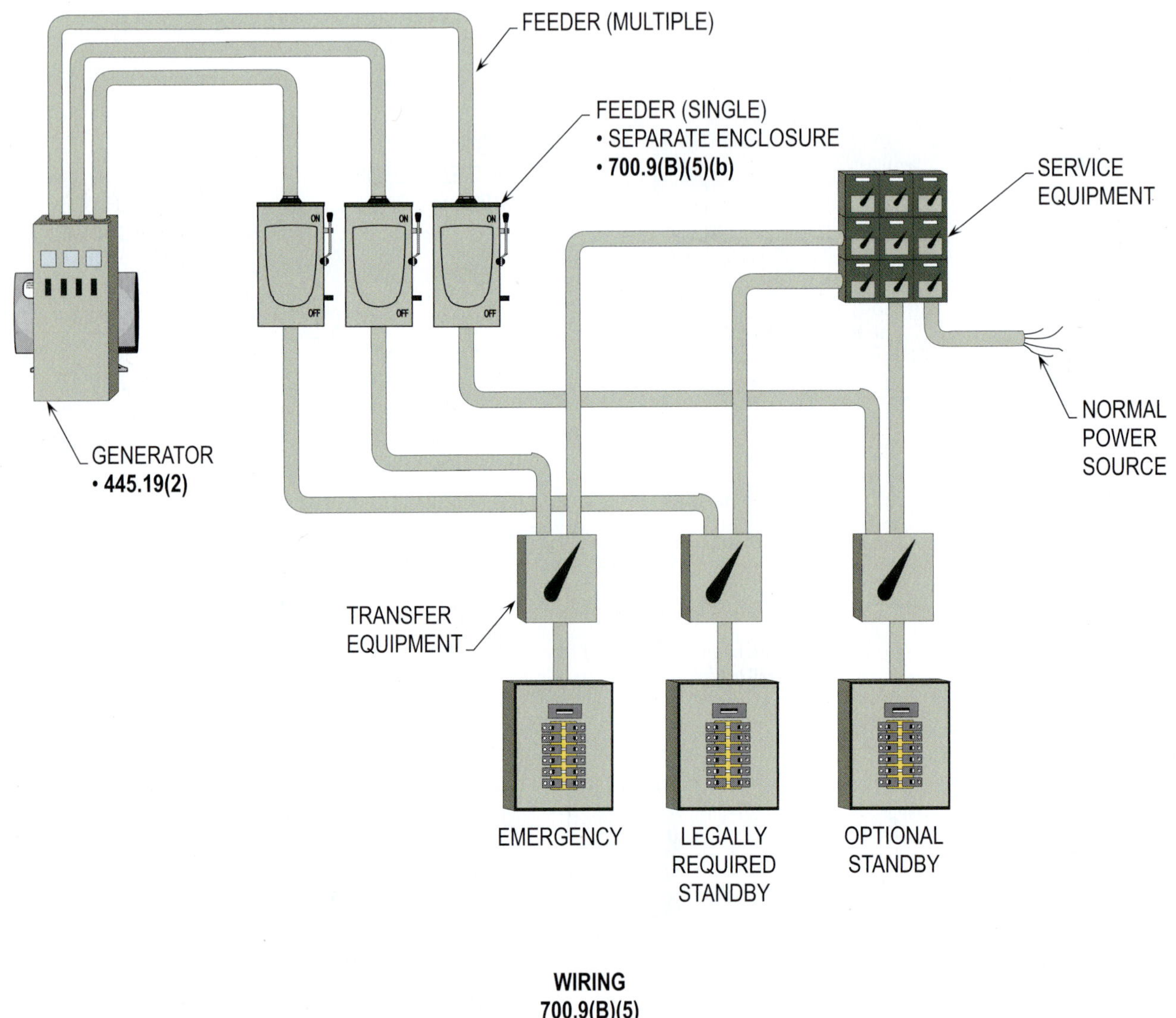

WIRING
700.9(B)(5)

Purpose of Change: A new subdivision has been added to provide wiring requirements for the emergency source.

NEC Ch. 7 - Article 700
Part II - 700.9(D)(3)

Type of Change		Panel Action		UL	UL 508	API 500 - 1997	API 505 - 1997	OSHA - 1994
New Subdivision		Accept in Principle		-	-	-	-	1910.308(b)(1)
ROP		ROC		NFPA 70E - 2004	NFPA 70B - 2006	NFPA 79 - 2007	NFPA	NEMA
pg. 770	# 13-126	pg. -	# -	450.2(A)	6.4.4.6(2)	-	-	-
log: 3015	CMP: 13	log: -	Submitter: James Conrad			2005 NEC: -		IEC: -

2008 NEC - 700.9 Wiring, Emergency System.

(D) Fire Protection. Emergency systems shall meet the additional requirements in 700.9(D)(1) and (D)(2) assembly occupancies for not less than 1000 persons or in buildings above 23 m (75 ft) in height with any of the following occupancy classes: assembly, educational, residential, detention and correctional, business, and mercantile.

(3) Generator Control Wiring. Control conductors installed between the transfer equipment and the emergency generator shall be kept entirely independent of all other wiring and shall meet the conditions of 700.9(D)(1).

Author's Comment: The generator start circuit is a critical component of the emergency system and, based on new subsection **(3)**, now has the same protection as the feeder-circuit wiring. This same requirement is currently in **695.14(F)** for fire pumps to ensure the starting of the standby generator where used for an alternate source of power.

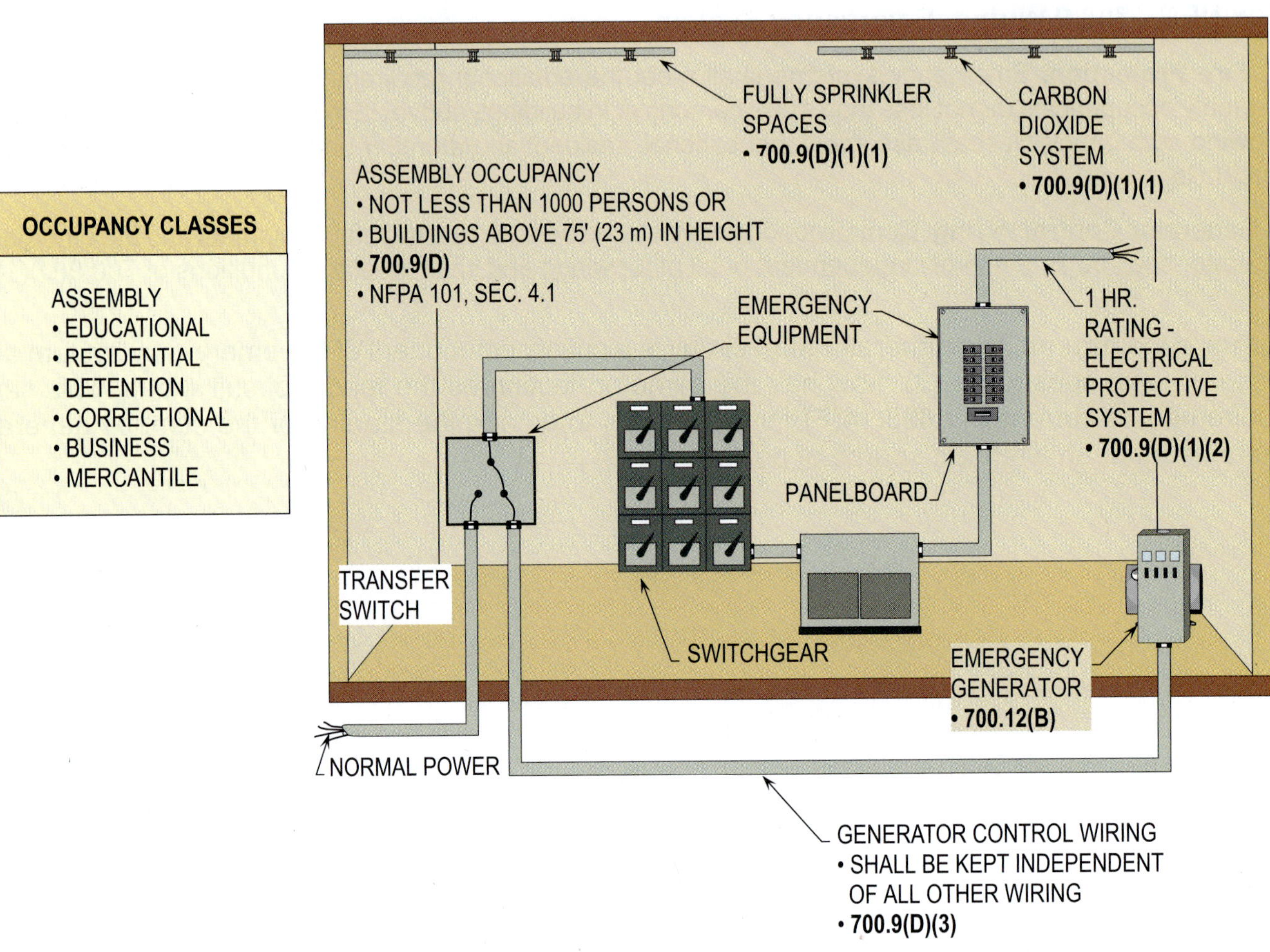

GENERATOR CONTROL WIRING
700.9(D)(3)

Purpose of Change: A new subdivision has been added to require generator control wiring installed between the transfer equipment and the emergency generator to be kept entirely independent of all other wiring.

NEC Ch. 7 - Article 700
Part V - 700.23

Type of Change	Panel Action	UL	UL 508	API 500 - 1997	API 505 - 1997	OSHA - 1994
New Section	Accept	924 - 1008	-	-	-	-
ROP	**ROC**	**NFPA 70E - 2004**	**NFPA 70B - 2006**	**NFPA 79 - 2007**	**NFPA**	**NEMA**
pg. 772 # 13-134	pg. 449 # 13-176	-	-	-	-	-
log: 3321 CMP: 13	log: 113	Submitter: Steven R. Terry		2005 NEC: -		IEC: -

2008 NEC - 700.23 Dimmer Systems.

A dimmer system containing more than one dimmer and listed for use in emergency systems shall be permitted to be used as a control device for energizing emergency lighting circuits. Upon failure of normal power, the dimmer system shall be permitted to selectively energize only those branch circuits required to provide minimum emergency illumination. All branch circuits supplied by the dimmer system cabinet shall comply with the wiring methods of Article 700.

Author's Comment: Dimmer systems listed for emergency use under UL 924 and containing more than one dimmer are common. These systems include a method to sense failure of normal power and selectively energize branch circuits fed from the dimmer cabinet using a reliable method, regardless of the setting of control switches or panels normally used to control the dimmer system. This is known as "Bypass Mode". Such systems are typically supplied by a feeder that is transferred from normal to emergency power by an upstream UL 1008 Transfer Switch. These dimmer systems often contain a greater number of branch circuits than those required to maintain minimum emergency illumination levels.

Clarification has been provided to permit such dimming systems to energize only those circuits required for minimum emergency illumination levels, rather than all circuits in the dimmer cabinet. The comment changed the text for compliance with the NEC Style Manual.

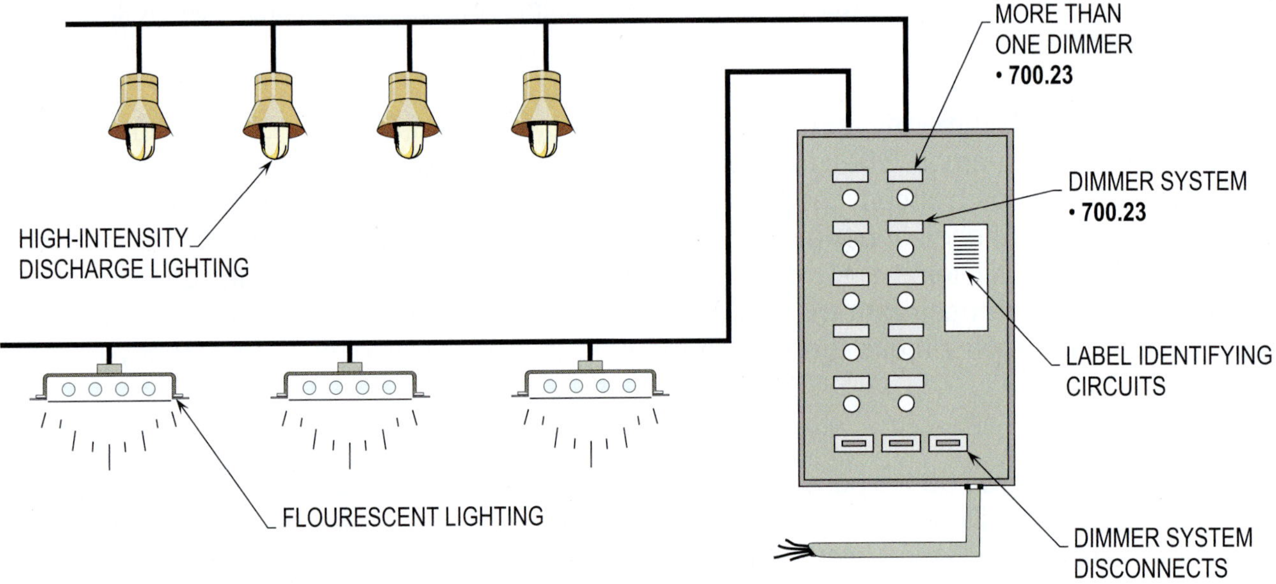

NOTE 2: IF NORMAL POWER FAILS, DIMMER SYSTEM SHALL SELECTIVELY ENERGIZE EMERGENCY BRANCH-CIRCUITS TO SUPPLY MINIMUM EMERGENCY ILLUMINATION.

DIMMER SYSTEMS
700.23

Purpose of Change: A new section has been added to address a dimmer system containing more than one dimmer and listed for use for emergency systems.

NEC Ch. 7 - Article 702
Part I - 702.5(B)

Type of Change		Panel Action		UL	UL 508	API 500 - 1997	API 505 - 1997	OSHA - 1994
Revision		Accept		1008	-	-	-	-
ROP		ROC		NFPA 70E - 2004	NFPA 70B - 2006	NFPA 79 - 2007	NFPA	NEMA
pg. 782	# 13-168	pg. 470	# 13-255	-	-	-	-	-
log: 2741	CMP: 13	log: 114	Submitter: Jim Pauley			2005 NEC: 702.5		IEC: -

2005 NEC - 702.5 Capacity and Rating.

~~An optional standby system shall have adequate capacity and rating for the supply of all equipment intended to be operated at one time.~~ Optional standby system equipment shall be suitable for the maximum available ~~fault~~ at its terminals. The user of the optional standby system shall be permitted to select the load connected to the system.

2008 NEC - 702.5 Capacity and Rating.

(A) Available Short Circuit Current. Optional standby system equipment shall be suitable for the maximum available short-circuit current at its terminals.

(B) System Capacity. The calculations of load on the standby source shall be made in accordance with Article 220 or by another approved method.

(1) Manual Transfer Equipment. Where manual transfer equipment is used, an optional standby system shall have adequate capacity and rating for the supply of all equipment intended to be operated at one time. The user of the optional standby system shall be permitted to select the load connected to the system.

(2) Automatic Transfer Equipment. Where automatic transfer equipment is used, an optional standby system shall comply with (2)(a) or (2)(b).

(a) Full Load. The standby source shall be capable of supplying the full load that is transferred by the automatic transfer equipment.

(b) Load Management. Where a system is employed that will automatically manage the connected load, the standby source shall have a capacity sufficient to supply the maximum load that will be connected by the load management system.

Author's Comment: The change in **702.5** has rearranged the existing text to split up the paragraph and provide headings that make it easier to use. It created an "Available Short Circuit Current" heading and moved the sentence about adequate ratings for fault current to this new heading. The term "fault current" was replaced with "short circuit current" to make it consistent with the rest of the code.

System capacity has been split into Manual Transfer and Automatic Transfer applications. For manual transfer, the existing text was used to require that power supply must be adequate to supply the equipment intended to be connected at one time. It retained the existing permission for the user to select the loads that can be connected.

For automatic transfer in **(B)(2)**, new **(a)** provides an option where standby supply must be able to pick up the entire load that is being transferred. The typical application is where a small generator is connected to a new subpanel with a set number of critical loads. Under normal power, the subpanel is supplied by the normal source. When power fails, the subpanel is transferred to the generator source. New **(b)** permits a system where some of the load in a larger panel may be shed in order to reduce the loading to an adequate level to be supplied by the standby source. The comment changed the text to comply with the NEC Style Manual.

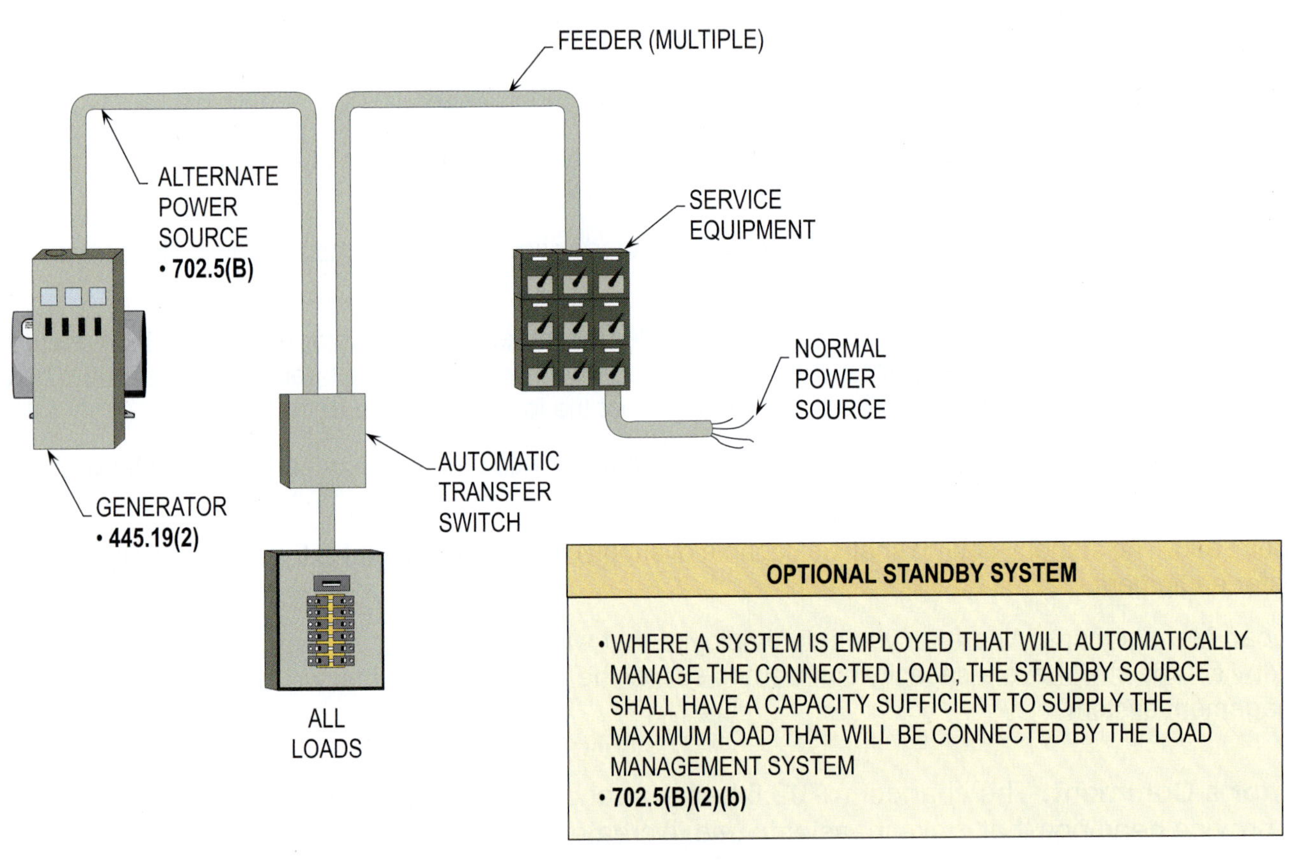

SYSTEM CAPACITY
702.5(B)

Purpose of Change: This revision clarifies the installation of manual transfer and automatic transfer applications for optional standby systems.

NEC Ch. 7 - Article 708
Parts I through V - 708.1 through 708.64

Type of Change		Panel Action		UL	UL 508	API 500 - 1997	API 505 - 1997	OSHA - 1994
New Article		Accept in Principle in Part		-	-	-	-	-
ROP		ROC		NFPA 70E - 2004	NFPA 70B - 2006	NFPA 79 - 2007	NFPA	NEMA
pg. 615	# 20-1	pg. 347	# 20-1a	-	-	-	-	-
log: 3497	CMP: 20	log: CC 2000	Submitter: Alan Manche			2005 NEC: -		IEC: -

2008 NEC - Article 708

Critical Operations Power Systems (COPS)

I. General.

II. Circuit Wiring and Equipment

III. Power Sources and Connection

IV. Overcurrent Protection

V. System Performance and Analysis.

Author's Comment: Article 708 is a new article dealing with mission critical power systems with detailed installation instructions for electrical power for vital infrastructures where interrupted power could disrupt national security, the economy, public health, or safety.

ARTICLE 708 - CRITICAL OPERATIONS POWER SYSTEMS (COPS)	
PART I - GENERAL	
SECTION	**HEADING**
708.1	SCOPE
708.2	DEFINITIONS
708.3	APPLICATION OF OTHER ARTICLES
708.4	RISK ASSESSMENT
708.5	PHYSICAL SECURITY
708.6	TESTING AND MAINTENANCE
708.8	COMMISSIONING
PART II - CIRCUIT WIRING AND EQUIPMENT	
708.10	FEEDER AND BRANCH-CIRCUIT WIRING
708.11	BRANCH CIRCUIT AND FEEDER DISTRIBUTION EQUIPMENT
708.12	FEEDERS AND BRANCH CIRCUITS SUPPLIED BY COPS
708.14	WIRING OF HVAC, FIRE ALARM, SECURITY, EMERGENCY COMMUNICATIONS AND SIGNALING SYSTEMS
PART III - POWER SOURCES AND CONNECTION	
708.20	SOURCES OF POWER
708.22	CAPACITY OF POWER SOURCES
708.24	TRANSFER EQUIPMENT
708.30	BRANCH CIRCUITS SUPPLIED BY COPS
PART IV - OVERCURRENT PROTECTION	
708.50	ACCESSIBILITY
708.52	GROUND-FAULT PROTECTION OF EQUIPMENT
708.54	COORDINATION
PART V - SYSTEM PERFORMANCE AND ANALYSIS	
708.64	EMERGENCY OPERATIONS PLAN

Purpose of Change: A new article has been added dealing with mission critical power systems with detailed installations for electrical power for vital infrastructures.

NEC Ch. 7 - Article 725
Parts I through IV - 725.1 through 725.179

Type of Change		Panel Action		UL	UL 508	API 500 - 1997	API 505 - 1997	OSHA - 1994
Renumbered		Accept		1310 - 1585 - 5085-3	-	-	-	1910.308(c)
ROP		ROC		NFPA 70E - 2004	NFPA 70B - 2006	NFPA 79 - 2007	NFPA	NEMA
pg. 773	# 3-137	pg. -	# -	450.3	-	-	-	-
log: 843	CMP: 3	log: -	Submitter: Vince Baclawski		2005 NEC: Article 725			IEC: -

2008 NEC - Article 725
Class 1, Class 2, and Class 3 Remote-Control, Signaling, and Power-Limited Circuits

I. General

II. Class 1 Circuits.

III. Class 2 and Class 3 Circuits

IV. Listing Requirements

Author's Comment: The number changes within **Articles 725** and **760** correlate the numbering of the two articles, as well as a similar numbering sequence for those related sections in Chapter 8 for overall correlation, while leaving adequate room between sections for future additions. It was impossible to renumber the two articles exactly, since some sections could not be moved without changing their intent. One example of this is newly renumbered **725.52** covering Class 1 circuits extending beyond one building and **760.32** covering both power limited fire alarm and non-power-limited fire alarm circuits extending beyond one building, both addressing the same issue, but the numbering could not be the same without changing the intent of these sections due to their location in their respective articles.

CROSS-REFRENCE		
PART I - GENERAL		
2008 NEC	**2005 NEC**	**SECTION HEADING**
725.1	725.1	SCOPE
725.2	725.2	DEFINITIONS
725.3	725.3	OTHER ARTICLES
725.21	725.7	ACCESS TO ELECTRICAL EQUIPMENT BEHIND PANELS DESIGNED TO ALLOW ACCESS
725.24	725.8	MECHANICAL EXECUTION OF WORK
725.25		ABANDONED CABLE
725.30	725.10	CLASS 1, CLASS 2, AND CLASS 3 CIRCUIT IDENTIFICATION
725.31	725.11	SAFETY-CONTROL EQUIPMENT
725.35	725.15	CLASS 1, CLASS 2, AND CLASS 3 CIRCUIT REQUIREMENTS
PART II - CLASS 1 CIRCUITS		
725.41	725.21	CLASS 1 CIRCUIT CLASSIFICATIONS AND POWER SOURCE REQUIREMENTS
725.43	725.23	CLASS 1 CIRCUIT OVERCURRENT PROTECTION
725.45	725.24	CLASS 1 CIRCUIT OVERCURRENT DEVICE LOCATION
725.46	725.25	CLASS 1 CIRCUIT WIRING METHODS
725.48	725.26	CONDUCTORS OF DIFFERENT CIRCUITS IN THE SAME CABLE, CABLE TRAY, ENCLOSURE, OR RACEWAY
725.49	725.27	CLASS 1 CIRCUIT CONDUCTORS
725.51	725.28	NUMBER OF CONDUCTORS IN CABLE TRAYS AND RACEWAY, AND DERATING
725.52	725.29	CIRCUITS EXTENDING BEYOND ONE BUILDING
PART III - CLASS 2 AND CLASS CIRCUITS		
725.121	725.41	POWER SOURCES FOR CLASS 2 AND CLASS 3 CIRCUITS
725.124	725.42	CIRCUIT MARKING
725.127	725.51	WIRING METHODS ON SUPPLY SIDE OF THE CLASS 2 OR CLASS 3 POWER SOURCE
725.130	725.52	WIRING METHODS AND MATERIALS ON LOAD SIDE OF THE CLASS 2 OR CLASS 3 POWER SOURCE
725.133	725.54	INSTALLATION OF CONDUCTORS AND EQUIPMENT IN CABLES, COMPARTMENTS, CABLE TRAYS, ENCLOSURES, MANHOLES, OUTLET BOXES, DEVICE BOXES AND RACEWAYS FOR CLASS 2 AND CLASS 3 CIRCUITS
725.136	725.55	SEPARATION FROM ELECTRIC LIGHT, POWER, CLASS 1, NON-POWER-LIMITED FIRE ALARM CIRCUIT CONDUCTORS AND MEDIUM POWER NETWORK-POWERED BROADBAND COMMUNICATIONS CABLES
725.139	725.56	INSTALLATION OF CONDUCTORS OF DIFFERENT CIRCUITS IN THE SAME CABLE, ENCLOSURE, OR RACEWAY
725.141	725.57	INSTALLATION OF CIRCUIT CONDUCTORS EXTENDING BEYOND ONE BUILDING
725.143	725.58	SUPPORT OF CONDUCTORS
725.154	725.61	APPLICATIONS OF LISTED CLASS 2, CLASS 3 AND PLTC CABLES
PART IV - LISTING REQUIREMENTS		
725.179	725.82	LISTING AND MARKING OF CLASS 2, CLASS 3 AND TYPE PLTC CABLES

CLASS 1, CLASS 2 AND CLASS 3 REMOTE-CONTROL, SIGNALING AND POWER-LIMITED CIRCUITS
725.1 through 725.179

Purpose of Change: This article has been renumbered to correlate with **Article 760**.

NEC Ch. 7 - Article 725
Part II - 725.48(B)(4)

Type of Change	Panel Action	UL	UL 508	API 500 - 1997	API 505 - 1997	OSHA - 1994
Revision	Accept in Principle	-	-	-	-	-
ROP	ROC	NFPA 70E - 2004	NFPA 70B - 2006	NFPA 79 - 2007	NFPA	NEMA
pg. 833 # 3-160	pg. 481 # 3-112	450.3(A)(3)	-	-	-	-
log: 821 CMP: 3	log: 622	Submitter: Richard P. Owen		2005 NEC: 725.26(B)(4)		IEC: -

2005 NEC - ~~725.26~~ Conductors of Different Circuits in the Same Cable, Cable Tray, Enclosure, or Raceway.

Class 1 circuits shall be permitted to be installed with other circuits as specified in ~~725.26~~(A) and 725.26(B).

(B) Class 1 Circuits with Power Supply Circuits. Class 1 circuits shall be permitted to be installed with power supply conductors as specified in ~~725.26~~(B)(1) through (B)(4).

(4) In Cable Trays. ~~In cable trays, where the Class 1 circuit conductors and power-supply conductors not functionally associated with them are separated by a solid fixed barrier of a material compatible with the cable tray, or where the power-supply or Class 1 circuit conductors are in a metal-enclosed cable.~~

2008 NEC - 725.48 Conductors of Different Circuits in the Same Cable, Cable Tray, Enclosure, or Raceway.

Class 1 circuits shall be permitted to be installed with other circuits as specified in 725.48(A) and (B).

(B) Class 1 Circuits with Power Supply Circuits. Class 1 circuits shall be permitted to be installed with power supply conductors as specified in 725.48(B)(1) through (B)(4).

(4) In Cable Trays. Installations in cable trays shall comply with 725.48(B)(4)(1) or (B)(4)(2).

(1) Class 1 circuit conductors and power-supply conductors not functionally associated with the Class 1 circuit conductors shall be separated by a solid fixed barrier of a material compatible with the cable tray.

(2) Class 1 circuit conductors and power-supply conductors not functionally associated with the Class 1 circuit conductors shall be permitted to be installed in a cable tray without barriers where all of the conductors are installed with separate multiconductor Type AC, Type MC, Type MI, or Type TC cables and all the conductors in the cables are insulated at 600 volts.

Author's Comment: In **725.48(B)**, Class 1 circuit conductors can be installed with power conductors where functionally associated, where installed in factory- or field-assembled control centers, in a manhole with various wiring methods. Class 1 circuit conductors and non-functionally associated power conductors installed in a cable tray must be separated by a solid fixed barrier of a material compatible with the cable tray, or the power-supply conductors or Class 1 circuit conductors must be installed in a metal-enclosed cable. The issue at hand was potential damage to power and Class 1 circuit conductors thus affecting the operation and safety of the other non-related circuits. In the newly added text, multiconductor Type AC, Type MC, Type MI or Type TC cables with 600 volt insulation can provide the separation of different wiring methods in a cable tray without making sweeping changes to the remainder of this section. In the Comment, the word "separate" was added after "with" and before "multiconductor" to ensure that where the Class 1 circuits and the power supply conductors are not functionally associated, separate cables will be installed, one for the Class 1 and one for the power-supply conductors. The text has been rearranged for clarity.

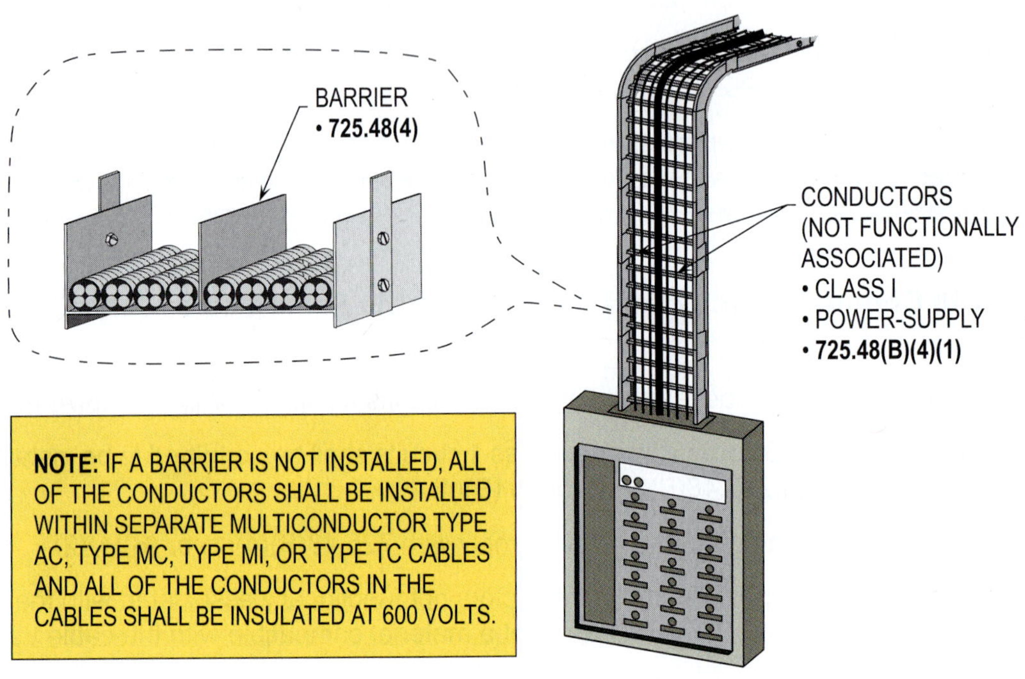

BARRIER
• 725.48(4)
CONDUCTORS
(NOT FUNCTIONALLY
ASSOCIATED)
• CLASS I
• POWER-SUPPLY
• 725.48(B)(4)(1)
NOTE: IF A BARRIER IS NOT INSTALLED, ALL OF THE CONDUCTORS SHALL BE INSTALLED WITHIN SEPARATE MULTICONDUCTOR TYPE AC, TYPE MC, TYPE MI, OR TYPE TC CABLES AND ALL OF THE CONDUCTORS IN THE CABLES SHALL BE INSULATED AT 600 VOLTS.

IN CABLE TRAYS
725.48(B)(4)

Purpose of Change: This revision clarifies the permitted methods of installation for Class I circuit conductors and power-supply conductors that are not functionally associated.

NEC Ch. 7 - Article 760
Parts I through IV - 760.1 through 760.179

Type of Change		Panel Action		UL	UL 508	API 500 - 1997	API 505 - 1997	OSHA - 1994
Renumbered		Accept		864	-	-	-	1910.308(d)
ROP		ROC		NFPA 70E - 2004	NFPA 70B - 2006	NFPA 79 - 2007	NFPA	NEMA - 2003
pg. 849	# 3-211	pg. -	# -	450.4	-	-	-	SB 28
log: 842	CMP: 3	log: -	Submitter: Richard P. Owen			2005 NEC: Article 760		IEC: -

2008 NEC - Article 760
Fire Alarm Systems

I. General

II. Non–Power-Limited Fire Alarm (NPLFA) Circuits

III. Power-Limited Fire Alarm (PLFA) Circuits

IV. Listing Requirements

Author's Comment: The number changes within **Articles 725** and **760** correlate to the numbering of the two articles, as well as a similar numbering sequence for those related sections in Chapter 8 for overall correlation, while leaving adequate room between sections for future additions. It was impossible to renumber the two articles exactly, since some sections could not be moved without changing their intent. One example of this is newly renumbered Sections **725.52** covering Class 1 circuits extending beyond one building and **760.32** covering both power limited fire alarm and non-power-limited fire alarm circuits extending beyond one building, both addressing the same issue, but the numbering could not be the same without changing the intent of these sections due to their location in their respective articles.

CROSS-REFRENCE		
PART I - GENERAL		
2008 NEC	**2005 NEC**	**SECTION HEADING**
760.1	**760.1**	SCOPE
760.2	**760.2**	DEFINITIONS
760.3	**760.3**	OTHER ARTICLES
760.21	**760.7**	ACCESS TO ELECTRICAL EQUIPMENT BEHIND PANELS DESIGNED TO ALLOW ACCESS
760.24	**760.8**	MECHANICAL EXECUTION OF WORK
760.25		ABANDONED CABLE
760.26	**760.9**	FIRE ALARM CIRCUIT AND EQUIPMENT GROUNDING
760.30	**760.10**	FIRE ALARM CIRCUIT IDENTIFICATION
760.32	**760.11**	FIRE ALARM CIRCUITS EXTENDING BEYOND ONE BUILDING
760.35	**760.15**	FIRE ALARM CIRCUIT REQUIREMENTS
PART II - NON-POWER-LIMITED FIRE ALARM (NPLFA) CIRCUITS		
760.41	**760.21**	NPLFA CIRCUIT POWER SOURCE REQUIREMENTS
760.43	**760.23**	NPLFA CIRCUIT OVERCURRENT PROTECTION
760.45	**760.24**	NPLFA CIRCUIT OVERCURRENT DEVICE LOCATION
760.46	**760.25**	NPLFA CIRCUIT WIRING
760.48	**760.26**	CONDUCTORS OF DIFFERENT CIRCUITS IN SAME CABLE, ENCLOSURE, OR RACEWAY
760.49	**760.27**	NPLFA CIRCUIT CONDUCTORS
760.51	**760.28**	NUMBER OF CONDUCTORS IN CABLE TRAYS AND RACEWAYS, AND DERATING
760.53	**760.30**	MULTICONDUCTOR NPLFA CABLES

FIRE ALARM SYSTEMS
ARTICLE 760

PART III - POWER-LIMITED FIRE ALARM (PLFA) CIRCUITS		
760.121	**760.41**	POWER SOURCES FOR PLFA CIRCUITS
760.124	**760.42**	CIRCUIT MARKING
760.127	**760.51**	WIRING METHODS ON SUPPLY SIDE OF THE PLFA POWER SOURCE
760.130	**760.52**	WIRING METHODS AND MATERIALS ON LOAD SIDE OF THE PLFA POWER SOURCE
760.133	**760.54**	INSTALLATION OF CONDUCTORS AND EQUIPMENT IN CABLES, COMPARTMENTS, CABLE TRAYS, ENCLOSURES, MANHOLES, OUTLET BOXES, AND RACEWAYS FOR POWER-LIMITED CIRCUITS
760.136	**760.55**	SEPARATION FROM ELECTRIC LIGHT, POWER, CLASS 1, NPLFA AND MEDIUM-POWER NETWORK-POWERED BROADBAND COMMUNICATIONS CIRCUIT CONDUCTORS
760.139	**760.56**	INSTALLATION OF CONDUCTORS OF DIFFERENT PLFA CIRCUITS, CLASS 2, CLASS 3 AND COMMUNICATIONS CIRCUITS IN THE SAME CABLE, CABLE TRAY, OR RACEWAY
760.142	**760.57**	CONDUCTOR SIZE
760.143	**760.58**	SUPPORT OF CONDUCTORS
760.145	**760.59**	CURRENT-CARRYING CONTINUOUS LINE-TYPE FIRE DETECTORS
760.154	**760.61**	APPLICATIONS OF LISTED PLFA CABLES
PART IV - LISTING REQUIREMENTS		
760.176	**760.81**	LISTING AND MARKING OF NPLFA CABLES
760.179	**760.82**	LISTING AND MARKING OF PLFA CABLES SND INSULATED CONTINUOUS LINE-TYPE FIRE DETECTORS

FIRE ALARM SYSTEMS
ARTICLE 760

Purpose of Change: This article has been renumbered to correlate with **Article 725**.

NEC Ch. 7 Article 770
Parts I through VI - 770.1 through 770.182

Type of Change		Panel Action		UL	UL 508	API 500 - 1997	API 505 - 1997	OSHA - 1994
Renumbered		Accept in Principle		1651 - 2024	-	-	-	-
ROP		ROC		NFPA 70E - 2004	NFPA 70B - 2006	NFPA 79 - 2007	NFPA	NEMA
pg. 868	# 16-2	pg. -	# -	-	-	-	-	-
log: 695	CMP: 16	log: -	Submitter: S. D. Kahn			2005 NEC: Article 770		IEC: -

2008 NEC - Article 770
Optical Fiber Cables and Raceways.

I. General

II. Cables Outside and Entering Buildings

III. Protection.

IV. Grounding Methods.

V. Installation Methods Within Buildings.

VI. Listing Requirements.

Author's Comment: The number of Parts within **Article 770** has been expanded in anticipation of having added sections.

ARTICLE 770 - OPTICAL FIBER CABLES AND RACEWAYS	
PART I - GENERAL	
SECTION	**HEADING**
770.1 **770.2** **770.3** **770.6** **770.12** **770.21** **770.24** **770.25** **770.26**	SCOPE DEFINITIONS OTHER ARTICLES OPTICAL FIBER CABLES INNERDUCT FOR OPTICAL FIBER CABLES ACCESS TO ELECTRICAL EQUIPMENT BEHIND PANELS DESIGNED TO ALLOW ACCESS MECHANICAL EXECUTION OF WORK ABANDONED CABLES SPREAD OF FIRE OR PRODUCTS OF COMBUSTION
PART II - CABLES OUTSIDE AND ENTERING BUILDINGS	
770.48	UNLISTED CABLES AND RACEWAYS ENTERING BUILDINGS
PART III - PROTECTION	
770.93	GROUNDING OR INTERRUPTION OF NON-CURRENT-CARRYING METALLIC MEMBERS OF OPTICAL FIBER CABLES
PART IV - GROUNDING METHODS	
770.100 **770.101** **770.106**	ENTRANCE CABLE GROUNDING GROUNDING GROUNDING OF ENTRANCE CABLES AT MOBILE HOMES
PART V - INSTALLATION METHODS WITHIN BUILDINGS	
770.110 **770.113** **770.733** **770.154**	RACEWAYS FOR OPTICAL FIBER CABLES INSTALLATION AND MARKING OF LISTED OPTICAL FIBER CABLES INSTALLATION OF OPTICAL FIBERS AND ELECTRICAL CONDUCTORS APPLICATIONS OF LISTED OPTICAL FIBER CABLES AND RACEWAYS
PART VI - LISTING REQUIREMENTS	
770.179 **770.182**	OPTICAL FIBER CABLES OPTICAL FIBER RACEWAYS

OPTICAL FIBER CABLES AND RACEWAYS
770.1 through 770.182

Purpose of Change: The number of parts within **Article 770** has been explained in anticipation of having added sectors.

8

Communications Systems

Chapter 8 of the NEC covers communications systems. Articles and sections of this chapter stand alone and are independent from chapters 1 through 7, except in specific cases where they are referenced. Communications systems include such systems as telephone, cable TV, radio, and broadband communications.

The rules and regulations in this article are mainly applied to those systems that are connected to a central station and operate as elements of such systems. When communications are involved, one of the Articles in the 800 series must be selected, based upon the system utilized.

Chapter 8 also includes **Article 810**, which deals with radio and television and receiving equipment and amateur radio transmitting and receiving equipment, but not equipment and antenna used for coupling carrier current to power line conductors.

Article 820 of **Chapter 8** covers coaxial cable distribution of radio frequency signals typically employed in community antenna television (CATV) systems.

Article 830 covers network-powered broadband commuications where a carrier frequency has multiple signals impressed on the carrier. At the network interface unit, the signals are converted into individual signals and then distributed for phone, TV burglar alarm, and other similar uses.

NEC Ch. 8 - Article 800
Part I - 800.2

Type of Change	Panel Action	UL	UL 508	API 500 - 1997	API 505 - 1997	OSHA - 1994
Revision	Accept	-	-	-	-	1910.308(e)
ROP	**ROC**	**NFPA 70E - 2004**	**NFPA 70B - 2006**	**NFPA 79 - 2007**	**NFPA**	**NEMA - 2003**
pg. 902 # 16-98	pg. 523 # 16-80	450.5	-	-	-	SB 28
log: 651 CMP: 16	log: 1534	Submitter: Technical Correlating Committee on NEC		2005 NEC: 800.1 and 800.2		IEC: -

2005 NEC - 800.2 Definitions.

See Article 100. For purposes of this article, the following additional definitions apply.

Communications Circuit. The circuit that extends voice, audio, video, interactive services, telegraph (except radio), ~~and~~ outside wiring for fire alarm and burglar alarm from the communications utility to the customer's communications equipment.

2008 NEC - 800.2 Definitions.

See Article 100. For purposes of this article, the following additional definitions apply.

Communications Circuit. The circuit that extends voice, audio, video, data, interactive services, telegraph (except radio), outside wiring for fire alarm and burglar alarm from the communications utility to the customer's communications equipment up to and including terminal equipment such as a telephone, fax machine or answering machine.

Author's Comment: The Scope of **Article 800** was changed to resolve a scope and correlation issue with **Article 725**. The term "telephone" implies a single, limited medium for the transmission of voice that is no longer valid in today's complex world of telecommunications. "Telephone" has evolved to the point where it is a communications system transporting information in various forms including voice, data, audio, video, and interactive services, and using varied technologies including copper wire, coaxial cable, optical fiber, and radio links, as well as high frequency carrier systems and advanced data processing and switching techniques. The revision is needed to convey the concept of modern-day telecommunications to the user of the NEC. The addition of the definition of "communications circuit" helps clarify the scope as covering communications services and equipment provided by a communications utility, including the associated services. Since the text in the 2005 NEC is unclear and may be construed as not including all premises communications wiring and equipment, such as a telephone or fax machine, this clarification has been added by the comment to the definition for communications circuit. For example, if a customer has a local telecommunications switch or PBX, the communications circuit would include not only the PBX, but all communications cabling and wiring connecting the customer's terminal equipment to the PBX.

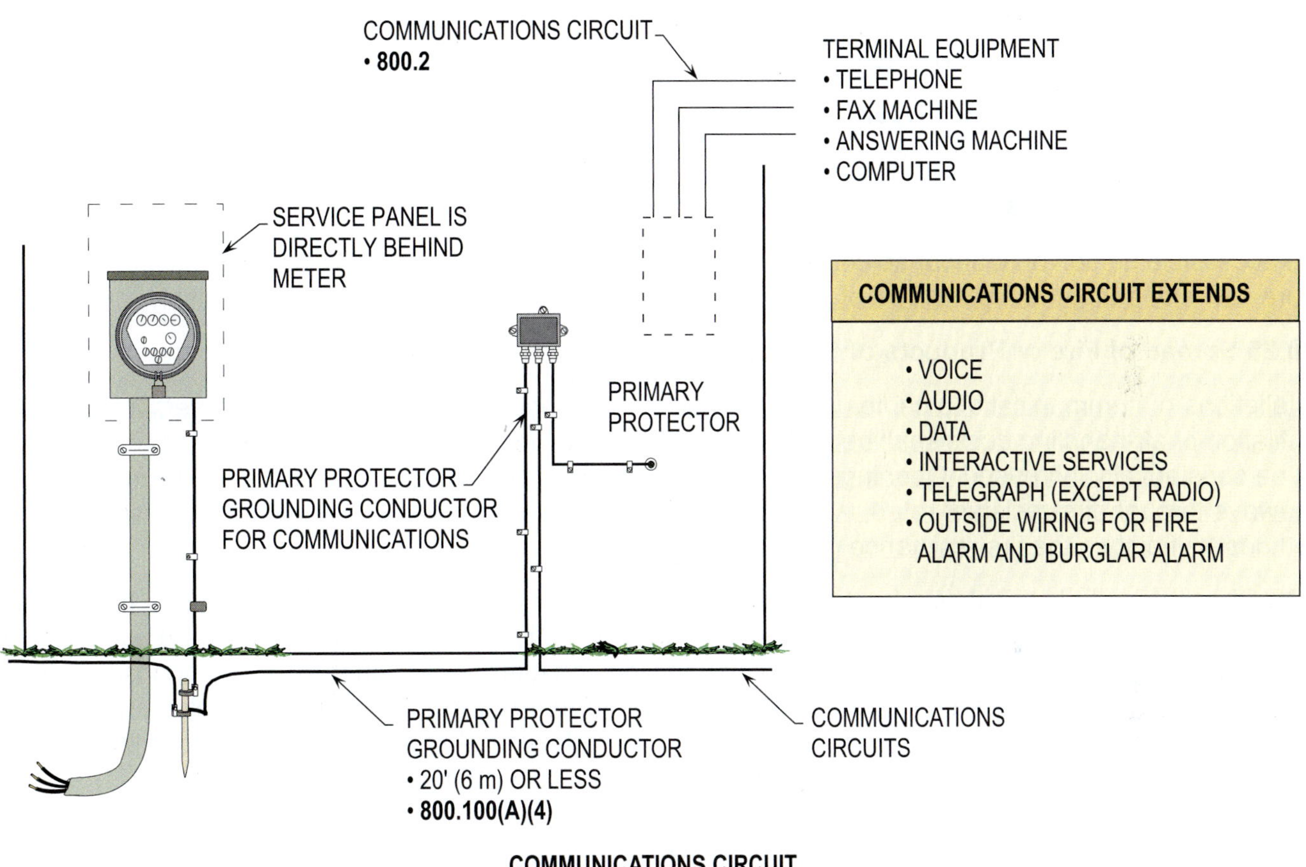

COMMUNICATIONS CIRCUIT
800.2

Purpose of Change: This revision clarifies that the definition of "communications circuits" covers communications services and equipment provided by a communications utility, including the associated services.

NEC Ch. 8 - Article 800
Part I - 800.25 and 800.26

Type of Change	Panel Action	UL	UL 508	API 500 - 1997	API 505 - 1997	OSHA - 1994
New Section	Accept in Principle	1863 - 494	-	-	-	-
ROP	**ROC**	**NFPA 70E - 2004**	**NFPA 70B - 2006**	**NFPA 79 - 2007**	**NFPA**	**NEMA**
pg. 908 # 16-128	pg. - # -	-	-	-	-	-
log: 728 CMP: 16	log: - Submitter: S.D. Kahn			2005 NEC: 800.3(C) and (D)		IEC: -

2005 NEC - 800.3 Other Articles

~~(C)~~ Spread of Fire or Products of Combustion. ~~Section 300.21 shall apply.~~ The accessible portion of abandoned communications cables shall ~~not be permitted to remain.~~

~~**(D) Equipment in Other Space Used for Environmental Air.** Section 300.22(C) shall apply.~~

2008 NEC - 800.25 Abandoned Cables.

The accessible portion of abandoned communications cables shall be removed. Where cables are identified for future use with a tag, the tag shall be of sufficient durability to withstand the environment involved.

800.26 Spread of Fire or Products of Combustion.

Installations of communications cables and communications raceways in hollow spaces, vertical shafts, and ventilation or air-handling ducts shall be made so that the possible spread of fire or products of combustion will not be substantially increased. Openings around penetrations of communications cables and communications raceways through fire-resistant-rated walls, partitions, floors, or ceilings shall be firestopped using approved methods to maintain the fire resistance rating.

FPN: Directories of electrical construction materials published by qualified testing laboratories contain many listing installation restrictions necessary to maintain the fire-resistive rating of assemblies where penetrations or openings are made. Building codes also contain restrictions on membrane penetrations on opposite sides of a fire-resistance-rated wall assembly. An example is the 600-mm (24-in.) minimum horizontal separation that usually applies between boxes installed on opposite sides of the wall. Assistance in complying with 800.26 can be found in building codes, fire resistance directories, and product listings.

Author's Comment: The title of **800.3** is "Other Articles." The requirement for the removal of abandoned cables is not in another article; it is in **Article 800**. It is out of place in **800.3** so it was moved it to a new section of **Article 800** and made editorially consistent with **Articles 770** and **820** by substituting "shall be removed" for "shall not be permitted to remain." Rather than refer to **300.21** for requirements for the prevention of the spread of fire, it is better to have the requirements in **Article 800**, which will be more familiar and readily accessible to communications installers. The text of new **800.26** is based on **300.21** but modified to apply to communications cables and raceways.

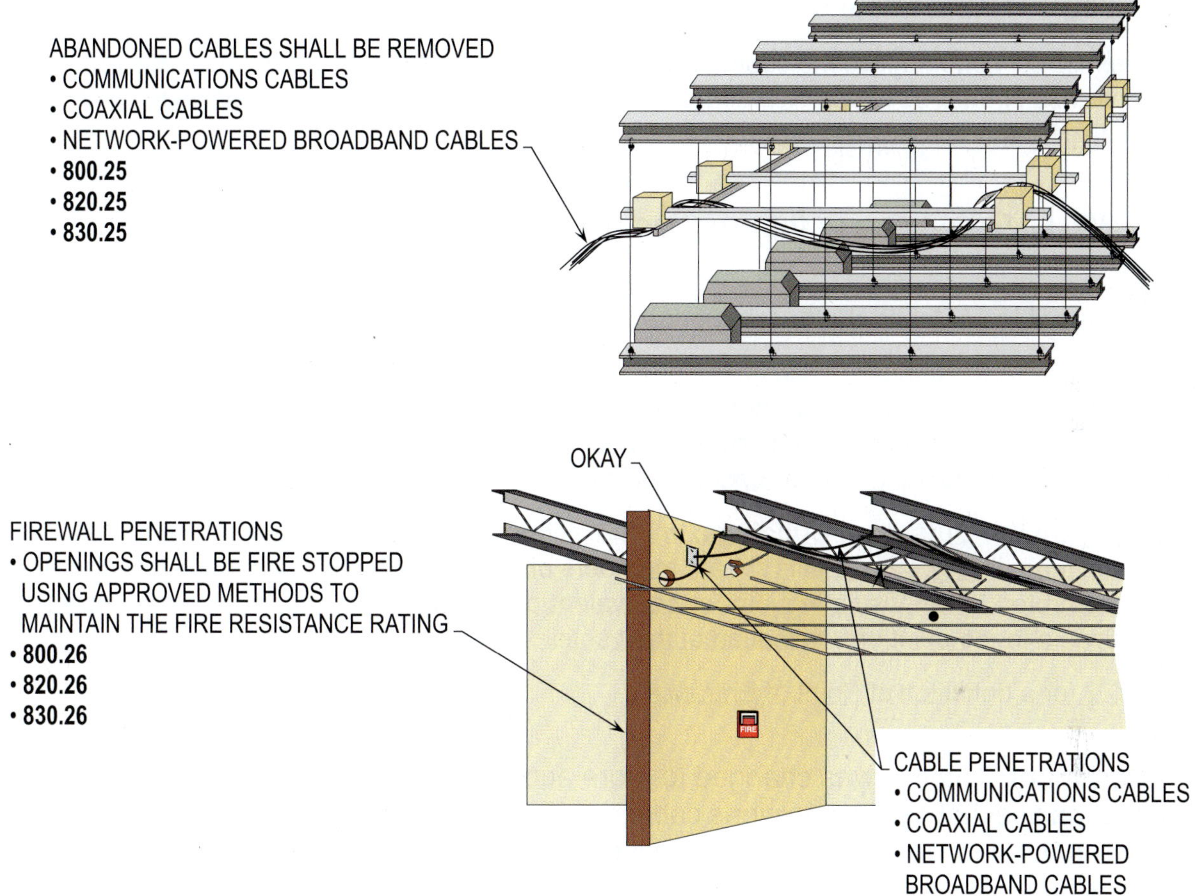

ABANDONED CABLES AND SPREAD OF FIRE OR PRODUCTS OF COMBUSTION 800.25 and 800.26

Purpose of Change: A new section has been added to address the requirements for the prevention of the spread of fire and one has been added dealing with abandoned cables.

NEC Ch. 8 - Article 800
Part III - 800.93

Type of Change	Panel Action	UL	UL 508	API 500 - 1997	API 505 - 1997	OSHA - 1994
Revision	Accept	-	-	-	-	1910.308(e)(5)
ROP	**ROC**	**NFPA 70E - 2004**	**NFPA 70B - 2006**	**NFPA 79 - 2007**	**NFPA**	**NEMA**
pg. 916 # 16-158	pg. 534 # 16-126	450.5(D)	-	-	-	-
log: 736 CMP: 16	log: 1541 Submitter: S.D. Kahn			2005 NEC: -		IEC: -

2005 NEC - 800.93 ~~Cable~~ Grounding.

~~The metallic sheath of communications cables entering buildings shall be grounded as close as practicable to the point of entrance or shall be interrupted as close to the point of entrance as practicable by an insulating joint or equivalent device.~~

2008 NEC - 800.93 Grounding or Interruption of Metallic Sheath Members of Communications Cables.

Communications cables entering the building or terminating on the outside of the building shall comply with 800.93(A) or (B).

(A) Entering Buildings. In installations where the communications cable enters a building, the metallic sheath members of the cable shall be either grounded as specified in 800.100 or interrupted by an insulating joint or equivalent device. The grounding or interruption shall be as close as practicable to the point of entrance.

(B) Terminating On the Outside of Buildings. In installations where the communications cable is terminated on the outside of the building, the metallic sheath members of the cable shall be either grounded as specified in 800.100, or interrupted by an insulating joint or equivalent device. The grounding or interruption shall be as close as practicable to the point of termination of the cable.

FPN: See 800.2 for a definition of Point of Entrance.

Author's Comment: The title was changed to more accurately reflect the text within the section. The Comment recognizes that communications cable does not always enter the building. In many cases, it is terminated on the outside of the building in a Network Interface Device (NID) or similar device. The Comment text provides editorial clarification by addressing the two scenarios, cables entering the building and cables terminating on the outside of the building, separately. Furthermore, the revised text eliminates any confusion that may result from the use of the terms "attached" and "point of attachment", and addresses the issue directly: cables that are terminated on the outside of the building.

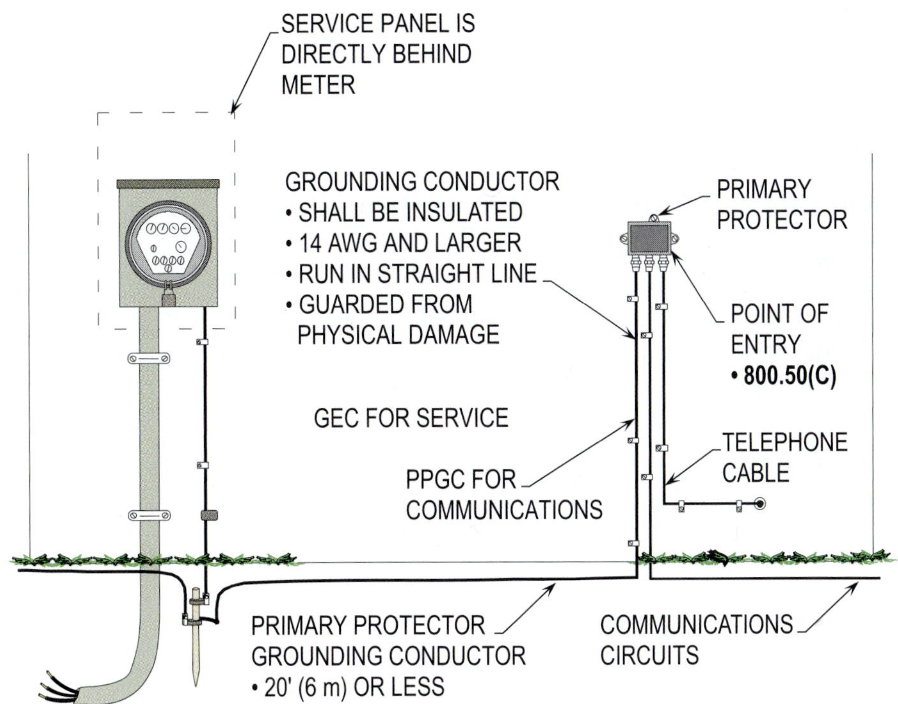

GROUNDING OR INTERRUPTION OF METALLIC SHEATH MEMBERS OF COMMUNICATIONS CABLES 800.93

Purpose of Change: This revision clarifies the requirements for grounding or interruption of metallic sheath members of communications cables entering buildings or terminating on the outside of the building.

NEC Ch. 8 - Article 800
Part IV - 800.100(A)(6)

Type of Change	Panel Action	UL	UL 508	API 500 - 1997	API 505 - 1997	OSHA - 1994
Revision	Accept	497	-	-	-	1910.308(e)(5)(iii)
ROP	**ROC**	**NFPA 70E - 2004**	**NFPA 70B - 2006**	**NFPA 79 - 2007**	**NFPA**	**NEMA**
pg. 918 # 16-166	pg. - # -	450.5(D)(1)(f)	-	-	-	-
log: 858 CMP: 16	log: - Submitter: Michael J. Johnston			2005 NEC: 800.100(A)(6)		IEC: -

2005 NEC - 800.100 Cable and Primary Protector Grounding.

(A) Grounding Conductor.

(6) Physical ~~Damage~~. ~~Where necessary,~~ the grounding conductor shall be ~~guarded from~~ physical damage. Where the grounding conductor is run in a metal raceway, both ends of the raceway shall be bonded to the grounding conductor or the same terminal or electrode to which the grounding conductor is connected.

2008 NEC - 800.100 Cable and Primary Protector Grounding.

(A) Grounding Conductor.

(6) Physical Protection. The grounding conductor shall be protected where exposed to physical damage. Where the grounding conductor is run in a metal raceway, both ends of the raceway shall be bonded to the grounding conductor or the same terminal or electrode to which the grounding conductor is connected.

Author's Comment: The metallic member(s) of the cable sheath, where required to be grounded by 800.93, and primary protectors shall be grounded as specified in **800.100(A) through 800.100(D)**.

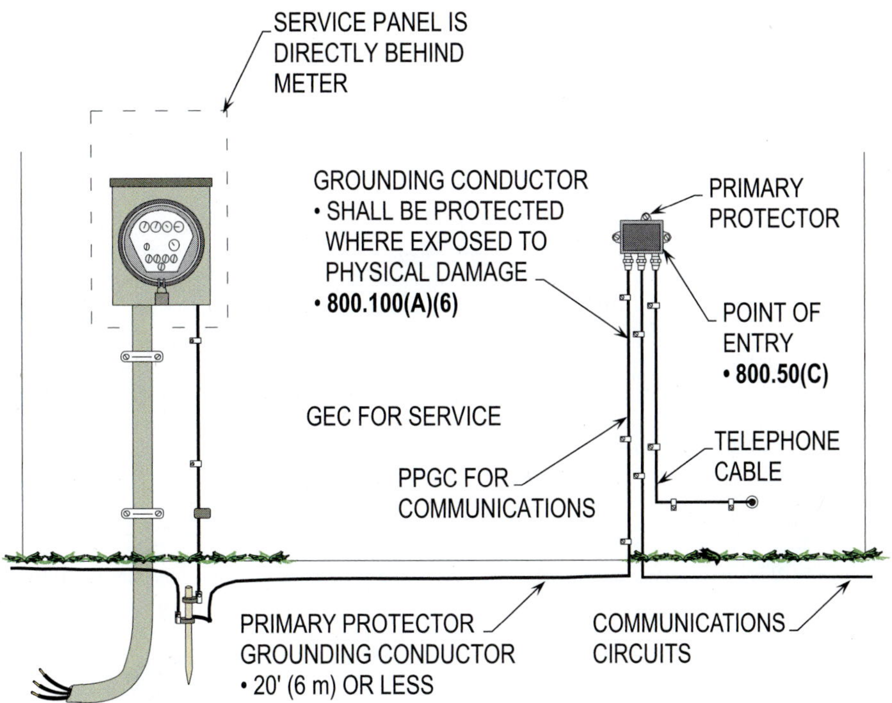

PHYSICAL PROTECTION
800.100(A)(6)

Purpose of Change: This revision clarifies that grounding conductors shall be protected where exposed to physical damage.

NEC Ch. 8 - Article 800
Part V - 800.156

Type of Change	Panel Action	UL	UL 508	API 500 - 1997	API 505 - 1997	OSHA - 1994
New Section	Accept in Principle in Part	1459	-	-	-	-
ROP	ROC	NFPA 70E - 2004	NFPA 70B - 2006	NFPA 79 - 2007	NFPA	NEMA
pg. 931 # 16-207	pg. - # -	-	-	-	-	-
log: 2655 CMP: 16	log: - Submitter: Robert W. Jensen			2005 NEC: -		IEC: -

2008 NEC - 800.156 Dwelling Unit Communications Outlet.

For new construction, a minimum of one communications outlet shall be installed within the dwelling and cabled to the service provider demarcation point.

Author's Comment: This change now requires a minimum of one communications outlet to be installed within a dwelling unit in new construction and cabled to the service provider demarcation point.

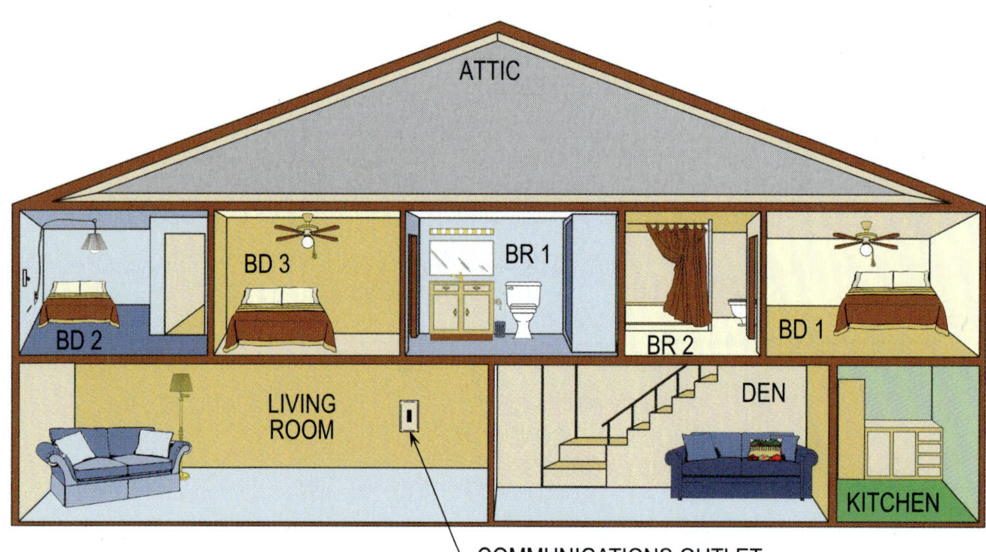

DWELLING UNIT COMMUNICATIONS OUTLET
800.156

Purpose of Change: A new section has been added to require a minimum of one communications outlet within a dwelling unit for new construction.

Name nec® Date

Questions

Section Answer

1. Branch circuit overcurrent devices are provided with interrupting ratings appropriate for the intended use. The interrupting ratings for the overcurrent protective devices shall not be less than how much ampacity? ________ ______

A. 10,000 amperes

B. 5,000 amperes

C. 25,000 amperes

D. 1,000 amperes

2. A branch-circuit overcurrent device is a device capable of providing protection for what type of circuits and equipment over the full range of overcurrents between its rated current and interrupting rating? ________ ______

A. Service, feeder, and branch circuits

B. Service and feeder circuits

C. Feeder and branch circuits

D. Branch circuits

3. A clothes closet is: ________ ______

A. A habitable room or space intended for storage of garments and apparel.

B. A non-habitable room or space intended only for storage of garments.

C. A habitable room or space intended only for storage of garments.

D. A non-habitable room or space intended primarily for storage of garments and apparel.

4. A qualified person is a person: ________ ______

A. Who has skills and knowledge related to the construction and operation of the electrical equipment and installations and has received safety training on the hazards involved.

B. Who has skills and knowledge related to the construction and operation of the electrical equipment and installations and has received safety training to recognize the hazards involved.

C. Who has skills and knowledge related to the construction and operation of the electrical equipment and installations and has received safety training to recognize and avoid the hazards involved.

D. Who has skills and knowledge related to the construction and operation of the electrical equipment and installations and has received safety training to avoid the hazards involved.

Questions

Section	Answer	

_______ _______ **5.** The prospective symmetrical fault current at a nominal voltage to which an apparatus or system can be connected without sustaining damage exceeding defined acceptance criteria is:

A. Short-circuit current rating.

B. Ground-fault current rating.

C. Arc-fault current rating.

D. Ground-fault and arc-fault current rating.

_______ _______ **6.** Short-Circuit Current Rating is the prospective symmetrical fault current at a nominal voltage:

A. To which an apparatus or system is able to be connected with sustaining damage exceeding defined acceptance criteria.

B. To which an apparatus or system is able to be connected without sustaining damage exceeding defined acceptance criteria.

C. To which an apparatus or system is able to be connected with sustaining damage.

D. To which an apparatus or system is able to be connected without sustaining damage exceeding defined acceptance criteria.

_______ _______ **7.** Based on 110.12(A), unused openings shall be closed to afford protection substantially equivalent to the wall of the equipment with the following exceptions (pick the most correct):

A. Those for the operation of equipment, those intended for mounting purposes or those permitted as part of the design for listed equipment.

B. Those for the operation of equipment, those intended for mounting purposes or those permitted as part of the design for approved equipment.

C. Those for the operation of equipment, those intended for cooling purposes or those permitted as part of the design for listed equipment.

D. Those for the operation of equipment, those intended for drainage purposes or those permitted as part of the design for listed equipment.

_______ _______ **8.** Based on 110.12(A), where metallic plugs or plates are used with nonmetallic enclosures, the metallic plugs or plates shall be recessed at least what distance from the outer surface of the enclosure?

A. 12.7 mm (1/2 in.)

B. 24.4 mm (1 in.)

C. 6 mm (1/4 in.)

D. 50.8 mm (2 in.)

Questions

Section Answer

9. What electrical equipment, in other than dwelling occupancies, that are likely required to be examined, adjusted, serviced, or maintained while energized, must be field marked to warn qualified persons of potential electric arc flash hazards? ________ ____

A. Switchboards

B. Motor control centers

C. Fusible switch

D. All the above.

10. Equipment subject to an electric arc flash hazard shall have marking: ________ ____

A. Located to be clearly visible to qualified persons before examination, adjustment, servicing, or maintenance of the equipment.

B. Located to be clearly visible to qualified persons after examination, adjustment, servicing, or maintenance of the equipment.

C. Located to be clearly visible to unqualified persons before examination, adjustment, servicing, or maintenance of the equipment.

D. Located to be clearly visible to all persons before examination, adjustment, servicing, or maintenance of the equipment.

11. An enclosure for switchboards, panelboards, industrial control panels, motor control centers, meter sockets, and motor controllers, rated not over 600 volts nominal, intended for an outdoor location and where subjected to prolonged submersion, shall be marked with what Enclosure Type number, based on Table 110.20? ________ ____

A. Type 6

B. Type 3RX

C. Type 6P

D. Type 3R

12. An enclosure for switchboards, panelboards, industrial control panels, motor control centers, meter sockets, and motor controllers, rated not over 600 volts nominal, intended for an outdoor location and where subjected to oil or coolant spraying or splashing, shall be marked with what Enclosure Type number, based on Table 110.20? ________ ____

A. Type 4X

B. Type 13

C. Type 12

D. Type 1

Questions

Section Answer

__________ _______ **13.** The term "raintight" is used in conjunction with what enclosure types?

A. Enclosure Types 3, 3S, 3SX, 3X, 4, 4X, 6, 6P

B. Enclosure Types 3, 3S, 3SX, 3X, 5, 12, 12K, 13

C. Enclosure Types 2, 5, 6, 6P, 3SX, 12, 12K, 13

D. Enclosure Types 2, 3, 3S, 3SX, 3X, 4, 4X, 6, 6P

__________ _______ **14.** The enclosures in Section 110.20 are not intended to protect against what conditions that may occur within the enclosure or enter via the conduit or unsealed openings?

A. Condensation and icing only.

B. Incidental contact with enclosed equipment.

C. Condensation, icing, corrosion, or contamination.

D. Rain, snow and sleet.

__________ _______ **15.** Which of the following is the most accurate statement?

A. Electrical equipment rated 1200 amperes or more that contain overcurrent devices, switching devices, or control devices, shall have one entrance to the required working space not less than 610 mm (24 in.) wide and 2.0 m (6 1/ 2 ft) high at each end of the working space.

B. Electrical equipment rated 1200 amperes or more and over 1.8 m (6 ft.) wide that contain overcurrent devices, switching devices, or control devices, shall have one entrance to the required working space not less than 610 mm (24 in.) wide and 2.0 m (6 1/ 2 ft) high at each end of the working space.

C. Electrical equipment rated 1000 amperes or more and over 1.8 m (6 ft.) wide that contain overcurrent devices, switching devices, or control devices, shall have one entrance to the required working space not less than 610 mm (24 in.) wide and 2.0 m (6 1/ 2 ft) high at each end of the working space.

D. Electrical equipment rated 1000 amperes or more and over 2.5 m (8 ft.) wide that contain overcurrent devices, switching devices, or control devices, shall have one entrance to the required working space not less than 610 mm (24 in.) wide and 2.0 m (6 1/ 2 ft) high at each end of the working space.

__________ _______ **16.** Personnel doors, intended for entrance to and egress from the working space and located within what distance from the working space, must have the door open in the direction of egress and must be equipped with a panic bar, pressure plate, or other device that is normally latched but opens under simple pressure?

A. 6.0 feet

B. 12.5 feet

C. 25.0 feet

D. 8.0 feet

Questions

	Section	Answer

17. The continuity of a grounded conductor shall: __________ ______

A. Depend upon the connection to a metallic enclosure, raceway, or cable armor.

B. Not depend upon the connection to a metallic enclosure, raceway, or cable armor.

C. Only depend upon the connection to a metallic enclosure in a service enclosure.

D. Depend upon the connection to a metallic enclosure of a panelboard supplied by a feeder circuit.

18. Where the premises wiring system has branch circuits supplied from more than one nominal voltage system, each ungrounded conductor of a branch circuit, where accessible, shall be identified by the following: __________ ______

A. Phase or line and system at all pull boxes.

B. Phase or line and system at all termination, connection, and splice points.

C. Phase or line and system at all termination, connection, splice points, and pull boxes.

D. Phase or line and system only at panelboards.

19. In a dwelling unit garage, a receptacle mounted in the ceiling of the garage for a garage door opener: __________ ______

A. Must be readily accessible and requires GFCI protection.

B. Must be accessible and does not require GFCI protection.

C. Must be accessible and requires GFCI protection.

D. Must not be accessible and requires GCI protection.

20. All 120-volt, single phase, 15- and 20-ampere branch circuits supplying outlets installed in dwelling unit in the following locations must be arc-fault circuit-interrupter protected: __________ ______

A. Family rooms, dining rooms, living rooms, parlors, libraries, dens, bedrooms, sun rooms, recreation rooms, closets, hallways, or similar rooms or areas.

B. Family rooms, dining rooms, kitchens, parlors, libraries, dens, bedrooms, sun rooms, recreation rooms, closets, hallways, or similar rooms or areas.

C. Family rooms, dining rooms, living rooms, parlors, libraries, dens, bedrooms, sun rooms, recreation rooms, hallways, or similar rooms or areas.

D. Family rooms, dining rooms, living rooms, parlors, libraries, dens, laundry rooms, sun rooms, recreation rooms, closets, hallways , or similar rooms or areas.

Questions

Section Answer

_______ _______ **21.** Grounded conductors that are not connected to an overcurrent device shall be permitted to be sized at what percentage of a continuous and non-continuous load?

A. 125 percent

B. 80 percent

C. 90 percent

D. 100 percent

_______ _______ **22.** Balconies, decks, and porches that are accessible from inside a dwelling unit:

A. Must have at least one receptacle outlet installed within the perimeter of the balcony, deck, or porch with a usable area of more than 1.86 m^2 (20 ft^2). The receptacle shall not be located more than 2.0 m (6 Ω ft) above the balcony, deck, or porch surface.

B. Must have at least two receptacle outlets installed within the perimeter of the balcony, deck, or porch with a usable area of more than 1.86 m^2 (20 ft^2). These receptacles shall not be located more than 2.0 m (6 Ω ft) above the balcony, deck, or porch surface.

C. Must have at least one receptacle outlet installed within the perimeter of the balcony, deck, or porch with a usable area of more than 1.86 m^2 (20 ft^2). The receptacle shall not be located more than 2.5 m (8 ft) above the balcony, deck, or porch surface.

D. Must have at least one receptacle outlet installed within the perimeter of the balcony, deck, or porch_with a usable area of less than 1.86 m^2 (20 ft^2). The receptacle shall not be located more than 2.0 m (6 Ω ft) above the balcony, deck, or porch surface.

_______ _______ **23.** For field installations where the length of transformer secondary conductors do not exceed 3 m (10 ft) and leave the enclosure or vault in which the supply connection is made, the rating of the overcurrent device protecting the primary of the transformer, multiplied by the primary to secondary transformer voltage ratio, shall not exceed:

A. Five times the ampacity of the secondary conductors.

B. Ten times the ampacity of the secondary conductors.

C. Fifteen times the ampacity of the secondary conductors.

D. Twenty times the ampacity of the secondary conductors.

Questions

Section | Answer

24. Grounding conductors and bonding jumpers shall be connected by one of the following means: ____________ ________

A. Listed pressure connectors; terminal bars; pressure connectors listed as grounding and bonding equipment; the exothermic welding process; machine screw-type fasteners that engage not less than two threads or are secured with a nut; thread-forming machine screws that engage not less than two threads in the enclosure; connections that are part of a listed assembly; or other listed means.

B. Listed pressure connectors; terminal bars; pressure connectors listed as grounding and bonding equipment; the exothermic welding process; machine screw-type fasteners that engage not less than two threads or are secured with a nut; thread-forming machine screws that engage not less than two threads in the enclosure; sheet metal screws; connections that are part of a listed assembly; or other listed means.

C. Approved pressure connectors; terminal bars; pressure connectors approved as grounding and bonding equipment; the exothermic welding process; machine screw-type fasteners that engage not less than two threads or are secured with a nut; thread-forming machine screws that engage not less than two threads in the enclosure; connections that are part of a approved assembly; or other approved means.

D. Listed pressure connectors; terminal bars; pressure connectors approved as grounding and bonding equipment; the exothermic welding process; machine screw-type fasteners that engage not less than two threads or are secured with a nut; thread-forming machine screws that engage not less than two threads in the enclosure; connections that are part of a listed assembly; or other listed means.

25. Alternating-current systems of less than 50 volts shall be grounded: ____________ ________

A. Where installed outside as underground conductors.

B. Where installed outside as overhead and underground conductors.

C. Where installed inside and outside as overhead conductors.

D. Where installed outside as overhead conductors.

26. A grounding electrode conductor for grounding service-supplied alternating-current systems shall be used to connect the equipment grounding conductors, the service-equipment enclosures, and, where the system is grounded, the grounded service conductor to the grounding electrode(s) required by Part III of this article. ____________ ________

A. This conductor shall be sized at 125% of the largest ungrounded conductor.

B. This conductor shall be sized at 100% of the largest ungrounded conductor.

C. This conductor shall be sized in accordance with 250.66.

D. This conductor shall be sized in accordance with 250.122.

Questions

Section	Answer

______ ______ **27.** Where multiple concrete-encased electrodes are present at a building or structure, it shall be permissible to bond only how many of the multiple concrete-encased electrodes into the grounding electrode system?

A. One

B. Two

C. Three

D. All of them

______ ______ **28.** Rod and pipe electrodes shall not be less than 2.44 m (8 ft) in length and shall consist of the following materials:

A. Grounding electrodes of stainless steel, copper or zinc coated steel shall be at least 15.87 mm (5/8 in.) in diameter, unless listed and not less than 19.1 mm (3/4 in.) in diameter.

B. Grounding electrodes of stainless steel, copper or zinc coated steel shall be at least 15.87 mm (5/8 in.) in diameter, unless listed and not less than 25.4 mm (1 in.) in diameter.

C. Grounding electrodes of stainless steel, copper or brass coated steel shall be at least 15.87 mm (5/8 in.) in diameter, unless listed and not less than 12.70 mm (1/2 in.) in diameter.

D. Grounding electrodes of stainless steel, copper or zinc coated steel shall be at least 15.87 mm (5/8 in.) in diameter, unless listed and not less than 12.70 mm (1/2 in.) in diameter.

______ ______ **29.** A single electrode consisting of a rod, pipe, or plate that does not have a resistance to ground of 25 ohms or less shall be augmented by one additional electrode consisting of

A. A concrete-encased electrode.

B. A ground ring.

C. Metal frame of the building.

D. Metal underground water pipe.

______ ______ **30.** Power-limited Class 2 or Class 3 circuit cables containing only circuits operating:

A. At greater than 50 volts shall be permitted to use a conductor with green insulation for other than equipment grounding purposes.

B. At less than 100 volts shall be permitted to use a conductor with green insulation for other than equipment grounding purposes.

C. At less than 50 volts shall be permitted to use a conductor with green insulation for other than equipment grounding purposes.

D. At greater than 75 volts shall be permitted to use a conductor with green insulation for other than equipment grounding purposes.

Questions

	Section	Answer

31. A cable- or raceway-type wiring method, installed in exposed or concealed locations under metal-corrugated sheet roof decking, shall be installed and supported so the nearest outside surface of the cable or raceway is what distance to the nearest surface of the roof decking. ______ ______

A. Not less than 31.8 mm (1 ° in.)

B. Not less than 38 mm (1 Ω in.)

C. Not less than 19.1 mm (3/4 in. in.)

D. Not less than 50.8 mm (2 in.)

32. The interior of enclosures or raceways installed underground shall be considered to be a: ______ ______

A. Wet location.

B. Damp location.

C. Dry location.

D. Damp or wet location.

33. Where raceways are installed in: ______ ______

A. Wet locations above grade, the interior of these raceways shall be considered to be a dry location where used with compression fittings.

B. Damp locations above grade, the interior of these raceways shall be considered to be a dry location where used with compression fittings.

C. Wet locations above grade, the interior of these raceways shall be considered to be a wet location.

D. Wet locations above grade, the exterior of these raceways shall be considered to be a wet location and the interior of the raceway shall be considered to be a dry location.

34. Raceways and cables installed into the bottom of open bottom equipment, such as switchboards, motor control centers, and floor or pad-mounted transformers, shall ______ ______

A. Be required to be mechanically and electrically secured to the equipment.

B. Not be required to be mechanically secured to the equipment.

C. Be required to be secured to the equipment.

D. Not be required to be electrically secured to the equipment where using metal raceways or cables.

Questions

Section **Answer**

______ ______ **35.** In industrial establishments, where conditions of maintenance and supervision ensure that qualified persons will service the installation, the minimum cover requirements, for other than rigid metal conduit and intermediate metal conduit, shall be permitted to be reduced:

A. 150 mm (6 inches) for each 50 mm (2 inches) of concrete or equivalent placed entirely within the trench over the underground installation.

B. 50 mm (2 inches) for each 150 mm (6 inches) of concrete or equivalent placed entirely within the trench over the underground installation.

C. 300 mm (12 inches) for each 100 mm (4 inches) of concrete or equivalent placed entirely within the trench over the underground installation.

D. 600 mm (24 in.) for each 50 mm (2 inches) of concrete or equivalent placed entirely within the trench over the underground installation.

______ ______ **36.** Where 6 AWG THWN copper conductors are installed in a conduit exposed to direct sunlight above a rooftop, the outside temperature is 90°F and the distance above the roof is Ω inch, the temperature and the derating factor is:

A. 150°F and 0.58

B. 130°F and 0.67

C. 150°F and 0.33

D. 140°F and 0.71

______ ______ **37.** Where determining box fill calculation, each conductor that originates outside the box and terminates or spliced within the box shall be counted once, and each conductor that passes through the box without splice or termination shall be counted once. In addition, the following applies:

A. Each loop or coil of unbroken conductor not less than the minimum length required for free conductors in 300.14 shall be counted twice.

B. Each loop or coil of unbroken conductor not less than twice the minimum length required for free conductors in 300.14 shall be counted twice.

C. Each loop or coil of unbroken conductor not less than twice the minimum length required for free conductors in 300.14 shall be counted once.

D. Each loop or coil of unbroken conductor not less than the minimum length required for free conductors in 300.14 shall be counted once.

Questions

	Section	Answer

38. Where determining box fill calculation, for each yoke or strap containing one or more devices or equipment, a double volume allowance in accordance with Table 314.16(B) shall be made for each yoke or strap based on the largest conductor connected to a device(s) or equipment supported by that yoke or strap. In addition, the following applies: ____________ ________

A. A device or utilization equipment wider than a single 100 mm (4 in.) device box as described in Table 314.16(A) shall have double volume allowances provided for each gang required for mounting.

B. A device or utilization equipment wider than a single 50 mm (2 in.) device box as described in Table 314.16(A) shall have quadruple volume allowances provided for each gang required for mounting.

C. A device or utilization equipment wider than a double 100 mm x 100 mm (4 in. x 4 in.) device box as described in Table 314.16(A) shall have double volume allowances provided for each gang required for mounting.

D. A device or utilization equipment wider than a single 50 mm (2 in.) device box as described in Table 314.16(A) shall have double volume allowances provided for each gang required for mounting.

39. Where oversized, concentric or eccentric knockouts are not encountered, Type IMC shall ____________ ________

A. Be permitted to be unsupported where the raceway is not more than 900 mm (3 ft) in length and remains in unbroken lengths (without coupling).

B. Be permitted to be unsupported where the raceway is not more than 450 mm (18 in.) in length and remains in unbroken lengths (without coupling).

C. Be required to be supported where the raceway is not more than 450 mm (18 in.) in length and remains in unbroken lengths (without coupling).

D. Be required to be supported where the raceway is not more than 900 mm (3 ft) in length and remains in unbroken lengths (without coupling).

40. A multipole, general use snap switch shall not be permitted to be fed from more than a single circuit unless it is listed and: ____________ ________

A. Marked as a one-circuit or two-circuit switch, or unless its voltage rating is not less than the nominal line-to-line voltage of the system supplying the circuits.

B. Marked as a two-pole switch, or unless its voltage rating is not less than the nominal line-to-line voltage of the system supplying the circuits.

C. Marked as a two-circuit or three-circuit switch, or unless its voltage rating is less than the nominal line-to-line voltage of the system supplying the circuits.

D. Marked as a two-circuit or three-circuit switch, or unless its voltage rating is not less than the nominal line-to-line voltage of the system supplying the circuits.

Questions

Section	Answer

______ ______ **41.** All 15- and 20-ampere, 125- and 250-volt receptacles installed in a wet location shall have an enclosure that is weatherproof whether or not the attachment plug cap is inserted. In addition, the following shall apply:

A. All 15- through 30-ampere, 125- and 250-volt non-locking receptacles shall be listed weather-resistant type.

B. All 15- and 20-ampere, 125- and 250-volt non-locking receptacles shall be listed weather-resistant type.

C. All 15- and 20-ampere, 125-volt non-locking receptacles shall be listed weather-resistant type.

D. All 15- and 20-ampere, 125- and 250-volt receptacles shall be listed weather-resistant type.

______ ______ **42.** In areas specified in 210.52 for dwelling unit receptacle outlets, what receptacle outlets shall be required to be listed tamper resistant receptacles?

A. Kitchens

B. Family rooms

C. Bedroom

D. All the above.

______ ______ **43.** Luminaires in bathtub or shower areas shall be located as follows:

A. Within a zone measured 900 mm (3 ft) horizontally and 2.5 m (8 ft) vertically from the top of the bathtub rim or shower stall threshold. This zone is all encompassing and includes the zone directly over the tub or shower stall. Luminaires (lighting fixtures) located in this zone shall be listed for damp locations, or listed for wet locations where subject to shower spray.

B. Within the actual outside dimension of the bathtub or shower to a height of 2.5 m (8 ft) vertically from the top of the bathtub rim or shower threshold and in a zone measured 900 mm (3 ft) horizontally and shall be marked for damp locations, or marked for wet locations where subject to shower spray.

C. Within the actual outside dimension of the bathtub or shower to a height of 2.5 m (8 ft) vertically from the top of the bathtub rim or shower threshold shall be marked for damp locations, or marked for wet locations where subject to shower spray.

D. Within the actual outside dimension of the bathtub or shower to a height of 2.5 m (8 ft) vertically from the top of the bathtub rim or shower threshold shall be listed for damp locations, or listed for wet locations where subject to shower spray.

Questions

Section Answer

44. The minimum clearance between luminaires in clothes closets and the nearest point of a storage space shall be as follows: ____________ ________

A. Surface-mounted fluorescent or LED luminaires shall be permitted to be installed within the storage space where identified for this use.

B. Surface-mounted fluorescent or LED luminaires shall be permitted to be installed with a clearance of a minimum clearance of 150 mm (6 in.) within the storage space where identified for this use.

C. Surface-mounted fluorescent or LED luminaires shall be permitted to be installed with a clearance of a minimum clearance of 300 mm (12 in.) within the storage space where identified for this use.

D. Surface-mounted fluorescent or LED luminaires shall be permitted to be installed with a clearance of a minimum clearance of 100 mm (4 in.) within the storage space where identified for this use.

45. In aircraft painting hangers, the following area shall be area classified as follows: ____________ ________

A. The area horizontally from aircraft surfaces between 3.0m (10ft) and 9.0m (30ft) from the floor to 9.0m (30ft) above the aircraft surface shall be classified as Class I, Division 1 or Class I, Zone 1

B. The area horizontally from aircraft surfaces between 9.0m (30ft) and 9.0m (30ft) from the floor to 9.0m (30ft) above the aircraft surface shall be classified as Class I, Division 2 or Class I, Zone 2.

C. The area horizontally from aircraft surfaces between 3.0m (10ft) and 9.0m (30ft) from the floor to 9.0m (30ft) above the aircraft surface shall be classified as Class I, Division 2 or Class I, Zone 2.

D. The area horizontally from aircraft surfaces between 3.0m (10ft) and 3.0m (10ft) from the floor to 9.0m (30ft) above the aircraft surface shall be classified as Class I, Division 2 or Class I, Zone 2.

46. Portable structures for fairs, carnivals, circuses, and similar events shall not be located: ____________ ________

A. Under or within 6.9 m (22.5 ft) horizontally of conductors operating in excess of 600 volts.

B. Under or within 4.5 m (15 ft) horizontally of conductors operating in excess of 600 volts.

C. Under or within 3.0 m (10 ft) horizontally of conductors operating in excess of 600 volts.

D. Under or within 3.7 m (12 ft) horizontally of conductors operating in excess of 600 volts.

Questions

Section **Answer**

________ ________ **47.** A sump pump or oil recovery pump located in the pit of an elevator hoistway shall be permitted to be cord connected. The cord shall be

A. An extra-hard usage oil resistant type, of a length not to exceed 1.8 m (6 ft), and shall be located to be protected from physical damage.

B. A hard usage oil resistant type, of a length not to exceed 900 mm (3 ft), and shall be located to be protected from physical damage.

C. A extra-hard usage oil resistant type, of a length not to exceed 2.5 m (8 ft), and shall be located to be protected from physical damage.

D. A hard usage oil resistant type, of a length not to exceed 1. 8 m (6 ft), and shall be located to be protected from physical damage.

________ ________ **48.** Each disconnecting means for all pool, spa, or hot tub utilization equipment shall be readily accessible and within sight from its equipment, and shall be located as follows:

A. At least 3.0 m (10 ft) horizontally from the inside walls of a pool, spa, or hot tub unless separated from the open water by a permanently installed barrier that provides a 1.5 m (5 ft) reach path or greater.

B. At least 1.5 m (5 ft) horizontally from the inside walls of a pool, spa, or hot tub unless separated from the open water by a permanently installed barrier that provides a 1.5 m (5 ft) reach path or greater.

C. At least 1.5 m (5 ft) horizontally from the inside walls of a pool, spa, or hot tub unless separated from the open water by a permanently installed barrier that provides a 900 mm (3 ft) reach path or greater.

D. At least 3.0 m (10 ft) horizontally from the inside walls of a pool, spa, or hot tub unless separated from the open water by a permanently installed barrier that provides a 2.5 m (8 ft) reach path or greater.

________ ________ **49.** All wet-niche luminaires shall be removable from the water for-inspection, relamping, or other maintenance. The forming shell location and length of cord in the forming shell shall permit personnel to place the removed luminaire on the deck or other dry location for such maintenance.

A. The luminaire maintenance location shall be readily accessible without entering or going in the pool water.

B. The luminaire maintenance location shall be accessible by entering or going in the pool water.

C. The luminaire maintenance location shall be readily accessible by entering or going in the pool water.

D. The luminaire maintenance location shall be accessible without entering or going in the pool water.

Questions

Section **Answer**

50. Swimming pool water in a permanently installed swimming pool shall be bonded in accordance with the following: ____________ ________

A. An intentional bond of a minimum conductive surface area of 2903 mm^2 (4.5 in^2) shall be installed in contact with the pool water.

B. An intentional bond of a minimum conductive surface area of 5806 mm^2 (9 in^2) shall be installed in contact with the pool water.

C. An intentional bond of a minimum conductive surface area of 11612 mm^2 (18 in^2) shall be installed in contact with the pool water.

D. An intentional bond of a minimum conductive surface area of 17418 mm^2 (27 in^2) shall be installed in contact with the pool water.

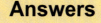

Answers

1. B Article 100
2. A Article 100
3. D Article 100
4. C Article 100
5. A Article 100
6. D Article 100
7. A 110.12
8. C 110.12
9. D 110.16
10. A 110.16
11. C 110.20
12. B 110.20
13. A 110.20
14. C 110.20
15. B 110.26(C)(2)
16. C 110.26(C)(3)
17. B 200.2(B)
18. B 210.5(C)
19. C 210.8(A)(2)
20. A 210.12(B)
21. D 210.19(A), Ex. 2
22. A 210.52(E)(3)
23. B 240.21(C)(4)
24. A 250.8(A)
25. D 250.20(A)(3)
26. C 250.24(D)
27. A 250.52(A)(3)
28. D 250.52(A)(5)
29. B 250.56
30. C 250.119
31. B 300.4(E)
32. A 300.5(B)
33. C 300.9
34. B 300.12
35. A Table 300.50, Note 3
36. C 310.15(B)(2)(c)
37. B 314.16(B)(1)
38. D 314.16(B)(4)
39. B 342.30(C)
40. D 404.8(C)
41. B 406.8(B)
42. D 406.11
43. C 410.10(D)
44. A 410.16
45. C 513.3(C)(2)
46. B 525.5(B)(2)
47. D 620.21(A)(1)(c)
48. B 680.12
49. D 680.23(B)(6)
50. B 680.26(C)

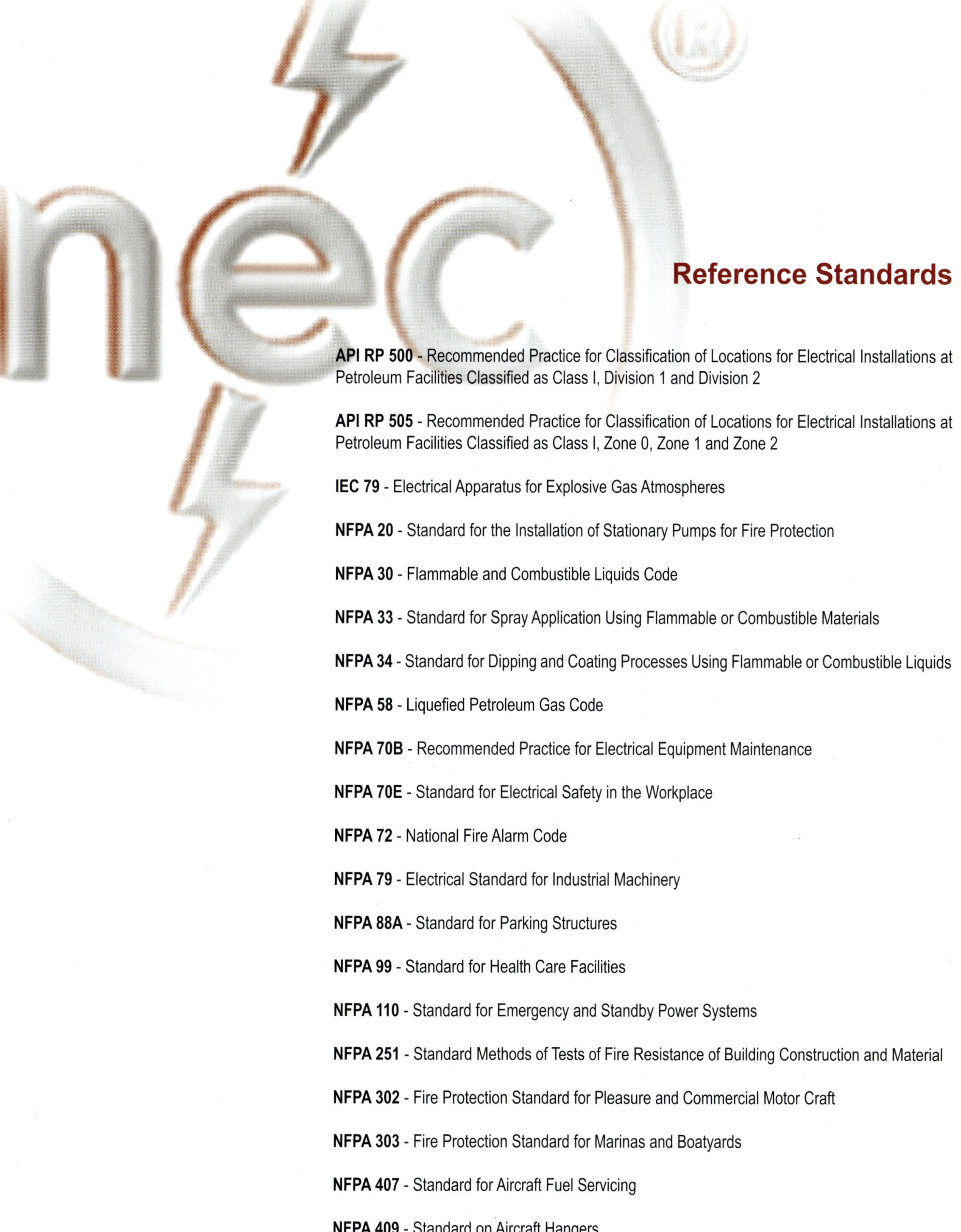

Reference Standards

API RP 500 - Recommended Practice for Classification of Locations for Electrical Installations at Petroleum Facilities Classified as Class I, Division 1 and Division 2

API RP 505 - Recommended Practice for Classification of Locations for Electrical Installations at Petroleum Facilities Classified as Class I, Zone 0, Zone 1 and Zone 2

IEC 79 - Electrical Apparatus for Explosive Gas Atmospheres

NFPA 20 - Standard for the Installation of Stationary Pumps for Fire Protection

NFPA 30 - Flammable and Combustible Liquids Code

NFPA 33 - Standard for Spray Application Using Flammable or Combustible Materials

NFPA 34 - Standard for Dipping and Coating Processes Using Flammable or Combustible Liquids

NFPA 58 - Liquefied Petroleum Gas Code

NFPA 70B - Recommended Practice for Electrical Equipment Maintenance

NFPA 70E - Standard for Electrical Safety in the Workplace

NFPA 72 - National Fire Alarm Code

NFPA 79 - Electrical Standard for Industrial Machinery

NFPA 88A - Standard for Parking Structures

NFPA 99 - Standard for Health Care Facilities

NFPA 110 - Standard for Emergency and Standby Power Systems

NFPA 251 - Standard Methods of Tests of Fire Resistance of Building Construction and Material

NFPA 302 - Fire Protection Standard for Pleasure and Commercial Motor Craft

NFPA 303 - Fire Protection Standard for Marinas and Boatyards

NFPA 407 - Standard for Aircraft Fuel Servicing

NFPA 409 - Standard on Aircraft Hangers

NFPA 410 - Standard on Aircraft Maintenance

NFPA 496 - Standard for Purged and Pressurized Enclosures for Electrical Equipment

NFPA 497 - Recommended Practice for the Classification of Flammable Liquids, Gases, or Vapors and of Hazardous (Classified) Locations for Electrical Installations in Chemical Process Areas

NFPA 499 - Recommended Practice for the Classification of Combustible Dusts and of Hazardous (Classified) Locations for Electrical Installations in Chemical Process Areas

NFPA 505 - Fire Safety Standard for Powered Industrial Trucks Including Type Designations, Areas of Use, Conversions, Maintenance and Operations